Gene Therapy in Neurodegenerative Disorders

Gene Therapy in Neurodegenerative Disorders

From Bench to Bedside

Edited by

Vahid Mansouri
Gene Therapy Research Center, Digestive Disease Research Institute; Shariati Hospital, Tehran University of Medical Sciences, Tehran, Iran

Maryam Bemanalizadeh
Gene Therapy Research Center, Digestive Disease Research Institute; Shariati Hospital, Tehran University of Medical Sciences, Tehran, Iran

ELSEVIER

ACADEMIC PRESS
An imprint of Elsevier

Academic Press is an imprint of Elsevier
125 London Wall, London EC2Y 5AS, United Kingdom
50 Hampshire Street, 5th Floor, Cambridge, MA 02139, United States

ISBN: 978-0-443-40467-2

For information on all Academic Press publications
visit our website at https://www.elsevier.com/books-and-journals

Publisher: Megan Ball
Acquisitions Editor: Ali Afzal-Khan
Editorial Project Manager: Stuti Das
Production Project Manager: M. Balakrishnan
Cover Designer: Mark Rogers

Typeset by Integra Software Services

To all patients and families affected by neurodegenerative and neuromuscular diseases.
Your courage, resilience, and hope are the reason science must continue to move from discovery to therapy, from possibility to treatment, and from the bench to the bedside.

Contents

Section I
Neuromuscular disorders

4 Gene therapy in spinal muscular atrophy (SMA)

Wladimir Bocca Vieira de Rezende Pinto and Paulo Sgobbi

5 Gene therapy in amyotrophic lateral sclerosis (ALS)

Negin Salemi and Arman Rahimmi

6 Gene therapy in Duchenne muscular dystrophy (DMD)

Arezou Azizsoltani, Zeinab Ghesmati, and Elnaz Rezaei

12 Gene therapy in Huntington's disease (HD)

Fatemeh Tirgar, Mina Hosseini, Sakineh Alijanpour, Mohamad-Reza Aghanoori, Sahar Alijanpour, and Hami Mahdavinataj

13 Gene therapy in spinocerebellar ataxia (SCA)

Feyza Sule Aslan and Enes Akyuz

Contributors

Mohamad-Reza Aghanoori Department of Molecular Medicine, National Institute of Genetic Engineering and Biotechnology (NIGEB), Tehran, Iran

Enes Akyuz Department of Pediatrics, School of Medicine, University of Wisconsin-Madison, Madison, WI, United States

Sahar Alijanpour Department of Genetics, Faculty of Medicine, Alborz University of Medical Sciences, Karaj, Iran

Sakineh Alijanpour Department of Biology, Faculty of Science, Gonbad Kavous University, Gonbad Kavous, Iran

Mitra Ashrafi School of Medicine, Zanjan University of Medical Sciences, Zanjan, Iran

Feyza Sule Aslan School of Medicine, Marmara University, Istanbul, Turkey

Arezou Azizsoltani Department of Medical Biotechnology and Molecular Medicine, School of Medicine, Arak University of Medical Sciences, Arak; Department of Medical Biotechnology, Faculty of Advanced Medical Sciences, Tabriz University of Medical Sciences, Tabriz, Iran

Maryam Bemanalizadeh Gene Therapy Research Center, Digestive Disease Research Institute; Shariati Hospital, Tehran University of Medical Sciences, Tehran, Iran

Wladimir Bocca Vieira de Rezende Pinto Division of Neuromuscular Diseases, Department of Neurology and Neurosurgery, Federal University of São Paulo (UNIFESP), São Paulo, Brazil

Mahdieh Eslami School of Medicine, Isfahan University of Medical Sciences, Isfahan, Iran

Zeinab Ghesmati Department of Medical Biotechnology, Faculty of Advanced Medical Sciences, Tabriz University of Medical Sciences, Tabriz, Iran

Zahra Heidari Shahid Chamran Hospital; Maternal-Fetal Medicine Research Center, Shiraz University of Medical Sciences, Shiraz, Iran

Mina Hosseini Department of Pharmaceutical Biotechnology, School of Pharmacy, Tehran University of Medical Sciences, Tehran, Iran

Asad Kazemi Department of Medical Biotechnology, School of Advanced Technologies in Medicine, Tehran University of Medical Sciences, Tehran, Iran

Iman Kiani Tehran University of Medical Sciences, Medicine Faculty, Tehran, Iran

Hami Mahdavinataj Department of Neuroscience and Addiction Studies, School of Advanced Technologies in Medicine, Tehran University of Medical Sciences, Tehran, Iran

Vahid Mansouri Gene Therapy Research Center, Digestive Disease Research Institute; Shariati Hospital, Tehran University of Medical Sciences, Tehran, Iran

Shiva Moein Department of Pathology, Brigham and Women's Hospital, Boston, MA, United States

Sayed Ali Reza Mousavi Department of Medical Biotechnology, School of Advanced Technologies in Medicine, Tehran University of Medical Sciences, Tehran, Iran

Parnaz Niazi Department of Medical Biotechnology, Zanjan University of Medical Sciences, Zanjan, Iran

Arman Rahimmi Neurosciences Research Center, Research Institute for Health Development, Kurdistan University of Medical Sciences, Sanandaj, Iran

Elnaz Rezaei Department of Agricultural Biotechnology, Faculty of Agriculture and Natural Resources, Imam Khomeini International University (IKIU), Qazvin, Iran

Zahra Oushyani Roudsari Department of Medical Biotechnology, Zanjan University of Medical Sciences, Zanjan, Iran

Fatemeh Saadatpour Pharmaceutical Biotechnology Lab, School of Biology, College of Science, University of Tehran, Tehran, Iran

Mojdeh Salehi Namini Stem Cell and Regenerative Medicine Research Center, Iran University of Medical Sciences; Department of Tissue Engineering, School of Advanced Technologies in Medicine, Tehran University of Medical Sciences, Tehran, Iran

Negin Salemi Isfahan University of Medical Sciences, Student Research Committee, Isfahan, Iran
Amir Savardashtaki Department of Medical Biotechnology, School of Advanced Medical Sciences and Technologies; Infertility Research Center, Shiraz University of Medical Sciences, Shiraz, Iran
Paulo Sgobbi Division of Neuromuscular Diseases, Department of Neurology and Neurosurgery, Federal University of São Paulo (UNIFESP), São Paulo, Brazil
Alireza Shoari Department of Cancer Biology, Mayo Clinic, Jacksonville, FL, United States
Fatemeh Tirgar Iranian National Center for Addiction Studies (INCAS); Department of Neuroscience and Addiction Studies, School of Advanced Technologies in Medicine, Tehran University of Medical Sciences, Tehran, Iran
Samaneh Toutounchian Tehran University of Medical Sciences, Medicine Faculty, Tehran, Iran
Asma Vafadar Student Research Committee; Department of Medical Biotechnology, School of Advanced Medical Sciences and Technologies, Shiraz University of Medical Sciences, Shiraz, Iran

Preface

The idea for this book grew out of several years of engagement with the rapidly evolving field of gene therapy and its translation into neurological and neuromuscular diseases. Over time, my work in this area repeatedly brought me back to the same question: How can a researcher, clinician, or translational scientist understand a disease deeply enough to design a rational gene therapy strategy for it?

This question became especially clear through my involvement in disease-focused reviews and in broader analyses of the global gene therapy clinical trial landscape. Examining clinical trials across different diseases, vectors, delivery routes, and therapeutic strategies provided a valuable view of where the field has been, where it is moving, and which scientific and translational gaps remain unresolved. It also showed that progress in gene therapy cannot be understood only by looking at a single technology or a single clinical trial. It must be interpreted through the biology of each disease, the suitability of therapeutic targets, the availability of relevant models, the feasibility of delivery, and the realities of clinical translation.

This perspective gradually shaped the framework of the present book. A meaningful disease-oriented review in gene therapy should begin with the disease itself: its clinical features, pathophysiology, genetic mechanisms, diagnostic considerations, and therapeutic targets. It should then move toward the gene therapy modalities that may be applicable, the delivery platforms that make these approaches possible, the experimental and animal models used to test them, the preclinical evidence supporting translation, and the clinical trials that reveal both promise and limitation. Such a structure does more than organize information; it helps reveal where the field is mature, where it is uncertain, and where future research may have the greatest impact.

This approach is particularly valuable for those entering a new disease area, as it can shorten the path from general familiarity to meaningful scientific contributions. It is also useful for investigators already working in one part of the translational pipeline, because it encourages a broader view: from molecular mechanism to therapeutic design, from preclinical modeling to clinical need, and from experimental possibility to patient-centered application.

This perspective was one of the main motivations behind this book. Gene therapy is no longer only a promise of future medicine. In some diseases, it has already changed clinical practice; in others, it is rapidly advancing through preclinical and clinical development. Yet the path from a genetic concept to a safe, effective, and accessible therapy remains complex. It requires not only technological innovation, but also disease-specific insight, appropriate models, rational vector selection, feasible manufacturing, careful clinical trial design, regulatory awareness, ethical reflection, and long-term safety considerations. These elements are most powerful when they are considered together.

This is especially true for neurodegenerative and neuromuscular diseases. Although these conditions are often discussed under shared scientific or clinical categories, each tells a different translational story. Duchenne muscular dystrophy, spinal muscular atrophy, Alzheimer's disease, Parkinson's disease, leukodystrophies, spinocerebellar ataxias, facioscapulohumeral muscular dystrophy, and other related disorders differ in their affected tissues, pathogenic mechanisms, therapeutic windows, delivery barriers, and clinical expectations. A strategy that appears suitable for one disease may be insufficient or even inappropriate for another. Therefore, progress in this field requires both a broad understanding of gene therapy platforms and a precise appreciation of each disease as its own biological and clinical problem.

My hope is that this structure will help readers see not only what has been achieved, but also why certain strategies have succeeded, why others have struggled, and where the next opportunities may emerge. For students and early-career researchers, I hope this book will provide a clear and comprehensive entry point into the field. For basic scientists, I hope it will help connect molecular and experimental work to clinical needs. For translational researchers and clinicians, I hope it will offer a disease-oriented map of the current landscape of gene therapy in neurodegenerative and neuromuscular disorders. More broadly, I hope it will encourage readers to think beyond isolated technologies and toward integrated therapeutic development.

At the heart of all these scientific efforts are patients and their families. Behind every genetic mechanism, animal model, vector platform, clinical trial, and regulatory pathway is the hope that one day a child may move more freely, a parent may preserve memory for longer, a patient may breathe more easily, or a family may be given more time, more certainty, and more joy. This human purpose is what gives translational science its deepest meaning. Gene therapy is technically demanding, scientifically complex, and often uncertain, but its ultimate value lies in its potential to change lives.

Editing this volume has also reminded me that scientific progress is rarely the result of a single perspective or a single community. This book was made possible through the generous contributions of authors and colleagues from different countries, institutions, and scientific backgrounds, each bringing a distinct view of gene therapy and neurodegenerative disease. I am deeply grateful to all chapter authors and contributors for sharing their knowledge, time, and insights. Their collective effort reflects the international and collaborative spirit that is essential for moving complex fields such as gene therapy forward.

I would also like to express my sincere gratitude to my coeditor, Dr. Maryam Bemanalizadeh, for her valuable collaboration and support throughout the development of this book. I am equally grateful to my colleagues and to the publishing team for their guidance and assistance in bringing this project to completion.

Gene therapy for neurodegenerative diseases remains an evolving field, but it is also one of the most hopeful frontiers in modern medicine. If this book helps even a small number of readers understand the field more clearly, identify an unanswered question, design a better experiment, or imagine a more effective path toward therapy, then it will have achieved its purpose.

Vahid Mansouri

Chapter 1

Introduction

Vahid Mansouri and Maryam Bemanalizadeh
Gene Therapy Research Center, Digestive Disease Research Institute, Tehran University of Medical Sciences, Tehran, Iran

Neurodegenerative and neurogenetic disorders, including Alzheimer's disease, Parkinson's disease, Huntington's disease, amyotrophic lateral sclerosis (ALS), spinal muscular atrophy (SMA), muscular dystrophies, spinocerebellar ataxias (SCAs), leukodystrophies, and lysosomal storage disorders, remain among the most devastating conditions in medicine. They are characterized by the progressive loss of neuronal or neuromuscular function, leading to cognitive decline, motor disability, and loss of independence across the lifespan. Despite major advances in diagnostics and symptomatic care, most of these conditions still lack therapies that durably modify the underlying disease process, and the cumulative burden on patients, families, and healthcare systems continues to grow worldwide.

Against this backdrop, gene therapy has moved from a theoretical concept to a realistic therapeutic modality. The approval of the first gene therapies and their rapid expansion, particularly as a life-changing treatment for otherwise untreatable rare monogenic diseases, have demonstrated that targeting the molecular root cause of disease can deliver sustained, clinically meaningful benefits for some patients. At the same time, advances in vector engineering, genome editing, antisense technologies, and delivery to the central and peripheral nervous systems have transformed what is technically possible. However, these successes coexist with ongoing scientific, clinical, ethical, and regulatory challenges that must be addressed if gene therapy is to become a broadly applicable strategy for neurodegenerative and neuromuscular disorders. Across indications and platforms, several translational bottlenecks repeatedly determine whether promising preclinical results can become safe, durable clinical benefits. These include delivery and biodistribution (achieving sufficient, cell-type-appropriate expression in the central nervous system (CNS) and peripheral tissues such as muscles while limiting off-target exposure), as well as immunogenicity and long-term durability (preexisting and treatment-induced immune responses, and sustained expression or editing effects over years).

Gene Therapy in Neurodegenerative Disorders: From Bench to Bedside was conceived to provide a coherent, translational overview of this rapidly evolving field. Rather than focusing exclusively on a single disease or technology, this volume brings together experts across basic neuroscience, molecular genetics, neuromuscular medicine, pediatric and adult neurology, and regulatory science to chart the path from mechanistic insight to clinical application. Our aim is to situate current gene therapy approaches within the broader biology of neurodegeneration, while providing readers with concrete examples of how these strategies are being developed, tested, and, in some cases, implemented in the clinic.

The opening chapters lay the conceptual and methodological groundwork for the book. Chapter 2 reviews contemporary gene-editing technologies and gene delivery modalities for neurodegenerative disorders, outlining the principles, strengths, and limitations of viral and nonviral vectors, genome-editing platforms, and emerging delivery strategies to the central nervous system. Chapter 3 addresses the ethical and regulatory dimensions of gene therapy, including trial design in rare diseases, long-term safety considerations, informed consent in vulnerable populations, and the evolving regulatory frameworks that govern advanced therapy medicinal products. Together, these chapters provide the scientific and societal context necessary for interpreting the disease-specific chapters that follow.

The remainder of the book is organized into three sections, each focusing on a major group of disorders in which gene-based approaches are particularly relevant. Section I (Neuromuscular disorders) examines gene therapy strategies for SMA, ALS, Duchenne muscular dystrophy (DMD), and limb-girdle muscular dystrophies (LGMD). These chapters discuss the genetic architecture of each disease, preclinical models, clinical trial experience where available, and practical issues such as timing of intervention, route of delivery, and outcome measures in neuromuscular populations.

Section II (Lysosomal storage disorders) focuses on the mucopolysaccharidoses (MPS), leukodystrophies such as metachromatic leukodystrophy (MLD), and neuronal ceroid lipofuscinosis (NCL). Here, the authors highlight how gene therapy

Gene Therapy in Neurodegenerative Disorders. https://doi.org/10.1016/B978-0-443-40467-2.00005-1

can address both systemic and central nervous system pathology, review current vector and promoter choices, and discuss lessons learned from early-phase clinical trials and long-term follow-up. Particular attention is paid to the challenges of treating pediatric-onset disorders, where early intervention must be balanced against uncertainties about durability and late adverse effects.

Section III (Neurodegenerative disorders) focuses on more common, often multifactorial conditions such as Alzheimer's disease, Parkinson's disease, Huntington's disease, and SCAs. In these chapters, contributors explore a spectrum of gene-based strategies, ranging from gene replacement and silencing to trophic factor delivery and modulation of protein aggregation, while also addressing the biological and methodological complexities of applying gene therapy to diseases that are not purely monogenic. Throughout, the emphasis is on how mechanistic insights are being translated into rational therapeutic design and where key gaps in knowledge remain.

A distinguishing feature of this volume is its cross-cutting focus on the practical and scientific requirements for successful delivery to the central nervous system and on the full translational pathway that connects discovery to real-world care. Across chapters, contributors trace how therapeutic concepts evolve from vector and construct design (including capsid selection, promoter choice, tropism, and dosing) through preclinical development (model selection, biodistribution, target engagement, safety, and durability), and into first-in-human and pivotal clinical trials (patient selection, endpoints and biomarkers, administration routes, immunogenicity management, and long-term follow-up). Importantly, the discussion extends beyond regulatory approval to address postmarketing considerations, including real-world effectiveness and late safety signals.

This volume is intended for a broad audience that includes basic and translational neuroscientists, clinical neurologists and neuromuscular specialists, pediatricians, geneticists, residents and fellows, PhD students, and colleagues in industry and regulatory agencies. Readers primarily interested in technology development may gravitate toward the early chapters on vectors and editing platforms, while clinicians may start with the disease-focused sections most relevant to their practice. We also hope the book will serve as a teaching resource for courses in neurogenetics, gene therapy, and translational neuroscience and as a reference for multidisciplinary teams planning or conducting gene therapy trials.

Ultimately, our motivation in assembling this book is both scientific and patient-centered. Scientifically, we are living through a period in which the tools to interrogate and modify the nervous system at the genetic level are advancing at an unprecedented speed. Clinically, there is an urgent need to translate these tools into safe, effective therapies for individuals living with neurodegenerative diseases, most of whom still have no disease-modifying options. By bringing together perspectives from bench and bedside, rare and common diseases, and early-stage discovery through to regulatory considerations, we hope this volume will help readers navigate the current landscape, recognize common challenges across disorders, and contribute to the next generation of gene-based therapies for neurodegenerative disorders.

Chapter 2

Gene-editing technologies and gene delivery modalities for neurodegenerative disorders: Current advances and challenges

Fatemeh Saadatpour
Pharmaceutical Biotechnology Lab, School of Biology, College of Science, University of Tehran, Tehran, Iran

1 Introduction

Neurological disorders (NDs) comprise a heterogeneous group of progressive and largely irreversible diseases, characterized by selective neuronal loss, synaptic dysfunction, and gradual collapse of neural networks (Gadhave et al., 2024). The intrinsic complexity of the central nervous system, combined with limited neuronal regenerative capacity and the simultaneous involvement of multiple molecular pathways, makes these disorders among the most challenging to treat in modern medicine. In most cases, diagnosis occurs at advanced stages, when widespread neuronal damage has already happened, significantly limiting the window for effective therapeutic intervention. Consequently, despite the growing personal, societal, and economic burden of NDs, currently available treatments are predominantly symptomatic and do not substantially modify the underlying disease processes (Gadhave et al., 2024; Liu and Song, 2025).

This profound gap between clinical need and the efficacy of conventional therapies has directed research toward novel therapeutic strategies that target the root genetic and molecular causes of disease, rather than merely alleviating secondary symptoms. Insights from the Human Genome Project and subsequent advances in genomic technologies have revealed that many neurodegenerative disorders, at least in specific patient subgroups, are associated with defined genetic mutations or dysregulated regulatory pathways. These discoveries have laid the groundwork for the emergence of gene therapy and gene manipulation as transformative approaches in modern neuromedicine (Shchaslyvyi et al., 2023; Liu and Song, 2025).

The conceptual foundation of gene therapy was established in 1972 with the proposal that targeted genetic modifications could be leveraged for therapeutic purposes (Devinsky et al., 2025). However, practical applications in NDs were limited for decades due to incomplete knowledge of disease-causing genes and technological constraints. A pivotal shift occurred in the 1990s following the Human Genome Project, which enabled the systematic identification of genes implicated in ND pathogenesis (Shchaslyvyi et al., 2023; Organization, 2025).

The first clinical gene therapy trial was conducted in 1990 for a nonneurological disease; however, initial applications relevant to NDs emerged through retinal degenerative disorders, often considered a subset of neurodegenerative conditions. Preclinical studies using adeno-associated virus (AAV) vectors, most notably the correction of RPE65 mutations in canine models of Leber congenital amaurosis (LCA) in 2001, provided critical proof-of-concept for gene replacement in the nervous system, leading to human trials by 2008 (Devinsky et al., 2025).

The early 2000s saw the introduction of programmable nucleases, beginning with zinc-finger nucleases (ZFNs), which combine sequence-specific DNA-binding domains with the FokI endonuclease to induce targeted double-strand breaks (DSBs) (Shchaslyvyi et al., 2023). ZFNs enabled precise gene disruption and correction applied to disease-relevant mutations such as *SNCA* A53T in induced pluripotent stem cells (iPSCs) modeling Parkinson's disease (PD) (Su et al., 2025). Building on this framework, transcription activator-like effector nucleases (TALENs) were introduced in 2011, offering

Gene Therapy in Neurodegenerative Disorders. https://doi.org/10.1016/B978-0-443-40467-2.00008-7

improved targeting flexibility through modular repeat-variable di-residues, though their large size limited efficient viral delivery (Khan et al., 2024).

A transformative advance occurred in 2013 with the adaptation of CRISPR–Cas9 for mammalian genome editing. Originating from bacterial adaptive immunity, CRISPR–Cas9 uses a single-guide RNA (sgRNA) to direct Cas9 to specific genomic loci adjacent to protospacer-adjacent motifs (PAMs), enabling efficient and programmable DNA cleavage (Khan et al., 2024; Mishra et al., 2025). Compared with ZFNs and TALENs, CRISPR–Cas9 offers unparalleled simplicity, scalability, and multiplexing capacity, rapidly establishing itself as the dominant genome-editing technology. Subsequent innovations, including base editors (BEs) (2017) and prime editors (PEs) (2019), have expanded the precision and versatility of CRISPR-based interventions by enabling single-nucleotide conversions or precise insertions/deletions without introducing DSBs (Sizova et al., 2013; Mishra et al., 2025). More recently, CRISPR-associated transposons (CASTs) have demonstrated the potential for large-scale genomic insertions without DSBs, representing a promising tool for complex genome engineering.

The 2010s saw a resurgence of gene-based therapeutic strategies for NDs, fueled by improved vector design, deeper mechanistic understanding, and accelerated clinical translation. Notable approvals include Spinraza (nusinersen) for spinal muscular atrophy (SMA) in 2016, Zolgensma (onasemnogene abeparvovec) in 2019, and Tofersen (Qalsody) for SOD1-linked amyotrophic lateral sclerosis (ALS) in 2023 (Finkel et al., 2016; Mendell et al., 2017; Everett and Bucelli, 2024). By mid-2024, four gene-based therapies had gained regulatory approval for neurodegenerative indications. Importantly, the first in vivo CRISPR-based clinical trial targeting RPE65 mutations in LCA began in 2019, marking a milestone in direct mutation correction in patients (Jo et al., 2019).

These advances underscore the importance of gene manipulation and delivery as central pillars of emerging disease-modifying strategies for NDs. The integration of increasingly precise genome engineering technologies with optimized delivery modalities offers a paradigm shift from symptomatic management to potentially curative interventions. This chapter provides a focused review of gene manipulation technologies and gene delivery systems for NDs, highlighting their mechanistic bases, translational advances, and current limitations. After summarizing shared pathophysiological mechanisms and major therapeutic targets, the chapter examines gene-centered, cellular, and metabolic intervention strategies. Molecular approaches for gene manipulation at the DNA and RNA levels are then discussed, followed by an overview of viral and nonviral delivery platforms optimized for central nervous system applications. Key challenges related to targeting specificity, off-target effects, and immunogenicity are critically evaluated, along with emerging strategies to improve safety and efficacy.

2 Common pathophysiological mechanisms in NDs

NDs span a heterogeneous array of conditions, including classic age-associated neurodegenerative diseases (NDDs), neuromuscular disorders (NMDs), and lysosomal storage disorders (LSDs) that, despite divergent etiologies and clinical presentations, share a set of fundamental pathogenic pathways. As shown in Fig. 2.1, these disorders converge on common intracellular vulnerabilities that compromise neuronal integrity, disrupt energy homeostasis, impair protein quality control, and undermine synaptic and axonal function (Fan et al., 2026). Such convergence provides a unified conceptual framework for disease progression and the identification of molecular targets that transcend traditional diagnostic boundaries (Nixon and Rubinsztein, 2024; Kapoor et al., 2025).

A fundamental hallmark of many NDs is the disruption of cellular proteostasis, leading to the accumulation of misfolded proteins and toxic aggregates. Key therapeutic targets include molecular chaperones, components of the ubiquitin–proteasome system, and regulators of autophagy–lysosomal pathways, all of which work to maintain protein quality control. Modulating these pathways through small molecules or biologics can enhance the clearance of pathogenic aggregates such as amyloid-β, α-synuclein, mutant huntingtin (mHTT), tau, and TAR DNA-binding protein 43 (TDP-43) (Settembre et al., 2008; Orr and Oddo, 2013).

In addition, the autophagy–lysosomal pathway dysfunction intersects with protein handling and cellular clearance mechanisms in many diseases (Orr and Oddo, 2013). Neurons and muscle cells rely on this pathway to remove damaged proteins and organelles (Nixon and Rubinsztein, 2024). When lysosomal degradation is impaired, autophagic vacuoles accumulate, and autophagic flux is blocked. In some NMDs, mutations that disrupt lysosomal acidification or pH regulation reduce degradative capacity (Malik et al., 2019). This leads to a buildup of autophagic vacuoles that cannot be cleared, contributing to muscle fiber degeneration and weakness, as seen in congenital myopathies caused by vacuolar ATPase assembly defects that elevate lysosomal pH and block autophagy (Du et al., 2025).

In LSDs, inherited deficiencies of lysosomal hydrolases prevent effective fusion between autophagosomes and lysosomes and impair substrate breakdown (Seranova et al., 2017; Darios and Stevanin, 2020). As a result, undegraded material

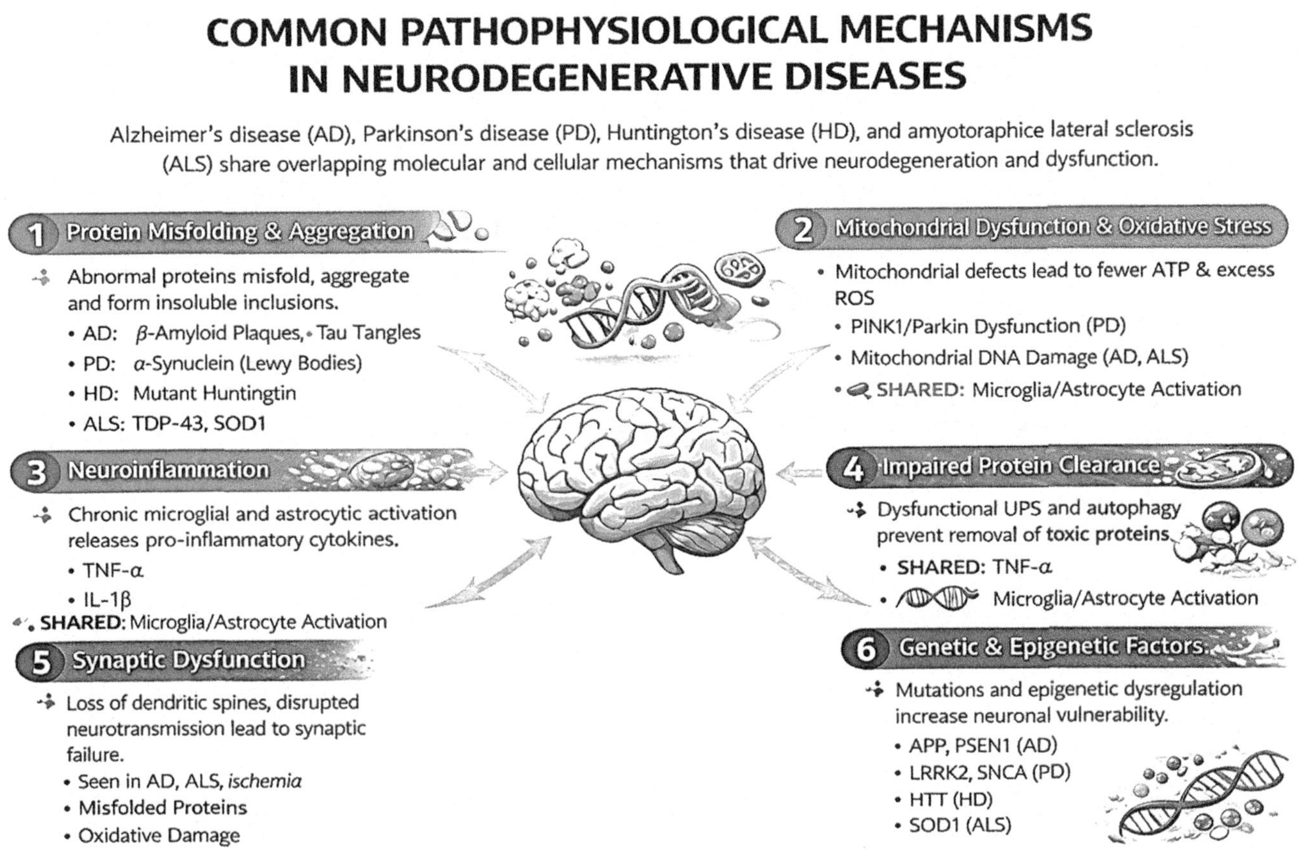

FIG. 2.1 Common pathophysiological mechanisms in NDs. Core processes include protein misfolding and aggregation, mitochondrial dysfunction and oxidative stress, neuroinflammation coupled with impaired protein clearance, synaptic dysfunction, and genetic and epigenetic dysregulation. Any mechanism listed as SHARED affects multiple NDs.

accumulates in lysosomes and autophagosomes. This defective clearance disrupts global protein homeostasis. It also interferes with mitochondrial quality control, increasing the presence of dysfunctional mitochondria. These combined failures in degradation and organelle turnover contribute to cell dysfunction and degeneration in affected tissues (Malik et al., 2019; Darios and Stevanin, 2020).

Energy imbalance arising from mitochondrial dysfunction and oxidative stress constitutes another shared pathogenic mechanism (Houldsworth, 2024). Neurons have among the highest metabolic demands of all cell types, relying extensively on oxidative phosphorylation to sustain ionic gradients, axonal transport, and synaptic transmission (Yusuf et al., 2025). Perturbations in electron transport chain activity, disrupted mitochondrial dynamics, defective mitophagy, and excessive reactive oxygen species (ROS) production are observed across Alzheimer's disease (AD), Huntington's disease (HD), PD, and ALS models, linking metabolic failure to neuron vulnerability and death (Rehman et al., 2023). Mitochondrial impairment not only limits ATP synthesis but also exacerbates oxidative damage and calcium dyshomeostasis, establishing vicious cycles of cellular stress that amplify proteostasis defects and trigger downstream degenerative cascades (Wadan et al., 2025).

A third unifying axis is neuroinflammation and excitotoxic stress, which contribute as both drivers and amplifiers of cellular injury (Kumar and Bhaskar, 2025). Persistent activation of glial cells, changes in the metabolic phenotypes of microglia, and elevated proinflammatory cytokine signaling (e.g., TNF-α, IL-1β) accelerate neuronal loss in NDDs, intensify motor neuron and muscle fiber damage in NMDs, and mediate secondary neurodegeneration in LSDs (Pandey, 2023; Du et al., 2025). Mitochondrial compromise can itself precipitate innate immune activation through the release of mitochondrial DNA and ROS, engaging inflammasome pathways that further propagate inflammatory cell death and microglial activation (Mahto et al., 2025; Miao et al., 2025).

Although these mechanisms are ubiquitously implicated, primary drivers differ among disorder classes. NDDs are dominated by toxic protein aggregation and early synaptic compromise; NMDs stem largely from structural, trafficking, or motor unit dysfunction; and LSDs originate from monogenic lysosomal enzyme deficiencies with profound effects on metabolic and degradative systems (Alkahtani et al., 2023; Fan et al., 2026). Nonetheless, the interconnectedness of proteostasis networks, mitochondrial quality control, and neuroimmune signaling suggests that therapeutic strategies targeting these core processes, such as gene-based modulation of proteostasis, small molecules that restore mitochondrial function and mitophagy, and interventions that rebalance inflammatory responses, may yield disease–agnostic benefits across these phenotypically distinct conditions (Džidić-Krivić et al., 2025; Kapoor et al., 2025; Kumar and Bhaskar, 2025). This integrated perspective not only clarifies the complex pathophysiological landscape of NDs but also supports the rational development of disease-modifying therapies aimed at shared molecular vulnerabilities rather than isolated clinical features.

3 Therapeutic targets and intervention strategies

The shared molecular and cellular mechanisms underlying neurodegenerative disorders not only define a complex, interconnected network of pathological processes but also expose fundamental vulnerabilities that can be exploited therapeutically. As illustrated in Fig. 2.2, therapeutic strategies have accordingly expanded from traditional small molecules to encompass precise gene-based interventions, targeted modulation of neuronal cell survival, strategic regulation of neuroimmune signaling, and cellular reprogramming and regeneration. In the sections that follow, the principal intervention strategies currently under investigation are reviewed in detail.

3.1 Gene-centered therapies

Gene-centered interventions aim to directly correct the molecular defects that underlie NDs by restoring or modulating disease-relevant gene function. These strategies leverage advances in viral vector engineering, RNA modulation technologies, and genome- and epigenome-editing platforms to achieve durable, precise therapeutic effects in the central nervous system (CNS) (Gitler et al., 2017).

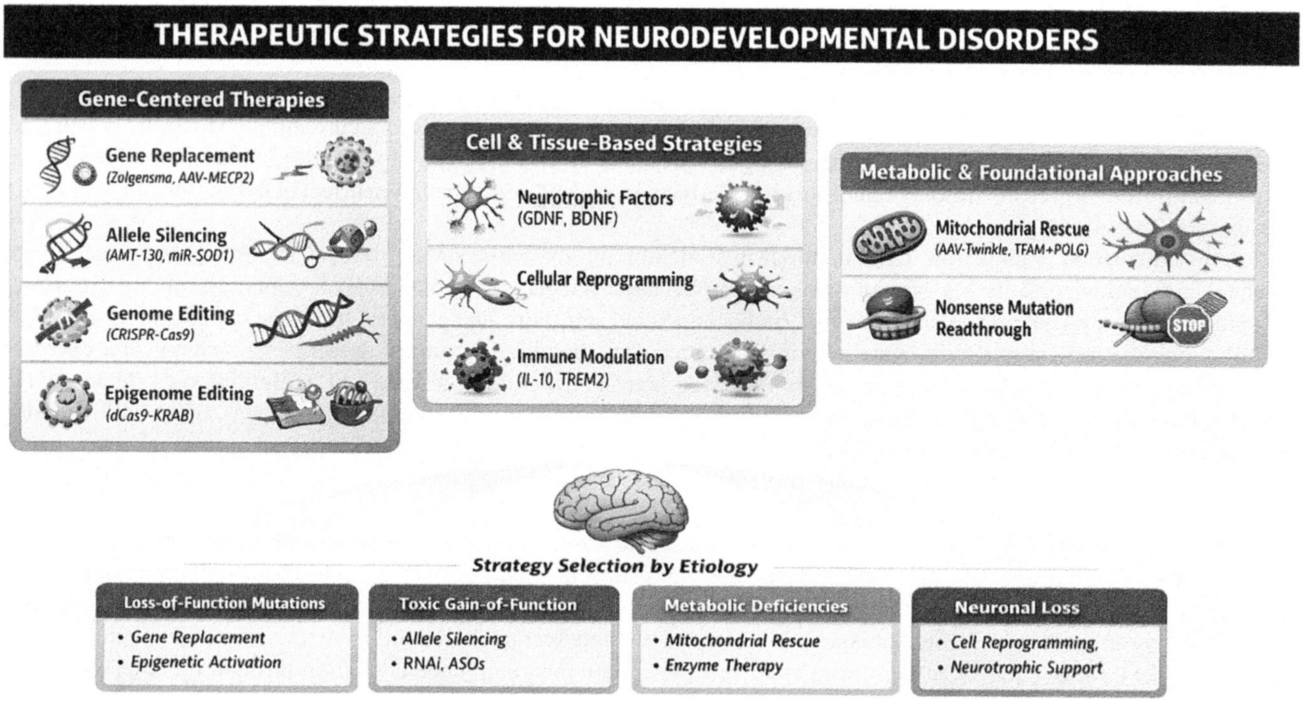

FIG. 2.2 Therapeutic strategies for NDs organized by mechanistic approach and disease etiology. The top panels summarize three major intervention categories: gene-centered therapies; cell- and tissue-based strategies; and metabolic and foundational approaches. The lower panel highlights strategy selection based on underlying etiology: loss-of-function mutations (gene replacement, epigenetic activation), toxic gain-of-function mutations (allele silencing, RNAi, antisense oligonucleotides), metabolic deficiencies (mitochondrial rescue, enzyme therapy), and neuronal loss (cell reprogramming, neurotrophic support).

3.1.1 *Gene augmentation and replacement*

Gene augmentation therapy is founded on a simple yet powerful premise: restoring physiological function by reintroducing intact copies of genes rendered nonfunctional by loss-of-function mutations. Among available delivery systems, AAV vectors, particularly serotypes such as AAV9, have become the platform of choice for targeting postmitotic neurons, enabling sustained episomal expression with an increasingly well-characterized safety profile (Nóbrega et al., 2020). The clinical success of systemic SMN1 gene replacement in SMA with onasemnogene abeparvovec (Zolgensma) represents a watershed moment for the field. The profound improvements in motor function and survival observed in treated infants have not only transformed the natural history of SMA but also demonstrated that effective genetic intervention in CNS disorders can be achieved through peripheral administration (Pascual-Morena et al., 2023; García-González et al., 2024).

This therapeutic framework has since expanded to a broad spectrum of monogenic neurodegenerative and neurometabolic diseases. In CLN2 Batten disease, delivery of TPP1 aims to restore lysosomal enzyme activity and prevent progressive neuronal loss driven by substrate accumulation. In giant axonal neuropathy (GAN), replacement of the GAN gene seeks to correct cytoskeletal disorganization and stabilize axonal integrity (Di Berardino et al., 2024). Likewise, in GM2 gangliosidoses, reconstitution of hexosaminidase subunit alpha (HEXA) and hexosaminidase subunit beta (HEXB) function is intended to recover hexosaminidase activity and mitigate the toxic buildup of gangliosides within neurons. Comparable principles guide therapeutic development in MLD, where restoration of arylsulfatase A (ARSA) activity targets the demyelinating cascade, and in Sanfilippo syndrome type IIIA, where augmentation of N-sulfoglucosamine sulfohydrolase (SGSH) addresses the enzymatic defect underlying heparan sulfate accumulation and progressive neurodegeneration (Amanat et al., 2022).

Beyond these recessive disorders, the paradigm is expanding to address more complex genetic scenarios (Roberts et al., 2024). For Rett syndrome, AAV-mediated *MECP2* replacement rescues phenotypes in preclinical models, though it requires precise expression control due to a narrow therapeutic window (Yang et al., 2025). In dominant disorders like SOD1-linked ALS and HD, hybrid knockdown-and-replace strategies are being explored, using AAV vectors to simultaneously silence mutant alleles and deliver protective transgenes. The approach also extends to diseases with multifactorial pathophysiology: experimental strategies for AD involve AAV-mediated delivery of neurotrophic factors (e.g., nerve growth factor [NGF], brain-derived neurotrophic factor [BDNF]) or protective alleles like *APOE2* (Ortega et al., 2025), while in PD, delivery of multienzyme cassettes (aromatic L-amino acid decarboxylase [AADC], tyrosine hydroxylase [TH], GTP cyclohydrolase 1 [GCH1]) to the striatum restores local dopamine synthesis, and in Dravet syndrome, *SCN1A* augmentation specifically in inhibitory interneurons aims to restore circuit balance (Kuang et al., 2024).

3.1.2 *Gene expression modulation*

Recent progress in gene-modulating therapies has expanded the therapeutic landscape for NDs by advancing both allele-specific suppression and expression enhancement strategies from preclinical validation toward clinical evaluation. These complementary approaches tackle genetic disorders from two angles: either by turning down harmful, overactive genes, or by precisely boosting the activity of genes that aren't producing enough protein.

Dominant-negative or toxic gain-of-function mutations present a compelling rationale for allele-specific suppression strategies that selectively silence pathogenic alleles while preserving wild-type function. Such approaches encompass single nucleotide polymorphism (SNP)-linked RNA interference (RNAi), exon-junction-targeting small interfering RNAs (siRNAs), and repeat-length-biased antisense oligonucleotides (ASOs) (Drongitis et al., 2025). Clinical evidence has emerged in HD, where next-generation ASOs, WVE-003, designed to reduce mHTT while sparing the normal protein, have demonstrated allele-selective lowering of mHTT in cerebrospinal fluid (CSF) with a favorable safety profile in Phase I/II (SELECT-HD) studies (Abigail et al., 2025; Hao and Ji, 2025). AMT-130 (AAV5-miHTT), a gene therapy delivering microRNA targeting *HTT* RNA via an AAV vector, has shown sustained reduction of mHTT and promising preliminary clinical biomarker and functional stabilization signals, highlighting the potential for long-lasting RNAi-based suppression in HD. Beyond ASOs and RNAi, emerging RNA-targeting CRISPR systems such as Cas13d–CAG^EX have also shown allele-selective knockdown of pathogenic *HTT* transcripts in preclinical models with corresponding phenotypic improvements, underscoring the expanding molecular toolkit for allele-specific suppression (Morelli et al., 2023; Bunting et al., 2025).

Complementary to suppression, expression enhancement strategies aim to correct disorders of insufficient gene activity by upregulating endogenous gene expression with high precision (Goyal et al., 2022). Cis-regulatory therapies fine-tune promoters or enhancers to elevate transcription without ectopic overexpression and can be delivered within a single AAV vector even for genes exceeding conventional cargo limits. Clinical-stage examples include ETX101, an AAV9-mediated gene regulation therapy in development for SCN1A + Dravet syndrome that uses engineered transcription factors under cell-selective regulatory elements to increase expression of the endogenous *SCN1A* gene in inhibitory interneurons, addressing the genetic basis of the disorder (de Lagrán et al., 2024; Sogorb-Gonzalez et al., 2024; Sullivan and Hood, 2025). Regulatory progress for ETX101, such as regenerative medicine advanced therapy (RMAT) and investigational new drug (IND) clearances and ongoing Phase I/II trials across multiple regions, reflects the translational momentum of cis-regulatory upregulation strategies in severe neurodevelopmental disorders (Isik et al., 2025).

3.1.3 *Genome- and epigenome-editing approaches*

Genome editing and epigenome editing represent two complementary strategies for modulating gene function in NDs.

Genome editing enables precise correction of disease-causing DNA changes. By leveraging CRISPR–Cas systems, researchers can directly correct pathogenic point mutations or small insertions or deletions in the genome without disrupting native regulatory architecture (Gao et al., 2025). For example, efforts in Rett syndrome models have used CRISPR approaches to correct *MECP2* mutations and restore normal gene function, demonstrating proof-of-principle that defective neuronal gene sequences can be repaired and endogenous protein expression rescued (Drongitis et al., 2025; Kumar et al., 2025). These tools are increasingly refined to reduce off-target effects and improve efficiency in postmitotic neurons, although challenges in delivery and immune responses remain significant hurdles for clinical translation (Zhang et al., 2024; Gao et al., 2025).

Epigenome editing, by contrast, modulates gene expression without altering the underlying DNA sequence. This approach typically uses catalytically inactive CRISPR effectors (e.g., dCas9) fused to epigenetic modifiers, such as Krüppel-associated box (KRAB) repressors, DNA methyltransferases, ten-eleven translocation methylcytosine dioxygenase 1 (TET1) demethylases, or histone acetyltransferases like p300, to reversibly repress or activate target loci (González Molina et al., 2025; Marei, 2026). In neurological contexts, epigenome editing has shown the ability to reactivate *MECP2* expression from the inactive X chromosome in Rett syndrome neurons and to demethylate the silenced *FMR1* promoter in Fragile X syndrome models, restoring gene expression and ameliorating cellular phenotypes (Mira-Bontenbal et al., 2022). These technologies provide a programmable means to adjust transcriptional states in disease models driven by epigenetic dysregulation, offering durable yet reversible control of gene activity without making permanent changes to the genome (Ueda et al., 2023).

3.2 Cell- and tissue-based strategies

Cell- and tissue-level interventions target the CNS microenvironment to enhance neuronal survival, promote circuit repair, and restore functional connectivity. Unlike gene-centered approaches, these strategies focus on supporting cellular resilience, modulating local signaling, and recruiting endogenous regenerative mechanisms to counteract neurodegeneration and neurodevelopmental deficits. Recent innovations leverage viral vector-mediated delivery, transcription factor-based reprogramming, and immunomodulatory cassettes to achieve durable, spatially restricted therapeutic effects (Tsurumaki et al., 2025).

3.2.1 *Enhancing neuronal survival*

Augmenting neurotrophic support is fundamental to sustaining vulnerable neuronal populations. Neurotrophic factors such as BDNF, glial cell line-derived neurotrophic factor (GDNF), NGF, and ciliary neurotrophic factor (CNTF) drive survival pathways, support axonal integrity, enhance synaptic plasticity, and mitigate apoptotic signaling in multiple neurodegenerative contexts (Kells et al., 2004; Numakawa and Kajihara, 2023). Sustained expression of these factors via AAV vectors has demonstrated preclinical efficacy. AAV-GDNF delivery to the putamen improves dopaminergic neuron survival in PD models, AAV-CNTF supports motor neuron viability in amyotrophic lateral sclerosis paradigms, and AAV-BDNF/NGF enhances hippocampal resilience in AD systems (Nicoletti et al., 2022; Stansberry and Pierchala, 2023). Clinical efforts to deliver neurotrophic genes directly to target CNS regions illustrate the translational momentum of this approach. Importantly, engineering advances such as activity-dependent promoters that restrict expression to active neural circuits, fusion constructs that enhance retrograde transport, and combinatorial CRISPRa codelivery to amplify endogenous factor production collectively optimize therapeutic outcomes while limiting off-target effects (Parambi et al., 2022; Szunyogh et al., 2025).

3.2.2 *Cellular reprogramming and neurogenesis*

In situ cellular reprogramming has emerged as a promising strategy to replenish lost neurons by converting endogenous glial cells into functional neurons within the injured or diseased CNS. Resident astrocytes and other glia can be redirected toward neuronal fates through the targeted delivery of fate-specifying transcription factors using viral vectors such as AAVs (Dokukin et al., 2025). For example, the expression of neurogenin-2 has been shown to promote direct astrocyte-to-neuron conversion in vitro and in vivo, driving lineage switching at the single-cell level and generating induced neurons with molecular signatures of their new identity (Clark et al., 2022; Qin et al., 2024).

Combinatorial transcription factor strategies have also proven effective: the classical binary alignment map (BAM) cocktail of Ascl1, Brn2, and Mytl1 can reprogram glial cells into induced neuronal subtypes, supporting the generation of neurons that exhibit key electrophysiological properties (Clark et al., 2022). In addition, miRNA-based approaches, particularly miR-124, can facilitate fate conversion by modulating posttranscriptional networks that favor neurogenic programs, thereby enhancing the efficiency of reprogramming toward immature neuron states (Papadimitriou et al., 2023).

RNA-binding protein modulation, such as polypyrimidine tract-binding protein 1 (PTBP1) knockdown has similarly been explored as a means to derepress neuron-specific gene networks and promote glia-to-neuron conversion in situ (Contardo et al., 2022). In ischemic stroke models, AAV-mediated PTBP1 suppression drives astrocyte conversion into neurons, improves neural tissue integrity, and reduces local inflammatory responses (Yuan et al., 2024). However, the efficacy of PTBP1 targeting in certain neurodegenerative contexts remains under active investigation and is subject to methodological debate (Nazim, 2024; Papadimitriou and Thomaidou, 2024).

Applications of these reprogramming approaches extend across multiple disease and injury paradigms. In PD models, glia-derived induced neurons can adopt a dopaminergic phenotype and integrate with host circuitry to restore dopaminergic signaling (Wei and Shetty, 2021; Yuan et al., 2024). In stroke, newly generated neurons, including GABAergic interneurons, can help normalize inhibitory tone and stabilize neural networks, promoting functional recovery (Chen et al., 2022b). Gene therapy vectors encoding proneural factors have also been used to convert astrocytes into region-appropriate neuronal subtypes in models of AD and HD (Yuan et al., 2024).

3.2.3 Immune modulation and neuroinflammation control

Chronic neuroinflammation, driven in part by sustained proinflammatory microglial activation, accelerates protein aggregation, synaptic loss, and neuronal death across many CNS disorders (Hours et al., 2025). Modulating this immune milieu toward a neuroprotective *state* is therefore a compelling therapeutic objective (Noh et al., 2025). Viral vector-mediated expression of immunomodulatory cassettes can shift microglial polarization, enhance clearance of pathological protein aggregates, and preserve synaptic integrity (Hours et al., 2025). Examples include AAV9 vectors expressing combinations of antiinflammatory cytokines (e.g., interleukin-10 (IL-10)) and receptors (e.g., triggering receptor expressed on myeloid cells 2 (TREM2)) in AD models, coexpression of ectonucleotidases such as CD39/CD73 to regulate purinergic signaling in PD, and AAV-IL-4/transforming growth factor beta 1 (TGF-β1) constructs that extend survival in ALS models by dampening deleterious inflammation (Bido et al., 2024; Zhou et al., 2024a). In addition to cytokine modulation, strategies that restore enzymatic homeostasis, such as AAV-aspartoacylase (ASPA) in Canavan disease, demonstrate the breadth of immune-centric and supportive interventions that preserve neuronal function (Leone et al., 2025). Parallel research on microglial biomarkers and targets such as TREM2 and progranulin underscores the growing appreciation of innate immune regulation as a driver of both pathology and repair in neurodegenerative conditions (Reifschneider et al., 2022).

3.3 Metabolic and foundational rescue strategies

Metabolic and foundational rescue strategies aim to restore essential cellular functions that are compromised in NMD, addressing systemic deficits in energy production, protein synthesis, and organelle integrity. By targeting fundamental cellular machinery, these approaches complement gene- and cell-based therapies, providing a foundational support system that enhances neuronal resilience and network stability (Thapa et al., 2025).

Mitochondrial dysfunction underlies a range of NDs, including mitochondrial encephalomyopathy, lactic acidosis, and stroke-like episodes (MELAS), myoclonic epilepsy with ragged red fibers (MERRF), Leigh syndrome, and requires organelle-specific interventions to restore energy homeostasis (Gammage et al., 2018). AAV-mediated strategies have been developed to target mitochondrial genome maintenance and dynamics (Thapa et al., 2025). For instance, AAV-mitochondrial targeting sequence (MTS)-Twinkle restores mtDNA helicase activity, AAVmiDys-tRNA-mito corrects pathogenic heteroplasmy, and multigene cassettes such as AAV9-[mitochondrial transcription factor A (TFAM) + DNA polymerase gamma (POLG) + single-stranded DNA-binding protein 1 (SSBP1) + optic atrophy 1 (OPA1)] simultaneously promote mitochondrial biogenesis, replication, and fusion. These interventions directly ameliorate oxidative phosphorylation deficits and complement nuclear gene replacement therapies by stabilizing energy metabolism, reducing ROS, and enhancing neuronal survival (Rehman et al., 2023; Barrera-Paez et al., 2025).

Moreover, premature stop codons caused by nonsense mutations result in truncated, nonfunctional proteins and contribute to a spectrum of NDs. Translational read-through strategies enable the bypass of these stop codons, restoring full-length protein expression (Howe and Patani, 2023; Tompa et al., 2025). Approaches include small-molecule read-through compounds that promote ribosomal incorporation of near-cognate tRNAs at premature termination sites and suppressor tRNAs engineered to selectively recognize nonsense codons (Wang et al., 2022; Peng et al., 2024b). These methods offer a broadly applicable platform for disorders in which single-nucleotide nonsense mutations disrupt critical proteins and can be combined with gene replacement or epigenetic therapies to maximize functional restoration (Devinsky et al., 2025).

4 Molecular gene manipulation approaches for NDs

Molecular gene manipulation strategies for NDs are engineered to precisely interrogate and modulate the molecular pathways driving disease pathogenesis, enabling targeted correction, suppression, or fine-tuning of gene expression across multiple regulatory layers. These approaches represent a cornerstone for the development of rational, disease-modifying therapies. The following section provides a systematic and integrative overview of key DNA- and RNA-based modalities, alongside emerging gene manipulation platforms, highlighting their potential to reshape and ameliorate the trajectory of NDs.

4.1 Molecular gene manipulation at the DNA level

4.1.1 Programmable DSB-based nucleases

Meganucleases

Meganucleases (MNs), also termed homing endonucleases, comprise an early generation of compact programmable nucleases, first identified in yeast mitochondria in 1985 (Flotte and Gessler, 2022). These enzymes recognize extended DNA sequences spanning 12–40 base pairs, conferring exceptional specificity and profoundly low off-target cleavage through tight β-strand interactions with DNA backbones and short 2–4 bp motifs. Predominantly from the LAGLIDADG family, such as I-CreI derived from the chloroplast genome of *Chlamydomonas reinhardtii* and I-SceI from the mitochondria of *Saccharomyces cerevisiae*, MNs are naturally encoded within introns or inteins (Dürrenberger and Rochaix, 1991; Kadyrov et al., 2007; Arnould et al., 2011). They facilitate the homing of genetic elements into homologous sites lacking these sequences, generating DSBs that trigger cellular repair pathways (Chevalier et al., 2005).

The compact molecular architecture and minimal cytotoxicity of MNs render them particularly amenable to AAV9 delivery, including intravenous routes, making them well-suited for postmitotic neurons in NDs. Precision is further augmented by the coadministration of DNA-protein kinase C substrate (PKcs) inhibitors, which bias repair toward high-fidelity homology-directed repair (HDR) or microhomology-mediated end joining over error-prone non-homologous end joining (NHEJ) (Silva et al., 2011; Datta et al., 2023).

Notwithstanding these strengths, MNs face challenges, including laborious redesign for novel targets, comparatively lower editing efficiency relative to ZFNs or CRISPR/Cas9, potential genotoxicity, and the requirement for preinsertion of cleavage sites (Datta et al., 2023). Off-target effects remain exceedingly rare owing to extended recognition lengths; mitigation strategies encompass fusion with transcription activator-like effector (TALE) domains, I-TevI hybrids, or transcription activator-like (TAL) arrays. Monomeric variants like I-DmoI further broaden applicability (Dupuy et al., 2013). Ultimately, MNs furnish a neuron-sparing, allele-selective platform poised for durable interventions in polygenic NDs, serving as a foundational antecedent to contemporary genome-editing paradigms (Flotte and Gessler, 2022).

Zinc-finger nucleases

ZFNs are the first generation of programmable nucleases for genome editing, enabling precise manipulation of specific DNA sequences. Structurally, ZFNs are chimeric proteins composed of a zinc finger DNA-binding domain and the nonspecific FokI nuclease catalytic domain from *Flavobacterium okeanokoites* (Carroll et al., 2006; Petolino, 2015). Each zinc finger typically recognizes a three-base-pair motif, and tandem arrays of three to six fingers expand recognition to 9–18 base pairs, enhancing specificity. FokI functions exclusively as a dimer, requiring two ZFNs to bind opposing DNA strands in a convergent orientation to induce a DSB, inherently limiting off-target cleavage and improving therapeutic safety (Ramalingam et al., 2011; Sizova et al., 2013).

The zinc finger motif is highly conserved, originally identified in transcription factor IIIA of *Xenopus laevis* (Carroll et al., 2006). A single finger coordinates a zinc ion via two cysteine and two histidine residues, enabling DNA triplet recognition. Arrays of zinc fingers allow targeting of longer, unique sequences, forming a robust framework for sequence-specific genome editing (Gersbach et al., 2014). Designing functional ZFNs is challenging due to context-dependent variability in zinc finger-DNA interactions. Approaches to address this include module assembly, which links precharacterized modules into target-specific arrays; oligomerized pool engineering, combining combinatorial libraries with iterative selection; bacterial one-hybrid systems, leveraging randomized DNA–protein interfaces; and context-dependent assembly, a rapid, publicly accessible platform enabling multifinger array construction within 1–2 weeks (Khan et al., 2024).

Based on recent studies, ZFNs demonstrate significant therapeutic potential in NDDs by targeting synuclein alpha (SNCA) mutations in iPSC-derived neurons and AAV-mediated suppression of mHTT expression in PDs and HDs, respectively. In Alzheimer's disease, ZFN-mediated editing of APP and PSEN genes models amyloid pathology and evaluates interventions (Raslan and Kee, 2013). In addition, ZFNs have been combined with microRNA modulation, such as targeting miR-34a in AD models, to reduce neuronal apoptosis and oxidative stress, illustrating their integration with endogenous gene-regulatory networks (Hussen et al., 2023).

Transcription activator-like effector nucleases

TALENs have emerged as highly versatile and precise genome-editing tools, offering unique advantages for investigating the molecular basis of neurodegenerative disorders. Their modular architecture enables specific recognition of target DNA sequences while minimizing off-target activity, making them particularly suitable for applications in sensitive neuronal tissues (Becker and Boch, 2021; Zaman et al., 2025).

A transformative application of TALENs in neurodegeneration is the creation of isogenic cell models harboring disease-relevant mutations. TALEN-mediated editing has been employed to introduce or correct mutations in *SNCA*, the gene encoding α-synuclein. These engineered iPSC-derived neural models provide a controlled system to investigate α-synuclein aggregation, propagation, and toxicity (McTague et al., 2021; Chen et al., 2022a). By allowing direct comparisons between genetically matched mutant and wild-type lines, TALENs have significantly advanced the understanding of early pathogenic events in PD (Arango et al., 2021). In HD, TALEN-directed correction of expanded CAG repeat tracts within the *HTT* gene in patient-derived iPSCs has demonstrated pronounced functional benefits, including enhanced neuronal viability, improved synaptic function, and partial restoration of cellular homeostasis (Hao and Ji, 2025). This strategy has also been extensively applied in AD, where they enable precise manipulation of genes central to amyloidogenic processing, including *APP* and *PSEN1*, allowing detailed analyses of amyloid precursor protein metabolism, γ-secretase activity, and the molecular determinants of Aβ production (Qin et al., 2022; Chen and Vogt, 2024). Beyond protein-coding genes, TALENs allow interrogation of regulatory networks by targeting microRNAs, such as miR-7 and miR-153, which regulate α-synuclein in PD neurons (Meza-Sosa et al., 2022; Lei et al., 2024), and miR-34a in AD models, enhancing cell survival and reducing oxidative stress (Teresa de Barros Viegas et al., 2017).

In addition to cellular systems, TALENs expand the toolkit for generating and refining animal models of NDDs. Precise genome modification allows the introduction or correction of pathogenic alleles in rodents and other models, reducing mutant protein expression and improving behavioral phenotypes (Nóbrega and Alves, 2018). TALENs also support transgenic line creation for mechanistic studies and high-throughput therapeutic screening, providing a robust platform for preclinical evaluation of gene-based interventions (Zhu et al., 2021).

CRISPR system

CRISPR and Cas proteins form a bacterial and archaeal adaptive immune system (Li et al., 2023b), repurposed for precise genome editing in eukaryotic cells. CRISPR arrays consist of direct repeats interspersed with spacer sequences derived from previous viral exposures, functioning as a molecular memory. Target recognition depends on the protospacer, a DNA segment complementary to the spacer, and the PAM, which ensures discrimination between self and foreign DNA (Sami et al., 2021; Klinkovskij et al., 2023).

CRISPR/Cas systems are phylogenetically diverse, categorized into two major classes, six types, and numerous subtypes, reflecting adaptation to a wide range of viral pressures. Class 1 systems (types I, III, IV) utilize multiprotein effector complexes, Cascade (type I) or Csm/Cmr complexes (type III), whereas Class 2 systems (types II, V, VI) rely on single, multidomain effector proteins, including Cas9 and Cas12. Types I–III are the most well-characterized and retain the conserved adaptation module consisting of Cas1 and Cas2 (Shivram et al., 2021). Type II systems uniquely employ tracrRNA in conjunction with crRNA to guide Cas9-mediated interference, and their single-effector architecture has rendered them particularly amenable to genome engineering (Kim et al., 2024; Nguyen et al., 2025). The discovery of type V Cas12 nucleases and compact type V–F Cas12f (Cas14) proteins has expanded the CRISPR toolkit, offering alternative PAM requirements, staggered DNA cleavage, and compatibility with AAV delivery due to their small size (Shareena et al., 2025a, Shareena et al. (2025b); Ali et al., 2026).

In NDD research, CRISPR enables gene knockouts, knockins, base editing, and prime editing to study or correct pathogenic mutations. Cas9-induced DSBs activate endogenous repair pathways, including NHEJ and HDR (Allemailem et al., 2024; Song and Bae, 2024). High-fidelity Cas9 variants, paired nickases, and PAM-flexible enzymes minimize off-target effects, expanding the range of editable loci. CRISPR's versatility and precision make it a transformative tool for investigating disease mechanisms, modeling neurodegenerative disorders, and developing potential gene-based therapies (Nureki et al., 2024).

4.1.2 DSB-free precision editors

Base editors

BEs enable precise single-nucleotide conversions without inducing DSBs, coupling catalytically impaired or nickase Cas proteins (e.g., nCas9 or dCas9) with DNA deaminase enzymes. Cytosine base editors (CBEs) use cytidine deaminases to convert C–G to T–A (via C→U deamination), while adenine base editors (ABEs) employ adenine deaminases for A–T to G–C, acting within a defined 4–8 nt editing window near a PAM site (Kantor et al., 2020; Bae and Song, 2023). The nickase creates a single-strand mismatch resolved by replication or excision repair, bypassing HDR in nondividing postmitotic neurons. In PD, ABEs revert LRRK2 G2019S (G→A transition), reducing toxic kinase activity and α-synuclein pathology in iPSC-derived dopaminergic neurons. BEs prioritize efficiency for common transitions (e.g., AD PSEN1 C-to-T, ALS SOD1 G-to-A). Numerous studies have demonstrated robust base editing in mammalian neuronal and sensory tissues, underpinning their potential for somatic gene correction in the nervous system (Lim et al., 2022). However, limitations include potential bystander editing within the activity window and restriction to transition mutations, thereby narrowing the spectrum of correctable pathogenic variants (Miskalis et al., 2024).

Prime editors

PEs offer versatile, high-precision search-and-replace genome rewriting without DSBs or HDR dependency. They fuse nCas9(H840A) to M-MLV reverse transcriptase (RT), guided by pegRNA containing a spacer (targeting sequence), reverse-transcriptase template, and nick guide RNA (nG) for a second-strand nick (Lee et al., 2023; Li et al., 2023a). The process nicks the PAM-distal strand, exposes a 3′ flap via displacement, copies the reverse transcriptase template (RTT) via RT into the flap, enabling all 12 substitutions, 1–80 nt indels/insertions, and resolves via excision repair at high efficiency. In familial AD models, optimized PE3/ePE variants installed protective APP A673T (T→C via RTT), reducing the Aβ42/40 ratio by inhibiting β-secretase cleavage in iPSC neurons (Rottner et al., 2025). For HD, PEs contract CAG repeats (e.g., 5–10 nt deletion in HTT exon 1), lowering mutant protein in striatal organoids (Hwang et al., 2024). PEs generally exhibit lower yields compared to BEs and face challenges such as pegRNA instability or degradation and a relatively large molecular size (~6.5 kb), which can complicate in vivo delivery for therapeutic applications (Böck et al., 2023). Nonetheless, their cell-cycle independence makes them particularly promising for editing postmitotic cells (BenDavid et al., 2024).

CRISPR-associated transposons

CASTs represent a next-generation genome engineering platform capable of inserting large DNA payloads (10–30 kb) into target genomes via RNA-guided mechanisms, without inducing DSBs. By bypassing the need for HDR (Rhiel et al., 2023), CASTs overcome a major limitation of conventional CRISPR–Cas nucleases, reducing indels, p53 activation (Peters et al., 2017; Song and Bae, 2024), and off-target effects while generating precise insertions flanked by transposon left and right ends (LE/RE) (Gao et al., 2025).

Initially discovered in bacterial metagenomes, CAST systems combine inactivated CRISPR effectors, such as Cas12k in type V–K or Cascade complexes in type I–F, with Tn7-like transposases (TnsA/B/C, TniQ) to achieve targeted integration approximately 60–66 base pairs downstream of a PAM site (Omura et al., 2015). Type I CASTs employ multisubunit Cascade complexes for target recognition, with TniQ bridging to the transposase machinery for donor excision and genome insertion. Type V–K CASTs simplify the process using a single Cas12k effector, with RNA-independent A/T-rich site selection and ATPase optimization, achieving high target specificity (Koonin and Makarova, 2024). Importantly, CAST-mediated integration occurs independently of the cell cycle, making it ideal for nondividing neurons in NDD models (Klompe et al., 2019).

CASTs are uniquely positioned for multiplex CNS applications, enabling the simultaneous delivery of complete protective allele cassettes that surpass the packaging constraints of conventional AAV vectors (~4.8 kb). For example, a single CAST vector can carry APP A673T, APOE3-Christchurch, and Klotho enhancer sequences, enabling coordinated protection against amyloidogenesis, tau propagation, and synaptic vulnerability. In human cells, evoCAST variants demonstrate high efficiency for gene-sized inserts and maintain low immunogenicity when delivered across the blood–brain barrier via AAVs. Unlike prime editing, CASTs can perform polygenic corrections without relying on HDR, making them particularly suitable for sporadic AD, PD, or mitochondrial-genome-associated neurodegenerative conditions (Hong et al., 2025; Witte et al., 2025).

4.2 Molecular gene manipulation at the RNA level

RNA-targeted interventions have emerged as a central therapeutic approach for NDDs because they modulate pathogenic transcripts directly, avoiding permanent genomic alterations such as DSBs or structural variants. This enables allele-selective and reversible regulation of gene expression, offering a precision advantage over DNA-level editing.

4.2.1 Adenosine deaminases acting on RNA-mediated RNA editing

Adenosine deaminases acting on RNA (ADARs) catalyze the site-specific conversion of adenosine to inosine (A-to-I) within double-stranded RNA (dsRNA), representing a posttranscriptional mechanism (Lu et al., 2024). Within the ADAR family, ADAR1 predominantly targets repetitive and noncoding regions, whereas ADAR2 preferentially edits coding sequences, including the GluA2 subunit of AMPA receptors (Jimeno et al., 2021).

Therapeutic adRNAs recruit endogenous ADARs to pathogenic transcripts via an antisense arm forming substrate-like dsRNA, augmented by recruitment domains from hyper-edited sites (GluR2 R/G or Alu elements) (Reautschnig et al., 2022). Bulged mismatches direct target A to the active site; 2'-O-methyl stabilization, circularization, or pseudouridine enhances neuronal persistence (Levanon et al., 2004; Tan et al., 2017). Delivery employs AAV for durable expression, lipid nanoparticles (LNPs) for transience, or stabilized oligos, correcting IDUA/Mecp2 stop codons with functional rescues in neurological models. Transcriptome RNA-seq refines designs by profiling off-targets (Sinnamon et al., 2017; Mendoza and Beal, 2023; Li et al., 2025).

Dysregulated ADAR-mediated RNA editing drives neurodegeneration across ALS, AD, and PD. In ALS, reduced ADAR2 activity at the GluA2 Q/R site yields Ca^{2+}-permeable AMPA receptors, fueling motor neuron excitotoxicity (Rainaldi et al., 2025). AD brains exhibit diminished ADAR2 and elevated ADAR3, disrupting the editing of SYT11 and ORAI2, while PD astrocytes show upregulated ADAR, modulating α-synuclein-triggered immune responses (Karagianni et al., 2024; Li et al., 2025). Editing QTLs (edQTLs) in ADAR, TOMM40, and FGD5-AS1 further influences disease via impaired neurotransmission, inflammation, and PI3K/Akt signaling (Mise et al., 2017).

Relative to CRISPR BEs, ADAR confers reversibility via transcript turnover, DSB-free neuronal safety, and endogenous enzyme recruitment for low immunogenicity (Li et al., 2022; Dredge et al., 2025). Emerging dCas9-ADAR hybrids enable programmable non-dsRNA targeting, while SPRING platforms with split ADAR domains deliver > 90% neuronal efficiency, curbing bystanders for precise ALS GluA2, PD SNCA, and AD APP corrections, heralding clinical translation in polygenic NDDs (Cuddleston et al., 2022; Rodriguez de Los Santos et al., 2024).

4.2.2 RNA-targeting Cas13 systems

Compact Class 2 CRISPR effectors, such as RfxCas13d (CasRx), enable programmable cleavage of single-stranded RNA (ssRNA) for transient degradation of toxic transcripts in NDs (Cox et al., 2017). This approach avoids genomic alterations, offering low neurotoxicity and reversibility, ideal for late-onset pathologies. crRNA-guided binding induces conformational activation of dual HEPN nuclease domains, cleaving targets at ~ 30-nucleotide intervals; high-fidelity hfCas13d variants minimize collateral bystander effects, preserving cellular homeostasis (Chen et al., 2025). Preclinical validation includes CasRx-mediated suppression of mutant HTT in HD mouse and pig models, slowing disease progression, alongside CRISPR-TO technology that localizes neuroprotective RNAs to ALS neurite tips, promoting regeneration (Kordyś et al., 2022; Wang et al., 2025). Engineered low-collateral Cas13d variants, coupled with RNA-seq-validated crRNAs, split/dimeric architectures, transient ribonucleoprotein (RNP) or LNP delivery, miRNA-responsive self-limiting AAVs, and Tet-On inducible promoters, ensure precise temporal control and reduced off-target activity (Granados-Riveron and Aquino-Jarquin, 2021).

4.2.3 Antisense oligonucleotides

ASOs are short, single-stranded, chemically modified nucleic acids designed to bind to complementary sequences in target pre-mRNA or mRNA. They act primarily through two mechanisms: RNase H-mediated cleavage, which degrades the RNA transcript, or steric blocking, which modulates RNA function, most commonly by altering splicing without degradation.

This versatility allows for either the reduction or increase of target protein levels, making ASOs a powerful tool for neurogenetic disorders (Bennett and Swayze, 2010; Adachi et al., 2021).

Contemporary ASOs incorporate advanced chemical modifications, such as phosphorothioate backbones, 2′-methoxyethyl (2′-MOE), locked nucleic acid (LNA), and 2′-fluoro (2′-F) substitutions, which enhance nuclease resistance, prolong tissue half-life, and fine-tune plasma protein binding (Glazier et al., 2020; Chakkyarath, 2025). These second- and third-generation chemistries are deliberately designed to mitigate class-specific adverse effects, including complement activation, thrombocytopenia, and renal accumulation. Additional safety is engineered at the sequence level by avoiding motifs that potently activate innate immune receptors (e.g., TLR7/8/9) or form G-quadruplex structures associated with off-target immune stimulation (Shi et al., 2024).

Because ASOs do not efficiently cross the blood-brain barrier (BBB), direct intrathecal (IT) injection into the CSF has become the standard route of administration for CNS indications. This approach localizes drug exposure to the CNS parenchyma, promotes broad distribution throughout the brain and spinal cord via diffusion, and limits systemic exposure, thereby reducing peripheral toxicity. Following CSF administration, ASOs are taken up widely by both neuronal and glial cells (Naeem et al., 2025). Their extended half-life in neural tissue supports prolonged dosing intervals. Patient safety is monitored through biomarkers such as CSF cell counts, neurofilament light chain, liver function tests, and platelet counts (Jorge et al., 2025).

Beyond conventional gene silencing, advanced allele-selective antisense oligonucleotide (ASO) designs are being developed for dominant disorders. These designs leverage genetic distinctions, such as SNPs, mutation-specific exon–intron junctions, or repeat-length biases, to selectively suppress mutant transcripts while preserving wild-type allele expression (Nguyen et al., 2024). This strategy achieves molecular precision without requiring cell-type-specific delivery vectors. Furthermore, deploying ASOs early in the disease course, including in presymptomatic carriers of familial mutations (as in planned trials for ALS and HD), may improve efficacy by intervening in less degenerated neural circuits, potentially allowing for lower doses and reduced local toxicity. While ASOs remain the pioneering RNA-targeting modality for the CNS, novel approaches, including chemically conjugated siRNAs, are now advancing, having surmounted earlier delivery challenges (Holm et al., 2022).

The clinical feasibility of IT ASO therapy is well-established, supported by over 15,000 patients treated, some for more than a decade (Younis et al., 2024). This platform has produced notable approved therapies, including nusinersen (Spinraza), approved in 2016 for SMA, which works via steric blocking to correct *SMN2* splicing and increase functional SMN protein, and tofersen (Qalsody), approved for *SOD1*-associated ALS, which acts through RNase H-mediated cleavage to degrade mutant *SOD1* mRNA and reduce toxic protein levels (Lovett et al., 2025). A robust pipeline of ASOs is now in clinical development for a range of other NDs.

4.2.4 RNAi constructs (shRNA/siRNA/miRNA)

RNAi constructs, delivered via AAV or lentivirus, achieve durable mRNA suppression in postmitotic CNS neurons but faced early challenges from shRNA overexpression saturating exportin-5/Dicer, causing microglial activation, neurodegeneration, and hepatotoxicity (Germain et al., 2023). Modern designs mitigate these issues through artificial miRNA scaffolds (e.g., pri-miR-30) mimicking endogenous processing to spare physiological miRNAs, moderate Pol II promoters (cell-specific over U6/H1), and precise vector dose titration below toxicity thresholds validated in rodents and primates (Ely et al., 2009).

Off-target silencing is curbed by bioinformatic seed filters avoiding 3′ UTR matches, mismatch-tolerant alleles (e.g., SNP-linked HTT), and validation via neuronal transcriptome/proteomics. Neuronal targeting combines tropism and promoters. Stereotactic AAV2/5 to the striatum with hSYN1/DARPP-32 confines miR-HTT to HD medium spiny neurons; AAV9/rh10 with ChAT/Hb9 targets ALS spinal motor neurons; AAV2-hSYN1-miR-SNCA focuses on PD nigrostriatal dopamine cells (Thakur et al., 2026).

Safety layers include miR-122/142-3p elements detargeting liver/immune cells, self-complementary AAVs reducing dose needs, CpG optimization minimizing TLR9, and corticosteroid preconditioning in large animals. These refinements deliver long-term, circuit-selective knockdown of HTT/SOD1/SNCA with favorable profiles for chronic NDs (He et al., 2025; Subasinghe et al., 2025).

4.2.5 Nonsense mutations and therapeutic tRNA strategies

Premature termination codons (PTCs) generated by nonsense mutations are a frequent cause of human disease, including neurodegeneration, as they lead to the truncation of essential proteins such as MeCP2, SOD1, or huntingtin (Lueck et al., 2019). These PTCs lack corresponding tRNAs for decoding, unlike natural stop codons (UGA, UAA, UAG), which are recognized by the release factor eRF1 to terminate translation. This results in nonfunctional, shortened proteins and underlying pathology (Porter et al., 2021; Xu et al., 2026). Engineered suppressor tRNAs (sup-tRNAs) represent a promising therapeutic strategy to correct these PTCs by decoding them, inserting the correct amino acid, and enabling the translation of a full-length, functional protein (Konermann et al., 2018).

The therapeutic appeal of tRNA-based approaches lies in several key features. Sup-tRNAs are exceptionally compact (~78 nucleotides), enabling their efficient packaging into AAV vectors for CNS delivery and facilitating coexpression with other small RNA therapeutics. Because they operate posttranscriptionally at the level of translation, sup-tRNAs avoid the risks associated with gene overexpression and systemic genomic alterations (Porter et al., 2021). This makes them particularly suitable for Goldilocks genes like *MECP2*, where expression must be maintained within a narrow physiological range (Coller and Ignatova, 2024) (Table 2.1).

TABLE 2.1 Mechanistic differences, therapeutic scope, and translational potential of DNA versus RNA interventions in NDDs therapy.

Strategy	Mechanism	Target cell type	Advantages	Limitations	Representative NDD examples
Molecular gene manipulation at the DNA level					
Meganucleases (MNs)	DSB-mediated gene insertion/correction	Postmitotic neurons	High specificity; low off-targets; compact for AAV	Labor-intensive redesign; preexisting site required; lower efficiency	Allele-specific correction of *HTT* in HD models
Zinc FINGER NUCLEASES (ZFNs)	DSB-mediated genome cleavage	Neurons, iPSC-derived neurons	High precision; dimerization limits off-targets	Complex design; context-dependent activity	*SNCA* editing in PD neurons; *HTT* suppression in HD
TALENs	Modular repeat-based DSB editing	Neurons, animal models	High specificity; flexible for regulatory targets	Large size; delivery challenges	Correction of CAG repeats in *HTT* (HD); *APP/PSEN1* editing in AD; *SNCA* mutations in PD
CRISPR/Cas9	sgRNA-guided DSB editing	Postmitotic and dividing cells	Versatile, multiplexable	Off-target risks; DSB-induced genotoxicity	*HTT* knockout in HD; *APP* or *PSEN1* editing in AD models
Base editors	DSB-free nucleotide conversion	Postmitotic neurons	Single-base precision; safe in nondividing cells	Limited to transition mutations; bystander edits	*LRRK2* G2019S correction in PD; *PSEN1* C→T editing in AD; *SOD1* G→A editing in ALS
Prime editors	DSB-free search-and-replace	Postmitotic neurons	All base substitutions and small indels; no HDR	Large size; pegRNA instability	*APP* A673T protective variant in AD; CAG contraction in *HTT* for HD
CAST	RNA-guided large DNA insertion	Nondividing neurons	Multi-kb insert; polygenic editing; cell cycle-independent	Novel technology; in vivo validation ongoing	Polygenic AD protection: *APP A673T*, *APOE3 Christchurch*, Klotho enhancer; PD protective alleles
Molecular gene manipulation at the RNA level					
ADAR-mediated editing	Site-specific A→I editing in dsRNA	Neurons, glia	Reversible; allele-selective; DSB-free	Editing window constraints; off-targets possible	*GluA2* Q/R site in ALS; *SNCA* correction in PD; *APP* in AD models
Cas13	Programmable ssRNA cleavage	Neurons	Reversible; low neurotoxicity; multiplexable	Collateral activity; transient effect	*HTT* knockdown in HD; *SOD1* knockdown in ALS; *SNCA* suppression in PD
Antisense oligonucleotides (ASOs)	RNase H degradation or steric block	Neurons, glia	Clinically validated; allele-selective	Requires repeated dosing; CNS delivery	*SMN2* splicing in SMA (Spinraza); *SOD1* in ALS; *C9orf72* repeat expansion in ALS
RNAi (shRNA/siRNA/miRNA)	RNAi-mediated transcript suppression	Postmitotic neurons	Durable; circuit-selective; flexible design	Early toxicity from overexpression; off-target silencing	*HTT* knockdown in HD models; *SOD1* suppression in ALS; *SNCA* in PD
tRNA-based readthrough	Suppression of premature stop codons	Neurons	Restores full-length protein; compact; DSB-free	Limited to nonsense mutations; partial efficiency	MECP2 nonsense mutations in Rett; SOD1 PTC correction in ALS; HTT PTC models

5 Gene delivery modalities

5.1 Viral vectors as gene therapy platforms in NDs

Viral vectors constitute the cornerstone of gene manipulation strategies for NDs, serving as engineered delivery vehicles that exploit natural viral entry mechanisms to introduce therapeutic transgenes, RNAi constructs, or genome-editing payloads into targeted CNS cell populations. These platforms are rendered replication-incompetent through targeted deletions of essential viral genes while preserving transduction efficiency, nuclear import, and, in select cases, long-term episomal or integrated expression in postmitotic neurons and glia (Iqubal et al., 2020). Vector selection is dictated by disease-specific pathology, encompassing anatomical distribution, cellular tropism, payload capacity, immunogenicity, and compatibility with clinically feasible administration routes, yielding a modular toolkit that complements molecular approaches such as CRISPR-based editing, RNAi, and epigenome modulation (Liu et al., 2024).

5.1.1 AAV vector platforms

AAVs represent the vanguard of CNS gene delivery platforms for NDs, with 13 natural serotypes and over 100 engineered variants enabling precise matching of vector tropism to disease-specific neuroanatomy and pathology. Serotype selection hinges on BBB penetrance, regional/cell-type tropism, delivery route feasibility, and safety, distinguishing AAVs from higher-capacity but more immunogenic alternatives like HSV-1 or lentiviruses. While all serotypes share episomal persistence and low pathogenicity, their utility in NDs varies markedly by circuit involvement and therapeutic modality (Mijanović et al., 2020).

AAV1 and AAV2 exemplify focal neuronal transduction paradigms suited to corticobasal ganglia disorders. AAV1's high-efficiency uptake in the sensorimotor cortex, corticospinal motor neurons, cerebellum, and brainstem (thousands of neurons per injection, moderate astrocytes) aligns with PD and AD motor circuit degeneration, outperforming AAV8's glial bias in neuronal-heavy pathologies. AAV2, with superlative local striatal/putamen transduction but negligible spread or BBB crossing, dominates stereotactic PD trials (glutamic acid decarboxylase (GAD), neurturin, AADC delivery), establishing unmatched safety for discrete nuclear targeting, superior to AAV6's retrograde spread for nonfocal applications (Géraudie et al., 2023).

Glia-neuron balanced serotypes, AAV5 and AAV6, address mixed pathologies. AAV5's oligodendrocyte/astrocyte preference alongside neuronal efficiency powers uniQure's AMT-130 (miRNA-mediated HTT knockdown in HD), where Phase I/II data signal progression slowing, a feat unattainable with neuron-exclusive AAV2. AAV6 extends this via robust striatal-to-SNpc retrograde transport for PD dopaminergic rescue, plus hippocampal transduction and microglial access, enabling dual neuronal/neuroinflammatory modulation absent in AAV1/2 (Martier and Konstantinova, 2020).

Astrocyte-centric AAV8 suits hippocampal neuroinflammation in AD/ALS but lags in corticospinal/retrograde potency versus AAV1/6, limiting broad-motor utility. AAV9's neonatal BBB traversal (Zolgensma for SMA motor neurons) shifts to IT adult delivery for ALS/AD widespread coverage, eclipsing (Chamakioti et al., 2022) spinal bias despite the latter's NHP-validated SOD1 shRNA efficacy in lumbar circuits (Lejman et al., 2023). AAV11's projection-specific retrograde labeling excels for circuit mapping (PD/HD/frontotemporal dementia (FTD)) but remains preclinical. AAV12's intranasal olfactory-hippocampal access (sialic acid-independent, low neutralizing antibodies (NAbs)) promises presymptomatic AD/PD intervention, contrasting AAV9's systemic hepatotoxicity (Fajardo-Serrano et al., 2021).

Across all serotypes, AAV vectors share an excellent safety profile, characterized by low inherent pathogenicity, minimal inflammation, and long-term stable transgene expression, primarily due to their episomal persistence in nondividing neurons. Nonetheless, they are limited by relatively small cargo capacity (~4.7 kb for most serotypes), the prevalence of preexisting neutralizing antibodies (nAbs) in humans (particularly for AAV1, AAV8, and AAV9), and dose-dependent hepatotoxicity following systemic administration (Wang et al., 2024b).

5.1.2 Engineered AAV variants

AAVrh.8

AAVrh.8, derived from a rhesus monkey isolate, exhibits limited CNS tropism with pronounced hepatic preference, making it less optimal for primary brain/spinal cord gene manipulation in NDs compared to CNS-optimized serotypes (Mietzsch et al., 2020). Preclinical studies highlight its utility in GM2 gangliosidosis feline models, where systemic delivery reduces CNS ganglioside storage through cross-correction by peripherally transduced cells secreting lysosomal enzymes that traffic across the BBB (Mietzsch et al., 2020). As a hybrid vector with moderate transduction efficiency (limited global CNS coverage in some models), it carries potential immune challenges from preexisting nAbs and variable promoter responsiveness, positioning it primarily for peripheral-focused or combination therapies rather than direct parenchymal CNS intervention in PD, HD, or ALS (Wang et al., 2024b).

AAV2/7 and AAV2/8

AAV2/7 and AAV2/8 are chimeric vectors combining the well-characterized AAV2 genome with capsids derived from AAV7 or AAV8, respectively. These hybrids retain the favorable safety and transcriptional control of AAV2 while exhibiting enhanced transduction efficiency and broader tissue distribution. AAV2/7 shows improved neuronal uptake and moderate BBB penetrance, whereas AAV2/8 exhibits strong astrocytic and hepatic tropism, making it suitable for applications requiring glial engagement or peripheral-to-CNS cross-correction (Gao et al., 2002; Wang et al., 2024b).

AAVrh.10

AAVrh.10 demonstrates high CNS gene expression in canine models, approaching AAV9 efficiency for spinal cord and brain transduction after IT or systemic administration, with tropism favoring neurons and oligodendrocytes. Clinically, it advances metachromatic leukodystrophy (MLD) therapy via ARSA enzyme replacement and shows preclinical promise in mucopolysaccharidosis VII (MPS VII) and globoid cell leukodystrophy (GLD), prolonging lifespan in canine models through robust myelination restoration (Albright et al., 2018). Dosing precision, promoter optimization (e.g., cytomegalovirus [CMV], cell-based assay for broad expression), and hybrid variability influence outcomes, yielding high transduction (global CNS coverage in canines/nonhuman primates via systemic routes) but moderate NAb prevalence, necessitating screening and immunosuppression protocols for translation to pediatric NDs like leukodystrophies with neurodegenerative components (Mietzsch et al., 2021; Sharma et al., 2025).

5.1.3 Lentiviral vector platforms

Lentiviral vectors (LVs), derived from replication-incompetent HIV-1, have established themselves as a crucial gene delivery platform for NDs, complementing AAV-based strategies with unique advantages. Their large cargo capacity (9–10 kb) accommodates full-length transgenes, multigene cassettes, or complex regulatory elements, WPRE and cPPT/CTS, far surpassing AAV's 4.7 kb limit. This capability shines in lysosomal storage leukodystrophies with ND features, enabling delivery of enzymes such as ARSA or galactocerebrosidase (GALC) under robust promoters. Unlike AAV episomes, LVs integrate into the host genome via integrase, ensuring stable, long-term expression in postmitotic neurons, glia, and progenitors. Such durability suits progressive NDs, where sustained therapeutic levels counter ongoing neuronal loss through single-dose administration rather than repeated injections (Blessing and Déglon, 2016; Humbel et al., 2021).

Ex vivo LV transduction of autologous hematopoietic stem cells (HSCs) with ARSA cassettes yields multilineage engraftment, supraphysiological enzyme activity in blood and CSF, and halts neurodegeneration in presymptomatic or early-symptomatic children, as evidenced by phase ½ trials (Meneghini et al., 2017). Secreted cross-correcting enzymes from engrafted cells cross the blood-brain barrier, eliminating the need for direct CNS injections; parallel approaches target GLD and related disorders. In vivo applications extend to complex NDs. In PD models, striatal or substantia nigra delivery of GDNF, neurturin, or dopamine enzymes (TH/AADC/GCH1) restores function for months. Integrase-deficient LVs (IDLVs) in spinocerebellar ataxia type 3 enhance cerebellar ubiquitin–proteasome activity, providing year-long benefits with reduced mutagenesis risk (Sahu et al., 2025).

LVs' broad tropism, low immunogenicity (fewer nAbs than AAVs), and pseudotyping versatility further enhance their ND utility. VSV-G pseudotypes yield high-titer production and efficient neuron/astrocyte transduction in rodent/primate CNS after intraparenchymal delivery (Sahu et al., 2025). Specialized envelopes, such as those containing RVG or rabies/vesicular stomatitis virus G (VSV-G) chimeras promote retrograde axonal transport, allowing them to reach distant projections, including ALS corticospinal motor neurons and PD nigrostriatal pathways. Chandipura viral glycoprotein (CNV-G) boosts neuronal efficiency beyond VSV-G, allowing tailored tropism for specific circuits (Arduini et al., 2025).

5.1.4 Herpes simplex virus type 1 vector platforms

Herpes simplex virus type 1 (HSV-1) vectors capitalize on the virus's innate neurotropism, accommodating payloads of up to 150 kb in bacterial artificial chromosome (BAC)-based amplicon systems or recombinant replication/capacity-deleted constructs, vastly exceeding AAV constraints of 4.7 kb. This capacity, coupled with glycoproteins gB/gD-driven retrograde axonal transport, affords unprecedented access to distal neuronal projections in NDs (Le Hars et al., 2025).

HSV-1 platforms encompass replication-conditional designs (e.g., ICP34.5/g34.5 mutants) and replication-defective architectures with deletions in essential immediate-early (IE) genes (ICP0, ICP4, ICP27, ICP47) (Joussain and Epstein, 2024). Helper-free amplicons, incorporating HSV origins (oriS, pac) and cis-acting nuclear retention elements on a BAC backbone, generate high-titer preparations devoid of helper virus, facilitating stereotactic or intranasal delivery in ND models. Predominant backbones for CNS applications include laboratory strains KOS, 17syn+, and F-strain and JS1 derivatives. Nonreplicative archetypes, helper-free amplicons for multitransgene cassettes in leukodystrophies, and multi-IE-deleted recombinants for protracted hippocampal/striatal/nigral expression dominate preclinical ND research (Ingusci et al., 2025).

HSV-1 surpasses AAV2-retro in retrograde transduction of projection neurons, including nigrostriatal dopamine cells from striatal infusions. Robust immunogenicity (anti-gB/gD IgG and CD8+ T-cell responses) precludes redosing but accommodates high-dose, single-administration paradigms with transient immunosuppression, yielding months-to-years episomal persistence via tet-off regulons (Cattaneo et al., 2025).

PrHSV-1 advantages for NDs comprise unrivaled capacity (>100 kb) for combinatorial cassettes, rabies virus-emulating retrograde efficiency, and glycoprotein engineering for neuronal/glial specificity. Drawbacks include cytotoxicity from vestigial viral genes (dose-dependent neurotoxicity), latency-associated reactivation (attenuated by ICP0 ablation), potent immunogenicity, and manufacturing intricacy relative to AAV scalability. Thus HSV-1 vectors emerge as a high-capacity adjunct to AAVs and lentiviruses, primed for NDs demanding retrograde traversal or multigenic intervention in PD, HD, ALS, and AD (Huang et al., 2022; Cattaneo et al., 2025; Miyagawa et al., 2025).

5.1.5 Adenovirus vector platforms

Adenovirus-based vectors represent a niche yet valuable platform for CNS gene delivery in NDs, complementing AAV and HSV-1 systems where large payloads (>36 kb), multicistronic constructs, or extensive parenchymal transduction are paramount (Ricobaraza et al., 2020).

Helper-dependent adenoviral vectors (HDAd and gutless designs) eliminate all viral coding sequences, retaining only inverted terminal repeats (ITRs) and packaging signals on a plasmid backbone. This minimizes immunogenicity while accommodating complex cassettes beyond the AAV limit (Cregan et al., 2000). Postintracerebral, IT, or convection-enhanced delivery (CED), HDAd efficiently transduces postmitotic neurons and glia across the cortex, hippocampus, and spinal cord, with retrograde transport from the striatum to the substantia nigra pars compacta (SNc) (Chen et al., 2023). In PD models, HDAd-encoded TH/AADC/GCH1 reconstitutes nigrostriatal dopamine synthesis, while GDNF bolsters dopaminergic survival. ALS benefits from neurotrophin-3 or insulin-like growth factor 1 delivery to spinal circuits, extending survival in SOD1^G93A mice. Lysosomal storage leukodystrophies with NDD features, such as MLD or GLD, leverage ARSA or GALC for cross-correcting enzyme secretion. HDAd's high titers, stable episomal persistence without integration, and broad tropism enable multilayered ND interventions, including CRISPR arrays with short-hairpin RNA (shRNA) and trophic factors (Peng et al., 2024a; Qia et al., 2025).

First-generation replication-deficient adenoviruses (FGAd), typically E1/E3-deleted Ad5 vectors, pioneered CNS gene transfer but are now largely historical. These transduced neurons/glia for TH, SOD1, and GDNF in PD; ASPA in Canavan disease; and neurotrophins in HD and ALS. Persistent viral gene expression, however, elicits potent innate and adaptive immunity, causing inflammation, cell loss, and transient (weeks-long) transgene expression, prompting supersession by safer HDAd (Hackett and Crystal, 2025).

Ad5 dominates neuronal transduction among serotypes, robustly targeting neurons, astrocytes, and oligodendrocytes in the cortex/striatum/hippocampus via intraparenchymal/intraventricular routes for GDNF, dopamine enzymes, or lysosomal payloads in PD/ALS/leukodystrophies. Limitations stem from high preexisting nAbs and inflammatory responses in FGAd; Ad2/Ad5 chimeras and fiber-modified Ad5 (altered knob/fiber) reduce liver tropism while preserving CNS efficacy (Bliss et al., 2024). Rare serotypes (Ad35/Ad11p) or fiber-pseudotyped Ad5 (e.g., Ad5/35) evade immunity but lack superior neuron tropism, serving mainly for immunological circumvention rather than enhanced ND targeting (Wang and Shao, 2023). Thus Ad5-derived HDAd and FGAd underpin validated ND applications, including PD dopaminergic restoration, ALS neuroprotection, HD modulation, and leukodystrophy enzyme replacement, positioning adenoviruses as a high-capacity adjunct for polygenic or combinatorial ND therapies (Mozharovskaya et al., 2023).

5.1.6 Hybrid viral systems

In recent years, increasing attention has been directed toward the development of hybrid and combinatorial gene delivery platforms designed to overcome the intrinsic constraints of single-vector systems and to meet the complex demands of neural gene therapy. These advanced architectures aim to balance transgene capacity, cell-type specificity, durability of expression, and safety. Among the most extensively explored strategies are hybrid vectors combining HSV and AAV. In these systems, the exceptionally large packaging capacity of HSV is leveraged to accommodate expansive or multigenic therapeutic payloads, while AAV-derived regulatory elements are incorporated to promote stable, long-term transgene expression in postmitotic neurons (Zou et al., 2022). Notably, such hybrid vectors can support substantially larger DNA cargos than conventional AAV and, in certain configurations, enable site-preferential genomic integration, a feature that is highly advantageous for sustained therapeutic intervention in the brain (Sahu et al., 2025).

A related approach involves AAV–adenovirus (Ad) hybrid vectors, which integrate the high transduction efficiency and robust packaging machinery of adenovirus with the genomic persistence properties of AAV. These vectors can accommodate transgenes of up to approximately 14 kb, rendering them suitable for the delivery of large neuronal genes, complex regulatory elements, or coordinated multigene expression systems required for circuit-level modulation (Słyk et al., 2024).

Beyond fully hybridized vectors, dual-vector and combinatorial delivery strategies have emerged as powerful tools for achieving spatial and circuit-specific gene transfer. In so-called double viral vector paradigms, two distinct vectors are administered concurrently to different neuronal populations or anatomical regions, enabling selective gene expression along defined neural pathways (Birchard, 2025). Preclinical studies have demonstrated that this approach enhances targeting precision within distributed neuronal circuits and improves functional outcomes, underscoring its potential for dissecting and therapeutically manipulating neural networks. Additional innovations combine viral vectors with nonviral delivery modalities to further optimize performance. For example, engineered virus-like particles (VLPs) or modified viral capsids integrated with lipid- or polymer-based nanoparticles can reduce immunogenicity and facilitate repeated administration (Mandalawatta et al., 2024).

5.2 Nonviral modalities

Nonviral delivery systems have emerged as promising alternatives to viral vectors in gene therapy for NDDs. Unlike viral platforms, nonviral approaches offer several distinct advantages, including higher safety profiles, tunable genetic payload capacity, reduced immunogenicity, and the potential for repeat dosing, which together address many of the limitations associated with viral vectors.

5.2.1 Chemical nanocarrier

Lipid nanoparticles

Lipid-based nanocarriers, including liposomes, LNPs, solid lipid nanoparticles (SLNs), and nanostructured lipid carriers (NLCs), represent a highly versatile and intensively studied class of nonviral delivery systems for gene therapy targeting the CNS (Satapathy et al., 2021; Cai et al., 2024). These carriers are constructed from biocompatible lipids that self-assemble into stable vesicular or particulate structures capable of encapsulating a broad array of nucleic acids, such as plasmid DNA, mRNA, and siRNA, thereby protecting therapeutic cargo from enzymatic degradation and facilitating cellular uptake. A critical challenge for CNS gene therapy is crossing the BBB, a tightly regulated endothelial interface that restricts entry of most macromolecules; lipid carriers address this through both physicochemical design and surface functionalization with targeting ligands (e.g., peptides, small molecules) that engage receptor-mediated transcytosis (RMT) pathways, enhancing transport across endothelial cells into brain tissue (Cai et al., 2024). Typical LNP formulations include ionizable cationic lipids, phospholipids, cholesterol, and PEGylated lipids that encapsulate nucleic acids via electrostatic interactions, balancing neutral surface charge at physiological pH for reduced immune activation via protonation in endosomes to promote efficient cytosolic release (Pardridge, 2023). LNP-mediated delivery of mRNA and siRNA has been explored as a therapeutic strategy for numerous NDs. In AD models, LNPs carrying siRNA targeting genes involved in Aβ production have been shown to reduce pathogenic protein levels and improve related cognitive deficits in animal studies, highlighting the potential of LNP-siRNA therapeutics for modifying disease progression (Rassu et al., 2017; Du et al., 2025). In PD, NLCs have been used to deliver neurotrophic factors such as GDNF via intranasal administration in rodent models, resulting in enhanced survival of dopaminergic neurons and improved motor function, underscoring their ability to support the delivery of biologically active proteins in addition to nucleic acids (Jagaran and Singh, 2022).

Recent advances in liposomal and LNP design also include brain-targeting lipid nanoparticles (BLNPs) that successfully cross the BBB and deliver mRNA to the brain following systemic injection, as demonstrated in mouse models with robust expression of therapeutic transgenes in multiple brain regions, paving the way for future treatments of diverse CNS disorders such as ALS and other NDDs. Furthermore, hybrid lipid carriers such as SLNs and NLCs with controlled release profiles offer enhanced cargo capacity and sustained therapeutic activity, making them particularly promising for chronic treatment paradigms required in NDD therapy (Kwak et al., 2024; Du et al., 2025).

Polymeric nanoparticles

Polymeric nanocarriers, composed of natural polymers such as chitosan, gelatin, and dextran, or synthetic polymers, including polyethyleneimine (PEI), poly(lactic-co-glycolic acid) (PLGA), and poly(butyl cyanoacrylate) (PBCA), provide highly adaptable platforms for condensing or encapsulating nucleic acids into nanoscale complexes. Cellular uptake of these carriers is predominantly mediated via endocytic pathways, and fine-tuning particle size, surface charge, and release kinetics enables them to overcome enzymatic degradation, limited cellular internalization, and the restrictive BBB, thereby facilitating efficient delivery to the CNS (Saini et al., 2024).

Cationic polymers such as branched PEI form compact polyplexes with DNA or RNA through electrostatic condensation and promote endosomal escape via the proton sponge effect, in which amine protonation buffers endosomal acidification and induces osmotic membrane disruption (Huang et al., 2025). Biodegradable matrices such as PLGA provide sustained release of genetic cargo through hydrolytic erosion, a feature particularly advantageous for chronic NDD therapy where prolonged gene expression is desired (Beach et al., 2024). Functionalization with targeting ligands further enhances RMT and selective delivery to neural cells (Caraway et al., 2022).

Preclinical studies underscore the therapeutic potential of polymeric nanocarriers in NDD models. In PD, PEI-mediated siRNA delivery targeting α-synuclein via intracerebroventricular (ICV) infusion substantially reduced pathogenic protein levels in the striatum (Feja et al., 2025). PLGA-based carriers have improved the uptake of miR-124 in dopaminergic stem cells, promoting differentiation and neuroprotective effects in toxin-induced PD models (Liu et al., 2024). Surface modifications, such as coating PBCA nanoparticles with polysorbate 80, leverage adsorptive and receptor-mediated pathways by recruiting plasma apolipoproteins, thereby facilitating uptake by brain endothelial cells. The incorporation of natural polymers like chitosan further enhances particle stability and CNS delivery efficiency (Mirzaei and Fatemizadeh, 2025).

Beyond PD, PLGA nanoparticles optimized for intranasal administration exploit the olfactory and trigeminal pathways to bypass the BBB, providing a noninvasive route for CNS delivery. These platforms have demonstrated efficacy across multiple ND models, delivering plasmid DNA, siRNA, and CRISPR/Cas components with controlled release, efficient BBB penetration, modular targeting, and the potential for repeated dosing (Zou et al., 2022).

5.2.2 Inorganic and hybrid nanocarriers

Inorganic and hybrid nanocarriers represent a distinct class of nonviral delivery systems that combine the structural advantages of inorganic cores with versatile surface chemistries to facilitate the transport of genetic cargo into the CNS. Unlike purely organic carriers, inorganic nanoparticles harness unique physicochemical properties, including tunable size, high surface area, and optical and magnetic responsiveness, which can be exploited for gene transport, imaging, and therapeutic modulation in NDD contexts. Common inorganic cores include gold, iron oxide, silica, and carbon-based nanomaterials, which can be modified with polymers, targeting ligands, or biomimetic coatings to enhance stability, biodistribution, and cellular uptake (Haque and Chowdhury, 2018).

Gold nanoparticles (AuNPs) have been widely studied for CNS delivery due to their small, tunable size, biocompatibility, and ability to cross the BBB when surface-functionalized appropriately. In vitro and in vivo studies have shown that galactose-coated AuNPs carrying oligonucleotides can cross human brain endothelial cells and penetrate astrocytes, potentially delivering therapeutic DNA into the brain parenchyma at significantly higher levels than free DNA. In these models, the oligonucleotide cargo remained attached during BBB transcytosis and reached target cells in functional amounts, highlighting the potential of AuNPs for gene transport across endothelial barriers (Zhang et al., 2022). Moreover, exosome-coated AuNPs have been shown to interact more effectively with brain cells under flow conditions and accumulate in mouse brains following systemic administration, illustrating how inorganic carriers can be paired with biological coatings to enhance CNS uptake via receptor-mediated transport (Nair et al., 2017; Sivamaruthi et al., 2025).

Mesoporous silica nanoparticles (MSNs) offer a high internal surface area and tunable pore structures that enable high loading capacities for nucleic acids and cocargo, such as small molecules or imaging agents. Core–shell MSNs functionalized with polymers or peptides have demonstrated the ability to protect plasmid DNA from nuclease degradation, facilitate endosomal escape, and transfect neuronal or progenitor cells in preclinical settings (Chen et al., 2024). These particles can be further modified with targeting ligands such as low-density lipoprotein receptor peptides to engage RMT mechanisms across the BBB, illustrating a powerful strategy for engineered CNS gene delivery (Huq et al., 2025).

Magnetic iron oxide nanoparticles and hybrid magnetic systems represent another inorganic class with relevance to neurodegeneration (Haro-Martínez et al., 2025). Superparamagnetic iron oxide nanoparticles can be guided by external magnetic fields and have been investigated for both drug and gene delivery applications; when coated with biocompatible polymers such as chitosan, they can ferry neuroprotective genes or cargoes through olfactory pathways or across BBB models, and support memory and cognitive improvement in AD models through enhanced delivery of therapeutic agents (Maheshwari et al., 2025). In addition to delivery, iron oxide and other inorganic cores can exert intrinsic therapeutic effects. For example, cerium oxide nanoparticles possess antioxidant properties that may reduce oxidative stress in neurodegenerative conditions like PD or AD, complementing gene therapy efforts (Haro-Martínez et al., 2025).

Emerging inorganic platforms include carbon dots and quantum dots, which combine delivery capability with imaging functionality. These materials are engineered to carry plasmid DNA, siRNA, or other gene modulators while providing fluorescent tracking, enabling simultaneous monitoring of biodistribution and therapeutic action (Bagherpour et al., 2025; Shareena et al., 2025a, Shareena et al. (2025b)). In addition, spherical nucleic acid architectures, which may incorporate an inorganic core such as gold and a dense shell of oligonucleotides, have demonstrated the capacity to cross the BBB and regulate gene expression in brain targets, including clinical trials assessing siRNA delivery for glioblastoma, underscoring the translational potential of inorganic-core gene nanocarriers (Xiao et al., 2025).

Hybrid nanocarriers that integrate inorganic cores with organic shells, for example, mesoporous silica cores coated with lipid or polymer layers, merge the advantageous payload capacity and structural stability of inorganic media with the biocompatibility and targeting flexibility of organic materials (Daksh et al., 2025). Such hybrids leverage modular surface chemistry to engage receptor-mediated or adsorptive transcytosis mechanisms across the BBB and minimize off-target interactions, making them promising candidates for gene therapy in complex NDs.

Hydrogels provide a complementary strategy to inorganic nanocarriers by serving as biomimetic, three- and four-dimensional delivery matrices that can be coengineered with inorganic nanoparticles or gene nanocarrier systems to enhance CNS targeting (Xu et al., 2024). Because hydrogels mimic the native extracellular matrix and can be formulated to gel in situ after injection, they offer a minimally invasive means to localize and sustain the release of therapeutic cargos directly within affected neural tissue (Hasanzadeh et al., 2023). In the context of NDs, hydrogels have been used to encapsulate nanocarriers and extracellular vesicles (EVs), protecting them from rapid clearance and controlling spatial release, which is critical for long-term modulation of gene expression in brain parenchyma (Pal et al., 2024). Moreover, injectable hydrogels have been studied as regenerative scaffolds that support extracellular matrix remodeling and neural tissue repair after injury, illustrating their dual utility in structural support and therapeutic delivery to CNS targets (Xu et al., 2024). Combined inorganic–hydrogel systems can therefore leverage high payload capacity, controlled release kinetics, and enhanced retention at pathological sites, making them particularly promising for sustained gene therapy in AD, PD, and other NDs (Santhamoorthy and Kim, 2025).

5.2.3 Physical nonviral strategies

Physical delivery approaches exploit biomechanical forces to transiently permeabilize cellular membranes or modulate the neurovascular unit, thereby facilitating the ingress of therapeutic nucleic acids into the CNS. These methods do not depend on viral vectors or chemical carriers alone; instead, they use external stimuli such as electric fields, acoustic energy, or magnetic forces to overcome biological barriers. When combined with LNPs, polymeric carriers, or other nonviral vectors, physical techniques can greatly enhance transduction efficiency in postmitotic neurons and other hard-to-transfect cell types, while minimizing systemic immunogenicity and bypassing certain biological constraints (Du et al., 2025).

Intracerebral electroporation remains a foundational physical gene delivery strategy in neuroscience research. Following direct parenchymal administration of naked DNA, RNA, or RNP complexes, precisely calibrated electrical pulses (typically in the range of 20–50 V amplitude, 10–20 ms duration, at ~ 1 Hz over multiple iterations) induce reversible dielectric membrane poration on the order of <100 nm (Fang et al., 2021). This transient poration permits efficient intracellular uptake of genetic cargo without requiring biological vectors and has been used to achieve substantial transfection rates in targeted neural populations. In HD models, for example, striatal electroporation yields high levels of medium spiny neuron transfection, enabling CAG repeat contraction or allele-specific ablation in a significant fraction of affected cells. Similarly, targeted electroporation in the substantia nigra of PD models has facilitated locus-specific editing of dopaminergic neuron genes such as LRRK2 (Murphy et al., 2024). Although effective, the invasiveness of direct electroporation can induce transient perilesional edema; this can be mitigated through optimization of pulse waveforms and delivery parameters to balance efficacy with tissue integrity.

Focused ultrasound (FUS)-mediated sonoporation is emerging as a leading noninvasive physical modality for CNS gene delivery. When low-intensity acoustic waves (commonly in the 1–5 MHz range with mechanical indices between ~ 0.5 and 2 MPa) are applied concomitantly with intravenous microbubbles, the resulting controlled cavitation disrupts tight junctions within the BBB reversibly, typically with recovery of barrier integrity within hours (Blesa et al., 2023). This transient opening allows systemically administered nanoparticles or gene vectors to extravasate into targeted brain regions. Preclinical evidence demonstrates that FUS can be harnessed to deliver gene-editing systems and oligonucleotides in models of major neurodegenerative disorders: in AD (APP/PS1) paradigms, FUS facilitates hippocampal delivery of CRISPR/Cas9 cargo, achieving measurable on-target editing and accompanying reductions in amyloid-β plaque pathology, indicating tangible disease-modifying effects in vivo (Gao, 2024). In ALS models, LNP-mediated silencing of SOD1 within motor neurons reaches 15%–25%

efficacy when paired with FUS-enhanced delivery, correlating with improved neuronal outcomes. Similarly, FUS has been used to open the BBB in HD model systems, enhancing the delivery and regional uptake of therapeutic AAV gene vectors across a range of age cohorts and demonstrating its broad utility for noninvasive CNS gene transport (Helal et al., 2025).

Other physical modalities, such as magnetofection and nanoinjection, offer complementary strategies for localized gene delivery. Magnetofection uses magnetic fields to guide nanoparticle–nucleic acid complexes into cells, improving localization and transfection efficacy in experimental CNS applications, while nanoinjection employs nanoscale probes to mechanically deliver DNA directly into individual cells with high precision. Each of these physical approaches presents distinct mechanistic advantages, from precise spatial control to minimized immunogenicity, and highlights the breadth of nonviral methods available for gene delivery in complex CNS environments (Aly et al., 2023).

Although many physical delivery methods remain predominantly in preclinical stages due to technical and translational challenges, recent advancements in waveform optimization, imaging guidance, and combinatorial strategies with nonviral carriers continue to enhance their effectiveness and clinical relevance. As these tools mature, they promise to expand the therapeutic landscape for NDDs by enabling precise, targeted, and minimally invasive gene delivery across previously formidable biological barriers (Fischell and Fishman, 2021; Dogbey et al., 2023).

5.2.4 Biological nonviral strategies

Biological nonviral strategies leverage naturally derived or biologically inspired carriers to deliver therapeutic nucleic acids into the CNS with high biocompatibility, low immunogenicity, and intrinsic cellular trafficking capabilities. These platforms exploit biological transport mechanisms, enabling them to navigate physiological barriers and engage directly with neural cells, including neurons, glia, and endothelial populations. Major classes of biological carriers include EVs (such as exosomes and microvesicles), engineered hybrid vesicles, and cell membrane-coated constructs, each offering unique mechanisms for gene delivery and therapeutic modulation in NDD contexts (Jiang et al., 2024).

EVs are naturally secreted membrane-enclosed particles that mediate intercellular communication by transporting proteins, lipids, and nucleic acids between cells. EVs are generally classified by size and biogenesis into exosomes (~30–150 nm), microvesicles (~50–1000 nm), and apoptotic bodies, with exosomes being the most widely studied for CNS applications due to their innate stability, favorable biodistribution, and capacity to cross the BBB via endogenous transport pathways. These vesicles protect their cargo from enzymatic degradation and can interact with recipient cells through receptor-mediated endocytosis, membrane fusion, or micropinocytosis, enabling targeted modulation of gene expression in the brain.

Exosomes have been engineered to deliver a spectrum of therapeutic nucleic acids and modulators in preclinical models of neurodegenerative disorders (Wu et al., 2021). For example, exosomes isolated from HEK293 cells overexpressing miR-124 were injected into the striatum of R6/2 transgenic HD mice, where they successfully reduced the expression of a disease-relevant target gene in the striatum, demonstrating that exosome-mediated miRNA delivery to the CNS is feasible even in the context of progressive degenerative disease processes (Choi et al., 2022). In ALS models, EVs released from adipose-derived mesenchymal stem cells (ADSCs) reduced the accumulation of mutant SOD1 protein in motor neurons, thereby protecting cells from oxidative damage and improving viability in vitro, illustrating the potential for naturally secreted vesicles to confer neuroprotection through cargo delivery (Ding et al., 2025).

In PD paradigms, engineered exosomes have been loaded with catalase mRNA, which preferentially delivered this therapeutic cargo to the brain in 6-hydroxydopamine (6-OHDA) PD models, reducing oxidative stress and neuroinflammation. Exosomes modified with the rabies virus glycoprotein (RVG) peptide and charged with DNA aptamers targeting pathological α-synuclein aggregates have efficiently transported these aptamers into the PD mouse brain, resulting in reduced α-synuclein pathology, protection of neuronal structures, and improved motor outcomes. Similarly, RVG-displaying exosomes loaded with siRNA against β-site amyloid precursor protein-cleaving enzyme 1 (BACE1) have been systemically administered in AD models, leading to significant BACE1 knockdown and associated reductions in amyloid-β pathology, highlighting the utility of engineered biological carriers for siRNA delivery in AD (Yu et al., 2021).

Beyond synthetic delivery platforms, biologically derived EVs from polarized immune cells have demonstrated significant neuroprotective potential. EVs released from M2-polarized microglia and enriched with miR-124-3p can cross the blood–brain barrier after intravenous administration and are efficiently internalized by hippocampal neurons (Yang et al., 2019). Once inside target cells, they attenuate neuroinflammation and enhance neurogenesis by modulating pathways such as Rela/ApoE, thereby promoting neuronal survival and functional recovery (Ke et al., 2026).

VLPs have emerged as highly attractive biomimetic nanocarriers due to their ability to replicate viral architecture while being devoid of infectious or replicative potential. These self-assembling protein nanostructures, typically ranging from 20 to 200 nm in diameter, preserve the capsid organization of the parent virus but lack viral nucleic acids. This combination confers high biosafety while maintaining the capacity for receptor-mediated endocytosis and effective translocation to neural targets (Mandalawatta et al., 2024). The physicochemical properties of VLPs, including surface charge, morphology, and cargo capacity, can be precisely tuned depending on the viral template, such as bacteriophage, adenovirus, or polyomavirus scaffolds. Notably, certain VLPs derived from neurotropic viruses, including JC polyomavirus, exhibit inherent tropism for oligodendrocytes and neural precursor cells, providing a platform for targeted CNS delivery (Daksh et al., 2025). Surface functionalization through chemical or genetic conjugation further enables the incorporation of peptides, antibodies, or neurotropic ligands (e.g., transferrin, ferritin, rabies virus glycoprotein peptide), promoting receptor-mediated uptake by neurons or glial cells (Lyu and Lu, 2022; Xiong et al., 2025). Owing to their biodegradability, low immunogenicity, and absence of replicative material, several VLP-based systems have received regulatory recognition, including Food and Drug Administration (FDA) approval for therapeutic or vaccine applications.

Hybrid and biologically inspired constructs further expand the delivery landscape. VLPs coated with EV membranes or decorated with targeting peptides combine the high cargo capacity of engineered vectors with the biocompatible surface architecture of biological systems to facilitate efficient uptake and trafficking in neural tissues. Cell membrane-coated nanoparticles and dendrosome-like vesicles similarly exploit membrane receptors and cellular recognition elements to enhance CNS targeting while minimizing immune activation.

Across these examples, key features emerge that underscore the promise of biological nonviral strategies: BBB penetration through inherent or engineered transport mechanisms; cargo diversity, ranging from miRNAs and siRNAs targeting disease drivers to therapeutic mRNAs encoding protective enzymes; and broad disease relevance, with experimental validation in models of HD, PD, AD, and ALS. Although challenges remain, continuous advances in EV engineering and hybrid biological carriers propel these strategies toward clinically viable gene therapy approaches for neurodegenerative disorders (Mandalawatta et al., 2024) (Table 2.2).

TABLE 2.2 Major gene delivery modalities of NDDs, with key properties, delivery mechanisms, and representative applications.

Category	Subclass/ platform	Key properties	Mechanism/delivery route	NDD relevance/examples
Viral vectors				
AAV vectors	AAV1, AAV2	High neuronal transduction; local stereotactic delivery; excellent safety with episomal persistence.	Local injection; focal regions; limited BBB crossing.	PD trials (GAD/neurturin/AADC); motor circuit targeting in PD/AD.
	AAV5, AAV6	Balanced glia–neuron tropism; moderate retrograde transport (AAV6).	Intracerebral/local; striatal-to-SNpc pathways.	HD HTT knockdown (AAV5-AMT-130); PD dopaminergic pathways.
	AAV8, AAV9	Astrocyte bias (AAV8); neonatal BBB penetration/IT delivery (AAV9).	Intrathecal/BBB crossing, systemic in neonates.	ALS/AD widespread CNS coverage; SMA.
	AAV11, AAV12	Projection-specific/olfactory routes, engineered for niche delivery.	Retrograde labeling; intranasal.	Preclinical PD/HD circuit mapping; presymptomatic AD/PD.
Engineered AAV variants	AAVrh.8	Moderate CNS tropism; hepatic bias.	Systemic; peripheral cross-correction models.	GM2 gangliosidosis (peripheral cross-correction).
	AAV2/7, AAV2/8	Chimeric capsids with enhanced uptake or glial engagement.	Enhanced transduction; BBB interactions.	CNS and glial engagement platforms.
	AAVrh.10	High CNS gene expression; neuronal/oligodendrocyte preference.	Systemic/intrathecal; global CNS coverage.	MLD enzyme replacement; other leukodystrophies.
Lentiviral vectors (LVs)	LV (HIV-1 derived)	Large cargo (≥9–10 kb); integration for persistent expression.	In vivo local or ex vivo HSC transduction.	MLD (ex vivo HSC-LV ARSA); PD GDNF/neurturin; IDLVs with reduced insertional risk.
	Pseudotypes (VSV-G, RVG, CNV-G)	Broad tropism; enhanced retrograde transport.	Intraparenchymal; circuit-specific delivery.	Motor neuron pathways in ALS/PD; broad CNS transduction.

Continued

Table 2.2 Major gene delivery modalities of NDDs, with key properties, delivery mechanisms, and representative applications.—cont'd

Category	Subclass/ platform	Key properties	Mechanism/delivery route	NDD relevance/examples
Herpes simplex virus (HSV-1)-based vectors	Replication-defective/ amplicon vectors	Very large capacity (≥100 kb); glycoprotein retrograde transport (gB/gD).	Stereotactic/intranasal; retrograde projection targeting.	High-capacity multigenic constructs; retrograde pathways in PD/HD/ALS/AD.
Adenovirus vectors	Helper-dependent adenovirus (HDAd)	Very large payloads (>36 kb); low immunogenicity for complex cassettes.	Intracerebral/intrathecal/ CED.	PD dopamine synthesis enzymes, GDNF; ALS neurotrophins; leukodystrophy enzyme secretion.
	First-generation Ad5/chimeras (FGAd)	Robust initial transduction; historical platforms.	Local CNS injection.	Early PD/ALS/HD delivery; high immunogenicity led to safer successors.
Hybrid viral systems	HSV–AAV hybrids	HSV cargo capacity + AAV long-term expression elements.	Combined high capacity and sustained expression.	Multigenic/CRISPR payloads for complex NDD therapy.
	AAV–adenovirus hybrids	Improved packaging and persistence properties.	Systemic/local with enhanced payloads (~14 kb).	Large gene or regulatory cassette delivery.
	Dual/ combinatorial vectors	Double vectors administered to distinct regions.	Pathway-specific targeting across circuits.	Enhanced circuit-specific modulation; improved outcomes in preclinical networks.
	VLPs with lipid/polymer integration	Viral capsid proteins and nonviral carriers.	Enhanced safety/repeat dosing.	Combination CRISPR/therapy delivery approaches.
Nonviral modalities				
Chemical nanocarriers	Lipid nanoparticles (LNPs)	Ionizable lipids, phospholipids, cholesterol, PEG; endosomal escape via ionization.	Systemic/intranasal/ local; BBB-targeted with ligands.	siRNA against amyloid- in AD; GDNF delivery in PD; BLNPs for broader CNS payloads.
	Polymeric Nanoparticles (PEI, PLGA, PBCA)	Electrostatic complexation and controlled release; polymer matrix degradation.	Endocytosis; receptor-mediated BBB uptake.	siRNA α-synuclein in PD; miR-124 delivery; intranasal routes to circumvent the BBB.
	Inorganic/Hybrid Nanocarriers	Gold, silica, magnetic cores with surface coatings; imaging and gene delivery.	Receptor-mediated transcytosis; multifunctional targeting.	siRNA delivery; functionalized pili for endothelial interaction.
Physical nonviral strategies	Electroporation	Electric pulses induce membrane poration for direct nucleic acid entry.	Intracerebral/stereotactic.	50%–70% medium spiny neuron transfection in HD; ~40 % LRRK2 correction in PD models.
	Focused ultrasound (FUS)	Microbubble–ultrasound cavitation transiently opens the BBB.	Systemic plus FUS for volumetric CNS uptake.	12%–18% Cas9 editing and 45 % amyloid- reduction in AD; 15%–25% SOD1 silencing in ALS.
	Magnetofection/ Gene Gun/ Nanoinjection	Magnetic field-guided or mechanical DNA delivery.	Localized transfection; precise targeting.	Research stages for various CNS applications.

Table 2.2 Major gene delivery modalities of NDDs, with key properties, delivery mechanisms, and representative applications.—cont'd

Category	Subclass/ platform	Key properties	Mechanism/delivery route	NDD relevance/examples
Biological nonviral strategies	Exosomes/EVs	Natural lipid vesicles with intrinsic BBB traversal and biological trafficking.	Systemic/stereotactic/ intranasal.	miR-124 in HD; catalase mRNA in PD; siRNA BACE1 in AD; ADSC-EVs in ALS; M2 microglial EVs mitigating inflammation.
	Engineered EVs (RVG-targeted)	Surface modification for receptor-mediated uptake.	Enhanced neuronal targeting.	PD α-synuclein aptamer delivery; AD amyloid-knockdown.
	VLP-EV hybrids	Viral-like cores with EV membranes.	Biological targeting plus engineered payloads.	Modular platforms for CRISPR and multicomponent therapeutics.
	Cell membrane-coated carriers	Native membrane surface proteins for homing.	Targeted uptake, immune evasion.	Emerging targeted strategies for neural subtypes.

6 Challenges and improvements of conventional gene delivery systems

Despite substantial advances, the delivery of therapeutic genes to the central nervous system remains constrained by formidable anatomical and physiological barriers, restricted vector payload capacity, and the need for precise, cell-specific targeting. In addition, off-target effects and vector-induced immune responses continue to impose significant limitations on both efficacy and safety. The subsequent sections examine these challenges comprehensively, encompassing the inherent limitations of conventional delivery systems and strategies devised to overcome them, methods to improve editing precision and minimize unintended modifications, and the immunogenicity of commonly used vectors alongside emerging approaches for immune modulation.

6.1 Delivery system challenges

The BBB represents the most formidable physiological obstacle to effective gene therapy for NDs, owing to its highly selective endothelial interface that rigorously limits the passage of systemically administered macromolecules and gene delivery vectors into the central nervous system. Under normal conditions, the BBB restricts paracellular and transcellular transport through tightly connected endothelial cells, active efflux systems, and minimal vesicular trafficking, thereby maintaining neural homeostasis but simultaneously excluding the vast majority of therapeutic agents from accessing brain parenchyma at therapeutic concentrations. This selectivity is a primary reason why fewer than 1 % of systemically delivered therapeutics reach their CNS targets, severely constraining the translational impact of many promising gene-editing and gene supplementation strategies. To address this barrier, substantial research has focused on engineered transport mechanisms, including RMT, nanocarrier-based delivery systems, and transient modulation of barrier integrity. In the subsequent sections, these approaches are examined in detail, highlighting innovations designed to enhance CNS bioavailability while minimizing off-target effects and systemic toxicity.

6.1.1 Protein-based modifications

Protein engineering leverages neurotropic viral proteins to endow synthetic carriers with BBB-traversing capability. Encephalitis virus membrane proteins integrated into liposomes generate virus-like nanoparticles that deliver siGL3 and siPLK1 across the BBB, silencing GL3/PLK1 oncogenes in glioblastoma models and suppressing tumor growth (Glass et al., 2018).

Advanced nanocapsules bearing angiopep-2 (LRP-1 ligand) encapsulate CRISPR–Cas9 RNPs with near-quantitative loading, enabling intravenous transcytosis and PLK1 editing in glioblastoma xenografts (Rodrigues et al., 2021). RVG29-functionalized nanoparticles target nAChRs for CNS entry in PD models, enhancing CRISPR RNP neuronal editing and α-synuclein knockdown (Yang et al., 2026). These strategies support CRISPR excision of expanded CAG repeats in HTT for HD, reducing neurodegeneration in mouse models, or allele-specific LRRK2 silencing in PD to rescue dopaminergic neurons (Vermilyea et al., 2020; Chung and Lee, 2022).

6.1.2 Aptamers and peptide-based modifications

Targeting BBB endothelial receptors via ligand-conjugated nanoparticles is another critical strategy. Aptamer-conjugated nanocarriers exploit RMT via BBB endothelial receptors, with TfR targeting predominance owing to microvascular abundance (Vermilyea et al., 2020). Ferritin nanocages coencapsulating iron/siGPX4 enable TfR-mediated glioblastoma silencing. Low-avidity anti-TfR liposomes minimize peripheral sequestration for parenchymal access, while acid-cleavable TfR-AuNPs enhance uptake two–fivefold via endosomal release (Monaco et al., 2017; Shen et al., 2025b). In NDs, TfR nanoparticles deliver CRISPR/Cas9 for SOD1 editing in ALS and LRRK2 knockout in PD (Boado, 2025).

Peptide ligands enable receptor-specific CNS homing. RVG-derived peptides bind nAChRs; RVG-decorated EVs (sEVs^RVG) loaded with anti-Zika siRNA avert fetal brain injury. RVG-poly(mannitol-co-PEI) vectors deliver siBACE1 to curb Aβ in AD; RVG29-LNPs facilitate Cas9 RNP/mRNA editing in PD neurons, slashing α-synuclein and safeguarding dopaminergic cells.

6.1.3 Biomimetic nanovesicles

Cell-membrane-derived nanovesicles confer biocompatibility, immune evasion, and fluidity for paracellular/fusion-based BBB traversal. Macrophage EVs with siSNCA/curcumin (REXO-C/ANP/S) penetrate PD brains, mitigating α-synuclein aggregation (Wang et al., 2024a). ROS-responsive exosome–liposome hybrids (TSEL; angiopep-2) codeliver siBACE1/pTREM2 in AD models, polarizing microglia to M2, boosting Aβ clearance, shrinking plaques, and restoring cognition in APP/PS1 mice (Dehghani et al., 2025). In NDs, these support CRISPRa/i for BDNF upregulation in AD or Cas13-mediated TDP-43 reduction in ALS (Lee, 2024). Furthermore, small EVs (50–150 nm) leverage endogenous proteins for RMT/fusion; Arc-sEVs encapsulate CRISPRi mRNA, transferring interneuronally to silence APP in AD iPSC models, reducing Aβ42 without lysosomal entrapment (Jiang et al., 2024).

VLPs offer versatile platforms for the encapsulation or surface presentation of small molecules, nucleic acids, or proteins relevant to NDs. They have been employed for the delivery of neuroprotective compounds (e.g., doxorubicin, resveratrol, curcumin), therapeutic enzymes (e.g., AADC), and genetic materials such as siRNA or mRNA targeting disease-associated transcripts (e.g., APP, α-synuclein, tau) (Xiong et al., 2025).

6.1.4 Synthetic lipid and polymeric nanoparticles

Synthetic nanocarriers, including lipid and polymeric nanoparticles, provide highly tunable platforms for gene and drug delivery. Ionizable lipid nanoparticles (iLNPs), composed of cationic lipids, helper phospholipids, cholesterol, and PEG-lipids, remain neutral at physiological pH but protonate in acidic endosomes to release cargo efficiently (Xiong et al., 2023). RVG29-LNPs have been used for prime editing of HTT CAG repeats in HD striatal slices, achieving 47% contraction and neuroprotection, while TfR-LNPs enable SOD1 base editing in ALS, clearing 60% of aggregates. PLGA/PEI nanoparticles functionalized with angiopep-2 facilitate CRISPR/Cas9-mediated C9orf72 repeat excision in ALS, reducing dipeptide repeats and restoring RAN translation (Zheng et al., 2024).

SLNs and NLCs enhance intranasal nose-to-brain delivery by improving mucoadhesion and avoiding ciliary clearance. For example, asenapine-loaded SLNs demonstrate enhanced CNS targeting in PD models, while donepezil-loaded NLCs show promising clinical effects for AD (Costa et al., 2021). Inorganic nanoparticles, such as MSNs with glutathione-sensitive linkers, enable CRISPR-sgNrf2 delivery in AD, restoring cognition in APP/PS1 mice, while gold nanoplates conjugated to Cas13 knock down TDP-43 in ALS, restoring autophagy in patient-derived neurons (Zheng et al., 2024).

6.1.5 Nanocarrier size optimization

Nanocarrier dimensions profoundly govern BBB traversal mechanisms in gene editing for NDD, balancing endothelial pore exploitation, transcytosis efficiency, renal clearance evasion, and brain extracellular matrix (ECM) diffusion (Fan et al., 2023). Optimal 10–100 nm carriers maximize paracellular diffusion through transient tight-junction gaps, RMT, and adsorptive endocytosis for CRISPR/Cas9 cargos, avoiding rapid hepatic/renal elimination of ultrasmall particles (<5 nm) or entrapment of supersized ones (>200 nm) (Song et al., 2021).

Nanoparticles smaller than 20 nm, including insulin receptor-targeted AuNPs, demonstrate high efficiency in RMT and are particularly well-suited for prime editing applications within the CNS (Habib and Singh, 2022). Their ultrasmall size facilitates endothelial transport and enhances tissue penetration. In this context, 5–10 nm nanoclusters capable of diffusing through the ECM have enabled effective delivery of Cas13 systems to ALS neurons, resulting in targeted knockdown of TDP-43 (Arduini et al., 2025). These findings underscore the critical influence of nanoscale dimensions on overcoming diffusion constraints within dense neural microenvironments. Within the intermediate size range (approximately 50–200 nm), nanoparticle platforms offer increased cargo capacity, accommodating larger genetic constructs such as CRISPRa plasmids (Wu et al., 2020). For example, 100-nm GLUT1-targeted quantum dots have been employed to enhance BDNF expression in AD models. Similarly, TfR-LNPs measuring 100–120 nm have optimized LRRK2 gene-editing efficiency in PD patient-derived hiPSCs, highlighting the importance of ligand–receptor specificity in improving targeting precision and intracellular delivery (Arora and Singh, 2021).

In addition, RVG29-decorated LNPs in the 15–50 nm range have successfully delivered CRISPR/Cas9 RNP complexes to dopaminergic neurons in PD models, enabling α-synuclein excision and resulting in significant motor improvement in A53T mice (Lenders et al., 2023). Likewise, ~ 30 nm TfR-targeted LNPs have facilitated SOD1 base editing in ALS models, reducing protein aggregation while effectively navigating the aged ECM. Furthermore, Angiopep-2-functionalized nanoparticles with size tuning between 20 and 80 nm have been applied for APP prime editing in AD hiPSCs, leading to substantial reductions in Aβ plaque burden with minimal glial off-target effects, findings subsequently validated in APP/PS1 murine models (Habib and Singh, 2022).

6.1.6 *External device development*

Physical methods that transiently and precisely modulate BBB permeability have emerged as crucial strategies for delivering gene-editing platforms to the central nervous system. These approaches enhance the delivery of CRISPR/Cas9, BEs, PEs, and RNA-targeting therapeutics, often complementing viral vectors such as AAV to enable targeted neuronal editing with improved efficiency.

MRI-guided focused ultrasound utilizes focused acoustic waves in combination with intravenously administered microbubbles to induce controlled cavitation at the vascular endothelium (Meng et al., 2024). This cavitation temporarily disrupts tight junctions between endothelial cells, creating a reversible opening in the BBB. The effect typically resolves within hours, allowing systemically administered nanoparticles, nucleic acids, or protein therapeutics to extravasate into targeted brain regions (Jolesz and McDannold, 2013; Nance et al., 2014). In preclinical models of NDD, FUS has facilitated the delivery of CRISPR/Cas9 to the putamen and substantia nigra in PD, enabled the editing of SOD1 in ALS motor neurons, and supported APP prime editing in AD. These interventions lead to reduced pathogenic protein aggregation, such as α-synuclein and Aβ, and improved neuronal function. FUS offers volumetric coverage of specific regions, providing spatially precise, noninvasive delivery. While generally safe, careful calibration of acoustic parameters is essential to minimize rare side effects, including microhemorrhage or endothelial injury (Meng et al., 2023; Wu et al., 2025).

In addition to high-resolution FUS, low-intensity pulsed ultrasound (LIPUS) represents an evolving variation that uses periodic acoustic waves with minimal thermal impact. When combined with microbubble injection, LIPUS can safely and reversibly disrupt the BBB to facilitate the delivery of large molecules and nanoparticles, including albumin-bound drugs, into brain tissue (Peng et al., 2024a). This approach offers flexibility in energy targeting and reduces the risk of tissue damage. Research also explores the use of acoustically activated nanodroplets as an alternative or supplement to microbubbles, functioning as contrast agents and mediators of BBB opening at specific therapeutic sites. These developments extend the physical toolbox for ultrasound-mediated BBB modulation beyond traditional microbubble systems (Géraudie et al., 2023; Dou et al., 2025).

Laser interstitial thermal therapy (LITT) provides another strategy to increase BBB permeability by thermally loosening tight junctions. Unlike FUS, LITT creates a more prolonged window of BBB modulation, sustaining permeability for several days to weeks (Bakrbaldawi et al., 2024). This extended period supports broader and more uniform distribution of gene-editing payloads in the target tissue. LITT has been applied to deliver BEs for the correction of HTT mutations in HD and to perform allele-specific editing of LRRK2 in PD models. Real-time imaging, such as MRI perfusion, facilitates monitoring of therapeutic spread and ensures targeted delivery (Kumar et al., 2024).

Noninvasive scalp electroporation is a technique in which mild electrical fields transiently reorganize endothelial cytoskeletons, increasing the uptake of lipid-based nanoparticles or gene-editing complexes across the BBB (Sharabi et al., 2021). This approach has demonstrated efficacy for allele-specific SNCA editing in PD and for the correction of PSEN1 mutations in AD models. When combined with high-fidelity Cas9 variants, electroporation can minimize off-target effects, making it particularly suitable for postmitotic neurons that are less amenable to viral delivery (Ulusoy et al., 2025).

CED bypasses the BBB entirely by directly infusing gene-editing agents into the brain parenchyma under controlled pressure. CED enables broad distribution of large molecules, including Cas13 complexes, across deep brain structures (Aquilina et al., 2022). In ALS models, this strategy allows effective depletion of TDP-43 in motor neurons, delaying disease progression and improving survival. Similarly, in PD models, CED-mediated delivery of α-synuclein-targeting agents reduces aggregation and preserves dopaminergic neuron function. MRI-guided CED further enables precise monitoring of infusion and sustained modulation of targets, such as MECP2 in Rett syndrome models (Elder and Lonser, 2025; Yang, 2025).

Coupling physical approaches with nonviral carriers also expands delivery flexibility (Dogbey et al., 2023). For example, FUS applied after intranasal administration of plasmid DNA nanoparticles can increase transgene expression in targeted regions many-fold compared with delivery alone, significantly improving plasmid penetration into deep brain structures and enhancing transfection of neurons and microglia in sonicated hemispheres. The additive effect of physical BBB modulation with advanced nanoparticle design underscores a powerful synergy that combines the noninvasive nature of physical methods with the high payload capacity and safety profile of nonviral gene vectors.

6.2 Precision and off-target effects and strategies for mitigation

Off-target effects remain a principal limitation in the clinical translation of gene-editing technologies, as unintended DNA modifications can generate unpredictable and potentially pathogenic consequences (Zhu et al., 2025). At the genomic level, such events may induce structural aberrations, including large deletions, chromosomal rearrangements, and loss of heterozygosity, all of which jeopardize genomic stability and cellular integrity. In severe cases, these alterations may disrupt essential regulatory networks, increasing the risk of oncogenic transformation. Therefore minimizing off-target activity represents a critical prerequisite for the development of safe and precise CRISPR-based therapeutics. Recent advances have introduced a spectrum of integrative strategies, ranging from refined nuclease engineering to algorithmic guide RNA optimization, that collectively achieve substantial reductions in off-target perturbations, particularly within neuronal systems (Kalter et al., 2025).

6.2.1 CNS and subcellular delivery strategies

Regional CNS delivery

Optimized delivery routes enhance therapeutic specificity in NDs by targeting early pathological epicenters while minimizing systemic exposure. ICV administration delivers therapeutics directly into CSF, achieving widespread CNS distribution for multifocal diseases. Intranasal delivery leverages olfactory and trigeminal nerve pathways to bypass the BBB, enabling rapid transport of biologics such as ASOs to the hippocampus and entorhinal cortex (Lee et al., 2025). FUS paired with intravenously administered microbubbles induces transient, reversible BBB permeability, allowing precise, noninvasive access to deep structures like the substantia nigra in PD (Blesa et al., 2023). In 1-Methyl-4-phenyl-1,2,3,6-tetrahydropyridine (MPTP)-induced PD nonhuman primates, FUS-mediated AAV2-GDNF delivery to the putamen/SNc increases striatal GDNF levels by threefold, yielding UPDRS motor score improvement and dopamine neuron preservation over 6 months, without liver toxicity (Mead et al., 2017).

Cellular and organelle-specific targeting

Subcellular precision addresses the root causes of NDs, such as organelle dysfunction and protein misfolding, by directing therapeutic cargos to specific intracellular compartments. Neuron-derived extracellular vesicles, derived from mesenchymal stem cells or engineered to carry siRNAs, recapitulate endogenous synaptic transfer mechanisms for circuit-restricted delivery, traversing the BBB while evading glial phagocytosis (Peng et al., 2021). Protein-based delivery platforms, encompassing antibody–drug conjugates and cell-penetrating peptides, such as TAT conjugated to CRISPR RNPs, enable selective cytosolic or nuclear translocation in vulnerable neuronal populations (Jafari et al., 2019; Du et al., 2025). Organelle-targeted nanocarriers incorporate motifs like triphenylphosphonium (TPP) for mitochondria and Lys-Asp-Glu-Leu signal (KDEL) sequences for the endoplasmic reticulum (ER), mitigating bioenergetic deficits in PD and ALS or proteostasis impairment in HD (Yu et al., 2021).

In SOD1 ALS mice, mitochondria-targeted AAV-PHP.eB bearing SOD1 shRNA with TPP-peptide conjugation diminishes mitochondrial reactive ROS accumulation and mutant SOD1 aggregates in spinal motor neuron mitochondria, thereby prolonging survival through restored ATP synthesis, enhanced mitochondrial membrane potential, and preserved axonal transport along microtubules (Mathiesen et al., 2020). Also, in progranulin (GRN)-deficient FTD models, ER-targeted LNPs delivering XBP1s mRNA under a neuron-specific promoter alleviate unfolded protein response activation in layer II/III pyramidal neurons of the frontal cortex; this intervention reduces insoluble TDP-43 phosphorylation and aggregation in the ER–Golgi intermediate compartment, concurrently preserving corticostriatal synaptic density and dendritic spine morphology over disease progression (Wauters et al., 2017; Zhou et al., 2021).

Neuron-specific promoters

Human synapsin 1 (hSyn1) and neuronal nuclei (NeuN) promoters ensure pan-neuronal restriction with high fidelity in postmitotic environments (McLean et al., 2014; Patzke et al., 2019). In HD, hSyn1-AAV9-HTT shRNA selectively silences mutant HTT in striatal medium spiny neurons (MSNs), achieving robust allele-specific knockdown while exhibiting negligible hepatic expression. TH promoters precisely limit AADC/GDNF cassettes to nigral dopaminergic neurons in PD, delivering motor improvement equivalent to UPDRS scores without off-target GDNF toxicity in nondopaminergic tissues. Synapsin-driven scAAV9–CRISPR ablates SNCA in PD dopaminergic neurons of the SNc, reducing α-synuclein overproduction, reactive microgliosis, neuronal loss, and motor deficits in 6-OHDA lesion models (Inoue et al., 2023). Additional examples include hSyn1-directed CRISPRa upregulation of GDNF in PD nigrostriatal projections, enhancing dopamine homeostasis and reversing akinesia, and NeuN promoter-driven prime editing of CAG repeats in HD cortical projectors, normalizing neuronal excitability without bystander edits (Inoue et al., 2023; Szunyogh et al., 2025).

Astrocyte-specific promoters

Truncated GFAP variants (gfa1405/GFAP-t) target reactive astrocytes surrounding AD amyloid plaques and PD Lewy bodies. In 5xFAD AD models, gfa1405-AAV9 upregulates TREM2 in astrocytes, enhancing microglial phagocytosis of amyloid-β plaques, suppressing neuroinflammation via reduced IL-1β/TNF-α secretion, and preserving synaptic integrity without hepatic involvement (Kaganovich, 2019). GFAP-t promoters also drive astrocytic SOD1 silencing in the ALS spinal cord, mitigating glutamate excitotoxicity through restored EAAT2 expression and prolonged motor neuron survival (Rosenblum and Trotti, 2017). Synapsin-directed BACE1 editing complements this by inhibiting amyloidogenesis in neurons, collectively ameliorating cognitive decline in APPswe/PS1 models (Maesako et al., 2019).

Microglia-specific promoters

CX3CR1 promoters selectively modulate neuroinflammation in microglia. CX3CR1-driven CRISPRi targeting glia maturation factor (GMF) inhibits p38 MAPK signaling and proinflammatory cytokines (IL-6, CCL2), thereby dampening AD progression and tau hyperphosphorylation in the microglial vicinity (Rosario et al., 2016; Zhou et al., 2024a). In PD α-synucleinopathy models, CX3CR1-LRRK2 knockout reduces microglial proliferation and dopaminergic synapse pruning, thereby preserving nigral circuitry (Kaganovich, 2019).

Oligodendrocyte-specific promoters

Myelin-associated promoters (myelin basic protein (MBP), myelin-associated glycoprotein (MAG), 2',3'-cyclic nucleotide 3'-phosphodiesterase (CNP)) target white matter pathology in ALS and progressive supranuclear palsy (Von Jonquieres et al., 2016). MBP-miR-SOD1 silences mutant transcripts in ALS spinal cord oligodendrocytes, preserving myelination, axonal conduction velocity, and hindlimb grip strength. CNP promoter-driven FUS editing in ALS models prevents oligodendroglial RNA foci, stabilizing RanBP1 translation and white matter integrity (Borel et al., 2016; Weiss et al., 2025).

6.2.2 Circuit-specific targeting in the CNS

Precision targeting of specific neuronal subtypes and neural circuits represents a cornerstone of advanced gene therapy for NDs, enabling selective modulation of disease-relevant cell populations while minimizing off-target effects in nonpathological tissues. This spatiotemporal control is paramount for disorders such as PD, AD, HD, and ALS, where pathology manifests within discrete circuits rather than uniformly across the CNS (Zhou et al., 2026).

Synapsin-guided CRISPRa elevates BDNF in hippocampal neurons of APP/PS1 mice, promoting synaptogenesis via TrkB activation, reducing Aβ plaque burden through enhanced clearance, and restoring spatial memory in Morris water maze assays without liver or kidney effects. DARPP-32 promoters refine MSN subtype targeting in HD R6/2 mice, selectively silencing indirect pathway mutant HTT to restore motor balance on the rotarod without direct pathway disruption (Zhang et al., 2025). Multiancestry genome-wide association studies identify novel loci (MTF2, PIK3CA, USP25) linked to inflammation and neurodegeneration, positioning them as prime CRISPR targets under CNS-restricted promoters for circuit-specific modulation in polygenic ND (Hallock et al., 2020; Zhou et al., 2026).

6.2.3 Circuit-level targeting using engineered capsids

Circuit-level targeting employs engineered AAV capsids, intersectional recombinase logic, cell-type-specific promoters, and microRNA-based detargeting to achieve projection-specific transgene transduction that transcends conventional cellular identity. This paradigm proves indispensable for NDDs, wherein neuropathology manifests within anatomically discrete neural circuits, including corticostriatal projections (Tervo et al., 2016).

Engineered capsids, including PHP.eB, AAV2-retro, AAV9-Retro, and SH10 variants, mediate retrograde axonal transport to enable precise circuit access. Dual-vector intersectional systems, such as AAV2-retro-FLEX-Cas9 coadministered with hSyn1-CRISPRa, restrict MAPT editing to corticostriatal projector neurons in FTD tauopathy models, thereby attenuating insoluble tau aggregate burden (Tervo et al., 2016; Liu et al., 2021, Challis et al., 2022). AAV-PHP.eB-miR142 incorporating the DARPP32 promoter achieves selective allele-specific silencing of mutant HTT in indirect-pathway medium spiny neurons (iMSNs) of R6/2 HD mice, reinstating motor coordination asymmetry without perturbing direct-pathway medium spiny neuron (dMSN) function (Chatterjee et al., 2022). AAV9-Retro-CRISPRi delivers dCas9-KRAB fusion proteins to suppress SOD1 expression in corticospinal motor neurons via intralumbar administration in ALS models, ameliorating ventral horn denervation while preserving sensory neuron integrity. SH10-SaCas9-nickase facilitates base editing of the SNCA A53T mutation within SNc dopaminergic projections in PD, enhancing akinesia phenotypes in M83 transgenic mice through targeted transduction (McCaffrey et al., 2016). AAV2-Retro-Flp recombinase paired with Frt-stop-GDNF cassettes elevates GDNF selectively in layer 2/3 cortical projectors synapsing onto FTD striatal afferents, thereby constraining trans-synaptic tau propagation (Ren et al., 2021).

Neuronal tropism in gene delivery is achieved by coupling optimized capsid platforms with cell-type-selective regulatory elements: engineered AAV9/PHP.eB vectors paired with hSyn1 or NeuN promoters restrict expression predominantly to telencephalic neurons; ChAT and Hb9 expression cassettes enable targeted transduction of cholinergic and motor neuron populations relevant to ALS and Alzheimer's models; TH and DAT1 promoters direct dopaminergic circuit specificity in PD contexts; and Cre/Flp-dependent intersectional constructs refine expression to defined corticostriatal projection neurons in HD (Ren et al., 2021; Chen, 2023). Promoter hierarchies rank ubiquitin C for maximal neocortical neuronal drive, phosphoglycerate kinase for robust neuronal tropism in VSV-G-pseudotyped lentiviruses, and hSyn1 for superior in vivo specificity and transduction efficiency; CMV and motor neuron disease (MND) promoters exhibit temporally delayed and comparatively subdued neuronal preference. Astrocyte-directed interventions leverage G1B3 promoters for glial-modulatory support in ND contexts (Jollé et al., 2019).

In addition, stereotaxic parenchymal infusion into disease epicenters curtails glial off-target transduction. MiRNA detargeting architectures, miR-122/142 for hepatic/glial suppression, miR-142-3p for microglial/macrophage silencing to avert innate immune activation, and miR-124T for nonneuronal exclusion, synergize with FuG-B2 or PHP.B2 pseudotyping to amplify retrograde transsynaptic efficiency. These multifaceted refinements ensure circuit fidelity, neuronal enrichment, and immunological evasion (Mseis-Jackson et al., 2024).

6.2.4 Self-complementary adeno-associated virus vectors

Conventional single-stranded AAV (ssAAV) vectors require host-cell-mediated second-strand DNA synthesis before transcription, a rate-limiting step that is particularly restrictive in postmitotic neurons and can delay transgene expression for several weeks. Self-complementary AAV (scAAV) vectors overcome this limitation by packaging a double-stranded DNA genome formed through ITR-mediated intramolecular base pairing upon nuclear entry. This configuration eliminates the need for second-strand synthesis, resulting in markedly accelerated and enhanced transgene expression (Dou et al., 2025; Zwi-Dantsis et al., 2025).

In neuronal tissues, scAAVs achieve faster onset and higher levels of transduction compared with ssAAVs, with detectable expression observed within days rather than weeks. The principal trade-off is a reduced packaging capacity, approximately half that of ssAAV vectors, which restricts payloads to compact genetic cargos. Consequently, scAAVs are particularly well-suited for delivering small CRISPR effectors such as Cas13d, shRNAs, or compact regulatory elements (Shen et al., 2025a).

In ND models, scAAV-mediated RNA targeting has demonstrated therapeutic benefits. For example, scAAV9 vectors expressing synapsin-driven Cas13d have effectively reduced pathological TDP-43 aggregation in ALS motor neurons, resulting in improved neuronal survival and extended lifespan in vivo (Shen et al., 2025a; Zwi-Dantsis et al., 2025). In addition, scAAV1 vectors administered via peripheral routes exhibit retrograde transport capabilities, enabling partial transduction of spinal motor neurons in SMA models.

6.2.5 Molecular determinants of editing precision

High-fidelity CRISPR nucleases in postmitotic neurons

High-fidelity Cas9 variants, including Cas9, eSpCas9 (enhanced specificity Cas9), SpCas9-HF1 (high-fidelity variant 1), and SpCas9-nCas9, demonstrate markedly reduced off-target activity through structural modifications that destabilize mismatched guide RNA–DNA interactions (Lv et al., 2022; Pedrazzoli et al., 2023). In addition, Cas9 nickase pairs and prime editing operate without creating DSBs, making them ideal for NDDs applications. Delivering CRISPR components as RNP complexes rather than plasmid DNA is an effective method to limit off-target activity. RNP delivery results in transient intracellular expression, reducing prolonged Cas9 activity and potential genotoxicity (Guo et al., 2023).

Optimization of sgRNA design

Optimizing sgRNA design is crucial for improving the success of gene editing and reducing off-target effects (Dang et al., 2015; Schubert et al., 2021). This involves selecting sgRNA sequences with the highest specificity for the target site and minimal complementarity to nontarget genomic regions, aided by advanced algorithms and software (Cui et al., 2018). GC content, nucleotide distribution, and the propensity for secondary RNA structures are considered to improve specificity. The use of chemically modified sgRNAs further enhances target specificity by reducing nonspecific binding, thereby markedly decreasing off-target activity (Schubert et al., 2021). Controlled delivery systems limit the cellular presence of both sgRNA and Cas9 to minimize side effects, with computational predictions guiding precise target site selection based on empirical evidence (Nasir et al., 2022). The integration of artificial intelligence (AI) and chemical modifications in sgRNA design significantly boosts the safety and efficacy of CRISPR-based gene editing, particularly in clinical applications where accuracy and reduced off-target effects are paramount (Cui et al., 2018).

Switchable gene-editing systems

Temporal regulation of CRISPR activity is critical for mitigating cumulative toxicity and unintended genomic effects, particularly in chronic neurodegenerative conditions. Inducible platforms, such as Tet-On/Tet-Off systems, enable precise control of editing initiation and termination through the administration of small-molecule inducers (Daksh et al., 2025; Pattanayak, 2025). These systems have been applied to modulate disease-associated alleles in neural organoids and animal models, allowing editing to be restricted to defined therapeutic windows.

Optogenetic CRISPR systems provide an additional layer of control by enabling light-dependent activation of Cas nucleases. Photoactivatable Cas9 constructs have been used to excise pathogenic repeat expansions in HD models with spatial and temporal precision, minimizing collateral genomic damage (Huang et al., 2025; Pulgarin et al., 2025). Such approaches are especially valuable for dissecting disease stage-specific gene function and for limiting editing to early or presymptomatic phases of neurodegeneration.

Multiplex and polygenic editing approaches

Most NDs arise from polygenic interactions and convergent pathogenic pathways rather than single-gene defects. Multiplex editing strategies address this complexity by enabling simultaneous modulation of multiple genomic targets (Daksh et al., 2025). PAM-relaxed nucleases such as Cas12a facilitate the use of compact gRNA arrays, allowing coordinated editing of several loci from a single construct (Cattaneo et al., 2025). RNA-targeting Cas13 systems further expand this toolkit by enabling posttranscriptional modulation without permanent genomic alteration (Rahman et al., 2022).

In ALS models, multiplex Cas12a-mediated editing has been utilized to simultaneously excise C9orf72 repeat expansions and suppress mutant SOD1, leading to enhanced motor neuron survival (Rocha et al., 2025). In Alzheimer's disease, triculture systems have been shown to reduce amyloid burden while reshaping microglial inflammatory phenotypes by concurrently editing APOE, APP, PSEN1, and TREM2 (El Kassem et al., 2025). Similarly, cotargeting *LRRK2* and *SNCA* in PD models has enhanced dopaminergic neuron resilience and restored synaptic function (Duan et al., 2023; Zhu et al., 2025).

Biochemical and kinetic control of editing activity

Regarding nuclease and sgRNA design, precise control of biochemical and kinetic parameters is essential for limiting unintended editing events. Physiological pH and temperature optimize Cas–gRNA binding kinetics and catalytic activity at intended loci, whereas deviations can weaken hydrogen bonding and promote off-target cleavage (Blaschke and Bajorath, 2022). In neuronal systems, even subtle alterations in intracellular conditions may amplify genotoxic stress, underscoring the need for controlled editing environments (Tyagi et al., 2021).

Temporal restriction of nuclease activity represents an additional layer of precision. Delivering CRISPR components as RNP complexes ensures a rapid onset of editing, followed by prompt degradation, minimizing prolonged exposure (Kalter et al., 2025). Nanoparticle-based delivery platforms further refine this process by enabling controlled intracellular release, effectively narrowing the editing window (Ocampo et al., 2025). Such kinetic regulation has been shown to reduce off-target mutagenesis in neuronal cultures while preserving therapeutic efficacy in vivo (Ichinose et al., 2025).

6.3 Immunogenicity and immune mitigation strategies

Immunogenicity, the activation of host immune responses against therapeutic vectors or transgene products, is a central challenge in gene therapy for NDs. Both innate and adaptive immune mechanisms can limit efficacy, reduce the durability of expression, provoke inflammation in the CNS, and prevent repeat dosing (Zaiss and Muruve, 2005). Understanding these responses and applying targeted mitigation strategies are essential for safe and effective translation.

6.3.1 Immune mechanisms activated by gene therapy

Innate immune responses represent the first line of defense and can be triggered by vector components themselves. Viral capsid proteins and vector DNA are recognized by pattern recognition receptors such as Toll-like receptors (TLRs) and cytosolic DNA sensors, inducing cytokine secretion and inflammation that can impair transduction (Arjomandnejad et al., 2023; Jie et al., 2026). For example, AAV genomes can stimulate TLR9, while capsid proteins engage TLR2, leading to proinflammatory signaling in antigen-presenting cells (APCs). This innate activation not only reduces vector uptake but can also initiate downstream adaptive responses (Zaiss and Muruve, 2005).

Adaptive immunity encompasses both humoral and cellular responses. Preexisting or therapy-induced nAbs against viral capsids block vector entry into target cells, reducing gene transfer efficiency (Fang et al., 2021; Han et al., 2025). A significant proportion of humans carry binding and nAbs to common AAV serotypes due to natural exposure, and even low-titer nAbs can prevent effective in vivo transduction. In parallel, capsid or transgene-derived peptides presented on MHC molecules can activate $CD8^+$ cytotoxic T lymphocytes, leading to clearance of transduced cells and loss of therapeutic expression. Memory B cells and $CD4^+$ helper T cells further amplify these adaptive responses, complicating repeated administration and long-term efficacy (Arjomandnejad et al., 2023).

6.3.2 Strategies to mitigate immune responses

To maximize therapeutic benefit while minimizing adverse immunogenicity, strategies fall into three broad categories: vector-oriented, transgene-oriented, and host immune-modulatory approaches.

Vector-oriented immune modulation

Engineering the vector itself to evade immune detection is a powerful strategy. Rational design and screening for novel AAV capsids with reduced antigenicity can decrease recognition by nAbs and innate receptors. For example, newer vector variants with enhanced retinal tropism have shown lower sensitivity to human nAbs and may thus be less immunogenic overall. Synthetic capsids designed with computational and machine learning approaches are also under development to evade preexisting immunity and enhance tissue specificity (Han et al., 2025; Huang et al., 2025).

Also, genome deimmunization by modifying the viral genome to reduce CpG motifs that trigger TLR9 can attenuate innate inflammatory signaling (Du et al., 2025). In early work, vectors with CpG-depleted genomes demonstrated reduced markers of innate immune activation and more stable transgene expression in animal models (Faust et al., 2013; Xiang et al., 2020). Proteasome inhibitors and inhibitory oligonucleotides targeting TLR signaling have also been explored to limit capsid antigen presentation on major histocompatibility complex class I (MHC-I) and prevent the cascade that leads to adaptive immune activation (Fang et al., 2021).

Serotype switching and empty capsid decoys are other approaches to improve efficacy by avoiding neutralization using AAV serotypes with lower prevalence of preexisting immunity (e.g., AAV5 or AAVrh74 versus AAV2). In addition, administering empty capsids as decoys before the therapeutic vector has been shown to absorb circulating nAbs in animal models, enhancing the subsequent delivery of full vectors (Słyk et al., 2024; Huang et al., 2025).

Transgene-oriented immune mitigation

Reducing immune responses against the therapeutic gene product itself is critical for the long-term success of gene therapy (Widner, 2024). Strategic codon optimization and humanization of the transgene sequence can align codon usage with the host's preferences and remove immunogenic motifs such as CpG islands, thereby diminishing the likelihood that the expressed protein is perceived as foreign and lowering cytotoxic T lymphocyte and antibody responses (Keeler et al., 2018; Prasad et al., 2022). Such optimization has been shown to reduce $CD8^+$ T-cell activation against introduced transgenes by minimizing immunogenic sequence features within the expression cassette. In addition to sequence design, targeted expression and detargeting strategies further mitigate immune visibility. Using tissue-specific promoters restricts transgene expression to intended cell types and prevents ectopic expression in professional APCs, which are key drivers of adaptive immune priming (Perez et al., 2020). Experimental evidence demonstrates that replacing ubiquitous promoters with tissue-specific ones can abolish transgene-specific humoral and cellular responses by avoiding APC expression (Prasad et al., 2022).

Complementing promoter specificity, miRNA detargeting employs endogenous miRNA recognition elements incorporated into the transgene's untranslated regions to suppress expression in cells that express the corresponding miRNA (Sahu et al., 2025). This posttranscriptional silencing approach effectively detargets expression from off-target cell populations, including APCs, further reducing immune activation while preserving expression in therapeutic target cells.

Host immune modulation

Direct modulation of the patient's immune system can significantly enhance the durability of therapeutic transgene expression in gene therapy. Pharmacological immunosuppression, including systemic corticosteroids administered perivector infusion, is routinely used to dampen both innate and adaptive immune responses against viral vectors and the encoded transgene, helping to reduce inflammation and preserve expression. Clinically relevant regimens also incorporate agents such as rapamycin, which inhibits T- and B-cell activation and promotes regulatory T-cell (Treg) activity, thereby mitigating antivector and antitransgene immune responses (Zhang et al., 2025). Rapamycin-loaded tolerogenic nanoparticles (ImmTOR) have demonstrated the ability to induce antigen-specific immune tolerance, suppress antibody formation against AAV capsids, and enable successful vector readministration in preclinical models when coadministered with gene therapy vectors at moderate doses (Cattaneo et al., 2025).

An additional immune-targeted approach focuses on the induction of antigen-specific immune tolerance, for example, by incorporating Treg-promoting elements into gene therapy protocols or by administering regulatory cell therapies (Keeler et al., 2023). Preclinical studies indicate that such strategies can achieve sustained transgene expression even in the presence of initial immune activation (Wagner et al., 2021; Prasad et al., 2022). Antibody modulation is also critical when preexisting nAbs threaten therapeutic efficacy. Approaches such as plasmapheresis can transiently reduce circulating anti-AAV antibodies, lowering neutralizing titers sufficiently to permit successful vector delivery and potentially support subsequent dosing (Choi et al., 2022).

7 Integration of AI with CNS gene manipulation systems

The fusion of AI with gene-editing systems has transcended incremental improvement to become a foundational, predictive science of genomic intervention in the human brain. Machine learning and deep learning models have added a computational and predictive layer to editing workflows, one that increases precision and uniformity of edits in neurons, anticipates molecular repair outcomes, identifies optimal strategies for correcting neurodegenerative mutations, minimizes off-target activity, enables editing in nondividing cells, and ultimately lays the foundation for personalized genomic therapies (Dai et al., 2024; Dixit et al., 2024). Through this integration, gene editing has evolved from a predominantly trial-and-error process into a predictive, safe, and ultra-precise form of genomic engineering.

7.1 Predictive engineering and optimization of DNA repair

AI has become a transformative force in gene editing, particularly for NDs, by predicting repair outcomes, optimizing editing tools, and improving precision. Platforms like Pythia analyze extensive repair datasets to design highly targeted DNA repair templates, enabling accurate correction of mutations such as CAG expansions in HD and C9orf72 repeats in ALS, which are difficult to repair in neurons (Naert et al., 2024).

Across editing modalities, including adenine and cytosine base editing and prime editing, BE-Hive, DeepBE, DeepPE, PE-Designer, and PRIDICT AI models forecast editing efficiency, reduce off-target risks, and design neuron-specific gRNAs and pegRNAs (Mathis et al., 2023; Alipanahi et al., 2024). AI further strengthens epigenetic editing by predicting enhancer activity and guiding expression-level modulation of genes like BDNF, which is therapeutically relevant for AD. AI also improves interventions that rely on transcriptional regulation rather than sequence modification. The Xpresso model predicts CRISPRa/i outcomes with high accuracy. Structural prediction tools inspired by AlphaFold2 and RNAfold enhance editing efficiency by identifying accessible target sites within complex DNA/RNA structures (Schneider et al., 2023; Mao et al., 2024).

Predictive gRNA-design platforms DeepCRISPR, SPROUT, and CRISPR-Net evaluate chromatin context and genomic accessibility to optimize specificity in sensitive neurological genes (e.g., HTT, SOD1, TARDBP, APP, LRRK2) (Tiwari). AI further accelerates the development of Cas9 variants tailored for brain delivery, as seen with DeepCas9. Finally, AI is advancing RNA-editing approaches using Cas13. DeepCas13 and Cas13Predict tools design optimal gRNAs, assess mRNA accessibility, and model neuronal network effects, supporting reversible correction of toxic proteins tau, huntingtin, and α-synuclein without altering DNA (Kim et al., 2025).

7.2 Predictive systems for diagnosis, modeling, and mechanistic insight

At a foundational level, AI-driven models reconstruct complex molecular landscapes underlying neurodegeneration. These algorithms simulate dynamic interactions between key pathological proteins, offering deeper insights into the temporal evolution and spatial propagation of disease. In diagnostics, multimodal AI systems integrate diverse inputs, including neuroimaging, electrophysiology, transcriptomics, and circulating biomarkers. Such frameworks enable not only early and ultra-accurate detection, but also delineation of discrete molecular subtypes within diseases.

Multiscale data integration is increasingly tractable through advanced computational architectures. These systems unify genomic, imaging, and clinical data into coherent, interpretable models. Meanwhile, dynamic disease modeling, previously limited by biological complexity, is being reshaped by deep neural networks and modern dynamical systems theory.

7.3 Intelligent design of next-generation therapeutics and precision medicine

AI has also redefined the landscape of drug discovery. Computational models leveraging 3D structural biology and molecular dynamics now design therapeutic molecules with multitarget capabilities, a necessity for the multifactorial nature of NDs (Roy et al., 2025). On the clinical front, AI-enabled precision medicine platforms integrate genomic, epigenetic, metabolic, and clinical data to generate individualized treatment strategies. These systems identify optimal therapeutic combinations, predict treatment responses, and tailor interventions to each patient's molecular profile, paving the way for fully personalized neurotherapeutics (Kaur et al., 2024; Zhou et al., 2024b).

8 Conclusion and future outlook

Gene-editing technologies are fundamentally reshaping both the conceptual framework and therapeutic possibilities for NDs. What was once a bold theoretical proposition has rapidly matured into a sophisticated and highly precise toolkit. CRISPR-based platforms, in combination with base and PEs and RNA-targeting systems, now enable researchers to systematically interrogate and correct the genetic, epigenetic, and transcriptomic perturbations that drive neuronal dysfunction. These tools offer the ability to intervene in processes previously considered immutable, including toxic protein accumulation, loss of gene dosage, repeat expansion instability, and widespread transcriptomic dysregulation. Consequently, they redefine not only the specific aspects of disease that can be targeted but also the timing and context in which interventions are likely to be most effective across the trajectory of disease progression.

Equally transformative are advances in delivery technologies. Innovations in AAV capsid engineering, programmable lipid-based nanoparticles, and physical strategies for transiently modulating the blood–brain barrier have markedly expanded access to neuronal populations and circuits. These delivery systems are not merely passive carriers; they actively influence intracellular trafficking, engage DNA and RNA repair machinery, and modulate the delicate balance between on-target efficacy and off-target risk. The therapeutic impact of gene editors depends as much on the interplay between biological context, editor modality, and delivery strategy as it does on the molecular tools themselves.

Despite these advances, considerable challenges remain. Achieving a broad, stable, and predictable distribution of editing agents within the structurally and pathologically heterogeneous human brain continues to be a formidable engineering and biological challenge. Long-term safety is a critical concern; durable and specific edits, potential immune activation, and delayed adverse effects all demand rigorous investigation. Defining the optimal therapeutic window and developing sensitive biomarkers to monitor efficacy are equally imperative, particularly in multifactorial diseases such as AD and sporadic PD. Moreover, delivery strategies must be optimized to mitigate off-target effects, reduce immunogenicity, and ensure efficient and controlled editing in relevant tissues.

Beyond their potential to directly repair disease-causing mutations, CRISPR/Cas systems serve as a powerful lens for interrogating the genome, revealing novel regulatory elements, epigenetic modifications, and transcriptomic networks, and offering unparalleled mechanistic insight into neuronal function and dysfunction. Early preclinical and clinical studies are evaluating CRISPR therapies in accessible cell types, yet CNS-targeted applications require further optimization of delivery. Integrating these interventions with conventional small molecules offers a strategy to improve efficacy and reduce risk.

The field has now entered a phase defined by precision, integration, and iterative refinement. Next-generation editors offer unprecedented accuracy, capable of modulating the genome, epigenome, and transcriptome in a cell-type-specific manner. Delivery platforms are becoming increasingly intelligent, selective, and responsive to the local neural microenvironment. This evolution relies on the convergence of molecular biology, computational neuroscience, bioengineering, and advanced neuroimaging, providing the mechanistic insight and resolution necessary to translate complex molecular interventions into effective therapies. Ethical and regulatory considerations are equally critical and must be addressed to ensure the safe and responsible application of these technologies in humans.

Looking forward, the outlook is optimistic. Coordinated advances across gene-editing and delivery technologies are beginning to delineate a feasible roadmap for interventions capable of halting, and in some cases reversing, the progression of NDDs. Translation from bench to bedside will require carefully designed, evidence-based steps, underpinned by robust preclinical models, advanced biomarker development, and innovative clinical trial design. Ultimately, the promise of gene editing in the central nervous system lies not merely in correcting pathogenic mutations but in enabling interventions that are precise, durable, and personalized, heralding a transformative era in the treatment of some of the most challenging neurological diseases.

References

Abigail, S., Brandon, L., Isabelle, T., Katherine, M., Daniel, C., Kelly, W., 2025. Huntington's disease clinical research congress, October 11-13, 2025, Nashville, TN. J. Huntington's Dis. 14 (1S), S1–S50.

Adachi, H., Hengesbach, M., Yu, Y.-T., Morais, P., 2021. From antisense RNA to RNA modification: therapeutic potential of RNA-based technologies. Biomedicines 9 (5), 550.

Albright, B.H., Storey, C.M., Murlidharan, G., Rivera, R.M.C., Berry, G.E., Madigan, V.J., Asokan, A., 2018. Mapping the structural determinants required for AAVrh. 10 transport across the blood-brain barrier. Mol. Ther. 26 (2), 510–523.

Ali, A., Malla, B.A., Bhat, M.S., Bhat, J.I.A., 2026. CRISPR-Cas systems in neurological diseases. In: Ali, A., Shafique, F. (Eds.), Gene Editing by CRISPR-Cas. CRC Press, pp. 209–235.

Alipanahi, R., Safari, L., Khanteymoori, A., 2024. DTMP-prime: a deep transformer-based model for predicting prime editing efficiency and PegRNA activity. Mol. Ther. Nucleic Acids 35 (4), 102370.

Alkahtani, S., Al-Johani, N.S., Alarifi, S., 2023. Mechanistic insights, treatment paradigms, and clinical progress in neurological disorders: current and future prospects. Int. J. Mol. Sci. 24 (2), 1340.

Allemailem, K.S., Almatroudi, A., Alrumaihi, F., Alradhi, A.E., Theyab, A., Algahtani, M., Alhawas, M.O., Dobie, G., Moawad, A.A., Rahmani, A.H., 2024. Current updates of CRISPR/Cas system and anti-CRISPR proteins: innovative applications to improve the genome editing strategies. Int. J. Nanomedicine 19, 10185–10212.

Aly, A.-E.-E., Sun, T., Zhang, Y., Li, Z., Kyada, M., Ma, Q., Padegimas, L., Sesenoglu-Laird, O., Cooper, M.J., McDannold, N.J., 2023. Focused ultrasound enhances transgene expression of intranasal hGDNF DNA nanoparticles in the sonicated brain regions. J. Control. Release 358, 498–509.

Amanat, M., Nemeth, C.L., Fine, A.S., Leung, D.G., Fatemi, A., 2022. Antisense oligonucleotide therapy for the nervous system: from bench to bedside with emphasis on pediatric neurology. Pharmaceutics 14 (11), 2389.

Aquilina, K., Chakrapani, A., Carr, L., Kurian, M., Hargrave, D., 2022. Convection-enhanced delivery in children: techniques and applications. Adv. Tech. Stand. Neurosurg. 45, 199–228.

Arango, D., Bittar, A., Esmeral, N.P., Ocasión, C., Muñoz-Camargo, C., Cruz, J.C., Reyes, L.H., Bloch, N.I., 2021. Understanding the potential of genome editing in Parkinson's disease. Int. J. Mol. Sci. 22 (17), 9241.

Arduini, A., Katiyar, H., Liang, C., 2025. Progress in pseudotyping lentiviral vectors towards cell-specific gene delivery in vivo. Viruses 17 (6), 802.

Arjomandnejad, M., Dasgupta, I., Flotte, T.R., Keeler, A.M., 2023. Immunogenicity of recombinant adeno-associated virus (AAV) vectors for gene transfer. BioDrugs 37 (3), 311.

Arnould, S., Delenda, C., Grizot, S., Desseaux, C., Pâques, F., Silva, G., Smith, J., 2011. The I-CreI meganuclease and its engineered derivatives: applications from cell modification to gene therapy. Protein Eng. Des. Sel. 24 (1-2), 27–31.

Arora, S., Singh, J., 2021. In vitro and in vivo optimization of liposomal nanoparticles based brain targeted vgf gene therapy. Int. J. Pharm. 608, 121095.

Bae, S., Song, B., 2023. Base Editors. Springer.

Bagherpour, I., Mozafari, M., Naghib, S.M., 2025. Inorganic nanoparticles-based drug delivery systems for neurodegenerative diseases therapy. Curr. Pharm. Des. 31 (25), 1998–2024.

Bakrbaldawi, A.A.A., Zhu, Z., Zheng, Z., Zhu, J., Jiang, H., 2024. An overview of MR-guided laser interstitial thermal therapy (MRg-LITT) in disrupting the blood-brain barrier: efficacy and duration. J. Integr. Neurosci. 23 (7), 130.

Barrera-Paez, J.D., Bacman, S.R., Balla, T., Van Booven, D., Gannamedi, D.P., Stewart, J.B., Mok, B., Liu, D.R., Lombard, D.B., Griswold, A.J., 2025. Correcting a pathogenic mitochondrial DNA mutation by base editing in mice. Sci. Transl. Med. 17(783), eadr0792.

Beach, M.A., Nayanathara, U., Gao, Y., Zhang, C., Xiong, Y., Wang, Y., Such, G.K., 2024. Polymeric nanoparticles for drug delivery. Chem. Rev. 124 (9), 5505–5616.

Becker, S., Boch, J., 2021. TALE and TALEN genome editing technologies. Gene Genome Ed. 2, 100007.

BenDavid, E., Ramezanian, S., Lu, Y., Rousseau, J., Schroeder, A., Lavertu, M., Tremblay, J.P., 2024. Emerging perspectives on prime editor delivery to the brain. Pharmaceuticals 17 (6), 763.

Bennett, C.F., Swayze, E.E., 2010. RNA targeting therapeutics: molecular mechanisms of antisense oligonucleotides as a therapeutic platform. Annu. Rev. Pharmacol. Toxicol. 50 (1), 259–293.

Bido, S., Nannoni, M., Muggeo, S., Gambarè, D., Ruffini, G., Bellini, E., Passeri, L., Iaia, S., Luoni, M., Provinciali, M., 2024. Microglia-specific IL-10 gene delivery inhibits neuroinflammation and neurodegeneration in a mouse model of Parkinson's disease. Sci. Transl. Med. 16(761), eadm8563.

Birchard, D., 2025. NVT-001: re-Mapping Memory with Dual-Vector Gene Therapy. Zenodo. doi:10.5281/zenodo.17831826.

Blaschke, T., Bajorath, J., 2022. Fine-tuning of a generative neural network for designing multi-target compounds. J. Comput. Aided Mol. Des. 36 (5), 363–371.

Blesa, J., Pineda-Pardo, J.A., Inoue, K.-I., Gasca-Salas, C., Balzano, T., Rey, N.L.-G.D., Reinares-Sebastián, A., Esteban-García, N., Rodríguez-Rojas, R., Márquez, R., 2023. BBB opening with focused ultrasound in nonhuman primates and Parkinson's disease patients: targeted AAV vector delivery and PET imaging. Sci. Adv. 9(16), eadf4888.

Blessing, D., Déglon, N., 2016. Adeno-associated virus and lentivirus vectors: a refined toolkit for the central nervous system. Curr. Opin. Virol. 21, 61–66.

Bliss, C.M., Hulin-Curtis, S.L., Williams, M., Marušková, M., Davies, J.A., Statkute, E., Baker, A.T., Stack, L., Kerstetter, L., Kerr-Jones, L.E., 2024. A pseudotyped adenovirus serotype 5 vector with serotype 49 fiber knob is an effective vector for vaccine and gene therapy applications. Mol. Ther. Methods Clin. Dev. 32 (3), 101308.

Boado, R.J., 2025. Brain delivery of biotherapeutics via receptor-mediated transcytosis across the blood–brain barrier. RSC Pharm. 2 (6), 1349–1375.

Böck, D., Tidecks, L., Wilhelm, M., Weber, Y., Ioannidi, E., Mumenthaler, J., Rothgangl, T., Schmidheini, L., Janjuha, S., Patriarchi, T., 2023. Prime editing of the β1 adrenoceptor in the brain reprograms mouse behavior. BioRxiv. doi:10.1101/2023.05.19.541410.

Borel, F., Gernoux, G., Cardozo, B., Metterville, J.P., Cabrera, G.T., Song, L., Su, Q., Gao, G.P., Elmallah, M.K., Brown Jr, R.H., 2016. Therapeutic rAAVrh10 mediated SOD1 silencing in adult SOD1 G93A mice and nonhuman primates. Hum. Gene Ther. 27 (1), 19–31.

Bunting, E.L., Donaldson, J., Cumming, S.A., Olive, J., Broom, E., Miclăuş, M., Hamilton, J., Tegtmeyer, M., Zhao, H.T., Brenton, J., 2025. Antisense oligonucleotide–mediated MSH3 suppression reduces somatic CAG repeat expansion in Huntington's disease iPSC–derived striatal neurons. Sci. Transl. Med. 17(785), eadn4600.

Cai, X., Drummond, C.J., Zhai, J., Tran, N., 2024. Lipid nanoparticles: versatile drug delivery vehicles for traversing the blood brain barrier to treat brain cancer. Adv. Funct. Mater. 34(41), 2404234.

Caraway, C.A., Gaitsch, H., Wicks, E.E., Kalluri, A., Kunadi, N., Tyler, B.M., 2022. Polymeric nanoparticles in brain cancer therapy: a review of current approaches. Polymers 14 (14), 2963.

Carroll, D., Morton, J.J., Beumer, K.J., Segal, D.J., 2006. Design, construction and in vitro testing of zinc finger nucleases. Nat. Protoc. 1 (3), 1329–1341.

Cattaneo, S., Bettegazzi, B., Ingusci, S., Verlengia, G., Sofia, T.A., Silvia, Z., Codazzi, F., Morelli, M.J., Marzulli, M., Glorioso, J.C., 2025. Genetic mutations in HSV-1 replication-defective vectors: implications for their safety in gene therapy applications. Gene Ther. 32 (6), 581–593.

Chakkyarath, V., 2025. Applications of advanced ASOs, RNA mimics, and others in RNA-based therapeutics. In: Sharma, N.K., Dakal, T.C., Maurya, P.K. (Eds.), RNA-Based Cancer Therapeutics: Techniques, Applications and Challenges in Prognosis. Springer Nature Singapore, pp. 35–62.

Challis, R.C., Kumar, S.R., Chen, X., Goertsen, D., Coughlin, G.M., Hori, A.M., Chuapoco, M.R., Otis, T.S., Miles, T.F., Gradinaru, V., 2022. Adeno-associated virus toolkit to target diverse brain cells. Annu. Rev. Neurosci. 45 (1), 447–469.

Chamakioti, M., Karantzelis, N., Taraviras, S., 2022. Advanced gene-targeting therapies for motor neuron diseases and muscular dystrophies. Int. J. Mol. Sci. 23 (9), 4824.

Chatterjee, D., Marmion, D.J., McBride, J.L., Manfredsson, F.P., Butler, D., Messer, A., Kordower, J.H., 2022. Enhanced CNS transduction from AAV. PHP. eB infusion into the cisterna magna of older adult rats compared to AAV9. Gene Ther. 29 (6), 390–397.

Chen, A.-P., Gao, P., Ashok, P., Ma, M.C., Zou, D., Chambers, A., Lin, L., Chang, H.-M., Arulanandam, A., Eyquem, J., 2022a. 323 TALEN®-based gene edited iPSC-derived NK (iNK) cells demonstrate enhanced antitumor activity. BMJ Spec. J.

Chen, J., Li, Z., Chen, L., 2022b. Reactive glia-to-GABAergic neuron reprogramming: a "golden touch" strategy to alleviate intractable seizures. Acta Epileptol. 4 (1), 35.

Chen, R.Z., Vogt, T.F., 2024. Gene editing for HD: therapeutic prospects. In: Yang, X.W., Thompson, L.M., Heiman, M. (Eds.), Huntington's Disease. Academic Press, pp. 551–570.

Chen, S.-H., He, B., Singh, S., Martin, N.P., 2023. Vector tropism. In: Eldridge, M.A.G., Galvan, A. (Eds.), Vectorology for Optogenetics and Chemogenetics, Neuromethods, vol. 195. Springer, pp. 105–123.

Chen, S., Liu, Z., Lo, C.-H., Wang, Q., Ning, K., Zhang, Q., Zhao, J., Shen, Y., Sun, Y., 2025. Gene therapy for ocular hypertension using hfCas13d-mediated mRNA targeting. PNAS Nexus 4(6), pgaf168.

Chen, X. (2023). Non-invasive functional gene delivery to the central and peripheral nervous system across species, California Institute of Technology.

Chen, Z.-A., Wu, C.-H., Wu, S.-H., Huang, C.-Y., Mou, C.-Y., Wei, K.-C., Yen, Y., Chien, I.-T., Runa, S., Chen, Y.-P., 2024. Receptor ligand-free mesoporous silica nanoparticles: a streamlined strategy for targeted drug delivery across the blood–brain barrier. ACS Nano 18 (20), 12716–12736.

Chevalier, B., Monnat Jr, R.J., Stoddard, B.L., 2005. The LAGLIDADG homing endonuclease family. In: Belfort, M., Derbyshire, V., Stoddard, B.L., Wood, D.L. (Eds.), Homing Endonucleases and Inteins. Springer, pp. 33–47.

Choi, H., Choi, K., Kim, D.-H., Oh, B.-K., Yim, H., Jo, S., Choi, C., 2022. Strategies for targeted delivery of exosomes to the brain: advantages and challenges. Pharmaceutics 14 (3), 672.

Chung, S.-K., Lee, S.-Y., 2022. Advances in gene therapy techniques to treat LRRK2 gene mutation. Biomolecules 12 (12), 1814.

Clark, I.H., Roman, A., Fellows, E., Radha, S., Var, S.R., Roushdy, Z., Borer, S.M., Johnson, S., Chen, O., Borgida, J.S., 2022. Cell reprogramming for regeneration and repair of the nervous system. Biomedicines 10 (10), 2598.

Coller, J., Ignatova, Z., 2024. tRNA therapeutics for genetic diseases. Nat. Rev. Drug Discov. 23 (2), 108–125.

Contardo, M., De Gioia, R., Gagliardi, D., Comi, G.P., Ottoboni, L., Nizzardo, M., Corti, S., 2022. Targeting PTB for glia-to-neuron reprogramming in vitro and in vivo for therapeutic development in neurological diseases. Biomedicines 10 (2), 399.

Costa, C.P., Moreira, J.N., Lobo, J.M.S., Silva, A.C., 2021. Intranasal delivery of nanostructured lipid carriers, solid lipid nanoparticles and nanoemulsions: a current overview of in vivo studies. Acta Pharm. Sin. B 11 (4), 925–940.

Cox, D.B., Gootenberg, J.S., Abudayyeh, O.O., Franklin, B., Kellner, M.J., Joung, J., Zhang, F., 2017. RNA editing with CRISPR-Cas13. Science 358 (6366), 1019–1027.

Cregan, S., MacLaurin, J., Gendron, T., Callaghan, S., Park, D., Parks, R., Graham, F., Morley, P., Slack, R., 2000. Helper-dependent adenovirus vectors: their use as a gene delivery system to neurons. Gene Ther. 7 (14), 1200–1209.

Cuddleston, W.H., Fan, X., Sloofman, L., Liang, L., Mossotto, E., Moore, K., Zipkowitz, S., Wang, M., Zhang, B., Wang, J., 2022. Spatiotemporal and genetic regulation of A-to-I editing throughout human brain development. Cell Rep. 41 (5), 111585.

Cui, Y., Xu, J., Cheng, M., Liao, X., Peng, S., 2018. Review of CRISPR/Cas9 sgRNA design tools. Interdiscip. Sci. Comput. Life Sci. 10 (2), 455–465.

Dai, M., Li, X., Zhang, Q., Liang, T., Huang, X., Fu, Q., 2024. Health research in the era of artificial intelligence: advances in gene-editing study. Medicine Plus 1(2), 100027.

Daksh, S., Bose, P., Kumar, N., Ojha, H., Deep, S., Datta, A., 2025. Mesoporous silica nanoparticles for targeted Aβ42 imaging and therapy in traumatic brain injury: enhanced blood–brain barrier penetration and sustained drug delivery. ACS Appl. Nano Mater. 8 (22), 11322–11337.

Dang, Y., Jia, G., Choi, J., Ma, H., Anaya, E., Ye, C., Shankar, P., Wu, H., 2015. Optimizing sgRNA structure to improve CRISPR-Cas9 knockout efficiency. Genome Biol. 16 (1), 280.

Darios, F., Stevanin, G., 2020. Impairment of lysosome function and autophagy in rare neurodegenerative diseases. J. Mol. Biol. 432 (8), 2714–2734.

Datta, A., Sarmah, D., Kaur, H., Chaudhary, A., Vadak, N., Borah, A., Shah, S., Wang, X., Bhattacharya, P., 2023. Advancement in CRISPR/Cas9 technology to better understand and treat neurological disorders. Cell. Mol. Neurobiol. 43 (3), 1019–1035.

de Lagrán, M.M., Bascón-Cardozo, K., Dierssen, M., 2024. Neurodevelopmental disorders: 2024 update. Free Neuropathol. 5, 20.

Dehghani, S., Ocakcı, O., Hatipoglu, P.T., Özalp, V.C., Tevlek, A., 2025. Exosomes as biomarkers and therapeutic agents in neurodegenerative diseases: current insights and future directions. Mol. Neurobiol. 62 (7), 9190–9215.

Devinsky, O., Coller, J., Ahrens-Nicklas, R., Liu, X.S., Ahituv, N., Davidson, B.L., Bishop, K.M., Weiss, Y., Mingorance, A., 2025. Gene therapies for neurogenetic disorders. Trends Mol. Med. 31 (9), 814–826.

Di Berardino, C., Massimino, L., Ungaro, F., Colasante, G., 2024. Gene therapy for Dravet syndrome: promises and impact on disease trigger and secondary modifications. Rare Dis. Orphan Drugs J. 3 (3), N/A.

Ding, F., Hou, R., Han, B., Fang, X., 2025. Cell membrane-and vesicle-based bionic nanodrugs: applications in central nervous system diseases and exploration of nasal–cerebral delivery. Gels 11 (11), 846.

Dixit, S., Kumar, A., Srinivasan, K., Vincent, P.D.R., Krishnan, N.R., 2024. Advancing genome editing with artificial intelligence: opportunities, challenges, and future directions. Front. Bioeng. Biotechnol. 11, 1335901.

Dogbey, D.M., Torres, V.E.S., Fajemisin, E., Mpondo, L., Ngwenya, T., Akinrinmade, O.A., Perriman, A.W., Barth, S., 2023. Technological advances in the use of viral and non-viral vectors for delivering genetic and non-genetic cargos for cancer therapy. Drug Deliv. Transl. Res. 13 (11), 2719–2738.

Dokukin, N.V., Chudakova, D.A., Shkap, M.O., Kovalchuk, A.M., Kibirsky, P.D., Baklaushev, V.P., 2025. Direct neural reprogramming in situ: existing approaches and their optimization. Biochemistry (Moscow) 90 (2), 214–230.

Dou, D., Lu, J., Chen, X., Zhang, X., 2025. Key challenges and optimization practices for non-clinical evaluation of AAV based gene therapy for central nervous system disease. Pharm. Res. 42 (10), 1661–1675.

Dredge, W.H., Kailash, B., Réal, A., Muller, B.Z., Jang, B., Fu, T., Seo, H., Won, -H.-H., Breen, M.S., Xiao, X., 2025. Meta-analysis of genetic regulation of RNA editing in the human brain identifies new genes underlying neurological disease. medRxiv. doi:10.1101/2025.09.30.25337026.

Drongitis, D., Verrillo, L., De Bellis, A., Miano, M.G., 2025. RNA-based therapies for neurodevelopmental disorders: innovative tools for molecular correction. Front. Mol. Biosci. 12, 1681647.

Du, M., Yu, Y., Wang, J., Ji, C., 2025. Autophagy–lysosome pathway dysfunction in neurodegeneration and cancer: mechanisms and therapeutic opportunities. Int. J. Mol. Sci. 27 (1), 366.

Duan, W., Urani, E., Mattson, M.P., 2023. The potential of gene editing for Huntington's disease. Trends Neurosci. 46 (5), 365–376.

Dupuy, A., Valton, J., Leduc, S., Armier, J., Galetto, R., Gouble, A., Lebuhotel, C., Stary, A., Pâques, F., Duchateau, P., 2013. Targeted gene therapy of xeroderma pigmentosum cells using meganuclease and TALEN™. PLoS ONE 8 (11), e78678.

Dürrenberger, F., Rochaix, J., 1991. Chloroplast ribosomal intron of *Chlamydomonas reinhardtii*: in vitro self-splicing, DNA endonuclease activity and in vivo mobility. EMBO J. 10 (11), 3495–3501.

Džidić-Krivić, A., Fajkić, A., Farhat, E.K., Lekić, L., Ejubović, A., Vukas, S.K., Ejubović, M., Lepara, O., Sher, E.K., 2025. Linking metabolic syndrome to neurodegeneration mechanisms and potential treatments. Mol. Neurobiol. 62 (11), 14344–14366.

El Kassem, G., Hillmer, J., Boettcher, M., 2025. Evaluation of Cas13d as a tool for genetic interaction mapping. Nat. Commun. 16 (1), 1631.

Elder, J.B., Lonser, R.R., 2025. Direct convective delivery for nervous system gene therapy. Neurosurg. Clin. 36 (1), 101–111.

Ely, A., Naidoo, T., Arbuthnot, P., 2009. Efficient silencing of gene expression with modular trimeric Pol II expression cassettes comprising microRNA shuttles. Nucleic Acids Res. 37 (13), e91.

Everett, W.H., Bucelli, R.C., 2024. Tofersen for SOD1 ALS. Neurodegener. Dis. Manag. 14 (5), 149–160.

Fajardo-Serrano, A., Rico, A.J., Roda, E., Honrubia, A., Arrieta, S., Ariznabarreta, G., Chocarro, J., Lorenzo-Ramos, E., Pejenaute, A., Vázquez, A., 2021. Adeno-associated viral vectors as versatile tools for Parkinson's research, both for disease modeling purposes and for therapeutic uses. Int. J. Mol. Sci. 22 (12), 6389.

Fan, T., Peng, J., Liang, H., Chen, W., Wang, J., Xu, R., 2026. Potential common pathogenesis of several neurodegenerative diseases. Neural Regen. Res. 21 (3), 972–988.

Fan, Y., Xu, C., Deng, N., Gao, Z., Jiang, Z., Li, X., Zhou, Y., Pei, H., Li, L., Tang, B., 2023. Understanding drug nanocarrier and blood–brain barrier interaction based on a microfluidic microphysiological model. Lab Chip 23 (7), 1935–1944.

Fang, Z., Chen, L., Moser, M.A., Zhang, W., Qin, Z., Zhang, B., 2021. Electroporation-based therapy for brain tumors: a review. J. Biomech. Eng. 143(10), 100802.

Faust, S.M., Bell, P., Cutler, B.J., Ashley, S.N., Zhu, Y., Rabinowitz, J.E., Wilson, J.M., 2013. CpG-depleted adeno-associated virus vectors evade immune detection. J. Clin. Invest. 123 (7), 2994–3001.

Feja, M., Drath, I., Weiß, S., Ewe, A., Gericke, B., Outeiro, T.F., Stefanis, L., Aigner, A., Richter, F., 2025. Nose-to-brain siRNA delivery by PEI/PPI-based nanoparticles reduces α-synuclein expression in a Parkinson's disease mouse model. Mol. Ther. Nucleic Acids 36 (3), 102671.

Finkel, R.S., Chiriboga, C.A., Vajsar, J., Day, J.W., Montes, J., De Vivo, D.C., Yamashita, M., Rigo, F., Hung, G., Schneider, E., 2016. Treatment of infantile-onset spinal muscular atrophy with nusinersen: a phase 2, open-label, dose-escalation study. The Lancet 388 (10063), 3017–3026.

Fischell, J.M., Fishman, P.S., 2021. A multifaceted approach to optimizing AAV delivery to the brain for the treatment of neurodegenerative diseases. Front. Neurosci. 15, 747726.

Flotte, T.R., Gessler, D.J., 2022. Gene therapy for rare neurological disorders. Clin. Pharmacol. Ther. 111 (4), 743–757.

Gadhave, D.G., Sugandhi, V.V., Jha, S.K., Nangare, S.N., Gupta, G., Singh, S.K., Dua, K., Cho, H., Hansbro, P.M., Paudel, K.R., 2024. Neurodegenerative disorders: mechanisms of degeneration and therapeutic approaches with their clinical relevance. Ageing Res. Rev. 99, 102357.

Gammage, P.A., Viscomi, C., Simard, M.-L., Costa, A.S., Gaude, E., Powell, C.A., Van Haute, L., McCann, B.J., Rebelo-Guiomar, P., Cerutti, R., 2018. Genome editing in mitochondria corrects a pathogenic mtDNA mutation in vivo. Nat. Med. 24 (11), 1691–1695.

Gao, G.-P., Alvira, M.R., Wang, L., Calcedo, R., Johnston, J., Wilson, J.M., 2002. Novel adeno-associated viruses from rhesus monkeys as vectors for human gene therapy. Proc. Natl. Acad. Sci. U. S. A. 99 (18), 11854–11859.

Gao, X., Zhou, C., Feng, Y., Ye, B., Zhao, Z., Qi, L., Hu, L., Deng, Y., Lin, C., Ding, Q., 2025. Research progress of gene editing technology in neurological diseases. Gene 962, 149534.

Gao, Z., 2024. Strategies for enhanced gene delivery to the central nervous system. Nanoscale Adv. 6 (12), 3009–3028.

García-González, N., Gonçalves-Sánchez, J., Gómez-Nieto, R., Gonçalves-Estella, J.M., López, D.E., 2024. Advances and challenges in gene therapy for neurodegenerative diseases: a systematic review. Int. J. Mol. Sci. 25(23), 12485.

Géraudie, A., Riche, M., Lestra, T., Trotier, A., Dupuis, L., Mathon, B., Carpentier, A., Delatour, B., 2023. Effects of low-intensity pulsed ultrasound-induced blood–brain barrier opening in P301S mice modeling Alzheimer's disease tauopathies. Int. J. Mol. Sci. 24(15), 12411.

Germain, N.D., Chung, W.K., Sarmiere, P.D., 2023. RNA interference (RNAi)-based therapeutics for treatment of rare neurologic diseases. Mol. Aspects Med. 91, 101148.

Gersbach, C.A., Gaj, T., Barbas III, C.F., 2014. Synthetic zinc finger proteins: the advent of targeted gene regulation and genome modification technologies. Acc. Chem. Res. 47 (8), 2309–2318.

Gitler, A.D., Dhillon, P., Shorter, J., 2017. Neurodegenerative disease: models, mechanisms, and a new hope. Dis. Model. Mech. 10 (5), 499–502.

Glass, Z., Lee, M., Li, Y., Xu, Q., 2018. Engineering the delivery system for CRISPR-based genome editing. Trends Biotechnol. 36 (2), 173–185.

Glazier, D.A., Liao, J., Roberts, B.L., Li, X., Yang, K., Stevens, C.M., Tang, W., 2020. Chemical synthesis and biological application of modified oligonucleotides. Bioconjug. Chem. 31 (5), 1213–1233.

González Molina, L.A., Dolga, A.M., Rots, M.G., Sarno, F., 2025. The promise of epigenetic editing for treating brain disorders. In: van Zundert, B., Montecino, M. (Eds.), Neuroepigenetics: Mechanisms in Health and Disease, Subcellular Biochemistry, vol. 108, pp. 111–190.

Goyal, A., Verma, A., Dubey, N., Raghav, J., Agrawal, A., 2022. Naringenin: a prospective therapeutic agent for Alzheimer's and Parkinson's disease. J. Food Biochem. 46(12), e14415.

Granados-Riveron, J.T., Aquino-Jarquin, G., 2021. CRISPR/Cas13-based approaches for ultrasensitive and specific detection of microRNAs. Cells 10 (7), 1655.

Guo, C., Ma, X., Gao, F., Guo, Y., 2023. Off-target effects in CRISPR/Cas9 gene editing. Front. Bioeng. Biotechnol. 11, 1143157.

Habib, S., Singh, M., 2022. Angiopep-2-modified nanoparticles for brain-directed delivery of therapeutics: a review. Polymers (Basel) 14 (4), 712.

Hackett, N.R., Crystal, R.G., 2025. Four decades of adenovirus gene transfer vectors: history and current use. Mol. Ther. 33 (5), 2192–2204.

Hallock, H.L., Quillian IV, H.M., Maynard, K.R., Mai, Y., Chen, H.-Y., Hamersky, G.R., Shin, J.H., Maher, B.J., Jaffe, A.E., Martinowich, K., 2020. Molecularly defined hippocampal inputs regulate population dynamics in the prelimbic cortex to suppress context fear memory retrieval. Biol. Psychiatry 88 (7), 554–565.

Han, J., Bai, H., Li, F., Zhang, Y., Zhou, Q., Li, W., 2025. Engineering a streamlined virus-like particle for programmable tissue-specific gene delivery. Nat. Commun. 16 (1), 9157.

Hao, C., Ji, H., 2025. Gene silencing therapies for Huntington's disease. Neurol. Lett. 4 (1), 69–75.

Haque, S.T., Chowdhury, E.H., 2018. Recent progress in delivery of therapeutic and imaging agents utilizing organic-inorganic hybrid nanoparticles. Current Drug Delivery 15 (4), 485–496.

Haro-Martínez, E., Muscolino, E., Moral, N., Duran, J., Fornaguera, C., 2025. Crossing the blood–brain barrier: nanoparticle-based strategies for neurodegenerative disease therapy. Drug Deliv. Transl. Res. 16 (3), 797–824.

Hasanzadeh, E., Seifalian, A., Mellati, A., Saremi, J., Asadpour, S., Enderami, S.E., Nekounam, H., Mahmoodi, N., 2023. Injectable hydrogels in central nervous system: unique and novel platforms for promoting extracellular matrix remodeling and tissue engineering. Mater. Today Bio 20, 100614.

He, Y.-N., Zhu, -H.-H., Zhou, Z.-H., Qu, -K.-K., 2025. Exosomal microRNAs in common mental disorders: mechanisms, biomarker potential and therapeutic implications. World J. Psychiatry 15(8), 108933.

Helal, M.M., Ibrahim, A.A., Beddor, A., Kashbour, M., 2025. Breaking barriers in Huntington's disease therapy: focused ultrasound for targeted drug delivery. Neurochem. Res. 50 (1), 68.

Holm, A., Hansen, S.N., Klitgaard, H., Kauppinen, S., 2022. Clinical advances of RNA therapeutics for treatment of neurological and neuromuscular diseases. RNA Biol. 19 (1), 594–608.

Hong, Y., Song, Y., Wang, W., Shi, J., Chen, X., 2025. Mitochondrial DNA editing: key to the treatment of neurodegenerative diseases. Genes Dis. 12(4), 101437.

Houldsworth, A., 2024. Role of oxidative stress in neurodegenerative disorders: a review of reactive oxygen species and prevention by antioxidants. Brain Commun. 6(1), fcad356.

Hours, C., Vayssière, P., Gressens, P., Laforge, M., 2025. Immunity in neuromodulation: probing neural and immune pathways in brain disorders. J. Neuroinflammation 22 (1), 1–38.

Howe, M.P., Patani, R., 2023. Nonsense-mediated mRNA decay in neuronal physiology and neurodegeneration. Trends Neurosci. 46 (10), 879–892.

Huang, H.-C., Wu, L.-F., Liu, K., Ma, B.-G., 2025. Opto-CRISPR: new prospects for gene editing and regulation. Trends Biotechnol. 43 (1), 10–23.

Huang, X., Li, X., Yang, L., Wang, P., Yan, J., Nie, Z., Gao, Y., Li, Z., Wen, J., Cao, X., 2022. Construction and optimization of herpes simplex virus vectors for central nervous system gene delivery based on CRISPR/Cas9-mediated genome editing. Curr. Gene Ther. 22 (1), 66–77.

Humbel, M., Ramosaj, M., Zimmer, V., Regio, S., Aeby, L., Moser, S., Boizot, A., Sipion, M., Rey, M., Déglon, N., 2021. Maximizing lentiviral vector gene transfer in the CNS. Gene Ther. 28 (1), 75–88.

Huq, T.B., Jeeja, P.A.K., Dam, S.K., Das, S.S., Vivero-Escoto, J.L., 2025. Recent applications of mesoporous silica nanoparticles in gene therapy. Adv. Healthc. Mater. 14, 2404781.

Hussen, B.M., Rasul, M.F., Abdullah, S.R., Hidayat, H.J., Faraj, G.S.H., Ali, F.A., Salihi, A., Baniahmad, A., Ghafouri-Fard, S., Rahman, M., 2023. Targeting miRNA by CRISPR/Cas in cancer: advantages and challenges. Mil. Med. Res. 10 (1), 32.

Hwang, H.-Y., Gim, D., Yi, H., Jung, H., Lee, J., Kim, D., 2024. Precise editing of pathogenic nucleotide repeat expansions in iPSCs using paired prime editor. Nucleic Acids Res. 52 (10), 5792–5803.

Ichinose, M., Teramoto, T., Nakamura, I., Shimajiri, Y., Yagi, Y., Gutmann, B., 2025. Fine-tuning of the PPR protein directs the RNA editing activity toward C-to-U or U-to-C conversion. Sci. Rep. 15 (1), 6288.

Ingusci, S., Goins, W.F., Cohen, J.B., Miyagawa, Y., Knipe, D.M., Glorioso, J.C., 2025. Next-generation replication-defective HSV vectors for delivery of large DNA payloads. Mol. Ther. 33 (5), 2205–2216.

Inoue, S., Nishimura, K., Gima, S., Nakano, M., Takata, K., 2023. CRISPR-Cas9-edited SNCA knockout human induced pluripotent stem cell-derived dopaminergic neurons and their vulnerability to neurotoxicity. Biol. Pharm. Bull. 46 (3), 517–522.

Iqubal, A., Iqubal, M.K., Khan, A., Ali, J., Baboota, S., Haque, S.E., 2020. Gene therapy, a novel therapeutic tool for neurological disorders: current progress, challenges and future prospective. Curr. Gene Ther. 20 (3), 184–194.

Isik, S., Osman, S., Yeman-Kiyak, B., Shamshir, S.R.M., Sanchez, N.M.E., 2025. Advances in neurodegenerative disease therapy: stem cell clinical trials and promise of engineered exosomes. CNS Neurosci. Ther. 31(9), e70577.

Jafari, B., Pourseif, M.M., Barar, J., Rafi, M.A., Omidi, Y., 2019. Peptide-mediated drug delivery across the blood-brain barrier for targeting brain tumors. Expert Opin. Drug Deliv. 16 (6), 583–605.

Jagaran, K., Singh, M., 2022. Lipid nanoparticles: promising treatment approach for Parkinson's disease. Int. J. Mol. Sci. 23 (16), 9361.

Jiang, S., Cai, G., Yang, Z., Shi, H., Zeng, H., Ye, Q., Hu, Z., Wang, Z., 2024. Biomimetic nanovesicles as a dual gene delivery system for the synergistic gene therapy of Alzheimer's disease. ACS Nano 18 (18), 11753–11768.

Jie, J., Yao, X., Deng, H., Zhou, Y., Jiang, X., Dai, X., Yang, Y., Yang, P., 2026. Regulatory T cells in neurological disorders and tissue regeneration: mechanisms of action and therapeutic potentials. Neural Regen. Res. 21 (4), 1277–1291.

Jimeno, S., Prados-Carvajal, R., Fernández-Ávila, M.J., Silva, S., Silvestris, D.A., Endara-Coll, M., Rodríguez-Real, G., Domingo-Prim, J., Mejías-Navarro, F., Romero-Franco, A., 2021. ADAR-mediated RNA editing of DNA: RNA hybrids is required for DNA double strand break repair. Nat. Commun. 12 (1), 5512.

Jo, D.H., Song, D.W., Cho, C.S., Kim, U.G., Lee, K.J., Lee, K., Park, S.W., Kim, D., Kim, J.H., Kim, J.-S., 2019. CRISPR-Cas9–mediated therapeutic editing of Rpe65 ameliorates the disease phenotypes in a mouse model of Leber congenital amaurosis. Sci. Adv. 5(10), eaax1210.

Jolesz, F.A., McDannold, N.J., 2013. MRI-guided focused ultrasound. In: Jolesz, F.A. (Ed.), Intraoperative Imaging and Image-Guided Therapy. pp. 403–412.

Jollé, C., Déglon, N., Pythoud, C., Bouzier-Sore, A.-K., Pellerin, L., 2019. Development of efficient AAV2/DJ-based viral vectors to selectively downregulate the expression of neuronal or astrocytic target proteins in the rat central nervous system. Front. Mol. Neurosci. 12, 201.

Jorge, A.F., Grijalvo, S., Aviñó, A., Fàbrega, C., Eritja, R., 2025. 7 Advances in therapeutic oligonucleotide. Nucleic Acids Chem. 7, 271.

Joussain, C., Epstein, A.L., 2024. Gene therapy of neurological and non-neurological diseases using herpes simplex virus vectors. Med. Res. Arch. 12 (7), 1–25.

Kadyrov, F.A., Holmes, S.F., Arana, M.E., Lukianova, O.A., O'Donnell, M., Kunkel, T.A., Modrich, P., 2007. Saccharomyces cerevisiae MutLα is a mismatch repair endonuclease. J. Biol. Chem. 282 (51), 37181–37190.

Kaganovich, A. (2019). "Contribution of microglia to the pathogenic mechanisms behind LRRK2-associated Parkinson's disease."

Kalter, N., Fuster-García, C., Silva, A., Ronco-Díaz, V., Roncelli, S., Turchiano, G., Gorodkin, J., Cathomen, T., Benabdellah, K., Lee, C., 2025. Off-target effects in CRISPR-Cas genome editing for human therapeutics: progress and challenges. Mol. Ther. Nucleic Acids 36 (1), 102433.

Kantor, A., McClements, M.E., MacLaren, R.E., 2020. CRISPR-Cas9 DNA base-editing and prime-editing. Int. J. Mol. Sci. 21 (17), 6240.

Kapoor, A.R., Rodríguez, M.F., Cheng, L.W., Granger, T.L., 2025. A comprehensive review on the pathophysiology, diagnosis, and emerging therapies in neurodegenerative diseases. Neurosci. Innov. Disord. 1 (1), 0002.

Karagianni, K., Bibi, A., Madé, A., Acharya, S., Parkkonen, M., Barbalata, T., Srivastava, P.K., de Gonzalo-calvo, D., Emanueli, C., Martelli, F., 2024. Recommendations for detection, validation, and evaluation of RNA editing events in cardiovascular and neurological/neurodegenerative diseases. Mol. Ther. Nucleic Acids 35 (1), 102133.

Kaur, A., Gupta, S., Kumar, D., 2024. AI-powered predictive modelling for disease diagnostics. In: Genomic Intelligence. CRC Press. pp. 154–170.

Ke, J., Ding, J., Xu, Y., Yu, C., Hong, Y., Li, S., Meng, T., Ping, Y., Yuan, H., Hu, F., 2026. Engineering microglial exosome-mediated microRNA-124-3p delivery for Alzheimer's disease combinational therapy. Biomater. Sci. 14 (1), 186–197.

Keeler, G.D., Gaddie, C.D., Sagadevan, A.S., Senior, K.G., Côté, I., Rechdan, M., Min, D., Mahan, D., Poma, B., Hoffman, B.E., 2023. Induction of antigen-specific tolerance by hepatic AAV immunotherapy regardless of T cell epitope usage or mouse strain background. Mol. Ther. Methods Clin. Dev. 28, 177–189.

Keeler, G.D., Kumar, S., Palaschak, B., Silverberg, E.L., Markusic, D.M., Jones, N.T., Hoffman, B.E., 2018. Gene therapy-induced antigen-specific tregs inhibit neuro-inflammation and reverse disease in a mouse model of multiple sclerosis. Mol. Ther. 26 (1), 173–183.

Kells, A.P., Fong, D.M., Dragunow, M., During, M.J., Young, D., Connor, B., 2004. AAV-mediated gene delivery of BDNF or GDNF is neuroprotective in a model of Huntington disease. Mol. Ther. 9 (5), 682–688.

Khan, M.Y., Khan, H., Maarfi, F., Khanam, A., Hasan, Z., Husain, A., 2024. Gene editing tool for neurodegenerative diseases. In: Khan, A., Rather, M.A., Ashraf, G.M. (Eds.), Mechanism and Genetic Susceptibility of Neurological Disorders. Springer, pp. 359–390.

Kim, D.Y., Lee, S.Y., Ha, H.J., Park, H.H., 2024. Structural basis of Cas3 activation in type IC CRISPR-Cas system. Nucleic Acids Res. 52 (17), 10563–10574.

Kim, M.-G., Go, M.-J., Kang, S.-H., Jeong, S.-H., Lim, K., 2025. Revolutionizing CRISPR technology with artificial intelligence. Exp. Mol. Med. 57 (7), 1419–1431.

Klinkovskij, A., Shepelev, M., Isaakyan, Y., Aniskin, D., Ulasov, I., 2023. Advances of genome editing with CRISPR/Cas9 in neurodegeneration: the right path towards therapy. Biomedicines 11 (12), 3333.

Klompe, S.E., Vo, P.L., Halpin-Healy, T.S., Sternberg, S.H., 2019. Transposon-encoded CRISPR–Cas systems direct RNA-guided DNA integration. Nature 571 (7764), 219–225.

Konermann, S., Lotfy, P., Brideau, N.J., Oki, J., Shokhirev, M.N., Hsu, P.D., 2018. Transcriptome engineering with RNA-targeting type VI-D CRISPR effectors. Cell 173 (3), 665–676.e614.

Koonin, E.V., Makarova, K.S., 2024. CRISPR in mobile genetic elements: counter-defense, inter-element competition and RNA-guided transposition. BMC Biol. 22 (1), 295.

Kordyś, M., Sen, R., Warkocki, Z., 2022. Applications of the versatile CRISPR-Cas13 RNA targeting system. Wiley Interdiscip. Rev. RNA 13 (3), e1694.

Kuang, G., Hou, J., Wang, Z., Zhao, L., Feng, Y., Wang, H., Cui, J., 2024. Spatial localization immobilization of metals and enzymes at Pickering emulsion droplet interfaces for chemo-enzymatic cascade catalysis. J. Colloid Interface Sci. 678, 139840.

Kumar, A., Gupta, A.K., Singh, P.K., 2024. Novel perspective of therapeutic modules to overcome motor and nonmotor symptoms in Parkinson's disease. AIMS Neurosci. 11 (3), 312.

Kumar, A., Pardhipa, N., Swami, S.G., Dwivedi, V., Singh, A.K., 2025. Therapeutic potential of genome editing tools in neurodegenerative diseases. In: Singh, S.K., et al., (Eds.), Genome Editing for Neurodegenerative Diseases. Elsevier, pp. 1–17.

Kumar, P., Bhaskar, K., 2025. Neuroinflammation and neurodegenerative diseases. Front. Media SA 19, 1561636.

Kwak, G., Grewal, A., Slika, H., Mess, G., Li, H., Kwatra, M., Poulopoulos, A., Woodworth, G.F., Eberhart, C.G., Ko, H.S., 2024. Brain nucleic acid delivery and genome editing via focused ultrasound-mediated blood–brain barrier opening and long-circulating nanoparticles. ACS Nano 18 (35), 24139–24153.

Le Hars, M., Joussain, C., Jégu, T., Epstein, A.L., 2025. Non-replicative herpes simplex virus genomic and amplicon vectors for gene therapy-an update. Gene Ther. 32 (3), 173–183.

Lee, C., 2024. Extracellular vesicle analysis and engineering for personalized medicine, Ulsan National Institute of Science and Technology.

Lee, G.Y., Glicksman, M.A., Patel, R., Malhotra, S., Moelis, N., Vanjani, N.N., Kumthekar, P., 2025. Retrospective review of intra-cerebrospinal fluid (CSF) drug delivery in CNS malignancies: safety, clinical efficacy and pharmacokinetic profiles of intracerebroventricular (ICV), lumbar intrathecal (LIT), and intra-cisterna magna (ICM) Injections. Cancers 17 (8), 1263.

Lee, J., Lim, K., Kim, A., Mok, Y.G., Chung, E., Cho, S.-I., Lee, J.M., Kim, J.-S., 2023. Prime editing with genuine Cas9 nickases minimizes unwanted indels. Nat. Commun. 14 (1), 1786.

Lei, S.-Y., Yang, Y.-Q., Liu, J.-C., Zhang, D.-H., Qu, Y., Sun, -Y.-Y., Zhu, H.-J., Zhou, S.-Y., Yang, Y., Guo, Z.-N., 2024. Nanodrug delivery systems for regulating microglial polarization in ischemic stroke treatment: a review. J. Tissue Eng. 15, 20417314241237052.

Lejman, J., Panuciak, K., Nowicka, E., Mastalerczyk, A., Wojciechowska, K., Lejman, M., 2023. Gene therapy in ALS and SMA: advances, challenges and perspectives. Int. J. Mol. Sci. 24 (2), 1130.

Lenders, V., Koutsoumpou, X., Phan, P., Soenen, S.J., Allegaert, K., de Vleeschouwer, S., Toelen, J., Zhao, Z., Manshian, B.B., 2023. Modulation of engineered nanomaterial interactions with organ barriers for enhanced drug transport. Chem. Soc. Rev. 52 (14), 4672–4724.

Leone, P., Lober, R.M., Francis, J., Flamini, O., Cecil, K.M., Shera, D., Janson, C.G., 2025. Oligodendrocyte-targeted adeno-associated virus gene therapy for Canavan disease in children: a phase 1/2 trial. Nat. Med. 31 (11), 3772–3779.

Levanon, E.Y., Eisenberg, E., Yelin, R., Nemzer, S., Hallegger, M., Shemesh, R., Fligelman, Z.Y., Shoshan, A., Pollock, S.R., Sztybel, D., 2004. Systematic identification of abundant A-to-I editing sites in the human transcriptome. Nat. Biotechnol. 22 (8), 1001–1005.

Li, J.M., Huang, J., Liao, Y., Hu, T., Wang, C.L., Zhang, W.Z., Huang, C.W., 2025. Gene and RNA editing: revolutionary approaches to treating diseases. MedComm 6(10), e70389.

Li, J., Zhang, C., He, Y., Li, S., Yan, L., Li, Y., Zhu, Z., Xia, L., 2023a. Plant base editing and prime editing: the current status and future perspectives. J. Integr. Plant Biol. 65 (2), 444–467.

Li, Q., Gloudemans, M.J., Geisinger, J.M., Fan, B., Aguet, F., Sun, T., Ramaswami, G., Li, Y.I., Ma, J.-B., Pritchard, J.K., 2022. RNA editing underlies genetic risk of common inflammatory diseases. Nature 608 (7923), 569–577.

Li, T., Yang, Y., Qi, H., Cui, W., Zhang, L., Fu, X., He, X., Liu, M., Li, P.-F., Yu, T., 2023b. CRISPR/Cas9 therapeutics: progress and prospects. Signal Transduct. Target. Ther. 8 (1), 36.

Lim, C.K., McCallister, T.X., Saporito-Magriña, C., McPheron, G.D., Krishnan, R., Powell, J.E., Clark, L.V., Perez-Pinera, P., Gaj, T., 2022. CRISPR base editing of cis-regulatory elements enables the perturbation of neurodegeneration-linked genes. Mol. Ther. 30 (12), 3619–3631.

Liu, D., Zhu, M., Zhang, Y., Diao, Y., 2021. Crossing the blood-brain barrier with AAV vectors. Metabolic Brain Disease 36 (1), 45–52.

Liu, W., Ma, C., Cao, J., Zhou, H., Guo, T., 2024. Tet1 peptide and zinc (II)-adenine multifunctional module functionalized polycations as efficient siRNA carriers for Parkinson's disease. J. Control. Release 367, 316–326.

Liu, Z., Song, S.-Y., 2025. Genomic and transcriptomic approaches advance the diagnosis and prognosis of neurodegenerative diseases. Genes 16 (2), 135.

Lovett, A., Chary, S., Babu, S., Bruneteau, G., Glass, J.D., Karlsborg, M., Ladha, S., Mayl, K., McDermott, C., Bucelli, R.C., 2025. Serious neurologic adverse events in tofersen clinical trials for amyotrophic lateral sclerosis. Muscle Nerve 71 (6), 1006–1015.

Lu, G., Shivalila, C., Monian, P., Yu, H., Harding, I., Briem, S., Byrne, M., Faraone, A., Friend, S., Huth, O., 2024. Rational design of base, sugar and backbone modifications improves ADAR-mediated RNA editing. Nucleic Acids Res. 52 (17), 10068–10084.

Lueck, J., Yoon, J., Perales-Puchalt, A., Mackey, A., Infield, D., Behlke, M., Pope, M., Weiner, D., Skach, W., McCray Jr, P., 2019. Engineered transfer RNAs for suppression of premature termination codons. Nat. Commun. 10, 822.

Lv, J., Xi, H., Lv, X., Zhou, Y., Wang, J., Chen, H., Yan, T., Jin, J., Zhao, J., Gu, F., 2022. Two high-fidelity variants: efSaCas9 and SaCas9-HF, which one is better?. Gene Ther. 29 (7), 458–463.

Lyu, P., Lu, B., 2022. New advances in using virus-like particles and related technologies for eukaryotic genome editing delivery. Int. J. Mol. Sci. 23 (15), 8750.

Maesako, M., Zoltowska, K.M., Berezovska, O., 2019. Synapsin 1 promotes Aβ generation via BACE1 modulation. PLoS One 14(12), e0226368.

Maheshwari, S., Singh, A., Verma, A., Shariq, M., Akhtar, J., Alsaidan, O.A., Prajapati, B.G., 2025. Superparamagnetic iron oxide nanoparticles (SPIONs) in targeting brain tumors: advances and challenges. Med. Oncol. 42 (8), 338.

Mahto, K., Kuwar, O.K., Maloo, A., Kalia, N., 2025. Therapeutic potential of luteolin in neurodegenerative disorders: targeting Nrf2, NFκB, MAPK, and JAK-STAT pathways to combat neuroinflammation and apoptosis. Inflammopharmacology 33 (9), 5011–5021.

Malik, B.R., Maddison, D.C., Smith, G.A., Peters, O.M., 2019. Autophagic and endo-lysosomal dysfunction in neurodegenerative disease. Mol. Brain 12 (1), 100.

Mandalawatta, H.P., Rajendra, K., Fairfax, K., Hewitt, A.W., 2024. Emerging trends in virus and virus-like particle gene therapy delivery to the brain. Mol. Ther. Nucleic Acids 35 (3), 102280.

Mao, F.Y., Tu, M.-J., Traber, G.M., Yu, A.-M., 2024. Comparison of three computational tools for the prediction of RNA tertiary structures. Non-Coding RNA 10 (6), 55.

Marei, H.E., 2026. Epigenetic editing in neurological and neuropsychiatric disorders: pioneering next-gen therapeutics for precision gene control. Mol. Neurobiol. 63 (1), 330.

Martier, R., Konstantinova, P., 2020. Gene therapy for neurodegenerative diseases: slowing down the ticking clock. Front. Neurosci. 14, 580179.

Mathiesen, S.N., Lock, J.L., Schoderboeck, L., Abraham, W.C., Hughes, S.M., 2020. CNS transduction benefits of AAV-PHP. eB over AAV9 are dependent on administration route and mouse strain. Mol. Ther. Methods Clin. Dev. 19, 447–458.

Mathis, N., Allam, A., Kissling, L., Marquart, K.F., Schmidheini, L., Solari, C., Balázs, Z., Krauthammer, M., Schwank, G., 2023. Predicting prime editing efficiency and product purity by deep learning. Nat. Biotechnol. 41 (8), 1151–1159.

McCaffrey, J., Sibert, J., Zhang, B., Zhang, Y., Hu, W., Riethman, H., Xiao, M., 2016. CRISPR-CAS9 D10A nickase target-specific fluorescent labeling of double strand DNA for whole genome mapping and structural variation analysis. Nucleic Acids Res. 44 (2), e11.

McLean, J.R., Smith, G.A., Rocha, E.M., Hayes, M.A., Beagan, J.A., Hallett, P.J., Isacson, O., 2014. Widespread neuron-specific transgene expression in brain and spinal cord following synapsin promoter-driven AAV9 neonatal intracerebroventricular injection. Neurosci. Lett. 576, 73–78.

McTague, A., Rossignoli, G., Ferrini, A., Barral, S., Kurian, M.A., 2021. Genome editing in iPSC-based neural systems: from disease models to future therapeutic strategies. Front. Genome Ed. 3, 630600.

Mead, B.P., Kim, N., Miller, G.W., Hodges, D., Mastorakos, P., Klibanov, A.L., Mandell, J.W., Hirsh, J., Suk, J.S., Hanes, J., 2017. Novel focused ultrasound gene therapy approach noninvasively restores dopaminergic neuron function in a rat Parkinson's disease model. Nano Lett. 17 (6), 3533–3542.

Mendell, J.R., Al-Zaidy, S., Shell, R., Arnold, W.D., Rodino-Klapac, L.R., Prior, T.W., Lowes, L., Alfano, L., Berry, K., Church, K., 2017. Single-dose gene-replacement therapy for spinal muscular atrophy. N. Engl. J. Med. 377 (18), 1713–1722.

Mendoza, H.G., Beal, P.A., 2023. Chemical modifications in RNA: elucidating the chemistry of dsRNA-specific adenosine deaminases (ADARs). Acc. Chem. Res. 56 (18), 2489–2499.

Meneghini, V., Frati, G., Sala, D., De Cicco, S., Luciani, M., Cavazzin, C., Paulis, M., Mentzen, W., Morena, F., Giannelli, S., 2017. Generation of human induced pluripotent stem cell-derived bona fide neural stem cells for ex vivo gene therapy of metachromatic leukodystrophy. Stem Cells Transl. Med. 6 (2), 352–368.

Meng, Y., Goubran, M., Rabin, J.S., McSweeney, M., Ottoy, J., Pople, C.B., Huang, Y., Storace, A., Ozzoude, M., Bethune, A., 2023. Blood–brain barrier opening of the default mode network in Alzheimer's disease with magnetic resonance-guided focused ultrasound. Brain 146 (3), 865–872.

Meng, Y., Kalia, L.V., Kalia, S.K., Hamani, C., Huang, Y., Hynynen, K., Lipsman, N., Davidson, B., 2024. Current progress in magnetic resonance-guided focused ultrasound to facilitate drug delivery across the blood-brain barrier. Pharmaceutics 16 (6), 719.

Meza-Sosa, K.F., Miao, R., Navarro, F., Zhang, Z., Zhang, Y., Hu, J.J., Hartford, C.C.R., Li, X.L., Pedraza-Alva, G., Pérez-Martínez, L., 2022. SPARCLE, a p53-induced lncRNA, controls apoptosis after genotoxic stress by promoting PARP-1 cleavage. Mol. Cell 82 (4), 785–802.e710.

Miao, G., Fortier, T.M., Liu, H., Schafer, D.P., Fitzgerald, K.A., Mao, J., Baehrecke, E.H., 2025. Microglia promote inflammatory cell death upon neuronal mitochondrial impairment during neurodegeneration. Nat. Struct. Mol. Biol. 32, 1–14.

Mietzsch, M., Barnes, C., Hull, J.A., Chipman, P., Xie, J., Bhattacharya, N., Sousa, D., McKenna, R., Gao, G., Agbandje-McKenna, M., 2020. Comparative analysis of the capsid structures of AAVrh. 10, AAVrh. 39, and AAV8. J. Virol. 94 (6). doi:10.1128/jvi.01769-01719.

Mietzsch, M., Yu, J.C., Hsi, J., Chipman, P., Broecker, F., Fuming, Z., Linhardt, R.J., Seeberger, P.H., Heilbronn, R., McKenna, R., 2021. Structural study of Aavrh. 10 receptor and antibody interactions. J. Virol. 95 (23). doi:10.1128/jvi.01249-01221.

Mijanović, O., Branković, A., Borovjagin, A.V., Butnaru, D.V., Bezrukov, E.A., Sukhanov, R.B., Shpichka, A., Timashev, P., Ulasov, I., 2020. Battling neurodegenerative diseases with adeno-associated virus-based approaches. Viruses 12 (4), 460.

Mira-Bontenbal, H., Tan, B., Gontan, C., Goossens, S., Boers, R., Boers, J., Dupont, C., van Royen, M., IJcken, W., French, P., 2022. Genetic and epigenetic determinants of reactivation of Mecp2 and the inactive X chromosome in neural stem cells. Stem Cell Rep. 17 (3), 693–706.

Mirzaei, S.Z., Fatemizadeh, F., 2025. Polysorbate emulsifier in drug delivery systems for cancer and neurodegenerative diseases: current status and emerging applications. Nano Micro Biosyst. 4 (2), 52–63.

Mise, A., Yoshino, Y., Yamazaki, K., Ozaki, Y., Sao, T., Yoshida, T., Mori, T., Mori, Y., Ochi, S., Iga, J.-I., 2017. TOMM40 and APOE gene expression and cognitive decline in Japanese Alzheimer's disease subjects. J. Alzheimers Dis. 60 (3), 1107–1117.

Mishra, S.K., Chang, H.-M., Obaid, A.A., Singh, S.K., 2025. Advances in CRISPER/Cas system and genome editing technologies for the treatment of neurodegenerative diseases. In: Singh, S.K., Yadav, A.K., Poeggeler, B., Perry, G. (Eds.), Genome Editing for Neurodegenerative Diseases. Elsevier, pp. 69–90.

Miskalis, A., Shirguppe, S., Winter, J., Elias, G., Swami, D., Nambiar, A., Stilger, M., Woods, W.S., Gosstola, N., Gapinske, M., 2024. SPLICER: a highly efficient base editing toolbox that enables in vivo therapeutic exon skipping. Nat. Commun. 15(1), 10354.

Miyagawa, Y., Maruyama, M., Sakai, A., Sato, Y., Shimizu, M., Kuroda, S., Hayashita-Kinoh, H., Yamamoto, M., Hashizume, R., Suzuki, H., 2025. Durable tissue-specific transgene expression in newborn mice following intraperitoneal delivery of non-cytotoxic HSV vectors. Mol. Ther. Methods Clin. Dev. 33 (3), 101573.

Monaco, I., Camorani, S., Colecchia, D., Locatelli, E., Calandro, P., Oudin, A., Niclou, S., Arra, C., Chiariello, M., Cerchia, L., 2017. Aptamer functionalization of nanosystems for glioblastoma targeting through the blood–brain barrier. J. Med. Chem. 60 (10), 4510–4516.

Morelli, K.H., Wu, Q., Gosztyla, M.L., Liu, H., Yao, M., Zhang, C., Chen, J., Marina, R.J., Lee, K., Jones, K.L., 2023. An RNA-targeting CRISPR–Cas13d system alleviates disease-related phenotypes in Huntington's disease models. Nat. Neurosci. 26 (1), 27–38.

Mozharovskaya, T., Popova, O., Zubkova, O., Vavilova, I., Pochtovyi, A., Shcherbakov, D., Gushchin, V., Logunov, D.Y., Gintsburg, A., 2023. Development and characterization of a vector system based on the simian adenovirus type 25. Bull. Russ. State Med. Univ. (1), 4–11. doi:10.24075/brsmu.2023.006.

Mseis-Jackson, N., Sharma, M., Li, H., 2024. Controlling the expression level of the neuronal reprogramming factors for a successful reprogramming outcome. Cells 13 (14), 1223.

Murphy, K.R., Aycock, K.N., Marsh, S., Hay, A.N., Athanasiadi, I., Bracha, S., Chang, C., Gourdie, R., Davalos, R.V., Rossmeisl, J.H., 2024. Tumor-derived extracellular vesicles disrupt the blood–brain barrier endothelium following high-frequency irreversible electroporation. Sci. Rep. 14(1), 28533.

Naeem, S., Zhang, J., Zhang, Y., Wang, Y., 2025. Nucleic acid therapeutics: past, present, and future. Mol. Ther. Nucleic Acids 36 (1), 102440.

Naert, T., et al., 2024. Pythia: Non-random DNA repair allows predictable CRISPR/Cas9 integration and gene editing. bioRxiv. doi:10.1101/2024.09.23.614424.

Nair, L., Nair, R., Shenoy, S., Thekkuveettil, A., Jayasree, R., 2017. Blood brain barrier permeable gold nanocluster for targeted brain imaging and therapy: an in vitro and in vivo study. J. Mater. Chem. B 5 (42), 8314–8321.

Nance, E., Timbie, K., Miller, G.W., Song, J., Louttit, C., Klibanov, A.L., Shih, T.-Y., Swaminathan, G., Tamargo, R.J., Woodworth, G.F., 2014. Non-invasive delivery of stealth, brain-penetrating nanoparticles across the blood– brain barrier using MRI-guided focused ultrasound. J. Control. Release 189, 123–132.

Nasir, M.A., Nawaz, S., Huang, J., 2022. A review: computational approaches to design sgRNA of CRISPR-Cas9. Curr. Bioinform. 17 (1), 2–18.

Nazim, M., 2024. Post-transcriptional regulation of the transcriptional apparatus in neuronal development. Front. Mol. Neurosci. 17, 1483901.

Nguyen, G.T., Raju, A., Schelling, M.A., Sashital, D.G., 2025. Rapid CRISPR–Cas9 target-strand nicking can provide phage resistance by reducing DNA abundance. Nucleic Acids Res. 53(18), gkaf900.

Nguyen, N.H., Nguyen-Thi, P.T., Nguyen, T.T., Bui, V.K.H., Nguyen, N.T.T., Vo, G.V., 2024. Applications and developments of gene therapy drug delivery systems for neurological disorders. Adv. Ther. 7(12), 2400269.

Nicoletti, V.G., Pajer, K., Calcagno, D., Pajenda, G., Nógrádi, A., 2022. The role of metals in the neuroregenerative action of BDNF, GDNF, NGF and other neurotrophic factors. Biomolecules 12 (8), 1015.

Nixon, R.A., Rubinsztein, D.C., 2024. Mechanisms of autophagy–lysosome dysfunction in neurodegenerative diseases. Nat. Rev. Mol. Cell Biol. 25 (11), 926–946.

Nóbrega, C., Alves, S., 2018. Gene silencing and editing strategies for neurodegenerative diseases. Front. Media SA 12, 425.

Nóbrega, C., Mendonça, L., Matos, C.A., 2020. Gene therapy strategies: gene augmentation. In: Nóbrega, C., Mendonça, L., Matos, C.A. (Eds.), A Handbook of Gene and Cell Therapy. Springer, pp. 117–126.

Noh, M.-Y., Kwon, H.S., Kwon, M.-S., Nahm, M., Jin, H.K., Bae, J.-S., Kim, S.H., 2025. Biomarkers and therapeutic strategies targeting microglia in neurodegenerative diseases: current status and future directions. Molecular Neurodegeneration 20 (1), 82.

Numakawa, T., Kajihara, R., 2023. Neurotrophins and other growth factors in the pathogenesis of Alzheimer's disease. Life 13 (3), 647.

Nureki, O., Nakagawa, R., Omura, S., Kajimoto, S., Okazaki, S., Hirano, H., Yamashita, K., Nishimasu, H., Ishiguro, S., Mori, H. (2024). "Molecular engineering and dynamic activation mechanism of a PAM-flexible compact CRISPR-Cas9."

Ocampo, R.F., Bravo, J.P., Dangerfield, T.L., Nocedal, I., Jirde, S.A., Alexander, L.M., Thomas, N.C., Das, A., Nielson, S., Johnson, K.A., 2025. DNA targeting by compact Cas9d and its resurrected ancestor. Nat. Commun. 16 (1), 457.

Omura, T., Omura, K., Tedeschi, A., Riva, P., Painter, M.W., Rojas, L., Martin, J., Lisi, V., Huebner, E.A., Latremoliere, A., 2015. Robust axonal regeneration occurs in the injured CAST/Ei mouse CNS. Neuron 86 (5), 1215–1227.

Organization, W.H., 2025. Global status report on neurology, World Health Organization.

Orr, M.E., Oddo, S., 2013. Autophagic/lysosomal dysfunction in Alzheimer's disease. Alzheimers Res. Ther. 5 (5), 53.

Ortega, A., Chernicki, B., Ou, G., Parmar, M.S., 2025. From lab bench to hope: emerging gene therapies in clinical trials for Alzheimer's disease. Mol. Neurobiol. 62 (1), 1112–1135.

Pal, P., Sharma, M., Bani, K.S., Gupta, S.K., 2024. Advances in neuronal regeneration: hydrogel-based delivery systems loaded with extracellular vesicles in modulating neural impulses and tissue repair. European Polymer Journal 220, 113457.

Pandey, M.K., 2023. Exploring pro-inflammatory immunological mediators: unraveling the mechanisms of neuroinflammation in lysosomal storage diseases. Biomedicines 11 (4), 1067.

Papadimitriou, E., Koutsoudaki, P.N., Thanou, I., Karagkouni, D., Karamitros, T., Chroni-Tzartou, D., Gaitanou, M., Gkemisis, C., Margariti, M., Xingi, E., 2023. A miR-124-mediated post-transcriptional mechanism controlling the cell fate switch of astrocytes to induced neurons. Stem Cell Rep. 18 (4), 915–935.
Papadimitriou, E., Thomaidou, D., 2024. Post-transcriptional mechanisms controlling neurogenesis and direct neuronal reprogramming. Neural Regen. Res. 19 (9), 1929–1939.
Parambi, D.G.T., Alharbi, K.S., Kumar, R., Harilal, S., Batiha, G.E.-S., Cruz-Martins, N., Magdy, O., Musa, A., Panda, D.S., Mathew, B., 2022. Gene therapy approach with an emphasis on growth factors: theoretical and clinical outcomes in neurodegenerative diseases. Mol. Neurobiol. 59 (1), 191–233.
Pardridge, W.M., 2023. Brain gene therapy with Trojan horse lipid nanoparticles. Trends Mol. Med. 29 (5), 343–353.
Pascual-Morena, C., Cavero-Redondo, I., Luceron-Lucas-Torres, M., Martinez-Garcia, I., Rodriguez-Gutierrez, E., Martinez-Vizcaino, V., 2023. Onasemnogene abeparvovec in type 1 spinal muscular atrophy: a systematic review and meta-analysis. Hum. Gene Ther. 34 (3-4), 129–138.
Pattanayak, S.P., 2025. CRISPR-guided epigenetic modulation as a next-generation therapeutic strategy in chronic inflammatory diseases. J. Pharmacol. Ther. Adv. 3117-6259. 1 (01), 41–50. 3117-6267.
Patzke, C., Brockmann, M.M., Dai, J., Gan, K.J., Grauel, M.K., Fenske, P., Liu, Y., Acuna, C., Rosenmund, C., Südhof, T.C., 2019. Neuromodulator signaling bidirectionally controls vesicle numbers in human synapses. Cell 179 (2), 498–513.e422.
Pedrazzoli, E., Bianchi, A., Umbach, A., Amistadi, S., Brusson, M., Frati, G., Ciciani, M., Badowska, K.A., Arosio, D., Miccio, A., 2023. An optimized SpCas9 high-fidelity variant for direct protein delivery. Mol. Ther. 31 (7), 2257–2265.
Peng, H., Harvey, B.T., Richards, C.I., Nixon, K., 2021. Neuron-derived extracellular vesicles modulate microglia activation and function. Biology 10 (10), 948.
Peng, J., Zou, -W.-W., Wang, X.-L., Zhang, Z.-G., Huo, R., Yang, L., 2024a. Viral-mediated gene therapy in pediatric neurological disorders. World J. Pediatr. 20 (6), 533–555.
Peng, Q., Cui, Y., Wu, J., Wu, L., Liu, J., Han, Y., Lu, G., 2024b. A c.726C>G (p. Tyr242Ter) nonsense mutation-associated with splicing alteration (NASA) of WDR45 gene underlies β-propeller protein-associated neurodegeneration (BPAN). Heliyon 10 (9), e30653.
Perez, B.A., Shutterly, A., Chan, Y.K., Byrne, B.J., Corti, M., 2020. Management of neuroinflammatory responses to AAV-mediated gene therapies for neurodegenerative diseases. Brain Sci. 10 (2), 119.
Peters, J.E., Makarova, K.S., Shmakov, S., Koonin, E.V., 2017. Recruitment of CRISPR-Cas systems by Tn7-like transposons. Proc. Natl. Acad. Sci. U.S.A. 114 (35), E7358–E7366 .
Petolino, J.F., 2015. Genome editing in plants via designed zinc finger nucleases. Vitro Cell. Dev. Biol. Plant 51 (1), 1–8.
Porter, J.J., Heil, C.S., Lueck, J.D., 2021. Therapeutic promise of engineered nonsense suppressor tRNAs. Wiley Interdiscip. Rev. RNA 12(4), e1641.
Prasad, S., Dimmock, D.P., Greenberg, B., Walia, J.S., Sadhu, C., Tavakkoli, F., Lipshutz, G.S., 2022. Immune responses and immunosuppressive strategies for adeno-associated virus-based gene therapy for treatment of central nervous system disorders: current knowledge and approaches. Hum. Gene Ther. 33 (23-24), 1228–1245.
Pulgarin, D.V., Pelo, N., Ferrandiz, L., Tršelič, T., Nyberg, W.A., Bowlin, G., Espinosa, A., 2025. Light-induced expression of gRNA allows for optogenetic gene editing of T lymphocytes in vivo. Nucleic Acids Res. 53(6), gkaf213.
Qia, Y., Culla, J.W., Tarbetb, E.B., Staerza, U.D., 2025 Systems of adenoviral gene transfer vectors moving into their fourth generation. In: Curiel, D.T., Parker, A.L. (Eds.), Adenoviral Vectors for Gene Therapy, third ed. pp. 403–432.
Qin, R., Zhang, Y., Yang, Y., Chen, J., Huang, L., Xu, W., Qin, Q., Liang, X., Lai, X., Huang, X., 2024. Decoding single-cell molecular mechanisms in astrocyte-to-iN reprogramming via Ngn2-and Pax6-mediated direct lineage switching. Eur. J. Med. Res. 29 (1), 390.
Qin, Y., Li, S., Li, X.-J., Yang, S., 2022. CRISPR-based genome-editing tools for Huntington's disease research and therapy. Neurosci. Bull. 38 (11), 1397–1408.
Rahman, M.U., Bilal, M., Shah, J.A., Kaushik, A., Teissedre, P.-L., Kujawska, M., 2022. CRISPR-Cas9-based technology and its relevance to gene editing in Parkinson's disease. Pharmaceutics 14 (6), 1252.
Rainaldi, J., Mali, P., Nourreddine, S., 2025. Emerging clinical applications of ADAR based RNA editing. Stem Cells Transl. Med. 14 (5), szaf016.
Ramalingam, S., Kandavelou, K., Rajenderan, R., Chandrasegaran, S., 2011. Creating designed zinc-finger nucleases with minimal cytotoxicity. J. Mol. Biol. 405 (3), 630–641.
Raslan, A.A., Kee, Y., 2013. Tackling neurodegenerative diseases: animal models of Alzheimer's disease and Parkinson's disease. Genes Genomics 35 (4), 425–440.
Rassu, G., Soddu, E., Posadino, A.M., Pintus, G., Sarmento, B., Giunchedi, P., Gavini, E., 2017. Nose-to-brain delivery of BACE1 siRNA loaded in solid lipid nanoparticles for Alzheimer's therapy. Colloids Surf. B Biointerfaces 152, 296–301.
Reautschnig, P., Wahn, N., Wettengel, J., Schulz, A.E., Latifi, N., Vogel, P., Kang, T.-W., Pfeiffer, L.S., Zarges, C., Naumann, U., 2022. CLUSTER guide RNAs enable precise and efficient RNA editing with endogenous ADAR enzymes in vivo. Nat. Biotechnol. 40 (5), 759–768.
Rehman, M.U., Sehar, N., Dar, N.J., Khan, A., Arafah, A., Rashid, S., Rashid, S.M., Ganaie, M.A., 2023. Mitochondrial dysfunctions, oxidative stress and neuroinflammation as therapeutic targets for neurodegenerative diseases: an update on current advances and impediments. Neurosci. Biobehav. Rev. 144, 104961.
Reifschneider, A., Robinson, S., van Lengerich, B., Gnörich, J., Logan, T., Heindl, S., Vogt, M.A., Weidinger, E., Riedl, L., Wind, K., 2022. Loss of TREM2 rescues hyperactivation of microglia, but not lysosomal deficits and neurotoxicity in models of progranulin deficiency. EMBO J. 41(4), e109108.
Ren, Y., Liu, Y., Luo, M., 2021. Gap junctions between striatal d1 neurons and cholinergic interneurons. Front. Cell. Neurosci. 15, 674399.
Rhiel, M., Geiger, K., Andrieux, G., Rositzka, J., Boerries, M., Cathomen, T., Cornu, T.I., 2023. T-CAST: an optimized CAST-Seq pipeline for TALEN confirms superior safety and efficacy of obligate-heterodimeric scaffolds. Front. Genome Ed. 5, 1130736.

Ricobaraza, A., Gonzalez-Aparicio, M., Mora-Jiménez, L., Lumbreras, S., Hernandez-Alcoceba, R., 2020. High-capacity adenoviral vectors: expanding the scope of gene therapy. Int. J. Mol. Sci. 21 (10), 3643.

Roberts, W.S., Price, S., Wu, M., Parmar, M.S., 2024. Emerging gene therapies for Alzheimer's and Parkinson's diseases: an overview of clinical trials and promising candidates. Cureus 16 (8), e67037.

Rocha, D.C., Omoregbee, M.O., Contiliani, D.F., Mandlik, R., Li, G., Mascoveto, J., Coleman, G., Culver, J.N., Leal, D.R., de Souza, A.A., 2025. Transgene-free genome editing in citrus and poplar trees using positive and negative selection markers. Plant Cell Rep. 44 (11), 1–17.

Rodrigues, J.P., Prajapati, N., DeCoster, M.A., Poh, S., Murray, T.A., 2021. Efficient LRP1-mediated uptake and low cytotoxicity of peptide L57 in vitro shows its promise as CNS drug delivery vector. J. Pharm. Sci. 110 (2), 824–832.

Rodriguez de Los Santos, M., Kopell, B.H., Grice, A.B., Ganesh, G., Yang, A., Amini, P., Liharska, L.E., Vornholt, E., Fullard, J.F., Dong, P., 2024. Divergent landscapes of A-to-I editing in postmortem and living human brain. Nat. Commun. 15 (1), 5366.

Rosario, A.M., Cruz, P.E., Ceballos-Diaz, C., Strickland, M.R., Siemienski, Z., Pardo, M., Schob, K.-L., Li, A., Aslanidi, G.V., Srivastava, A., 2016. Microglia-specific targeting by novel capsid-modified AAV6 vectors. Mol. Ther. Methods Clin. Dev. 3, 16026.

Rosenblum, L.T., Trotti, D., 2017. EAAT2 and the molecular signature of amyotrophic lateral sclerosis. In: Ortega, A. (Ed.), Glial Amino Acid Transporters, Advances in Neurobiology, vol. 16, pp. 117–136.

Rottner, A.K., Lundin, A., Li, S., Firth, M., Maresca, M., Sienski, G., 2025. Optimized prime editing of the Alzheimer's disease-associated APOE4 mutation. Stem Cell Rep. 20 (1), 102372.

Roy, P.D., Chowdhory, U.G., Dey, A., Sagor, D.H., 2025. AI and machine learning in healthcare: advancing diagnostics, personalized treatment, and predictive modeling. Preprints.org. doi:10.20944/preprints202504.0007.v1.

Sahu, B., Behera, B., Kumar, U., 2025. Transgene delivery system: viral, nonviral, and other methods for central nervous system. In: Singh, S.K., Yadav, A.K., Poeggeler, B., Perry, G. (Eds.), Genome Editing for Neurodegenerative Diseases. Elsevier, pp. 135–155.

Saini, L., Dubey, A., Pal, R., Pandey, P., Mandal, R.K., 2024. Synthetic and natural polymers enhancing drug delivery and their treatment: a comprehensive review. J. Drug Deliv. Ther. 14 (10), 153–165.

Sami, A., Xue, Z., Tazein, S., Arshad, A., Zhu, Z.H., Chen, Y.P., Hong, Y., Zhu, X.T., Zhou, K.J., 2021. CRISPR–Cas9-based genetic engineering for crop improvement under drought stress. Bioengineered 12 (1), 5814–5829.

Santhamoorthy, M., Kim, S.-C., 2025. A review of the development of biopolymer hydrogel-based scaffold materials for drug delivery and tissue engineering applications. Gels 11 (3), 178.

Satapathy, M.K., Yen, T.-L., Jan, J.-S., Tang, R.-D., Wang, J.-Y., Taliyan, R., Yang, C.-H., 2021. Solid lipid nanoparticles (SLNs): an advanced drug delivery system targeting brain through BBB. Pharmaceutics 13 (8), 1183.

Schneider, B., Sweeney, B.A., Bateman, A., Cerny, J., Zok, T., Szachniuk, M., 2023. When will RNA get its AlphaFold moment?. Nucleic Acids Res. 51 (18), 9522–9532.

Schubert, M.S., Thommandru, B., Woodley, J., Turk, R., Yan, S., Kurgan, G., McNeill, M.S., Rettig, G.R., 2021. Optimized design parameters for CRISPR Cas9 and Cas12a homology-directed repair. Sci. Rep. 11(1), 19482.

Seranova, E., Connolly, K.J., Zatyka, M., Rosenstock, T.R., Barrett, T., Tuxworth, R.I., Sarkar, S., 2017. Dysregulation of autophagy as a common mechanism in lysosomal storage diseases. Essays Biochem. 61 (6), 733–749.

Settembre, C., Fraldi, A., Rubinsztein, D.C., Ballabio, A., 2008. Lysosomal storage diseases as disorders of autophagy. Autophagy 4 (1), 113–114.

Sharabi, S., Last, D., Daniels, D., Fabian, I., Atrakchi, D., Bresler, Y., Liraz-Zaltsman, S., Cooper, I., Mardor, Y., 2021. Non-invasive low pulsed electrical fields for inducing BBB disruption in mice—feasibility demonstration. Pharmaceutics 13, 169.

Shareena, G., Viswalingam, V., Kumar, D., 2025a. Carbon dots as versatile nano-architectures for the treatment of neurological disorders. In: Patel, V., Rajput, M.S. (Eds.), Targeted Therapy for the Central Nervous System. Elsevier, pp. 349–368.

Shareena, G., Viswalingam, V., Rajput, M.S., Kumar, D., 2025b. CRISPR/Cas9 gene editing: a new hope for Alzheimer's disease. In: Patel, V., Rajput, M.S. (Eds.), Targeted Therapy for the Central Nervous System. Elsevier, pp. 301–323.

Sharma, S., Joshi, V., Kumar, V., 2025. Implications of AAV serotypes in neurological disorders: current clinical applications and challenges. Clin. Transl. Neurosci. 9 (3), 32.

Shchaslyvyi, A.Y., Antonenko, S.V., Tesliuk, M.G., Telegeev, G.D., 2023. Current state of human gene therapy: approved products and vectors. Pharmaceuticals 16 (10), 1416.

Shen, W., Chen, S., Li, Z., Yu, W., 2025a. Significant enhancement of adeno-associated virus vector yield for gene therapy Using multidimensional optimization strategies. FASEB J. 39(23), e71305.

Shen, X., Li, H., Zhang, B., Li, Y., Zhu, Z., 2025b. Targeting transferrin receptor 1 for enhancing drug delivery through the blood–brain barrier for Alzheimer's disease. Int. J. Mol. Sci. 26 (19), 9793.

Shi, Y., Zhen, X., Zhang, Y., Li, Y., Koo, S., Saiding, Q., Kong, N., Liu, G., Chen, W., Tao, W., 2024. Chemically modified platforms for better RNA therapeutics. Chem. Rev. 124 (3), 929–1033.

Shivram, H., Cress, B.F., Knott, G.J., Doudna, J.A., 2021. Controlling and enhancing CRISPR systems. Nat. Chem. Biol. 17 (1), 10–19.

Silva, G., Poirot, L., Galetto, R., Smith, J., Montoya, G., Duchateau, P., Paques, F., 2011. Meganucleases and other tools for targeted genome engineering: perspectives and challenges for gene therapy. Curr. Gene Ther. 11 (1), 11–27.

Sinnamon, J.R., Kim, S.Y., Corson, G.M., Song, Z., Nakai, H., Adelman, J.P., Mandel, G., 2017. Site-directed RNA repair of endogenous Mecp2 RNA in neurons. Proc. Natl. Acad. Sci. U. S. A. 114 (44), E9395–E9402.

Sivamaruthi, B.S., Kesika, P., Sisubalan, N., Chaiyasut, C., 2025. Advances in gold nanoparticles for the diagnosis and management of Alzheimer's disease. Pharmaceutics 17 (9), 1158.

Sizova, I., Greiner, A., Awasthi, M., Kateriya, S., Hegemann, P., 2013. Nuclear gene targeting in C hlamydomonas using engineered zinc-finger nucleases. Plant J. 73 (5), 873–882.

Słyk, Ż., Stachowiak, N., Małecki, M., 2024. Recombinant adeno-associated virus vectors for gene therapy of the central nervous system: delivery routes and clinical aspects. Biomedicines 12 (7), 1523.

Sogorb-Gonzalez, M., Landles, C., Caron, N.S., Stam, A., Osborne, G., Hayden, M.R., Howland, D., van Deventer, S., Bates, G.P., Vallès, A., 2024. Exon 1-targeting miRNA reduces the pathogenic exon 1 HTT protein in Huntington's disease models. Brain 147 (12), 4043–4055.

Song, B., Bae, S., 2024. Genome editing using CRISPR, CAST, and Fanzor systems. Mol. Cells 47(7), 100086.

Song, J., Lu, C., Leszek, J., Zhang, J., 2021. Design and development of nanomaterial-based drug carriers to overcome the blood–brain barrier by using different transport mechanisms. Int. J. Mol. Sci. 22(18), 10118.

Stansberry, W.M., Pierchala, B.A., 2023. Neurotrophic factors in the physiology of motor neurons and their role in the pathobiology and therapeutic approach to amyotrophic lateral sclerosis. Front. Mol. Neurosci. 16, 1238453.

Su, D., Cui, Y., He, C., Yin, P., Bai, R., Zhu, J., Lam, J.S., Zhang, J., Yan, R., Zheng, X., 2025. Projections for prevalence of Parkinson's disease and its driving factors in 195 countries and territories to 2050: modelling study of global burden of disease study 2021. BMJ 388, e080952.

Subasinghe, K., Hall, C., Rowe, M., Zhou, Z., Barber, R., Phillips, N., 2025. Neuronal enriched extracellular vesicle miR-122-5p as a potential biomarker for Alzheimer's disease. Cells 14 (22), 1784.

Sullivan, J., Hood, V., 2025. Emerging genetic therapies in epilepsy. Semin. Neurol. 45 (2), 298–302.

Szunyogh, S., Carroll, E., Wade-Martins, R., 2025. Recent developments in gene therapy for Parkinson's disease. Mol. Ther. 33 (5), 2052–2064.

Tan, M.H., Li, Q., Shanmugam, R., Piskol, R., Kohler, J., Young, A.N., Liu, K.I., Zhang, R., Ramaswami, G., Ariyoshi, K., 2017. Dynamic landscape and regulation of RNA editing in mammals. Nature 550 (7675), 249–254.

Teresa de Barros Viegas, A., Guedes, J.R., Oliveira, A.R., Cardoso, A.M.S., Cardoso, A.L.C., 2017. miRNAs: new biomarkers and therapeutic targets in dementia. Curr. Pharm. Des. 23 (5), 669–692.

Tervo, D.G.R., Hwang, B.-Y., Viswanathan, S., Gaj, T., Lavzin, M., Ritola, K.D., Lindo, S., Michael, S., Kuleshova, E., Ojala, D., 2016. A designer AAV variant permits efficient retrograde access to projection neurons. Neuron 92 (2), 372–382.

Thakur, A., Chowdhury, K.R., Kumar, A., Sharma, V.V., Bhatia, R., 2026. Targeting non-coding RNAs in neurodegeneration: advances in therapeutic RNA modalities and next-gen delivery technologies. Curr. Alzheimer Res. doi:10.2174/0115672050421604251108045622.

Thapa, K., Khan, H., Chahuan, S., Dhankhar, S., Kaur, A., Garg, N., Saini, M., Singh, T.G., 2025. Insights into therapeutic approaches for the treatment of neurodegenerative diseases targeting metabolic syndrome. Mol. Biol. Rep. 52 (1), 260.

Tompa, M., Sinko, G., Mally, J., Karteszi, J., Kalman, B., 2025. using whole exome sequencing to identify genetic causes of neurodevelopmental disorders in a cohort of 11 patients: a single center experience. Int. J. Mol. Sci. 26(20), 10176.

Tsurumaki, Y., Kagawa, Y., Hayashi, S., Mizoguchi, H., Miyano, M., Fujii, M., Shinozaki, R., Kellathur, S.N., Zhang, A.J., Suzuki, K., 2025. Regulatory insights on gene therapies, adeno-associated virus-based gene therapies, cell/tissue-based products, and medical care/practice in cell and gene therapies: report from the 7th Asia partnership conference-April 25, 2024. Regen. Ther. 29, 404–418.

Tyagi, B.S., Mamrutha, H., Sheoran, S., Gupta, O.P., 2021. Mechanism of CRISPER/Cas9-mediated genome editing: scope and opportunities. In: Khan, M.S., Altakh, M.A., Alghamdi, A.A. (Eds.), Genome Editing in Plants: Principles and Applications. CRC Press, pp. 37–54.

Ueda, J., Yamazaki, T., Funakoshi, H., 2023. Toward the development of epigenome editing-based therapeutics: potentials and challenges. Int. J. Mol. Sci. 24 (5), 4778.

Ulusoy, A., Asci, H., Taner, R., Tepebasi, M.Y., Ilhan, I., Karabacak, P., Comlekci, S., Ozmen, O., 2025. Acute exposure to 27.12 MHz radiofrequency electromagnetic field disrupts blood-brain barrier integrity via eNOS activation and occludin down-regulation. Iran. J. Basic Med. Sci. 28 (12), 1691.

Vermilyea, S.C., Babinski, A., Tran, N., To, S., Guthrie, S., Kluss, J.H., Schmidt, J.K., Wiepz, G.J., Meyer, M.G., Murphy, M.E., 2020. In vitro CRISPR/Cas9-directed gene editing to model LRRK2 G2019S Parkinson's disease in common marmosets. Sci. Rep. 10 (1), 3447.

Von Jonquieres, G., Fröhlich, D., Klugmann, C.B., Wen, X., Harasta, A.E., Ramkumar, R., Spencer, Z.H., Housley, G.D., Klugmann, M., 2016. Recombinant human myelin-associated glycoprotein promoter drives selective AAV-mediated transgene expression in oligodendrocytes. Front. Mol. Neurosci. 9, 13.

Wadan, A.-H.S., Shaaban, A.H., El-Sadek, M.Z., Mostafa, S.A., Moshref, A.S., El-Hussein, A., Ellakwa, D.E.-S., Mehanny, S.S., 2025. Mitochondrial-based therapies for neurodegenerative diseases: a review of the current literature. Naunyn Schmiedebergs Arch. Pharmacol. 398, 1–30.

Wagner, D.L., Peter, L., Schmueck-Henneresse, M., 2021. Cas9-directed immune tolerance in humans—a model to evaluate regulatory T cells in gene therapy?. Gene Ther. 28 (9), 549–559.

Wang, J., Zhang, Y., Mendonca, C.A., Yukselen, O., Muneeruddin, K., Ren, L., Liang, J., Zhou, C., Xie, J., Li, J., 2022. AAV-delivered suppressor tRNA overcomes a nonsense mutation in mice. Nature 604 (7905), 343–348.

Wang, R., Wang, X., Zhao, H., Li, N., Li, J., Zhang, H., Di, L., 2024a. Targeted delivery of hybrid nanovesicles for enhanced brain penetration to achieve synergistic therapy of glioma. J. Control. Release 365, 331–347.

Wang, S., Xu, L., Ma, F., Xu, M., Chen, G., Wang, D., Wu, X., Wang, P., Yu, J., Shi, L., 2025. Recent advances and clinical prospects of non-viral brain-targeted gene delivery systems. Appl. Phys. Rev. 12 (4), 041301.

Wang, X., Klann, P.J., Wiedtke, E., Sano, Y., Fischer, N., Schiller, L., Elfert, A., Güttsches, A.-K., Weyen, U., Grimm, D., 2024b. Seroprevalence of binding and neutralizing antibodies against 18 adeno-associated virus types in patients with neuromuscular disorders. Front. Immunol. 15, 1450858.

Wang, Y., Shao, W., 2023. Innate immune response to viral vectors in gene therapy. Viruses 15 (9), 1801.

Wauters, E., Van Mossevelde, S., Van der Zee, J., Cruts, M., Van Broeckhoven, C., 2017. Modifiers of GRN-associated frontotemporal lobar degeneration. Trends Mol. Med. 23 (10), 962–979.

Wei, Z.-Y.D., Shetty, A.K., 2021. Treating Parkinson's disease by astrocyte reprogramming: progress and challenges. Sci. Adv. 7(26), eabg3198.

Weiss, A., Gilbert, J.W., Flores, I.V.R., Belgrad, J., Ferguson, C., Dogan, E.O., Wightman, N., Mocarski, K., Echeverria, D., Harkins, A.L., 2025. RNAi-mediated silencing of SOD1 profoundly extends survival and functional outcomes in ALS mice. Mol. Ther. 33, 102450.

Widner, H., 2024. Immunology of cell and gene therapy approaches for neurologic diseases. Handb. Clin. Neurol. 205, 135–144.

Witte, I.P., Lampe, G.D., Eitzinger, S., Miller, S.M., Berríos, K.N., McElroy, A.N., King, R.T., Stringham, O.G., Gelsinger, D.R., Vo, P.L.H., 2025. Programmable gene insertion in human cells with a laboratory-evolved CRISPR-associated transposase. Science 388 (6748), eadt5199.

Wu, J.-Y., Li, Y.-J., Hu, X.-B., Huang, S., Luo, S., Tang, T., Xiang, D.-X., 2021. Exosomes and biomimetic nanovesicles-mediated anti-glioblastoma therapy: a head-to-head comparison. J. Control. Release 336, 510–521.

Wu, S.-K., Tsai, C.-L., Mir, A., Hynynen, K., 2025. MRI-guided focused ultrasound for treating Parkinson's disease with human mesenchymal stem cells. Sci. Rep. 15 (1), 2029.

Wu, S., Fu, J., Liu, D., Chen, D., Hu, H., 2020. The blood–brain barrier cell-targeted gene delivery system to enhance nerve growth factor protein secretion in the brain. ACS Biomater. Sci. Eng. 6 (11), 6207–6216.

Xiang, Z., Kurupati, R.K., Li, Y., Kuranda, K., Zhou, X., Mingozzi, F., High, K.A., Ertl, H.C., 2020. The effect of CpG sequences on capsid-specific CD8+ T cell responses to AAV vector gene transfer. Mol. Ther. 28 (3), 771–783.

Xiao, Y., Liang, Z., Shyngys, M., Baekova, A., Cheung, S., Muljadi, M.B., Bai, Q., Zeng, L., Choi, C.H.J., 2025. In vivo interactions of nucleic acid nanostructures with cells. Adv. Mater. 37(2), 2314232.

Xiong, K., Wang, X., Feng, C., Zhang, K., Chen, D., Yang, S., 2025. Vectors in CRISPR gene editing for neurological disorders: challenges and opportunities. Adv. Biol. 9(3), 2400374.

Xiong, R., Ling, G., Zhang, Y., Guan, J., Zhang, P., 2023. Nucleic acid delivery by ionizable nanocarriers for brain disease treatment. Brain-X 1 (1), e7.

Xu, K., Zhang, Q., Zhu, D., Jiang, Z., 2024. Hydrogels in gene delivery techniques for regenerative medicine and tissue engineering. Macromol. Biosci. 24(6), 2300577.

Xu, M., Liu, H., Wang, J., Vieira, A.F.C., Zhou, X., Liu, N., Liang, J., Du, A., Chen, X., Xing, R. Yang Y. Gonzalez-Perez M.P. Kumar V. Wang D., 2026. An engineered UGA suppressor tRNA gene for disease-agnostic AAV delivery. Nat. Biotechnol. 44. doi:10.1038/s41587-025-02982-5.

Yang, D., Wu, X., Yao, Y., Duan, M., Wang, X., Li, G., Guo, A., Wu, M., Liu, Y., Zheng, J., 2025. An RNA editing strategy rescues gene duplication in a mouse model of MECP2 duplication syndrome and nonhuman primates. Nat. Neurosci. 28 (1), 72–83.

Yang, H.-M., 2025. Overcoming the blood–brain barrier: advanced strategies in targeted drug delivery for neurodegenerative diseases. Pharmaceutics 17 (8), 1041.

Yang, Q., Sang, Y., Geng, N., Liu, Y., Zhang, D., Zou, Y., Zheng, M., 2026. Non-invasive CRISPR/Cas9 nanocapsules specifically edit α-synuclein for effective Parkinson's disease treatment. Nano Today 66, 102903.

Yang, Y., Ye, Y., Kong, C., Su, X., Zhang, X., Bai, W., He, X., 2019. MiR-124 enriched exosomes promoted the M2 polarization of microglia and enhanced hippocampus neurogenesis after traumatic brain injury by inhibiting TLR4 pathway. Neurochem. Res. 44 (4), 811–828.

Younis, H.S., Templin, M., Whiteley, L.O., Kornbrust, D., Kim, T.-W., Levin, A.A., Henry, S.P., 2024. Overview of the nonclinical development strategies and class effects of oligonucleotide-based therapeutics. In: Faqi, A.S. (Ed.), A Comprehensive Guide to Toxicology in Nonclinical Drug Development. Elsevier, pp. 769–790.

Yu, Y., Li, W., Mao, L., Peng, W., Long, D., Li, D., Zhou, R., Dang, X., 2021. Genetically engineered exosomes display RVG peptide and selectively enrich a neprilysin variant: a potential formulation for the treatment of Alzheimer's disease. J. Drug Target. 29 (10), 1128–1138.

Yuan, M., Tang, Y., Huang, T., Ke, L., Huang, E., 2024. In situ direct reprogramming of astrocytes to neurons via polypyrimidine tract-binding protein 1 knockdown in a mouse model of ischemic stroke. Neural Regen. Res. 19 (10), 2240–2248.

Yusuf, J.A., Akanbi, S.T., Ijibadejo, M.S., Sulaiman, S.E., Olorunlowu, D.R., Akinola, A.O., Abe, A.P., Opoola, E.K., Adejobi, E.O., Adigun, T.D., 2025. A review of mitochondrial dysfunction, pathophysiology and therapeutic prospects in neurodegenerative diseases. Discov. Neurosci. 20 (1), 25.

Zaiss, A.K., Muruve, D.A., 2005. Immune responses to adeno-associated virus vectors. Curr. Gene Ther. 5 (3), 323–331.

Zaman, W., Khalil, A.A.K., Amin, A., 2025. TALEN-interceded genome editing in plants: unveiling new frontiers in secondary metabolite improvement and genetic diversity. Plants 14 (19), 3024.

Zhang, J., Yang, T., Huang, W., Yu, Y., Sun, T., 2022. Applications of gold nanoparticles in brain diseases across the blood-brain barrier. Current Medicinal Chemistry 29 (39), 6063–6083.

Zhang, M.-L., Li, H.-B., Jin, Y., 2024. Application and perspective of CRISPR/Cas9 genome editing technology in human diseases modeling and gene therapy. Front. Genet. 15, 1364742.

Zhang, Y., Liao, Q., Wen, X., Fan, J., Yuan, T., Tong, X., Jia, R., Chai, P., Fan, X., 2025. Hijacking of the nervous system in cancer: mechanism and therapeutic targets. Molecular Cancer 24 (1), 44.

Zheng, Y., Cui, L., Lu, H., Liu, Z., Zhai, Z., Wang, H., Shao, L., Lu, Z., Song, X., Zhang, Y., 2024. Nose to brain: exploring the progress of intranasal delivery of solid lipid nanoparticles and nanostructured lipid carriers. Int. J. Nanomedicine 19, 12343–12368.

Zhou, L., Wang, Y., Xu, Y., Zhang, Y., Zhu, C., 2024a. A comprehensive review of AAV-mediated strategies targeting microglia for therapeutic intervention of neurodegenerative diseases. J. Neuroinflammation 21 (1), 232.

Zhou, W., Wang, Z., Wang, W., Yang, J., Qi, W., Xu, H., Xue, J. (2026). Transporter× Circuit Specificity Paradigm: integration Pathway of Cross-BBB Transporter Heterogeneity and Neural Circuit Precision Delivery. Available at SSRN 6076076.

Zhou, X., Kukar, T., Rademakers, R., 2021. Lysosomal dysfunction and other pathomechanisms in FTLD: evidence from progranulin genetics and biology. In: Ghetti, B., Buratti, E., Boeve, B., Rademakers, R. (Eds.), Frontotemporal Dementias: Emerging Milestones of the 21st Century. Springer, pp. 219–242.

Zhou, Y., Shen, X., He, Z., Weng, H., Chen, W., 2024b. Utilizing AI-enhanced multi-omics integration for predictive modeling of disease susceptibility in functional phenotypes. J. Theory Pract. Eng. Sci. 4 (02), 45–51.

Zhu, M., Xu, R., Yuan, J., Wang, J., Ren, X., Cong, T., You, Y., Ju, A., Xu, L., Wang, H., 2025. Tracking-seq reveals the heterogeneity of off-target effects in CRISPR–Cas9-mediated genome editing. Nat. Biotechnol. 43 (5), 799–810.

Zhu, X., Zhang, Y., Yang, X., Hao, C., Duan, H., 2021. Gene therapy for neurodegenerative disease: clinical potential and directions. Front. Mol. Neurosci. 14, 618171.

Zou, Y., Sun, X., Yang, Q., Zheng, M., Shimoni, O., Ruan, W., Wang, Y., Zhang, D., Yin, J., Huang, X., 2022. Blood-brain barrier–penetrating single CRISPR-Cas9 nanocapsules for effective and safe glioblastoma gene therapy. Sci. Adv. 8(16), eabm8011.

Zwi-Dantsis, L., Mohamed, S., Massaro, G., Moeendarbary, E., 2025. Adeno-associated virus vectors: principles, practices, and prospects in gene therapy. Viruses 17 (2), 239.

Chapter 3

Ethical and regulatory aspects of gene therapy in neurodegenerative disorders

Alireza Shoari
Department of Cancer Biology, Mayo Clinic, Jacksonville, FL, United States

1 Introduction

Gene therapy represents a transformative advance in biomedical science, reshaping treatment by directly addressing the molecular and genetic causes of disease rather than merely alleviating symptoms (Gonçalves and Paiva, 2017). This paradigm shift is particularly consequential for neurodegenerative disorders, which is an area of medicine historically constrained by the limited efficacy of available treatments and the irreversible nature of neuronal degeneration (Chen et al., 2020).

Neurodegenerative diseases such as Alzheimer's disease, Parkinson's disease, amyotrophic lateral sclerosis (ALS), and Huntington's disease are among the most debilitating and complex disorders affecting the human nervous system. They are characterized by the progressive loss of structure or function of neurons, often culminating in neuronal death; these diseases impose a heavy burden not only on patients but also on caregivers and healthcare systems worldwide (Lamptey et al., 2022). Cognitive decline, motor dysfunction, and neuropsychiatric disturbances contribute to a cascade of impairments that severely compromise quality of life. Importantly, current treatments primarily aim to delay symptom progression and provide temporary relief, offering little to no influence over the underlying neurodegenerative processes (Wareham et al., 2022). Within this therapeutic void, gene therapy presents a transformative opportunity where it offers a mechanism-based approach to intervene at the genetic and molecular level, aiming to correct or compensate for pathogenic mutations, upregulate protective factors, or suppress harmful proteins (Rittiner et al., 2022). By altering the genetic program of affected cells, gene therapy opens new avenues to potentially modify disease trajectories in a durable, and in some cases, curative manner. In contrast to symptomatic therapies, gene-based strategies seek to reestablish neural homeostasis, restore functional protein expression, and mitigate toxic cascades that drive neurodegeneration (Maguire et al., 2014).

The feasibility of gene therapy has been significantly advanced by decades of research in molecular genetics, virology, and cellular biology and also breakthroughs in vector design, especially viral vectors such as adeno-associated viruses (AAVs) and lentiviruses, have enhanced the safety, efficiency, and specificity of gene delivery to neural tissues (Bulcha et al., 2021). Concurrently, genome-editing technologies like CRISPR–Cas9, zinc finger nucleases, and transcription activator-like effector nucleases (TALENs) have enabled unprecedented precision in correcting genetic anomalies. These tools allow for tailored interventions that can be adapted to individual genetic profiles, marking the dawn of personalized neurotherapeutics (Singh et al., 2024). Furthermore, innovations in targeting strategies and delivery systems have facilitated the traversal of previously impermeable barriers such as the blood–brain barrier (BBB), which has long hindered therapeutic access to the central nervous system (Upadhyay, 2014). Coupled with advances in bioinformatics (Bono, 2025), synthetic biology (Kitada et al., 2018), and neuroimaging (Daci et al., 2024), the deployment of gene therapy in the nervous system is becoming increasingly feasible and precise.

Given the multifactorial nature of neurodegenerative diseases, encompassing genetic, epigenetic, environmental, and lifestyle factors, gene therapy must often act on multiple fronts (Firdaus and Li, 2024). For example, in addition to correcting mutations, it may also need to suppress aberrant protein aggregation, modulate immune responses, and enhance neurotrophic support (Chen et al., 2024). The multifaceted nature of gene therapies elevates them beyond conventional treatment approaches, establishing them as a versatile platform that can integrate with and strengthen other strategies, such as pharmacologic interventions, stem cell–based treatments, and neurorehabilitation programs. Their use in neurodegenerative

Gene Therapy in Neurodegenerative Disorders. https://doi.org/10.1016/B978-0-443-40467-2.00007-5

disease represents a significant step forward in modern therapeutics. Ongoing progress in delivery technologies, safety profiling, and disease-specific biology is steadily moving the field toward the possibility of modifying the course of neurodegeneration itself. This opening section sets the stage for a more detailed discussion of the ethical dimensions, regulatory pathways, clinical progress, and scientific advances that collectively drive the transition of gene therapy from experimental studies to practical applications in the clinic.

1.1 Overview of gene therapy and its significance

Gene therapy represents a sophisticated and evolving therapeutic modality that intervenes at the genomic level to correct or mitigate disease, and at its essence, gene therapy entails the deliberate alteration of genetic information within somatic cells to elicit therapeutic effects (Addison, 2017). This may be achieved through various strategies, including the replacement of defective genes, the introduction of functional genes capable of producing therapeutic proteins, the silencing of deleterious gene expression, or the precise correction of mutations using genome-editing technologies such as CRISPR–Cas9 (Cox et al., 2015). Each of these approaches leverages the molecular underpinnings of disease to generate a more direct and often more durable therapeutic outcome than conventional drug therapies.

Two principal modalities define the method of gene delivery: ex vivo and in vivo. In the ex vivo approach, cells are harvested from the patient, genetically engineered outside the body under controlled laboratory conditions, and then reintroduced into the patient and this method is particularly advantageous for ensuring high precision and quality control of genetic modifications and has found success in hematologic disorders and certain cancers (Wu et al., 2022). In contrast, in vivo gene therapy involves the direct delivery of genetic constructs into the patient's body, typically through viral vectors or other carrier systems where this approach is particularly relevant for neurodegenerative diseases, where direct access to neural tissues is critical and cell harvesting may be impractical or impossible (Mendell et al., 2021). A central development in the maturation of gene therapy is the introduction of precise gene-editing platforms, particularly the CRISPR–Cas9 system. This technology allows for the direct and specific modification of DNA sequences at targeted genomic loci, enabling the repair of pathogenic mutations or the disruption of harmful genes with high fidelity (Uddin et al., 2020). The advent of such tools has transformed the landscape of gene therapy from one of general gene augmentation to one of tailored, patient-specific genomic correction.

Crucially, current gene therapy efforts focus on somatic cells, thereby limiting genetic modifications to the individual receiving the treatment and avoiding transgenerational transmission, and this distinction is of profound ethical significance (Chen and Zhang, 2022). Somatic gene therapy is widely considered more ethically acceptable as it does not interfere with the germline, nor does it raise the specter of altering human heredity (Ayanoğlu et al., 2020). In contrast, germline gene therapy, which involves editing reproductive cells or early embryos, introduces the possibility of inheritable genetic changes, and despite its theoretical potential to eliminate hereditary diseases, this approach remains highly controversial due to unresolved concerns over long-term safety, equity, and the potential for misuse in nontherapeutic enhancement (Rubeis and Steger, 2018). Consequently, it remains largely confined to basic research and is subject to stringent regulatory and ethical oversight in most countries. In the domain of neurodegenerative diseases, gene therapy holds exceptional promise due to its capacity to address the molecular and cellular mechanisms that drive disease progression (Martier and Konstantinova, 2020).

Traditional treatments for conditions such as Alzheimer's, Parkinson's, ALS, and Huntington's disease are often palliative, aimed at mitigating symptoms without halting or reversing the underlying neurodegenerative process. Moreover, the efficacy of these treatments tends to diminish over time, and their ability to cross the BBB, a major anatomical and physiological obstacle in neurology, is frequently limited (Hussain et al., 2018). Gene therapy offers the potential to overcome these limitations by enabling sustained and targeted interventions within the central nervous system, where it can be designed to deliver trophic support to neurons, thereby enhancing survival and functional integrity (Dooley et al., 2024). In addition, it allows for the suppression of pathogenic proteins, the correction of disease-causing mutations, and the modulation of neuroinflammatory and apoptotic pathways that exacerbate neuronal death (Luo et al., 2022). The specificity with which gene therapy can be targeted to particular cell types or regions of the brain not only enhances therapeutic efficacy but also minimizes systemic toxicity and off-target effects, thereby improving overall safety profiles (Morris et al., 2021). Ultimately, the significance of gene therapy lies not only in its therapeutic potential but also in its conceptual innovation, and it represents a shift from reactive to proactive medicine, from managing disease to fundamentally altering its course at the molecular level (Dunbar et al., 2018). For neurodegenerative diseases, where conventional pharmacological strategies have often failed to achieve meaningful outcomes, gene therapy offers a hopeful and scientifically grounded alternative. As research continues to refine delivery systems, increase targeting accuracy, and expand our understanding of neurogenetics, the role of gene therapy in transforming the treatment of these devastating conditions is likely to become increasingly central.

1.2 Hope versus hype: Ethical and regulatory imperatives

Gene therapy, hailed by many as a revolutionary advancement in modern medicine, carries with it both the profound promise of transformative healing and the sobering potential for ethical missteps and unintended harm where this dual-edged nature, a tension between hope and hype, defines much of the public discourse surrounding gene-based treatments and underscores the urgency of carefully navigating the ethical and regulatory terrain (Arjmand et al., 2020). As gene therapy moves from experimental innovation to clinical reality, particularly in the treatment of neurodegenerative disorders, the stakes grow ever higher. While the potential to alter the course of debilitating conditions such as Alzheimer's disease, Parkinson's disease, ALS, and Huntington's disease is undeniably compelling, this potential must be weighed against the profound ethical responsibilities and regulatory challenges such interventions entail (Garcia-Gonzalez et al., 2024). The ethical landscape of gene therapy is both complex and evolving, and one of the most pressing issues is the adequacy of informed consent, particularly in the context of neurodegenerative diseases that affect cognition and decision-making capacity. Patients with progressive cognitive decline may be unable to fully comprehend the intricacies of gene-based treatments, including their experimental nature, long-term implications, and associated risks (Karlawish et al., 2013; Kim, 2011). This raises difficult questions about autonomy, surrogate decision-making, and the ethical obligations of clinicians and researchers to ensure that consent is not only legally valid but also morally meaningful. The challenge lies not only in simplifying complex scientific information but also in ensuring that patients or their legal representatives are making decisions that align with the patients' values and best interests (Chaet, 2019).

Beyond informed consent, gene therapy prompts a reevaluation of our societal obligations regarding justice, equity, and access. The high cost of approved gene therapies, often exceeding one or two million dollars per treatment, has ignited global debate about affordability and distributive fairness (Kimmelman, 2024). Although these therapies hold curative potential, access remains restricted to a narrow group of patients, most often those within well-funded healthcare systems or supported by comprehensive insurance coverage. This financial exclusivity raises the risk of widening existing health disparities and fostering a tiered model of care in which advanced medical technologies are available only to the most privileged (Wong et al., 2023). Ethical stewardship demands that developers, policymakers, and payers work collaboratively to establish pricing models, subsidy mechanisms, and global access strategies that prioritize inclusion and sustainability over market-driven exclusivity (Lee and Sawai, 2024).

From a regulatory perspective, gene therapy occupies a unique position at the intersection of innovation and oversight. Agencies such as the Food and Drug Administration (FDA), European Medicines Agency (EMA), and regulatory bodies in other countries have developed dedicated pathways for the evaluation of gene-based treatments (Drago et al., 2021). These frameworks encompass a wide range of considerations, including vector safety, manufacturing standards, dosing consistency, clinical trial design, and postmarketing surveillance (Gonin et al., 2005; Zhao et al., 2017; Koutsilieri et al., 2020). However, the rapidly evolving nature of gene therapy, fueled by advances in genome editing, delivery systems, and personalized medicine, often outpaces existing regulations. This dynamic landscape necessitates the development of adaptive regulatory models that can respond to new scientific realities while maintaining rigorous standards of safety and efficacy. Key regulatory challenges include the need to manage individual variability in response to gene therapy, ensure reproducibility of outcomes across diverse populations, and address uncertainties related to long-term follow-up (LTFU). Furthermore, regulators must grapple with the implications of compassionate use programs, expanded access protocols, and accelerated approval mechanisms, which may introduce therapies into the market based on limited or surrogate endpoints. While such pathways can expedite access to life-saving treatments, they also require careful balancing of urgency with thorough risk assessment to preserve patient safety and public trust.

2 Ethical considerations in gene therapy

As gene therapy advances from theoretical possibility to clinical application, it presents not only scientific and medical breakthroughs but also profound ethical challenges. Unlike conventional treatments, gene therapy intervenes at the level of human DNA, introducing changes that can be long-lasting, irreversible, and, in some cases, heritable. This unique nature of gene therapy raises questions that transcend the laboratory and clinic, forcing society to confront dilemmas related to justice, consent, identity, access, and the appropriate boundaries of human intervention in biology. In the context of neurodegenerative diseases, where patients are often vulnerable, cognitively impaired, or in desperate need of options, these ethical considerations become even more urgent and complex.

This section examines the complex ethical dimensions of gene therapy through several thematic perspectives. It opens with a discussion of the societal and moral implications of applying genetic technologies to delay aging or extend human lifespan, raising fundamental questions about enhancement, fairness, and the meaning of mortality. Attention then shifts to the cultural and religious contexts that shape acceptance of gene therapy, underscoring the diversity of values influencing public attitudes. The issue of informed consent is considered next, with particular focus on patients experiencing progressive cognitive decline. Additional themes include the pursuit of justice and equitable access to therapies with high financial barriers, as well as the long-term responsibility to monitor safety across generations. Collectively, these considerations outline the ethical frameworks needed to guide the development of gene therapy in ways that honor human dignity, promote social equity, and safeguard collective responsibility.

2.1 Genetic modification and societal concerns: Implications of delaying aging or achieving immortality through gene therapy

As gene therapy continues to evolve, its application is no longer limited to treating life-threatening genetic disorders or rare diseases and, increasingly, conversations are emerging around the use of gene-editing technologies to influence the human aging process and potentially extend human lifespan well beyond current biological limits (Kitaeva et al., 2024). While such possibilities remain largely speculative, they introduce a provocative set of ethical and societal concerns that extend far beyond the clinical domain and the theoretical capacity to delay aging or achieve biological immortality through genetic intervention forces a reckoning not only with scientific feasibility but also with broader philosophical, cultural, and moral implications (Mackey, 2003). At the heart of this discourse is the question of intent: whether gene therapy should be used strictly to treat disease or also to enhance and extend life beyond its natural trajectory (Penticuff, 1994). The line between therapeutic intervention and human enhancement becomes increasingly blurred when genetic tools are applied to slow cellular senescence, increase resistance to oxidative stress, or repair age-related DNA damage (Coller, 2019). Early experimental models have already demonstrated that manipulation of specific genes, such as those regulating telomerase activity, DNA repair enzymes, or insulin/IGF-1 signaling, can extend lifespan in various organisms and if such findings are eventually translatable to humans, the implications could be transformative, triggering a paradigm shift in how society conceives of aging, health, and mortality (Bernardes de Jesus et al., 2012; De Luca et al., 2013; Berryman et al., 2008; Kim and Lee, 2019).

The pursuit of longevity through gene therapy raises profound ethical dilemmas. Foremost among these is the question of equity and access. If antiaging interventions become feasible, who will be able to benefit from them? There is a real danger that such therapies would be limited to the wealthy and well-insured, thereby widening existing health and social inequalities. In this context, life-extending treatments risk becoming luxury goods, fostering a biologically advantaged class and intensifying social divisions in unprecedented ways. Such a scenario would directly challenge core principles of justice and fairness in healthcare, while also provoking difficult debates about the commodification of human life (Cornetta et al., 2018). Second, the notion of significantly extending human life, or potentially achieving immortality, disrupts long-standing social and cultural norms. Concepts such as generational turnover, retirement, reproductive cycles, and resource allocation are built on the assumption of a finite lifespan. A world in which individuals live well beyond a century could strain economic systems, exacerbate environmental degradation, and generate intergenerational tensions (Bogojević et al., 2008). Third, there are profound existential and psychological concerns associated with delaying or escaping death. For many individuals and cultures, the awareness of mortality is a fundamental aspect of the human condition, shaping personal meaning, ethical behavior, and spiritual belief. Efforts to overcome aging could disrupt these existential frameworks, creating new forms of alienation, psychological distress, or identity crises, and the prospect of indefinite life may not necessarily lead to enhanced well-being; rather, it could produce novel forms of anxiety about the purpose and trajectory of human existence (Hayry, 2011). Philosophers and ethicists argue that the finitude of life imbues it with urgency and meaning, and that a world without death could lead to ethical apathy or existential fatigue.

In conclusion, while the potential to delay aging or extend life through gene therapy captivates scientific imagination and public curiosity, it also compels a deep ethical interrogation. As the boundaries of what is biologically possible expand, so too must our capacity to engage in thoughtful, inclusive, and foresighted dialog. The future of gene therapy is not only a question of molecular mastery, but also a question of how we choose to define and pursue the good life, both individually and collectively.

2.2 Cultural and religious dimensions of accepting gene therapy

The acceptance of gene therapy is not governed by scientific validation alone; it is deeply influenced by cultural beliefs, religious values, and moral worldviews which these dimensions play a crucial role in shaping how individuals and communities perceive the legitimacy, desirability, and ethical acceptability of biomedical innovations, especially those that intervene at the level of human genetics (Delhove et al., 2020). While gene therapy offers unprecedented hope for curing debilitating illnesses, particularly neurodegenerative disorders, its reception varies widely across different cultural and religious contexts. This variation is not merely academic; it has profound implications for policy, clinical implementation, public trust, and global equity in healthcare (Hunt, 2008).

Cultural perspectives strongly influence how people interpret health, illness, and the human body. In many communities, genetic disorders are understood not only in biomedical terms but also through frameworks such as fate, karma, ancestral imbalance, or spiritual testing. These beliefs shape whether interventions rooted in genetics are welcomed as appropriate solutions or regarded as intrusive disruptions to the natural or spiritual order (Koloi-Keaikitse et al., 2025). For example, in collectivist cultures where identity is closely linked to family or community, altering one's genetic makeup might be perceived as disrupting lineage or heritage, raising concerns about identity and continuity (Paniagua and Taylor, 2008). Similarly, in societies that emphasize harmony with nature, gene editing may be viewed as an unnatural interference in the course of life, even if it is intended for therapeutic purposes (Gaydarska et al., 2024). Religious traditions, meanwhile, offer structured moral frameworks that guide decisions about medical treatment and human intervention in the natural world, and some religious groups welcome gene therapy as a manifestation of divine providence, an extension of human ingenuity granted by a higher power to alleviate suffering and heal disease (Cassidy and Pellegrino, 1993; Rosner, 1998; Nor, 2010). Within such frameworks, using gene therapy to treat fatal or severely debilitating conditions may be not only permissible but ethically imperative, aligning with core values of compassion, stewardship, and the sanctity of life (Peters et al., 2008). Culturally sensitive outreach, transparent communication, and community engagement are essential to rebuilding trust and facilitating informed decision-making in these contexts.

It is important to recognize that religious and cultural responses to gene therapy are far from uniform. Within every tradition, there exists a range of perspectives, with some leaders urging cautious acceptance, others rejecting such interventions outright, and still others offering strong support. Religious bioethics continues to evolve as communities engage with new biotechnologies and formulate positions that balance theological commitments with contemporary realities. Far from being peripheral, these cultural and religious dimensions are central to the legitimacy and success of gene therapy. Approaching them with humility and respect helps ensure that its benefits are pursued in ways that honor human diversity, secure genuine consent, and strengthen the relationship between science and society.

2.3 Patient autonomy and informed consent in the context of neurodegenerative disorders

At the foundation of modern medical ethics lies the principle of respect for patient autonomy: the right of individuals to make informed and voluntary decisions about their healthcare (Varkey, 2021). In the context of gene therapy for neurodegenerative diseases, however, the application of this principle becomes ethically complex and operationally challenging and neurodegenerative disorders such as Alzheimer's disease, Parkinson's disease, Huntington's disease, and ALS often involve progressive cognitive decline, impaired judgment, and loss of decision-making capacity (Sun and Roy, 2021). These conditions not only compromise the individual's ability to comprehend the risks and benefits of novel treatments but also raise critical questions about how to ethically obtain consent when autonomy is in flux (Meaney et al., 1996). Gene therapy differs from many conventional treatments in that its effects can be long-lasting or even permanent, with potential consequences that extend beyond what is currently measurable (Salzman et al., 2018). This heightened level of uncertainty and irreversibility amplifies the ethical significance of the informed consent process. For patients in the early stages of neurodegenerative disease, consent may still be viable if conducted with sensitivity and support. However, as the disease progresses, individuals may lose the cognitive capacity required to engage meaningfully in such decisions, and this progression introduces a temporal dimension to consent: What is understood and agreed to at one stage may no longer be valid or fully comprehensible later (High, 1992).

In these cases, the role of advanced directives and surrogate decision-makers becomes central. Advance directives allow individuals to articulate their preferences for future medical interventions while they are still cognitively competent. However, the implementation of these directives in experimental treatments like gene therapy can be fraught with ambiguity (Sedini et al., 2022). A patient's directive may not explicitly anticipate gene-based interventions, leaving room for uncertainty in how such instructions should be applied. In some cases, the directive can be reasonably interpreted as supportive of innovative yet potentially risky therapies, but such judgments remain open to debate. These interpretive ambiguities place a significant ethical burden on caregivers and healthcare professionals, who must weigh respect for the patient's prior wishes against the realities of current clinical practice (Evans et al., 2020). When consent must be obtained from legally authorized representatives, spouses, adult children, or guardians, additional ethical considerations emerge, and surrogates may face emotional pressure, fear of loss, or desperation, all of which can influence their decisions (Gong et al., 2010). Moreover, there may be discrepancies between what a surrogate believes is best and what the patient would have wanted. In such instances, the ideal of substituted judgment, where the surrogate makes decisions aligned with the patient's known values, may conflict with the surrogate's own perspectives or hopes (Torke et al., 2008). Healthcare providers must be vigilant in facilitating conversations that clarify these distinctions and ensure that surrogate consent reflects, as closely as possible, the values and preferences of the patient rather than the wishes of the decision-maker (McSwiggan et al., 2016).

Complicating matters further is the issue of therapeutic misconception, wherein patients or their families mistakenly believe that participation in a gene therapy trial guarantees personal benefit (Heynemann et al., 2023). In the setting of progressive neurodegenerative disease, where few effective treatments exist and hope is in short supply, this misconception can be especially powerful. It is ethically imperative that researchers and clinicians make a clear distinction between research and therapy, emphasizing the experimental nature of gene therapy, the possibility of receiving a placebo, and the realistic range of outcomes, including potential harm or no effect at all (Joseph et al., 2022). Ethical consent in these cases also involves more than a one-time signature on a form, and it should be treated as a process of engagement, tailored to the patient's evolving cognitive state and communicative ability (Kadam, 2017). Strategies such as using simplified language, visual decision aids, and multiple counseling sessions over time can enhance understanding and retention (Muscat et al., 2021). Consent discussions should be revisited periodically, especially in long-term trials or when new information becomes available about risks or benefits, and this iterative approach respects patient autonomy even as their cognitive condition changes and fosters trust in the research process (Verheggen and van Wijmen, 1996). Respecting patient autonomy in the context of neurodegenerative disease and gene therapy demands an ethically sensitive and adaptive approach. It requires acknowledging the challenges posed by cognitive decline, implementing mechanisms to support decision-making, and ensuring that consent is genuinely informed, voluntary, and reflective of the patient's values (Entwistle et al., 2010). In doing so, clinicians and researchers affirm the fundamental dignity of individuals living with neurodegenerative conditions and uphold the ethical integrity of one of the most promising frontiers in medicine.

2.4 Justice and accessibility in healthcare, and the societal impact of expensive gene therapies

As gene therapy transitions from experimental treatment to clinical reality, its high cost and limited accessibility have raised urgent ethical questions surrounding justice and equity in healthcare. These concerns are particularly acute in the context of neurodegenerative diseases, where the promise of potentially life-altering interventions is counterbalanced by structural and financial barriers that determine who can access these therapies and who cannot (Cornetta et al., 2018). The issue is not merely about affordability; it touches on deeper societal values regarding fairness, prioritization of healthcare resources, and the social obligations of governments, industries, and healthcare institutions. Justice in healthcare refers to the principle that all individuals should have equitable access to medically necessary interventions, regardless of socioeconomic status, geographic location, or demographic background (Pratt et al., 2020). Gene therapy, however, currently challenges this principle. Treatments such as Zolgensma for SMA or Luxturna for inherited retinal disease come with price tags that can exceed one or two million dollars per patient. These costs are not incidental, and they are often tied to the complexity of the research, manufacturing processes, small patient populations, and lack of competition (Ng et al., 2024). While such pricing may be justified by developers in terms of research recovery and value-based pricing, the result is a system in which access is effectively rationed by wealth, insurance status, or national healthcare capacity. In low- and middle-income countries, where public health budgets are already constrained, the introduction of high-cost gene therapies presents an almost insurmountable barrier (Cornetta et al., 2022). Even in high-income nations, many public healthcare systems struggle to accommodate the sudden budgetary impact of these treatments, and insurance coverage may be inconsistent, reimbursement policies uncertain, and cost-effectiveness thresholds difficult to meet for therapies with limited long-term outcome data (Kruk et al., 2018). Consequently, patients are often left in a liminal state, aware that a potentially curative treatment exists, yet unable to access it due to financial or bureaucratic limitations. This disparity creates a two-tiered healthcare landscape in which gene therapy becomes an option only for the fortunate few (Dabbous et al., 2022).

The societal implications of this inequity are far-reaching. As gene therapy becomes more integrated into clinical practice, questions will arise about how public funds should be allocated (Hanna et al., 2017). Decisions about whether healthcare systems should allocate substantial portions of their budgets to fund costly gene therapies for a small group of patients inevitably carry consequences for other critical services, including primary care, vaccination programs, and chronic disease management. Such choices are not solely economic in nature but also moral, as they define the values a society chooses to uphold and determine who is prioritized within its healthcare system (Wong et al., 2023). Moreover, unequal access to gene therapy can exacerbate existing social determinants of health. Communities that already experience healthcare disparities, whether due to race, income, disability, or geographic remoteness, are likely to be the last to benefit from advanced genetic treatments (Cornetta et al., 2018). These dynamics risk entrenching a cycle of exclusion, in which disadvantaged populations fall further behind not only in health outcomes but in their capacity to participate in emerging medical innovations.

The danger is not only one of biological inequity but also of deepening social injustice and intergenerational disadvantage (Kruk et al., 2018).

Healthcare professionals and institutions must also prepare for the integration of gene therapy by ensuring that patients are not excluded due to logistical or systemic barriers, and this includes building infrastructure for delivery, training clinicians in genetic medicine, and developing referral systems that identify eligible patients early (Bhatia et al., 2024). Advocacy groups, patient organizations, and community leaders must continue to play a critical role in raising awareness, shaping policy, and ensuring that the voices of those most affected are not marginalized in decision-making processes (Holden et al., 2016). In summary, the development of gene therapy represents a monumental scientific achievement, but its ethical legitimacy depends on the extent to which it is accessible, affordable, and aligned with principles of social justice. If these therapies are to fulfill their transformative potential, they must not be reserved for a privileged few but instead be integrated into a healthcare model that recognizes health as a human right rather than a market commodity, so that society can realize the full value of gene therapy not just as a clinical breakthrough but as a catalyst for a more just and inclusive future.

2.5 Long-term safety and ethical implications

The long-term safety of gene therapy remains one of the most pressing ethical and scientific concerns in the field, and unlike traditional pharmacological treatments, which are typically reversible and short-acting, gene therapy introduces genetic changes that may persist indefinitely (Goswami et al., 2019). This permanence offers the promise of durable therapeutic benefit but also magnifies the potential risks, especially when those risks may not manifest until years or even decades after treatment. The ethical implications of introducing long-lasting or irreversible modifications into the human body are profound, particularly when the full spectrum of outcomes remains uncertain (Niemiec and Howard, 2020). One of the central challenges in evaluating long-term safety is the unpredictability of biological responses to genetic modification (Resnik and Vorhaus, 2006). Even with advances in precision tools like CRISPR–Cas9 and improved viral vectors, the possibility of off-target effects, unintended changes to regions of the genome outside the intended target, remains a significant concern, and such alterations can lead to harmful consequences, including oncogenesis, immune reactions, or unanticipated disruptions in cellular function (Guo et al., 2023). These effects may not become apparent in the short term, particularly in neurodegenerative diseases that progress slowly and may initially mask adverse responses.

Another layer of complexity arises from the potential immunological consequences of gene therapy. Many gene delivery systems rely on viral vectors, such as AAVs, which, despite being engineered for safety, can still provoke immune responses (Shirley et al., 2020). These reactions may limit the effectiveness of the therapy, require immunosuppression, or cause harmful inflammation. In some cases, patients may become sensitized to the vector, rendering them ineligible for future treatments or readministration, and from an ethical standpoint, the need to manage such risks carefully, especially in already vulnerable populations, is paramount (Wang and Shao, 2023). Furthermore, the irreversibility of gene therapy raises critical ethical questions about consent and risk tolerance which once a genetic alteration has been introduced into somatic cells, especially in the central nervous system, there is often no way to remove or undo the change (Buedo et al., 2023). This permanence means that both known and unknown effects must be weighed carefully at the time of consent, a particularly difficult task when long-term outcomes are uncertain or under-researched. Patients, or their surrogates, must be made fully aware not only of the potential benefits but also of the possibility that the therapy could have latent adverse effects that may not be preventable or reversible (Powell, 2021).

For neurodegenerative diseases, where patients experience progressive decline and available treatments are scarce, the ethical considerations become especially complex. Offering a therapy with uncertain risks to individuals facing terminal conditions may be seen as justified, since long-term side effects could be less relevant in the context of limited life expectancy. At the same time, the unpredictability of those risks can be viewed as undermining the ethical basis for intervention. These are not solely clinical judgments but moral dilemmas that demand reflection and input from ethicists, patients, families, and the wider public (Watt et al., 2019; Kohlschutter, 2021). The obligation for long-term monitoring of patients treated with gene therapy is another ethical priority, and researchers and developers must commit to robust follow-up protocols that extend beyond the traditional scope of clinical trials which these should include regular health assessments, neurocognitive evaluations, and monitoring for late-onset adverse events (Mizoguchi et al., 2024). Ethical responsibility also demands that patients be supported over the course of this follow-up, not only medically but also logistically and emotionally, and this means ensuring access to care, continuity of data collection, and transparency in communicating new findings that may affect their health or future treatment options (Olejarczyk and Young, 2024). There is also a moral duty to contribute to the broader scientific community by sharing long-term safety data, and ethical research practice must therefore include commitments to data sharing, publication of negative results, and participation in global registries and collaborative oversight frameworks (Ohmann et al., 2017).

TABLE 3.1 Ethical challenges in gene therapy for neurodegenerative disorders.

Ethical challenge	Description	Proposed ethical strategies
Informed consent in cognitive decline	Patients with neurodegenerative diseases may have impaired decision-making capacity, complicating consent.	Advance directives, surrogate consent, iterative consent processes.
Justice and access	High costs and unequal access risk exacerbating healthcare disparities.	Equitable pricing models, global access frameworks, advocacy group engagement.
Long-term safety	Unknown long-term risks due to the durable and irreversible nature of gene therapies.	15+ year follow-ups, transparent risk communication, global safety registries.
Use for enhancement (e.g., antiaging)	Raises questions about natural lifespan limits, fairness, and societal impact.	Robust public discourse, regulation limiting nontherapeutic enhancement.
Cultural/religious acceptance	Varied acceptance across cultures and religions depending on beliefs about medical intervention.	Inclusive policymaking, religious dialog, cultural sensitivity in implementation.

Finally, the possibility of transgenerational consequences, though currently limited to hypothetical germline interventions, warrants ethical reflection. If gene editing were ever applied to germline cells or embryos, the risks would extend not only to the individual being treated but also to future generations who cannot consent. This raises questions about intergenerational justice, precautionary ethics, and the long-term stewardship of human genetic integrity. Even in somatic therapy, unintended germline transmission cannot be entirely ruled out, particularly in treatments involving the reproductive system, which highlights the need for strict regulatory safeguards and transparent scientific practices. These ethical issues, ranging from informed consent to long-term safety, are summarized in Table 3.1.

3 Regulatory frameworks for gene therapy

As gene therapy moves from a pioneering scientific breakthrough to a practical clinical option, regulation has emerged as one of the most dynamic and influential dimensions of biomedical governance. The novelty, complexity, and potential durability of these interventions call for regulatory frameworks that are both scientifically rigorous and ethically responsive, while remaining flexible enough to keep pace with rapidly advancing technologies. This section considers how regulatory authorities worldwide are addressing the distinctive demands of gene therapy through adaptations in approval pathways, expanded postmarket surveillance, and the development of tools for assessing personalized treatments. It also highlights both the diversity and areas of convergence in regional approaches, examines the difficulties of achieving standardization in precision medicine, and evaluates innovative mechanisms such as adaptive licensing, orphan drug programs, and compassionate use initiatives. Together, these discussions aim to map the current regulatory environment while identifying opportunities for reform and areas where continued evolution is essential.

3.1 Overview of global regulatory systems

The regulatory oversight of gene therapy has undergone a profound evolution in recent decades, driven by the accelerating pace of scientific breakthroughs and the pressing need to balance innovation with safety and public accountability. Although regulatory bodies worldwide share core goals, such as ensuring patient safety, product efficacy, and ethical integrity, the mechanisms they use to achieve these objectives vary significantly (Donaldson et al., 2000). These differences reflect not only institutional priorities but also the broader sociopolitical, economic, and cultural contexts in which these regulatory systems are embedded. In the United States, the FDA, through the Center for Biologics Evaluation and Research (CBER), has been at the forefront of gene therapy regulation (Bailey et al., 2015). The FDA has developed a suite of specialized guidance documents specifically for human gene therapy, addressing everything from preclinical study design to vector characterization, patient monitoring, and LTFU (Takefman and Bryan, 2012). Central to the FDA's regulatory framework is the Investigational New Drug application process, which allows companies and researchers to begin clinical trials once sufficient safety and scientific rationale are demonstrated (Holbein, 2009). In addition, the FDA requires detailed Chemistry, Manufacturing, and Controls documentation to ensure that gene therapy products are produced under conditions of high quality, consistency, and traceability (Cauchon et al., 2019). A hallmark of the FDA's approach is its strong emphasis on data-driven, phased evaluation and the necessity of long-term patient follow-up, often extending for 15 years, to assess delayed adverse effects and therapeutic durability (Umscheid et al., 2011).

Across the Atlantic, the EMA regulates gene therapy products under the Advanced Therapy Medicinal Products (ATMP) classification. This category includes gene therapy, somatic-cell therapy, and tissue-engineered products, and it provides a harmonized framework for review and authorization across all EU member states (Iglesias-Lopez et al., 2019). Approval via the EMA's centralized procedure enables a product to be marketed throughout the European Union with a single application, and EMA incorporates a comprehensive risk management strategy, which includes an emphasis on pharmacovigilance, postmarketing studies, and patient engagement in the review process (Detela and Lodge, 2019). It also operates the Committee for Advanced Therapies (CAT), a multidisciplinary body tasked with evaluating the scientific, technical, and ethical aspects of ATMPs (Aguilera-Cobos et al., 2022). The EMA places strong importance on early scientific advice and dialog with developers, aiming to facilitate innovation while upholding rigorous regulatory standards.

Japan has emerged as a distinct leader in regenerative medicine and gene therapy regulation, thanks in part to the progressive stance of the Pharmaceuticals and Medical Devices Agency (PMDA) (Ikka et al., 2023). The PMDA has pioneered a conditional and time-limited approval system, introduced under the 2014 Regenerative Medicine Law, which allows promising gene therapy products to reach the market based on early-phase data, provided that confirmatory trials continue postapproval (Nagai, 2019). This framework is particularly appealing to patients with severe or life-threatening conditions who have limited treatment options. The Japanese model prioritizes patient access and therapeutic innovation while maintaining regulatory oversight through mandatory postmarket surveillance, including stringent efficacy and safety reporting requirements. Japan's approach exemplifies a regulatory culture that values speed, adaptability, and proactive risk management (Kramer et al., 2013).

Other countries are also adapting their regulatory landscapes to accommodate gene therapy. Health Canada and Australia's Therapeutic Goods Administration have established pathways similar to those in the United States and EU, emphasizing rigorous evaluation and long-term oversight (Yoffe et al., 2023). In China, the National Medical Products Administration (NMPA) has begun to implement a more structured and transparent framework for gene therapies, building on lessons from past controversies to increase regulatory credibility and attract global biotech investment (Lu et al., 2023). In conclusion, global regulatory systems for gene therapy are diverse but are converging in their goal to facilitate safe, effective, and ethically sound therapeutic innovation. As gene therapies continue to move from bench to bedside, ongoing international dialog and policy alignment will be critical to ensuring that regulatory systems support both scientific progress and public interest across diverse healthcare environments. A comparative overview of these regional regulatory frameworks is provided in Table 3.2.

3.2 Challenges in regulating personalized medicine and the implications for standardization and quality control

Personalized medicine, particularly in gene therapy, marks a transformative shift in healthcare by tailoring treatments to an individual's genetic profile and disease characteristics (Mathur and Sutton, 2017). While this approach enhances therapeutic precision, it challenges traditional regulatory frameworks designed for standardized, population-wide treatments.

TABLE 3.2 Global regulatory approaches for gene therapy.

Region	Key regulatory body	Distinct features	Focus areas
United States	FDA (CBER)	Breakthrough Therapy, RMAT, Fast Track, 15-year follow-ups.	Safety, efficacy, adaptive licensing.
European Union	EMA (CAT)	Advanced therapy medicinal products (ATMPs), centralized approval, strong postmarketing surveillance.	Pharmacovigilance, ethical evaluation, multinational access.
Japan	PMDA	Conditional approval based on early data, with a strong emphasis on postmarket studies.	Rapid access, patient safety, innovation.
China	NMPA	Emerging structured system, aiming for international credibility and investment.	Infrastructure development, regulatory maturity.
Global Harmonization	ICH/WHO	Guideline convergence, shared data platforms, mutual recognition agreements.	Standardization, capacity-building, ethical alignment.

EMA (CAT), European Medicines Agency, Committee for Advanced Therapies; *FDA (CBER)*, US Food and Drug Administration, Center for Biologics Evaluation and Research; *ICH/WHO*, International Council for Harmonisation/World Health Organization; *NMPA*, National Medical Products Administration (China); *PMDA*, Pharmaceuticals and Medical Devices Agency (Japan).

Personalized gene therapies, often bespoke and produced in small batches, defy conventional norms of uniformity in dosing, manufacturing, and reproducibility, complicating assessments of safety, efficacy, and quality (Ho et al., 2020). Standard regulatory models depend on comparability across large patient groups, but personalized therapies target unique genetic mutations and often involve patient-derived materials, complicating reproducibility and standardization (Molla and Bitew, 2024). The traditional reliance on randomized controlled trials and fixed quality specifications becomes less feasible, as outcomes may vary significantly between patients, and regulators face difficulties in ensuring product consistency when each therapy is tailored, manufactured under different conditions, and potentially used only once (Young and Smith, 2025). To address these issues, regulatory bodies are exploring adaptive approaches. One promising strategy is the validation of platform technologies, which are standardized delivery systems that support interchangeable therapeutic components (Rosa et al., 2021). The FDA's acceptance of N-of-1 trials, where detailed data from a single patient supports regulatory evaluation, exemplifies this shift, and these methods introduce greater variability and uncertainty, requiring regulators to balance flexibility with rigor (Selker et al., 2023).

Manufacturing quality presents another hurdle. Personalized therapies often involve intricate processes like viral vector engineering and autologous cell modification, all under Good Manufacturing Practice (GMP) conditions (Ramamurthy et al., 2024). However, decentralized production and limited infrastructure in clinical settings pose compliance challenges. Ensuring consistent quality across diverse sites and individualized raw materials calls for centralized facilities or public-private partnerships to uphold regulatory standards (Shore et al., 2022). The rapid pace of innovation further strains regulatory systems, and emerging tools like base and prime editing demand swift yet thorough evaluation (Xu et al., 2024). To remain effective, regulatory science must evolve in parallel with technological advances, adopting tools such as predictive analytics, bioinformatics, and systems biology (Anklam et al., 2022). The evolution of personalized therapies demands continuous training, cross-sector collaboration, and agile governance, while also presenting ethical challenges to informed consent and risk communication, as the unique profile of each therapy can obscure potential risks, particularly in urgent or compassionate-use scenarios, increasing the likelihood of therapeutic misconception and emphasizing the need for regulatory frameworks that ensure transparent, individualized consent to protect patient autonomy (Kruk et al., 2018).

Access and equity remain major concerns, as the high costs and technical complexity of personalized therapies risk restricting their availability to affluent or well-resourced institutions, and without proactive regulatory policies that promote accessibility and affordability, such advancements may deepen existing health disparities (MacDonald et al., 2022). Ensuring equitable distribution and fostering inclusive innovation is a critical responsibility for regulatory agencies, while the increasing reliance on real-world evidence in evaluating personalized therapies highlights the need for robust, secure, and interoperable data systems, along with clear standards for data governance to support responsible monitoring of long-term efficacy and safety (Abernethy et al., 2022). Regulating personalized gene therapies necessitates a paradigm shift. Regulatory systems must adopt novel, flexible approaches while upholding public trust, safety, and equity in this evolving landscape.

3.3 Approval process, adaptive licensing, and compassionate use programs

The clinical development and approval of gene therapies require a nuanced balance between rigorous scientific validation and the ethical imperative to provide timely access to potentially life-saving treatments (Lapteva et al., 2020). Unlike conventional drugs, gene therapies often target rare or severe diseases with limited therapeutic options, small patient populations, and rapid progression, necessitating flexible regulatory approaches (Berry et al., 2024). Traditional drug development follows a phased model, such as preclinical studies, followed by Phase I (safety), Phase II (preliminary efficacy), and Phase III (large-scale trials) (Umscheid et al., 2011). However, this framework is often unsuitable for gene therapies, which are usually single-dose treatments with durable effects, and consequently, conventional trial designs, emphasizing repeated dosing and short-term outcomes, may not apply (Phares et al., 2024).

In response, regulatory bodies have introduced adaptive licensing and conditional approval mechanisms which these allow therapies to enter the market based on preliminary data, with ongoing evidence collection postapproval (Franco et al., 2023). This staged approach supports early access while maintaining long-term oversight. For example, the EMA has integrated adaptive licensing into its conditional marketing authorization framework (Nicotera et al., 2019). In the United States, the FDA offers pathways such as Breakthrough Therapy, Fast Track, and Regenerative Medicine Advanced Therapy (RMAT), which expedite development and review for promising therapies, sometimes based on surrogate endpoints (Tandulje et al., 2025). Japan's PMDA further facilitates early access by permitting conditional approval after Phase II data, contingent on robust postmarketing surveillance (Ali et al., 2024). These pathways underscore the growing recognition of the need for regulatory flexibility, especially in addressing urgent medical needs.

However, accelerated approvals carry inherent risks, as limited early-stage data heightens uncertainty about long-term safety, efficacy, and cost-effectiveness, which are especially critical concerns for irreversible and high-cost gene therapies, making it essential for regulatory agencies to enforce stringent postmarketing commitments, such as long-term registries, adverse event monitoring, and ongoing benefit–risk reassessments (Escandon and Lantos, 2023). Compassionate use or expanded access programs provide a pathway to unapproved therapies for patients who are ineligible for clinical trials, which is especially important in rapidly progressing diseases where treatment delays can be life-threatening, and in the United States, such programs are overseen by the FDA, contingent upon manufacturer consent and a favorable benefit–risk assessment (Jarow et al., 2017). The 2018 "Right to Try" Act relaxed access requirements further, though it has raised concerns about circumventing FDA oversight (Fountzilas et al., 2018). The EU and other countries, including Japan and Canada, have implemented similar frameworks.

Despite their ethical importance, compassionate use programs face practical and methodological challenges, as limited product availability can strain manufacturing capacity, and data collected from noncontrolled settings is often inadequate for regulatory decision-making (Goyal et al., 2020). In addition, access to therapies outside of clinical trials may discourage patient participation in formal studies, thereby delaying comprehensive data collection, and to address these concerns, clear criteria for compassionate access should be established to promote fairness and transparency, while institutions must provide thorough counseling to patients and families to help manage expectations and ensure understanding of potential risks (Goyal et al., 2020). As these therapies redefine modern medicine, regulatory systems must ensure access while maintaining rigorous standards of safety, efficacy, and equity.

3.4 Postapproval monitoring and safety reporting

Postapproval monitoring is a cornerstone of gene therapy regulation due to the long-lasting, often irreversible nature of these treatments, and while clinical trials provide valuable short-term safety and efficacy data, they cannot capture the full scope of long-term or rare adverse events (Crawford and Sumner, 2021). Regulatory oversight thus extends beyond initial approval, entering a phase of real-world pharmacovigilance designed to track delayed effects, assess durability, and monitor unforeseen outcomes (Donzanti, 2018). Gene therapies pose unique risks, including insertional mutagenesis, immune responses, and off-target effects, especially when using viral vectors or gene-editing tools, and because these risks may not appear until years after treatment, LTFU protocols are essential (Haart, 2025). Agencies like the FDA and EMA often mandate up to 15 years of posttreatment surveillance, including clinical assessments, laboratory tests, imaging, and patient-reported outcomes (Almeida, 2024).

Central to this effort are patient registries, which aggregate long-term data from gene therapy recipients. These databases facilitate the identification of safety signals, support real-world effectiveness assessments, and enable continued learning in small, diverse populations. Adverse event reporting systems, such as the FDA's MedWatch and the EMA's EudraVigilance, are essential for the timely detection of emerging risks, with reporting responsibilities shared by providers, patients, and manufacturers, while developers must also submit Periodic Safety Update Reports and Annual Safety Reports to provide comprehensive reviews of safety and benefit–risk profiles, which may prompt regulatory actions such as label changes or market withdrawal if concerns are identified (Ruggiero et al., 2024; Klepper, 2004).

Digital health technologies, including wearables and mobile applications, present new opportunities for real-time monitoring of patient health, treatment adherence, and quality of life, but they also introduce challenges related to data security, regulatory validation, and the integration of evidence into clinical and regulatory decision-making (Canali et al., 2022). Agencies must develop clear frameworks to ensure that digital surveillance strengthens, rather than undermines, safety oversight (Mumtaz et al., 2023). Beyond biological outcomes, the psychosocial impacts of gene therapy must also be considered, as these treatments can influence mental health, self-image, and social experiences, particularly in pediatric or neurodegenerative conditions, making it essential for monitoring systems to include assessments of emotional well-being and social functioning to support holistic patient care (Cunliffe, 2016). Transparency in postapproval monitoring is critical, especially as more therapies are approved through expedited pathways, since public trust relies on clear communication about ongoing risks and real-world outcomes, and open access to surveillance data promotes informed decision-making and accountability (Mennella et al., 2024). Ultimately, postapproval surveillance reflects a scientific and ethical commitment to patients. It ensures that gene therapy development remains responsible, responsive, and patient-centered. Robust monitoring frameworks, transparent risk communication, and long-term engagement are essential to sustaining the promise of gene-based medicine.

3.5 Fast-tracking and orphan drug designation

As gene therapies increasingly address rare and severe diseases lacking effective treatments, regulatory agencies worldwide have implemented specialized mechanisms to accelerate their development and approval (Berry et al., 2024).

Two of the most impactful tools are expedited review pathways and orphan drug designations, which aim to lower regulatory barriers, provide incentives, and ensure faster patient access to transformative therapies (Michaeli et al., 2024). As mentioned, in the United States, the FDA offers several fast-track options such as Fast Track, Breakthrough Therapy, Accelerated Approval, and RMAT designation. These pathways cater to drugs that treat serious conditions and meet unmet medical needs. Fast Track enables rolling review and increased FDA interaction, while Breakthrough Therapy status offers intensive guidance based on early evidence of substantial improvement over existing therapies (Jeong et al., 2024). Accelerated Approval permits conditional marketing based on surrogate endpoints, with the requirement for post-approval confirmation of clinical benefit, while RMAT status, designed specifically for cell and gene therapies, combines these advantages to facilitate the development and review of innovative biologics (Chhina et al., 2022). The EMA has similar programs, and the PRIME (PRIority MEdicines) initiative offers early, enhanced regulatory guidance to developers of therapies addressing unmet needs, often in conjunction with Conditional Marketing Authorization, which allows earlier market entry under commitments to provide additional data (Hanaizi et al., 2023). Japan's PMDA employs the Sakigake designation to support first-in-class, high-impact therapies with shortened review times and proactive regulatory engagement (Tanaka et al., 2021).

Complementing these fast-track mechanisms, Orphan Drug Designation provides economic and regulatory incentives for therapies targeting rare diseases, which are defined as affecting fewer than 200,000 people in the United States or 5 in 10,000 in the European Union (Belančić et al., 2025). Given the limited commercial viability of treatments for rare diseases, orphan designations promote development by offering benefits such as market exclusivity for 7 years in the United States, along with tax credits, waived regulatory fees, and access to research support, which is especially important in gene therapy due to high development costs and the often one-time nature of treatment (Seoane-Vazquez et al., 2008). Many gene therapies receive both expedited and orphan designations, benefiting from accelerated development support and extended market protection, as exemplified by Zolgensma, which was approved for SMA and granted Breakthrough Therapy, Fast Track, Priority Review, and Orphan status, enabling rapid approval and establishing a precedent for treating rare pediatric conditions (Kerpel-Fronius et al., 2020).

Despite their benefits, these pathways raise concerns where expedited approval based on limited data can lead to uncertainties around long-term safety and efficacy, particularly for irreversible and high-cost gene therapies (Escandon and Lantos, 2023). In addition, orphan designations may be misused by artificially narrowing the target population to qualify for incentives, a practice sometimes referred to as salami slicing, before later expanding the market (Franco et al., 2023). To address these issues, regulators enforce rigorous postmarketing obligations, including confirmatory trials, long-term safety reporting, and real-world evidence submissions (Alipour-Haris et al., 2024). Ongoing oversight and transparent implementation are critical to ensuring that accelerated access mechanisms do not undermine scientific rigor or patient safety. Fast-track pathways and orphan drug designations have already played a pivotal role in advancing gene therapy, bringing new possibilities to conditions once considered untreatable. Sustaining this progress will require a careful balance between fostering innovation and upholding strong regulatory and ethical standards, thereby safeguarding public trust and long-term health outcomes.

4 Case studies and real-world examples

While theoretical frameworks and regulatory guidelines offer important blueprints for how gene therapy should be developed and implemented, it is real-world clinical experience that reveals how these systems function in practice. Case studies serve as critical lenses through which the ethical complexities, regulatory tensions, and societal impacts of gene therapy are brought into sharp focus. They illuminate not only the scientific promise of these therapies but also the challenges of equitable access, public trust, and long-term accountability. This section draws from a diverse set of real-world examples to examine how ethical and regulatory considerations intersect in actual gene therapy deployment, offering lessons that are both cautionary and aspirational.

The section begins by exploring high-profile success stories, such as the use of Zolgensma for SMA, which has redefined expectations for regulatory agility, ethical consent, and posttreatment stewardship. It then examines more contentious cases, like the approval of gene therapies for Duchenne muscular dystrophy (DMD), where questions about evidence standards, cost-effectiveness, and regulatory transparency have sparked significant debate. Furthermore, it considers the ethical and logistical challenges of conducting gene therapy trials across international borders, especially in the context of rare diseases, and highlights the indispensable roles of patient advocacy groups and national registries in shaping policy, guiding research priorities, and protecting patient interests. Together, these case studies underscore the importance of context-sensitive governance and ethically grounded innovation in realizing the full potential of gene therapy.

4.1 Ethical and regulatory insights from gene therapy applications

Real-world applications of gene therapy reveal how ethical principles and regulatory frameworks are tested in practice. A key example is onasemnogene abeparvovec, commercially known as Zolgensma, a gene therapy for SMA, a rare and life-threatening condition caused by mutations in the *SMN1* gene; approved by the FDA in 2019, Zolgensma uses an AAV9 vector to deliver a functional copy of the *SMN1* gene through a one-time intravenous infusion (Blair, 2022). Clinical trials demonstrated significant improvements in motor function and survival, establishing Zolgensma as a transformative intervention for affected families and highlighting how fast-tracking and orphan drug designations can accelerate access to critical therapies, as evidenced by its receipt of Breakthrough Therapy, Priority Review, and Orphan Drug status, which expedited its regulatory review (Rhee et al., 2024). However, this rapid approval based on limited data raised concerns about long-term safety and efficacy. Regulators had to balance urgent clinical need with scientific uncertainty, which is a recurring challenge in gene therapy oversight (Weiß et al., 2024).

The case also underscores pressing ethical concerns, as informed consent becomes particularly complex when therapies are administered to infants, requiring parents to make high-stakes decisions under emotional stress and with limited understanding of the underlying science, making clarity, neutrality, and transparency in communication essential to ensuring ethically sound consent processes (Kretzschmar et al., 2024). Zolgensma's price, which exceeds two million dollars, has sparked global debate over affordability and equitable access, with some health systems negotiating outcomes-based reimbursement models, while others were unable to cover the cost, leaving families to pursue alternative funding through donations or advocacy efforts (Nuijten, 2022; Garrison et al., 2021). This disparity highlights ethical concerns related to justice and the potential for widening healthcare inequities through high-cost, high-impact therapies, while LTFU remains essential, as patients treated with Zolgensma are enrolled in monitoring programs lasting up to fifteen years to evaluate durability, safety, and late-emerging effects (Yeo et al., 2022). This reflects an ethical commitment to patient stewardship, reinforcing that gene therapy is not a one-time transaction but a long-term partnership requiring continued oversight and care.

Other gene therapy applications raise distinct ethical and regulatory challenges, as seen with Luxturna, or voretigene neparvovec, which restored partial vision in patients with Leber congenital amaurosis, an inherited form of blindness, prompting questions about how to assess outcomes such as vision restoration, which are inherently subjective and influenced by cultural and individual perceptions (Gerhardt et al., 2023). In sickle cell disease, where affected populations have historically faced medical neglect, ethical engagement is paramount. Clinical trials must emphasize community trust, inclusive recruitment, and culturally sensitive communication to ensure equitable participation and access. Adenosine deaminase severe combined immunodeficiency, one of the earliest conditions treated successfully with ex vivo gene therapy, demonstrated the importance of integrating diagnostics, genetic counseling, and multidisciplinary care (Ferrua and Aiuti, 2017). The long-term monitoring protocols established for this therapy have helped shape broader gene therapy surveillance practices, and together, these case studies demonstrate that the development of gene therapies is as ethically complex as it is scientifically advanced, underscoring the need for transparent pricing, robust informed consent processes, sustained patient support, and equitable access (Kumar et al., 2016). As gene therapies evolve, these early experiences will be critical for shaping future policies and ensuring that progress in precision medicine is guided by social responsibility and ethical rigor.

4.2 Case studies on regulatory failures or controversies: The case of DMD and beyond

Gene therapy has brought transformative successes, but it has also provoked significant regulatory controversies that underscore the tensions between innovation, evidence, and public trust (Olaghere et al., 2025). A notable example is the case of DMD, a fatal X-linked disorder caused by mutations in the *DMD* gene, leading to progressive muscle degeneration and premature death (Angulski, 2023). In June 2023, the FDA granted accelerated approval to delandistrogene moxeparvovec, marketed as Elevidys, a gene therapy for DMD developed by Sarepta Therapeutics, with the decision based on the expression of microdystrophin protein as a surrogate biomarker rather than direct evidence of improved motor function, which remains the most clinically relevant endpoint for patients. The supporting trial involved a small cohort and lacked rigorous placebo controls in some arms, prompting considerable debate, as proponents, including advocacy groups, supported early access to the therapy for a devastating disease with limited treatment options, arguing that earlier intervention could prevent irreversible damage and that delaying approval would deny patients a potentially life-saving treatment (Zaidman et al., 2023; Mendell et al., 2025; Mullard, 2023). Critics, however, questioned the scientific rigor of using an unvalidated surrogate marker and raised concerns about methodological weaknesses, including small sample size and partial unblinding (Boehler et al., 2023). They warned that lowering evidentiary standards for approval risks undermining the credibility of regulatory agencies and could discourage participation in future controlled trials (Mendell et al., 2024). The three-million-dollar price tag further intensified ethical concerns, underscoring the importance of cost-effectiveness and equitable resource allocation, particularly when therapeutic benefits remain uncertain, and this controversy reflects broader ethical tensions in the regulation of rare diseases, especially within accelerated approval pathways, which aim to expedite access to promising therapies but also carry the risk of premature approvals without sufficient long-term safety and efficacy data, a concern that is particularly significant for irreversible and high-cost interventions such as gene therapies (Klimchak et al., 2023).

Other gene therapy cases have also raised serious concerns. A 2020 trial for an AAV-based therapy for X-linked myotubular myopathy was halted following severe liver toxicity and multiple patient deaths, which subsequent investigations pointed to inadequate trial design, improper dosing, and insufficient preclinical modeling. The incident emphasized the need for rigorous screening, dose optimization, and transparent risk disclosure in clinical trials (Shieh, 2020; Duan, 2023). Commercial pressures introduce an additional layer of complexity, as the high financial stakes create a risk of regulatory capture, in which agencies may align more closely with industry interests than with the public good, potentially leading to accelerated approvals of inadequately vetted therapies and undermining public trust in regulatory systems (Shapiro, 2012). In addition, inconsistent global regulatory standards have contributed to the rise of gene therapy tourism, where patients pursue unapproved treatments in countries with weaker oversight, often facing high costs, unproven benefits, and inadequate follow-up care, which raises ethical concerns about exploitation and safety and highlights the need for international regulatory harmonization and enforcement to prevent misuse and protect patients (Cornetta et al., 2022).

Regulatory controversies in gene therapy, especially in high-profile cases like DMD, highlight the delicate balance between hope and evidence. While the urgency to treat severe diseases is morally compelling, regulatory integrity requires adherence to rigorous scientific standards, transparent processes, and postmarketing vigilance. Moving forward, collaboration among regulators, clinicians, and developers is essential to refine approval criteria, enhance long-term surveillance, and maintain public engagement. Only with such a balanced, ethically grounded approach can gene therapy fulfill its promise responsibly and sustainably.

4.3 Cross-border clinical trials, international collaborations, and ethical and regulatory hurdles

Gene therapy research increasingly relies on international collaboration, especially for rare diseases with globally dispersed patient populations, and these partnerships are critical to overcoming the scientific, logistical, and financial barriers inherent in developing advanced therapies (Julkowska et al., 2017). However, global clinical trials introduce complex ethical and regulatory challenges that must be carefully addressed to ensure patient safety, scientific integrity, and equitable access (Olopade et al., 2012). A major driver of international collaboration is the need for sufficient and diverse patient enrollment, as rare diseases often affect only a few individuals per million, making multicountry recruitment essential for assembling meaningful trial cohorts, while also promoting genetic and ethnic diversity to reduce bias and improve the generalizability of results (Marshall, 2017). However, the lack of regulatory harmonization complicates these efforts, as countries maintain distinct requirements for trial approval, informed consent, manufacturing, and safety reporting, resulting in logistical and financial burdens, while misaligned expectations can lead to inconsistent data, protocol modifications, and delays in trial execution (Ndebele et al., 2014). Initiatives such as the International Council for Harmonisation (ICH), World Health Organization (WHO) guidelines, the EU Clinical Trials Regulation, and FDA–EMA coordination aim to align standards, but full harmonization, especially around ethics and data protection, remains a work in progress (Kahn et al., 2024).

Ethical challenges are also significant, as informed consent must be adapted to local linguistic, cultural, and educational contexts, and in low-resource settings, patients and families may have limited understanding of gene therapy, making it more difficult to ensure truly informed decision-making (Halkoaho et al., 2016). Collaborating with local ethics committees, community leaders, and advocacy groups is vital to ensure consent is clear, transparent, and voluntary (Strauss et al., 2001). Exploitation risks in low- and middle-income countries remain a major concern, as lower regulatory barriers and reduced costs may encourage sponsors to conduct trials in these regions. However, without strong oversight and equitable benefit-sharing, such practices risk reinforcing historical inequities (Alemayehu et al., 2018). Key issues include the lack of posttrial care, insufficient long-term monitoring, and the absence of guarantees for future access to successful therapies, making it essential for ethical standards to ensure that participation is accompanied by fair treatment and tangible benefits, while concerns about equity also extend to the affordability and accessibility of these therapies (Doval et al., 2015). When clinical trials are conducted in low- and middle-income countries, developers are not always required to market the resulting therapies locally or at accessible prices, raising fundamental questions about justice, including whether participants in resource-limited settings should be guaranteed access if the therapy proves successful, while additional challenges arise related to data sharing and privacy (Osei, 2012). Regulations such as the EU's General Data Protection Regulation (GDPR) place strict limits on genetic data use and transfer, potentially hindering international collaboration, and balancing data protection with scientific openness requires standardized protocols, robust governance frameworks, and trust among institutions (Shabani and Borry, 2018).

Operational hurdles further complicate global trials. These include shipping genetically modified products, ensuring cold-chain storage, coordinating central laboratory testing, and meeting biosafety requirements (van den Bos et al., 2024). Gene therapies are especially sensitive, demanding high standards for logistics and compliance. Despite these challenges, successful international trials for SMA demonstrate the feasibility and impact of cross-border collaboration (Finkel et al., 2020). These efforts facilitate diverse recruitment, accelerate data collection, and support regulatory convergence, and to ensure that international trials are both ethical and effective, it is essential to implement global ethical guidelines, such as those from UNESCO or CIOMS, at the local level, while also advancing capacity-building initiatives to strengthen regulatory systems in low- and middle-income countries (Li et al., 2017). International collaboration is crucial for advancing gene therapy, particularly for rare and diverse diseases. However, this must be pursued with strong ethical safeguards, regulatory alignment, and a commitment to global equity. If responsibly managed, cross-border trials can drive innovation while reinforcing the ethical foundations of gene therapy research.

4.4 Patient advocacy groups and national registries in regulatory and ethical governance

In the dynamic landscape of gene therapy, patient advocacy groups and national disease registries have emerged as influential stakeholders and no longer passive recipients of biomedical innovation, patients and their advocates actively shape the research agenda, influence regulatory policy, and ensure that ethical considerations reflect lived experiences (Flotte, 2015). Patient advocacy organizations often originate as grassroots movements and evolve into structured entities capable of funding research, lobbying for policy change, and contributing to regulatory consultations (Rose et al., 2017). In the context of gene therapy, particularly for rare and life-threatening neurodegenerative diseases, patient advocacy groups have played pivotal roles by raising public awareness, supporting early-stage research, facilitating clinical trial recruitment, and advocating for expedited review of promising therapies (Nori et al., 2022). A prime example is the development of Zolgensma for SMA. Advocacy groups like Cure SMA and SMA Europe worked closely with researchers and regulators, contributing to trial design, patient-reported outcome measures, and discussions on access, with their engagement helping to frame SMA as a condition warranting accelerated regulatory pathways and ensuring that patient perspectives were central to the process (Balaji et al., 2025). Ethically, advocacy groups provide essential insights into issues of access, affordability, and risk tolerance, challenging restrictive eligibility criteria and pricing models while advocating for broader inclusion and equitable access, and by highlighting the daily realities of living with disease, they bring a sense of moral urgency to policy and regulatory discussions (Boyer et al., 2024).

Advocacy groups also enhance informed consent and communication, particularly for pediatric or cognitively impaired patients, by developing educational resources and hosting forums that help patients and caregivers navigate complex decisions related to gene therapy, where the outcomes may have lifelong implications (Flotte, 2015). Complementing the efforts of advocacy groups, national disease registries serve as critical tools in the context of rare diseases and gene therapy by compiling longitudinal clinical, genetic, and demographic data that support trial recruitment, epidemiological research, and postmarketing surveillance (Patterson et al., 2023). For regulators, these databases provide essential real-world evidence on treatment outcomes and safety, particularly when initial trials are small or exclude certain populations, and registries also serve an ethical function by promoting transparency and accountability through tracking who receives therapies, how effective they are, and what adverse effects occur (Dang, 2023). This information helps ensure that gene therapy is used equitably and responsibly, and when linked with biobanks or genomic databases, registries further support precision medicine while upholding data privacy and informed consent (Olson et al., 2014).

The synergy between advocacy organizations and registries amplifies their impact, as advocacy groups often contribute to the creation or maintenance of registries, helping to ensure that they remain patient-centered and sustain active participation (Kwame and Petrucka, 2021). In return, registries supply the data needed for informed advocacy, funding decisions, and policy interventions. However, despite their importance, both advocacy groups and registries face challenges, as some groups may not fully represent the diversity of affected communities and may encounter conflicts of interest when funded by therapy developers, while registries often struggle with issues such as data standardization, interoperability, and long-term sustainability, particularly in multinational settings (Chiu et al., 2021). To address these issues, transparent governance frameworks are essential, stakeholder engagement must be inclusive, and mechanisms should protect against undue influence (Maurer et al., 2022). Regulators should institutionalize patient input at every stage, from preclinical development to postapproval monitoring, and ethics committees must recognize the legitimacy of patient perspectives when evaluating trial design, acceptable risk thresholds, and access criteria (Faulkner et al., 2023).

In conclusion, patient advocacy groups and disease registries are not peripheral; they are foundational to the scientific and ethical integrity of gene therapy. Their collaboration ensures that innovation is driven not only by technical advances but by compassion, justice, and the lived experiences of patients. As gene therapy evolves, these organizations will remain vital in guiding its responsible and inclusive future.

5 Challenges and future directions

Gene therapy for neurodegenerative disorders is rapidly evolving, bringing along a host of challenges that span scientific, regulatory, ethical, and societal domains. As technologies mature and treatments enter clinical practice, stakeholders must anticipate and address these emerging issues to sustain progress and public trust. This section provides a comprehensive overview of the key obstacles and future trajectories shaping the field.

5.1 Gaps in ethical and regulatory guidelines: Challenges in neurodegenerative disease gene therapy

Despite progress in gene therapy, ethical and regulatory frameworks remain insufficient, particularly in addressing the complexities of neurodegenerative diseases like Alzheimer's, Parkinson's, Huntington's, and ALS which these conditions are biologically complex, clinically diverse, and marked by irreversible neuronal decline (Garcia-Gonzalez et al., 2024). Current governance structures often fall short in addressing the unique challenges posed by these diseases, including issues such as endpoint validation and patient vulnerability, with a central regulatory difficulty being the lack of reliable biomarkers and well-defined clinical endpoints (Mulberg et al., 2019). Unlike monogenic diseases with well-defined genetic markers, most neurodegenerative disorders lack validated biomarkers for diagnosis, monitoring disease progression, or assessing therapeutic efficacy, as illustrated by the ongoing debate over the use of amyloid-beta and tau as surrogate endpoints in Alzheimer's disease (Cummings, 2019). Parkinson's disease also lacks definitive biomarkers for tracking symptoms, creating ambiguity that complicates clinical trial design and delays regulatory approvals, since agencies generally require evidence of functional benefit, such as improvements in cognition or motor function, which may take years to demonstrate (Li and Le, 2020). Determining the optimal therapeutic window adds further complexity to regulation, as many neurodegenerative diseases begin with a prolonged asymptomatic phase, making early treatment potentially risky and late intervention potentially ineffective, requiring regulators to navigate this uncertainty and develop frameworks that responsibly balance potential benefits and risks across the entire disease continuum (Hampel et al., 2016).

Informed consent presents a significant ethical challenge in neurodegenerative diseases, as cognitive impairment can compromise patients' capacity to consent to participation and LTFU, and as cognitive decline progresses, maintaining the validity of consent and ensuring ethical patient engagement becomes increasingly difficult (Karneboge et al., 2025; Soofi, 2022). Nevertheless, current regulatory guidelines provide limited tailored guidance on managing decision-making capacity, surrogate consent, or ongoing consent renewal in populations with neurodegenerative diseases, even though long-term safety monitoring is essential due to the irreversible nature of gene therapy and the typically slow progression of these conditions (Hanna et al., 2017). However, existing postmarketing surveillance systems are often insufficient for detecting cognitive or neuropsychiatric side effects, highlighting the need to expand monitoring efforts to include not only physical but also emotional and cognitive outcomes, which are essential to evaluating patient quality of life (Alomar et al., 2020). Personalized gene therapies targeting specific mutations in conditions such as ALS or Parkinson's disease challenge existing regulatory norms, as they often involve small, genetically distinct patient populations that make large-scale trials with standard controls impractical, and current regulatory frameworks lack the flexibility to accommodate these ultra-personalized interventions, thereby slowing approval pathways and raising ethical concerns about equitable access (Xie et al., 2025).

In conclusion, the current ethical and regulatory frameworks for neurodegenerative gene therapy are ill-equipped to meet the field's unique demands. Moving toward tailored, ethically informed governance will ensure that innovation in this space proceeds responsibly, with full consideration of patient needs, scientific rigor, and global equity.

5.2 The intersection of artificial intelligence and gene therapy: Optimization and ethical implications

The integration of artificial intelligence (AI) into gene therapy is reshaping biomedical innovation by enhancing the development, delivery, and personalization of treatments, helping to address challenges in vector design, genome editing, and patient stratification, while also introducing ethical and regulatory complexities that require careful consideration (Dixit et al., 2023). AI significantly enhances delivery mechanisms, particularly for hard-to-reach tissues such as the central nervous system, by enabling researchers to use deep learning and systems biology models to predict viral vector behavior, optimize tissue targeting, and minimize off-target effects, drawing on extensive datasets from experiments and clinical trials to design more efficient and safer delivery systems (Wang et al., 2025). Genome editing also benefits from AI. Algorithms can select optimal guide RNAs, anticipate off-target activity, and simulate molecular consequences, which is critical for minimizing risks such as tumorigenesis (Erdoğan, 2024). AI also expedites therapeutic discovery by predicting gene targets, identifying suitable patients, and prioritizing candidates, streamlining development and lowering costs (Serrano et al., 2024).

In clinical care, AI supports personalized gene therapy by integrating genomic, clinical, and biometric data to help clinicians identify suitable candidates, determine optimal intervention timing, and select appropriate delivery routes, which is especially valuable for complex conditions such as neurodegenerative diseases (Parekh et al., 2023).

AI holds great potential in advancing gene therapy, yet it introduces significant ethical concerns, chief among them being algorithmic opacity. Numerous studies have highlighted that many AI models function as "black boxes," with decision-making processes that are difficult to interpret. In the context of gene therapy, where interventions can be irreversible, this lack of transparency raises concerns about patient trust, informed consent, and accountability. Consequently, the development and implementation of explainable AI frameworks have been widely advocated to support the ethical and clinically responsible deployment of AI-driven tools in medicine (Vaassen, 2022; Barredo Arrieta et al., 2020). Bias is another critical concern, as AI models trained on unrepresentative data may marginalize certain populations and exacerbate existing healthcare disparities, making fairness audits, diverse datasets, and inclusive design processes essential to ensuring equitable access and outcomes (Cross et al., 2024). Data privacy is also critical, and AI-driven gene therapy requires large-scale data sharing across borders. Legal compliance with frameworks like the GDPR and the Health Insurance Portability and Accountability Act is necessary, but ethical governance must ensure patients understand and control how their data is used, stored, and repurposed (Mennella et al., 2024). Accountability in AI-informed decisions remains unclear, as harm resulting from AI guidance raises questions about liability, including whether responsibility lies with developers, clinicians, or institutions, highlighting the urgent need for clear legal frameworks and robust oversight systems (Habli et al., 2020).

Lastly, AI risks displacing clinical judgment. While it should support, not replace, medical expertise, safeguards and training are needed to preserve the human aspects of care. AI can revolutionize gene therapy, but its ethical integration requires transparency, fairness, and accountability to ensure socially responsible innovation.

5.3 Future trends in gene therapy regulation and ethics

As gene therapy becomes a mainstream medical approach, its ethical and regulatory frameworks must evolve to meet new scientific, clinical, and societal challenges. Several key trends are poised to shape this evolution, such as emphasizing adaptability, equity, and foresight. One major development will be the adoption of dynamic regulatory frameworks. Traditional fixed-phase drug approvals are ill-suited to the complexity of gene therapies (Escandon and Lantos, 2023). Future regulatory models are expected to incorporate adaptive licensing, enabling iterative approvals based on real-world data and ongoing risk–benefit reassessment, while platform-based evaluations that preapprove vectors and manufacturing systems may streamline the approval process for customized therapies by concentrating review on the gene-specific payloads (Innes et al., 2025).

Ethics will become more embedded in the innovation lifecycle. "Ethics by design" approaches will proactively address potential dilemmas such as germline editing, neuroenhancement, and access disparities (Brey and Dainow, 2024). Ethics committees and regulators are expected to adopt ongoing ethical audits, stakeholder consultations, and community engagement as core components of oversight. Global regulatory harmonization will be essential as gene therapy continues to expand internationally, since inconsistent standards can lead to fragmentation, delays, and inequities, and greater coordination through organizations such as the WHO and ICH can facilitate shared trial protocols, mutual recognition of data, and unified approval pathways, which are particularly important for rare diseases that depend on international patient recruitment (Drago et al., 2021; Arcidiacono, 2023). As the use of data-intensive technologies such as AI and genomics expands, the ethical issues surrounding consent, privacy, and data governance grow increasingly complex. Addressing these concerns will require regulatory frameworks to evolve, incorporating dynamic consent models and cross-border data protections that safeguard transparency, security, and patient autonomy in the handling of health information (Oliva et al., 2014).

Affordability and access will continue to be central ethical concerns, as gene therapies often cost over one to two million dollars per treatment, and future policies may require developers to submit affordability plans, implement tiered pricing structures, and engage in public-private partnerships, while support for local manufacturing and technology transfer in low- and middle-income countries will be essential to promoting global equity (Salzman et al., 2018). Attention will also shift toward the long-term psychosocial impacts of gene therapy, as patients, particularly those treated early, live longer postintervention, requiring care systems to address mental health, identity, and social integration, with future evaluations increasingly considering well-being and quality of life alongside clinical efficacy (Oliveri et al., 2018; Hardy et al., 2023). Finally, the growing potential for genetic enhancement will force a clearer boundary between therapy and nontherapeutic use. Regulations will need enforceable limits, backed by public engagement and global consensus, to ensure gene editing serves health, not inequality or speculation.

5.4 Building public trust through transparency and communication

The success of gene therapy relies not only on scientific and regulatory progress but also on public trust, which is essential for its acceptance, ethical sustainability, and societal legitimacy in an era marked by misinformation and skepticism toward scientific institutions, and building this trust requires consistent transparency, clear communication, and inclusive public engagement (Robillard et al., 2014). Transparency involves more than disclosing clinical data which it requires making complex science accessible to diverse audiences. Gene therapy stakeholders must explain how treatments work, what risks and uncertainties exist, and what evidence supports their use. Avoiding hype and acknowledging limitations are essential. Overstated claims or underplayed risks damage credibility. Instead, trust grows when stakeholders are candid about both benefits and challenges (Elwy et al., 2022).

Effective communication is inherently two-way, as public trust is not achieved through top-down messaging but through meaningful dialog. Engaging patients, families, advocacy groups, and marginalized communities early in the research and policy design process fosters mutual understanding and shared responsibility (Lansing et al., 2023). Forums, consultations, and participatory panels provide opportunities for the public to express their values, concerns, and questions, helping to align innovation with societal needs. Trust-building is especially critical in communities that have been historically harmed by medical systems, where past exploitation or systemic exclusion has fostered deep mistrust of biomedical research (Davila et al., 2023). Building trust in these contexts requires cultural competence, sustained engagement, and equitable access to both therapies and opportunities for research participation, while transparency must extend across the entire process, including clinical trials and postmarket data (Bodicoat et al., 2021).

The digital age introduces new challenges. Misinformation spreads quickly online, often conflating gene therapy with controversial technologies like germline editing. Proactive science communication, using digital platforms, trusted messengers, and fact-checking collaborations, is essential to correct misconceptions and promote informed discourse (Traberg et al., 2023). Economic transparency is also essential, as public trust can be undermined when gene therapies are seen as primarily profit-driven, making it important for companies to communicate openly about pricing, funding sources, and commitments to equity, while value-based pricing and access models can demonstrate a broader commitment to social responsibility (Garrison et al., 2019). Ultimately, trust is built through consistency and ethical leadership. Regulators, researchers, and companies must act with integrity and openness. By embedding transparency and inclusive communication throughout the gene therapy ecosystem, the field can gain public confidence and support for its transformative potential.

5.5 Harmonizing international regulatory standards

As gene therapy expands globally, the urgency of harmonizing international regulatory standards has grown, since what was once a niche technology now spans multiple therapeutic areas and regions, with clinical trials and approvals extending across continents (Drago et al., 2021). However, a fragmented landscape of national regulations hampers scientific progress, equitable access, and patient safety. Currently, agencies like the FDA (United States), EMA (Europe), PMDA (Japan), and NMPA (China) regulate gene therapy independently, each with unique requirements for clinical trials, manufacturing, safety monitoring, and postmarketing commitments (Tang et al., 2021). This divergence leads to delays, duplicative processes, and inconsistent access, particularly problematic for global trials in rare diseases that require multinational recruitment. To address these inefficiencies, stakeholders increasingly advocate for regulatory convergence. Harmonization does not require identical systems but calls for alignment on core principles, mutual recognition of data, and streamlined procedures. The ICH sets global standards for clinical trial design and manufacturing, while the WHO and organizations like International Society for Cell & Gene Therapy (ISCT) and Alliance for Regenerative Medicine (ARM) offer guidance tailored to gene and cell therapies (Hoffmann et al., 2023). Mutual recognition agreements represent a practical step toward regulatory alignment, as demonstrated by existing cooperation between the FDA and EMA in recognizing certain GMP inspections, and extending such agreements to gene therapies could reduce redundancy and accelerate multinational product launches, while shared data platforms and interoperable registries would enhance postmarket surveillance by enabling cross-border tracking of long-term safety and efficacy (Gaspar Ernesto and Dias Loureiro, 2020).

However, regulatory harmonization faces significant challenges, as disparities between high- and low-income countries risk widening access gaps, making it essential for harmonization strategies to include capacity-building efforts such as training, infrastructure development, and institutional support to ensure equitable participation across all nations (Arcidiacono, 2023). Developing shared ethical benchmarks grounded in human rights and inclusive dialog will help sustain trust and accountability, and, looking ahead, transnational regulatory consortia modeled after the EMA's centralized review could enable simultaneous multicountry approvals, reducing delays, expanding access, and accelerating innovation while maintaining rigorous oversight (Cavaleri et al., 2025).

In conclusion, international regulatory harmonization is essential for realizing the full potential of gene therapy. It must be guided by principles of safety, equity, and collaboration, ensuring that all populations, regardless of geography, can benefit from the next generation of genetic medicine.

6 Conclusion

Gene therapy represents a transformative frontier in modern medicine, particularly in the context of neurodegenerative diseases such as Alzheimer's, Parkinson's, Huntington's, and ALS. These disorders, long marked by progressive decline and limited therapeutic options, are now entering a new phase of treatment possibilities that aim to modify their underlying causes rather than merely manage symptoms. However, the promise of gene therapy is accompanied by significant ethical responsibilities and complex regulatory demands that must evolve to match the scientific pace.

Traditional ethical and regulatory frameworks often fall short when applied to gene therapy in neurodegenerative contexts. The irreversible nature of these interventions, the vulnerability of the patient populations involved, and the uncertainties surrounding long-term outcomes call for a more nuanced and adaptive approach. Ethical engagement must be ongoing and responsive to changes in patient cognition and capacity. Regulatory systems must move beyond rigid models and embrace adaptive pathways, extended postmarket surveillance, and context-specific endpoints that reflect the realities of slowly progressing, multifactorial diseases. Accessibility and equity remain central challenges. The extraordinary cost of gene therapies, coupled with variable health infrastructure and policy across countries, risks entrenching existing disparities unless proactively addressed. Global harmonization of regulatory standards, shared ethical principles, and cooperative frameworks for data sharing and trial oversight are essential to ensure that gene therapy does not become a privilege for the few but a shared benefit for all.

Public trust in gene therapy will be a defining factor in its long-term success. Building and sustaining that trust will require transparency in clinical outcomes, clarity in communication, meaningful inclusion of patient voices, and responsible management of emerging technologies such as AI. Trust is not a static resource; it must be earned through ethical conduct, sustained through accountability, and nurtured through genuine engagement with the public. Looking ahead, researchers must design inclusive and flexible trials, remain transparent in sharing outcomes, and collaborate across disciplines and borders. Policymakers must invest in long-term monitoring infrastructure, develop targeted guidance for neurodegenerative conditions, and embed ethical oversight into every stage of innovation. Clinicians, meanwhile, will continue to play a vital role as mediators between patients and these new technologies, providing compassionate guidance and ensuring that clinical decisions are informed, equitable, and person-centered.

Ultimately, the advancement of gene therapy for neurodegenerative diseases will depend not only on the ingenuity of science, but on the integrity of the systems that govern its use. It is not enough to pursue cures; we must also ensure that they are delivered justly, safely, and transparently. The collective challenge for the global medical and regulatory community is to guide this powerful technology with foresight and responsibility, so that its impact is not only transformative, but profoundly ethical.

References

Abernethy, A., Adams, L., Barrett, M., Bechtel, C., Brennan, P., Butte, A., Faulkner, J., Fontaine, E., Friedhoff, S., Halamka, J., Howell, M., Johnson, K., Long, P., McGraw, D., Miller, R., Lee, P., Perlin, J., Rucker, D., Sandy, L., Savage, L., Stump, L., Tang, P., Topol, E., Tuckson, R., Valdes, K., 2022. The promise of digital health: then, now, and the future. NAM Perspect. 2578-6865. 6 (22). doi:10.31478/202206e.

Addison, C., 2017. Bench, bedside, boardroom: negotiating translational gene therapy. New Genet. Soc. 1463-6778. 36 (1), 22–42. doi:10.1080/14636778.2017.1289468.

Aguilera-Cobos, L., Piedad Rosario-Lozano, M., Ponce-Polo, A., Blasco-Amaro, J.A., Epstein, D., 2022. Barriers for the evaluation of advanced therapy medicines and their translation to clinical practice: umbrella review. Health Policy. 0168-8510. 126 (12), 1248–1255. doi:10.1016/j.healthpol.2022.10.007.

Alemayehu, C., Mitchell, G., Nikles, J., 2018. Barriers for conducting clinical trials in developing countries- a systematic review. Int. J. Equity Health. 1475-9276. 17 (1). doi:10.1186/s12939-018-0748-6.

Ali, F., Ilyas, A., Ahmadeen, S., 2024. Regulations in Japan. In: Global Regulations of Medicinal, Pharmaceutical, and Food Products. CRC Press, Botswana, pp. 153–167. doi:10.1201/9781003296492-11.

Alipour-Haris, G., Liu, X., Acha, V., Winterstein, A.G., Burcu, M., 2024. Real-world evidence to support regulatory submissions: a landscape review and assessment of use cases. Clin. Transl. Sci. 1752-8062. 17 (8). doi:10.1111/cts.13903.

Almeida, D., 2024. Leveraging patient experience data to guide medicines development, regulation, access decisions and clinical care in the EU. Front. Med. (Lausanne).

Alomar, M., Tawfiq, A.M., Hassan, N., Palaian, S., 2020. Post marketing surveillance of suspected adverse drug reactions through spontaneous reporting: current status, challenges and the future. Ther. Adv. Drug Saf. 2042-0986. 11. doi:10.1177/2042098620938595.

Angulski, B.B., 2023. Duchenne muscular dystrophy: disease mechanism and therapeutic strategies. Front. Physiol. 14, 1183101.

Anklam, E., Bahl, M.I., Ball, R., Beger, R.D., Cohen, J., Fitzpatrick, S., Girard, P., Halamoda-Kenzaoui, B., Hinton, D., Hirose, A., Hoeveler, A., Honma, M., Hugas, M., Ishida, S., Kass, G.E.N., Kojima, H., Krefting, I., Liachenko, S., Liu, Y., Masters, S., Marx, U., McCarthy, T., Mercer, T., Patri, A., Pelaez, C., Pirmohamed, M., Platz, S., Ribeiro, A.J.S., Rodricks, J.V., Rusyn, I., Salek, R.M., Schoonjans, R., Silva, P., Svendsen, C.N., Sumner, S., Sung, K., Tagle, D., Tong, L., Tong, W., van den Eijnden-van-raaij, J., Vary, N., Wang, T., Waterton, J., Wang, M., Wen, H., Wishart, D., Yuan, Y., Slikker Jr, W., 2022. Emerging technologies and their impact on regulatory science. Exp. Biol. Med. 1535-3702. 247 (1), 1–75. doi:10.1177/15353702211052280.

Arcidiacono, J., 2023. International harmonization for cell and gene therapy products. Adv. Exp. Med. Biol. 1430, 235–240. doi:10.1007/978-3-031-34567-8_14.

Arjmand, B., Larijani, B., Hosseini, M.S., Payab, M., Gilany, K., Goodarzi, P., Roudsari, P.P., Baharvand, M.A., Mohammadi, N.S.H., 2020. The horizon of gene therapy in modern medicine: advances and challenges. Adv. Exp. Med. Biol. 1247, 33–64. doi:10.1007/5584_2019_463.

Ayanoğlu, F.B., Elçin, A.E., Elçin, Y.M., 2020. Bioethical issues in genome editing by CRISPR-Cas9 technology. Turk. J. Biol. 1303-6092. 44 (2), 110–120. doi:10.3906/biy-1912-52.

Bailey, A.M., Arcidiacono, J., Benton, K.A., Taraporewala, Z., Winitsky, S., 2015. United States food and drug administration regulation of gene and cell therapies. Adv. Exp. Med. Biol. 871, 1–29. doi:10.1007/978-3-319-18618-4_1.

Balaji, L., Farrar, M.A., Yiu, E.M., Kariyawasam, D., 2025. A state-of-the-art review of registries in spinal muscular atrophy: a valuable resource for clinical research. J. Neuromuscul. Dis. 2214–3599. doi:10.1177/22143602241313113.

Barredo Arrieta, A., Díaz-Rodríguez, N., Ser, J.D., Bennetot, A., Tabik, S., Barbado, A., Garcia, S., Gil-Lopez, S., Molina, D., Benjamins, R., Chatila, R., Herrera, F., 2020. Explainable artificial intelligence (XAI): concepts, taxonomies, opportunities and challenges toward responsible AI. Inf. Fusion. 1566-2535. 58, 82–115. doi:10.1016/j.inffus.2019.12.012.

Belančić, A., Gkrinia, E.M.M., Likić, R., Vitezić, D., 2025. Navigating challenges and opportunities in orphan medicines: a spotlight commentary on rare diseases. Br. J. Clin. Pharmacol. 0306-5251. 91 (4), 1084–1087. doi:10.1002/bcp.70013.

Bernardes de Jesus, B., Vera, E., Schneeberger, K., Tejera, A.M., Ayuso, E., Bosch, F., Blasco, M.A., 2012. Telomerase gene therapy in adult and old mice delays aging and increases longevity without increasing cancer. EMBO Mol. Med. 1757-4676. 4 (8), 691–704. doi:10.1002/emmm.201200245.

Berry, D., Donigan, K., Kahlman, L., Long, J., Markus, C., McCombs, C.K., 2024. Optimizing regulatory frameworks for gene therapies in rare diseases: challenges and solutions. Mol. Ther. Methods Clin. Dev. 2329-0501. 32 (4), 101386. doi:10.1016/j.omtm.2024.101386.

Berryman, D.E., Christiansen, J.S., Johannsson, G., Thorner, M.O., Kopchick, J.J., 2008. Role of the GH/IGF-1 axis in lifespan and healthspan: lessons from animal models. Growth Horm. IGF Res. 1532-2238. 18 (6), 455–471. doi:10.1016/j.ghir.2008.05.005.

Bhatia, S., Cam, Y.L., Carrion, J., Diamond, L., Fennessy, P., Gassman, S., Gutzwiller, F., Kagan, S., Pankevich, D., Maloney, J.Y., Mahadev, N., Schulz, M., Wong-Rieger, D., Morgese, P., 2024. Strengthening health systems for access to gene therapy in rare genetic disorders. Mol. Ther. Methods Clin. Dev. 2329-0501. 32 (2), 101220. doi:10.1016/j.omtm.2024.101220.

Blair, H.A., 2022. Onasemnogene abeparvovec: a review in spinal muscular atrophy. CNS Drugs. 1179-1934. 36 (9), 995–1005. doi:10.1007/s40263-022-00941-1.

Bodicoat, D.H., Routen, A.C., Willis, A., Ekezie, W., Gillies, C., Lawson, C., Yates, T., Zaccardi, F., Davies, M.J., Khunti, K., 2021. Promoting inclusion in clinical trials—a rapid review of the literature and recommendations for action. Trials. 1745-6215. 22 (1). doi:10.1186/s13063-021-05849-7.

Boehler, J.F., Brown, K.J., Beatka, M., Gonzalez, J.P., Dreghici, R.D., Soustek-Kramer, M., McGonigle, S., Ganot, A., Palmer, T., Lowie, C., Chamberlain, J.S., Lawlor, M.W., Morris, C.A., 2023. Clinical potential of microdystrophin as a surrogate endpoint. Neuromuscul. Disord. 1873-2364. 33 (1), 40–49. doi:10.1016/j.nmd.2022.12.007.

Bogojević, A., Balaž, A., Karapandža, R., 2008. Consequences of increased longevity for wealth, fertility, and population growth. Phys. A Stat. Mech. Appl. 0378-4371. 387 (2-3), 543–550. doi:10.1016/j.physa.2007.09.004.

Bono, H., 2025. Recent advances in genome editing and bioinformatics: addressing challenges in genome editing implementation and genome sequencing. Int. J. Mol. Sci. 1422-0067. 26 (7), 3442. doi:10.3390/ijms26073442.

Boyer, B., Huber, K., Zimlichman, E., Saunders, R., McClellan, M., Kahn, C., Noach, R., Salzberg, C., 2024. Advancing the future of equitable access to health care: recommendations from international health care leaders. Health Aff. Scholar. 2976-5390. 2 (8). doi:10.1093/haschl/qxae094.

Brey, P., Dainow, B., 2024. Ethics by design for artificial intelligence. AI Ethics. 2730-5953. 4 (4), 1265–1277. doi:10.1007/s43681-023-00330-4.

Buedo, P., Bianchini, A., Klas, K., Waligora, M., 2023. Bioethics of somatic gene therapy: what do we know so far?. Curr. Med. Res. Opin. 0300-7995. 39 (10), 1355–1365. doi:10.1080/03007995.2023.2257600.

Bulcha, J.T., Wang, Y., Ma, H., Tai, P.W.L., Gao, G., 2021. Viral vector platforms within the gene therapy landscape. Signal Transduct. Target. Ther. 2059-3635. 6 (1). doi:10.1038/s41392-021-00487-6.

Canali, S., Schiaffonati, V., Aliverti, A., Mulvaney, S., 2022. Challenges and recommendations for wearable devices in digital health: data quality, interoperability, health equity, fairness. PLoS Digit. Health. 2767-3170. 1 (10), e0000104. doi:10.1371/journal.pdig.0000104.

Cassidy, J.D., Pellegrino, E.D., 1993. A catholic perspective on human gene therapy. J. Int. Bioethique = Int. J. Bioethics 4 (1), 11–18.

Cauchon, N.S., Oghamian, S., Hassanpour, S., Abernathy, M., 2019. Innovation in chemistry, manufacturing, and controls—A regulatory perspective from industry. J. Pharm. Sci. 1520-6017. 108 (7), 2207–2237. doi:10.1016/j.xphs.2019.02.007.

Cavaleri, M., Alves de Sousa, C.M., Hacker, A., Higgs, E.S., Lumpkin, M.M., Maia, C.S., Mathur, R., Fimbo, A.M., Reis, A., Shin, K.S., Vaughn, D.W., Zhang, W., Moorthy, V., 2025. A roadmap for fostering timely regulatory and ethics approvals of international clinical trials in support of global health research systems. Lancet Glob. Health. 2214-109X. 13 (4), e769. doi:10.1016/s2214-109x(24)00515-1.

Chaet, D.H., 2019. AMA code of medical ethics' opinions related to unrepresented patients. AMA J. Ethics. 2376-6980. 21 (7), 600–602. doi:10.1001/amajethics.2019.600.

Chen, A.A., Zhang, X., 2022. Rethinking the knowledge-attitudes model and introducing belief in human evolution: examining antecedents of public acceptability of human gene editing. Health, Risk Soc. 1469-8331. 24 (7-8), 297–316. doi:10.1080/13698575.2022.2123903.

Chen, K.S., Koubek, E.J., Sakowski, S.A., Feldman, E.L., 2024. Stem cell therapeutics and gene therapy for neurologic disorders. Neurotherapeutics. 1878-7479. 21 (4), e00427. doi:10.1016/j.neurot.2024.e00427.

Chen, W., Hu, Y., Ju, D., 2020. Gene therapy for neurodegenerative disorders: advances, insights and prospects. Acta Pharm. Sin. B. 2211-3843. 10 (8), 1347–1359. doi:10.1016/j.apsb.2020.01.015.

Chhina, M., Drago, D., Ndu, A., 2022. Planning for progress: a US regulatory approach to advancing the clinical development of gene therapies. Mol. Ther. 1525-0024. 30 (7), 2397–2400. doi:10.1016/j.ymthe.2022.05.005.

Chiu, P., Cummings, G.G., Thorne, S., Schick-Makaroff, K., 2021. Policy advocacy and nursing organizations: a scoping review. Policy Polit. Nurs. Pract. 1552-7468. 22 (4), 271–291. doi:10.1177/15271544211050611.

Coller, B.S., 2019. Ethics of human genome editing. Annu. Rev. Med. 1545-326X. 70, 289–305. doi:10.1146/annurev-med-112717-094629.

Cornetta, K., Bonamino, M., Mahlangu, J., Mingozzi, F., Rangarajan, S., Rao, J., 2022. Gene therapy access: global challenges, opportunities, and views from Brazil, South Africa, and India. Mol. Ther. 1525-0016. 30 (6), 2122–2129. doi:10.1016/j.ymthe.2022.04.002.

Cornetta, K., Patel, K., Wanjiku, C.M., Busakhala, N., 2018. Equitable access to gene therapy: a call to action for the American society of gene and cell therapy. Mol. Ther. 1525-0016. 26 (12), 2715–2716. doi:10.1016/j.ymthe.2018.11.002.

Cox, D.B.T., Platt, R.J., Zhang, F., 2015. Therapeutic genome editing: prospects and challenges. Nat. Med. 1546-170X. 21 (2), 121–131. doi:10.1038/nm.3793.

Crawford, T.O., Sumner, C.J., 2021. Assuring long-term safety of highly effective gene-modulating therapeutics for rare diseases. J. Clin. Invest. 1558-8238. 131 (15). doi:10.1172/JCI152817.

Cross, J.L., Choma, M.A., Onofrey, J.A., 2024. Bias in medical AI: implications for clinical decision-making. PLoS Digit. Health. 2767-3170. 3 (11). doi:10.1371/journal.pdig.0000651.

Cummings, 2019. The role of biomarkers in Alzheimer's disease drug development. Adv. Exp. Med. Biol. 1118 (2), 29–61.

Cunliffe, V.T., 2016. The epigenetic impacts of social stress: how does social adversity become biologically embedded?. Epigenomics. 1750-192X. 8 (12), 1653–1669. doi:10.2217/epi-2016-0075.

Dabbous, M., Toumi, M., Simoens, S., Wasem, J., Saal, G., Wang, Y., Luis Huerta Osuna, J., François, C., Annemans, L., von der Schulenburg, J.-M.G., Sola-Morales, O., Malone, D., Garrison, L.P., 2022. Amortization of gene replacement therapies: a health policy analysis exploring a mechanism for mitigating budget impact of high-cost treatments. Health Policy. 0168-8510. 126 (1), 49–59. doi:10.1016/j.healthpol.2021.11.005.

Daci, R., Gray-Edwards, H., Shazeeb, M.S., Vardar, Z., Vachha, B., Cataltepe, O.I., Flotte, T.R., 2024. Neuroimaging applications for the delivery and monitoring of gene therapy for central nervous system diseases. Hum. Gene Ther. 1043-0342. 35 (21-22), 886–895. doi:10.1089/hum.2024.057.

Dang, A., 2023. Real-world evidence: a primer. Pharm. Med. 1178-2595. 37 (1), 25–36. doi:10.1007/s40290-022-00456-6.

Davila, C., Chan, S.H., Gosline, A., Arenas, Z., Kavanagh, J., Feltz, B., McCarthy, E., Pitts, T., Ritchie, C., 2023. Online forums as a tool for broader inclusion of voices on health care communication experiences and serious illness care: mixed methods study. J. Med. Internet Res. 1438-8871. 25 (1), e48550. doi:10.2196/48550.

De Luca, G., Ventura, I., Sanghez, V., Russo, M.T., Ajmone-Cat, M.A., Cacci, E., Martire, A., Popoli, P., Falcone, G., Michelini, F., Crescenzi, M., Degan, P., Minghetti, L., Bignami, M., Calamandrei, G., 2013. Prolonged lifespan with enhanced exploratory behavior in mice overexpressing the oxidized nucleoside triphosphatase hMTH1. Aging Cell. 1474-9718. 12 (4), 695–705. doi:10.1111/acel.12094.

Delhove, J., Osenk, I., Prichard, I., Donnelley, M., 2020. Public acceptability of gene therapy and gene editing for human use: a systematic review. Hum. Gene Ther. 1043-0342. 31 (1-2), 20–46. doi:10.1089/hum.2019.197.

Detela, G., Lodge, A., 2019. EU regulatory pathways for ATMPs: standard, accelerated and adaptive pathways to marketing authorisation. Mol. Ther. Methods Clin. Dev. 2329-0501. 13, 205–232. doi:10.1016/j.omtm.2019.01.010.

Dixit, S., Kumar, A., Srinivasan, K., Vincent, P.M.D.R., Krishnan, N.R., 2023. Advancing genome editing with artificial intelligence: opportunities, challenges, and future directions. Front. Bioeng. Biotechnol. 2296-4185. 11. doi:10.3389/fbioe.2023.1335901.

Donaldson, M.S., Corrigan, J.M., Kohn, L.T., 2000. To Err Is Human: Building a Safer Health System. National Academies Press (US), Washington (DC).

Donzanti, B.A., 2018. Pharmacovigilance is everyone's concern: let's work it out together. Clin. Ther. 1879-114X. 40 (12), 1967–1972. doi:10.1016/j.clinthera.2018.09.016.

Dooley, J., Hughes, J.G., Needham, E.J., Palios, K.A., Liston, A., 2024. The potential of gene delivery for the treatment of traumatic brain injury. J. Neuroinflammation. 1742-2094. 21 (1). doi:10.1186/s12974-024-03156-x.

Doval, D., Shirali, R., Sinha, R., 2015. Post-trial access to treatment for patients participating in clinical trials. Perspect. Clin. Res. 2229-3485. 6 (2), 82. doi:10.4103/2229-3485.154003.

Drago, D., Foss-Campbell, B., Wonnacott, K., Barrett, D., Ndu, A., 2021. Global regulatory progress in delivering on the promise of gene therapies for unmet medical needs. Mol. Ther. Methods Clin. Dev. 2329-0501. 21, 524–529. doi:10.1016/j.omtm.2021.04.001.

Duan, D., 2023. Lethal immunotoxicity in high-dose systemic AAV therapy. Mol. Ther. 1525-0016. 31 (11), 3123–3126. doi:10.1016/j.ymthe.2023.10.015.

Dunbar, C.E., High, K.A., Joung, J.K., Kohn, D.B., Ozawa, K., Sadelain, M., 2018. Gene therapy comes of age. Science. 1095-9203. 359 (6372). doi:10.1126/science.aan4672.

Elwy, A.R., Maguire, E.M., Kim, B., West, G.S., 2022. Involving stakeholders as communication partners in research dissemination efforts. J. Gen. Intern. Med. 1525-1497. 37, 123–127. doi:10.1007/s11606-021-07127-3.

Entwistle, V.A., Carter, S.M., Cribb, A., McCaffery, K., 2010. Supporting patient autonomy: the importance of clinician-patient relationships. J. Gen. Intern. Med. 1525-1497. 25 (7), 741–745. doi:10.1007/s11606-010-1292-2.

Erdoğan, S., 2024. Integration of artificial intelligence and genome editing system for determining the treatment of genetic disorders. Balkan Med. J. 2146-3123. 41 (6), 419–420. doi:10.4274/balkanmedj.galenos.2024.2024-080824.

Escandon, R., Lantos, J., 2023. Challenges in accelerated approvals for gene therapies. Mol. Ther. 1525-0024. 31 (9), 2572–2574. doi:10.1016/j.ymthe.2023.08.004.

Evans, C.J., Yorganci, E., Lewis, P., Koffman, J., Stone, K., Tunnard, I., Wee, B., Bernal, W., Hotopf, M., Higginson, I.J., Tanner, D., Henry, C., Grande, G., Dewar, S., Owen, G., Burman, R., Adamis, D., Dunn, M., Kim, S., Woods, S., Vohora, R., 2020. Processes of consent in research for adults with impaired mental capacity nearing the end of life: systematic review and transparent expert consultation (MORECare_Capacity statement). BMC Med. 1741-7015. 18 (1). doi:10.1186/s12916-020-01654-2.

Faulkner, S.D., Somers, F., Boudes, M., Nafria, B., Robinson, P., 2023. Using patient perspectives to inform better clinical trial design and conduct: current trends and future directions. Pharm. Med. 1179-1993. 37 (2), 129–138. doi:10.1007/s40290-022-00458-4.

Ferrua, F., Aiuti, A., 2017. Twenty-five years of gene therapy for ADA-SCID: from bubble babies to an approved drug. Hum. Gene Ther. 1557-7422. 28 (11), 972–981. doi:10.1089/hum.2017.175.

Finkel, R.S., Day, J.W., De Vivo, D.C., Kirschner, J., Mercuri, E., Muntoni, F., Shieh, P.B., Tizzano, E., Desguerre, I., Quijano-Roy, S., Saito, K., Droege, M., Dabbous, O., Khan, F., Renault, L., Anderson, F.A., Servais, L., 2020. RESTORE: a prospective multinational registry of patients with genetically confirmed spinal muscular atrophy—rationale and study design. J. Neuromuscul. Dis. 2214-3602. 7 (2), 145–152. doi:10.3233/JND-190451.

Firdaus, Z., Li, X., 2024. Unraveling the genetic landscape of neurological disorders: insights into pathogenesis, techniques for variant identification, and therapeutic approaches. Int. J. Mol. Sci. 1422-0067. 25 (4). doi:10.3390/ijms25042320.

Flotte, T.R., 2015. The role of patient advocacy organizations in advancing human gene therapy. Hum. Gene Ther. 1557-7422. 26 (12), 782. doi:10.1089/hum.2015.29011.trf.

Fountzilas, E., Said, R., Tsimberidou, A.M., 2018. Expanded access to investigational drugs: balancing patient safety with potential therapeutic benefits. Expert Opin. Investig. Drugs. 1744-7658. 27 (2), 155–162. doi:10.1080/13543784.2018.1430137.

Franco, P., Jain, R., Rosenkrands-Lange, E., Hey, C., Koban, M.U., 2023. Regulatory pathways supporting expedited drug development and approval in ICH member countries. Ther. Innov. Regul. Sci. 2168-4790. 57 (3), 484–514. doi:10.1007/s43441-022-00480-3.

Garcia-Gonzalez, N., Gonçalves-Sánchez, J., Gómez-Nieto, R., Gonçalves-Estella, J.M., López, D.E., 2024. Advances and challenges in gene therapy for neurodegenerative diseases: a systematic review. Int. J. Mol. Sci. 25 (23), 12485.

Garrison, L.P., Jackson, T., Paul, D., Kenston, M., 2019. Value-based pricing for emerging gene therapies: the economic case for a higher cost-effectiveness threshold. J. Manag. Care Spec. Pharm. 2376-1032. 25 (7), 793–799. doi:10.18553/jmcp.2019.18378.

Garrison, L.P., Jiao, B., Dabbous, O., 2021. Gene therapy may not be as expensive as people think: challenges in assessing the value of single and short-term therapies. J. Manag. Care Spec. Pharm. 2376-1032. 27 (5), 674–681. doi:10.18553/jmcp.2021.27.5.674.

Gaspar Ernesto, L.M., Dias Loureiro, R.M., 2020. Pharmaceutical manufacturing in the EU and the EU-US mutual recognition agreement. PDA J. Pharm. Sci. Technol. 1948-2124. 74 (6), 617–626. doi:10.5731/pdajpst.2020.011437.

Gaydarska, H., Takashima, K., Shahrier, S., Raz, A., Minari, J., 2024. The interplay of ethics and genetic technologies in balancing the social valuation of the human genome in UNESCO declarations. Eur. J. Hum. Genet. 1018-4813. 32 (6), 725–730. doi:10.1038/s41431-024-01549-3.

Gerhardt, M.J., Priglinger, C.S., Rudolph, G., Hufendiek, K., Framme, C., Jägle, H., Salchow, D.J., Anschütz, A., Michalakis, S., Priglinger, S.G., 2023. Gene therapy with voretigene neparvovec improves vision and partially restores electrophysiological function in pre-school children with leber congenital amaurosis. Biomedicines. 2227-9059. 11 (1). doi:10.3390/biomedicines11010103.

Gonçalves, G.A.R., Paiva, R.D.M.A., 2017. Gene therapy: advances, challenges and perspectives. Einstein (São Paulo). 1679-4508. 15 (3), 369–375. doi:10.1590/s1679-45082017rb4024.

Gong, M.N., Winkel, G., Rhodes, R., Richardson, L.D., Silverstein, J.H., 2010. Surrogate consent for research involving adults with impaired decision making: survey of Institutional review board practices. Crit. Care Med. 1530-0293. 38 (11), 2146–2154. doi:10.1097/CCM.0b013e3181f26fe6.

Gonin, P., Buchholz, C.J., Pallardy, M., Mezzina, M., 2005. Gene therapy bio-safety: scientific and regulatory issues. Gene Ther. 0969-7128. 12 (S1), S146. doi:10.1038/sj.gt.3302629.

Goswami, R., Subramanian, G., Silayeva, L., Newkirk, I., Doctor, D., Chawla, K., Chattopadhyay, S., Chandra, D., Chilukuri, N., Betapudi, V., 2019. Gene therapy leaves a vicious cycle. Front. Oncol. 2234-943X. 9. doi:10.3389/fonc.2019.00297.

Goyal, P., Mathur, R., Medhi, B., 2020. Understanding the challenges and ethical aspects of compassionate use of drugs in emergency situations. Indian J. Pharmacol. 1998-3751. 52 (3), 163–171. doi:10.4103/ijp.IJP_665_20.

Guo, C., Ma, X., Gao, F., Guo, Y., 2023. Off-target effects in CRISPR/Cas9 gene editing. Front. Bioeng. Biotechnol. 2296-4185. 11. doi:10.3389/fbioe.2023.1143157.

Haart, 2025. Long-term follow-up after authorization of gene therapy: leveraging real-world data. Drug Discov. Today 30 (5), 104337.

Habli, I., Lawton, T., Porter, Z., 2020. Artificial intelligence in health care: accountability and safety. Bull. World Health Organ. 1564-0604. 98 (4), 251–256. doi:10.2471/BLT.19.237487.

Halkoaho, A., Pietilä, A.M., Ebbesen, M., Karki, S., Kangasniemi, M., 2016. Cultural aspects related to informed consent in health research: a systematic review. Nurs. Ethics. 1477-0989. 23 (6), 698–712. doi:10.1177/0969733015579312.

Hampel, H., O'Bryant, S.E., Castrillo, J.I., Ritchie, C., Rojkova, K., Broich, K., Benda, N., Nisticò, R., Frank, R.A., Dubois, B., Escott-Price, V., Lista, S., 2016. Precision medicine—the golden gate for detection, treatment and prevention of Alzheimer's disease. J. Prev. Alzheimers Dis. 2274-5807. 3 (4), 1–17. doi:10.14283/jpad.2016.112.

Hanaizi, Z., Kweder, S., Thor, S., Ribeiro, S., Marcal, A., 2023. Considering global development? Insights from applications for FDA breakthrough therapy and EMA PRIME designations. Ther. Innov. Regul. Sci. 2168-4790. 57 (2), 321–328. doi:10.1007/s43441-022-00475-0.

Hanna, E., Rémuzat, C., Auquier, P., Toumi, M., 2017. Gene therapies development: slow progress and promising prospect. J. Mark. Access Health Policy. 2001-6689. 5 (1), 1265293. doi:10.1080/20016689.2017.1265293.

Hardy, S.J., Crosby, L.E., Porter, J.S., Sil, S., Valrie, C.R., Jonassaint, C.R., Bediako, S.M., Andrews, C., Rivera, M., Woolford, T., Coleman-Cowger, V.H., 2023. Assessing psychosocial readiness for gene therapy in sickle cell disease: a consensus statement. Blood. 0006-4971. 142 (Supplement 1), 371. doi:10.1182/blood-2023-190398.

Hayry, M., 2011. Considerable life extension and three views on the meaning of life. Camb. Q. Healthc. Ethics 20 (1), 21–29.

Heynemann, S., Lipworth, W., McLachlan, S., Philip, J., John, T., Kerridge, I., 2023. When research becomes practice: the concept of the therapeutic misconception and challenges to consent in clinical trials. Intern. Med. J. 1444-0903. 53 (2), 271–274. doi:10.1111/imj.16015.

High, D.M., 1992. Research with Alzheimer's disease subjects: informed consent and proxy decision making. J. Am. Geriatr. Soc. 1532-5415. 40 (9), 950–957. doi:10.1111/j.1532-5415.1992.tb01995.x.

Ho, D., Quake, S.R., McCabe, E.R.B., Chng, W.J., Chow, E.K., Ding, X., Gelb, B.D., Ginsburg, G.S., Hassenstab, J., Ho, C.-M., Mobley, W.C., Nolan, G.P., Rosen, S.T., Tan, P., Yen, Y., Zarrinpar, A., 2020. Enabling technologies for personalized and precision medicine. Trends Biotechnol. 0167-7799. 38 (5), 497–518. doi:10.1016/j.tibtech.2019.12.021.

Hoffmann, J.M., Blümle, A., Grossmann, R., Yau, H., Lang, B., Bradbury, C., 2023. Toward a global harmonization of service infrastructure in academic clinical trial units: an international survey. Front. Med. 2296-858X. 10. Available from: doi:10.3389/fmed.2023.1252352.

Holbein, M.E.B., 2009. Understanding FDA regulatory requirements for investigational new drug applications for sponsor-investigators. J. Investig. Med. 1081-5589. 57 (6), 688–694. doi:10.2310/jim.0b013e3181afdb26.

Holden, K., Akintobi, T., Hopkins, J., Belton, A., McGregor, B., Blanks, S., Wrenn, G., 2016. Community engaged leadership to advance health equity and build healthier communities. Soc. Sci. 2076-0760. 5 (1), 2. doi:10.3390/socsci5010002.

Hunt, S., 2008. Controversies in treatment approaches: gene therapy, IVF, stem cells, and pharmacogenomics. Nat. Educ. 1, 222.

Hussain, R., Zubair, H., Pursell, S., Shahab, M., 2018. Neurodegenerative diseases: regenerative mechanisms and novel therapeutic approaches. Brain Sci. 2076-3425. 8 (9), 177. doi:10.3390/brainsci8090177.

Iglesias-Lopez, C., Agustí, A., Obach, M., Vallano, A., 2019. Regulatory framework for advanced therapy medicinal products in Europe and United States. Front. Pharmacol. 1663-9812. 10. doi:10.3389/fphar.2019.00921.

Ikka, T., Hatta, T., Saito, Y., Fujita, M., 2023. Does the act on the safety of regenerative medicine in Japan ensure "safety"?: implications of low adverse event reporting. Stem Cell Rep. 2213-6711. 18 (12), 2297–2299. doi:10.1016/j.stemcr.2023.10.012.

Innes, G.K., Smith, K.A., Kuzucan, A., Li, J., Rivera, D., Panagiotou, O.A., Concato, J., 2025. Real-world evidence in new drug and biologics license application approvals during fiscal years 2020–2022. Clin. Pharmacol. Ther. 1532-6535. doi:10.1002/cpt.3688.

Jarow, J.P., Lurie, P., Ikenberry, S.C., Lemery, S., 2017. Overview of FDA's expanded access program for investigational drugs. Ther. Innov. Regul. Sci. 2168-4804. 51 (2), 177–179. doi:10.1177/2168479017694850.

Jeong, H., Purja, S., Kim, E., 2024. Analysis and comparative evaluation of expedited programs for gene therapy products: insights from the United States, the European Union. Gene Ther. 1476-5462. 31 (5-6), 242–254. doi:10.1038/s41434-023-00437-7.

Joseph, A.M., Karas, M., Ramadan, Y., Joubran, E., Jacobs, R.J., 2022. Ethical perspectives of therapeutic human genome editing from multiple and diverse viewpoints: a scoping review. Cureus. 2168-8184. 14 (11). doi:10.7759/cureus.31927.

Julkowska, D., Austin, C.P., Cutillo, C.M., Gancberg, D., Hager, C., Halftermeyer, J., Jonker, A.H., Lau, L.P.L., Norstedt, I., Rath, A., Schuster, R., Simelyte, E., van Weely, S., 2017. The importance of international collaboration for rare diseases research: a European perspective. Gene Ther. 0969-7128. 24 (9), 562–571. doi:10.1038/gt.2017.29.

Kadam, R., 2017. Informed consent process: a step further towards making it meaningful!. Perspect. Clin. Res. 2229-5488. 8 (3), 107–112. doi:10.4103/picr.PICR_147_16.

Kahn, J.P., Smith, C.W., Worku, T.L., Shore, C.K., 2024. Regulatory processes for rare disease drugs in the United States and European Union: flexibilities and collaborative opportunities. National Academies Press. doi:10.17226/27968.

Karlawish, J., Cary, M., Moelter, S.T., Siderowf, A., Sullo, E., Xie, S., Weintraub, D., 2013. Cognitive impairment and PD patients' capacity to consent to research. Neurology. 0028-3878. 81 (9), 801–807. doi:10.1212/WNL.0b013e3182a05ba5.

Karneboge, J., Haberstroh, J., Geschke, K., Perry, J., Radenbach, K., Jessen, F., Rostamzadeh, A., 2025. Facing the new diagnostic and treatment options of Alzheimer's disease: the necessity of informed consent. Alzheimers Dement. 1552-5260. 21 (1). doi:10.1002/alz.14204.

Kerpel-Fronius, S., Baroutsou, V., Becker, S., Carlesi, R., Collia, L., Franke-Bray, B., Kleist, P., Kurihara, C., Laranjeira, L.F., Matsuyama, K., Naseem, S., Schenk, J., Silva, H., 2020. Development and use of gene therapy orphan drugs—Ethical needs for a broader cooperation between the pharmaceutical industry and society. Front. Med. 2296-858X. 7. doi:10.3389/fmed.2020.608249.

Kim, S.S., Lee, C.K., 2019. Growth signaling and longevity in mouse models. BMB Rep. 1976-670X. 52 (1), 70–85. doi:10.5483/BMBRep.2019.52.1.299.

Kim, S.Y.H., 2011. The ethics of informed consent in Alzheimer disease research. Nat. Rev. Neurol. 1759-4758. 7 (7), 410–414. doi:10.1038/nrneurol.2011.76.

Kimmelman, J., 2024. Ethics of gene and cell therapy development for neurologic disorders. Handb. Clin. Neurol. 205, 111–121. doi:10.1016/B978-0-323-90120-8.00002-2.

Kitada, T., DiAndreth, B., Teague, B., Weiss, R., 2018. Programming gene and engineered-cell therapies with synthetic biology. Science. 0036-8075. 359 (6376). doi:10.1126/science.aad1067.

Kitaeva, K.V., Solovyeva, V.V., Blatt, N.L., Rizvanov, A.A., 2024. Eternal youth: a comprehensive exploration of gene, cellular, and pharmacological anti-aging strategies. Int. J. Mol. Sci. 1422-0067. 25 (1). doi:10.3390/ijms25010643.

Klepper, M.J., 2004. The periodic safety update report as a pharmacovigilance tool. Drug Saf. 27 (8), 569–578. doi:10.2165/00002018-200427080-00008.

Klimchak, A.C., Sedita, L.E., Rodino-Klapac, L.R., Mendell, J.R., McDonald, C.M., Gooch, K.L., Malone, D.C., 2023. Assessing the value of delandistrogene moxeparvovec (SRP-9001) gene therapy in patients with Duchenne muscular dystrophy in the United States. J. Mark. Access Health Policy. 2001-6689. 11 (1). doi:10.1080/20016689.2023.2216518.

Kohlschutter, A., 2021. Ethical issues in care and treatment of neuronal ceroid lipofuscinoses (NCL)-A personal view. Front. Neurol. 12, 692527.

Koloi-Keaikitse, S., Kasule, M., Kwape, I., Jankie, D., Ralefala, D., Ntseane, D.M., Mokone, G.G., 2025. Understanding cultural values, norms and beliefs that may impact participation in genome-editing related research: perspectives of local communities in Botswana. Dev. World Bioeth. 1471-8731. 25 (1), 24–34. doi:10.1111/dewb.12440.

Koutsilieri, S., Tzioufa, F., Sismanoglou, D.-C., Patrinos, G.P., 2020. Unveiling the guidance heterogeneity for genome-informed drug treatment interventions among regulatory bodies and research consortia. Pharmacol. Res. 1043-6618. 153, 104590. doi:10.1016/j.phrs.2019.104590.

Kramer, D.B., Tan, Y.T., Sato, C., Kesselheim, A.S., 2013. Postmarket surveillance of medical devices: a comparison of strategies in the US, EU, Japan, and China. PLoS Med. 1549-1676. 10 (9), e1001519. doi:10.1371/journal.pmed.1001519.

Kretzschmar, A.K.M., Teixeira, E., Galato, D., da Silva, E.N., 2024. Judicialization of zolgensma in the ministry of health: costs and clinical profile of patients. Revis. Saude Public. 0034-8910. 58. doi:10.11606/s1518-8787.2024058005899.

Kruk, M.E., Gage, A.D., Arsenault, C., Jordan, K., Leslie, H.H., Roder-DeWan, S., Adeyi, O., Barker, P., Daelmans, B., Doubova, S.V., English, M., Elorrio, E.G., Guanais, F., Gureje, O., Hirschhorn, L.R., Jiang, L., Kelley, E., Lemango, E.T., Liljestrand, J., Malata, A., Marchant, T., Matsoso, M.P., Meara, J.G., Mohanan, M., Ndiaye, Y., Norheim, O.F., Reddy, K.S., Rowe, A.K., Salomon, J.A., Thapa, G., Twum-Danso, N.A.Y., Pate, M., 2018. High-quality health systems in the sustainable development goals era: time for a revolution. Lancet Glob. Health. 2214-109X. 6 (11), e1196. doi:10.1016/S2214-109X(18)30386-3.

Kumar, S.R., Markusic, D.M., Biswas, M., High, K.A., Herzog, R.W., 2016. Clinical development of gene therapy: results and lessons from recent successes. Mol. Ther. Methods Clin. Dev. 2329-0501. 3, 16034. doi:10.1038/mtm.2016.34.

Kwame, A., Petrucka, P.M., 2021. A literature-based study of patient-centered care and communication in nurse-patient interactions: barriers, facilitators, and the way forward. BMC Nurs. 1472-6955. 20 (1). doi:10.1186/s12912-021-00684-2.

Lamptey, R.N.L., Chaulagain, B., Trivedi, R., Gothwal, A., Layek, B., Singh, J., 2022. A review of the common neurodegenerative disorders: current therapeutic approaches and the potential role of nanotherapeutics. Int. J. Mol. Sci. 1422-0067. 23 (3). doi:10.3390/ijms23031851.

Lansing, A.E., Romero, N.J., Siantz, E., Silva, V., Center, K., Casteel, D., Gilmer, T., 2023. Building trust: leadership reflections on community empowerment and engagement in a large urban initiative. BMC Public Health. 1471-2458. 23 (1). doi:10.1186/s12889-023-15860-z.

Lapteva, L., Purohit-Sheth, T., Serabian, M., Puri, R.K., 2020. Clinical development of gene therapies: the first three decades and counting. Mol. Ther. Methods Clin. Dev. 2329-0501. 19, 387–397. doi:10.1016/j.omtm.2020.10.004.

Lee, T.L., Sawai, T., 2024. Navigating equity in global access to genome therapy expanding access to potentially transformative therapies and benefiting those in need requires global policy changes. Front. Genet. 1664-8021. 15. doi:10.3389/fgene.2024.1381172.

Li, R., Ruiz, F., Culyer, A.J., Chalkidou, K., Hofman, K.J., 2017. Evidence-informed capacity building for setting health priorities in low- and middle-income countries: a framework and recommendations for further research. F1000Research. 2046-1402. 6, 231. doi:10.12688/f1000research.10966.1.

Li, T., Le, W., 2020. Biomarkers for Parkinson's disease: how good are they?. Neurosci. Bull. 1995-8218. 36 (2), 183–194. doi:10.1007/s12264-019-00433-1.

Lu, J., Xu, L., Wei, W., He, W., 2023. Advanced therapy medicinal products in China: regulation and development. MedComm 4 (3), e251.

Luo, M., Lee, L.K.C., Peng, B., Choi, C.H.J., Tong, W.Y., Voelcker, N.H., 2022. Delivering the promise of gene therapy with nanomedicines in treating central nervous system diseases. Adv. Sci. 2198-3844. 9 (26). doi:10.1002/advs.202201740.

MacDonald, J., Demiris, G., Shevin, M., Thadaney-Israni, S., Carney, T.J., Cupito, A., 2022. Health technology for all: an equity-based paradigm shift opportunity. NAM Perspect. 2578-6865. 12. doi:10.31478/202212a.

Mackey, T., 2003. An ethical assessment of anti-aging medicine. J. Anti-Aging Med. 1094-5458. 6 (3), 187–204. doi:10.1089/109454503322733045.

Maguire, C.A., Ramirez, S.H., Merkel, S.F., Sena-Esteves, M., Breakefield, X.O., 2014. Gene therapy for the nervous system: challenges and new strategies. Neurotherapeutics. 1878-7479. 11 (4), 817–839. doi:10.1007/s13311-014-0299-5.

Marshall, J.C., 2017. Global collaboration in acute care clinical research: opportunities, challenges, and needs. Crit. Care Med. 1530-0293. 45 (2), 311–320. doi:10.1097/CCM.0000000000002211.

Martier, R., Konstantinova, P., 2020. Gene therapy for neurodegenerative diseases: slowing down the ticking clock. Front. Neurosci. 1662453X. 14. doi:10.3389/fnins.2020.580179.

Mathur, S., Sutton, J., 2017. Personalized medicine could transform healthcare (Review). Biomed. Rep. 2049-9442. 7 (1), 3–5. doi:10.3892/br.2017.922.

Maurer, M., Mangrum, R., Hilliard-Boone, T., Amolegbe, A., Carman, K.L., Forsythe, L., Mosbacher, R., Lesch, J.K., Woodward, K., 2022. Understanding the influence and impact of stakeholder engagement in patient-centered outcomes research: a qualitative study. J. Gen. Intern. Med. 0884-8734. 37 (S1), 6–13. doi:10.1007/s11606-021-07104-w.

McSwiggan, S., Meares, S., Porter, M., 2016. Decision-making capacity evaluation in adult guardianship: a systematic review. Int. Psychogeriatr. 1741-203X. 28 (3), 373–384. doi:10.1017/S1041610215001490.

Meaney, F.J., Kinney, S., Kling, S., Landenburger, G., Panny, S., Schwartz, M., Heidenreich, R., Therrell, B.L., Brokopp, C., Eckman, J., Glass, M., Hannon, W.H., Lorey, F., Pass, K.A., Shapira, E., 1996. Assessing genetic risks—implications for health and social policy: response from the newborn screening committee of the council of regional networks for genetic services. Screening. 0925-6164. 4 (4), 247–249. doi:10.1016/0925-6164(96)00133-8.

Mendell, J.R., Al-Zaidy, S.A., Rodino-Klapac, L.R., Goodspeed, K., Gray, S.J., Kay, C.N., Boye, S.L., Boye, S.E., George, L.A., Salabarria, S., Corti, M., Byrne, B.J., Tremblay, J.P., 2021. Current clinical applications of in vivo gene therapy with AAVs. Mol. Ther. 1525-0024. 29 (2), 464–488. doi:10.1016/j.ymthe.2020.12.007.

Mendell, J.R., Muntoni, F., McDonald, C.M., Mercuri, E.M., Ciafaloni, E., Komaki, H., Leon-Astudillo, C., Nascimento, A., Proud, C., Schara-Schmidt, U., Veerapandiyan, A., Zaidman, C.M., Guridi, M., Murphy, A.P., Reid, C., Wandel, C., Asher, D.R., Darton, E., Mason, S., Potter, R.A., Singh, T., Zhang, W., Fontoura, P., Elkins, J.S., Rodino-Klapac, L.R., 2025. AAV gene therapy for Duchenne muscular dystrophy: the EMBARK phase 3 randomized trial. Nat. Med. 1546-170X. 31 (1), 332–341. doi:10.1038/s41591-024-03304-z.

Mendell, J.R., Sahenk, Z., Lehman, K.J., Lowes, L.P., Reash, N.F., Iammarino, M.A., Alfano, L.N., Lewis, S., Church, K., Shell, R., Potter, R.A., Griffin, D.A., Hogan, M., Wang, S., Mason, S., Darton, E., Rodino-Klapac, L.R., 2024. Long-term safety and functional outcomes of delandistrogene moxeparvovec gene therapy in patients with Duchenne muscular dystrophy: a phase 1/2a nonrandomized trial. Muscle Nerve. 1097-4598. 69 (1), 93–98. doi:10.1002/mus.27955.

Mennella, C., Maniscalco, U., De Pietro, G., Esposito, M., 2024. Ethical and regulatory challenges of AI technologies in healthcare: a narrative review. Heliyon. 2405-8440. 10 (4), e26297. doi:10.1016/j.heliyon.2024.e26297.

Michaeli, D.T., Michaeli, T., Albers, S., Boch, T., Michaeli, J.C., 2024. Special FDA designations for drug development: orphan, fast track, accelerated approval, priority review, and breakthrough therapy. Eur. J. Health Econ. 1618-7601. 25 (6), 979–997. doi:10.1007/s10198-023-01639-x.

Mizoguchi, H., Zhang, A.J., Gupta, P.K., Komuro, M., Cheun, W.K., Chiu, C.W., Choi, B., 2024. Regulatory systems and requirements for clinical trials of AAV-based gene therapies—perspectives from six Asian countries or regions: report from the 6th Asia Partnership Conference of Regenerative Medicine—April 20, 2023. Regen. Ther. 2352-3204. 26, 334–345. doi:10.1016/j.reth.2024.06.005.

Molla, G., Bitew, M., 2024. Revolutionizing personalized medicine: synergy with multi-omics data generation, main hurdles, and future perspectives. Biomedicines. 2227-9059. 12 (12). doi:10.3390/biomedicines12122750.

Morris, J.A., Boshoff, C.H., Schor, N.F., Wong, L.M., Gao, G., Davidson, B.L., 2021. Next-generation strategies for gene-targeted therapies of central nervous system disorders: a workshop summary. Mol. Ther. 1525-0024. 29 (12), 3332–3344. doi:10.1016/j.ymthe.2021.09.010.

Mulberg, A.E., Bucci-Rechtweg, C., Giuliano, J., Jacoby, D., Johnson, F.K., Liu, Q., Marsden, D., McGoohan, S., Nelson, R., Patel, N., Romero, K., 2019. Regulatory strategies for rare diseases under current global regulatory statutes: a discussion with stakeholders. Orphanet J of Rare Dis. 14 (1), 36.

Mullard, A., 2023. FDA approves first gene therapy for Duchenne muscular dystrophy, despite internal objections. Nat. Rev. Drug Discov. 1474-1776. 22 (8), 610. doi:10.1038/d41573-023-00103-y.

Mumtaz, H., Riaz, M.H., Wajid, H., Saqib, M., Zeeshan, M.H., Khan, S.E., Chauhan, Y.R., Sohail, H., Vohra, L.I., 2023. Current challenges and potential solutions to the use of digital health technologies in evidence generation: a narrative review. Front. Digit. Health. 2673-253X. 5. doi:10.3389/fdgth.2023.1203945.

Muscat, D.M., Smith, J., Mac, O., Cadet, T., Giguere, A., Housten, A.J., Langford, A.T., Smith, S.K., Durand, M.A., McCaffery, K., 2021. Addressing health literacy in patient decision aids: an update from the international patient decision aid standards. Med. Decis. Mak. 1552-681X. 41 (7), 848–869. doi:10.1177/0272989X211011101.

Nagai, S., 2019. Flexible and expedited regulatory review processes for innovative medicines and regenerative medical products in the US, the EU, and Japan. Int. J. Mol. Sci. 1422-0067. 20 (15), 3801. doi:10.3390/ijms20153801.

Ndebele, P., Blanchard-Horan, C., Shahkolahi, A., Sanne, I., 2014. Regulatory challenges associated with conducting multicountry clinical trials in resource-limited settings. J. Acquir. Immune Defic. Syndr. 1525-4135. 65 (Supplement 1), S29. doi:10.1097/qai.0000000000000037.

Ng, Q.X., Ong, C., Yaow, C.Y.L., Chan, H.W., Thumboo, J., Wang, Y., Koh, G.C.H., 2024. Cost-of-illness studies of inherited retinal diseases: a systematic review. Orphanet J. Rare Dis. 1750-1172. 19 (1). doi:10.1186/s13023-024-03099-9.

Nicotera, G., Sferrazza, G., Serafino, A., Pierimarchi, P., 2019. The iterative development of medicines through the european medicine agency's adaptive pathway approach. Front. Med. 2296-858X. 6. doi:10.3389/fmed.2019.00148.

Niemiec, E., Howard, H.C., 2020. Ethical issues related to research on genome editing in human embryos. Comput. Struct. Biotechnol. J. 2001-0370. 18, 887–896. doi:10.1016/j.csbj.2020.03.014.

Nor, S.N.M., 2010. Human genetic technologies and islamic bioethics. In: GenEthics and Religion. S. Karger AG, Malaysia, pp. 129–137. doi:10.1159/000315608.

Nori, M., Fisher-Vance, D., Wuerth, L., Colenso, R., Donovan, D.J., 2022. The global role of patients, advocates and caregivers in rare diseases. Future Rare Dis. 2399-5270. 2 (2). doi:10.2217/frd-2022-0003.

Nuijten, 2022. Pricing Zolgensma—the world's most expensive drug. J. Mark. Access Health Policy 10 (1), 2022353.

Ohmann, C., Banzi, R., Canham, S., Battaglia, S., Matei, M., Ariyo, C., Becnel, L., Bierer, B., Bowers, S., Clivio, L., Dias, M., Druml, C., Faure, H., Fenner, M., Galvez, J., Ghersi, D., Gluud, C., Groves, T., Houston, P., Karam, G., Kalra, D., Knowles, R.L., Krleža-Jerić, K., Kubiak, C., Kuchinke, W., Kush, R., Lukkarinen, A., Marques, P.S., Newbigging, A., O'Callaghan, J., Ravaud, P., Schlünder, I., Shanahan, D., Sitter, H., Spalding, D., Tudur-Smith, C., van Reusel, P., van Veen, E.-B., Visser, G.R., Wilson, J., Demotes-Mainard, J., 2017. Sharing and reuse of individual participant data from clinical trials: principles and recommendations. BMJ Open. 2044-6055. 7 (12), e018647. doi:10.1136/bmjopen-2017-018647.

Olaghere, J., Williams, D.A., Farrar, J., Büning, H., Calhoun, C., Ho, T., Inamdar, M.S., Liu, D., Makani, J., Nyarko, K., Ruiz, S., Tisdale, J., McCune, J.M., Boadi, E., 2025. Scientific advancements in gene therapies: opportunities for global regulatory convergence. Biomedicines. 2227-9059. 13 (3), 758. doi:10.3390/biomedicines13030758.

Olejarczyk, J.P., Young, M., 2024. Patient rights and ethics. In: StatPearls [Internet].

Oliva, A., Kaphle, A., Reguant, R., Sng, L.M., Twine, N.A., Malakar, Y., Wickramarachchi, A., Keller, M., Ranbaduge, T., Chan, E.K., Breen, J.. 2014. Future-proofing genomic data and consent management: a comprehensive review of technology innovations. GigaScience. 13, giae021.

Oliveri, S., Ferrari, F., Manfrinati, A., Pravettoni, G., 2018. A systematic review of the psychological implications of genetic testing: a comparative analysis among cardiovascular, neurodegenerative and cancer diseases. Front. Genet. 1664-8021. 9. doi:10.3389/fgene.2018.00624.

Olopade, C.O., Olugbile, S., Olopade, O.I., 2012. Issues and challenges for clinical research in international settings. In: Principles and Practice of Clinical Research. Elsevier Inc., United States, pp. 689–699. doi:10.1016/B978-0-12-382167-6.00047-3.

Olson, J.E., Bielinski, S.J., Ryu, E., Winkler, E.M., Takahashi, P.Y., Pathak, J., Cerhan, J.R., 2014. Biobanks and personalized medicine. Clin. Genet. 0009-9163. 86 (1), 50–55. doi:10.1111/cge.12370.

Osei, A., 2012. Challenges of clinical trials in low- and middle-income countries. Int. Psychiatry. 1749-3676. 9 (2), 30–31. doi:10.1192/s1749367600003039.

Paniagua, C.T., Taylor, R.E., 2008. The cultural lens of genomics. Online J. Issues Nurs. 1091-3734. 13 (1). Available from: http://www.nursingworld.org/MainMenuCategories/ANAMarketplace/ANAPeriodicals/OJIN/TableofContents/vol132008/No1Jan08/CulturalLens.aspx.

Parekh, A.-D.E., Shaikh, O.A., Simran, Manan, S., Hasibuzzaman, M.A., 2023. Artificial intelligence (AI) in personalized medicine: AI-generated personalized therapy regimens based on genetic and medical history: short communication. Ann. Med. Surg. 2049-0801. 85 (11), 5831–5833. doi:10.1097/ms9.0000000000001320.

Patterson, A.M., O'Boyle, M., VanNoy, G.E., Dies, K.A., 2023. Emerging roles and opportunities for rare disease patient advocacy groups. Ther. Adv. Rare Dis. 2633-0040. 4. doi:10.1177/26330040231164425.

Penticuff, J., 1994. Ethical issues in genetic therapy. J. Obstet. Gynecol. Neonatal Nur. 0884-2175. 23 (6), 498–501. doi:10.1111/j.1552-6909.1994.tb01911.x.

Peters, T., Aguilar-Cordova, E., Crawford, C., Lebacqz, K., 2008. Religious Traditions and Genetic Enhancement, Vol. 98. Springer Science and Business Media LLC, pp. 109–159. doi:10.1007/978-1-4020-6923-9_4.

Phares, S., Trusheim, M., Emond, S.K., Pearson, S.D., 2024. Managing the challenges of paying for gene therapy: strategies for market action and policy reform in the United States. J. Comp. Eff. Res. 2042-6305. 13 (12). doi:10.57264/cer-2024-0118.

Powell, 2021. Discussing treatment burden. Breathe (Sheff) 17, 200284.

Pratt, B., Wild, V., Barasa, E., Kamuya, D., Gilson, L., Hendl, T., Molyneux, S., 2020. Justice: a key consideration in health policy and systems research ethics. BMJ Glob. Health. 2059-7908. 5 (4), e001942. doi:10.1136/bmjgh-2019-001942.

Ramamurthy, A., Tommasi, A., Saha, K., 2024. Advances in manufacturing chimeric antigen receptor immune cell therapies. Semin. Immunopathol. 1863-2300. 46 (5). doi:10.1007/s00281-024-01019-4.

Resnik, D.B., Vorhaus, D.B., 2006. Genetic modification and genetic determinism. Philos. Ethics, Humanit. Med. 1747-5341. 1 (1). doi:10.1186/1747-5341-1-9.

Rhee, J., Kang, J., Jo, Y., Yoo, K., Kim, Y.L., Hann, S., Kim, Y., Kim, H., Kim, J., Kong, Y., 2024. Improved therapeutic approach for spinal muscular atrophy via ubiquitination-resistant survival motor neuron variant. J. Cachexia, Sarcopenia Muscle. 2190-5991. 15 (4), 1404–1417. doi:10.1002/jcsm.13486.

Rittiner, J., Cumaran, M., Malhotra, S., Kantor, B., 2022. Therapeutic modulation of gene expression in the disease state: treatment strategies and approaches for the development of next-generation of the epigenetic drugs. Front. Bioeng. Biotechnol. 2296-4185. 10. doi:10.3389/fbioe.2022.1035543.

Robillard, J.M., Roskams-Edris, D., Kuzeljevic, B., Illes, J., 2014. Prevailing public perceptions of the ethics of gene therapy. Hum. Gene Ther. 1557-7422. 25 (8), 740–746. doi:10.1089/hum.2014.030.

Rosa, C., Marsch, L.A., Winstanley, E.L., Brunner, M., Campbell, A.N.C., 2021. Using digital technologies in clinical trials: current and future applications. Contemp. Clin. Trials. 1551-7144. 100, 106219. doi:10.1016/j.cct.2020.106219.

Rose, S.L., Highland, J., Karafa, M.T., Joffe, S., 2017. Patient advocacy organizations, industry funding, and conflicts of interest. JAMA Intern. Med. 2168-6106. 177 (3), 344–350. doi:10.1001/jamainternmed.2016.8443.

Rosner, F., Judaism, 1998. Genetic screening and genetic therapy. Mt. Sinai J. Med. 65 (5-6), 406–413.

Rubeis, G., Steger, F., 2018. Risks and benefits of human germline genome editing: an ethical analysis. Asian Bioeth. Rev. 1793-9453. 10 (2), 133–141. doi:10.1007/s41649-018-0056-x.

Ruggiero, R., Balzano, N., Nicoletti, M.M., Di Mauro, G., Fraenza, F., Campitiello, M.R., Rossi, F., Capuano, A., 2024. Real-world safety data of the orphan drug onasemnogene abeparvovec (Zolgensma®) for the SMA rare disease: a pharmacovigilance study based on the EMA adverse event reporting system. Pharmaceuticals. 17 (3), 394.

Salzman, R., Cook, F., Hunt, T., Malech, H.L., Reilly, P., Foss-Campbell, B., Barrett, D., 2018. Addressing the value of gene therapy and enhancing patient access to transformative treatments. Mol. Ther. 1525-0016. 26 (12), 2717–2726. doi:10.1016/j.ymthe.2018.10.017.

Sedini, C., Biotto, M., Crespi Bel'skij, L.M., Grandini, R.E.M., Cesari, M., 2022. Advance care planning and advance directives: an overview of the main critical issues. Aging Clin. Exp. Res. 1720-8319. 34 (2), 325–330. doi:10.1007/s40520-021-02001-y.

Selker, H.P., Dulko, D., Greenblatt, D.J., Palm, M., Trinquart, L., 2023. The use of N-of-1 trials to generate real-world evidence for optimal treatment of individuals and populations. J. Clin. Transl. Sci. 2059-8661. 7 (1). doi:10.1017/cts.2023.604.

Seoane-Vazquez, E., Rodriguez-Monguio, R., Szeinbach, S.L., Visaria, J., 2008. Incentives for orphan drug research and development in the United States. Orphanet J. Rare Dis. 1750-1172. 3 (1). doi:10.1186/1750-1172-3-33.

Serrano, D.R., Luciano, F.C., Anaya, B.J., Ongoren, B., Kara, A., Molina, G., Ramirez, B.I., Sánchez-Guirales, S.A., Simon, J.A., Tomietto, G., Rapti, C., Ruiz, H.K., Rawat, S., Kumar, D., Lalatsa, A., 2024. Artificial intelligence (AI) applications in drug discovery and drug delivery: revolutionizing personalized medicine. Pharmaceutics. 1999-4923. 16 (10). doi:10.3390/pharmaceutics16101328.

Shabani, M., Borry, P., 2018. Rules for processing genetic data for research purposes in view of the new EU general data protection regulation. Eur. J. Hum. Genet. 1018-4813. 26 (2), 149–156. doi:10.1038/s41431-017-0045-7.

Shapiro, S.A., 2012. The complexity of regulatory capture: diagnosis, causality, and remediation. Roger Williams U.L. Rev. 17, 11.

Shieh, P.B., 2020. Moving forward after two deaths in a gene therapy trial of myotubular myopathy" by Wilson and Flotte. Hum. Gene Ther. 31, 695–696.

Shirley, J.L., de Jong, Y.P., Terhorst, C., Herzog, R.W., 2020. Immune responses to viral gene therapy vectors. Mol. Ther. 1525-0024. 28 (3), 709–722. doi:10.1016/j.ymthe.2020.01.001.

Shore, C., Brown, L., Hopp, W.J., 2022. National Academies of Sciences, Engineering, and Medicine. Causes and consequences of medical product supply chain failures. InBuilding Resilience into the Nation's Medical Product Supply Chains. National Academies Press (US).

Singh, K., Bhushan, B., Kumar, S., Singh, S., Macadangdang, R.R., Pandey, E., Varma, A.K., Kumar, S., 2024. Precision genome editing techniques in gene therapy: current state and future prospects. Curr. Gene Ther. 1566-5232. 24 (5), 377–394. doi:10.2174/0115665232279528240115075352.

Soofi, H., 2022. Ethical aspects of facilitating the recruitment of people with dementia for clinical trials: a call for further debate. Br. J. Clin. Pharmacol. 0306-5251. 88 (1), 22–26. doi:10.1111/bcp.14968.

Strauss, R.P., Sengupta, S., Quinn, S.C., Goeppinger, J., Spaulding, C., Kegeles, S.M., Millett, G., 2001. The role of community advisory boards: involving communities in the informed consent process. Am. J. Public Health. 0090-0036. 91 (12), 1938–1943. doi:10.2105/AJPH.91.12.1938.

Sun, J., Roy, S., 2021. Gene-based therapies for neurodegenerative diseases. Nat. Neurosci. 1546-1726. 24 (3), 297–311. doi:10.1038/s41593-020-00778-1.

Takefman, D., Bryan, W., 2012. The state of gene therapies: the FDA perspective. Mol. Ther. 1525-0024. 20 (5), 877–878. doi:10.1038/mt.2012.51.

Tanaka, M., Idei, M., Sakaguchi, H., Kato, R., Sato, D., Sawanobori, K., Kawarasaki, S., Hata, T., Yoshizaki, A., Nakamura, M., Ikuma, M., 2021. Achievements and challenges of the Sakigake designation system in Japan. Br. J. Clin. Pharmacol. 0306-5251. 87 (10), 4027–4035. doi:10.1111/bcp.14807.

Tandulje, A.A., Varpe, P.C., Chaugule, P.D., Saji, S., Nizar, F., Raghuvanshi, R.S., Srivastava, S., 2025. Cell and gene therapy products: navigating the regulatory landscape of paradigm approvals in the US (2020 to 2024). J. Pharm. Innov. 1939-8042. 20 (2). doi:10.1007/s12247-025-09973-8.

Tang, W., Huang, Y., Zhou, D., Huang, Y., Chen, Y., Ren, S., Li, Y., Wu, S., Zhao, X., Song, X., Wang, H., Jin, Y., Hongtao, Y., Zhang, L., Li, Y., Boulton, D., Shen, K., 2021. Evolving drug regulatory landscape in China: a clinical pharmacology perspective. Clin. Transl. Sci. 1752-8054. 14 (4), 1222–1230. doi:10.1111/cts.12987.

Torke, A.M., Alexander, G.C., Lantos, J., 2008. Substituted judgment: the limitations of autonomy in surrogate decision making. J. Gen. Intern. Med. 1525-1497. 23 (9), 1514–1517. doi:10.1007/s11606-008-0688-8.

Traberg, C.S., Harjani, T., Basol, M., Biddlestone, M., Maertens, R., Roozenbeek, J., van der Linden, S., 2023. Prebunking against misinformation in the modern digital age. In: Managing Infodemics in the 21st Century: Addressing New Public Health Challenges in the Information Ecosystem. Springer International Publishing, United Kingdom, pp. 99–111. doi:10.1007/978-3-031-27789-4_8.

Uddin, F., Rudin, C.M., Sen, T., 2020. CRISPR gene therapy: applications, limitations, and implications for the future. Front. Oncol. 2234-943X. 10. doi:10.3389/fonc.2020.01387.

Umscheid, C.A., Margolis, D.J., Grossman, C.E., 2011. Key concepts of clinical trials: a narrative review. Postgrad. Med. 1941-9260. 123 (5), 194–204. doi:10.3810/pgm.2011.09.2475.

Upadhyay, R.K., 2014. Drug delivery systems, CNS protection, and the blood brain barrier. Biomed. Res. Int. 2314-6141. 2014. doi:10.1155/2014/869269.

Vaassen, B., 2022. AI, opacity, and personal autonomy. Philos. Technol. 2210-5433. 35 (4). doi:10.1007/s13347-022-00577-5.

van den Bos, C., Keefe, R., Keville, M., McCaman, M., 2024. Manufacturing and quality of advanced therapeutics: cellular therapies, CAR-T and gene therapies. In: Cell Engineering. Springer Science and Business Media LLC, pp. 453–488. doi:10.1007/978-3-031-45669-5_15.

Varkey, B., 2021. Principles of clinical ethics and their application to practice. Med. Princ. Pract. 1011-7571. 30 (1), 17–28. doi:10.1159/000509119.

Verheggen, F.W.S.M., van Wijmen, F.C.B., 1996. Informed consent in clinical trials. Health Policy. 0168-8510. 36 (2), 131–153. doi:10.1016/0168-8510(95)00805-5.

Wang, B., Zhang, T., Liu, Q., Sutcharitchan, C., Zhou, Z., Zhang, D., Li, S., 2025. Elucidating the role of artificial intelligence in drug development from the perspective of drug-target interactions. J. Pharm. Anal. 2095-1779. 15 (3), 101144. doi:10.1016/j.jpha.2024.101144.

Wang, Y., Shao, W., 2023. Innate immune response to viral vectors in gene therapy. China Viruses. 1999-4915. 15 (9). doi:10.3390/v15091801.

Wareham, L.K., Liddelow, S.A., Temple, S., Benowitz, L.I., Di Polo, A., Wellington, C., Goldberg, J.L., He, Z., Duan, X., Bu, G., Davis, A.A., Shekhar, K., Torre, A.L., Chan, D.C., Canto-Soler, M.V., Flanagan, J.G., Subramanian, P., Rossi, S., Brunner, T., Bovenkamp, D.E., Calkins, D.J., 2022. Solving neurodegeneration: common mechanisms and strategies for new treatments. Mol. Neurodegener. 1750-1326. 17 (1). doi:10.1186/s13024-022-00524-0.

Watt, A.D., Jenkins, N.L., McColl, G., Collins, S., Desmond, P.M., 2019. Ethical issues in the treatment of late-stage Alzheimer's disease. J. Alzheimers Dis. 1875-8908. 68 (4), 1311–1316. doi:10.3233/JAD-180865.

Weiß, C., Becker, L.L., Friese, J., Blaschek, A., Hahn, A., Illsinger, S., Schwartz, O., Bernert, G., von der Hagen, M., Husain, R.A., Goldhahn, K., 2024. Efficacy and safety of gene therapy with onasemnogene abeparvovec in children with spinal muscular atrophy in the DA-CH-region: a population-based observational study. The Lancet Regional Health–Europe. 1, 47.

Wong, C.H., Li, D., Wang, N., Gruber, J., Lo, A.W., Conti, R.M., 2023. The estimated annual financial impact of gene therapy in the United States. Gene Ther. 1476-5462. 30 (10-11), 761–773. doi:10.1038/s41434-023-00419-9.

Wu, X., He, X., Liu, F., Jiang, X., Wang, P., Zhang, J., Jiang, J., 2022. Development and clinical translation of ex vivo gene therapy. Comput. Struct. Biotechnol. J. 2001-0370. 20, 2986–3003. doi:10.1016/j.csbj.2022.06.015.

Xie, Q., Li, K., Chen, Y., Li, Y., Jiang, W., Cao, W., Yu, H., Fan, D., Deng, B., 2025. Gene therapy breakthroughs in ALS: a beacon of hope for 20% of ALS patients. Transl. Neurodegener. 2047-9158. 14 (1). doi:10.1186/s40035-025-00477-6.

Xu, W., Zhang, S., Qin, H., Yao, K., 2024. From bench to bedside: cutting-edge applications of base editing and prime editing in precision medicine. J. Transl. Med. 1479-5876. 22 (1). doi:10.1186/s12967-024-05957-3.

Yeo, C.J.J., Simmons, Z., De Vivo, D.C., Darras, B.T., 2022. Ethical perspectives on treatment options with spinal muscular atrophy patients. Ann. Neurol. 1531-8249. 91 (3), 305–316. doi:10.1002/ana.26299.

Yoffe, A., Liu, J., Smith, G., Chisholm, O., 2023. Regulatory reform outcomes and accelerated regulatory pathways for new prescription medicines in Australia. Ther. Innov. Regul. Sci. 2168-4790. 57 (2), 271–286. doi:10.1007/s43441-022-00465-2.

Young, M., Smith, M.A., 2025. Standards and Evaluation of Healthcare Quality, Safety, and Person-Centered Care. In: StatPearls. StatPearls Publishing, Treasure Island (FL).

Zaidman, C.M., Proud, C.M., McDonald, C.M., Lehman, K.J., Goedeker, N.L., Mason, S., Murphy, A.P., Guridi, M., Wang, S., Reid, C., Darton, E., Wandel, C., Lewis, S., Malhotra, J., Griffin, D.A., Potter, R.A., Rodino-Klapac, L.R., Mendell, J.R., 2023. Delandistrogene moxeparvovec gene therapy in ambulatory patients (aged ≥4 to <8 years) with Duchenne muscular dystrophy: 1-year interim results from study SRP-9001-103 (ENDEAVOR). Annals of Neurology 94 (5), 955–968. doi:10.1002/ana.26755.

Zhao, Y., Stepto, H., Schneider, C.K., 2017. Development of the first World Health Organization lentiviral vector standard: toward the production control and standardization of lentivirus-based gene therapy products. Hum. Gene Ther. Methods. 1946-6544. 28 (4), 205–214. doi:10.1089/hgtb.2017.078.

Section I

Neuromuscular disorders

Chapter 4

Gene therapy in spinal muscular atrophy (SMA)

Wladimir Bocca Vieira de Rezende Pinto and Paulo Sgobbi
Division of Neuromuscular Diseases, Department of Neurology and Neurosurgery, Federal University of São Paulo (UNIFESP), São Paulo, Brazil

1 Introduction

5q Spinal muscular atrophy (5q SMA) represents one of the most important autosomal recessive inherited neuromuscular disorders, constituting the main genetically determined presentation of lower motor neuron disease. 5q SMA can cause clinical impairment from the prenatal period to late-onset forms, including in adults and the elderly, with the onset in infancy and preschool children being the most characteristic clinical context (Smith, 2020; Nishio et al., 2023).

A population frequency of asymptomatic heterozygotes is attributed to one individual with a single heterozygous deletion for every 50 individuals in the general population, with regional and country variations in the frequency of carriers, and it is more common in population clusters and in groups with higher rates of consanguinity. 5q SMA is estimated to have a prevalence of 1–2 cases for every 100,000 individuals and an incidence of 1 case for every 10,000 live newborns per year (Verhaart et al., 2017; Smith, 2020; Nishio et al., 2023).

5q SMA can present clinically from the prenatal period to the elderly, with predominant manifestations and consequences related to isolated lower motor neuron dysfunction. These include muscle weakness (appendicular, axial, and bulbar), amyotrophy, fasciculations, decreased deep tendon reflexes, and reduced muscle tone. In the earlier presentations, the occurrence of floppy baby syndrome is common, as well as generalized hypotonia, delayed motor development, and early and progressive ventilatory and bulbar involvement. In the later presentations of 5q SMA, there is typically predominant involvement of the pelvic and shoulder girdles, with slowly progressive limb-girdle muscle weakness, amyotrophy, and reduced deep tendon reflexes. Complications in these scenarios include weakness of the axial muscle groups (paravertebral and abdominal), progressive scoliosis, joint contractures, and significant ventilatory restriction. The different clinical forms of 5q SMA are summarized in Table 4.1 (Serra-Juhe and Tizzano, 2019; Smith, 2020; Nishio et al., 2023). Electroneuromyography typically shows evidence of chronic, preganglionic, neurogenic involvement diffusely, with markedly enlarged motor unit action potentials. Needle electromyography may also reveal signs of acute or ongoing active denervation, such as fibrillation potentials or positive sharp waves. Other techniques such as motor unit number estimation and motor unit number index may be used to support neurophysiological monitoring in conjunction with clinical follow-up of disease progression (Günther et al., 2019).

The diagnosis of 5q SMA is established based on the results of specific genetic testing, through the detection of deletions in the *SMN1* gene and the quantification of *SMN2* gene copies using the molecular technique of multiplex ligation-dependent probe amplification. Detection of homozygosity for a deletion in the *SMN1* gene can confirm the diagnosis in over 95% of cases, while in the remaining cases, a heterozygous deletion is usually accompanied by a point variant in the other allele, which can only be identified through complementary *SMN1* gene sequencing (de Holanda Mendonça et al., 2020; Smith, 2020). It is important to note that the genetic diagnosis can be established in some individuals in the absence of clinical, neurophysiological, or biomarker abnormalities, characterizing presymptomatic individuals (considered by some authors as 5q SMA type 5) (Serra-Juhe and Tizzano, 2019).

The prognosis in 5q SMA depends fundamentally on the clinical form presented by the patient, with types I and II usually presenting higher rates of ventilatory, motor, and bulbar complications earlier, requiring invasive support measures more frequently. In type I forms, life expectancy is usually short, averaging < 2 years of age, despite the changes that have occurred since the advent of disease-modifying therapies and advances in ventilatory and multidisciplinary management.

Gene Therapy in Neurodegenerative Disorders. https://doi.org/10.1016/B978-0-443-40467-2.00009-9

TABLE 4.1 Summary of the main clinical and genetic aspects related to the different types of 5q SMA (Serra-Juhe and Tizzano, 2019; Smith, 2020; Nishio et al., 2023).

Type of 5q SMA (frequency)	Definition and maximum achieved motor milestone	Age at onset and *SMN2* copy number	Natural history and clinical and neuromuscular complications
5q SMA type I (Werdnig-Hoffmann's disease) (40%–60% of cases)	Subtypes: 0: severe compromise with antenatal onset; very early bulbar, respiratory, and motor compromise; early signs identified by prenatal imaging methods Ia: neonatal onset of symptoms; no motor developmental milestones achieved Ib: absent or insufficient head control Ic: normal head control; supported sitter	Age at onset for subtypes: 0: antenatal onset (1 *SMN2* copy) Ia: neonatal or first weeks of life (1 *SMN2* copy) Ib: before age 3 months (1–2 *SMN2* copies) Ic: after age 3 months and commonly before 6 months (1–3 *SMN2* copies)	Type 0: Antenatal (fetal) death or early neonatal death (first weeks). Potential association with arthrogryposis multiplex congenita, akinesia deformation sequence, severe congenital cardiopathy, and brain malformations Ia: Severe respiratory compromise; failure to thrive; severe bulbar involvement; no achievement of motor milestones; early death during the first year of life. Ib/Ic: Early death commonly occurs during the first 2 years, sometimes with short periods of plateau; early feeding complications; hypoventilation and respiratory failure with the need for noninvasive and invasive ventilation support
5q SMA type II (intermediary) (30%–40% of cases)	Maximum motor achievement of unassisted sitter; nonwalker; may stand with support (IIb)	Subtypes: IIa: most commonly after 6 months (2–3 copies) IIb: generally, after 12 months (2–4 copies)	Progressive severe scoliosis; restrictive chronic respiratory failure due to hypoventilation, with the need for noninvasive (or rarely invasive) ventilatory support; severe joint contractures; commonly accompanied by loss of motor milestones; normal to above-average cognitive performance (intellectual ability)
5q SMA type III (Kugelberg–Welander disease) (10% of cases)	Generally normal motor milestones, with a loss of motor function during the disease course	IIIa: onset commonly occurs between 18 months and 36 months (3 copies) IIIb: onset after 3 months and up to 18/21 years (3–4 copies)	Progressive severe scoliosis; slowly progressive restrictive respiratory failure with the need for noninvasive ventilatory support; progressive joint contractures; loss of motor milestones (walking ability is frequently lost)
5q SMA type IV (adult) (< 1%–2% of cases)	Normal motor development, walker; symptom-onset after 18/21 years old	Adult onset; 3–8 copies	Normal life expectancy; rarely with loss of walking ability; generally with mild motor compromise; rarely with late respiratory involvement
5q SMA type V or presymptomatic (unknown)	Unknown maximum motor milestone, depending on age at diagnosis and treatment availability (commonly after newborn screening or family screening in asymptomatic cases)	Any age (asymptomatic); any *SMN2* copy number	Variable motor compromise depending on age at diagnosis, *SMN2* copy number, and treatment availability

In type II form, the prognosis is usually better than in type I, but ventilatory and orthopedic complications may bring additional factors for greater severity, especially after years of disease in adolescents and young adults. In type III, orthopedic and ventilatory complications tend to occur later and are less severe than in type II form, while in type IV, the best prognosis is generally observed, with life expectancy close to that of the general population (Smith, 2020; Nishio et al., 2023).

2 Disease pathogenesis and genetic patterns in 5q SMA

In healthy individuals, the *SMN1* gene represents a telomeric copy of the survival motor neuron (SMN) located in 5q13.2, while *SMN2* gene represents a centromeric copy with > 99% nucleotide identity and a single-nucleotide variation (c.840C > T), leading to transcriptional changes and distinct protein products. Both *SMN1* and *SMN2* are located in a 5q region with a classic inverted duplication. Most healthy individuals have a characteristic configuration with two *SMN1* copies and four *SMN2* copies; however, a single deletion of *SMN1* gene may also be observed. The *SMN1* gene copies are responsible for the production of normal amounts of functional SMN protein, while the *SMN2* copies and their transcripts produce only around 10% of functional SMN protein due to an alternative splicing mechanism. The remaining 90% of transcripts from *SMN2* give rise to abnormal, truncated, and nonfunctional SMN protein (Fig. 4.1) (Farrar and Kiernan, 2015).

SMN protein is a predominantly cytoplasmic protein, which is involved in several intracellular functions, including transcriptional regulation (i.e., including other transcription factors), RNA processing, neuronal differentiation, cellular trafficking, neuronal migration, telomerase regeneration, and small nuclear ribonucleoprotein biogenesis. To a lesser extent, SMN protein is also present in the nuclear gems (or gemini of the coiled bodies) in the nucleus associated with different ribonucleoproteins and the spliceosome (Mourelatos et al., 2001; Baccon et al., 2002; Giavazzi et al., 2006).

5q SMA results most commonly from a biallelic deletion involving at exon 7 of the *SMN1* gene. A minor proportion of cases, most commonly in 5q SMA types II, III, and IV, results from a deletion at exon 7 of one allele of the *SMN1* gene and a point mutation in the other allele of the *SMN1* gene. It is estimated that < 4%–5% of cases result from "de novo" genetic mechanisms, without the occurrence of maternal or paternal inheritance of one of the homozygous deletions or the point

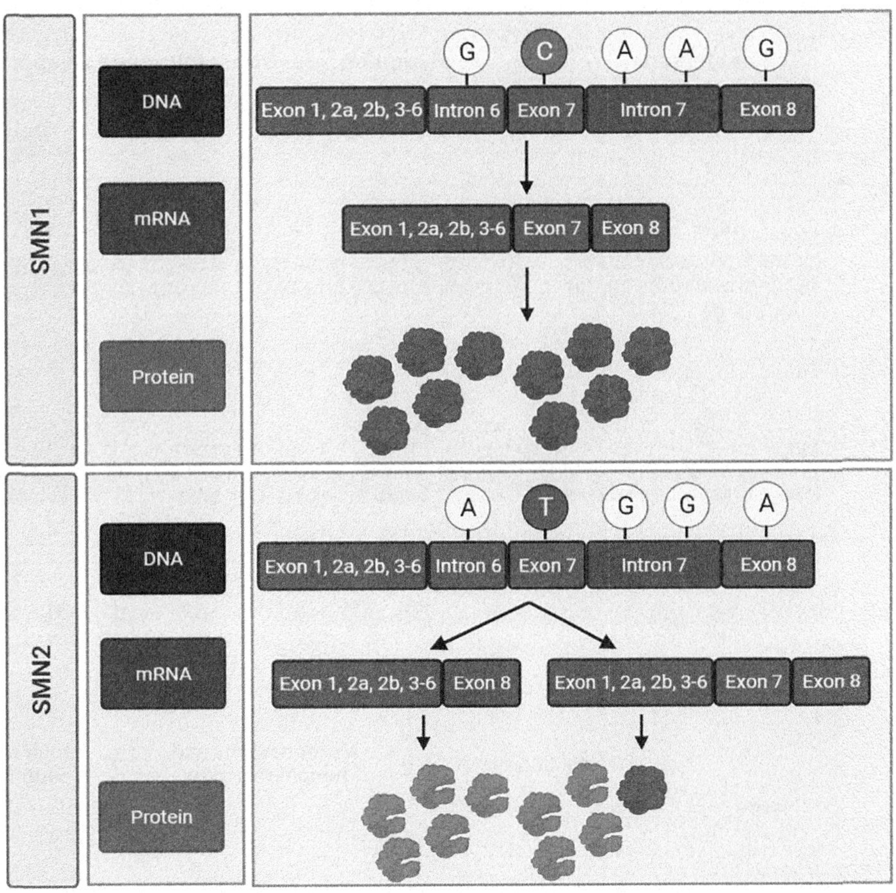

FIG. 4.1 SMN protein coding, *SMN1* and *SMN2* genes, and the splicing mechanism. *Figure content designed and distributed by Dr. Maryam Bemanalizadeh.*

mutation in the case of compound heterozygotes. Depending on *SMN2* number of copies and the residual amount of normal and functional SMN protein produced, patients may develop one of the 5q SMA types. Specific *SMN2* copy numbers do not specifically determine one type of 5q SMA, as two copies of *SMN2* may be observed, for example, in patients with SMA types II, III, and I and three copies of *SMN2* may be identified, for example, in individuals with SMA types I, II, and III. Thus *SMN2* copy numbers are implicated in the modulation of phenotype severity but do not represent the single genetic factor involved in the determination of clinical manifestations and phenotypes. Other genes are also involved in the pathophysiology of 5q SMA and in the occurrence of clinical heterogeneity (Farrar and Kiernan, 2015).

3 Currently approved therapeutic approaches for 5q SMA

The therapeutic approach to 5q SMA is based on the integration of multidisciplinary and supportive therapies combined with the availability of disease-modifying pharmacological interventions. Nonpharmacological management of 5q SMA requires the structured implementation of care by a specialized multidisciplinary team, including nutrition, motor and respiratory physiotherapy, speech therapy, psychology, occupational therapy, and various related medical specialties. Motor physiotherapy and occupational therapy are essential pillars for patients across all age groups, often complemented by orthopedic and physiatric interventions. Prevention of joint contractures is a priority to avoid functional decline, and specialized individualized follow-up is necessary for the management of congenital hip dislocation. Respiratory physiotherapy is equally indispensable for patients of all ages, including the appropriate indication and management of noninvasive ventilation with bilevel positive airway pressure and, in cases of permanent dependency, the use of invasive mechanical ventilation via tracheostomy. Nutritional and speech therapy follow-up is essential, particularly in the presence of severe oropharyngeal dysphagia and high risk of bronchoaspiration, with early implementation of percutaneous gastrostomy feeding representing an important measure (Arnold et al., 2015); Mercuri et al., 2018; Finkel et al., 2018; Waldrop and Elsheikh, 2020).

Currently, three different drugs (Nusinersen, Risdiplam, and Onasemnogene abeparvovec) have been approved for the treatment of 5q SMA in most countries, following extensive clinical trials and evaluation by international regulatory agencies. Fig. 4.2 summarizes the mechanism of action involved with the three drugs. The main aspects related to the three drugs used in

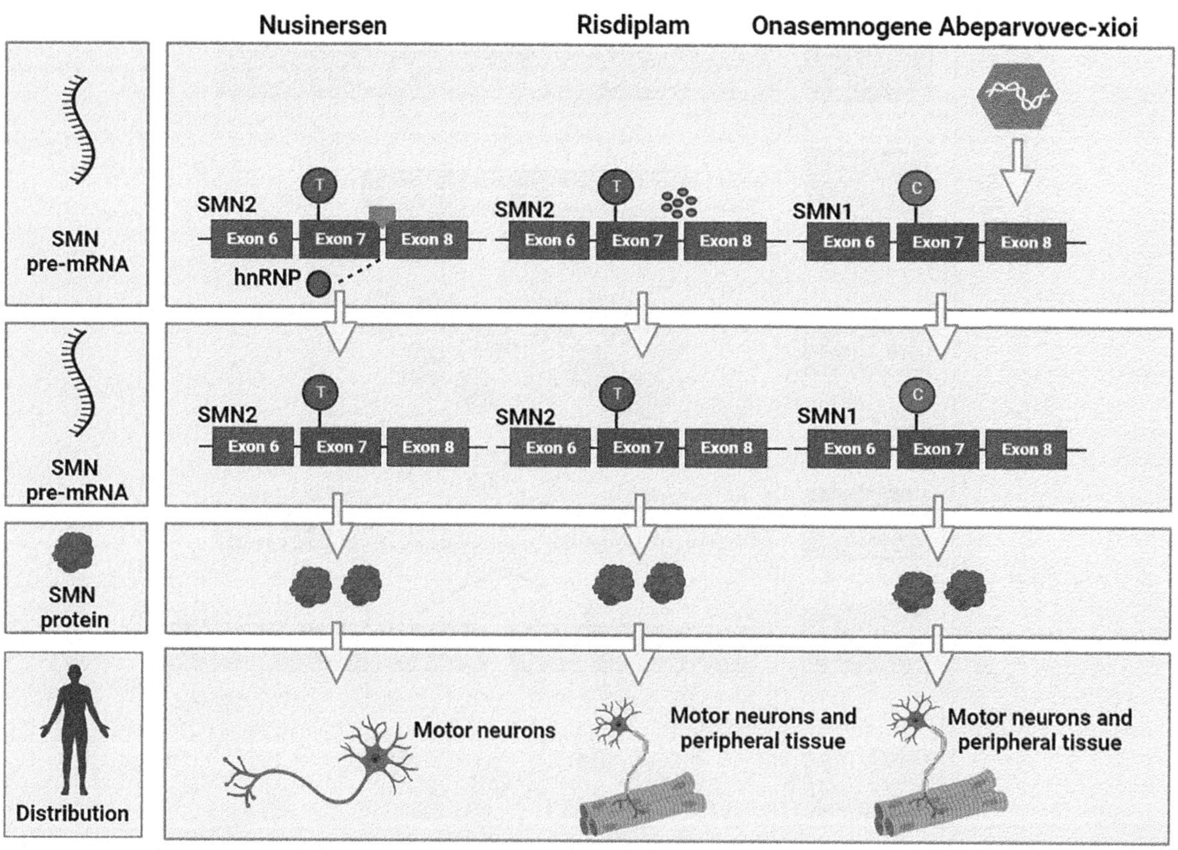

FIG. 4.2 Mechanism of action related to the treatment with Nusinersen, Risdiplam, and Onasemnogene abeparvovec. *Figure content designed and distributed by Dr. Maryam Bemanalizadeh*

TABLE 4.2 Summary of the main characteristics of the three approved drugs used in the treatment of 5q SMA (Hagenacker and Schara-Schmidt, 2024; Hatanaka et al., 2024; Waldrop, 2024; Hjartarson et al., 2022; Nishio et al., 2023; De Waele and Servais, 2025; McPheron and Felker, 2024; Ramdas et al., 2024).

Disease-modifying therapy	Nusinersen	Risdiplam	Onasemnogene abeparvovec
Route of administration and dose frequency	Intrathecal injection: loading phase (4 doses in 2 months) and maintenance phase (every 4 months)	Oral solution; once daily	Single-dose intravenous infusion
Dosing	12 mg per dose (both pediatric and adult patients)	Up to 20 kg: 0.2 mg/kg per day; 20 kg or above: 5 mg per day	1.1 x 10^{14} vector genomes per kg
Main clinical trials and studies related to drug development and approval	ENDEAR (Phase 3), CHERISH (Phase 3), NURTURE, EMBRACE (Phase 2), CS2/CS12, SHINE	FIREFISH (Phase ½), SUNFISH (Phases 2 and 3), JEWELFISH, RAINBOWFISH	START (Phase 1), STR1VE-US (Phase 3), STR1VE-EU (Phase 3), SPR1NT (Phase 3)
Class of genetic therapy	Antisense oligonucleotide (binder of *SMN2* pre-mRNA and promotion of exon 7 inclusion)	Small-molecule *SMN2* splicing modifier (inclusion of exon 7 in SMN2 mRNA)	Gene replacement therapy via AAV9-based gene therapy (delivery of a functional *SMN1* copy to lower motor neurons)
Potential adverse events	Back pain, postlumbar puncture syndrome, headache, fever, upper respiratory tract infection, meningitis, hydrocephalus, aseptic meningitis	Fever, elevated liver enzymes, rash, upper respiratory tract infection, gastroenteritis, vomiting, diarrhea	Fever, liver failure, thrombocytopenia, nausea, vomiting, thrombotic microangiopathy, bronchiolitis, pneumonia
Recommended routine monitoring	Renal function, coagulation profile, platelet counts, proteinuria evaluation	Liver function tests, complete blood count tests, ophthalmologic examination	Liver function tests, cardiac biomarkers (troponin I/T), platelet counts
Observations	Need for intrathecal access and its potential limitations in patients with severe scoliosis, high cost	Chronic daily use, high cost	Variable frequency of positive high titers to AAV9 before treatment, evidence of benefits related to early therapy use, high cost

the treatment of 5q SMA are summarized in Table 4.2 (Hagenacker and Schara-Schmidt, 2024; Waldrop, 2024; Hjartarson et al., 2022; Nishio et al., 2023; De Waele and Servais, 2025; McPheron and Felker, 2024; Hatanaka et al., 2024; Ramdas et al., 2024).

4 Gene therapy in 5q SMA

The use of gene therapy has become a reality in the treatment of 5q SMA, especially after international regulatory agencies approved the use of Onasemnogene abeparvovec (Chongmelaxme et al., 2025). Other alternatives to this therapy are currently being developed, given the challenges related primarily to the high costs of therapy and access to therapy in different countries. The development of therapies that share similar mechanisms of action but have a lower frequency of adverse events and serious adverse events, similar or higher efficacy rates, and even include other routes of gene therapy administration is also underway. The main clinical studies conducted to date with Onasemnogene abeparvovec are summarized in Table 4.3.

Onasemnogene abeparvovec (Zolgensma or OAV101; formerly AVXS-101) represents the first in vivo viral vector-based gene therapy approved for the treatment of 5q SMA. This therapy received Food and Drug Administration approval in May 2019 for patients up to 2 years of age, including presymptomatic individuals identified through genetic neonatal screening. European Medicines Agency approval in May 2020 was extended to patients with SMA type I or any type with up to three copies of the *SMN2* gene (Pascual-Morena et al., 2023; Kichula et al., 2021; Blair, 2022; Waldrop, 2024).

Treatment consists of a single intravenous dose using a nonreplicating recombinant adeno-associated viral vector serotype 9 (AAV9) to deliver a functional copy of the *SMN1* transgene, resulting in the production of functional SMN protein. The gene therapy promoter is a hybrid of the CMV enhancer and the chicken beta-actin gene (Pascual-Morena et al., 2023; Kichula et al., 2021; Blair, 2022).

TABLE 4.3 Summary of the main characteristics of the clinical studies related to Onasemnogene abeparvovec in the treatment of 5q SMA.

Clinical trial (study name) (references)	Phase and NCT identifier	Study summary and purpose	Inclusion criteria and sample	Results
AVXS-101-CL-101 (Al-Zaidy et al., 2019)	Phase 1 (NCT02122952)	Open-label, dose escalation study, with symptomatic 5q SMA type I infants, aged up to 6 months	Symptomatic individuals with genetically confirmed 5q SMA type I, with up to 2 *SMN2* copies and motor symptom onset up to 6 months	Prolonged survival without permanent ventilation; rapid and significant improvement in motor milestones
START (Mendell et al., 2021)	Phase 1 (part of protocol: NCT02122952)	Open-label, dose escalation study, with symptomatic 5q SMA type I infants, aged up to 6 months	Symptomatic individuals with 5q SMA type I, up to 2 *SMN2* copies, and symptom-onset before 6 months	Marked improvement in survival and motor function; sustained benefits during follow-up
STR1VE-US (AVXS-101-CL-303) (Day et al., 2021)	Phase 3 (NCT03306277)	Multicenter, open-label study, with symptomatic 5q SMA type I infants, aged up to 6 months and 2 *SMN2* copies	Individuals aged up to 6 months with symptom onset before 6 months, genetically confirmed 5q SMA and 2 *SMN2* copies	Marked improvement in motor developmental milestones: 60% sat unassisted for > 30 s by 18 months; 91% became free from permanent ventilation by 14 months
STRONG (Finkel et al., 2023)	Phase ½ (NCT03381729)	Multicenter, open-label study with intrathecal evaluation of Onasemnogene abeparvovec for nonambulatory 5q SMA type II children, aged > 6 months and < 5 years	Nonambulatory individuals aged > 6 months and up to 5 years, genetically confirmed 5q SMA, 3 *SMN2* copies; no prior use of Risdiplam or Nusinersen	92% of individuals achieved 3 or more points of HFMSE improvement; favorable safety profile (consistent with safety data from previous studies)
STR1VE-EU (Mercuri et al., 2021)	Phase 3 (NCT03461289)	Multicenter, open-label study, with symptomatic 5q SMA type I infants	Age ≤ 6 months, genetically confirmed 5q SMA, with onset before 6 months, and 2 *SMN2* copies	Confirmed efficacy and safety; data consistent with results from STR1VE-US
SPR1NT (Strauss et al., 2022a, 2022b)	Phase 3 (NCT03505099)	Multicenter, open-label study, with presymptomatic infants with 2 or 3 *SMN2* copies	Genetically confirmed 5q SMA, presymptomatic individual with 2–3 *SMN2* copies, and ≤ 6 weeks at dosing	Presymptomatic treatment prevented disease progression with most children achieving normal motor milestones
OFELIA (Saute et al., 2025)	Phase 4 (NCT05073133)	Multicenter, open-label study, with symptomatic 5q SMA infants ≤ 24 months old and ≤ 17 kg	Age ≤ 24 months, ≤ 17 kg, and symptomatic genetically confirmed 5q SMA	Sustained or improved motor function up to 18 months posttreatment; consistent safety
STEER (Press Release Data in text)	Phase 3 (NCT05089656)	Multicenter, randomized, sham-controlled, double-blinded study with intrathecal Onasemnogene abeparvovec, for treatment-naïve 5q SMA type II patients, sitters, aged 2–<18 years	Treatment-naïve 5q SMA type II individuals aged between 2 and 18 years	Reported data disclosing statistically significant HFMSE improvement with favorable safety compared to sham

The approved intravenous dose is 1.1×10^{14} vector genomes/kg, with a standard formulation of 2.0×10^{13} vector genomes/mL in vials of 5.5 mL or 8.3 mL. Infusion should be administered slowly over approximately 60 min. The frozen product must be received and stored at 2°C–8°C for up to 14 days, and infusion must be completed within 8 h after syringe preparation. Specific procedures for handling and disposal of materials that may have come into contact with the drug are required (Pascual-Morena et al., 2023; Kichula et al., 2021; Blair, 2022; Kirschner et al., 2024).

In the Phase 1 AVXS-101-CL-101 study (NCT02122952), an initial cohort of 15 infants with type I SMA, aged 0.9–7.9 months, was evaluated, including a low-dose cohort (n = 3) and a high-dose cohort (n = 12). At 24 months, all children in the high-dose cohort remained free of serious adverse events and did not require mechanical ventilation. Among the 12 patients with 5q SMA weighing 3.6–8.4 kg, all remained independent of mechanical ventilation at both 14 and 24 months. At 24 months, 10 patients achieved independent sitting for ≥ 10 s, 9 for ≥ 30 s, and 2 achieved independent walking. In the START extension study (NCT03421977), which included 10 of the original patients (total n = 13) with follow-up exceeding 5.7 years, all participants remained free of mechanical ventilation. All patients maintained or attained additional motor milestones, with five receiving monotherapy exclusively from study initiation, including two who achieved supported standing (Mendell et al., 2017, 2021; Pascual-Morena et al., 2023).

The Phase 3b SMART study (NCT04851873), a multicenter and open-label trial, aimed to evaluate both the safety and efficacy of Onasemnogene abeparvovec in patients with SMA weighing between 8.5 and 21 kg. The study included symptomatic patients with various types of SMA, who were divided into three cohorts based on body weight: 8.5–13 kg, > 13–17 kg, and > 17–21 kg. A total of 24 patients were enrolled, including both treatment-naive individuals and those who had previously discontinued pharmacological treatment. Patients were followed for 52 weeks under the same standard corticosteroid regimen used in other clinical trials of this therapy. The primary objective was to assess safety, while secondary outcomes included motor function and the attainment of developmental motor milestones. Within the study population, most patients had type II SMA (n = 11), had received prior pharmacological treatment (n = 11), and carried three copies of the *SMN2* gene (n = 18). The overall safety profile of the drug was similar across the weight-based groups, with frequent occurrences of asymptomatic thrombocytopenia (71%) and elevated liver transaminases (83%). No deaths occurred during the study. Serious treatment-related adverse events were reported in 29% of participants, and 96% experienced one or more adverse events of special interest. At the end of the 52-week period, 33% of participants were still receiving corticosteroids. No cases of ganglionopathy or thrombotic microangiopathy were observed. Most participants showed stabilization or improvement in motor function across the population weighing up to 21 kg. After 52 weeks, 16 out of 18 patients demonstrated improvement or stability in Hammersmith Functional Motor Scale Expanded (HFMSE) scores, and 15 out of 17 in RULM scores, with four participants carrying three *SMN2* copies achieving gains in developmental motor milestones. Thus the SMART study demonstrated good safety of Onasemnogene abeparvovec in the SMA population weighing between 8.5 and 21 kg (McMillan et al., 2024). Other studies have also evaluated Onasemnogene abeparvovec in the SMA population over 8.5 kg (Chand et al., 2022).

The STRONG study (AVXS-101-CL-102; NCT03381729) evaluated the safety and clinical impact of intrathecally administered Onasemnogene abeparvovec using an ascending-dose regimen in patients with 5q SMA who were able to sit but not walk. Eligible participants had homozygous deletions of the *SMN1* gene and three copies of *SMN2*, aged 6 months to 5 years. Primary endpoints included the ability to stand independently for ≥ 3 s and changes in HFMSE scores. Among younger patients (6–24 months), 7.7% achieved independent standing following treatment with intermediate doses. In the older cohort (24–60 months), a statistically significant improvement in HFMSE scores was observed at 12 months with the intermediate dose (Finkel et al., 2023). These findings support the potential expansion of Onasemnogene abeparvovec use to populations with later-onset symptoms, in conjunction with evidence from subsequent clinical studies (René and Parks, 2023).

The Phase 3 STR1VE-US study (AVXS-101-CL-303; NCT03306277), an open-label, single-arm, nonplacebo-controlled trial, evaluated 22 patients with infantile-onset 5q SMA carrying two copies of the *SMN2* gene. At study entry, none of the participants required mechanical ventilation, and the mean age was 3.7 months. All received a single dose of 1.1×10^{14} vector genomes/kg. Following treatment, 21 patients remained free from permanent mechanical ventilation for ≥ 10.5 months, 20 survived for ≥ 14 months, and 20 survived without serious adverse events through 18 months. At 18 months of age, 18 participants remained independent of any ventilatory support. Regarding motor developmental milestones, 17 patients achieved head control (mean age: 6.8 months), and 14 achieved independent sitting for ≥ 30 s during the study (mean age: 12.6 months), with 13 reaching this milestone by 18 months. One patient achieved supported ambulation at 12.9 months. In terms of motor function, outcomes exceeded those reported in historical cohorts of type I 5q SMA, which typically fail to reach CHOP-INTEND scores ≥ 40:21 patients achieved scores ≥ 40, 14 achieved ≥ 50, and 9 achieved ≥ 58 (Day et al., 2021; Yang et al., 2023).

The Phase 3 STR1VE-EU study (AVXS-101-CL-302; NCT03461289), an open-label, single-dose, nonplacebo-controlled trial, enrolled 33 patients with infantile-onset 5q SMA carrying two copies of the *SMN2* gene, treated with the same dosage regimen as in the CL-303 study. At baseline, 9 participants required ventilatory support and 9 required nutritional support. The mean age at study entry was 4.1 months, and the mean CHOP-INTEND score was 33. Thirty-two patients completed the trial within the protocol-defined age range, with one death reported, directly attributable to 5q SMA. Regarding motor developmental milestones, 23 patients achieved head control (mean age: 8 months), 14 achieved independent sitting for ≥ 10 s by 18 months (mean age: 15.9 months), and 1 patient achieved independent walking within the same period. CHOP-INTEND scores of ≥ 40, ≥ 50, and ≥ 58 points were reached by 24, 14, and 3 patients, respectively. In addition, 31 patients remained free from mechanical ventilation for ≥ 14 months (Mercuri et al., 2021; Pascual-Morena et al., 2023; Yang et al., 2023).

Similarly, the Phase 3 STR1VE-AP trial is ongoing in Asia-Pacific populations, with baseline characteristics comparable to those enrolled in the STR1VE-US and STR1VE-EU studies. In Latin America (Brazil and Argentina), the open-label, multicenter, Phase 4 OFELIA study (NCT05073133) was conducted as a single-dose, single-arm trial enrolling patients with 5q SMA, irrespective of *SMN2* copy number, weighing < 17 kg and aged ≤ 24 months. After 18 months of follow-up, 83.3% of participants achieved the motor milestone of independent sitting, 58.3% could stand with support, 25% could stand without support, 16.7% walked with assistance, and 8.3% walked independently. The overall safety profile and incidence of adverse events were consistent with previous Onasemnogene abeparvovec studies; however, two deaths occurred, one of which was attributed to thrombotic microangiopathy (Saute et al., 2025). Other real-world multicenter and observational studies of Onasemnogene abeparvovec monotherapy have reported efficacy and safety outcomes consistent with the pivotal clinical trials (Servais, 2024; Bitetti et al., 2023; Bitetti et al., 2024; Stettner et al., 2023; Gowda et al., 2024; Desguerre et al., 2024; Weiß et al., 2024; Mendonça et al., 2024). In another study, conducted across 18 neuromuscular centers in Germany and Austria, patients with 5q SMA type I or II aged ≤ 24 months and weighing ≤ 15 kg were included, the majority having received prior Nusinersen treatment. Statistically significant improvements in CHOP-INTEND scores were observed, with mean increases of 13.8 points in children < 8 months of age and 7.7 points in those aged 8–24 months (Weiß et al., 2022).

In the multicenter, Phase 3 SPR1NT study (AVXS-101-CL-304; NCT03505099), an open-label, nonplacebo-controlled trial, two cohorts of presymptomatic patients with genetically confirmed 5q SMA were evaluated. Cohort 1 included 14 patients with two *SMN2* copies, and Cohort 2 included 15 patients with three SMN2 copies. Primary and secondary endpoints comprised attainment of motor developmental milestones, ventilatory support requirements, survival, and performance on the CHOP-INTEND and Bayley-III scales. In Cohort 1 (two *SMN2* copies), all 14 patients remained free from mechanical ventilation and nutritional support at 14 months. All achieved independent sitting for ≥ 30 s between 5.7 and 11.8 months of age, with 11 reaching this milestone before day 279. Nine patients achieved independent walking, and all attained CHOP-INTEND scores ≥ 58 by 18 months. In Cohort 2 (three *SMN2* copies), all 15 patients remained free from mechanical ventilation and nutritional support at 24 months. All achieved independent sitting between 9.5 and 18.3 months, 14 achieved unassisted standing by day 514, and 14 achieved independent walking. All patients scored within two standard deviations of the normative mean on the Bayley-III by 24 months (Strauss et al., 2022a, 2022b; Pascual-Morena et al., 2023). Furthermore, additional case reports have demonstrated remarkable benefits in premature infants born at 30-week gestational age with one *SMN2* copy and the genetic modifier variant c.835–44A > G, who developed normal motor developmental milestones after treatment at 3.5 weeks of life (Brown et al., 2024).

In the multicenter Phase 3 STEER study (NCT05089656), the efficacy and safety of intrathecal administration of Onasemnogene abeparvovec (OAV101 IT) were evaluated in treatment-naïve patients with SMA type II (with symptom-onset after 6 months of age) and aged over 2 years and under 18 years. This was a randomized, double-blind study with a sham control group. Patients were excluded if they had high serum anti-AAV9 antibody titers (> 1:50), active infections near the period of screening or drug infusion, liver dysfunction (based on laboratory criteria), significant sensory abnormalities on neurological examination, scoliosis or hip fixation surgeries within 12 months prior to screening or planned within 64 weeks after treatment initiation, clinical complications that could interfere with motor performance assessments (including a Cobb angle > 40° in a seated position or severe joint contractures), or ventilatory needs such as invasive ventilation, noninvasive ventilation for > 12 h per day, or the need for tracheostomy. In the STEER study, the intrathecal treatment group met the primary endpoint, showing a significant increase in HFMSE scores at 52 weeks compared to baseline. There was no significant difference in the profile of adverse events or serious adverse events at 52 weeks compared to the sham control group, with fever, vomiting, and upper respiratory tract infections being the most commonly observed events. To date, data from the STEER study have only been made available through a Novartis Pharmaceuticals press release (available at: URL: https://www.novartis.com/news/media-releases/novartis-intrathecal-onasemnogene-abeparvovec-phase-iii-study-meets-primary-endpoint-children-and-young-adults-sma) (last accessed on August 8, 2025). Another study related to the intrathecal use of Onasemnogene abeparvovec is the Phase 3 STRENGTH study (NCT05386680), targeting patients with SMA aged 2–18 years who previously discontinued use of Nusinersen or Risdiplam. However, final data from this study have not yet been published.

The most frequently observed adverse events include gastrointestinal symptoms (primarily nausea and vomiting), elevations in liver transaminases, and transient hepatopathy (toxic or immune-mediated in nature), fever, thrombocytopenia (particularly within the first 2 weeks of treatment), transient increases in troponin I, respiratory infections (such as bronchiolitis, upper respiratory tract infections, and aspiration pneumonia), chronic respiratory failure, and, in presentations with severe complications, the occurrence of fulminant and fatal liver failure, acute kidney injury, communicating hydrocephalus, hemolytic anemia, and systemic thrombotic microangiopathy, generally within the first 2 weeks after intravenous infusion. The safety and efficacy profile in patients aged > 2 years or weighing > 13.5 kg has not yet been completely established. Routine preinfusion screening for anti-AAV9 antibodies is recommended due to the estimated seroprevalence (based on some cohorts, e.g., in the United States), with titers ideally < 150. Oral prednisolone (1 mg per kg per day) is recommended for at least 30 days pre- and postinfusion, with gradual tapering during clinical follow-up. Owing to the variable onset of liver dysfunction, monitoring of liver enzymes, bilirubin, and coagulation parameters is advised before infusion and for up to 3 months thereafter. Platelet counts and coagulation profiles should be assessed weekly during the first month after therapy and biweekly for the next 2 months until normalization. In preclinical juvenile primates, mild acute nonprogressive mononuclear inflammatory processes of the dorsal root and trigeminal sensory ganglia were identified, leading to partial recovery after 6 months of follow-up (Kichula et al., 2021; Blair, 2022; Yang et al., 2023; Zhuang et al., 2023; Pascual-Morena et al., 2023; Chand et al., 2021b; 2021a; Aharoni et al., 2023; Day et al., 2023; Braun et al., 2024; Guillou et al., 2022; Ogbonmide et al., 2023; Tukov et al., 2022).

5 Recent and ongoing clinical trials on gene therapy in 5q SMA

There are several clinical trials that are currently ongoing with promising data related to early clinical phases or experimental findings. The main drugs and their associated trials are summarized in Tables 4.4 and 4.5. In addition, there are several other studies evaluating the potential use of combination therapies in the treatment of 5q SMA, including dual and triple therapies. In this context, add-on approaches, switching therapies, and switching plus add-on therapy are being explored. Currently, the RESPOND (Phase 4) study (NCT04488133) is evaluating the combined use of Nusinersen following initial treatment with Onasemnogene abeparvovec in patients aged 2–36 months. The Phase 3 STRENGTH study (NCT05386680) is assessing the combined use of Onasemnogene abeparvovec after initial treatment with Risdiplam or Nusinersen. Furthermore, there are two additional single-group Phase 4 studies evaluating the use of Risdiplam in patients previously treated with Onasemnogene abeparvovec (NCT05861986, NCT05861999) (Bemanalizadeh et al., 2025).

6 Current challenges and limitations in gene therapy for 5q SMA

Major challenges remain regarding the development of gene therapies for the treatment of 5q SMA. One of the major current challenges is the difficulty in accessing available therapies, such as Onasemnogene abeparvovec, given the high cost of therapy, which is often not covered by both public and private healthcare systems. Also challenging is the fact that the potential role of this drug in the treatment of older patients (such as adolescents) or even adults remains unknown, scenarios for which there are no specific clinical studies or specific approval for the drug's use.

As evidenced by several reports and case series in the current medical literature, there are significant doubts regarding the need for supplemental or even combined genetic therapies in some patients, as well as whether a single-dose treatment strategy is sufficient from a neuromuscular clinical perspective for patient management, especially in cases with a long treatment history. It is also unknown whether any type of readministration of therapy may be necessary during long-term monitoring or follow-up.

It is also a matter of great interest to study the development of novel drug strategies, such as preclinical studies with gene-editing approaches such as base editing and CRISPR/Cas9 (*Clustered Regularly Interspaced Short Palindromic Repeats/CRISPR-associated protein 9*) (Arbab et al., 2023; Li et al., 2020; Alves et al., 2024). There are currently novel approaches under development by a multicenter collaboration using fibroblasts derived from individuals with 5q SMA and a mouse model of 5q SMA: adenosine base editors targeting exon 7 of *SMN2* gene and its regulatory elements, attempting to normalize SMN levels (Alves et al., 2024). After evaluating > 100 distinct RNAs and base editors and using Cas9 with high editing properties, the authors achieved almost 99% reversion of the point mutation in *SMN2*, resulting in increased exon 7 transcripts and elevated SMN protein levels (Alves et al., 2024).

Another genome-editing strategy previously used Cas9 and targeted splicing regulatory elements. In this attempt, after disrupting intronic splicing silencers using Cas9 and sgRNA (single guide RNA), the authors increased exon 7 inclusion of *SMN2* gene for full-length SMN protein production in SMA induced pluripotent stem cells. In zygotes from a mouse

TABLE 4.4 Summary of the main ongoing clinical studies on gene therapies in the treatment of 5q SMA.

ClinicalTrial ID	NCT05901987	NCT06421831	NCT05824169	NCT06288230	NCT06191354
5q SMA	SMA type II	SMA type III	SMA type I	No specific types	SMA type I
Sponsor	GeneCradle Inc.	GeneCradle Inc.	GeneCradle Inc.	Lantu Biopharma	Skyline Therapeutics; Xinhua Hospital
Molecule/Drug	GC101	GC101	GC101	AAV-hSMN1 (Vesemnogene Lantuparvovec)	SKG0201
Phase	Phase ½	Phase ½	Phase ½	Phase ½	Phase 1
Route	Intrathecal	Intrathecal	Intrathecal	Intravenous	Intrathecal
Vector/Gene construct	Self-complementary AAV9 (scAAV9) carrying codon-optimized *SMN1* driven by the CMV enhancer + chicken β-actin promoter	scAAV9 + codon-optimized *SMN1* + chicken β-actin promoter	scAAV9 + codon-optimized *SMN1* + chicken β-actin promoter	rAAV9 with full-length noncodon-optimized *SMN1* + regulatory elements + optimized promoter, bovine growth hormone polyadenylation sequence	AAV vector with codon-optimized *SMN1* and CNS-selective promoter
Participants	33	21	18	6	12
Study design and dosing	Nonrandomized; three doses: low (1.2×10^{14} vg), medium (2.4×10^{14} vg), high (4.8×10^{14} vg); single intrathecal dose	Nonrandomized; single cohort (1.2×10^{14} vg); single intrathecal dose	Nonrandomized; low (2.4×10^{14} vg) and high (4.8×10^{14} vg); single intrathecal dose	Nonrandomized; single 1.5×10^{14} vg dose; 24-month follow-up	Dose escalation; single intrathecal dose
Primary outcomes	AEs/SAEs (52 weeks); ability to stand unassisted ≥ 3 s or HFMSE change (52 weeks)	AEs/SAEs (52 weeks); HFMSE change (52 weeks)	AEs/SAEs (18 months); sitting unassisted ≥ 10 s at 18 months	AEs/SAEs (24 months)	Dose-limiting toxicity (4 weeks); AEs/SAEs at 18 months
Secondary and other outcomes	Walk 5 steps; HFMSE + 3 points; ventilatory support time	HFMSE improvement of 3 or more points; RULM; CGI; MFM; FVC, FEV1, MIP, MEP; SMAIS; 6MWT	CHOP-INTEND; thrive ability; event-free survival at 14 months; percentage of individuals achieving milestones on MGRS	Ambulation; change in HFMSE score; changes in ventilatory support hours; motor milestones, at 24 months	Survival at 14 months; CHOP-INTEND change (24 weeks); new milestones on BSID-III (24 weeks)
Inclusion criteria	**(i)** Male or female child with SMA type II presenting between 6 and 60 months when parents/legal guardians sign consent; **(ii)** symptom-onset between 6 and 18 months in participants who cannot stand or walk alone but can sit; **(iii)** biallelic homozygous deletion or compound heterozygous deletion + point mutation in *SMN1* with 2 *SMN2* copies; **(iv)** parents or legal guardians signing informed consent	**(i)** Male or female with SMA type III, aged ≥2 years when consent is signed; **(ii)** biallelic homozygous deletion in *SMN1*; **(iii)** parents or legal guardians providing signed consent; **(iv)** use of effective contraception for 6 months posttreatment; **(v)** HFMSE score of 10–54 at screening	**(i)** Male or female child with SMA type I, ≤ 6 months old at consent; **(ii)** biallelic *SMN1* deletion or compound heterozygous (deletion + point mutation) with ≤ 2 *SMN2* copies; **(iii)** parents or legal guardians signing consent	**(i)** Male or female with SMA; **(ii)** biallelic *SMN1* deletion or compound heterozygous (deletion + point mutation); **(iii)** participant/guardian signs consent	**(i)** Male or female child with SMA type I, aged ≤ 6 months at the time of infusion; **(ii)** clinical history and signs consistent with SMA type I; **(iii)** parents/legal guardians sign consent

Exclusion criteria	**(i)** Previous gene therapy trials; **(ii)** Risdiplam within 15 days or Nusinersen within 4 months; **(iii)** AAV9 neutralizing antibody ≥ 1:200; **(iv)** SMN2 gene mutation c.859G > C; **(v)** need for noninvasive ventilation ≥ 12 h/day, invasive ventilation, or O_2 saturation < 95%; **(vi)** positive *serology for HIV, Treponema pallidum*, HCV, or HBsAg; **(vii)** glucocorticoid allergy; **(viii)** contraindications to lumbar puncture; **(ix)** active infection requiring systemic antiviral therapy; **(x)** vaccination ≤ 2 weeks before infusion; **(xi)** major clinical diseases preventing participation; **(xii)** abnormal laboratories, including AST, ALT, bilirubin, GGT, hemoglobin > 15 or < 11 g/dL, low platelets, or class IV modified Ross heart failure	**(i)** Previous SMA gene therapy trials; **(ii)** Risdiplam for 15 days or Nusinersen for 4 months prior; **(iii)** AAV9 nAbs ≥ 1:200; **(iv)** > 4 *SMN2* copies; **(v)** need for ventilatory support ≥ 16 h/day; **(vi)** positive serology for HIV, *Treponema pallidum*, HCV, HBsAg; **(vii)** allergy/hypersensitivity to prednisolone or glucocorticoids; **(viii)** severe joint contractures preventing intrathecal access or motor testing; **(ix)** unresolved surgical lesions; **(x)** vaccination ≤ 2 weeks before infusion; **(xi)** major clinical comorbidities preventing participation	**(i)** Previous gene therapy trials; **(ii)** previous Risdiplam or Nusinersen; **(iii)** AAV9 nAbs ≥ 1:200; **(iv)***SMN2* c.859G > C mutation; **(v)** need for noninvasive ventilation ≥ 16 h/day, invasive ventilation, or O_2 saturation < 95%; **(vi)** positive serology HIV, *Treponema pallidum*, HCV, HBsAg; **(vii)** allergy to prednisolone or glucocorticosteroids; **(viii)** laboratory abnormalities (AST, ALT, bilirubin, hemoglobin > 15 or < 11 g/dL, platelet count abnormalities, class IV modified Ross heart failure); **(ix)** lumbar puncture contraindications; **(x)** active infection requiring antiviral therapy; **(xi)** vaccination ≤ 2 weeks; **(xii)** major diseases preventing participation	**(i)** Recent SMA trials (gene therapy); **(ii)** invasive ventilation or O_2 saturation < 95%; **(iii)** active viral infection, including HIV/HCV/HBV; **(iv)** significant laboratory abnormalities; **(v)** swallowing test showing *aspiration risk and inability/* unwillingness to use alternative feeding; **(vi)** major clinical diseases adding unnecessary risk	**(i)** Allergy to prednisolone or glucocorticoids; **(ii)** active viral infection requiring hospitalization; **(iii)** severe infections/diseases; **(iv)** weight < third percentile for age/sex; **(v)** SpO_2 < 96% without support; **(vi)** previous Spinraza, *Evrysdi, Zolgensma, or* other SMA trials; **(vii)** major scheduled surgeries during the study
Safety and efficacy data	No safety/efficacy data available	No safety/efficacy data available	No safety/efficacy data available	No safety/efficacy data available	No safety/efficacy data available

AAV, adeno-associated virus; *AE*, adverse events; *ALT*, alanine aminotransferase; *AST*, aspartate aminotransferase; *BSID-III*, Bayley Scales of Infant and Toddler Development Third Edition; *CGI*, Clinical Global Impression; *CHOP-INTEND*, Children's Hospital of Philadelphia Infant Test of Neuromuscular Disorders; *FEV1*, Forced Expiratory Volume in 1 s; *GGT*, gamma-glutamyltransferase; *HFMSE*, Hammersmith Functional Motor Scale Expanded; *HINE-2*, Hammersmith Infant Neurological Examination Module 2; *FVC*, forced vital capacity; *MEP*, maximal expiratory pressure; *MFM*, motor function measure; *MGRS*, Multicenter Growth Reference Study (World Health Organization); *MIP*, maximal inspiratory pressure; *nAb*, neutralizing antibody; *RULM*, revised upper limb module; *SAE*, serious adverse events; *sc*, self-complementary; *SMA*, spinal muscular atrophy; *SMAIS*, SMA Independence Scale; *6MWT*, six-minute walk test.

TABLE 4.5 Summary of the main ongoing clinical studies on gene therapies in the treatment of 5q SMA.

ClinicalTrial ID	NCT06971094	NCT06888661	NCT05614531	NCT05747261(Bluebell)	ChiCTR2500095667/ NCT06772402	ChiCTR2200066399
5q SMA	SMA type II	SMA types I and II	SMA type I	SMA type I/presymptomatic*	SMA type II	SMA type I
Sponsor	GeneCradle Inc.	Hangzhou Jiayin Biotech Ltd. /Exegenesis Bio	Hangzhou Jiayin Biotech Ltd./Exegenesis Bio	Biocad	The Children's Hospital of Zhejiang University School of Medicine/ Genecombio Ltd.	GeneCradle Inc.
Molecule/drug	GC101	EXG001-307	EXG001-307	ANB-004	GCB-001	GC101
Phase	Phase 3	Phase 1	Phase ½	Phase ½	Phase 1	Phase 1/2a
Route	Intrathecal	Intrathecal	Intravenous	Intravenous	Intrathecal	Intrathecal
Vector/gene construct	Self-complementary AAV9 (scAAV9) carrying codon-optimized *SMN1* driven by the CMV enhancer + chicken β-actin promoter	Nonreplicating rAAV9 carrying a human SMN cDNA cassette + human synapsin promoter + CMV enhancer	Nonreplicating rAAV9 carrying a human SMN cDNA cassette + human synapsin promoter + CMV enhancer	Self-complementary AAV9 carrying codon-optimized *SMN1* driven by the CMV enhancer + chicken β-actin promoter	Self-complementary AAV9 (scAAV9) carrying a full-length human SMN transgenic product	Self-complementary AAV9 (scAAV9) carrying codon-optimized *SMN1* driven by the CMV enhancer + chicken β-actin promoter
Participants	50 (SMA type II; aged 2–12 years; previous use of Nusinersen)	4 (up to 2 years old)	12 (up to 6 months)	40 (up to 240 days)	6 (SMA type II; 2–12 years)	15
Study design and dosing	Randomized, open-label; 1.2×10^{14} vg; single intrathecal dose; 2 groups (one receives GC101 and discontinues Nusinersen; the other continues Nusinersen)	Open-label, dose escalation	Open-label, dose escalation	Nonrandomized, dose escalation (3 cohorts)	Nonrandomized, open-label; 2 groups (low and high dose); single intrathecal dose	Single-arm, open-label; single intrathecal dose
Primary outcomes	HFMSE score change at 52 weeks.	Safety and tolerability (AE/SAE up to 52 weeks)	Safety and tolerability	At 12 months: AE/SAE; change in HINE score; motor development score	0–12 months: AEs; AAV viral load, viral shedding, and immunogenicity after infusion; change in HFMSE score; change in RULM score	Phase I: safety and tolerance; stage IIa: efficacy
Secondary and other outcomes	HFMSE score change at 26 weeks; individuals with 3 or more points of improvement in HFMSE at 26/52 weeks; changes in MGRS at 26/52 weeks; RULM score change at 26/52 weeks; SMAIS score change at 26/52 weeks; AE/SAE at 52 weeks	New milestones on BSID-III (52 weeks); improvement on CHOP-INTEND (52 weeks); proportion of CHOP-INTEND > 40; improvement on HINE-2 (52 weeks)	Number of survivors at 14 months: individuals who sit unsupported for at least 30 s	At 12 months: event occurrence (fatal event, tracheostomy, noninvasive respiratory support for at least 16 h for 14 days or more); change in motor abilities; IgM/IgG antibody to AAV9; IgM/IgG to SMN1; T cells specific to AAV9 or SMN transgene; AAV9 vector in body fluids	Not reported	AAV load and immunogenicity

Inclusion criteria	SMA type II; treatment with Nusinersen for at least 1 year; no treatment with Risdiplam for 2 months prior to or up to 12 months after enrollment; HFMSE score of 10 points or more; consent from patients or legal guardians	SMA type I with age 180 days or less; type II with 180 days or more (and up to 2 years); type I with up to 2 copies of *SMN2*, type II with up to 3 copies of *SMN2*; consent form from legal guardians; in accordance with general standard care measures for 5q SMA	SMA type I with 2 *SMN2* copies; up to 6 months of age; consent form from legal guardians; in accordance with general standard care measures	SMA with 2 or 3 *SMN2* copies; age up to 240 days; consent from a legal representative; presymptomatic with 2 *SMN2* copies; disease onset up to 180 days	SMA type II, onset from 6 to 18 months; age 2–12 years; 2–4 *SMN2* copies; consent from guardians; compliance with protocol procedures	SMA type I with 2 *SMN2* copies; onset before 6 months; age at infusion up to 6 months; consent from legal guardian
Exclusion criteria	Serum anti-AAV9 nAbs > 1:50 (screening); Nusinersen within 2 months prior; severe liver injury; patients with contraindications to glucocorticoid use; contraindications to lumbar puncture or intrathecal therapy; individuals with medical conditions that affect the assessment of motor function or pose a risk to safety; positive HBsAg, HIV, HCV, or syphilis antibodies; vaccinations within 2 weeks prior; other prior gene therapies; other trials within 3 months prior	Contraindications to lumbar puncture: other intrathecal therapies; previous implantable shunts or cannulas to drain cerebrospinal fluid; severe scoliosis (50 degrees or more); gestational age at birth < 35 weeks; oxygen saturation < 95%; weight < third percentile; tracheostomy, noninvasive ventilatory support for > 12 months; active viral infections (HIV, hepatitis B or C, Epstein-Barr virus), syphilis; upper or lower respiratory tract infection within 4 weeks; severe infections within 4 weeks; bacterial meningitis; cardiopathy; hypersensitivity to prednisolone or other glucocorticoids; immunosuppressive therapy within 3 months; current use of immunomodulatory drugs; prior use of Risdiplam, Nusinersen, or Onasemnogene abeparvovec; previous clinical trials related to SMA; type 2 with prior scoliosis surgery within 12 months or scheduled in up to 52 weeks; major surgeries programmed	Gestational age at birth < 35 weeks; oxygen saturation < 96%; tracheostomy or noninvasive ventilatory support for 16 h or more; weight < third percentile; active viral infections (HIV, hepatitis B or C, Epstein-Barr virus), syphilis; upper or lower respiratory tract infection within 4 weeks; severe infections within 2 weeks or currently under treatment; cardiopathy; hypersensitivity to prednisolone or other glucocorticoids; immunosuppressive therapy within 3 months; serum anti-AAV9 nAbs > 1:50; clinically significant abnormal laboratory values (liver function, renal function, platelet counts); prior use of Risdiplam, Nusinersen, or Onasemnogene abeparvovec; previous clinical trials related to SMA; major surgeries programmed	Diagnosis of HIV, hepatitis B or C, or congenital syphilis; maternal HIV infection; anti-AAV9 antibody > 1:50; agreement of legal representatives with the possibility of alternative feeding methods; tracheostomy or respiratory support for 16 h or more per day; previous treatment with Nusinersen, Risdiplam, Branaplam, or Onasemnogene abeparvovec (or planned within 12 months after infusion); need for glucocorticoids or immunosuppressive drugs; abnormal laboratory tests (renal and liver function, anemia, leukocytosis, raised troponin I levels); acute/chronic liver failure; allergy or hypersensitivity to components; participation in other clinical trials	Current clinical trial with other SMA drugs; previous AAV gene therapy; risks of gene replacement therapy higher than benefits; prior Nordenafil sodium injection within 4 months; prior use of beta2 receptor agonists within 30 days; prior risperidone use within 15 days; allergy or hypersensitivity to components; noninvasive ventilatory support for at least 12 h daily; serum anti-AAV9 nAbs > 1:200	Participating in SMA drug trials or other AAV gene therapies; prior treatment with Nosenazone Sodium and Lispiran; use of oral beta-2-receptor agonists within 30 days; immunosuppressive treatments up to 3 months after infusion; invasive ventilation or noninvasive ventilation for 16 h daily or more; serum anti-AAV9 nAbs > 1:200; heterozygous variant c.859G > C in *SMN2* gene; hypersensitivity or allergy to components; contraindications to lumbar puncture or intrathecal infusion; concomitant severe diseases related to gene therapy; abnormal laboratory testing (liver and renal function, anemia, low platelet count); active infection (HIV, hepatitis B or C, syphilis); severe infections within 2 weeks; weight below 2 standard deviations; vaccination <2 weeks
Safety and efficacy data	No safety/efficacy data available	No safety/efficacy data available	No safety/efficacy data available	No safety/efficacy data available	No safety/efficacy data available	No safety/efficacy data available

AAV, adeno-associated virus; *AE*, adverse events; *ALT*, alanine aminotransferase; *AST*, aspartate aminotransferase; *BSID-III*, Bayley Scales of Infant and Toddler Development Third Edition; *CGI*, Clinical Global Impression; *CHOP-INTEND*, Children's Hospital of Philadelphia Infant Test of Neuromuscular Disorders; *FEV1*, Forced Expiratory Volume in 1 s; *GGT*, gamma-glutamyltransferase; *HFMSE*, Hammersmith Functional Motor Scale Expanded; *HINE-2*, Hammersmith Infant Neurological Examination Module 2; *FVC*, forced vital capacity; *MEP*, maximal expiratory pressure; *MFM*, motor function measure; *MGRS*, Multicenter Growth Reference Study (World Health Organization); *MIP*, maximal inspiratory pressure; *nAb*, neutralizing antibody; *RULM*, revised upper limb module; *SAE*, serious adverse events; *sc*, self-complementary; *SMA*, spinal muscular atrophy; *SMAIS*, SMA Independence Scale; *6MWT*, six-minute walk test.

model with severe SMA, Cas9 injection together with sgRNA markedly increased SMN protein levels, resulting in a raised median survival of the mouse model (Li et al., 2020).

7 Conclusion

Disease-modifying therapies based on genetic approaches have brought about marked changes in the natural history of the different types of 5q SMA. The current treatment of choice for 5q SMA includes multidisciplinary supportive measures combined with one of the disease-modifying drugs, such as Onasemnogene abeparvovec, Nusinersen, and Risdiplam, with several aspects considered in the treatment decision-making process. Gene therapy with Onasemnogene abeparvovec, as well as new molecules in clinical trials and under development, represents a major advance in changing prognostic aspects of motor function, ventilation, and life expectancy in both symptomatic and presymptomatic individuals diagnosed with 5q SMA.

References

Aharoni, S., Bistritzer, J., Levine, H., Sagi, L., Fattal-Valevski, A., Ginzberg, M., Noyman, I., Cohen, R., Nevo, Y., 2023. Adeno-associated virus serotype 9 antibody titers in patients with SMA pre-screened for treatment with onasemnogene abeparvovec –routine care evidence. Gene Ther. 30 (1-2), 101–106. doi:10.1038/s41434-022-00339-0.

Al-Zaidy, S., Pickard, A.S., Kotha, K., Alfano, L.N., Lowes, L., Paul, G., Church, K., Lehman, K., Sproule, D.M., Dabbous, O., Maru, B., Berry, K., Arnold, W.D., Kissel, J.T., Mendell, J.R., Shell, R., 2019. Health outcomes in spinal muscular atrophy type 1 following AVXS-101 gene replacement therapy. Pediatr. Pulmonol. 54 (2), 179–185. doi:10.1002/ppul.24203.

Alves, C.R.R., Ha, L.L., Yaworski, R., Sutton, E.R., Lazzarotto, C.R., Christie, K.A., Reilly, A., Beauvais, A., Doll, R.M., de la Cruz, D., Maguire, C.A., Swoboda, K.J., Tsai, S.Q., Kothary, R., Kleinstiver, B.P., 2024. Optimization of base editors for the functional correction of SMN2 as a treatment for spinal muscular atrophy. Nat. Biomed. Eng. 8 (2), 118–131. doi:10.1038/s41551-023-01132-z.

Arbab, M., Matuszek, Z., Kray, K.M., Du, A., Newby, G.A., Blatnik, A.J., Raguram, A., Richter, M.F., Zhao, K.T., Levy, J.M., Shen, M.W., Arnold, W.D., Wang, D., Xie, J., Gao, G., Burghes, A.H.M., Liu, D.R., 2023. Base editing rescue of spinal muscular atrophy in cells and in mice. Science 380 (6642). doi:10.1126/science.adg6518.

Arnold, W.D, Kassar, D., Kissel, J.T., 2015. Spinal muscular atrophy: diagnosis and management in a new therapeutic era. Muscle Nerve 51 (2), 157–167. doi:10.1002/mus.24497.

Baccon, J., Pellizzoni, L., Rappsilber, J., Mann, M., Dreyfuss, G., 2002. Identification and characterization of Gemin7, a novel component of the survival of motor neuron complex. J. Biol. Chem. 277 (35), 31957–31962. doi:10.1074/jbc.M203478200.

Bemanalizadeh, M., Heidary, L., Dakkali, M.S., Hadizadeh, S., Ahmadbeigi, N., Heidari, M., Zamani, G., Ashrafi, M., Mansouri, V., 2025. Combination therapies in spinal muscular atrophy: a systematic review. Eur. J. Pediatr. 184 (9), 583. doi: 10.1007/s00431-025-06386-0.

Bitetti, I., Lanzara, V., Margiotta, G., Varone, A., 2023. Onasemnogene abeparvovec gene replacement therapy for the treatment of spinal muscular atrophy: a real-world observational study. Gene Ther. 30 (7-8), 592–597. doi:10.1038/s41434-022-00341-6.

Bitetti, I., Manna, M.R., Stella, R., Varone, A., 2024. Motor and neurocognitive profiles of children with symptomatic spinal muscular atrophy type 1 with two copies of SMN2 before and after treatment: a longitudinal observational study. Front. Neurol. 15. doi:10.3389/fneur.2024.1326528.

Blair, H.A., 2022. Onasemnogene abeparvovec: a review in spinal muscular atrophy. CNS Drugs 36 (9), 995–1005. doi:10.1007/s40263-022-00941-1.

Braun, M., Lange, C., Schatz, P., Long, B., Stanta, J., Gorovits, B., Tarcsa, E., Jawa, V., Yang, T.-Y., Lembke, W., Miller, N., McBlane, F., Christodoulou, L., Yuill, D., Milton, M., 2024. Preexisting antibody assays for gene therapy: considerations on patient selection cutoffs and companion diagnostic requirements. Mol. Ther. Methods Clin. Dev. 32 (1). doi:10.1016/j.omtm.2024.101217.

Brown, S.M., Ajjarapu, A.S., Ramachandra, D., Blasco-Pérez, L., Costa-Roger, M., Tizzano, E.F., Sumner, C.J., Mathews, K.D., 2024. Onasemnogene-abeparvovec administration to premature infants with spinal muscular atrophy. Ann. Clin. Transl. Neurol. 11 (11), 3042–3046. doi:10.1002/acn3.52213.

Chand, D.H., Mitchell, S., Sun, R., LaMarca, N., Reyna, S.P., Sutter, T., 2022. Safety of onasemnogene abeparvovec for patients with spinal muscular atrophy 8.5 kg or heavier in a global managed access program. Pediatr. Neurol. 132, 27–32. doi:10.1016/j.pediatrneurol.2022.05.001.

Chand, D.H., Zaidman, C., Arya, K., Millner, R., Farrar, M.A., Mackie, F.E., Goedeker, N.L., Dharnidharka, V.R., Dandamudi, R., Reyna, S.P., 2021b. Thrombotic microangiopathy following onasemnogene abeparvovec for spinal muscular atrophy: a case series. J. Pediatr. 231, 265–268. doi:10.1016/j.jpeds.2020.11.054.

Chand, D., Mohr, F., McMillan, H., Tukov, F.F., Montgomery, K., Kleyn, A., Sun, R., Tauscher-Wisniewski, S., Kaufmann, P., Kullak-Ublick, G., 2021a. Hepatotoxicity following administration of onasemnogene abeparvovec (AVXS-101) for the treatment of spinal muscular atrophy. J. Hepatol. 74 (3), 560–566. doi:10.1016/j.jhep.2020.11.001.

Chongmelaxme, B., Yodsurang, V., Vichayachaipat, P., Srimatimanon, T., Sanmaneechai, O., 2025. Gene-based therapy for the treatment of spinal muscular atrophy types 1 and 2: a systematic review and meta-analysis. Gene Ther. 32 (4), 301–330. doi:10.1038/s41434-024-00503-8.

Day, J.W., Finkel, R.S., Chiriboga, C.A., Connolly, A.M., Crawford, T.O., Darras, B.T., Iannaccone, S.T., Kuntz, N.L., Peña, L.D.M., Shieh, P.B., Smith, E.C., Kwon, J.M., Zaidman, C.M., Schultz, M., Feltner, D.E., Tauscher-Wisniewski, S., Ouyang, H., Chand, D.H., Sproule, D.M., Macek, T.A., Mendell, J.R., 2021. Onasemnogene abeparvovec gene therapy for symptomatic infantile-onset spinal muscular atrophy in patients with two copies of SMN2 (STR1VE): an open-label, single-arm, multicentre, phase 3 trial. Lancet Neurol. 20 (4), 284–293. doi:10.1016/S1474-4422(21)00001-6.

Day, J.W., Mendell, J.R., Burghes, A.H.M., van Olden, R.W., Adhikary, R.R., Dilly, K.W., 2023. Adeno-associated virus serotype 9 antibody seroprevalence for patients in the United States with spinal muscular atrophy. Mol. Ther. Methods Clin. Dev. 31. doi:10.1016/j.omtm.2023.101117.

de Holanda Mendonça, R., Matsui, C., Polido, G.J., Macedo Serafim Silva, A., Kulikowski, L., Dias, A.T., Zanardo, E.A., Solla, D.J.F., Gurgel-Giannetti, J., de Moura, A.C.M.L., Sampaio, G.P.C., Oliveira, A.S.B., de Souza, P.V.S., de Rezende Pinto, W.B.V., Gonçalves, E.A., Farias, I.B., Nardes, F., de Queiroz Campos Araújo, A.P., Marques, W., José Tomaselli, P., Ribeiro, M.D.O., Paulo Kitajima, J., Paoli Monteiro, F., Saute, J.A.M., Becker, M.M., Saraiva-Pereira, M.L., Brusius-Facchin, A.C., van der Linden, V., Florêncio, R.N., Vinícius Soares Barbosa, A., Machado-Costa, M.C., Luiz Santos Pessoa, A., Souza, L.S., Franca, M.C., Kok, F., Reed, U.C., Zanoteli, E., 2020. Intragenic variants in the SMN1 gene determine the clinical phenotype in 5q spinal muscular atrophy. Neurol. Genet. 6 (5). doi:10.1212/NXG.0000000000000505.

De Waele, L., Servais, L., 2025. Treatment strategies for patients with spinal muscular atrophy. Expert Rev. Neurother. 25 (3), 259–261. doi:10.1080/14737175.2024.2439486.

Desguerre, I., Barrois, R., Audic, F., Barnerias, C., Chabrol, B., Davion, J.B., Durigneux, J., Espil-Taris, C., de la Banda, M.G.-G., Guichard, M., Isapof, A., Nougues, M.C., Laugel, V., Goff, L.L., Mercier, S., Pervillé, A., Richelme, C., Thibaud, M., Sarret, C., Schweitzer, C., Testard, H., Trommsdorff, V., Vanhulle, C., Walther-Louvier, U., Altuzarra, C., Chouchane, M., Ropars, J., Quijano-Roy, S., Cances, C., 2024. Real-world multidisciplinary outcomes of onasemnogene abeparvovec monotherapy in patients with spinal muscular atrophy type 1: experience of the French cohort in the first three years of treatment. Orphanet J. Rare Dis. 19 (1). doi:10.1186/s13023-024-03326-3.

Farrar, M.A., Kiernan, M.C., 2015. The genetics of spinal muscular atrophy: progress and challenges. Neurotherapeutics 12 (2), 290–302. doi:10.1007/s13311-014-0314-x.

Finkel, R.S., Darras, B.T., Mendell, J.R., Day, J.W., Kuntz, N.L., Connolly, A.M., Zaidman, C.M., Crawford, T.O., Butterfield, R.J., Shieh, P.B., Tennekoon, G., Brandsema, J.F., Iannaccone, S.T., Shoffner, J., Kavanagh, S., Macek, T.A., Tauscher-Wisniewski, S., 2023. Intrathecal onasemnogene abeparvovec for sitting, nonambulatory patients with spinal muscular atrophy: phase i ascending-dose study (STRONG). J. Neuromuscul. Dis. 10 (3), 389–404. doi:10.3233/JND-221560.

Finkel, R.S., Mercuri, E., Meyer, O.H., Simonds, A.K., Schroth, M.K., Graham, R.J., Kirschner, J., Iannaccone, S.T., Crawford, T.O., Woods, S., Muntoni, F., Wirth, B., Montes, J., Main, M., Mazzone, E.S., Vitale, M., Snyder, B., Quijano-Roy, S., Bertini, E., Davis, R.H., Qian, Y., Sejersen, T., 2018. Diagnosis and management of spinal muscular atrophy: part 2: pulmonary and acute care; medications, supplements and immunizations; other organ systems; and ethics. Neuromuscul. Disord. 28 (3), 197–207. doi:10.1016/j.nmd.2017.11.004.

Giavazzi, A., Setola, V., Simonati, A., Battaglia, G., 2006. Neuronal-specific roles of the survival motor neuron protein: evidence from survival motor neuron expression patterns in the developing human central nervous system. J. Neuropathol. Exp. Neurol. 65 (3), 267–277. doi:10.1097/01.jnen.0000205144.54457.a3.

Gowda, V., Atherton, M., Murugan, A., Servais, L., Sheehan, J., Standing, E., Manzur, A., Scoto, M., Baranello, G., Munot, P., McCullagh, G., Willis, T., Tirupathi, S., Horrocks, I., Dhawan, A., Eyre, M., Vanegas, M., Fernandez-Garcia, M.A., Wolfe, A., Pinches, L., Illingworth, M., Main, M., Abbott, L., Smith, H., Milton, E., D'Urso, S., Vijayakumar, K., Marco, S.S., Warner, S., Reading, E., Douglas, I., Muntoni, F., Ong, M., Majumdar, A., Hughes, I., Jungbluth, H., E. Wraige, 2024. Efficacy and safety of onasemnogene abeparvovec in children with spinal muscular atrophy type 1: real-world evidence from 6 infusion centres in the United Kingdom. Lancet Reg. Health Eur. 37. doi:10.1016/j.lanepe.2023.100817.

Guillou, J., de Pellegars, A., Porcheret, F., Frémeaux-Bacchi, V., Allain-Launay, E., Debord, C., Denis, M., Péréon, Y., Barnérias, C., Desguerre, I., Roussey, G., Mercier, S., 2022. Fatal thrombotic microangiopathy case following adeno-associated viral SMN gene therapy. Blood Adv. 6 (14), 4266–4270. doi:10.1182/bloodadvances.2021006419.

Günther, R., Neuwirth, C., Koch, J.C., Lingor, P., Braun, N., Untucht, R., Petzold, D., Weber, M., Hermann, A., 2019. Motor unit number index (MUNIX) of hand muscles is a disease biomarker for adult spinal muscular atrophy. Clin. Neurophysiol. 130 (2), 315–319. doi:10.1016/j.clinph.2018.11.009.

Hagenacker, T., Schara-Schmidt, U., 2024. Gene replacement therapy in spinal muscular atrophy: filling the data gaps. Lancet Reg. Health Eur. 37. doi:10.1016/j.lanepe.2023.100822.

Hatanaka, F., Suzuki, K., Shojima, K., Yu, J., Takahashi, Y., Sakamoto, A., Prieto, J., Shokhirev, M., Nuñez Delicado, E., Esteban, C.R., Belmonte, J.C.I., 2024. Therapeutic strategy for spinal muscular atrophy by combining gene supplementation and genome editing. Nat. Commun. 15 (1). doi:10.1038/s41467-024-50095-5.

Hjartarson, H.T., Nathorst-Böös, K., Sejersen, T., 2022. Disease modifying therapies for the management of children with spinal muscular atrophy (5q SMA): an update on the emerging evidence. Drug Des., Dev. Ther. 16, 1865–1883. doi:10.2147/DDDT.S214174.

Kichula, E.A., Proud, C.M., Farrar, M.A., Kwon, J.M., Kayoko Saito, Desguerre, I., McMillan, H.J., 2021. Expert recommendations and clinical considerations in the use of onasemnogene abeparvovec gene therapy for spinal muscular atrophy. Muscle Nerve 64 (4), 413–427. doi:10.1002/mus.27363.

Kirschner, J., Bernert, G., Butoianu, N., De Waele, L., Fattal-Valevski, A., Haberlova, J., Moreno, T., Klein, A., Kostera-Pruszczyk, A., Mercuri, E., Quijano-Roy, S., Sejersen, T., Eduardo, F.T., Ludo van der Pol, W., Wallace, S., Zafeiriou, D., Ziegler, A., Muntoni, F., Servais, L., 2024. 2024 update: european consensus statement on gene therapy for spinal muscular atrophy. Eur. J. Paediatr. Neurol. 51, 73–78. doi:10.1016/j.ejpn.2024.06.001.

Li, -J.-J., Lin, X., Tang, C., Lu, Y.-Q., Hu, X., Zuo, E., Li, H., Ying, W., Sun, Y., Lai, -L.-L., Chen, H.-Z., Guo, -X.-X., Zhang, Q.-J., Wu, S., Zhou, C., Shen, X., Wang, Q., Lin, M.-T., Ma, L.-X., Wang, N., Krainer, A.R., Shi, L., Yang, H., Chen, W.-J., 2020. Disruption of splicing-regulatory elements using CRISPR/Cas9 to rescue spinal muscular atrophy in human iPSCs and mice. Natl. Sci. Rev. 7 (1), 92–101. doi:10.1093/nsr/nwz131.

McMillan, H.J., Baranello, G., Farrar, M.A., Zaidman, C.M., Teresa Moreno, De Waele, L., Jong, Y.-J., Laugel, V., Quijano-Roy, S., Mercuri, E., Chien, Y.-H., Straub, V., Darras, B.T., Seibert, J., Escudero, R.B., Alecu, I., Freischläger, F., Muntoni, F., 2024. Safety and efficacy of IV onasemnogene abeparvovec for pediatric patients with spinal muscular atrophy: the phase 3b SMART study. Neurology 104 (2). doi:10.1212/WNL.0000000000210268.

McPheron, M.A., Felker, M.V., 2024. Clinical perspectives: treating spinal muscular atrophy. Mol. Ther. 32 (8), 2489–2504. doi:10.1016/j.ymthe.2024.06.020.

Mendell, J.R., Al-Zaidy, S.A., Lehman, K.J., McColly, M., Lowes, L.P., Alfano, L.N., Reash, N.F., Iammarino, M.A., Church, K.R., Kleyn, A., Meriggioli, M.N., Shell, R., 2021. Five-year extension results of the phase 1 START trial of onasemnogene abeparvovec in spinal muscular atrophy. JAMA Neurol. 78 (7), 834–841. doi:10.1001/jamaneurol.2021.1272.

Mendell, J.R., Al-Zaidy, Shell, Arnold, Rodino-Klapac, Prior, Lowes, Alfano, Berry, Church, Kissel, Nagendran, L'Italien, Sproule, Wells, Cardenas, Heitzer, Kaspar, Corcoran, Braun, Likhite, Miranda, Meyer, Foust, Burghes, Kaspar, 2017. Single-dose gene-replacement therapy for spinal muscular atrophy. N. Engl. J. Med. 377 (18), 1713–1722. doi:10.1056/NEJMoa1706198.

Mendonça, R.H., Ortega, A.B., Matsui Jr, C., van der Linden, V., Kerstenetzky, M., Grossklauss, L.F., Silveira-Lucas, E.L., Polido, G.J., Zanoteli, E., 2024. Gene replacement therapy for spinal muscular atrophy: safety and preliminary efficacy in a Brazilian cohort. Gene Ther. 31 (7-8), 391–399. doi:10.1038/s41434-024-00456-y.

Mercuri, E., Finkel, R.S., Muntoni, F., Wirth, B., Montes, J., Main, M., Mazzone, E., Vitale, M., Snyder, B., Quijano-Roy, S., Bertini, E., Davis, R.H., Meyer, O.H., Simonds, A.K., Schroth, M.K., Graham, R.J., Kirschner, J., Iannaccone, S.T., Crawford, T., Woods, S., Qian, Y., Sejersen, T., Tiziano, F.D., Tizzano, E., Swoboda, K., Swoboda, K., Laing, N., Kayoko, S., Prior, T., Chung, W.K., Wu, S.-M., Coleman, C., Gee, R., Glanzman, A., Kroksmark, A.-K., Krosschell, K., Nelson, L., Rose, K., Stępień, A., Vuillerot, C., Dubousset, J., Farrington, D., Flynn, J., Halanski, M., Hasler, C., Miladi, L., Reilly, C., Roye, B., Sponseller, P., Hurst, R., Tarrant, S., Barja, S., Foust, K., Kyle, B., Rodan, L., Roper, H., Seffrood, E., Szlagatys-Sidorkiewicz, A., 2018. Diagnosis and management of spinal muscular atrophy: part 1: recommendations for diagnosis, rehabilitation, orthopedic and nutritional care. Neuromuscul. Disord. 28 (2), 103–115. doi:10.1016/j.nmd.2017.11.005.

Mercuri, E., Muntoni, F., Baranello, G., Masson, R., Boespflug-Tanguy, O., Bruno, C., Corti, S., Daron, A., Deconinck, N., Servais, L., Straub, V., Ouyang, H., Chand, D., Tauscher-Wisniewski, S., Mendonca, N., Lavrov, A., Seferian, De Lucia, Tachibana, Jollet, Mouffak, Pedemonte, Brolatti, Morando, Vanlander, De Vos, Tahon, Govoni, Magri, Comi, Foa, Parente, Buscemi,, Dal Farra, Schneider, Jonas, Defeldre, Pagliano, Zanin, Arnoldi, Schembri, Del Sole, Mandelli, Pera, Antonaci, Coratti, de Sanctis, Pane,, Scoto, Groves, Edel, Abel, Van Ruiten, Lofra, Thompson, 2021. Onasemnogene abeparvovec gene therapy for symptomatic infantile-onset spinal muscular atrophy type 1 (STR1VE-EU): an open-label, single-arm, multicentre, phase 3 trial. Lancet Neurol. 20 (10), 832–841. doi:10.1016/S1474-4422(21)00251-9.

Mourelatos, Z., Abel, L., Yong, J., Kataoka, N., Dreyfuss, G., 2001. SMN interacts with a novel family of hnRNP and spliceosomal proteins. EMBO J. 20 (19), 5443–5452. doi:10.1093/emboj/20.19.5443.

Nishio, H., Niba, E.T.E., Saito, T., Okamoto, K., Takeshima, Y., Awano, H., 2023. Spinal muscular atrophy: the past, present, and future of diagnosis and treatment. Int. J. Mol. Sci. 24 (15). doi:10.3390/ijms241511939.

Ogbonmide, T., Rathore, R., Rangrej, S.B., Hutchinson, S., Lewis, M., Ojilere, S., Carvalho, V., Kelly, I., 2023. Gene therapy for spinal muscular atrophy (SMA): a review of current challenges and safety considerations for onasemnogene abeparvovec (Zolgensma). Cureus 15 (3). doi:10.7759/cureus.36197.

Pascual-Morena, C., Cavero-Redondo, I., Lucerón-Lucas-Torres, M., Martínez-García, I., Rodríguez-Gutiérrez, E., Martínez-Vizcaíno, V., 2023. Onasemnogene abeparvovec in type 1 spinal muscular atrophy: a systematic review and meta-analysis. Hum. Gene Ther. 34 (3-4), 129–138. doi:10.1089/hum.2022.161.

Ramdas, S., Oskoui, M., Servais, L., 2024. Treatment options in spinal muscular atrophy: a pragmatic approach for clinicians. Drugs 84 (7), 747–762. doi:10.1007/s40265-024-02051-2.

René, C.A., Parks, R.J., 2023. Expanding the availability of onasemnogene abeparvovec to older patients: the evolving treatment landscape for spinal muscular atrophy. Pharmaceutics 15 (6). doi:10.3390/pharmaceutics15061764.

Saute, J.A., Muntadas, J., Gurgel-Giannetti, J., Monges, S., Aliberti, P., Mendonça, R.H., Alecu, I., Ritter, S., de Lana, J.M., Mumneh, N., Zanoteli, E., 2025. Safety and tolerability of onasemnogene abeparvovec for patients with spinal muscular atrophy weighing ≤17 kg and ≤24 months old from OFELIA, a phase 4, open-label, multicenter, non-randomised, interventional study. Lancet Reg. Health Am. 49. doi:10.1016/j.lana.2025.101193.

Serra-Juhe, C., Tizzano, E.F., 2019. Perspectives in genetic counseling for spinal muscular atrophy in the new therapeutic era: early pre-symptomatic intervention and test in minors. Eur. J. Hum. Genet. 27 (12), 1774–1782. doi:10.1038/s41431-019-0415-4.

Servais, L., 2024. Gene therapy for spinal muscular atrophy: timing is key. Lancet Reg. Health Eur. 47. doi:10.1016/j.lanepe.2024.101112.

Smith, A.G., 2020. Spinal Muscular Atrophy. Continuum Lifelong Learn. Neurol. 26 (5), 1348–1368. doi:10.1212/CON.0000000000000918.

Stettner, G.M., Hasselmann, O., Tscherter, A., Galiart, E., Jacquier, D., Klein, A., 2023. Treatment of spinal muscular atrophy with onasemnogene abeparvovec in Switzerland: a prospective observational case series study. BMC Neurol. 23 (1). doi:10.1186/s12883-023-03133-6.

Strauss, K.A., Farrar, M.A., Muntoni, F., Saito, K., Mendell, J.R., Servais, L., McMillan, H.J., Finkel, R.S., Swoboda, K.J., Kwon, J.M., Zaidman, C.M., Chiriboga, C.A., Iannaccone, S.T., Krueger, J.M., Parsons, J.A., Shieh, P.B., Kavanagh, S., Tauscher-Wisniewski, S., McGill, B.E., Macek, T.A., 2022b. Onasemnogene abeparvovec for presymptomatic infants with two copies of SMN2 at risk for spinal muscular atrophy type 1: the Phase III SPR1NT trial. Nat. Med. 28 (7), 1381–1389. doi:10.1038/s41591-022-01866-4.

Strauss, K.A., Farrar, M.A., Muntoni, F., Saito, K., Mendell, J.R., Servais, L., McMillan, H.J., Finkel, R.S., Swoboda, K.J., Kwon, J.M., Zaidman, C.M., Chiriboga, C.A., Iannaccone, S.T., Krueger, J.M., Parsons, J.A., Shieh, P.B., Kavanagh, S., Wigderson, M., Tauscher-Wisniewski, S., McGill, B.E., Macek, T.A., 2022a. Onasemnogene abeparvovec for presymptomatic infants with three copies of SMN2 at risk for spinal muscular atrophy: the Phase III SPR1NT trial. Nat. Med. 28 (7), 1390–1397. doi:10.1038/s41591-022-01867-3.

Tukov, F.F., Mansfield, K., Milton, M., Meseck, E., Penraat, K., Chand, D., Hartmann, A., 2022. Single-dose intrathecal dorsal root ganglia toxicity of onasemnogene abeparvovec in cynomolgus monkeys. Hum. Gene Ther. 33 (13-14), 740–756. doi:10.1089/hum.2021.255.

Verhaart, I.E.C., Robertson, A., Wilson, I.J., Aartsma-Rus, A., Cameron, S., Jones, C.C., Cook, S.F., Lochmüller, H., 2017. Prevalence, incidence and carrier frequency of 5q-linked spinal muscular atrophy—A literature review. Orphanet J. Rare Dis. 12 (1). doi:10.1186/s13023-017-0671-8.

Waldrop, M.A., 2024. Clinical decision making around commercial use of gene and genetic therapies for spinal muscular atrophy. Neurotherapeutics 21 (4). doi:10.1016/j.neurot.2024.e00437.

Waldrop, M.A., Elsheikh, B.H., 2020. Spinal muscular atrophy in the treatment era. Neurol. Clin. 38 (3), 505–518. doi:10.1016/j.ncl.2020.03.002.

Weiß, C., Becker, L.-L., Friese, J., Blaschek, A., Hahn, A., Illsinger, S., Schwartz, O., Bernert, G., von der Hagen, M., Husain, R.A., Goldhahn, K., Kirschner, J., Pechmann, A., Flotats-Bastardas, M., Schreiber, G., Schara, U., Plecko, B., Trollmann, R., Horber, V., Wilichowski, E., Baumann, M., Klein, A., Eisenkölbl, A., Köhler, C., Stettner, G.M., Cirak, S., Hasselmann, O., Kaindl, A.M., Garbade, S.F., Johannsen, J., Ziegler, A., Baum, P., Baumgartner, M., Bertsche, A., Blankenburg, M., Denecke, J., Deschauer, M., Eckenweiler, M., Geis, T., Groß, M., Günther, R., Hagenacker, T., Hamelmann, E., Kamm, C., Kauffmann, B., Koch, J.C., Löscher, W., Ludolph, A.C., Martin, P., Mensch, A., Zu Hörste, G.M., Neuwirth, C., Petri, S., Pühringer, M., Rathmann, I., Schäfer, D., Schimmel, M., Schrank, B., Schreiber-Katz, O., Schwerin-Nagel, A., Smitka, M., Steinbach, M., Steiner, E., Stoffels, J., Theophil, M., Topakian, R., Türk, M., Vorgerd, M., Walter, M.C., Weiler, M., Wiegand, G., Wunderlich, G., Wurster, C.D., Zeller, D., Metelmann, M., Zeiner, F., Pilshofer, V., Rappold, M., Pauschek, J., Reihle, C., Homma, A.K., Lingor, P., Henzi, B., Reinhardt, T., Holzwarth, D., Wittmann, W., Kappel, S., Freigang, M., Stolte, B., Martakis, K., Classen, G., Roland-Schäfer, D., Steuernagel, D., Hartmann, H., Fischer, S., Wermuth, M., Muhandes, M.T., Hotter, A., Uzelac, Z., Naegel, S., Wiethoff, S., Braun, N., Bjelica, B., Kölbel, H., Angelova-Toshkina, D., Wilken, B., Osmanovic, A., Fiedler, B., Tomforde, M., Voelkl, T., von Moers, A., Müller, P., Behring, B., Güttsches, A., Reilich, P., Wick, W., Stoltenburg, C., Witzel, S., Bellut, J., Hoffmann, G.F., Mörtlbauer, K., Ille, A., Schroth, M., Driemeyer, J., Semmler, L., Müller, C., Dörnbrack, K., Zemlin, M., Geitmann, S., Lapp, H.S., Brakemeier, S., Gehrke, T., Ntemiris, K., Kaiser, N., Borowski, S., Ramadan, B., Hustedt, U., Baum, T., Schneider, I., Akova-Oztürk, E., Vill, K., Dibrani, Z., Wohnrade, C., Della-Marina, A., Jung, L., Deba, T., Zobel, J., Schallner, J., Kraut, C., Vollmann, P., Schüssler, S., Roeder, M., Hiebeler, M., Berberich, N., Schneider, J., Brauner, B., Kölker, S., Pernegger, E., Gosk-Tomek, M., Braun, S., Weiss, D., Machetanz, G., Langer, T., Saier, C., Baumann, S., Hettrich, S., Dworschak, G., Müller-Kaempffer, K., Dittes, I., Thimm, A., Quinten, L., Albers, K., Bevot, A., Bretschneider, C., Dorst, J., Kendzierski, T., Hannibal, I., Bischofberger, J., Riesmeier, T., Gangfuß, A., to Settel, E.J., Grässl, M., Fiebig, S., Hollerauer, C., Seeber, L., Krahwinkler, I., Lange, I., Montagnese, F., Mann-Richter, M., Wagner, A., Leypold, C., Saffari, A., Anna, E., Wiesenhofer, A., Wendel, E.-M., Steffens, P.-S., Wider, S., Tassoni, A., Dall, A., Busch, F., Zeisler, D., Wessel, M., Lipka, J., Hackemer, A., Plugge, L., Jansen, E., Roth, E., Schuster, J., Koelsch, A., Warken-Madelung, B., Schwippert, M., Holtkamp, B., Köbbing, K., Claeys, S., Foerster, S., Thiele, S., Rochau-Trumpp, H., George, A., Niesert, M., Neimair, T., Vettori, K., Haverkamp, J., Taherpour, J., Hug, J., Wenzel, F., Bant, C., Baur, U., Bühner, K., Schlag, M., Ruß, L., Küpper, H., Müller, A., Wollinsky, K., Well, T., Leinert, A., Andres, B., Omran, H., Claus, N., Hagenmeyer, A., Schnurr, M., Dukic, V., Ludolph, A.C., Specht, S., Angermair, V., Hüpper, A., Banholzer, D., Stein, S., Kampowski, T., Richmann, M., Nicolai, S., Atta, O., Meßmer, B., de Vries, H., Rotenfusser, E., Oscmanovic, A., Renger, I., Guillemot, H., Lehnert, I., Grünwedel, M., Grimm, L., Stocker, G., Hoevel, A., Stadler, T., Fischer, M., Vogt, S., Gebert, A., Goldbach, S., Lochmüller, H., Müller-Felber, W., Schara-Schmidt, U., Probst-Schendzielorz, K., Lang, A., Nitzsche, M., Hammer, J., Müller-Kaempfer, K., Wirner-Piotrowski, C., van der Stam, L., Bongartz, A., Enzmann, C., Fluss, J., Galiart, E., Jacquier, D., Metzler, D.B., Tscherter, A., 2024. Efficacy and safety of gene therapy with onasemnogene abeparvovec in children with spinal muscular atrophy in the D-A-CH-region: a population-based observational study. Lancet Reg. Health Eur. 47. doi:10.1016/j.lanepe.2024.101092.

Weiß, C., Ziegler, A., Becker, L.-L., Johannsen, J., Brennenstuhl, H., Schreiber, G., Flotats-Bastardas, M., Stoltenburg, C., Hartmann, H., Illsinger, S., Denecke, J., Pechmann, A., Müller-Felber, W., Vill, K., Blaschek, A., Smitka, M., van der Stam, L., Weiss, K., Winter, B., Goldhahn, K., Plecko, B., Horber, V., Bernert, G., Husain, R.A., Rauscher, C., Trollmann, R., Garbade, S.F., Hahn, A., von der Hagen, M., Kaindl, A.M., 2022. Gene replacement therapy with onasemnogene abeparvovec in children with spinal muscular atrophy aged 24 months or younger and bodyweight up to 15 kg: an observational cohort study. Lancet Child Adolesc. Health 6 (1), 17–27. doi:10.1016/S2352-4642(21)00287-X.

Yang, D., Ruan, Y., Chen, Y., 2023. Safety and efficacy of gene therapy with onasemnogene abeparvovec in the treatment of spinal muscular atrophy: a systematic review and meta-analysis. J. Paediatr. Child Health 59 (3), 431–438. doi:10.1111/jpc.16340.

Zhuang, W., Lu, M., Wu, Y., Chen, Z., Wang, M., Wang, X., Guan, S., Lin, W., 2023. Safety concerns with nusinersen, risdiplam, and onasemnogene abeparvovec in spinal muscular atrophy: a real-world pharmacovigilance study. Clin. Drug Investig. 43 (12), 949–962. doi:10.1007/s40261-023-01320-4.

Chapter 5

Gene therapy in amyotrophic lateral sclerosis (ALS)

Negin Salemi[a] and Arman Rahimmi[b]
[a]*Isfahan University of Medical Sciences, Student Research Committee, Isfahan, Iran*
[b]*Neurosciences Research Center, Research Institute for Health Development, Kurdistan University of Medical Sciences, Sanandaj, Iran*

1 Introduction

Amyotrophic lateral sclerosis (ALS) is a progressive neurodegenerative disorder that poses significant challenges to healthcare systems worldwide. The disease is characterized by the degeneration of motor neurons in the brain and spinal cord, leading to muscle weakness, loss of voluntary movement, muscular atrophy, and ultimately, respiratory failure. The incidence and prevalence of ALS vary across different populations. Estimates indicate that incidence ranges from 0.26 per 100,000 person-years in Ecuador to 23.46 per 100,000 person-years in Japan, while point prevalence spans from 1.57 per 100,000 in Iran to an estimated 10.5 per 100,000 in the United States in 2022, with projections suggesting a rise to over 11.2 per 100,000 by 2030 due to an aging population. Variations based on geographical location, ethnicity, and gender further shape these trends, with higher prevalence observed in North America and Europe compared to Asian populations, prompting investigations into genetic predispositions and environmental influences that may drive these differences (Mehta et al., 2025).

The clinical presentation of ALS is multifaceted and changes over time, varying significantly among individuals. Initially, patients may experience subtle symptoms such as muscle weakness, twitching (fasciculations), or cramping, typically in the hands, legs, or bulbar area. As the disease progresses, these symptoms escalate, often resulting in significant functional impairment and a wide range of difficulties, including speaking (dysarthria) and swallowing (dysphagia) (Engelberg-Cook et al., 2024). Eventually, the relentless progression may lead to complete paralysis, emphasizing the urgent need for effective therapeutic interventions that not only alleviate symptoms but also enhance the patient's overall quality of life.

Despite advancements in understanding the nature of ALS, the prognosis for individuals diagnosed with it remains complex and variable. A recent study analyzed demographic and clinical variables in a cohort of 1442 patients, revealing a median survival of 17.2 months after diagnosis and a median survival of 30 months after the onset of symptoms. Older age at the time of diagnosis was correlated with shorter survival (Engelberg-Cook et al., 2024). The unpredictability of ALS progression further complicates prognostic assessments, leaving both patients and caregivers in a state of uncertainty and anxiety.

Given these pressing challenges, researching novel treatment modalities, including gene therapy, is increasingly critical for the future of ALS treatment. This chapter delves into these developments, emphasizing the potential that gene therapy holds for altering the course of ALS while also examining the associated scientific and ethical considerations. By providing a comprehensive overview of ALS and its implications, we aim to underscore the importance of innovative therapeutic approaches that can significantly impact the lives of individuals afflicted with this devastating disorder. Through ongoing research and collaboration, there is hope that new breakthroughs will emerge, fostering a brighter future for ALS patients and their families.

2 Disease pathogenesis and genetic patterns

ALS is considered an idiopathic disorder, and its causes seem to be multifactorial and are only partially understood. In general, these factors can be divided into two categories: genetic and environmental. From the genetic point of view, the studies conducted so far have suggested the possibility of >20 genes being associated with ALS, and it is expected that more genes

Gene Therapy in Neurodegenerative Disorders. https://doi.org/10.1016/B978-0-443-40467-2.00012-9

will be discovered in the future. Among these genes, *SOD1*, *tardbptrim72*, and *C9orf72* can be mentioned. Also, among the environmental risk factors, age, male sex, smoking, body mass index, exercise, long-term exposure (due to occupational or living environment) to metals and insecticides, head trauma, and viral infections can be mentioned (Masrori and Van Damme, 2020).

By assessing an ALS patient's autopsies with the unaided eye, some characteristics are observable, including atrophy in skeletal muscles and the motor cortex, hardening of the corticospinal and corticobulbar tracts, reduced thickness in cranial nerve XII, and the anterior spinal cord roots (van Es et al., 2017). One of the unique features of ALS is that both upper and lower motor neurons are damaged. Upper motor neurons are located in the motor cortex of the brain, and lower motor neurons are located in the brain stem and spinal cord (Brown and Al-Chalabi, 2017). Aggregation of inclusion bodies within the cytoplasm of motor neurons is another feature of ALS. These inclusion bodies are known as Bunina bodies, and their main component is the TDP-43 protein (van Es et al., 2017). However, in ALS patients with mutations in the *c9orf72*, *SOD1*, or *FUS* genes, the protein products of these genes are also present (Gendron et al., 2017). In the following, the molecular mechanisms of ALS and its genetic patterns are discussed.

2.1 Molecular mechanisms and genes associated with ALS

The cause of motor neuron degeneration in ALS is not yet fully understood, but several molecular pathways are thought to be involved. Studies indicate that many cellular processes are disrupted during ALS, such as protein degradation, cytoskeleton maintenance, RNA processing, and DNA repair (Alsultan et al., 2016; Konopka and Atkin, 2022). Mutant SOD1 protein forms aggregates that inhibit protein degradation, and similar aggregations are seen in sporadic ALS with wild-type SOD1, which suggests other factors such as oxidative stress or coaggregations with other proteins such as TDP-43, FUS, and C9orf72 (Bosco et al., 2010). Other genes involved in protein degradation whose mutations may lead to ALS include *vcp*, *optn*, *tbk1*, and *sqstm1*. In addition, genes important for cytoskeleton maintenance and axonal transport, *such* as *dctn1*, *pfn1*, and *tuba4a*, may play a role in the development of ALS (Brown and Al-Chalabi, 2017). Furthermore, RNA-binding proteins like TDP-43 and FUS are also implicated in ALS, and mutations in their coding genes cause them to aggregate in the cytoplasm, leading to decreased levels in the nucleus and improper RNA processing. However, TDP-43 aggregates are also present in almost all cases of ALS, even in the absence of a mutation in its coding gene. Other RNA metabolism genes associated with ALS include *ang*, *setx*, and *matr3* (van Es et al., 2017).

The most commonly mutated gene in ALS is *c9orf72*, which causes motor neuron death through mechanisms like RNA transcript deposition, toxic dipeptide repeat (DPR) protein translation, and decreased levels of normal C9orf72 protein. Mitochondrial dysfunction plays a significant role in the progression of ALS and frontotemporal dementia (FTD), particularly in cases linked to mutations in the *c9orf72* gene. Mitochondria, as the main energy supplier of cells, are of great importance, especially in energy-hungry cells like motor neurons, which are heavily affected in ALS (Brown and Al-Chalabi, 2017). Indeed, 40% of familial and 8% of sporadic ALS cases are due to the hexanucleotide (GGGGCC) repeat expansions in the first intron of the *c9orf72* gene (Konopka and Atkin, 2022).

Excitotoxicity is believed to play an axial role in the death of motor neurons during ALS. Indeed, excitotoxicity is common in all forms of ALS, and it is characterized by high intracellular calcium levels due to excessive glutamate stimulation. Motor neurons are particularly sensitive to excitotoxicity because of their lower calcium-buffering capacity and more permeable α-amino-3-hydroxy-5-methyl-4-isoxazolepropionic acid (AMPA) receptors (van Es et al., 2017).

ALS develops through the molecular events triggered by genetic mutations and protein dysfunction mentioned above. The following explanations elucidate the order of these events and the mechanistic progression of ALS step-by-step:

Mutations in genes like *SOD1*, *c9orf72*, *TARDBP* (encoding TDP-43), and *FUS* disrupt protein homeostasis, leading to toxic aggregates of misfolded proteins that overwhelm cellular cleanup systems like the ubiquitin–proteasome pathway and autophagy. These aggregates—including mutant SOD1, cytoplasmic TDP-43 clumps, and DPRs from *c9orf72* expansions—interfere with RNA processing, impair mitochondrial function, and generate oxidative stress (Taylor et al., 2016). Simultaneously, defects in cytoskeletal genes (*dctn1*, *pfn1*, *tuba4a*) destabilize microtubules and disrupt axonal transport, starving neurons of vital organelles and nutrients. As damaged mitochondria fail to meet energy demands, calcium dysregulation and excitotoxicity further stress motor neurons. Surrounding glial cells, activated by neuronal distress, amplify damage through chronic inflammation, releasing toxic cytokines and losing their supportive roles. Over time, synaptic connections at neuromuscular junctions degenerate, axons retract, and motor neurons die, culminating in progressive paralysis. This vicious cycle of protein aggregation, transport failure, metabolic collapse, and neuroinflammation ultimately leads to the irreversible motor deficits characteristic of ALS (Peters et al., 2015).

2.2 Genetic patterns

Approximately 10% of ALS cases appear to be hereditary, and the rest are sporadic. Over 30 genes, including *SOD1*, *c9orf72*, *tdp-43*, and FUS, have been linked to ALS, contributing to its genetic complexity (Suzuki et al., 2023). For family members, the risk is small but notable: first-degree relatives of ALS patients have a 1.1% chance of developing the disease, compared to about 0.05% in the general population (Wingo et al., 2011). This increased risk highlights the genetic component in hereditary cases. Most hereditary ALS forms follow Mendelian inheritance patterns, primarily autosomal dominant, where a single mutated gene copy from one parent is enough to increase disease risk. For example, mutations in SOD*1* or *c9orf72* are often inherited this way, leading to ALS in about 50% of carriers over time. Less commonly, autosomal recessive patterns occur, such as in some *aladin* gene mutations, where both gene copies must be mutated for the disease to manifest (van Es et al., 2017). Recent studies, however, suggest that ALS may not always follow simple Mendelian rules. The oligogenic inheritance hypothesis proposes that ALS can result from the combined effects of multiple genetic variants, each contributing a small risk. For instance, a patient might carry variants in *c9orf72* and *tbk1*, which together increase disease likelihood more than either alone. This model explains why some patients with known ALS gene mutations do not develop symptoms, while others with no single dominant mutation do (Renton et al., 2014). To confirm the role of oligogenic inheritance, researchers need comprehensive genome-wide data from thousands of ALS patients across diverse populations. Making these datasets publicly available, such as through initiatives like the Project MinE database, would allow scientists to study gene interactions more thoroughly and uncover new insights into ALS's genetic architecture.

3 Current approved therapeutic approaches for ALS

With a complex pathophysiology involving glutamate excitotoxicity, oxidative stress, mitochondrial dysfunction, protein aggregation, and inflammation, treatment approaches are challenging. Currently, only a few pharmacological therapies are approved, offering modest benefits.

3.1 Pharmacological treatments

Riluzole was the first Food and Drug Administration (FDA)-approved drug for ALS, introduced in 1995. It is a benzothiazole compound, a type of thiazole derivative originally developed as an anticonvulsant before its application in ALS was discovered. It functions by inhibiting glutamate release and reducing excitotoxicity, a process in which excessive glutamate damages motor neurons. Although its exact mechanism is still under investigation, riluzole may also indirectly antagonize glutamate receptors and inactivate voltage-gated sodium channels, contributing to neuroprotection. This mechanism slightly extends survival, typically by a few months, but does not significantly improve quality of life (Miller et al., 2003).

Edaravone was approved by the FDA in 2017. It is a pyrazolone derivative, a class of compounds with antioxidant properties, initially developed in Japan for stroke treatment before being repurposed for ALS. It works by scavenging free radicals and reducing oxidative stress, a key contributor to neuronal damage in ALS. Clinical trials have shown that edaravone slows functional decline in patients, particularly those in the early stages of the disease, though its impact on survival is minimal. It may be an optimal choice for individuals with recent-onset ALS (Sawada, 2017).

Tofersen was approved by the FDA in 2023. It is an antisense oligonucleotide (ASO), a synthetic strand of nucleic acid designed to target genetic mutations, specifically for ALS patients with SOD1 gene mutations (about 2% of cases). It binds to SOD*1* mRNA, reducing the production of a toxic misfolded protein that drives motor neuron degeneration. Clinical trials have demonstrated that tofersen slows disease progression in these patients, though its benefits are limited to this genetic subgroup and it does not broadly extend survival (Miller et al., 2022).

3.2 Symptomatic or supportive treatments for ALS

3.2.1 Respiration

Noninvasive ventilation has been shown to extend survival and improve quality of life, especially for patients without severe bulbar impairment, by addressing respiratory insufficiency as a common complication in ALS (Kim et al., 2020).

3.2.2 Sialorrhea management

Excessive salivation, or sialorrhea, can be managed with anticholinergic medications, botulinum toxin injections, or radiation therapy on the salivary glands. Botulinum toxin injections offer temporary relief and are often effective (Bjelica and Petri, 2024).

3.2.3 Muscle cramps and spasticity

Baclofen or Tizanidine can help reduce spasticity, a common cause of discomfort in ALS patients. For cramps, quinine sulfate and other muscle relaxants are prescribed; their effectiveness is moderate (Kwak, 2022).

3.2.4 Pseudobulbar affect

This condition involves involuntary emotional expressions (e.g., uncontrollable laughter or crying). Nuedexta, a combination of dextromethorphan and quinidine, is FDA-approved for managing pseudobulbar affect (PBA) in ALS patients. Clinical studies indicate that Nuedexta not only addresses PBA but also shows improvements in bulbar functions such as speech, swallowing, and salivation (Trojsi et al., 2023).

3.3 Clinical trials and research

Clinical research is exploring diverse approaches, including neuroprotective agents, gene therapy, and stem cell transplantation. These clinical trial approaches and investigational therapies reflect an expanding landscape in ALS treatment, bringing hope for slowing the progression of the disease and improving patients' quality of life.

3.3.1 Under investigation compounds

Beyond approved treatments, ALS therapy research explores agents targeting its multifactorial nature. For example, protein kinase inhibitors, such as those targeting CK-1, show potential in preclinical studies. These compounds reduce TDP-43 phosphorylation, a hallmark of ALS pathology seen in nearly all cases, by inhibiting kinases that add phosphate groups to TDP-43, which can lead to toxic protein clumps. In animal models, CK-1 inhibitors have provided neuroprotection and extended lifespan, but they are not yet approved for human use and require further clinical trials to confirm safety and efficacy (Richardson et al., 2023). Another investigational approach involves Beta-2 adrenoceptor agonists, drugs originally developed for asthma, like salbutamol. In ALS preclinical models, these agonists protect neuromuscular junctions—the connections between motor neurons and muscles—and reduce protein degradation, potentially slowing muscle wasting. While early studies in animal models are encouraging, these drugs are not FDA-approved for ALS and are still under investigation to determine their effectiveness in humans (Inchiosa, 2024).

The search for ALS treatments is like piecing together a complex puzzle, with each therapy targeting a different aspect of the disease's multifactorial nature. While tofersen is a step forward for SOD1-related ALS, the unapproved status of therapies like protein kinase inhibitors and beta-2 adrenoceptor agonists highlights the need for more research to bring effective treatments to all ALS patients.

3.3.2 Gene therapy

Advances in genetic research have highlighted ALS-related mutations, such as *SOD1* and *c9orf72*, opening new chapters for gene therapy. This approach includes ASOs that target mutant genes, RNA interference (RNAi), and CRISPR-based editing. Current clinical trials focus on delivering these therapies via viral vectors, particularly adeno-associated virus (AAV) platforms, which have shown potential in early trials (Amado and Davidson, 2021).

3.3.3 Neuroprotective agents

Several neuroprotective medications have shown early promise, including ceftriaxone and minocycline, which target inflammation and excitotoxicity in ALS. Although these agents have demonstrated efficacy in animal models, translating results to human trials remains challenging. Efforts to optimize dosage and delivery are ongoing (Maragakis et al., 2023).

3.3.4 Stem cell therapy

Stem cell-based therapies aim to replace lost motor neurons or provide neuroprotective support through mesenchymal and neural stem cells. Recent trials include the transplantation of neural stem cells directly into the spinal cord to promote motor neuron survival and reduce disease progression. Early results suggest safety and feasibility, with limited but encouraging efficacy indications (Sironi et al., 2023).

3.3.5 Adaptive platform trials

The HEALEY ALS Platform Trial is transforming ALS treatment research by using an adaptive platform design to test multiple therapies simultaneously under a unified framework, accelerating the identification of effective drugs. Unlike traditional single-drug trials, it shares a common placebo group, employs biomarkers like neurofilament light chain (NfL) to track progression, and adapts in real time, adding, modifying, or dropping treatments based on interim data. This approach offers significant advantages, including faster results, reduced costs, and greater flexibility, but faces challenges such as statistical complexity, potential bias, and questions about generalizability across diverse ALS populations. Since its 2020 launch, the trial has evaluated drugs like zilucoplan and verdiperstat, aiming to overcome historical bottlenecks in ALS therapeutic development (Quintana et al., 2023; Paganoni et al., 2020).

3.3.6 Drug repurposing

Drugs originally designed for other diseases are being investigated for ALS. For instance, masitinib is in advanced trials. It has demonstrated potential to slow ALS progression and improve functional outcomes (Duan et al., 2024).

3.3.7 Combination therapies

Combination therapies for ALS represent a promising multitarget approach to address the disease's complex pathophysiology, involving protein misfolding, neuroinflammation, and oxidative stress. These regimens typically pair neuroprotective agents (e.g., FDA-approved riluzole) with investigational drugs targeting complementary pathways, such as anti-inflammatories or mitochondrial stabilizers like CNM-Au8, to achieve synergistic effects. While no multidrug combinations are currently FDA-approved, platform trials like HEALEY are actively evaluating combination strategies in Phases II–III, utilizing biomarkers such as NfL to monitor treatment response. Although approved monotherapies (e.g., riluzole, edaravone) are sometimes studied in experimental combinations, these approaches still remain investigational, with researchers working to establish optimal drug pairings and demonstrate clinical benefits beyond single-agent therapies (Paganoni et al., 2020; Kiernan et al., 2021).

Current approved pharmacological treatments.

Therapeutic (references)	Chemical nature and history	Mechanism of action	Efficacy	Side effects/cautions
Riluzole (Miller et al., 2003)	Benzothiazole compound (a thiazole derivative), originally developed as an anticonvulsant	Reduces excitotoxicity by inhibiting glutamate release; may antagonize glutamate receptors and inactivate voltage-gated sodium channels	Slightly extends survival (by a few months); no significant impact on quality of life	Mild-to-moderate adverse events (e.g., nausea, fatigue); monitor liver function
Edaravone (Sawada, 2017)	Pyrazolone derivative, initially developed in Japan for stroke treatment	Scavenges free radicals and reduces oxidative stress	Slows functional decline, particularly in early-stage patients; minimal impact on survival	Mild-to-moderate adverse events (e.g., bruising, gait disturbance); contraindicated in severe renal impairment
Tofersen (Miller et al., 2022)	Antisense oligonucleotide (ASO), a synthetic nucleic acid targeting genetic mutations, developed from genetic therapy advances	Binds to SOD1 mRNA, reducing the production of toxic misfolded protein in SOD1-mutated ALS (2% of cases)	Slows disease progression in SOD1-mutated patients; limited to this subgroup, no broad survival benefit	Mild-to-moderate adverse events (e.g., headache, pain at the injection site); risk of serious neurological events (e.g., myelitis)

4 Gene therapy in ALS

4.1 Targets for gene-therapy-based approaches and their mechanisms of action in ALS

In the early 1990s, scientists discovered that the cause of familial ALS, at least in some cases, is mutations in the *superoxide dismutase 1* (*SOD1*) gene. Thereafter, they came up with the idea of treating ALS as a genetic disorder using gene therapy. The first gene therapy trial for ALS dates back to the late 1990s, when researchers used a combination of cell therapy and gene therapy to genetically manipulate stem cells to secrete neuroprotective factors such as brain-derived neurotrophic factor (BDNF) and ciliary neurotrophic factor (CNTF) and then placed these cells in polymer capsules and transferred them subcutaneously or into the cerebrospinal fluid to secrete factors that protect motor neurons. Of course, most of these studies were preclinical and conducted on animal models. However, most of these attempts were not successful and at best extended the survival of the mice by only 2–3 weeks (Aebischer and Kato, 1995). Although these efforts were promising as the first attempts, many technical challenges lay ahead, including the difficulty of delivering the genes effectively and precisely to target cells, ensuring long-term efficacy, and avoiding the induction of an immune response.

In the following years, researchers studied various gene therapy-based approaches to improve their efficacy and solve the challenges. Gene therapy for ALS is divided into three types based on the nature of the action: reducing the contents of a toxic protein via inhibition of its mutant coding gene expression, applying genomic modifications, or providing healthy copies of a gene that is malfunctioning. However, most studies focus on gene silencing of toxic mutant genes (Giovannelli et al., 2023). From another perspective, gene therapies being studied for ALS can be categorized into two groups: viral-based and nonviral (Amado and Davidson, 2021). In virus-based strategies, AAVs, and especially AAV9, are among the best and most popular choices for gene therapy applications in neurological diseases because they can cross the blood–brain barrier (BBB) and deliver engineered genetic material to target cells with high efficiency (Giovannelli et al., 2023). Another advantage of viral vectors is that after a single treatment, the desired gene is stably expressed in the transduced cells. However, viral vectors are not only used for gene transfer, and in some cases, they are also used for gene inhibition and gene knockout (Amado and Davidson, 2021).

In addition to providing healthy copies of defective genes or neuroprotective factors for motor neurons, which were discussed earlier, silencing and inhibiting harmful genes is another approach that is being explored in ALS gene therapy studies. ASOs, RNAi, or gene-editing technologies such as CRISPR are occasionally recruited for this purpose. These materials may be delivered to the target cells naked, via viral vectors, or in the form of nanoparticles (Cappella et al., 2019).

Several clinical trials are currently underway to evaluate the safety and efficacy of various gene therapy approaches for ALS. These trials are testing different delivery methods, target genes, and patient populations. While gene therapy holds great promise for the treatment of ALS, it is important to note that it is still a relatively new field. Further research is needed to optimize delivery methods, improve target cell specificity, and enhance long-term efficacy. In addition, careful monitoring of potential side effects and long-term outcomes will be essential to ensure the safety and effectiveness of these therapies.

4.2 Current role of gene therapy in the treatment of ALS

Currently, there is no official, approved gene therapy for ALS, and all gene therapies are in the preclinical or clinical trial stages. In the following sections, there are some of the most important clinical trials on gene therapies for ALS that have been conducted or are currently underway, and the results of each of them. In addition, some of the preclinical gene therapy studies on animal models of ALS are discussed.

5 Gene therapy clinical trials for ALS

Gene therapy represents a promising precision medicine approach for ALS, targeting the specific genetic drivers of the disease. The landscape is dynamic, marked by both significant breakthroughs and challenging setbacks. Therapies primarily utilize two modalities: ASOs, which degrade target mRNA, and AAV vectors, which deliver genetic material to cells to suppress harmful genes (Table 5.1).

5.1 Approved or discontinued clinical trials

These trials have provided critical insights, whether leading to a landmark approval or highlighting the complexities of specific targets.

5.1.1 BIIB067 (tofersen): Approved

BIIB067 (tofersen) is an ASO targeting SOD1 mRNA for the approximately 2% of ALS cases caused by SOD1 mutations. Administered intrathecally, it was evaluated in the Phase III VALOR study (NCT02623699) and its open-label extension. The trial achieved significant biomarker success, demonstrating reductions in cerebrospinal fluid SOD1 protein and plasma NfL, a marker of neuronal injury. Although the primary clinical endpoint (change in ALSFRS-R) in the initial placebo-controlled period was not statistically significant, long-term data showed that early initiation of tofersen led to a slower functional decline. The therapy was generally well-tolerated, with most adverse events being mild to moderate. Serious adverse events, including myelitis in approximately 2% of participants, were managed with treatment interruption and immunosuppressive therapy. Based on this evidence, tofersen received FDA approval in April 2023, establishing itself as the first approved gene-targeting therapy for SOD1-ALS and a foundation for precision neurology.

TABLE 5.1 A summary of important gene therapy clinical trials for ALS.

Agent	Targeted genes and vectors	Clinical trial phase and NCT number	Enrolled patients	Dosage regimen	Trial design	Findings	Final status
BIIB078 (NCT03626012)	Targets *c9orf72* sense RNA using an antisense oligonucleotide (ASO)	Phase I (NCT03626012)	106	Intrathecal; multiple ascending doses (10–120 mg) every 4 weeks	Randomized, placebo-controlled, dose-escalating	Mild-to-moderate adverse events (e.g., headache, procedural pain); no reduction in neurofilament light chain (NfL) or clinical benefits (ALSFRS-R, respiratory function)	Discontinued (2022) due to lack of efficacy
BIIB067 (NCT02623699) (tofersen)	Targets *sod1* mRNA using an ASO	Phase I/II (NCT02623699); Phase III (VALOR, NCT02623699); completed 2021	183 (Phase I/II: 50; Phase III: 133)	Intrathecal; 100 mg loading doses (3 doses, 2 weeks apart), then 100 mg every 4 weeks	Randomized, placebo-controlled (Phase III); open-label extension	Reduced CSF SOD1 protein and NfL levels; slowed ALSFRS-R decline in SOD1-ALS patients; mild-to-moderate adverse events (e.g., myelitis in 2%)	FDA-approved (2023) for SOD1-ALS
AAV-miR-SOD1 (NCT06100276) (AMT-162)	*sod1* mRNA delivered via AAV vector	Phase I/II (NCT03268603); ongoing	20 (estimated)	Intrathecal; single-dose AAV vector (dose not publicly specified)	Open-label, single-arm	Generally well-tolerated; transient muscle strength improvements in one participant; disease progression continued; managed with immunosuppression	Under investigation
Ulefnersen (NCT04768972) (ION363)	Targets FUS mRNA using an ASO	Phase III (FUSION, NCT04768972); ongoing	77 (estimated)	Intrathecal; multiple doses (specifics not fully disclosed)	Randomized, double-blind, placebo-controlled	Preliminary data suggest tolerability; no severe adverse events; ongoing evaluation of ALSFRS-R, ventilation-free survival	Under investigation
BIIB105 (NCT04494256) (ION541)	Targets *atxn2* mRNA using an ASO	Phase I/II (ALSpire, NCT04494256); completed 2024	99	Intrathecal; multiple ascending doses (10–100 mg) over 3–6 months	Randomized, placebo-controlled, dose-escalating; open-label extension	Significant reduction in CSF ATXN2 protein; no reduction in NfL or clinical benefits (ALSFRS-R, strength, breathing)	Discontinued (2024) due to a lack of efficacy
WVE-004 (NCT04931862)	Targets *c9orf72* RNA using an ASO	Phase I/II (FOCUS-C9, NCT04931862); terminated 2023	42 (estimated)	Intrathecal; multiple ascending doses (3–30 mg)	Randomized, double-blind, placebo-controlled	Reduced poly(GP) DPRs in CSF; no significant clinical benefits reported yet; generally well-tolerated	Discontinued (2023) due to a lack of efficacy

5.1.2 *BIIB078: Discontinued*

BIIB078 was an ASO targeting C9orf72 sense RNA, intended for the approximately 10% of ALS cases linked to C9orf72 hexanucleotide expansions. It was administered intrathecally in a Phase I trial (NCT03626012). The trial found the therapy to be generally safe, with mild-to-moderate adverse events such as headache and procedural pain (van den Berg et al., 2024). However, it showed no reduction in NfL levels or clinical benefit on the ALSFRS-R scale compared to placebo. Due to this lack of efficacy, development was discontinued in March 2022 (Biogen, 2022). This outcome underscores the significant challenge of targeting C9orf72-ALS, which involves complex gain-of-function and loss-of-function mechanisms.

5.1.3 *BIIB105/ION541: Discontinued*

BIIB105/ION541 was an ASO targeting ATXN2 mRNA, a genetic risk factor for ALS. It was evaluated in a Phase I/II study (NCT04494256) via intrathecal administration. While the therapy successfully reduced ATXN2 protein levels, it failed to demonstrate any clinical benefits on measures such as the ALSFRS-R or respiratory function. Despite an acceptable safety profile, development was discontinued in May 2024 due to the absence of clinical efficacy (Biogen, 2020).

5.1.4 *WVE-004: Discontinued for ALS*

WVE-004 was an ASO targeting C9orf72 RNA. In its Phase I/II trial (NCT04931862), which involved intrathecal administration, the therapy successfully reduced poly(GP) DPR proteins, a key toxic species in C9orf72-ALS. However, it failed to demonstrate significant clinical improvements in patients. Following these results, Wave Life Sciences halted its ALS program for WVE-004 in 2024, though development continues for Huntington's disease (Sciences and Multicenter, 2021).

5.2 Ongoing clinical trials

These investigational therapies are currently being evaluated for safety and efficacy.

5.2.1 *ION363 (Ulefnersen)*

ION363 (Ulefnersen) is an ASO targeting FUS mRNA, which is responsible for a rare and often aggressive form of ALS, accounting for about 1% of cases. The therapy is being evaluated in the ongoing Phase III FUSION trial (NCT04768972) with intrathecal administration. Preliminary data reported by the developer, Ionis Pharmaceuticals, suggest the therapy is well-tolerated, with no severe adverse events reported thus far. The trial is assessing efficacy through outcomes such as changes in the ALSFRS-R and ventilation assistance-free survival (Korobeynikov et al., 2022).

5.2.2 *APB-102 (AAV-miR-SOD1)*

APB-102 is an AAV vector designed to deliver a microRNA that targets and suppresses SOD1 mRNA. This therapy for SOD1-ALS is under investigation in an ongoing Phase I/II trial (NCT03268603) involving intrathecal administration. Preliminary data indicate the therapy is generally well-tolerated. One participant was reported to have transient improvements in muscle strength, though disease progression continued. Adverse events such as back pain have been managed with immunosuppression, with no major safety concerns raised to date (Mueller et al., 2020). The trial continues to monitor biomarkers and clinical outcomes.

5.2.3 *AMT-162*

AMT-162 is an AAV vector-based therapy targeting SOD1 mRNA, developed by uniQure. The Phase I/II trial (NCT05455736) is evaluating its safety and tolerability following a single intrathecal injection. The study design is open-label and will measure biomarker reduction of SOD1 and changes in ALSFRS-R scores (Biopharma, 2023).

5.2.4 *SNUG01*

SNUG01 is an investigational therapy with an undisclosed gene target, utilizing an ASO-based platform. A Phase I/II trial (NCT06645197) is underway, administering the therapy intrathecally. The study is designed as an open-label, single-arm trial, primarily assessing safety and tolerability, with secondary outcomes, including ALSFRS-R and NfL levels. No data has been made public yet (Dongsheng, 2024).

6 Current advances in preclinical gene therapy for ALS

Recent progress in gene therapy for ALS has shown remarkable promise in animal models, bringing renewed optimism for treating this fatal neurodegenerative disease. Researchers are employing viral vectors (AAV, lentivirus), CRISPR–Cas9 gene editing, and ASOs to silence toxic genes, deliver neuroprotective factors, and restore motor neuron function in SOD1, C9orf72, and FUS rodent models (Mueller et al., 2020; Jiang et al., 2016). The studies have demonstrated extended survival, delayed symptom onset, and improved motor function, highlighting the therapeutic potential of these approaches. This section examines the most compelling preclinical breakthroughs, discusses key challenges in translation, and explores how these findings could shape future human clinical trials for ALS.

6.1 Animal models used for preclinical studies

To investigate preclinical gene therapy approaches in ALS, a comprehensive understanding of established laboratory models is essential. Animal models are crucial for studying ALS, as they help researchers understand the disease's mechanisms and test potential treatments. For a model to be considered valid, it must be similar to the actual disease in humans in three aspects: face, structure, and predictive performance.

These models include a wide range of living organisms, from invertebrates such as worms and fruit flies to vertebrates such as zebrafish, mice, and large mammals such as dogs, pigs, and nonhuman primates (NHPs). In most cases, ALS models are created by producing animals with specific mutations in the *SOD1* gene (e.g., SOD1-G93A mouse). These models are primarily applied for therapeutic testing (e.g., ASOs, gene therapy), mechanistic studies (motor neuron degeneration, glial interactions), and biomarker discovery. Key advantages include a well-characterized, reproducible phenotype with predictable disease progression and translational relevance, as evidenced by their use in developing FDA-approved therapies like tofersen. However, limitations include low genetic relevance (only ~ 2% of ALS patients carry *sod1* mutations), artifacts from *sod1* overexpression, and a focus on motor neuron pathology rather than TDP-43 aggregation, a hallmark of most ALS cases. Despite these drawbacks, SOD1 models remain indispensable for preclinical ALS research due to their reliability and standardized protocols (Bonifacino et al., 2021).

Although in some cases, animals with mutations in other ALS-related genes, such as *c9orf72*, *FUS*, or *TDP-43*, are also used (Gordon, 2013). C9orf72 models, including BAC transgenic mice and knock-in/knockout strains, are critical for studying GGGGCC repeat-mediated toxicity, including RNA foci and DPR protein aggregates, as well as for testing ASO therapies targeting repeat expansions. These models are highly clinically relevant, as *c9orf72* mutations account for ~40% of familial ALS cases, and they recapitulate key disease features like *TDP-43* mislocalization and neuroinflammation (Balendra and Isaacs, 2018). However, challenges include inconsistent phenotypes (e.g., mild or late-onset symptoms in some strains), limited motor neuron degeneration in many models, and genetic complexity, making standardization difficult (Zhu et al., 2023). Despite these limitations, C9orf72 models remain essential for understanding ALS mechanisms linked to repeat expansions.

FUS-ALS mouse models, such as those expressing the ALS-associated FUSR521G or FUSP525L mutations, are critical for investigating pathologies related to RNA-binding protein dysfunction, including nuclear loss of function and cytoplasmic aggregation. These models have been instrumental in evaluating FUS-targeting therapies, such as the ASO ON363, which aims to reduce mutant FUS expression and mitigate toxicity. Their key advantage lies in directly modeling mutations responsible for a small but significant subset of aggressive, often early-onset familial ALS cases (Pelaez et al., 2023).

TDP-43 and FUS models, including TDP-43 transgenic mice/rats (overexpression or knock-in) and FUS mutant mice (e.g., FUSR521C), are pivotal for studying protein aggregation and RNA metabolism defects in ALS. These models are particularly valuable for testing TDP-43-targeting therapies, such as ASOs and immunotherapies. A key advantage is their high clinical relevance, as TDP-43 pathology is observed in > 97% of ALS cases, including sporadic forms, making them broadly applicable. However, limitations include reliance on nonphysiological overexpression in severe models, which may distort disease mechanisms, incomplete motor neuron loss in some strains, and short survival times in aggressive models, which restrict long-term therapeutic evaluation (Palomo et al., 2019; Nolan et al., 2016). Despite these challenges, TDP-43 and FUS models remain indispensable for advancing ALS research, particularly for sporadic disease mechanisms.

TABLE 5.2 ALS laboratory models and their applications in research.

Research goal	Best model	Why?	Application in drug/study
Therapeutic screening	SOD1G93A mice	Predictable, FDA-validated phenotype	Tofersen (BIIB067)
C9orf72 mechanisms	C9orf72 BAC mice	Recapitulates repeat toxicity	C9orf72 ASOs
TDP-43/ FUS pathology	TDP-43 knock-in/ FUS mutant mice	High relevance to sporadic ALS; protein aggregation	TDP-43 ASOs, ON363 (FUS)
ATXN2-targeting	ATXN2 knockout/repeat expansion mice	Models modifier of TDP-43 toxicity	BIIB105 (ASO)
High-throughput screening	Zebrafish/Drosophila	Fast, cost-effective	Early-stage drug discovery
Patient-specific studies	iPSC motor neurons	Human genetic context	Patient-derived mutation studies
Translational safety/ efficacy	Nonhuman primates (NHPs)	Clinically relevant anatomy and physiology; required for regulatory approval	SNUG01, KLTO-202

Models targeting genetic modifiers of ALS, such as ATXN2, are also gaining prominence. The polyglutamine (polyQ) expansion in the ATXN2 gene is a known risk factor and disease modifier, with intermediate-length repeats associated with increased ALS risk. Knockout or expanded repeat models of ATXN2 are used to evaluate therapies that aim to reduce ataxin-2 protein levels, as lowering ataxin-2 has been shown to mitigate TDP-43 toxicity and pathology. This approach has been instrumental in the preclinical development of ASOs like BIIB105, which directly targets ATXN2 RNA to reduce ataxin-2 expression, demonstrating the potential of targeting genetic modifiers for therapeutic benefit (Vieira de Sá et al., 2024).

Nonmammalian models, such as zebrafish and Drosophila, are also widely used for high-throughput drug screening and genetic modifier studies due to their low cost, rapid results, and ease of genetic manipulation (e.g., CRISPR, morpholinos). These models enable efficient large-scale testing of potential therapeutics and identification of disease-modifying genes. However, their simpler nervous systems limit translational relevance to human ALS, and they do not fully replicate the complex motor neuron degeneration seen in patients (Liguori et al., 2021; Patten et al., 2016). Despite these drawbacks, zebrafish and Drosophila remain valuable tools for early-stage ALS research, particularly for mechanistic insights and preliminary drug discovery.

Another useful model that is becoming increasingly popular is human-induced pluripotent stem cell (iPSC)-derived motor neurons taken from patients with hereditary ALS. Indeed, they are taken from individuals with mutations in ALS-related genes such as *SOD1*, *c9orf72*, and *TDP-43*. They enable patient-specific disease modeling and personalized drug screening, offering a unique platform with human genetic fidelity and applicability to sporadic ALS studies. A key advantage is the elimination of species-specific differences, improving translational relevance. However, these models often fail to fully replicate disease phenotypes, particularly neuromuscular junction degeneration, and exhibit high variability between cell lines, which can complicate data interpretation (Fujimori et al., 2018). Despite these limitations, iPSC-derived motor neurons provide valuable insights into ALS mechanisms and patient-specific therapeutic responses.

The transition from small to large animal models is a critical final step in the preclinical pipeline, with NHPs representing the gold standard for translational validation. Due to their close phylogenetic relationship to humans, NHPs possess a comparable neuroanatomical scale and complexity that directly impact therapy delivery and distribution. Studies in NHPs are therefore essential to bridge the gap between rodents and humans, providing definitive data on the safety, biodistribution, and optimal dosing of intrathecally delivered gene therapies such as SNUG01 and KLTO-202. This includes quantifying vector delivery throughout the human-sized spinal cord and brain, which is crucial for ensuring effective target engagement in clinical trials. Despite significant challenges related to cost, logistics, and ethical oversight, NHP studies generate a dataset that is often considered indispensable by regulatory agencies for approving first-in-human studies (Yang et al., 2021).

A comprehensive summary of the experimental models discussed in Section 6.1 is presented in Table 5.2.

6.2 Targeted genes and vectors

In fact, so far, very extensive efforts have been made in the field of gene therapy on laboratory models of ALS, and it is not possible to mention all of them in this chapter. However, some of the most important studies in this field will be briefly described below.

Neurotrophic factors (NTFs) have emerged as key candidates for gene therapy in ALS, with preclinical studies demonstrating their potential to modulate motor neuron survival. These factors are involved in a wide range of physiological processes of central nervous system (CNS) neurons, including differentiation, survival, and homeostasis. Numerous studies have demonstrated their protective role against neurodegeneration in in vitro and in vivo models. Among these NTFs, we can mention factors such as BDNF, CNTF, and pituitary adenylate cyclase-activating polypeptide (PACAP) (Volonté et al., 2020).

AAV-mediated delivery of BDNF has shown mixed results: while AAV9-BDNF initially enhanced neuronal survival in SOD1-G93A mice, chronic overexpression led to excitotoxicity via TrkB-mediated calcium dysregulation, suggesting the need for tightly regulated expression systems (Chang and Wang, 2021). Conversely, PACAP delivered via AAVrh10 in ALS models exhibited consistent neuroprotection by reducing glutamate excitotoxicity through PAC1 receptor activation, with intracerebroventricular administration prolonging survival (Cherait et al., 2023).

Another research demonstrated that CNS-targeted delivery of insulin-like growth factor-1 (IGF-1) via AAV vectors exerts potent neuroprotective effects in SOD1G93A ALS mouse models through dual mechanisms: direct motor neuron protection and modulation of pathogenic glial activity. Stereotaxic injection of AAV-IGF-1 into the deep cerebellar nuclei (DCN) of symptomatic mice enabled widespread distribution of IGF-1 throughout brainstem and spinal cord networks via axonal transport, resulting in significantly delayed disease progression (29% increase), extended lifespan (15–22 days), improved motor function, and reduced neuropathology. Complementary in vitro studies revealed IGF-1's capacity to attenuate neurotoxic glial responses, specifically suppressing TNF-α and nitric oxide release while activating prosurvival AKT pathways. These findings collectively establish that single-site CNS delivery of IGF-1 can simultaneously target both neuronal and nonneuronal disease mechanisms, with the DCN serving as an effective gateway for broad neurotrophic distribution. The results validate IGF-1's multimodal therapeutic potential in ALS and support the development of clinical translation strategies combining neuroprotection with glial modulation, while highlighting the utility of coculture systems for screening future neuroprotective therapies targeting the ALS microenvironment (Dodge et al., 2008).

Another approach being considered in gene therapy for ALS is the use of ASOs. In one important preclinical study, which later led to clinical trials, researchers used a 20-nucleotide ASO to target the SOD1 gene in a mouse model of ALS (*SOD1*^{G93A}). This process reduced the expression of SOD mRNA and protein, attenuated the development of ALS, and elongated the animal's survival (Smith et al., 2006). The clinical trial started later based on this study was named BIIB067/tofersen.

Hexanucleotide repeat expansions in *c9orf72*, the most common genetic cause of ALS and FTD, were shown to induce neurodegeneration through a gain-of-toxicity mechanism involving RNA foci and dipeptide-repeat protein (DPR) accumulation, rather than through loss of C9ORF72 function. Transgenic mouse studies demonstrated that while complete *c9orf72* knockout caused only mild immune abnormalities without motor deficits, expression of expanded GGGGCC repeats (up to 450 repeats) led to age- and repeat length-dependent formation of RNA foci, DPR proteins via RAN translation, hippocampal neuron loss, and cognitive/behavioral impairments. Critically, a single dose of ASOs targeting repeat-containing RNAs—while preserving normal *c9orf72* mRNA levels—produced sustained reductions in both RNA foci and DPR proteins (poly-GP/GA), along with significant behavioral improvements lasting up to 6 months posttreatment. These findings not only establish repeat RNA-mediated toxicity as central to C9ALS/FTD pathogenesis but also validate the therapeutic potential of selective ASO-mediated suppression of pathogenic repeat transcripts without disrupting normal C9ORF72 function, providing a promising targeted treatment strategy (Jiang et al., 2016). Later, a clinical trial was designed based on these studies, which was named BIIB078.

As mentioned previously, many other preclinical studies have been conducted on similar targets in ALS models, but with different approaches, such as using miRNA and siRNA. In these studies, as in the studies mentioned above, the approaches were either silencing mutated genes or transferring and amplifying neuroprotective genes (Amado and Davidson, 2021). In most of these studies, the vectors used for gene transfer were AAV9 and AAV10, as they have an inherent tendency to bind to motor neurons. However, lentiviruses have also been used successfully in some cases. In addition, ASOs were delivered via direct intrathecal injection (Amado and Davidson, 2021). A summary of Section 6.2 is visualized in Fig. 5.1.

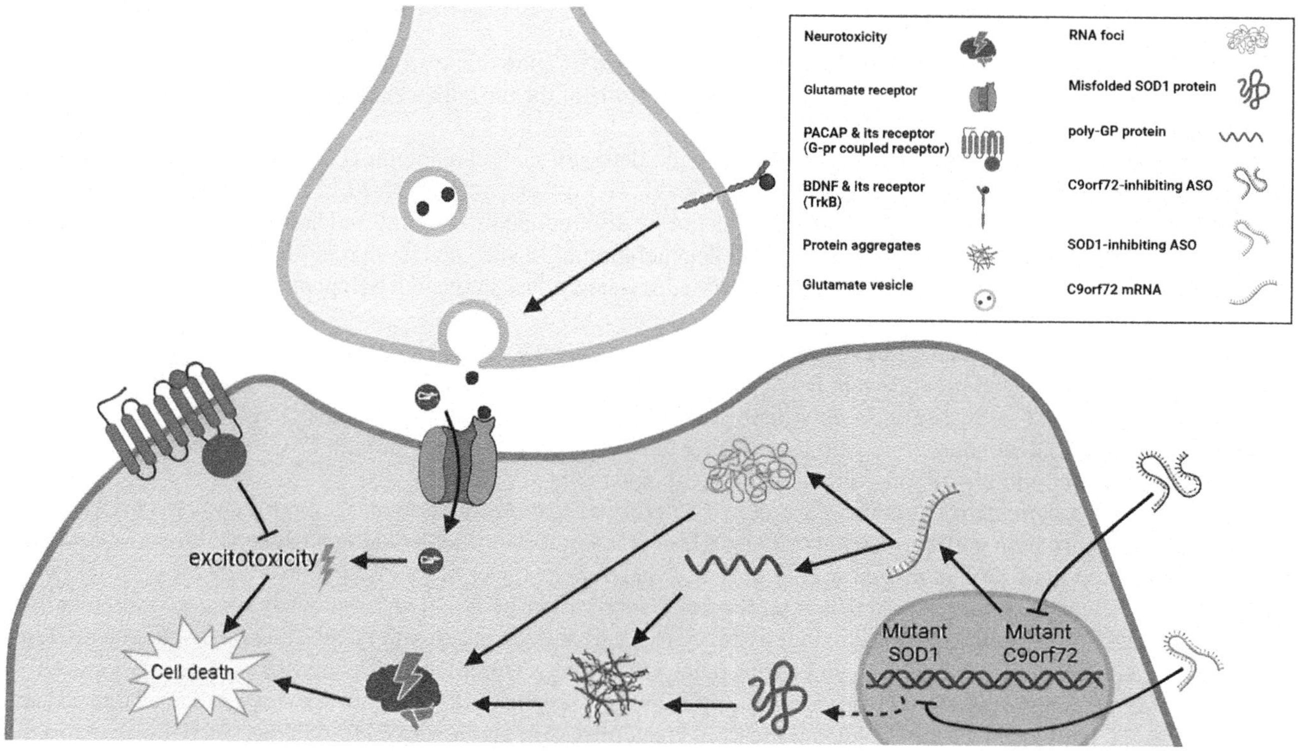

FIG. 5.1 A summary of protective and deleterious factors and pathways in motor neurons. The figure illustrates how various gene therapy approaches inhibit deleterious factors. Designed using the BioRender Application.

7 Current challenges and limitations in gene therapy

While gene therapy offers promising avenues for treating ALS, several challenges and limitations still impede its clinical application:

7.1 Delivery challenges

Efficient delivery of therapeutic agents through the BBB to target motor neurons in the CNS remains a significant hurdle. The BBB's selective permeability restricts the passage of many therapeutic molecules, complicating treatment strategies (Lejman et al., 2023).

7.2 Genetic heterogeneity

ALS presents with considerable genetic diversity, with numerous mutations implicated in its pathogenesis. This can complicate the development of gene therapy, necessitating personalized approaches tailored to specific genetic profiles (Meadows et al., 2024).

7.3 Safety concerns

Prospective adverse effects, for example, immune responses to viral vectors or off-target genetic modifications, carry major safety concerns. Guaranteeing the long-term safety of gene therapies is a principal concern, requiring extensive preclinical and clinical evaluations (García-González et al., 2024).

7.4 Ethical and regulatory hurdles

Gene therapy for ALS holds critical issues, with treatments like tofersen being costly and limited to those specific to one genetic mutation. However, ethical and regulatory challenges, specifically as ALS has an aggressive course and complex pathology, significantly impact treatment progress and drug delivery. These challenges complicate informed consent. Indeed, patients need long-term follow-up for gene therapy safety, but ALS's short survival time makes this challenging. These boundaries, distinct from general ethical aspects in gene therapy, directly affect patient access and trial outcomes (Lejman et al., 2023; Fang et al., 2022).

7.5 Limited understanding of disease mechanisms

The limited understanding of ALS pathophysiology affects the identification of effective therapeutic targets, hindering the development of comprehensive gene therapy strategies. Addressing these challenges requires continued research to develop delivery methods, enhance understanding of ALS genetics, ensure safety, and address ethical considerations, ultimately advancing gene therapy as a viable treatment for ALS (Amado and Davidson, 2021).

8 Conclusion

In summary, gene therapy has emerged as a transformative approach for treating ALS, offering targeted strategies to address the disease's complex genetic and molecular underpinnings. Gene therapy holds immense potential to revolutionize ALS treatment, but its success will depend on interdisciplinary collaboration, continued innovation in delivery technologies, and rigorous clinical validation. By addressing current limitations and building on preclinical successes, the field moves closer to delivering meaningful therapies for this devastating disease.

ALS pathogenesis involves diverse, intersecting pathways, including toxic protein aggregation (such as SOD1, C9ORF72, and TDP-43), glutamate-mediated excitotoxicity, and neuroinflammation. Gene therapy strategies have been designed to intervene at these mechanistic levels by silencing mutant genes, delivering neuroprotective factors (for example, IGF-1, BDNF, and CNTF), or modulating glial cell activity to preserve motor neuron function.

Several therapeutic platforms are particularly promising. ASOs have advanced into clinical testing, with trials such as BIIB067 (tofersen) for SOD1-associated ALS demonstrating the potential to reduce toxic protein levels and slow disease progression. Viral vector-based approaches, especially those using AAV serotypes like AAV9 and AAVrh10, enable efficient targeting of the CNS and have shown preclinical success in delivering NTFs (such as IGF-1 and PACAP) and mediating gene silencing. In parallel, gene-editing and gene-silencing technologies, including CRISPR and RNAi, are emerging as powerful tools in preclinical models for precise editing or knockdown of pathogenic mutations.

Clinical translation has, however, been complex. Trials such as BIIB078 (targeting C9ORF72) and APB-102 (AAV-miR-SOD1) have faced significant challenges, including limited efficacy and safety concerns. Nonetheless, these studies have yielded valuable lessons on how best to optimize delivery routes, such as intrathecal or intracerebroventricular administration, and refine dosing regimens to balance therapeutic benefit with acceptable risk.

Despite this progress, major challenges remain. Overcoming the BBB, addressing the underlying genetic and clinical heterogeneity of ALS, and ensuring the long-term safety of gene-based interventions are central hurdles. Moreover, ethical and regulatory considerations continue to shape the pace and design of translational efforts. Given the multifactorial nature of ALS, there is increasing recognition that effective treatment will likely require combination approaches that target multiple pathogenic pathways simultaneously.

Looking ahead, advances in biomarker development, such as the use of NfL as a disease activity and treatment-response marker, are expected to facilitate earlier diagnosis and more sensitive monitoring in clinical trials. Innovative trial designs, including adaptive platform trials like the HEALEY Platform Trial, aim to accelerate the evaluation of multiple candidate therapies. In addition, patient-specific models, particularly iPSC-derived motor neurons, will be crucial for dissecting disease mechanisms and tailoring therapies, ultimately helping to accelerate meaningful therapeutic breakthroughs in ALS gene therapy.

References

Aebischer, P., Kato, A.C., 1995. Treatment of amyotrophic lateral sclerosis using a gene therapy approach. Eur. Neurol. 1421-9913. 35 (2), 65–68. doi:10.1159/000117095.

Alsultan, et al., 2016. The genetics of amyotrophic lateral sclerosis: current insights. Degener. Neurol. Neuromuscul. Dis. 6, 49–64.

Amado, D.A., Davidson, B.L., 2021. Gene therapy for ALS: a review. Mol. Ther. 1525-0024. 29 (12), 3345–3358. Available from: https://www.journals.elsevier.com/molecular-therapy. doi:10.1016/j.ymthe.2021.04.008.

Balendra, R., Isaacs, A.M., 2018. C9orf72-mediated ALS and FTD: multiple pathways to disease. Nat. Rev. Neurol. 1759-4766. 14 (9), 544–558. Available from: http://www.nature.com/nrneurol/archive/index.html. doi:10.1038/s41582-018-0047-2.

Biogen, 2020. A Phase 1/2 Multiple-Ascending-Dose Study With a Long-Term Open-Label Extension to Assess the Safety, Tolerability, Pharmacokinetics, Pharmacodynamics, and Effect on Disease Progression of BIIB105 Administered Intrathecally to Adults With Amyotrophic Lateral Sclerosis With or Without Poly-CAG Expansion in the ATXN2 Gene.

Biogen. 2022. Available from: https://investors.biogen.com/news-releases/news-release-details/biogen-and-ionis-announce-topline-phase-1-study-results.

Biopharma, U., 2023. A Phase 1/2, Multicenter, Single Ascending Dose Study to Evaluate the Safety, Tolerability, and Exploratory Efficacy of Intrathecally Administered Gene Therapy AMT-162 in Adult Participants with SOD1 Amyotrophic Lateral Sclerosis (SOD1-ALS).

Bjelica, B., Petri, S., 2024. Narrative review of diagnosis, management and treatment of dysphagia and sialorrhea in amyotrophic lateral sclerosis. J. Neurol. 1432-1459. 271 (10), 6508–6513. Available from: https://www.springer.com/journal/415. doi:10.1007/s00415-024-12657-x.

Bonifacino, T., Zerbo, R.A., Balbi, M., Torazza, C., Frumento, G., Fedele, E., Bonanno, G., Milanese, M., 2021. Nearly 30 years of animal models to study amyotrophic lateral sclerosis: a historical overview and future perspectives. Int. J. Mol. Sci. 1422-0067. 22 (22), 12236. doi:10.3390/ijms222212236.

Bosco, D.A., Morfini, G., Karabacak, N.M., Song, Y., Gros-Louis, F., Pasinelli, P., Goolsby, H., Fontaine, B.A., Lemay, N., McKenna-Yasek, D., Frosch, M.P., Agar, J.N., Julien, J.P., Brady, S.T., Brown, R.H., 2010. Wild-type and mutant SOD1 share an aberrant conformation and a common pathogenic pathway in ALS. Nat. Neurosci. 1097-6256. 13 (11), 1396–1403. doi:10.1038/nn.2660.

Brown, R.H., Al-Chalabi, A., 2017. Amyotrophic lateral sclerosis. N. Engl. J. Med. 1533-4406. 377 (2), 162–172. Available from: http://www.nejm.org/doi/pdf/10.1056/NEJMra1603471. doi:10.1056/NEJMra1603471.

Cappella, M., Ciotti, C., Cohen-Tannoudji, M., Biferi, M.G., 2019. Gene therapy for ALS-A perspective. Int. J. Mol. Sci. 1422-0067. 20 (18). Available from: https://www.mdpi.com/1422-0067/20/18/4388/pdf. doi:10.3390/ijms20184388.

Chang, E., Wang, J., 2021. Brain-derived neurotrophic factor attenuates cognitive impairment and motor deficits in a mouse model of Parkinson's disease. Brain Behav. 2162-3279. 11 (8). Available from: http://onlinelibrary.wiley.com/journal/10.1002/(ISSN)2157-9032. doi:10.1002/brb3.2251.

Cherait, A., Banks, W.A., Vaudry, D., 2023. The potential of the nose-to-brain delivery of PACAP for the treatment of neuronal disease. Pharmaceutics. 1999-4923. 15 (8). Available from: http://www.mdpi.com/journal/pharmaceutics. doi:10.3390/pharmaceutics15082032.

Dodge, J.C., Haidet, A.M., Yang, W., Passini, M.A., Hester, M., Clarke, J., Roskelley, E.M., Treleaven, C.M., Rizo, L., Martin, H., Kim, S.H., Kaspar, R., Taksir, T.V., Griffiths, D.A., Cheng, S.H., Shihabuddin, L.S., Kaspar, B.K., 2008. Delivery of AAV-IGF-1 to the CNS extends survival in ALS mice through modification of aberrant glial cell activity. Mol. Ther. 1525-0024. 16 (6), 1056–1064. Available from: https://www.journals.elsevier.com/molecular-therapy. doi:10.1038/mt.2008.60.

Dongsheng, F., 2024. Available from: https://www.clinicaltrials.gov/study/NCT06645197.

Duan, Q.Q., Wang, H., Su, W.M., Gu, X.J., Shen, X.F., Jiang, Z., Ren, Y.L., Cao, B., Li, G.B., Wang, Y., Chen, Y.P., 2024. TBK1, a prioritized drug repurposing target for amyotrophic lateral sclerosis: evidence from druggable genome Mendelian randomization and pharmacological verification in vitro. BMC Med. 1741-7015. 22 (1). Available from: https://bmcmedicine.biomedcentral.com/. doi:10.1186/s12916-024-03314-1.

Engelberg-Cook, E., Shah, J.S., Hucke, A.T.D.S., Vera-Garcia, D.V., Dagher, J.E., Donahue, M.H., Belzil, V.V., Oskarsson, B., 2024. Prognostic factors and epidemiology of amyotrophic lateral sclerosis in Southeastern United States. Mayo Clinic Proceedings: Innovations, Quality & Outcomes. 2542-4548. 8 (5), 482–492. doi:10.1016/j.mayocpiqo.2024.07.008.

Fang, T., Je, G., Pacut, P., Keyhanian, K., Gao, J., Ghasemi, M., 2022. Gene therapy in amyotrophic lateral sclerosis. Cells. 2073-4409. 11 (13), 2066. doi:10.3390/cells11132066.

Fujimori, K., Ishikawa, M., Otomo, A., Atsuta, N., Nakamura, R., Akiyama, T., Hadano, S., Aoki, M., Saya, H., Sobue, G., Okano, H., 2018. Modeling sporadic ALS in iPSC-derived motor neurons identifies a potential therapeutic agent. Nat. Med. 1078-8956. 24 (10), 1579–1589. doi:10.1038/s41591-018-0140-5.

García-González, N., Gonçalves-Sánchez, J., Gómez-Nieto, R., Gonçalves-Estella, J.M., López, D.E., 2024. Advances and challenges in gene therapy for neurodegenerative diseases: a systematic review. Int. J. Mol. Sci. 1422-0067. 25 (23), 12485. doi:10.3390/ijms252312485.

Gendron, T.F., Chew, J., Stankowski, J.N., Hayes, L.R., Zhang, Y.J., Prudencio, M., Carlomagno, Y., Daughrity, L.M., Jansen-West, K., Perkerson, E.A., O'Raw, A., Cook, C., Pregent, L., Belzil, V., Van Blitterswijk, M., Tabassian, L.J., Lee, C.W., Yue, M., Tong, J., Song, Y., Castanedes-Casey, M., Rousseau, L., Phillips, V., Dickson, D.W., Rademakers, R., Fryer, J.D., Rush, B.K., Pedraza, O., Caputo, A.M., Desaro, P., Palmucci, C., Robertson, A., Heckman, M.G., Diehl, N.N., Wiggs, E., Tierney, M., Braun, L., Farren, J., Lacomis, D., Ladha, S., Fournier, C.N., McCluskey, L.F., Elman, L.B., Toledo, J.B., McBride, J.D., Tiloca, C., Morelli, C., Poletti, B., Solca, F., Prelle, A., Wuu, J., Jockel-Balsarotti, J., Rigo, F., Ambrose, C., Datta, A., Yang, W., Raitcheva, D., Antognetti, G., McCampbell, A., Van Swieten, J.C., Miller, B.L., Boxer, A.L., Brown, R.H., Bowser, R., Miller, T.M., Trojanowski, J.Q., Grossman, M., Berry, J.D., Hu, W.T., Ratti, A., Traynor, B.J., Disney, M.D., Benatar, M., Silani, V., Glass, J.D., Floeter, M.K., Rothstein, J.D., Boylan, K.B., Petrucelli, L., 2017. Poly(GP) proteins are a useful pharmacodynamic marker for C9ORF72-associated amyotrophic lateral sclerosis. Sci. Transl. Med. 1946-6242. 9 (383). Available from: http://stm.sciencemag.org/content/scitransmed/9/383/eaai7866.full.pdf. doi:10.1126/scitranslmed.aai7866.

Giovannelli, I., Higginbottom, A., Kirby, J., Azzouz, M., Shaw, P.J., 2023. Prospects for gene replacement therapies in amyotrophic lateral sclerosis. Nat. Rev. Neurol. 1759-4758. 19 (1), 39–52. doi:10.1038/s41582-022-00751-5.

Gordon, P.H., 2013. Amyotrophic lateral sclerosis: an update for 2013 clinical features, pathophysiology, management and therapeutic trials. Aging and Dis.. 2152-5250. 4 (5), 295–310. Available from: http://www.aginganddisease.org/AD-2013-Gordon.pdf. doi:10.14336/AD.2013.0400295.

Inchiosa, M.A., 2024. Beta2-adrenergic suppression of neuroinflammation in treatment of parkinsonism, with relevance for neurodegenerative and neoplastic disorders. Biomedicines. 2227-9059. 12 (8), 1720. doi:10.3390/biomedicines12081720.

Jiang, J., Zhu, Q., Gendron, T.F., Saberi, S., McAlonis-Downes, M., Seelman, A., Stauffer, J.E., Jafar-nejad, P., Drenner, K., Schulte, D., Chun, S., Sun, S., Ling, S.-C., Myers, B., Engelhardt, J., Katz, M., Baughn, M., Platoshyn, O., Marsala, M., Watt, A., Heyser, C.J., Ard, M.C., De Muynck, L., Daughrity, L.M., Swing, D.A., Tessarollo, L., Jung, C.J., Delpoux, A., Utzschneider, D.T., Hedrick, S.M., de Jong, P.J., Edbauer, D., Van Damme, P., Petrucelli, L., Shaw, C.E., Frank Bennett, C., Cruz, S.D., Ravits, J., Rigo, F., Cleveland, D.W., Lagier-Tourenne, C., 2016. Gain of toxicity from ALS/

FTD-linked repeat expansions in C9ORF72 is alleviated by antisense oligonucleotides targeting GGGGCC-containing RNAs. Neuron. 0896-6273. 90 (3), 535–550. doi:10.1016/j.neuron.2016.04.006.

Kiernan, M.C., Vucic, S., Talbot, K., McDermott, C.J., Hardiman, O., Shefner, J.M., Al-Chalabi, A., Huynh, W., Cudkowicz, M., Talman, P., Van den Berg, L.H., Dharmadasa, T., Wicks, P., Reilly, C., Turner, M.R., 2021. Improving clinical trial outcomes in amyotrophic lateral sclerosis. Nat. Rev. Neurol. 1759-4766. 17 (2), 104–118. Available from: http://www.nature.com/nrneurol/archive/index.html. doi:10.1038/s41582-020-00434-z.

Kim, J., Bireley, J.D., Hayat, G., Kafaie, J., 2020. Effects of noninvasive ventilation in amyotrophic lateral sclerosis: the complication of bulbar impairment. Clin. Transl. Neurosci. 2514-183X. 4 (1), 2514183X2091418. doi:10.1177/2514183x20914183.

Konopka, A., Atkin, J.D., 2022. DNA damage, defective DNA repair, and neurodegeneration in amyotrophic lateral sclerosis. Front. Aging Neurosci. 1663-4365. 14. Available from: https://www.frontiersin.org/journals/aging-neuroscience#. doi:10.3389/fnagi.2022.786420.

Korobeynikov, V.A., Lyashchenko, A.K., Blanco-Redondo, B., Jafar-Nejad, P., Shneider, N.A., 2022. Antisense oligonucleotide silencing of FUS expression as a therapeutic approach in amyotrophic lateral sclerosis. Nat. Med. 1546-170X. 28 (1), 104–116. Available from: http://www.nature.com/nm/index.html. doi:10.1038/s41591-021-01615-z.

Kwak, S., 2022. Pain in amyotrophic lateral sclerosis: a narrative review. J. Yeungnam Med. Sci. 2799-8010. 39 (3), 181–189. doi:10.12701/jyms.2022.00332.

Lejman, J., Panuciak, K., Nowicka, E., Mastalerczyk, A., Wojciechowska, K., Lejman, M., 2023. Gene therapy in ALS and SMA: advances, challenges and perspectives. Int. J. Mol. Sci. 1422-0067. 24 (2), 1130. doi:10.3390/ijms24021130.

Liguori, F., Amadio, S., Volonté, C., 2021. Fly for ALS: drosophila modeling on the route to amyotrophic lateral sclerosis modifiers. Cell. Mol. Life Sci. 1420-9071. 78 (17-18), 6143–6160. Available from: http://link.springer.de/link/service/journals/00018/index.htm. doi:10.1007/s00018-021-03905-8.

Maragakis, N.J., de Carvalho, M., Weiss, M.D., 2023. Therapeutic targeting of ALS pathways: refocusing an incomplete picture. Ann. Clin. Transl. Neurol. 2328-9503. 10 (11), 1948–1971. Available from: http://onlinelibrary.wiley.com/journal/10.1002/(ISSN)2328-9503. doi:10.1002/acn3.51887.

Masrori, P., Van Damme, P., 2020. Amyotrophic lateral sclerosis: a clinical review. Eur. J. Neurol. 1351-5101. 27 (10), 1918–1929. doi:10.1111/ene.14393.

Meadows, C., Rau, N.A., Faridi, W., Ly, C.V., 2024. Translating the ALS genetic revolution into therapies: a review. Curr. Treat. Options Neurol. 1092-8480. 26 (3), 35–49. doi:10.1007/s11940-024-00781-y.

Mehta, P., Raymond, J., Nair, T., Han, M., Berry, J., Punjani, R., Larson, T., Mohidul, S., Horton, D.K., 2025. Amyotrophic lateral sclerosis estimated prevalence cases from 2022 to 2030, data from the national ALS registry. Amyotroph. Lateral Scler. Frontotemporal Degener. 2167-9223. 26 (3-4), 290–295. Available from: http://www.tandfonline.com/loi/iafd20. doi:10.1080/21678421.2024.2447919.

Miller, R.G., Mitchell, J.D., Lyon, M., Moore, D.H., 2003. Riluzole for amyotrophic lateral sclerosis (ALS)/motor neuron disease (MND). Amyotroph. Lateral Scler. Other Motor Neuron Disord. 1466-0822. 4 (3), 191–206. doi:10.1080/14660820310002601.

Miller, T.M., Cudkowicz, M.E., Genge, A., Shaw, P.J., Sobue, G., Bucelli, R.C., Chiò, A., Van Damme, P., Ludolph, A.C., Glass, J.D., Andrews, J.A., Babu, S., Benatar, M., McDermott, C.J., Cochrane, T., Chary, S., Chew, S., Zhu, H., Wu, F., Nestorov, I., Graham, D., Sun, P., McNeill, M., Fanning, L., Ferguson, T.A., Fradette, S., 2022. Trial of antisense oligonucleotide tofersen for SOD1 ALS. N. Engl. J. Med. 1533-4406. 387 (12), 1099–1110. Available from: http://www.nejm.org/medical-index. doi:10.1056/NEJMoa2204705.

Mueller, C., Berry, J.D., McKenna-Yasek, D.M., Gernoux, G., Owegi, M.A., Pothier, L.M., Douthwright, C.L., Gelevski, D., Luppino, S.D., Blackwood, M., Wightman, N.S., Oakley, D.H., Frosch, M.P., Flotte, T.R., Cudkowicz, M.E., Brown, R.H., 2020. SOD1 suppression with adeno-associated virus and MicroRNA in familial ALS. N. Engl. J. Med. 0028-4793. 383 (2), 151–158. doi:10.1056/nejmoa2005056.

Nolan, M., Talbot, K., Ansorge, O., 2016. Pathogenesis of FUS-associated ALS and FTD: insights from rodent models. Acta Neuropathol. Commun. 2051-5960. 4 (1), 99. doi:10.1186/s40478-016-0358-8.

Paganoni, S., Macklin, E.A., Hendrix, S., Berry, J.D., Elliott, M.A., Maiser, S., Karam, C., Caress, J.B., Owegi, M.A., Quick, A., Wymer, J., Goutman, S.A., Heitzman, D., Heiman-Patterson, T., Jackson, C.E., Quinn, C., Rothstein, J.D., Kasarskis, E.J., Katz, J., Jenkins, L., Ladha, S., Miller, T.M., Scelsa, S.N., Vu, T.H., Fournier, C.N., Glass, J.D., Johnson, K.M., Swenson, A., Goyal, N.A., Pattee, G.L., Andres, P.L., Babu, S., Chase, M., Dagostino, D., Dickson, S.P., Ellison, N., Hall, M., Hendrix, K., Kittle, G., McGovern, M., Ostrow, J., Pothier, L., Randall, R., Shefner, J.M., Sherman, A.V., Tustison, E., Vigneswaran, P., Walker, J., Yu, H., Chan, J., Wittes, J., Cohen, J., Klee, J., Leslie, K., Tanzi, R.E., Gilbert, W., Yeramian, P.D., Schoenfeld, D., Cudkowicz, M.E., 2020. Trial of sodium phenylbutyrate–taurursodiol for amyotrophic lateral sclerosis. N. Engl. J. Med. 0028-4793. 383 (10), 919–930. doi:10.1056/nejmoa1916945.

Palomo, V., Tosat-Bitrian, C., Nozal, V., Nagaraj, S., Martin-Requero, A., Martinez, A., 2019. TDP-43: a key therapeutic target beyond amyotrophic lateral sclerosis. ACS Chem. Neurosci. 1948-7193. 10 (3), 1183–1196. doi:10.1021/acschemneuro.9b00026.

Patten, S.A., Parker, J.A., Wen, X.Y., Drapeau, P., 2016. Simple animal models for amyotrophic lateral sclerosis drug discovery. Expert Opin. Drug Discov. 1746-045X. 11 (8), 797–804. doi:10.1080/17460441.2016.1196183.

Pelaez, M.C., Desmeules, A., Gelon, P.A., Glasson, B., Marcadet, L., Rodgers, A., Phaneuf, D., Pozzi, S., Dutchak, P.A., Julien, J.P., Sephton, C.F., 2023. Neuronal dysfunction caused by FUSR521G promotes ALS-associated phenotypes that are attenuated by NF-κB inhibition. Acta Neuropathol. Commun. 2051-5960. 11 (1). Available from: https://actaneurocomms.biomedcentral.com. doi:10.1186/s40478-023-01671-1.

Peters, O.M., Ghasemi, M., Brown, R.H., 2015. Emerging mechanisms of molecular pathology in ALS. J. Clin. Invest. 1558-8238. 125 (5), 1767–1779. Available from: http://www.jci.org/articles/view/71601/pdf. doi:10.1172/JCI71601.

Quintana, M., Saville, B.R., Vestrucci, M., Detry, M.A., Chibnik, L., Shefner, J., Berry, J.D., Chase, M., Andrews, J., Sherman, A.V., Yu, H., Drake, K., Cudkowicz, M., Paganoni, S., Macklin, E.A., 2023. Design and statistical innovations in a platform trial for amyotrophic lateral sclerosis. Ann. Neurol. 0364-5134. 94 (3), 547–560. doi:10.1002/ana.26714.

Renton, A.E., Chiò, A., Traynor, B.J., 2014. State of play in amyotrophic lateral sclerosis genetics. Nat. Neurosci. 1546-1726. 17 (1), 17–23. doi:10.1038/nn.3584.

Richardson, P.J., Smith, D.P., de Giorgio, A., Snetkov, X., Almond-Thynne, J., Cronin, S., Mead, R.J., McDermott, C.J., Shaw, P.J., 2023. Janus kinase inhibitors are potential therapeutics for amyotrophic lateral sclerosis. Transl. Neurodegener. 2047-9158. 12 (1). Available from: https://translational-neurodegeneration.biomedcentral.com/. doi:10.1186/s40035-023-00380-y.

Sawada, H., 2017. Clinical efficacy of edaravone for the treatment of amyotrophic lateral sclerosis. Expert Opin. Pharmacother. 1465-6566. 18 (7), 735–738. doi:10.1080/14656566.2017.1319937.

Sciences, W.L., Multicenter, A., 2021. Randomized, Double-blind, Placebo-controlled, Phase 1b/2a Study of WVE-004 Administered Intrathecally to Patients With C9orf72-associated Amyotrophic Lateral Sclerosis (ALS) or Frontotemporal Dementia (FTD).

Sironi, F., De Marchi, F., Mazzini, L., Bendotti, C., 2023. Cell therapy in ALS: an update on preclinical and clinical studies. Brain Res. Bull. 0361-9230. 194, 64–81. doi:10.1016/j.brainresbull.2023.01.008.

Smith, R.A., Miller, T.M., Yamanaka, K., Monia, B.P., Condon, T.P., Hung, G., Lobsiger, C.S., Ward, C.M., McAlonis-Downes, M., Wei, H., Wancewicz, E.V., Bennett, C.F., Cleveland, D.W., 2006. Antisense oligonucleotide therapy for neurodegenerative disease. J. Clin. Invest. 1558-8238. 116(8), 2290–2296. Available from: http://www.jci.org/cgi/reprint/116/8/2290. doi: 10.1172/JCI25424United States . :.

Suzuki, N., Nishiyama, A., Warita, H., Aoki, M., 2023. Genetics of amyotrophic lateral sclerosis: seeking therapeutic targets in the era of gene therapy. J. Hum. Genet. 1434-5161. 68 (3), 131–152. doi:10.1038/s10038-022-01055-8.

Taylor, J.P., Brown, R.H., Cleveland, D.W., 2016. Decoding ALS: from genes to mechanism. Nature. 1476-4687. 539 (7628), 197–206. Available from: http://www.nature.com/nature/index.html. doi:10.1038/nature20413.

Trojsi, F., Di Nardo, F., D'Alvano, G., Caiazzo, G., Passaniti, C., Mangione, A., Sharbafshaaer, M., Russo, A., Silvestro, M., Siciliano, M., Cirillo, M., Tedeschi, G., Esposito, F., 2023. Resting state fMRI analysis of pseudobulbar affect in amyotrophic lateral sclerosis (ALS): motor dysfunction of emotional expression. Brain Imaging Behav. 1931-7557. 17 (1), 77–89. doi:10.1007/s11682-022-00744-4.

van den Berg, L.H., Rothstein, J.D., Shaw, P.J., Babu, S., Benatar, M., Bucelli, R.C., Genge, A., Glass, J.D., Hardiman, O., Libri, V., Mobach, T., Oskarsson, B., Pattee, G.L., Ravits, J., Shaw, C.E., Weber, M., Zinman, L., Jafar-nejad, P., Rigo, F., Lin, L., Ferguson, T.A., Gotter, A.L., Graham, D., Monine, M., Inra, J., Sinks, S., Eraly, S., Garafalo, S., Fradette, S., 2024. Safety, tolerability, and pharmacokinetics of antisense oligonucleotide BIIB078 in adults with C9orf72-associated amyotrophic lateral sclerosis: a phase 1, randomised, double blinded, placebo-controlled, multiple ascending dose study. Lancet Neurol. 1474-4465. 23 (9), 901–912. Available from: https://www.sciencedirect.com/science/journal/14744422. doi:10.1016/S1474-4422(24)00216-3.

van Es, M.A., Hardiman, O., Chio, A., Al-Chalabi, A., Pasterkamp, R.J., Veldink, J.H., van den Berg, L.H., 2017. Amyotrophic lateral sclerosis. The Lancet. 1474-547X. 390 (10107), 2084–2098. Available from: http://www.journals.elsevier.com/the-lancet/. doi:10.1016/S0140-6736(17)31287-4.

Vieira de Sá, R., Sudria-Lopez, E., Cañizares Luna, M., Harschnitz, O., van den Heuvel, D.M.A., Kling, S., Vonk, D., Westeneng, H.-J., Karst, H., Bloemenkamp, L., Varderidou-Minasian, S., Schlegel, D.K., Mars, M., Broekhoven, M.H., van Kronenburg, N.C.H., Adolfs, Y., Vangoor, V.R., de Jongh, R., Ljubikj, T., Peeters, L., Seeler, S., Mocholi, E., Basak, O., Gordon, D., Giuliani, F., Verhoeff, T., Korsten, G., Pla, T.C., Venø, M.T., Kjems, J., Talbot, K., Michael A., V.E., Veldink, J.H., van den Berg, L.H., Zelina, P., Pasterkamp, R.J., 2024. ATAXIN-2 intermediate-length polyglutamine expansions elicit ALS-associated metabolic and immune phenotypes. Nat. Commun. 2041-1723. 15 (1). doi:10.1038/s41467-024-51676-0.

Volonté, C., Morello, G., Spampinato, A.G., Amadio, S., Apolloni, S., D'Agata, V., Cavallaro, S., 2020. Omics-based exploration and functional validation of neurotrophic factors and histamine as therapeutic targets in ALS. Ageing Res. Rev. 1568-1637. 62, 101121. doi:10.1016/j.arr.2020.101121.

Wingo, T.S., Cutler, D.J., Yarab, N., Kelly, C.M., Glass, J.D., Cookson, M.R., 2011. The heritability of amyotrophic lateral sclerosis in a clinically ascertained United States research registry. PLoS ONE. 1932-6203. 6 (11), e27985. doi:10.1371/journal.pone.0027985.

Yang, W., Chen, X., Li, S., Li, X.-J., 2021. Genetically modified large animal models for investigating neurodegenerative diseases. Cell Biosci. 2045-3701. 11 (1). doi:10.1186/s13578-021-00729-8.

Zhu, L., Li, S., Li, X.-J., Yin, P., 2023. Pathological insights from amyotrophic lateral sclerosis animal models: comparisons, limitations, and challenges. Transl. Neurodegener. 2047-9158. 12 (1). doi:10.1186/s40035-023-00377-7.

Chapter 6

Gene therapy in Duchenne muscular dystrophy (DMD)

Arezou Azizsoltani[a,b], Zeinab Ghesmati[b], and Elnaz Rezaei[c]
[a]Department of Medical Biotechnology and Molecular Medicine, School of Medicine, Arak University of Medical Sciences, Arak, Iran
[b]Department of Medical Biotechnology, Faculty of Advanced Medical Sciences, Tabriz University of Medical Sciences, Tabriz, Iran
[c]Department of Agricultural Biotechnology, Faculty of Agriculture and Natural Resources, Imam Khomeini International University (IKIU), Qazvin, Iran

1 Introduction

Duchenne muscular dystrophy (DMD) is an X-linked recessive neuromuscular disorder and constitutes both the most prevalent and the most severe pediatric form of muscular dystrophy. The condition predominantly affects males and typically manifests in early childhood. DMD arises from pathogenic variants in the DMD gene located on the X chromosome, which encodes dystrophin, a critical cytoskeletal protein required for stabilizing muscle fibers and preserving sarcolemmal integrity during cycles of contraction and relaxation (Salari et al., 2022; Mah et al., 2011).

Although DMD is primarily classified as a progressive neuromuscular disorder, it is also associated with important neurodevelopmental and neurobehavioral manifestation due to its impacts on central nervous system (CNS) development and function. CNS involvement in DMD is well-documented, with a substantial proportion of patients exhibiting cognitive impairment, learning disabilities, autism spectrum features, and neurobehavioral disorders linked to the loss of brain-expressed dystrophin isoforms including Dp140 and Dp71 (Bresolin et al., 1994; Vaillend et al., 2025; Darmahkasih et al., 2020). From a translational perspective, DMD has become a paradigmatic disease for addressing gene therapy platform challenges that are shared across neurological indications, including systemic adeno-associated virus (AAV) delivery, dose-limiting immune responses, biodistribution constraints, and long-term safety monitoring (Zanoteli et al., 2024; High and Roncarolo, 2019). Consequently, insights gained from preclinical and clinical gene therapy development in DMD are directly informative for therapeutic strategies targeting neurodegenerative disorders, particularly those requiring widespread CNS or neuromuscular gene delivery.

The global birth prevalence is estimated to range from approximately 1 in 3500–6000 live male births (Crisafulli et al., 2020; Kariyawasam et al., 2022; Emery, 1991; Stark, 2015). Beyond its public health implications, understanding the epidemiology and genetic transmission patterns of DMD is essential for gene therapy research, as mutation prevalence, carrier frequency, and genotype distribution directly influence patient stratification, trial design, and the long-term feasibility of curative genetic interventions. According to epidemiological analyses, the global prevalence of DMD has been estimated at approximately 4.8 per 100,000 individuals (95% CI: 3.6–6.3 per 100,000 people) (Salari et al., 2022; Crisafulli et al., 2020). The relatively high global prevalence of DMD underscores the urgent need to strengthen prenatal screening programs and provide comprehensive genetic counseling for families with a positive history (Crisafulli et al., 2020). In genetic counseling prior to pregnancy, recurrence-risk assessment primarily requires consideration of maternal carrier status, the familial pathogenic variant, fetal sex, family history, and the possibility of germline mosaicism (Manzur et al., 2008). Although global analyses of DMD prevalence are available, substantial between study heterogeneity limits the precision and comparability of regional prevalence estimates (Crisafulli et al., 2020; Salari et al., 2022).

There are no significant ethnic or geographic differences in the incidence of DMD, as the disorder arises from spontaneous mutations in the *DMD* gene that occur at a relatively constant rate worldwide. Reported variations in incidence primarily reflect disparities in diagnostic availability, access to genetic testing, and the quality of healthcare reporting systems. In countries with advanced genetic screening and early diagnostic programs, DMD is typically detected more frequently

Gene Therapy in Neurodegenerative Disorders. https://doi.org/10.1016/B978-0-443-40467-2.00006-3

and at earlier stages compared to regions with limited healthcare infrastructure. Therefore patients with DMD in developing countries with lower socioeconomic status often face significant challenges in accessing subspecialty care and genetic testing (Counterman et al., 2020).

The high frequency of de novo mutations in the *DMD* gene continues to pose a major obstacle to the complete eradication of the disease (Vaillend et al., 2025). While *DMD* gene mutations mainly occur in males, reports indicate that approximately 2.5%–7.8% of female carriers can present with symptoms, thus being identified as symptomatic carriers (Lee et al., 2015).

2 Common clinical signs and symptoms

The characteristic early signs and symptoms indicative of DMD that prompt clinical evaluation typically appear in most cases at about 2–5 years of age, including delayed motor milestones, difficulty climbing stairs, toe walking, frequent falls, clumsiness, and the presence of Gowers' sign (difficulty rising from the floor) (Birnkrant et al., 2018; Messina and Vita, 2018; Lee et al., 2022). Affected boys often experience difficulty running and climbing stairs, frequent falls, and generally lose the ability to walk independently by 10–13 years (Boland et al., 1996; Brooke et al., 1989; McDonald et al., 1995; Davidson et al., 2025). In DMD, rapidly progressive muscle weakness typically results in wheelchair dependence by around 12 years of age and death before the third decade of life. The most common causes of death are respiratory complications and heart failure from progressive cardiomyopathy (Salari et al., 2022; Birnkrant et al., 2018). Patients with DMD can be categorized into five stages based on their clinical presentation (Birnkrant et al., 2018; Suthar and Sankhyan, 2018; Nascimento Osorio et al., 2019). The main clinical characteristics identified in each stage are summarized in Table 6.1.

Currently, no definitive cure is available for DMD, and the overall prognosis remains poor (Venugopal and Pavlakis, 2023). The disorder significantly impairs both the quantity and quality of life and imposes considerable healthcare and socioeconomic burdens on affected families and society; therefore preventive strategies are of paramount importance (Manzur et al., 2008; Theadom et al., 2015).

2.1 Prognosis of DMD

The prognosis of DMD reflects the relentless progression of skeletal, respiratory, and cardiac muscle degeneration resulting from dystrophin deficiency. Although contemporary multidisciplinary care has substantially improved survival, the overall outlook remains poor. The disease trajectory is characterized by early motor decline, progressive respiratory insufficiency, and cardiomyopathy, each contributing significantly to morbidity and mortality (Venugopal and Pavlakis, 2023).

TABLE 6.1 The main clinical signs and symptoms in different stages of DMD.

Clinical stage	Age	Clinical signs and symptoms
Presymptomatic	Birth–2 years	Possible delay in gait development
Early ambulatory	2–6 years	• Lower limb weakness • Abnormal gait (Trendelenburg gait) • Gowers' maneuver • Toe walking • Difficulty in climbing stairs • Inability to jump • Learning difficulties • Behavioral problems
Late ambulatory	6–10 years	• Increasing difficulty in walking • Inability to climb stairs • Inability to get off the floor • Early signs of scoliosis
Early nonambulatory	11–15 years	• Loss of independent ambulation • Able to stand with support • Onset and progression of scoliosis
Late nonambulatory	>15 years	• Progressive weakness of the upper limbs • Inability to remain standing • Respiratory complications • Cardiac complications

Loss of independent ambulation (loss of ambulation, LOA) remains one of the most robust clinical milestones in DMD prognosis. Most patients lose the ability to walk independently between 10 and 13 years of age, and this transition strongly correlates with subsequent declines in respiratory and cardiovascular function (Ferizovic et al., 2022; Weber et al., 2022). Following LOA, forced vital capacity (FVC) typically decreases at a rate of 5%–10% annually, predisposing patients to nocturnal hypoventilation, recurrent respiratory infections, and increased reliance on noninvasive ventilation during late adolescence and early adulthood (Sheehan et al., 2018).

Cardiac involvement is a major predictor of survival in DMD. Dystrophin deficiency causes progressive dilated cardiomyopathy, and cardiac MRI markers reliably forecast heart failure progression and cardiac-related mortality (Magrath et al., 2018). Without cardiac-specific therapy, heart failure becomes a leading cause of death alongside respiratory failure. With optimized respiratory care, corticosteroid therapy, angiotensin-converting enzyme (ACE) inhibitors, and β-blockers, life expectancy has risen from the late teens to the late 20s–30s, with some patients in specialized care centers now surviving into their fourth decade (Eagle et al., 2002).

Underweight and poor pulmonary function have been associated with increased mortality in DMD (Cheeran et al., 2017; Phillips et al., 2001); however, additional prognostic factors may help predict disease progression, optimize treatment, and enhance both quality of life and survival. Prognostic factors in DMD provide valuable insights into disease progression and can aid in the personalization of patient management.

Key prognostic factors that guide treatment decisions include younger age at symptom onset, lower baseline motor scores, earlier decline in pulmonary function, more severe cardiac dysfunction, and low body mass index or malnutrition. These indicators help clinicians determine the optimal timing and strategy for therapeutic interventions. Among motor assessments, longitudinal trajectories of the North Star Ambulatory Assessment (NSAA) and 6-minute walk distance (6MWD) are among the strongest predictors of time to LOA and subsequent functional decline (Ferizovic et al., 2022; Weber et al., 2022; Hibma et al., 2023).

Biomarkers have emerged as a crucial component of modern prognostic modeling. As reviewed by Fortunato and colleagues, serum biomarkers (e.g., myomiRs, titin-fragments), inflammatory mediators, and proteomic signatures correlate with disease severity and may predict the rate of muscle degeneration (Fortunato and Ferlini, 2023). Quantitative muscle MRI—particularly fat fraction measurements using Dixon techniques or T2 mapping—outperforms many clinical measures in predicting disease progression and future LOA, offering an objective and highly reproducible prognostic tool (Naarding et al., 2020). Similarly, cardiac MRI fibrosis indices predict the timing and severity of future cardiomyopathy, underscoring their relevance in prognostic stratification (Magrath et al., 2018).

Genetic modifiers represent an emerging layer of prognostic understanding. Despite identical DMD mutations, patients may display markedly different disease trajectories, suggesting a modulatory role for non-DMD genes. Variants in ***SPP1 (osteopontin)***, latent transforming growth factor-β binding protein **(LTBP4)**, and α-actinin-3 **(*ACTN3*)** have been shown to influence inflammation, fibrosis, regenerative capacity, and overall disease severity (Bello et al., 2023). According to Bello et al., the integration of modifier genotypes with DMD mutation type may soon allow for precision-based prognostic modeling, although routine clinical implementation requires further validation (Bello et al., 2023).

Recent systematic reviews and metaanalyses show that accurate prognostication in DMD cannot rely on a single biomarker or clinical measure. Instead, the most robust predictive models integrate multiple domains, including genotype (both mutation type and modifier genes), longitudinal motor assessments such as NSAA and 6MWD, respiratory decline measured by FVC trajectories, quantitative muscle and cardiac MRI, serum molecular biomarkers, and treatment exposure to steroids, cardiac medications, and ventilator support. These multimodal frameworks are essential for personalized patient management and are increasingly employed to stratify participants in clinical trials, thereby enhancing the sensitivity and interpretability of therapeutic outcomes (Ferizovic et al., 2022; Weber et al., 2022; Sheehan et al., 2018).

Overall, while DMD remains a life-limiting disorder, advances in prognostic markers, genetic characterization, imaging techniques, and supportive care have markedly improved the ability to anticipate disease trajectory and guide personalized interventions aimed at maximizing both survival and quality of life.

3 Molecular mechanisms and pathogenesis

DMD is a severe, progressive neuromuscular disorder characterized by the degeneration of skeletal and cardiac muscle, caused by the absence or profound deficiency of dystrophin, a large cytoskeletal protein encoded by the DMD gene on the X chromosome (Hoffman et al., 1987). Dystrophin plays a central role in maintaining sarcolemmal integrity by linking the intracellular actin cytoskeleton to the extracellular matrix through the dystrophin-associated glycoprotein complex (DGC) (Birnkrant et al., 2018). The loss of dystrophin destabilizes the sarcolemma, resulting in membrane fragility, abnormal calcium influx, mitochondrial dysfunction, oxidative stress, and chronic inflammation. These pathological processes establish a self-amplifying cycle of myofiber necrosis and impaired regeneration, ultimately leading to progressive muscle wasting and fibrosis (Bonilla et al., 1988; Duan et al., 2021; Allen et al., 2015). Beyond its mechanical function, dystrophin also serves as a critical signaling scaffold that organizes key molecules involved in calcium homeostasis, redox balance, and local blood flow regulation, highlighting its multifaceted role in muscle physiology and disease pathogenesis (Allen et al., 2015; Le et al., 2018; Amann et al., 1998).

3.1 Membrane fragility and calcium influx: A central pathogenic mechanism

One of the earliest and most critical events in DMD pathogenesis is sarcolemmal damage induced by mechanical stress during muscle contraction. In dystrophin-deficient myofibers, the loss of dystrophin-mediated mechanical reinforcement renders the sarcolemma highly susceptible to repeated microinjuries. These disruptions allow uncontrolled Ca^{2+} entry through multiple pathways, including stretch-activated channels, store-operated calcium channels, transient receptor potential channels (TRPC1/3/6), and nonspecific membrane leak pathways (Gailly et al., 1993; Vandebrouck et al., 2002; Ducret et al., 2006). The resulting elevation in cytosolic calcium concentration ($[Ca^{2+}]_i$) activates Ca^{2+}-dependent proteases such as calpains, as well as phospholipases, leading to progressive proteolysis of cytoskeletal and membrane-associated proteins (Gailly et al., 2007). Excess cytosolic Ca^{2+} causes mitochondrial Ca^{2+} overload and disrupts SR Ca^{2+} handling. Mitochondrial Ca^{2+} overload compromises organelle integrity, promotes permeability transition and initiates apoptotic or necrotic cell death pathways, thereby amplifying muscle fiber degeneration (Mareedu et al., 2021).

3.2 Mitochondrial dysfunction and oxidative stress

Mitochondria play a central role in energy production and intracellular Ca^{2+} buffering. In DMD, sustained Ca^{2+} influx into mitochondria via the mitochondrial Ca^{2+} uniporter results in mitochondrial Ca^{2+} overload. This condition promotes the opening of the mitochondrial permeability transition pore, loss of mitochondrial membrane potential, and excessive generation of reactive oxygen species (ROS) (Mareedu et al., 2021; Zabłocka et al., 2021). Elevated ROS levels oxidize proteins, lipids, and nucleic acids, thereby exacerbating cellular injury. Importantly, ROS further potentiate Ca^{2+} dysregulation by enhancing the activity of Ca^{2+}-permeable channels, creating a deleterious positive feedback loop between oxidative stress and calcium overload (Escobar-Huertas et al., 2024). Concurrently, ATP depletion resulting from mitochondrial dysfunction impairs Ca^{2+} reuptake into the SR via SERCA pumps, accelerating sarcolemmal instability and cellular collapse. Under physiological conditions, dystrophin anchors neuronal nitric oxide synthase (nNOS) to the sarcolemma through interactions with syntrophins, enabling nitric oxide (NO)-mediated regulation of vasodilation and redox signaling. In DMD, mislocalization of nNOS reduces NO bioavailability, leading to functional ischemia, impaired blood flow during muscle activity, and further accumulation of ROS (Allen et al., 2015; Escobar-Huertas et al., 2024).

3.3 Reactive oxygen species and Ca^{2+} crosstalk

A hallmark of DMD pathology is the pathological interplay between Ca^{2+} dysregulation and oxidative stress, mediated largely through ROS signaling. In dystrophin-deficient muscle, ROS are generated by both the mitochondrial electron transport chain and NADPH oxidases (NOX). Among these, NOX2 is predominantly upregulated in skeletal muscle, whereas NOX4 expression is increased in cardiac tissue.

Enhanced NOX2-derived ROS promotes sarcolemmal Ca^{2+} influx, further elevating intracellular Ca^{2+} levels and accelerating myofiber damage. In parallel, NOX4-generated ROS induces ryanodine receptor 1 (RyR1)-mediated Ca^{2+} leak from the SR, particularly at SR–T-tubule junctions. This localized Ca^{2+} release stimulates additional NOX2 activation, establishing a self-amplifying feedback loop that intensifies oxidative stress, Ca^{2+} overload, and structural membrane damage. At the molecular level, NOX-dependent ROS signaling alters ion channel function, destabilizes Ca^{2+} homeostasis, and exacerbates membrane fragility, thereby promoting myofiber necrosis. Collectively, ROS–Ca^{2+} signaling crosstalk driven by NOX enzymes represents a central mechanism linking dystrophin loss to progressive muscle degeneration in DMD (Allen et al., 2015; Cully and Rodney, 2020).

3.4 Cellular degeneration, necrosis, and regeneration failure

The combined effects of sarcolemmal instability, sustained Ca^{2+} overload, mitochondrial dysfunction, and oxidative stress culminate in widespread myofiber necrosis. Necrotic fibers release intracellular damage-associated molecular patterns that activate inflammatory signaling pathways, including NF-κB, TGF-β, TNF-α, and IL-6 cascades. This inflammatory milieu promotes infiltration of immune cells, particularly macrophages, which initially support tissue repair but ultimately contribute to chronic inflammation and tissue damage. Repeated cycles of degeneration and incomplete regeneration progressively exhaust the muscle stem cell pool, leading to impaired regenerative capacity. Over time, functional muscle tissue is replaced by fibrotic and adipose tissue, a defining pathological feature of advanced-stage DMD (Allen et al., 2015). This irreversible remodeling underlies the progressive loss of muscle function and highlights the interconnected nature of molecular, metabolic, and inflammatory processes driving disease progression.

4 Genetic pattern and associated genes

DMD is an X-linked neuromuscular genetic disorder caused by mutations in the DMD gene, located on Xp21. The DMD gene spans approximately 2.4 megabases (Mb) of genomic DNA and was first identified in 1986 (Kunkel, 2005; Monaco et al., 1986). It is one of the largest genes in the human genome, comprising 79 exons and accounting for nearly 0.1% of the entire genome (Duan et al., 2021; Koenig et al., 1987). Through the use of alternative promoters and splicing, the gene gives rise to multiple tissue-specific dystrophin isoforms. The full-length muscle isoform Dp427 (427 kDa) is predominantly expressed in skeletal and cardiac muscle, whereas shorter isoforms—including Dp260, Dp140, Dp116, and Dp71—are expressed in tissues such as the retina, brain, kidney, and Schwann cells (Flanigan et al., 2009a, 2009b). Loss or severe reduction of dystrophin expression results in progressive muscle weakness, myofiber degeneration, atrophy, and replacement of muscle tissue by fibrotic and adipose tissue. Because DMD follows an X-linked recessive inheritance pattern, it predominantly affects males, while heterozygous females may exhibit mild or subclinical manifestations. Although the DMD gene exhibits a high rate of de novo mutations, approximately two-thirds of affected individuals inherit the mutation from carrier mothers (Salari et al., 2022). In each pregnancy of a carrier mother, there is a 50% probability of transmitting the pathogenic allele. Male offspring who inherit the mutation typically develop the disease, whereas clinical manifestations in carrier females depend largely on patterns of X-chromosome inactivation (Salari et al., 2022). The absence of dystrophin in DMD arises from diverse genetic alterations that disrupt the stability, transcription, or translation of dystrophin mRNA, leading to a marked deficiency or complete absence of the muscle-specific dystrophin isoform (Dp427m). Mutations within the same gene can also cause Becker muscular dystrophy (BMD), a milder allelic disorder characterized by later onset and slower disease progression (Flanigan et al., 2009a, 2009b; Monaco et al., 1988). Owing to its large size and complex genomic architecture, the DMD gene is particularly susceptible to a wide spectrum of mutational events, including deletions, duplications, insertions, point mutations, and splice-site alterations (Gatto et al., 2024; Aartsma-Rus et al., 2006).

4.1 The DMD gene and its functional domains

Comprehensive insight into the global distribution and mutational spectrum of DMD has been provided by the TREAT-NMD DMD Global Database, which integrates data from regional and national registries worldwide. Analysis of 7149 genetically confirmed cases demonstrated that large structural variants spanning one or more exons constitute the majority of pathogenic mutations (~80%), whereas small sequence variants account for approximately 20% of cases (Bladen et al., 2015). This mutational distribution has important implications for genotype–phenotype correlations and for the development of mutation-specific and domain-targeted therapeutic strategies.

Dystrophin functions as the central structural component of the DGC, which links the intracellular cytoskeleton to the extracellular matrix. The DGC consists of intracellular proteins (dystrophin, syntrophins, and dystrobrevins), transmembrane components (β-dystroglycan; α-, β-, γ-, and δ-sarcoglycans; and sarcospan), and extracellular ligands, including α-dystroglycan and laminin (Gao and McNally, 2015; Wilson et al., 2022; Ervasti, 2007).

Structurally, dystrophin is organized into four major functional domains that collectively mediate its mechanical anchoring and signaling functions within muscle fibers (Zhao et al., 2016) (Fig. 6.1).

These domains include:

1. N-terminal (NT) actin-binding domain anchors dystrophin to filamentous actin;
2. Central rod domain, containing 24 spectrin-like repeats separated by four flexible hinge regions (H1–H4) that confer elasticity and structural resilience;
3. Cysteine-rich (CR) domain interacts directly with β-dystroglycan, forming a crucial link between the cytoskeleton and the sarcolemma; and
4. C-terminal (CT) domain associates with several signaling and scaffolding proteins such as syntrophins and dystrobrevins that regulate cellular communication and membrane integrity (Flanigan et al., 2009a, 2009b; Ahn and Kunkel, 1993).

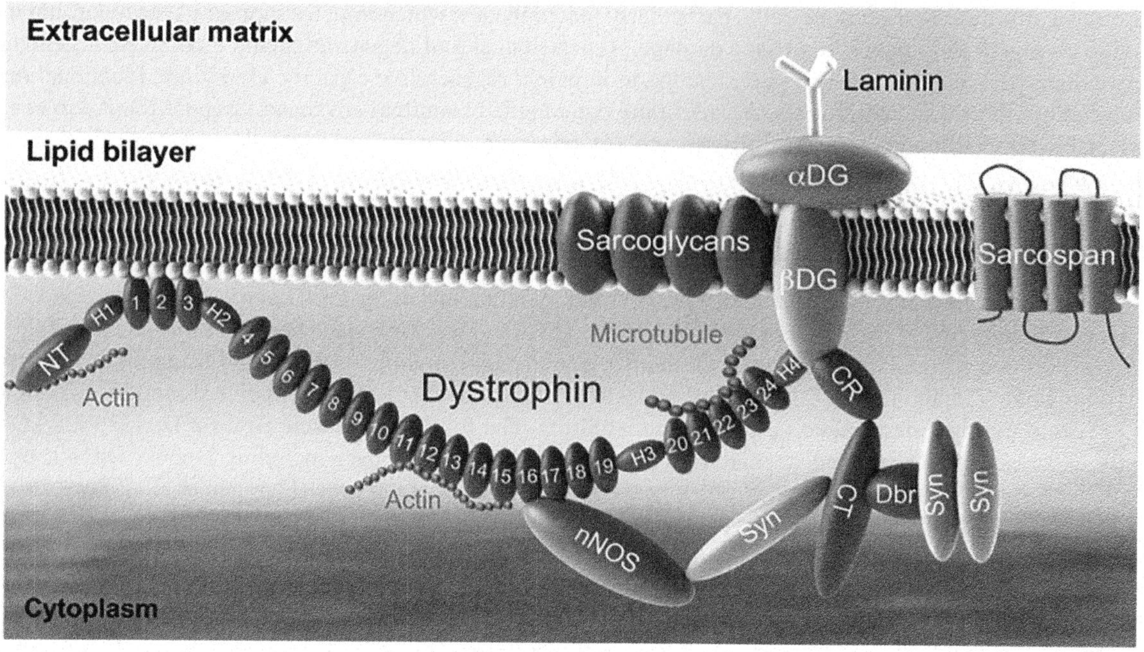

FIG. 6.1 This schematic illustrates the structural organization of dystrophin and its interactions within the dystrophin-associated glycoprotein complex (DGC). Dystrophin is composed of four major domains: The N-terminal (NT) actin-binding domain, a central rod domain, a cysteine-rich (CR) domain, and a C-terminal (CT) domain. The central rod domain contains 24 spectrin-like repeats (R1–R24) interspersed with four hinge regions (H1–H4), which confer flexibility to the protein; positively charged repeats are indicated. Dystrophin harbors two actin-binding sites located within the NT domain and spectrin repeats 11–15. Spectrin repeats 1–3 interact with the negatively charged sarcolemmal lipid bilayer, whereas repeats 16 and 17 form the binding interface for neuronal nitric oxide synthase (nNOS). Interaction with microtubules is mediated by spectrin repeats 20–23. The cysteine-rich domain, together with part of hinge 4, binds the β-subunit of dystroglycan, thereby anchoring dystrophin to the sarcolemma. The C-terminal domain associates with adaptor proteins such as syntrophins and dystrobrevins. Through these coordinated interactions, dystrophin mechanically links intracellular cytoskeletal elements, including actin filaments and microtubules, to laminin in the extracellular matrix via the DGC, which comprises dystrophin, dystroglycans, sarcoglycans, sarcospan, syntrophins, dystrobrevins, and nNOS (McGreevy et al., 2015). *(Reproduced from McGreevy, J.W., Hakim, C.H., McIntosh, M.A., Duan, D. 2015. Animal models of Duchenne muscular dystrophy: from basic mechanisms to gene therapy. Dis. Model. Mech. 8 (3), 195–213. https://doi.org/10.1242/dmm.018424. ©2015 The Company of Biologists Ltd. Licensed under CC BY 3.0.)*

Specific spectrin repeats within the rod domain mediate additional functional interactions. Repeats 1–3 interact with the sarcolemmal lipid bilayer, repeats 11–15 provide a secondary actin-binding site, repeats 16–17 form the docking interface for nNOS, and repeats 20–23 contribute to microtubule binding. Through these coordinated interactions, dystrophin integrates actin filaments and microtubules into a unified cytoskeletal network that is mechanically coupled to laminin in the extracellular matrix via the DGC.

Mutations disrupting the NT actin-binding domain, CR domain, or CT region typically abolish dystrophin–DGC interactions and result in the severe Duchenne phenotype. In contrast, mutations confined to the central rod domain often preserve partial dystrophin functionality and are more frequently associated with the milder BMD phenotype. These domain-specific structure–function relationships form the biological basis for current gene replacement and micro-/mini-dystrophin therapeutic designs (Fig. 6.1).

4.2 Spectrum and frequency of mutations

4.2.1 Large mutations

Large genomic rearrangements constitute the majority of DMD mutations (~80%), consisting mainly of deletions (~68%) and duplications (~11%) (Bladen et al., 2015; Monaco et al., 1986). These mutations frequently disrupt the translational reading frame, resulting in the complete loss of functional dystrophin. Large mutations exhibit a nonrandom distribution and cluster within two well-defined hotspots: a proximal hotspot encompassing exons 2–20, which affects the NT actin-binding domain, and a distal hotspot spanning exons 45–55, involving the central rod domain. Exon 45 is the most frequently deleted single exon, accounting for approximately 4% of all deletions, whereas exon 2 represents the most commonly duplicated exon, comprising about 11% of all duplications. The predominance of deletions within exons 45–55 is consistent with data from the Leiden and UMD-DMD databases, confirming the intrinsic instability of this genomic region (Aartsma-Rus et al., 2006; Tuffery-Giraud et al., 2009).

4.2.2 Small mutations

Small mutations (<1 exon) accounted for 20% of all cases in the TREAT-NMD dataset. These included small deletions (25% of small variants), small insertions (9%), splice-site mutations (14%), and point mutations (52%, of which 50% were nonsense and 2% missense variants) (Bladen et al., 2015). The predominance of nonsense mutations highlights the high mutational susceptibility of CpG dinucleotides, which undergo spontaneous C→T transitions due to deamination of methylated cytosine residues (Bladen et al., 2015; Krawczak et al., 1998). Approximately 10% of all DMD cases result from premature stop codons (TGA, TAG, or TAA), which lead to truncated dystrophin and are targets for read-through therapies such as ataluren (Bladen et al., 2015; Howard et al., 2000).

4.2.3 Reading frame rule and genotype–phenotype correlation

A defining feature of DMD genetics is the reading-frame rule (Monaco et al., 1988). This model explains that frame-shift mutations disrupt translation, producing unstable mRNA degraded by nonsense-mediated mRNA decay (NMD), resulting in the absence of dystrophin and the DMD phenotype. In-frame mutations preserve the reading frame, allowing the synthesis of truncated but partially functional dystrophin, characteristic of BMD (Monaco et al., 1988; Hoffman and Kunkel, 1989). According to the TREAT-NMD data, 93% of mutations followed the reading-frame rule, whereas 7% represented exceptions, often involving in-frame deletions in the actin-binding domain or CR region (Bladen et al., 2015). This correlation underscores the fundamental role of the type and frame consequence of mutation, rather than its size or location alone, in determining disease severity (Flanigan et al., 2009a, 2009b; Bladen et al., 2015).

4.3 Geographic distribution and genetic diversity

Across all major geographic regions, large exon deletions remain the predominant class of DMD mutations, consistently accounting for the majority of cases worldwide, with frequencies ranging from approximately 64% in Oceania to as high as 88% in Africa. Large duplications constitute the second most common category, representing between 5% of mutations in African cohorts and up to 12% in European cohorts. In contrast, the proportion of small sequence variants—including point mutations, splice-site alterations, and small insertions and deletions (indels)—shows greater regional variability (7% in Africa to 22% in Oceania). This variation is unlikely to reflect true biological differences across populations; rather, it is largely attributable to disparities in diagnostic capacity. Many low- and middle-income countries still rely primarily on Multiplex Ligation-dependent Probe Amplification (MLPA) or other exon-level assays, which efficiently detect large deletions and duplications but do not routinely screen for smaller sequence changes (Hwa et al., 2007). Consequently, small mutations tend to be underreported in regions where next-generation sequencing (NGS) technologies are not widely accessible. In addition, several continents, especially Africa, have substantially smaller sample sizes in mutation registries, further amplifying apparent variability through sampling bias. Recent global genomic studies confirm that when comprehensive sequencing approaches such as NGS or whole-exome sequencing are uniformly applied, the distribution of small DMD mutations becomes more consistent across populations, underscoring that observed geographical differences primarily reflect diagnostic and reporting limitations rather than true mutational heterogeneity (Efthymiou et al., 2016; O'Brien and Kunkel, 2001).

4.4 Modifier genes and genetic interactions

Although *DMD* is the primary gene responsible for disease onset, genetic modifiers influence the severity and progression of muscular degeneration. Variants in SPP1, LTBP4, and ACTN3 have been shown to affect inflammation, fibrosis, and regenerative capacity (Gatto et al., 2024). Such modifier loci contribute to inter-individual variability in disease course, even among patients carrying identical *DMD* mutations, illustrating the complex polygenic nature of dystrophinopathy phenotypes (Duan et al., 2021).

5 Current therapeutic strategies

Multiple therapeutic interventions have now gained regulatory approval and constitute the current standard of care for DMD. Although these therapies are not curative, they provide an essential clinical baseline for evaluating the efficacy, safety, and outcome measures of emerging gene therapy approaches. These approved treatments are designed to slow disease progression, preserve muscle function, improve cardiac and respiratory outcomes, and ultimately prolong survival. Current evidence-based management of DMD combines corticosteroid-based therapy, including prednisone, deflazacort, and vamorolone with mutation-specific exon-skipping agents; givinostat; AAV-mediated microdystrophin gene transfer, and standardized cardiopulmonary supportive care (D'Ambrosio and Mendell, 2023).

Corticosteroids represent the earliest and most widely adopted pharmacological intervention for DMD and remain the global standard of care. They are the only class of drugs consistently shown to delay loss of ambulation, preserve upper-limb function, reduce scoliosis progression, and slow pulmonary and cardiac decline. Prednisone is commonly administered at a dose of 0.75 mg/kg/day, either as a continuous daily regimen or in intermittent schedules, and has demonstrated significant benefits in motor performance and disease stabilization. However, long-term corticosteroid therapy is associated with substantial adverse effects, including weight gain, behavioral changes, hypertension, growth suppression, osteoporosis, and increased fracture risk, which often limit long-term tolerability. Deflazacort (DFZ; Emflaza), a prednisolone derivative, was introduced as an alternative glucocorticoid with a potentially improved safety profile and has been shown to provide comparable functional benefits with a reduced impact on weight gain and bone health in some patients (Mesa et al., 1991). Overall, corticosteroids remain the first-line therapy in DMD and continue to serve as the therapeutic foundation upon which newer disease-modifying strategies are layered. Exon-skipping therapies utilize antisense oligonucleotides (AONs) to restore the disrupted reading frame of the *DMD* gene, thereby enabling production of a truncated but partially functional dystrophin protein. These therapies are inherently mutation-specific and therefore applicable only to patients whose genotypes are amenable to skipping of selected exons. To date, four exon-skipping agents have received Food and Drug Administration (FDA) approval.

Eteplirsen (Exondys 51) was the first approved exon-skipping therapy and is indicated for patients eligible for exon 51 skipping. Although it induces relatively low levels of dystrophin expression, modest stabilization of functional outcomes, including 6MWD, has been reported in selected cohorts. Eteplirsen is administered as a weekly intravenous (IV) infusion (Syed, 2016).

Peptide-conjugated phosphorodiamidate morpholino oligomers (PPMOs) were developed to enhance cellular uptake and exon-skipping efficiency compared with conventional phosphorodiamidate morpholino oligomers (PMOs). SRP-5051 (vesleteplirsen) demonstrated approximately 1.6-fold higher exon-skipping activity and a 5-fold increase in dystrophin production following monthly IV administration in early clinical studies. Despite encouraging efficacy signals, clinical development was temporarily paused in 2022 due to safety concerns, including electrolyte abnormalities and muscle cramps; following regulatory review, clinical evaluation has since resumed (D'Ambrosio and Mendell, 2023). Golodirsen (Vyondys 53) targets exon 53 and has demonstrated increased dystrophin expression in treated patients, thereby expanding therapeutic coverage for individuals with exon 53-amenable mutations (Servais et al., 2022; Aartsma-Rus and Corey, 2020). Viltolarsen (Viltepso) is another PMO designed for exon 53 skipping. In a clinical study involving 16 boys with DMD, weekly IV administration at doses of 40 or 80 mg/kg resulted in a 5.7%–5.9% increase in dystrophin expression, accompanied by meaningful functional stabilization compared with natural history controls. Improvements were particularly evident in timed motor assessments and 6MWD, and the therapy demonstrated a favorable safety profile with predominantly mild respiratory adverse events. Long-term extension data further supported sustained motor benefit, justifying the approved dose of 80 mg/kg administered weekly (Clemens et al., 2023).

Casimersen (Amondys 45) received FDA accelerated approval in 2021 for patients eligible for exon 45 skipping. Weekly administration at 30 mg/kg produced a modest increase in dystrophin expression and was generally well-tolerated, with adverse events limited mainly to mild infections and headaches (Clemens et al., 2020). Collectively, exon-skipping therapies represent the largest group of mutation-targeted FDA-approved treatments for DMD, offering personalized therapeutic approaches based on a patient's specific genetic mutation.

Building upon these mutation-specific approaches, AAV-mediated microdystrophin gene transfer has emerged as an important strategy aimed at partially addressing the underlying dystrophin deficiency. The approval of Elevidys (delandistrogene moxeparvovec-rokl; SRP-9001) marks a major milestone in DMD therapeutics. This therapy employs AAVrh74 to deliver a microdystrophin transgene capable of providing partial dystrophin function in muscle fibers. Following an FDA labeling revision in 2025, Elevidys is indicated only for ambulatory patients aged 4 years and older with a confirmed mutation in the DMD gene; the indication for nonambulatory patients was removed the updated prescribing information also includes a boxed warning for serious liver injury and acute liver failure, including fatal outcomes. Clinical trials have demonstrated robust microdystrophin expression; however, the large randomized trial did not meet its primary NSAA endpoint, although secondary and exploratory outcomes suggested potential function benefit. Long-term durability and safety continue to be evaluated. As such, Elevidys provides a critical reference point for future gene therapy development and optimization in DMD (Mullard, 2023; Mendell et al., 2025; U.S. Food and Drug Administration, 2025) (Table 6.2).

TABLE 6.2 Summary of standard-of-care, approved region-specific, and investigational therapies for DMD.

Drug	Dose	Mechanism	Target	Dystrophin restoration	Regulatory status	Year	Ref
Eteplirsen (Exondys 51)	30 mg/kg Weekly	Antisense exon skipping (PMO)	Exon 51	~1%	FDA-approved	2016	(D'Ambrosio and Mendell, 2023)
Golodirsen (Vyondys 53®)	30 mg/kg weekly	Antisense exon skipping (PMO)	Exon 53	~1%	FDA-approved	2019	(Frank et al., 2020)
Viltolarsen (Viltepso)	80 mg/kg weekly	Antisense exon skipping (PMO)	Exon 53	~5%–6%	FDA-approved	2020	(Clemens et al., 2020)
Casimersen (Amondys 45)	30 mg/kg weekly	Antisense exon skipping (PMO)	Exon 45	~1%–2%	FDA-approved	2021	(D'Ambrosio and Mendell, 2023)
Deflazacort (Emflaza)	0.9 mg/kg/day	Glucocorticoid antiinflammatory	Glucocorticoid receptor (GR)	Not applicable	FDA-approved corticosteroid	2017	(Griggs et al., 2016)
Prednisone	0.75 mg/kg/day	Glucocorticoid antiinflammatory	Glucocorticoid receptor (GR)	Not applicable	Standard of care	–	(Griggs et al., 2016)
Vamorolone Agamree	0.25–6.0 mg/kg/day	Dissociative steroid (antiinflammatory)	NF-Kb modulation	Not applicable	FDA-approved	2023	(Keam, 2024)
Vesleteplirsen (SRP-5051)	20 or 30 mg/kg/monthly	PPMO exon skipping	Exon 51	~2%–3%	Phase II	2022	(D'Ambrosio and Mendell, 2023)
Ataluren (Translarna)	40 mg/kg/day	Nonsense mutation read through	Stop codons		Conditional authorization (EU)	2015	(Mercuri et al., 2023)
Deramiocel (CAP-1002)	25 million cells	Paracrine effect	cardiosphere-derived cells therapy	Not applicable	Phase I/II	–	(Taylor et al., 2019)

Note: Ataluren (Translarna) received conditional marketing authorization in the EU for ambulatory patients with Duchenne muscular dystrophy caused by a nonsense mutation in the dystrophin gene. On 28 March 2025, the European Commission decided not to renew the conditional marketing authorization because its effectiveness had not been confirmed. Regulatory status may vary by region and over time. *PMO*, phosphorodiamidate morpholino oligomer; *PPMO*, peptide-conjugated phosphorodiamidate morpholino oligomer; *GR*, glucocorticoid receptor; *EU*, European Union.

Finally, comprehensive cardiopulmonary supportive care remains an indispensable component of DMD management. Early initiation of cardioprotective therapies, including angiotensin-converting enzyme inhibitors, angiotensin receptor blockers, β-blockers, and mineralocorticoid receptor antagonists, delays the progression of cardiomyopathy, while structured respiratory interventions reduce morbidity and mortality associated with respiratory failure. When integrated with disease-modifying approaches such as corticosteroids, exon-skipping therapies, and AAV-based gene replacement, these supportive measures define the contemporary standard of care and have substantially improved both survival and quality of life in patients with DMD.

6 Gene therapy strategies and mechanisms in DMD

Gene therapy has emerged as a central disease-modifying strategy for DMD, shifting therapeutic development from symptomatic management toward direct correction or bypass of the underlying genetic defect (Mendell et al., 2012). Because full-length dystrophin cannot be packaged into a single AAV vector (see AAV-mediated gene therapy section), microdystrophin gene replacement has become the leading clinical platform (Le Guiner et al., 2017; Chamberlain and Chamberlain, 2017; Mendell et al., 2020). These engineered constructs function as functional dystrophin surrogates, preserving key domains required for sarcolemmal stabilization and force transmission rather than restoring the full-length dystrophin protein (Koenig et al., 1987; Harper et al., 2002).

Typically, microdystrophins retain the NT actin-binding domain, selected spectrin-like repeats critical for mechanical integrity and nNOS anchoring, and the CR dystroglycan-binding domain, while omitting large portions of the central rod and CT regions that are dispensable for basic mechanical function (Koenig et al., 1987; Lai et al., 2009). In parallel, additional gene-based strategies—including exon-skipping technologies, viral splice modulation, and genome editing—have expanded the therapeutic landscape by targeting dystrophin expression at the RNA or DNA level (Aartsma-Rus and Krieg, 2017; Chemello et al., 2020; Amoasii et al., 2018). Together, these approaches define gene therapy in DMD as a multifaceted effort aimed at restoring dystrophin-mediated structural stability, rather than broadly suppressing downstream disease manifestations (Shimizu-Motohashi, 2025).

6.1 Targets for gene therapy and mechanisms of action

DMD is caused by mutations in the *DMD* gene that disrupt the open reading frame, resulting in the absence of functional dystrophin. Loss of dystrophin compromises sarcolemmal integrity and initiates a cascade of secondary pathological processes, including chronic activation of NF-κB-mediated inflammatory signaling, increased cytokine and chemokine release, calcium dysregulation, mitochondrial dysfunction, oxidative stress, and progressive muscle degeneration and fibrosis. While these downstream pathways contribute substantially to disease progression, they arise primarily as consequences of dystrophin deficiency rather than as independent initiating events (Akshaya et al., 2024; Budzinska et al., 2021).

Accordingly, gene therapy strategies in DMD are principally designed to restore dystrophin expression, thereby addressing the primary molecular defect and indirectly attenuating many secondary pathological cascades. In contrast, pathways such as inflammation, oxidative stress, and fibrosis are more effectively targeted by adjunctive pharmacological or gene modifier approaches, rather than by dystrophin replacement alone. This distinction is critical for therapeutic targeting and trial interpretation.

For conceptual clarity, therapeutic strategies discussed in this section can be divided into (1) dystrophin-restoring gene therapy approaches—comprising viral gene replacement, viral exon skipping, and genome editing—and (2) non–gene therapy or adjunctive approaches, such as pharmacologic nonsense readthrough and muscle-enhancing gene transfer, which do not directly modify the DMD gene (Stephenson and Flanigan, 2021).

6.2 Dystrophin-restoring gene therapy approaches

At the RNA level, AON-mediated exon-skipping aim to restore the mRNA reading frame by selectively omitting specific exons, enabling the synthesis of a truncated yet partially functional dystrophin protein. By contrast, CRISPR/Cas9-based exon excision act at the DNA level (Anthony et al., 2012). An alternative platform involves viral exon skipping mediated by U7 small nuclear ribonucleoprotein (snRNP), in which modified U7 snRNPs carrying exon-targeting antisense sequences are delivered via AAV vectors to enable sustained exon skipping within muscle tissue (Monceau et al., 2022). Clinically approved exon-skipping agents—including eteplirsen (exon 51), golodirsen and viltolarsen (exon 53), and casimersen (exon 45)—illustrate the translational viability of this mechanism, each achieving partial dystrophin restoration in mutation-specific patient subgroups (Aartsma-Rus and Goemans, 2019; Shirley, 2021).

Gene replacement via AAV-mediated delivery of microdystrophin constructs constitutes the most advanced gene therapy modality in DMD. Because full-length dystrophin exceeds AAV packaging limits, microdystrophins are engineered to retain essential functional domains while fitting within the vector. Systemic administration enables widespread skeletal muscle transduction and has demonstrated robust microdystrophin expression in early clinical trials. Although immune activation and hepatic toxicity, including serious liver injury and acute liver failure, have been reported, AAV-mediated microdystrophin gene transfer remains one of the most promising strategies for reestablishing dystrophin-dependent mechanical stability (Duan et al., 2021; U.S. Food and Drug Administration, 2025).

Genome-editing approaches, including CRISPR/Cas9- and base-editing platforms, aim to permanently correct or bypass pathogenic *DMD* mutations at the DNA level. By excising mutation-bearing exons or restoring the reading frame, these strategies offer the theoretical potential for durable dystrophin expression following a single intervention, although challenges related to delivery efficiency, off-target effects, and long-term safety remain under active investigation (Young et al., 2017).

6.3 Nongene therapy and adjunctive approaches

A distinct, nongene therapy approach involves the pharmacologic read-through of premature stop codons, which does not alter the DMD gene or transcript but instead modulates ribosomal translation. Agents such as gentamicin promote the insertion of an amino acid at the premature termination site, allowing continued translation of dystrophin. Although this strategy restored dystrophin expression in a substantial proportion of fibers in mdx mouse models, clinical efficacy in humans has been variable, and it is not considered gene therapy per se (Pichavant et al., 2011; Palma and Lejeune, 2021). Adjunctive muscle-enhancing gene transfer strategies, such as the delivery of follistatin or GALGT2, aim to increase muscle mass, modify glycosylation pathways, or enhance muscle resilience rather than restore dystrophin expression. These approaches are best viewed as supportive modifiers that may complement dystrophin-restoring gene therapies but do not correct the primary molecular defect (Flanigan et al., 2022). Beyond these approaches, regulatory noncoding RNAs (ncRNAs), including microRNAs and long ncRNAs, have emerged as modulators of muscle degeneration and regeneration. MyomiRs such as miR-1, miR-133, and miR-206 are consistently dysregulated in dystrophin-deficient muscle and may serve as biomarkers or adjunctive therapeutic targets rather than primary gene replacement strategies (Brusa et al., 2020).

Collectively, gene therapy in DMD is primarily directed toward restoring dystrophin expression through viral gene replacement, viral exon skipping, or genome editing, thereby addressing the root genetic defect. Modulation of downstream inflammatory, oxidative, and fibrotic pathways represents a complementary strategy that may enhance clinical benefit when combined with dystrophin-restoring therapies but does not substitute for correction of the primary molecular deficiency.

7 Current advances in preclinical gene therapy for DMD

Preclinical gene therapy for DMD has advanced significantly, progressing from initial proof-of-concept studies in mice to large-scale and translational investigations in canine models. Among the leading approaches are AAV-mediated micro- and mini-dystrophin replacement and AON-based exon skipping. Additional emerging strategies include dual- and tri-AAV systems designed to deliver larger inserts, modulation of dystrophin-independent pathways (such as utrophin upregulation, α7-integrin, and myostatin inhibition), and early-stage DNA-level repair methods (McGreevy et al., 2015).

7.1 Animal models used for preclinical studies in DMD

Muscular dystrophies are caused by mutations in various genes, prompting the development of numerous animal models to study their underlying mechanisms and potential treatments. Among them, DMD and BMD are the most common X-linked forms, both resulting from mutations in the *DMD* gene, which encodes dystrophin, a key structural protein essential for normal muscle function. In DMD, out-of-frame mutations disrupt the reading frame, leading to a severe loss of functional dystrophin and producing <5% of the normal protein level in skeletal muscle (Hoffman et al., 1987). In contrast, BMD is caused by in-frame mutations that permit the production of a shorter but partially functional dystrophin protein, resulting in milder symptoms and a longer life expectancy (Fig. 6.2; Hoffman and Kunkel, 1989).

A diverse range of naturally occurring and genetically engineered animal models—now numbering >60—has been developed to study dystrophin deficiency and advance therapeutic strategies for DMD. These models are essential for understanding the molecular and pathological basis of the disease and for evaluating the safety and efficacy of gene therapies before clinical use. Although none fully replicates all features of human DMD, each provides specific advantages for particular research goals. Importantly, CRISPR-based genome editing has greatly accelerated the creation of more precise DMD models, strengthening the foundation for preclinical gene therapy development and translational progress (Young et al., 2017).

The mdx mouse remains the most widely used and best-characterized animal model for DMD. First described by Bulfield et al. (1984), this model carries a spontaneous nonsense mutation in exon 23 of the *DMD* gene, leading to a premature stop codon and a near-complete loss of dystrophin expression (Bulfield et al., 1984).

Despite the absence of dystrophin, mdx mice exhibit a relatively mild phenotype compared to humans, with minimal muscle weakness, limited fibrosis, and a near-normal lifespan, mainly due to compensatory upregulation of utrophin, a dystrophin-homologous protein (Allen et al., 2015; Chamberlain et al., 2007; Sicinski et al., 1989).

Despite these limitations, due to its well-characterized genetics and manageable size, the mdx mouse remains the primary model for preclinical DMD research and gene therapy development (Aartsma-Rus et al., 2006; Willmann et al., 2012).

In 1989 Chapman and colleagues introduced four chemically induced mdx variant strains. These mutants—designated mdx2cv, mdx3cv, mdx4cv, and mdx5cv—were created on a C57BL/6 genetic background through mutagenesis using N-ethyl-N-nitrosourea (ENU). Each strain harbors a distinct point mutation within the *Dmd* gene (Chapman et al., 1989; Im et al., 1996).

Several additional dystrophin-deficient mouse lines—such as Dup2 and MD-null—have been generated using genetic engineering to complement traditional mdx models. To enable cell and gene therapy research without immune interference,

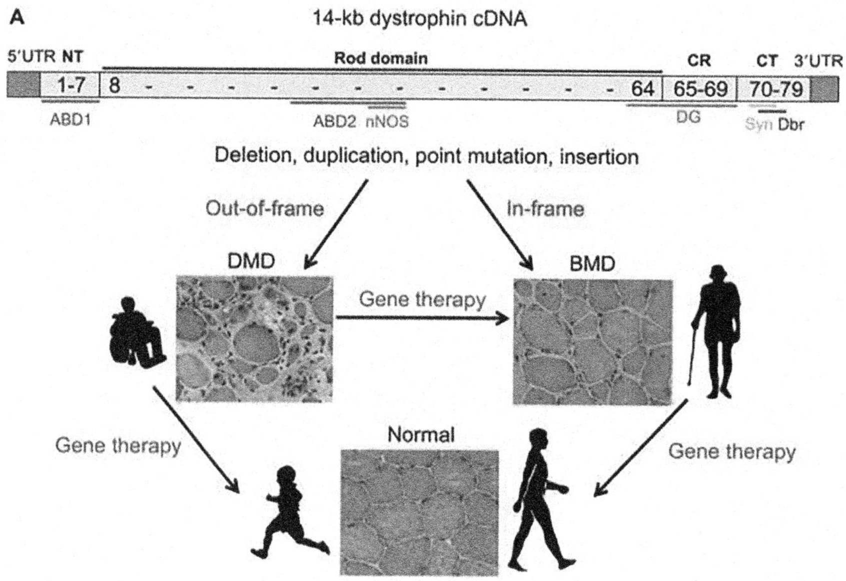

FIG. 6.2 Structure of the dystrophin gene and genotype–phenotype relationship in dystrophinopathies. A schematic representation of the full-length dystrophin cDNA (~14 kb), illustrating its major functional domains, including the N-terminal actin-binding domain (NT), central rod domain containing spectrin-like repeats and hinge regions, cysteine-rich (CR) domain, and C-terminal (CT) domain. The locations of key interaction sites, including actin-binding domains (ABD1 and ABD2), the neuronal nitric oxide synthase (nNOS) binding region, dystroglycan (DG), syntrophin (Syn), and dystrobrevin (Dbr), are indicated. Various mutation types—deletions, duplications, point mutations, and insertions—can disrupt the dystrophin reading frame. Out-of-frame mutations lead to DMD, whereas in-frame mutations result in BMD, a milder phenotype with partial dystrophin expression. Gene therapy approaches aim to restore dystrophin expression or reading frame integrity, thereby improving muscle histology and functional outcomes and potentially shifting the disease phenotype toward a milder or near-normal state (McGreevy et al., 2015). *(Reproduced from Fig. 2A in McGreevy JW, Hakim CH, McIntosh MA, Duan D. Animal models of Duchenne muscular dystrophy: from basic mechanisms to gene therapy. Dis Model Mech. 2015;8(3):195–213. https://doi.org/10.1242/dmm.018424. © 2015 The Company of Biologists Ltd. Licensed under CC BY 3.0)*

immunodeficient mdx strains have also been developed by crossing dystrophin-null mice with immune-deficient backgrounds. Classic examples include nude-mdx and scid-mdx, while newer models like NSG-mdx4cv feature deeper immune suppression, offering improved systems for preclinical evaluation of therapeutic approaches (Farini et al., 2007; Arpke et al., 2013; Walsh et al., 2011). Recent CRISPR/Cas9 studies in mdx mice have successfully restored the reading frame of the *Dmd* gene, resulting in partial dystrophin reexpression and improved muscle function (Long et al., 2014).

To overcome the mild phenotype of the classic mdx mouse and better mimic human disease severity, several modified strains have been developed with deletions such as ΔEx43, ΔEx45 (Min et al., 2020), ΔEx50 (Amoasii et al., 2017), ΔEx52 (Min et al., 2020), and ΔEx8–34 using newer approaches (Egorova et al., 2019). Several CRISPR- and base-editing-derived exon 20 nonsense or frameshift models have also been created, showing mdx-like disease characteristics but lacking compatibility with human sequence-targeted therapies (Kim et al., 2017; Koo et al., 2018).

Despite the absence of dystrophin, mdx mice display a relatively mild phenotype due to the upregulation of compensatory proteins such as utrophin and $\alpha7\beta1$-integrin, which stabilize the sarcolemma. Genetic elimination of these compensatory pathways or humanization of the mdx background yields mouse models that more faithfully recapitulate the severe muscle pathology characteristic of human DMD. Double-knockout models, such as mdx/utrn$^{-/-}$, lacking both dystrophin and utrophin, develop severe muscular dystrophy and shortened lifespans, more closely reflecting the human DMD phenotype. However, these animals are challenging to breed and maintain, and they often die earlier than single-knockout strains (Rooney et al., 2006; Deconinck et al., 1997; Grady et al., 1997). More recent work indicates that utrophin-heterozygous mdx mice may serve as a useful intermediate model, exhibiting a phenotype between the mild mdx mouse and the extremely severe double-knockout models (van Putten et al., 2012). In addition, humanized DMD mouse strains, including hDMDΔ52/mdx, hDMDΔ45/mdx, and hEx45KI-mdx44, were generated to support the preclinical evaluation of exon-skipping agents and gene-editing approaches (Young et al., 2017; Pickar-Oliver et al., 2021; Kenjo et al., 2021).

Compared to mice, rat models more closely reproduce the pathological features of human DMD, enabling behavioral assessments with greater statistical robustness and offering a strong preclinical alternative. The earliest Dmdmdx rat, created using TALEN technology, showed marked muscle and cardiac defects that closely resembled the structural damage and functional impairments observed in DMD patients (Larcher et al., 2014).

Additional rat DMD models, such as ΔEx52, ΔEx3–16, and ΔEx45, also present with pronounced muscle and cardiac pathology, closely reflecting the human disease. Their larger body size enables more robust behavioral and physiological assessments, further enhancing their utility in translational research (Taglietti et al., 2022; Nakamura et al., 2014; Gallagher, 2025). However, because their predictive value for human outcomes is limited, key findings often require confirmation in large animal models, such as canine (GRMD) or porcine DMD models, before advancing to clinical trials.

The DMD rabbit model (ΔEx51), created by CRISPR, shows loss of dystrophin, muscle weakness, cardiomyopathy, and early death, bridging small and large animal models (Sui et al., 2018).

Porcine DMD models (ΔEx52, ΔEx27, mEx13) share a strong anatomical and physiological similarity with humans but are costly and short-lived (Xie et al., 2019).

Dogs are among the most clinically relevant and well-established large-animal models of DMD, as their body size, muscle architecture, lifespan, and disease progression more closely resemble those of human patients than rodent models. The best-studied strain is the golden retriever muscular dystrophy (GRMD) dog, which carries a naturally occurring mutation in the *DMD* gene that eliminates dystrophin expression.

Canine DMD models, including GRMD and related strains, faithfully reproduce key clinical features of human disease, such as progressive muscle degeneration, severe weakness, gait abnormalities, respiratory compromise, and cardiomyopathy. Their systemic muscle involvement, cardiac pathology, and immune responses make them indispensable for evaluating whole-body gene therapies, including AAV-microdystrophin, exon-skipping compounds, and CRISPR-based genome editing (Sharp et al., 1992).

CRISPR/Cas9 technologies have also enabled the development of additional canine variants, such as exon-6-deleted dogs, to expand modeling capabilities. However, despite their strong translational relevance, these models do not fully mirror certain human hallmarks, most notably the loss of ambulation during early adolescence, because affected dogs develop severe symptoms at a younger age (Oh et al., 2022).

Monkey models of DMD represent the closest physiological and genetic approximation to humans among all preclinical DMD models. Due to their high genomic similarity, immune system resemblance, and comparable body size and muscle physiology, nonhuman primates (NHPs) provide an essential platform for evaluating safety, immunogenicity, biodistribution, and long-term effects of gene therapies before human clinical trials (Chen et al., 2015). These mutations were introduced by injecting CRISPR components into zygotes, similar to the process used for generating CRISPR pigs and rabbits. Rhesus monkeys with CRISPR-induced exon 4 and 46 mutations have been developed for preliminary pharmacological and safety assessments (Chen et al., 2015). Despite over 60 DMD animal models, none perfectly reproduces all human disease features; each offers unique strengths for specific study aims. The current and main animal models of DMD are summarized in Table 6.3.

Overall, CRISPR technology has revolutionized DMD model creation, accelerating preclinical gene therapy research and translation to human treatment.

7.2 Targeted genes and vectors

7.2.1 Dystrophin replacement strategies

Efforts to restore dystrophin expression in DMD have explored multiple vector platforms. Early work attempted full-length DMD cDNA (large cargo) delivery using high-capacity viral systems such as gutted adenovirus and herpes simplex virus, which can accommodate the ~14-kb dystrophin transcript but were ultimately limited by significant immunogenicity and toxicity. More recently, tri-AAV approaches divide the full-length cDNA into three AAV vectors for in vivo reconstitution; however, these strategies still require substantial improvements in efficiency and reliability (Happi Mbakam et al., 2022).

7.2.2 Mini- and microdystrophin constructs

Consistent with AAV packaging constraints discussed above, vector design is critical; optimized constructs incorporate functional domains that bind syntrophin and dystrobrevin and include motifs that restore nNOS localization, such as the R16/17 spectrin repeats, which are essential for optimal physiological rescue (Wang et al., 2000; Chwalenia et al., 2025). Conversely, certain hinge regions, including the poly-proline-rich sequence in hinge 2, have been shown to compromise stability or performance and are generally avoided in modern therapeutic designs (Banks et al., 2010; Ramos et al., 2019).

7.3 Administration process in DMD gene therapy

Gene therapy approaches for DMD, including AAV-microdystrophin, exon skipping, and CRISPR-based gene editing, require efficient delivery to skeletal and cardiac muscle. Preclinical studies use several administration routes depending on the tool, animal model, and therapeutic goal.

TABLE 6.3 The current and main animal models of DMD.

Model/ref	Genetic mutation/ mechanism	Phenotypic severity and key features	Typical research applications
Rodent model			
mdx (classic) (Bulfield et al., 1984)	Spontaneous, nonsense point mutation in exon 23	Mild phenotype; initial muscle necrosis at ~ 3 weeks, peak necrosis at ~ 9 weeks; predominant skeletal muscle involvement with a near-normal lifespan	Most widely used baseline model; exon-skipping, gene therapy screening, antiinflammatory/ antifibrotic testing, basic pathophysiology
Mdx4cv (Chapman et al., 1989; Im et al., 1996)	ENU-induced point mutation in exon 53	Moderate-to-severe phenotype; persistent degeneration; fewer revertant fibers; pronounced diaphragm pathology	Fibrosis studies, antisense testing, reproducible therapeutic outcomes
Mdx5cv (Maguire et al., 2009; Danko et al., 1992)	Splice site mutation affecting exon 10	Mild to moderate; defective splicing; variant of classic mdx	Alternative splicing studies
mdx52 (Araki et al., 1997)	Targeted deletion of mouse exon 52; loss of Dp427/ Dp260/Dp140, with preservation of Dp71	Moderate severity; fewer revertant fibers than mdx	CNS involvement, genotype–phenotype correlations, exon 51–53 skipping, gene therapy validation
Dmd-null (full knockout) (Yucel et al., 2018)	Complete deletion of the entire Dmd locus using Cre-loxP	Very severe, no revertant fibers, rapid degeneration, early cardiopulmonary involvement, shortened lifespan	High-studies stringency mechanistic studies, isoform-specific analysis, therapeutic stress testing
D2-mdx (DBA/2 background) (Hammers et al., 2020)	Exon 23 mutation on the DBA/2 strain; severity modified by genes such as (e.g., LTBP4, ANXA6)	Severe phenotype; early fibrosis, profound weakness, impaired regeneration, mitochondrial dysfunction	Testing high-potency therapies, fibrosis, and mitochondrial dysfunction studies
Immunodeficient and compound mouse models			
Nude-mdx (Farini et al., 2007)	mdx mutation + Foxn1 deficiency (T-cell deficiency)	Dystrophin deficiency with impaired adaptive immunity	Xenogeneic cell transplantation; immune-attenuated gene therapy testing
SCID-mdx (Farini et al., 2007)	mdx crossed onto SCID background (T-cell and B-cell deficiency)	Severe immunodeficiency with dystrophin deficiency	Human cell engraftment; evaluation of gene/cell therapies without adaptive immune interference
NSG-mdx4cv (Arpke et al., 2013)	mdx4cv mutation on an NSG background (T-, B-, and NK-cell deficiency)	Profound immunodeficiency; stable dystrophin-null phenotype	Gold-standard model for long-term human stem cell and gene-edited cell engraftment
Utrophin/ dystrophin double knockout (dKO) (Deconinck et al., 1997)	Combined knockout of dystrophin (Dmd) and utrophin (Utrn)	Extremely severe; early-onset weakness, extensive fibrosis, kyphosis, growth retardation, premature death	Studying utrophin compensation, severe degeneration mechanisms, utrophin-targeted therapies
α7-Integrin/ dystrophin dKO (Rooney et al., 2006)	Loss of dystrophin and α7-integrin	Severe dystrophy with impaired regeneration and early mortality; milder than utrophin/dystrophin dKO	ECM–cytoskeleton linkage studies; adhesion-targeted therapies
Mdx4cv/mTR (Sacco et al., 2010; Mourkioti et al., 2013)	Dystrophin loss and telomerase RNA component deficiency	Severe muscle wasting, cardiac defects, stem cell depletion; closely mimics human DMD	Satellite cell exhaustion, gene therapy testing, fibrosis, and aging studies
Humanized and mutation-specific mouse models			
hDMD (full-length human DMD transgenic) ('t Hoen et al., 2008)	BAC transgene carrying complete human DMD; endogenous mouse dystrophin intact	Normal phenotype; expression of human dystrophin	Baseline control for human-specific CRISPR editing; off-target analysis

Model/ref	Genetic mutation/ mechanism	Phenotypic severity and key features	Typical research applications
hDMDΔ52/mdx (Pickar-Oliver et al., 2021)	Human DMD transgene with exon 52 deletion; mouse Dmd inactivated	Moderate-to-severe; human-like Δ52 defect	Exon 51/53 skipping, AAV-antisense therapy, CRISPR frame restoration
hDMDΔ45/mdx (Young et al., 2017)	Human DMD exon 45 deletion; mdx background	Moderate severity; reliable dystrophin rescue assay	Casimersen testing, exon 44/46 skipping, gene editing
hEx45KI-mdx44 (Kenjo et al., 2021)	Human exon 45 knock-in with mouse exon 44 deletion	Precise humanized exon–intron structure	CRISPR/base editing, AAV-U7 snRNA exon correction
Rat model			
Dmdmdx rat (Larcher et al., 2014)	11bp deletion in exon 23 frameshift, no dystrophin	Severe progressive dystrophy; fibrosis, cardiomyopathy, reduced lifespan	AAV microdystrophin, cardiac-targeted therapy, biomarker discovery
W-Dmdem1Kykn (cDMDR) (Nakamura et al., 2014)	CRISPR/Cas9-induced out-of-frame mutations in exon 3 or exon 16	Severe muscle wasting; satellite cell senescence; cardiomyopathy; bone defects; reduced muscle regeneration; reduced lifespan	Stem cell senescence studies; senolytic therapies; metabolic interventions; cardiac dysfunction mechanisms
Dmd-KO (exons 22–26) (Iyer et al., 2020)	CRISPR/Cas9-induced out-of-frame deletion of exons 22–26	Severe dystrophic phenotype with muscle degeneration, inflammation, fibrosis, impaired regeneration, and early cardiomyopathy	Gene replacement and exon-skipping evaluation; studies of muscle regeneration failure and cardiac involvement
R-DMDdel52 (Taglietti et al., 2022)	Deletion of exon 52 (human hotspot)	Severe skeletal and cardiac pathology; median survival: 11–12 months	Fibrosis biomarkers, gene therapy validation
R-DMDdup10–17 (Masson et al., 2025)	Duplication of exons 10–17 (~125 kb)	Severe cardiopulmonary involvement	Duplication-type mutation modeling
DMD-KO (exon 3) (Nakamura et al., 2025)	14-bp deletion in exon 3, causing an out-of-frame mutation and dystrophin loss	Severe muscle weakness with inflammation and fibrosis; reduced strength	Pharmacological screening (e.g., selective androgen receptor modulators)
DmdΔ45 (Wang et al., 2026)	Deletion of exon 45 with spontaneous exon 44 skipping	Moderately severe phenotype; partial dystrophin restoration; delayed cardiomyopathy; longer survival	Exon-skipping mechanisms; revertant fiber biology; genotype–phenotype correlation
DMD-IF (Teramoto et al., 2020)	In-frame deletion of exons 3–16, producing truncated dystrophin	Becker-like mild phenotype with delayed muscle and cardiac involvement	Becker muscular dystrophy modeling; in-frame mutation studies
Large animal models			
GRMD: canine model (Golden Retriever) (Sharp et al., 1992)	Intron 6 mutation causing exon 7 skipping	Severe progressive dystrophy closely mimicking human DMD	Gold-standard large-animal model for systemic AAV delivery and dose scaling
CXMD: canine (Beagle) (Shimatsu et al., 2003; Valentine et al., 1988)	GRMD mutation on Beagle background	Milder than GRMD; suitable for longitudinal studies	Long-term safety and efficacy assessment
CKCS-MD: canine (Walmsley et al., 2010)	Point mutation in intron 50 causing exon 50 skipping; hotspot region (exons 45–53)	Severe progressive muscular dystrophy with early-onset muscle weakness, skeletal muscle degeneration, and cardiomyopathy	Evaluation of exon-skipping strategies (e.g., exon 51 skipping); genotype–phenotype correlation; preclinical testing of gene and RNA-based therapies
GSHP MD: canine (Valentine et al., 1988)	Large genomic deletion of the entire DMD gene	Severe dystrophy; complete loss of dystrophin; early-onset progressive muscle degeneration	Studying full dystrophin absence; gene replacement feasibility; preclinical testing of AAV-based gene therapy in severe DMD

Continued

TABLE 6.3 The current and main animal models of DMD—cont'd

Model/ref	Genetic mutation/ mechanism	Phenotypic severity and key features	Typical research applications
Tibetan Terrier MD: canine(Kornegay et al., 2012)	Large deletion of exons 8–29resulting in an out-of-frame transcript	Severe progressive muscular dystrophy with early-onset muscle weakness, marked skeletal muscle degeneration, and cardiomyopathy	Structural gene deletion modeling; evaluation of gene replacement and exon-skipping strategies; preclinical testing of AAV-based gene therapy
Labrador Retriever MD: canine (Smith et al., 2007)	Repetitive element insertion in intron 19	Progressive muscular dystrophy with skeletal muscle weakness, degeneration, and variable cardiac involvement	Preclinical gene therapy studies; evaluation of AAV vector biodistribution and transgene expression; modeling splice-disrupting DMD mutations
Japanese Spitz MD: canine (Atencia-Fernandez et al., 2015)	Chromosomal inversion disrupting the DMD gene	Variable-severity muscular dystrophy with progressive muscle weakness and degeneration; dystrophin truncation	Structural rearrangement modeling; complex mutation analysis
DMD Rabbit (CRISPR/Cas9 (Sui et al., 2018)	CRISPR/Cas9-mediated knockout of the DMD gene (frameshift or exon disruption)	Severe progressive muscular dystrophy; marked muscle weakness, fibrosis, elevated CK; cardiomyopathy and respiratory impairment; reduced lifespan; closer in severity to human DMD than mdx mice	Translational gene therapy testing, AAV dose optimization, cardiac and respiratory outcome studies, pharmacological screening
DMD pig (exon 52 deletion) (Klymiuk et al., 2013)	Targeted deletion of DMD exon 52, resulting in a frameshift mutation	Severe early-onset muscular dystrophy with rapid skeletal muscle degeneration, cardiomyopathy, respiratory failure, and shortened survival	Translational large-animal model for gene and RNA-based therapies; evaluation of systemic delivery, safety, and dose scaling prior to human clinical trials
MDM (DMDex52del miniature pig) (Echigoya et al., 2021)	rAAV-mediated homologous recombination and somatic cell nuclear transfer, resulting in exon 52 deletion and a dystrophin-null genotype	Severe early-onset dystrophy with marked muscle degeneration, fibrosis	Preclinical evaluation of ASO-mediated exon skipping (PMO/PPMO), dystrophin restoration, DAP complex reassembly, and translational safety
Becker-like pig (Nonneman et al., 2012)	In-frame DMD exon deletion or missense mutation producing truncated dystrophin	Mild-to-moderate Becker-like phenotype with partial dystrophin expression, delayed skeletal muscle involvement, and stress-induced cardiac dysfunction	Modeling Becker muscular dystrophy; genotype–phenotype correlation; evaluation of cardiac-targeted and gene stabilization therapies
Nonhuman primate (Rhesus macaque Ex50) (Bai et al., 2025)	CRISPR/Cas9-mediated targeting of DMD exon 50, resulting in a frameshift mutation and the complete loss of dystrophin	Severe early-onset Duchenne-like phenotype with progressive skeletal muscle degeneration, fibrosis, elevated serum creatine kinase, impaired gait, and declining motor function	High-fidelity translational model for DMD; evaluation of systemic gene-editing and AAV-based gene therapies; immune response, biodistribution, and long-term safety assessment prior to clinical translation

AAV, adeno-associated virus; *ANXA6*, annexin A6; *ASO*, antisense oligonucleotide; *BAC*, bacterial artificial chromosome; *CK*, creatine kinase; *CKCS-MD*, Cavalier King Charles Spaniel muscular dystrophy; *CNS*, central nervous system; *CRISPR/Cas9*, clustered regularly interspaced short palindromic repeats/CRISPR-associated protein 9; *CXMD*, canine X-linked muscular dystrophy; *DAP*, dystrophin-associated protein; *dKO*, double knockout; *Dp*, dystrophin protein isoform; *ECM*, extracellular matrix; *ENU*, N-ethyl-N-nitrosourea; *GRMD*, Golden Retriever muscular dystrophy; *GSHP*, German Shorthaired Pointer; *KI*, knock-in; *KO*, knockout; *LTBP4*, latent transforming growth factor beta-binding protein 4; mdx, murine dystrophin-deficient X-linked; *MDM*, Duchenne muscular dystrophy exon 52 deletion miniature pig; *mTR*, mouse telomerase RNA component; NK cells, natural killer cells; *NSG*, NOD scid gamma; *PMO*, phosphorodiamidate morpholino oligomer; *PPMO*, peptide-conjugated phosphorodiamidate morpholino oligomer; *RPGR*, retinitis pigmentosa GTPase regulator; *SCID*, severe combined immunodeficiency; *U7 snRNA*, U7 small nuclear ribonucleic acid.

Systemic IV delivery is the most common route for DMD gene therapy, used to deliver vectors widely to skeletal, diaphragm, and cardiac muscle: systemic AAV-microdystrophin in GRMD dogs (Oh et al., 2022), systemic CRISPR/Cas9 AAV9 in mdx mice (Min et al., 2020; Pickar-Oliver et al., 2021), and systemic PPMO in DMDex52del pigs (Echigoya et al., 2021).

Local intramuscular (IM) injection proved the concept and enabled construct optimization in mice but is not suitable for whole-body therapy; it enables local histology and functional assessment, including IM AAV-CRISPR in mdx mice and IM mini-dystrophin AAV8 (Pickar-Oliver et al., 2021).

7.4 Follow-up and monitoring of gene therapy

In preclinical DMD gene therapy studies, follow-up relies on long-term evaluation of molecular, histological, biochemical, immunological, and functional parameters to assess therapeutic durability, safety, and biodistribution. In mouse models specifically, monitoring includes measuring dystrophin recovery, examining skeletal and cardiac muscle pathology, and conducting functional performance tests at scheduled intervals after treatment (Min et al., 2019).

After the administration of AAV-microdystrophin, exon-skipping agents, or CRISPR/Cas9-based therapies, animals are monitored over weeks to months using repeated blood sampling for serum CK as a marker of ongoing muscle damage, Western blot and immunofluorescence to quantify dystrophin protein restoration and localization, and RT-PCR or deep sequencing to evaluate mRNA correction or genome-editing efficiency (Min et al., 2020; Egorova et al., 2019). Histopathological monitoring includes serial muscle biopsies or postmortem tissue collection to examine fiber morphology, fibrosis, necrosis, and inflammation (Larcher et al., 2014; Taglietti et al., 2022). Functional follow-up—critical for evaluating therapeutic benefit—consists of grip-strength testing, treadmill endurance, in vivo muscle force measurements, echocardiography, and respiratory function evaluation, depending on the species (Young et al., 2017; Larcher et al., 2014).

Biodistribution and vector persistence are monitored through quantitative PCR to assess AAV genome copies in muscle, heart, liver, and other organs, while immune monitoring evaluates antibody responses against AAV capsid, Cas9, or dystrophin using ELISA and T-cell assays (Kenjo et al., 2021; Min et al., 2019).

In CRISPR-based approaches, long-term monitoring additionally involves sequencing off-target sites and assessing genomic rearrangements to ensure editing precision and minimize genotoxicity (Chemello et al., 2020; Min et al., 2019).

Together, these multimodal follow-up strategies allow comprehensive evaluation of the safety, efficacy, durability, and systemic impact of emerging gene therapy approaches for DMD.

7.5 Complications and management of complications in DMD gene therapy

Gene therapy for DMD has demonstrated significant therapeutic promise; however, a number of complications have been documented across preclinical and clinical studies, particularly in the context of systemic AAV delivery. The most frequently observed adverse events involve immune- and inflammation-mediated toxicities, including transient hepatic enzyme elevations (Mendell et al., 2020), flu-like systemic reactions, and, in more severe cases, AAV-associated hepatotoxicity, complement activation, and thrombotic microangiopathy following high-dose vector infusion (Hordeaux et al., 2024; Schwotzer et al., 2024). Reports of acute liver failure and multiorgan dysfunction after high-dose AAV administration underscore the importance of careful dose selection and early safety monitoring (Byrne et al., 2025b).

Exon-skipping therapies generally display a more favorable safety profile; however, certain antisense chemistries, particularly phosphorothioate-modified oligonucleotides, have been associated with renal toxicity and injection-site reactions (Leckie et al., 2024). Preclinical CRISPR/Cas9-based strategies introduce a distinct set of risks, including off-target editing, unintended large deletions or inversions at the cut site, and potential immune responses against bacterial Cas9 proteins (Erkut and Yokota, 2022; Padmaswari et al., 2023).

Management of these complications relies on a combination of risk mitigation, close clinical surveillance, and rapid therapeutic intervention. In AAV-based trials, standard practice includes pretreatment screening for anti-AAV neutralizing antibodies (NAbs), detailed baseline evaluation of hepatic, renal, and cardiac function, and the initiation of prophylactic corticosteroids to suppress early immune activation (Chwalenia et al., 2025; Duan, 2018a).

Following vector administration, patients undergo frequent laboratory monitoring, including serial measurements of alanine aminotransferase, aspartate aminotransferase, platelet counts, complement markers (e.g., C3, C5b-9), renal function, and cardiac biomarkers such as troponin (Duan, 2018a).

Overall, the safe translation of DMD gene therapy requires an integrated strategy of preventive screening, dose optimization, and robust posttreatment surveillance, ensuring a balance between therapeutic benefit and patient safety.

8 Recent clinical gene therapy trials in DMD

DMD is caused by a variant in the *DMD* gene located on Xp21 (Monaco et al., 1986). Recent progress in drug development has led to the approval of novel therapies utilizing diverse therapeutic approaches, such as exon skipping, antiinflammatory approaches, read-through mechanisms, and gene therapies. Nevertheless, these medications primarily act as disease-modifying agents, and their approval status differs among countries (Shimizu-Motohashi, 2025).

As of June 22, 2023, delandistrogene moxeparvovec (SRP-9001; Elevidys) received US FDA accelerated approval as the first AAV-based gene replacement therapy for DMD. The product consists of a recombinant AAVrh74 vector encoding a microdystrophin transgene under the control of the muscle-specific MHCK7 promoter, administered as a single IV infusion to enable systemic skeletal (and cardiac) muscle transduction. The recommended dose is 1.33×10^{14} vector genomes (vg)/kg for patients weighing < 70 kg with a fixed total dose of 9.31×10^{15} vg for those weighing ≥ 70 kg, delivered as a one-time systemic infusion.

The phase 3 EMBARK (SRP-9001-301) trial is a multinational, randomized, placebo-controlled, two-part crossover study evaluating safety and efficacy in ambulatory boys aged ≥4 to <8 years with a confirmed DMD mutation. The primary analysis at week 52 did not demonstrate a statistically significant improvement in the NSAA total score versus placebo. However, the treatment was associated with robust microdystrophin expression and favorable trends in several secondary functional endpoints (Mendell et al., 2025).

On November 14, 2025, the FDA added a boxed warning for serious liver injury, including cases of acute liver failure (with fatal outcomes), and restricted the indication to ambulatory patients ≥4 years of age with a confirmed DMD mutation; the prior nonambulatory indication was removed. These regulatory updates underscore the evolving benefit–risk profile of systemic AAV gene therapy and the importance of long-term safety monitoring. Beyond microdystrophin gene replacement, additional gene-based strategies under clinical investigation include AAV-mediated splice modulation and genome-editing platforms such as CRISPR/Cas systems. Collectively, these approaches reflect the ongoing transition of DMD therapy from supportive management toward molecularly targeted, disease-modifying interventions (Shimizu-Motohashi, 2025).

8.1 AAV-mediated gene therapy

Gene therapy has emerged as a promising approach to modify DMD progression. The use of AAV vectors for in vivo gene delivery represents a major advancement in gene therapy and offers a promising approach for treating disorders such as DMD. Extensively studied and recently approved vector-based gene therapies, such as delandistrogene moxeparvovec, utilize AAV vectors to deliver a truncated yet functional version of the *DMD* gene (delivery of microdystrophin constructs), thereby aiming to restore dystrophin production. These truncated genes retain essential functional domains, enabling the restoration of dystrophin expression within the packaging limits of AAV vectors (Khan and Hussain, 2026). Several viral vectors, including lentiviruses and AAVs, have been explored; however, AAVs are preferred due to their low immunogenicity, defective replication, and long-term transgene expression in nondividing cells (Shimizu-Motohashi, 2025). The main limitation of AAV vectors is their packaging capacity (~4.7–5 kb), which is insufficient to accommodate the approximately 14-kb full-length dystrophin cDNA (Shimizu-Motohashi, 2025). The discovery of exceptionally mild DMD cases involving deletions of nearly half of the DMD gene (England et al., 1990) prompted the development of micro- and mini-dystrophin constructs, which are shortened yet retain essential functional domains. The lengths of mini- and microdystrophins were determined to be around 6–8 kb and below 4 kb, respectively (Duan, 2018b). Each AAV serotype exhibits distinct organ-specific tropism. Recombinant AAV (rAAV) serotypes 1, 6, 7, 8, and 9 have been shown to preferentially target skeletal muscle tissue (Muraine et al., 2020; Zincarelli et al., 2008). rAAV rh74, derived from rhesus monkeys, demonstrates high skeletal muscle tropism (Chicoine et al., 2014a; 2014b). Among individuals with DMD, seroprevalence of rAAV9 and rAAV rh74 is comparatively low (Zygmunt et al., 2017; Verma et al., 2023). Appropriate selection of a muscle-specific promoter is essential to reduce off-target immunogenicity and improve expression efficiency. Promoters such as CK8 and MHCK7 have been tested clinically, with MHCK7 used in delandistrogene moxeparvovec (Mendell et al., 2025).

Adjunctive gene therapy strategies for DMD do not act on the dystrophin gene itself; rather, they are designed to improve muscle function through alternative molecular pathways. For example, rAAV1.CMV.huFollistatin344 promotes muscle growth through follistatin overexpression by inhibiting myostatin/activin signaling. rAAVrh74.MCK.GALGT2 delivers the human GALGT2 transgene to skeletal and cardiac muscle, where GALGT2 overexpression increases glycosylation of alpha-dystroglycan and is intended to enhance sarcolemmal stability through a dystrophin-independent mechanism (Flanigan et al., 2022).

Gene therapy for DMD extends beyond dystrophin restoration (microdystrophin replacement via AAV vectors), incorporating approaches such as exon-skipping approaches via AAV vectors or base editing, endogenous dystrophin upregulation via CRISPR, and adjunctive muscle-enhancement strategies designed to compensate for the lack of functional dystrophin (Shimizu-Motohashi, 2025). Table 6.4 summarizes major recent or ongoing AAV-mediated gene therapy clinical programs for DMD. Only one AAV DMD therapy is FDA-approved (Elevidys). All other AAV programs remain investigational.

TABLE 6.4 Major recent or ongoing AAV-mediated gene therapy clinical programs for DMD (as of February 2026).

Product	Study start/ completion	AAV serotype/ promoter/construct	Trial status	Reported dose (vg/kg)	Route of administration	Age range/key inclusion/exclusion	Phase	Sponsor / developer	Trial ID	Last updated	Ref
Delandistrogene moxeparvovec (SRP-9001-101) Elevidys	Start: 2018-01-04 Completion: 2023-04-25	rAAVrh74; MHCK7 muscle-specific hybrid promoter; SRP-9001 microdystrophin cassette (ΔR4–R23/ΔC)	Completed	2.0×10^{14} vg/kg by the supercoiled qPCR method (equivalent to ≈ 1.33×10^{14} vg/kg by the linear standard qPCR)	Single-dose intravenous (IV) infusion	3 months to 7 years; **Inclusion**: confirmed DMD, ambulatory, steroid-stable; **Exclusion**: preexisting anti-AAVrh74 above threshold, some mutation restrictions used in later trials.	Phase 1/2a	Sarepta/ Roche	NCT03375164	2024-11-14	(Mendell et al., 2020, 2024; Zaidman et al., 2023)
Delandistrogene moxeparvovec (SRP-9001-102)	Start: 2018-12-05 Completion: 2023-08-16	AAVrh74; MHCK7 promoter; SRP-9001 microdystrophin cassette	Completed	~1.33×10^{14} vg/kg (some subjects received a lower lot dose of ~ 6.29×10^{13} vg/kg)	Single IV infusion	4–7 years; **Inclusion:** clinically diagnosed DMD with a qualifying mutation, ambulatory, stable corticosteroid regimen for ≥ 12 weeks; **Exclusion:** high anti-AAVrh74 Ab titer, prior gene therapy, certain DMD genotypes in some parts.	Phase ½	Sarepta	NCT03769116	2024-11-14	(Zaidman et al., 2023; Mendell et al., 2023)
Delandistrogene moxeparvovec (SRP-9001-103, ENDEAVOR)	Start: 2020-11-23 Completion: 2028-02-29	rAAVrh74; MHCK7 promoter; microdystrophin cassette	Recruiting	1.33×10^{14} vg/kg	Single IV infusion	2 years and older; **Inclusion:** previously untreated or previously dosed participants, ambulatory status varied by cohort; **Exclusion**: anti-AAVrh74 above the threshold unless eligible for specific cohorts.	Phase 1	Sarepta	NCT04626674	2025-12-29	Zaidman et al., 2023; Potter et al., 2025)
Delandistrogene moxeparvovec (SRP-9001-301, EMBARK)	Start: 2021-10-27 Completion: 2024-10-25	rAAVrh74; MHCK7 promoter; microdystrophin cassette	Completed	1.33×10^{14} vg/kg	Single IV infusion	≥4 to <8 years; **Inclusion**: ambulatory males with genetically confirmed DMD, NSAA window, rAAVrh74 Ab titer below threshold, stable steroids ≥ 12 weeks; **Exclusion**: certain DMD genotypes, high anti-AAVrh74 titers, prior gene therapy.	Phase 3	Sarepta/ Roche	NCT05096221	2025-07-08	(Mendell et al., 2025, Mendell et al., 2026; Audhya et al., 2026; Vandenborne et al., 2025; McDonald et al., 2025)

Continued

Table 6.4 Major recent or ongoing AAV-mediated gene therapy clinical programs for DMD (as of February 2026)—cont'd

Product	Study start/ completion	AAV serotype/ promoter/con- struct	Trial status	Reported dose (vg/kg)	Route of adminis- tration	Age range/key inclu- sion/exclusion	Phase	Sponsor / developer	Trial ID	Last up- dated	Ref
Delandistrogene moxeparvovec (EXPEDITION)	Start: 2023-09-27 Completion: 2030-11-30	n/a	Enrolling	n/a (observa- tional)	n/a	Age range varies; **Inclusion**: partici- pants who previously received SRP-9001 in a clinical study; **Exclu- sion**: per protocol.	Phase 3	Sarepta/ Roche	NCT05967351	2026- 02-06	(ClinicalTrials. gov, 2026b)
Delandistrogene moxeparvovec (SRP-9001-302/EN- VOL)	Start: 2023-11-29 Completion: 2034-01-30	rAAVrh74; MHCK7 promoter; mi- crodystrophin cassette	Recruit- ing	1.33×10^{14} vg/kg	Single IV infusion	Infant/<4 years; **Inclusion**: younger children/infants with confirmed DMD; **Ex- clusion**: per protocol.	Phase 2	Sarepta	NCT06128564	2025- 12-24	(ClinicalTrials. gov, 2026f)
Delandistrogene moxeparvovec (Imlifidase + SRP- 9001 study)	Start: 2024-01-29Com- pletion: 2025- 10-10	rAAVrh74 (SRP- 9001) + imlifi- dase; MHCK7 promoter; microdystro- phin cassette	Termi- nated	Dose per proto- col: 1.33×10^{14} vg/kg; imlifidase dose per protocol	Imlifidase IV infu- sion, fol- lowed by SRP-9001 IV infu- sion, per protocol	4–9 years; **Inclusion**: participants with pre- existing anti-AAVrh74 antibody (Ab) who otherwise meet clinical eligibility; **Exclusion**: contraindi- cations to imlifidase or SRP-9001, activeinfec- tion.	Phase 1	Sarepta	NCT06241950	2025- 11-20	(ClinicalTrials. gov, 2026h)
Delandistrogene moxeparvovec (ENDURE)	Start: 2024-02-07 Completion: 2038-12-31	n/a	Enrolling	n/a	n/a	Varies (4 years and older); **Inclusion**: treat- ed participants and matched standard- of-care patients per protocol; **Exclusion** per registry.		Sarepta/ Roche	NCT06270719	2025- 12-02	(NCT06270719, C.g.T.R., 2026)
Delandistrogene moxeparvovec (HORIZON-plasma- pheresis + SRP-9001)	Start: 2024-09-18Com- pletion: 2025- 08-05	rAAVrh74; MHCK7 promoter; mi- crodystrophin cassette with plasmapheresis as precondi- tioning	Termi- nated	$\approx 1.33 \times 10^{14}$ vg/kg	Precon- ditioning plasma- pheresis procedures followed by IV SRP-9001 infusion	4–8 years; **Inclusion:** participants with pre- existing anti-AAVrh74 Abs intended for plas- mapheresis to reduce titers; **Exclusion**: stan- dard safety exclusions and inability to toler- ate plasmapheresis.	Phase 1	Sarepta	NCT06597656	2025- 09-04	(ClinicalTrials. gov, 2026j)

Delandistrogene moxeparvovec (SRP-9001-303, ENVISION)	Start: 2023-05-31 Completion: 2028-06-30	rAAVrh74; MHCK7 promoter; microdystrophin cassette	Active, not recruiting	1.33×10^{14} vg/kg	IV infusion	Broader (older ambulatory and nonambulatory groups); **Inclusion**: ambulatory and nonambulatory males with DMD, rAAVrh74 Ab titer limits, stable background care; **Exclusion**: prior AAV gene therapy, high antibody titers, specific genotypes per label/contraindications.	Phase 3	Sarepta/Roche	NCT05881408	2025-06-22	(ClinicalTrials.gov, 2026e)
Fordadistrogene Movaparvovec (PF-06939926)	Start: 2018-01-23 Completion: 2025-07-28	AAV9; AAV9-mini/micro-dystrophin transgene under a muscle-selective promoter	Terminated	$\sim 1.0 \times 10^{14}$ vg/kg (low) and $\sim 2.0 \times 10^{14}$ vg/kg (high)	Single systemic IV infusion	~6–12 years; **Inclusion**: genetically confirmed DMD, ambulatory and/or cohort-defined ambulatory status, stable corticosteroid regimen per protocol; **Exclusion**: high anti-AAV9 neutralizing antibodies (NAbs), prior AAV gene therapy, major uncontrolled illness.	Phase 1b	Pfizer	NCT03362502	2025-09-15	(Walsh et al., 2025; Byrne et al., 2025a; Sherlock et al., 2025; Butterfield et al., 2025)
Fordadistrogene Movaparvovec (CIFFREO)	Start: 2020-11-05 Completion: 2039-04-15	AAV9; AAV9-mini-dystrophin with muscle-preferential promoter	Active, not recruiting	$\approx 2.0 \times 10^{14}$ vg/kg	Single IV infusion	4–7 years; **Inclusion**: genetically confirmed DMD, ambulatory males aged 4–7 years on stable glucocorticoids, negative for AAV9 neutralizing antibodies; **Exclusion**: prior systemic AAV gene therapy, significant comorbidities, high antibody titers.	Phase 3	Pfizer	NCT04281485	2026-01-02	(Philippidis, 2022, Philippidis, 2024; Dystrophy, 2024)
Fordadistrogene Movaparvovec (DAYLIGHT)	Start: 2022-08-08 Completion: 2025-10-03	AAV9; AAV9-mini-dystrophin cassette	Terminated	$\sim 2.0 \times 10^{14}$ vg/kg	Single IV infusion	~2–3 years; **Inclusion**: very young boys with genetically confirmed DMD; **Exclusion**: AAV9 NAbs above cutoff, major comorbidities, prior AAV therapy.	Phase 2	Pfizer	NCT05429372	2025-10-21	(ClinicalTrials.gov, 2026a)

Continued

Table 6.4 Major recent or ongoing AAV-mediated gene therapy clinical programs for DMD (as of February 2026)—cont'd

Product	Study start/ completion	AAV serotype/ promoter/construct	Trial status	Reported dose (vg/kg)	Route of administration	Age range/key inclusion/exclusion	Phase	Sponsor / developer	Trial ID	Last updated	Ref
Long-term follow-up (LTFU) entries	Start: 2023-03-13 Completion: 2025-09-24	n/a; n/a	Terminated	n/a	Observational	Ages vary; **Inclusion**: participants who previously received PF-06939926 in a preceding Pfizer interventional study; **Exclusion**: investigator site staff involved in the study and their family members.	Phase 3	Pfizer	NCT05689164	2025-10-21	(ClinicalTrials.gov, 2026c)
RGX-202 AFFINITY DUCHENNE	Start: 2023-01-04 Completion: 2028-08	AAV8; Spc5-12 muscle/cardiac promoter; RGX-202 microdystrophin cassette	Recruiting	~1×10^{13} to 3×10^{13} vg/kg	Systemically via a one-time IV infusion	1 year and older; **Inclusion:** genotype confirmed DMD, antibody/preexisting anti-AAV8 screening per protocol, pediatric ambulatory cohorts included; **Exclusions**: liver disease, serious comorbidity.	Phase 2/3	REGENX-BIO	NCT05693142	2025-12-17	(ClinicalTrials.gov, 2026d)
AFFINITY BEYOND	Start: 2022-12-20 Completion: 2026-09	n/a; n/a	Recruiting	n/a	n/a	0–25 years; **Inclusion**: people with DMD who consent to antibody testing; **Exclusion**: per registry.		REGENX-BIO	NCT05683379	2025-11-25	(ClinicalTrials.gov, 2026b)
LTFU for RGX-202	Start: 2024-05-08 Completion: 2029-12-01	n/a; n/a	Enrolling	n/a	Observational	Pediatric to adolescent; **Inclusion**: participants who previously received RGX-202 in an interventional trial; **Exclusion**: standard per registry.		REGENX-BIO	NCT06491927	2025-05-13	(ClinicalTrials.gov, 2026i)
SGT-003	Start: 2024-05-06 Completion: 2031-05-06	AAV (engineered capsid: AAV-SLB101); muscle-specific promoter; next-generation microdystrophin cassette	Recruiting	~1×10^{14} vg/kg	Systemically via IV infusion	0–17 years; **Inclusion:** pediatric ambulatory cohorts; **Exclusions**: include prior AAV exposure/high anti-AAV titers, active severe organ disease.	Phase ½	Solid Biosciences Inc.	NCT06138639	2026-02-03	(ClinicalTrials.gov, 2026g)

SGT-001	Start: 2017-12-06 Completion: 2026-10-15	**AAV9** carrying a microdystrophin transgene; promoter CK8/CK7	Active, not recruiting	5×10^{13} vg/kg and 2×10^{14} vg/kg	Systemically via IV infusion	4–17 years; **Inclusion**: genetically confirmed DMD, ambulatory, adequate organ function, negative for anti-AAV9 neutralizing antibodies; **Exclusions**: nonambulatory, anti-AAV9 antibodies, patients at higher risk for immune or organ-related complications.	Phase ½	Solid Biosciences Inc.	NCT03368742	2026-02-03	(Boehler et al., 2024)
scAAV9.U7.ACCA (AT702/exon-2 Dup2 program)	Start: 2020-01-15 Completion: 2025-07-01	Self-complementary AAV9 (scAAV9); U7snRNA splice-modulator (not microdystrophin)	Completed	3×10^{13} vg/kg	Systemically via IV infusion	≥6 months to < 14 years; **Inclusion**: genetically confirmed DMD exon-2 duplication (Dup2), ambulatory boys, adequate liver, kidney, and cardiac function, negative AAV9 NAbs, stable corticosteroid regimen; **Exclusions**: Nonambulatory patients, positive anti-AAV9 antibodies, significant cardiac, hepatic, or renal disease, prior AAV gene therapy exposure.	Phase ½	Ohio State/Nationwide Children's Hospital	NCT04240314	2025-09-11	(Wein et al., 2014; Vulin et al., 2015; Gurvich et al., 2009; Flanigan et al., 2009b; Vulin et al., 2012)
GNT0004	Start: 2020-11-30 Completion:2025-07-28	AAV8; Spc5-12 muscle/cardiac-specific promoter; hMD1 microdystrophin transgene	Recruiting	3×10^{13} vg/kg	Single-dose IV infusion	~6–10 years; **Inclusion:** ambulatory boys, genotype confirmation; **Exclusions:** immune or liver issues.	Phase ½/3	Généthon	EudraCT: 2020-002093-27A, Sponsor protocol code: GNT-016-MDYF	2025-12-12	(Agency, 2026)

Continued

Table 6.4 Major recent or ongoing AAV-mediated gene therapy clinical programs for DMD (as of February 2026)—cont'd

Product	Study start/ completion	AAV serotype/ promoter/con-struct	Trial status	Reported dose (vg/kg)	Route of adminis-tration	Age range/key inclu-sion/exclusion	Phase	Sponsor / developer	Trial ID	Last up-dated	Ref
GEN6050X	Start: 2024-07-05 Completion: 2027-12	AAV9; muscle-specific promoter (e.g., MCK, tMCK, or CK8); microdystrophin/sequence tailored for patients amenable to exon-50 skipping	Active, not recruiting	$\sim 5 \times 10^{13}$ vg/kg	Single-dose IV infusion	4–10 years; **Inclusion:** genotype-defined, adequate liver, kidney, and heart function, negative for anti-AAV antibodies; **Exclusions**: nontarget mutation types, severe organ dysfunction, risk of interfering with assessments, anti-AAV antibodies present.	Early Phase 1	Peking Union Medical College Hospital	NCT06392724	2025-07-29	(Yuan et al., 2018)
rAAV1.CMV. huFollistatin344	Start: 2015-01 Completion: 2017-11	AAV1; CMV (cytomegalovirus) promoter	Completed	$\sim 1 \times 10^{12}$ vg/kg	Intramuscular (IM) injection	7 years and older; **Inclusion:** diagnosis of muscle-wasting disorder, adequate liver, kidney, and cardiac function; **Exclusions**: severe cardiac, hepatic, or renal disease, prior AAV1 exposure in the same muscle groups.	Phase ½	Jerry R. Mendell	NCT02354781	2023-10-11	(Mendell et al., 2015; Kota et al., 2009; Rodino-Klapac et al., 2009; Miller et al., 2006; Bian et al., 2025)
rAAVrh74.MCK. GALGT2	Start: 2017-11-06Completion: 2023-12-31	Recombinant AAVrh74; MCK (muscle-creatine kinase) muscle-specific promoter; GALGT2 gene transfer	Completed	2.5×10^{13} to 5×10^{13} vg/kg	Single-dose intravascular limb infusion (ILI) in early studies; IV infusion in later strategies	4 years and older; **Inclusion**: genetic requirement, ambulatory boys, stable steroid regimen, low/negative anti-AAVrh74 NAbs; **Exclusions:** ambulation loss, cardiac and pulmonary disease, high anti-AAVrh74 antibody titers, any previous AAV exposure.	Phase ½	Kevin Flanigan, Nationwide Children's Hospital	NCT03333590	2025-03-13	(Chicoine et al., 2014b; Xu et al., 2009)

https://clinicaltrials.gov/ accessed 2026/8/2 with the search term DMD or Duchenne muscular dystrophy.rAAV, recombinant adeno-associated virus; scAAV, self-complementary adeno-associated virus; AAVrh74, adeno-associated virus serotype rh74; AAV1, adeno-associated virus serotype 1; AAV8, adeno-associated virus serotype 8; AAV9, adeno-associated virus serotype 9; CMV, cytomegalovirus; MCK, muscle creatine kinase (promoter); tMCK, truncated muscle creatine kinase (promoter); MHCK7, muscle-specific hybrid creatine kinase promoter (MHCK7); Spc5-12, synthetic muscle/cardiac promoter (Spc5-12); U7snRNA, U7 small nuclear RNA; IV, intravenous; IM, intramuscular; ILI, intravascular limb infusion; vg/kg, vector genomes per kilogram; qPCR, quantitative polymerase chain reaction; Ab, antibody; NAb, neutralizing antibody; NSAA, North Star Ambulatory Assessment; LTFU, long-term follow-up.

8.2 CRISPR-mediated gene therapy

Genome engineering approaches can be broadly divided by how they alter DNA or gene expression. CRISPR activation (CRISPRa) primarily utilizes a deactivated Cas9 protein fused with transcriptional activators to upregulate the gene target's expression without altering its DNA sequence (McLaughlin et al., 2025; Yao et al., 2023; Pan et al., 2022). This precise, targeted approach offers significant advantages over traditional methods of random mutagenesis or transgene-based overexpression (McLaughlin et al., 2025).

Nuclease editing (e.g., Cas9 or Cas12a) introduces double-strand breaks that are repaired by nonhomologous end joining (NHEJ) or homology-directed repair (HDR), enabling knockouts, insertions, or precise sequence replacement (Liao et al., 2024). In contrast to the HDR pathway, which is generally more precise but not error-free, the NHEJ pathway is error-prone, and small insertions or deletions (indels) occur frequently at the cleavage site (Song et al., 2021).

Base editing is a newer genome-editing approach that uses components from CRISPR systems, together with other enzymes, to directly install point mutations into cellular DNA or RNA without making double-stranded DNA breaks (DSBs). DNA base editors comprise a catalytically disabled nuclease fused to a nucleobase deaminase enzyme and, in some cases, a DNA glycosylase inhibitor (Rees and Liu, 2018).

Prime editing (PE) holds great potential in CRISPR therapy as it mediates targeted insertions, deletions, and all 12 possible base-to-base conversions without introducing a double-strand DNA break. PE typically employs a Cas9 nickase along with a tethered or untethered Moloney murine leukemia virus (M-MLV) reverse transcriptase (RT), and a PE guide RNA (pegRNA). epegRNAs enhance the efficiency of the installation or correction of disease-relevant mutations (Nelson et al., 2022; Li et al., 2025).

CRISPR and CRISPR-associated (Cas) protein systems represent a new generation of precise and versatile gene-editing technologies, several of which have recently progressed into clinical trial phases. Table 6.5 focuses on registered human trials. The CRISPR/Cas system uses a guide RNA to target Cas9 to specific genomic loci, generating double-strand breaks. These breaks are repaired by either NHEJ, which may create indels, or HDR, enabling precise correction with a donor DNA template (Mali et al., 2013; Chemello et al., 2023). Ongoing CRISPR-based approaches for DMD aim to restore the open reading frame through techniques such as exon deletion, splice-site modification, and frame-shifting indel introduction (Chemello et al., 2023).

Here is a current table summarizing the recent clinical trials and new/ongoing clinical trials for CRISPR-mediated gene (or genome) editing therapies in DMD. Because CRISPR-based therapy for DMD is still in its very early stages, the number of trials is small; the table includes all publicly registered or announced in vivo editing trials (single-patient, early-phase, or first-in-human).

As of February 2026, there are only two publicly known in vivo CRISPR-based human gene-editing trials for DMD. Both are first-in-human (one single-patient, one small cohort). Both trials target specific subsets of DMD-HG302 targets patients amenable to exon 51 skipping/reframing; CRD-TMH-001 targets a rare promoter/exon 1 mutation. This reflects a core limitation of CRISPR editing: for now, many therapies are tailored to particular genotypes, not universal.

The CRD-TMH-001 therapeutic approach involved systemic administration of an rAAV9 vector encoding a CRISPRa system, in which a catalytically inactive Cas9 from *Staphylococcus aureus* (dSaCas9) was fused to VP64, a synthetic transcriptional activator. This system was designed to enhance endogenous DMD transcription by recruiting RNA polymerase II and coactivators to the promoter or enhancer regions, with the goal of partially restoring dystrophin expression via the upregulation of an alternative isoform, such as the cortical dystrophin variant (Dp427c). In the CRD-TMH-001 trial, a fatal acute respiratory distress syndrome event was reported after high-dose systemic rAAV9 delivery of a dSaCas9-VP64 CRISPR transactivator in an N-of-1 DMD trial; the report implicated an innate immune reaction consistent with dose-limiting AAV toxicity (Lek et al., 2023).

The HG302 program uses a high-fidelity Cas12 (hfCas12Max) rather than the standard Cas9; this is aimed at minimizing off-target editing and, hopefully, allowing lower vector doses. The first patient was dosed in December 2024 (Therapeutics, 2024). Early preclinical data are promising (dystrophin restoration in mice and NHPs, good safety). CRISPR-based therapies are among the most promising emerging approaches but face challenges in delivery (muscle tropism, immune response) and genotypic restrictions in each treatment. Many CRISPR/genome-editing preclinical programs are still in the animal or cell stage. For example, companies like Precision BioSciences announced a preclinical in vivo editing candidate (PBGENE-DMD) in 2025, but as of now, there is no public IND or clinical trial registration.

8.3 Complications and management of complications

Early-phase clinical trials have demonstrated encouraging safety profiles and transgene expression; however, challenges such as immune responses, efficacy, and long-term durability remain (Khan and Hussain, 2026). Immune response-related adverse events observed following delandistrogene moxeparvovec administration include vomiting, liver injury,

TABLE 6.5 Recent clinical trials for CRISPR-mediated gene therapies in DMD (as of February 2026).

Product	Study start/completion	Reported dose (vg/kg)	Route of administration	CRISPR agent	Mechanism of action	Current status	Age range/key inclusion/exclusion	Phase	Sponsor/developer	Trial ID	Last updated
HG302 (M.U.S.C.L.E.)	Start: 2024-11-06 Completion: 2026-09-30	~1×10^{13} vg/kg	IV single injection	CRISPR–hf-Cas12Max in single AAV vector	Exon 51 skipping	Recruiting	4–8 years; **Inclusion**: male, ≥ 4 and ≤ 8 years at enrollment with a confirmed DMD diagnosis, DMD gene mutation: deletions in exons 52, 52–61, or 52–63, able to walk ≥ 10 m independently; **Exclusion**: active infection, DMD-associated cardiomyopathy, respiratory insufficiency, or use of any investigational drugs or exon-skipping therapy within 6 months.	Not applicable	HuidaGene Therapeutics Co., Ltd	NCT06594094	2024-11-25
CRD-TMH-001	Start: 2022-08-31 Completion: 2023-09	~1×10^{14} vg/kg	IV injection	CRISPR–Cas9 delivered via rAAV	Upregulates dystrophin	Unknown status	18–28 years**; Inclusion**: completion of informed consent, confirmation of genetic mutation, confirmation of absence of elevated AAV9 NAbs; **Exclusion**: any significant medical issue(s) that would, in the opinion of the Principal Investigator, prevent this patient from being dosed.	Phase 1	Cure Rare Disease, Inc.	NCT05514249	2022-09-01

https://clinicaltrials.gov/ accessed 2026/8/2 with the search term DMD or Duchenne muscular dystrophy.*vg/kg*, vector genomes per kilogram; *IV*, intravenous; *AAV*, adeno-associated virus; *hfCas12Max*, high-fidelity Cas12Max nuclease; *AAV9*, adeno-associated virus serotype 9; NAb, neutralizing antibody.

myocarditis, myositis, and rhabdomyolysis, which highlight the importance of rigorous patient selection, immunosuppression protocols, and ongoing monitoring (Shimizu-Motohashi, 2025). Despite the recent approval of microdystrophin gene therapy, the long-term efficacy and stability of transgene expression remain significant clinical challenges.

Future advancements in gene therapy for DMD may address current challenges related to efficacy, long-term transgene durability, and immune response-related accessibility of gene therapy. A key challenge is the limited packaging capacity of AAV vectors, which precludes delivery of the full-length DMD gene and necessitates the use of truncated microdystrophin constructs. Although these truncated constructs have shown functional benefits in clinical studies, complete restoration of full-length dystrophin remains a major objective, potentially achievable through dual- or triple-vector systems (Duan, 2024) or alternative nonviral delivery platforms (Chulanova et al., 2025).

Recent studies have demonstrated the feasibility of dual- and triple-AAV vector systems combined with split-intein technology for in vivo delivery of full-length or large dystrophin proteins (Zhou et al., 2024; Tasfaout et al., 2024). Enhancing transduction efficiency across skeletal and cardiac muscles remains a critical objective. Progress in capsid engineering and promoter refinement has enhanced muscle specificity and minimized the required vector dosage (Zhou et al., 2024). Another key objective is to ensure the durability of dystrophin expression. Although sustained expression has been observed for up to 1 year in preclinical models (Baine et al., 2024) and functional benefits have persisted for 4 years in early-phase trials (Mendell et al., 2024), definitive evidence of long-term expression is still lacking. Finally, immune barriers continue to pose a significant challenge. Preexisting NAbs directed against AAV capsids currently preclude treatment in a proportion of otherwise eligible patients (Shimizu-Motohashi, 2025).

9 Functional outcome measures

In DMD clinical trials, ambulatory functional measures—such as 6MWD: distance walked in 6 min; NSAA: 17-item motor scale evaluating walking, climbing, rising, and running; timed function tests (TFTs): time to rise from the floor (TTSTAND), time to climb 4 stairs (TTCLIMB), and time to run or walk 10 m (TTRW)—have been widely used to assess disease progression and quantify therapeutic benefit in boys who are still able to walk. These measures are sensitive to subtle changes in motor performance and can detect deviations from the predictable decline observed in natural history studies (Todd et al., 2025).

As individuals transition to later stages of DMD and lose ambulation, additional outcome measures become crucial to continue capturing meaningful clinical change. Nonambulatory and secondary endpoints, such as the Performance of Upper Limb score, pulmonary function tests, including FVC (% predicted), and cardiac assessments (e.g., ejection fraction or myocardial fibrosis on MRI), extend the evaluable window of clinical trials across the disease spectrum. Biomarkers and imaging measures—including dystrophin expression, CK levels, and quantitative muscle MRI—along with patient-reported outcomes, provide further supportive evidence and help build a comprehensive picture of therapeutic impact beyond lower-limb function (Mcdonald et al., 2013).

Long-term evaluation of microdystrophin expression on ambulatory and lower-limb function is conducted using validated clinical measures, including NSAA (Pane et al., 2014; Ricotti et al., 2016). NSAA, developed to track disease progression, is currently recognized by regulatory agencies, including the FDA and European Medicines Agency, as valid for the assessment of functional outcomes in DMD. It is widely used in clinical trials, natural history studies, and clinical practice. NSAA is clinically meaningful as a disease-specific, functionally interpretable measure that is sensitive to disease progression and treatment effects (Gillman et al., 2025; Ayyar Gupta et al., 2023). Individual NSAA items are being analyzed to determine their sensitivity in detecting changes in ambulatory and lower-limb function relative to the overall composite score. Accumulating data from treated and control cohorts enable assessment of the functional impact of microdystrophin gene transfer using validated outcome measures (Asher et al., 2020).

While microdystrophin expression confirms successful gene delivery and translation, expression levels do not consistently translate into proportional functional improvement. Therefore functional outcome measures, such as NSAA and TFTs, remain essential for determining clinical benefit in patients treated with AAV-based gene therapies (Mendell et al., 2020; Bushby and Connor, 2011).

10 Safety and current challenges in gene therapy for DMD

AAV vectors have demonstrated a comparatively favorable safety profile in clinical studies, due in part to their lower immunogenicity relative to adenoviral vectors. Clinical experience with AAV9 and related capsids has nonetheless revealed risks—including complement activation, thrombocytopenia, and immune-mediated liver toxicity—highlighting that adverse events remain clinically meaningful even with next-generation vectors (Asher et al., 2020; Hinderer et al., 2018; Binks and Early, 2019). In contrast, preclinical assessments of AAVrh74 in mice and NHPs have shown no toxicity at doses up to 6×10^{14} vg/kg—approximately threefold higher than current clinical dosing—supporting its potential tolerability

(Ehrhardt et al., 2003). This is consistent with its mechanism of maintaining delivered genes as stable, nonintegrating episomes, with integration into the host genome occurring at very low frequency and therefore posing minimal oncogenic risk (Ehrhardt et al., 2003). Early human studies further corroborate the acceptable safety profile of AAVrh74, although continued evaluation across multiple ongoing clinical programs remains essential (Mendell et al., 2019; Rodino-Klapac et al., 2019).

Despite these advantages, AAV-mediated gene therapy for DMD faces major obstacles that constrain long-term benefits. Because current AAV-based approaches rely on microdystrophin surrogates (see Mechanism section), clinical efficacy is further influenced by immune barriers and tissue-specific transduction efficiency. Systemic AAV delivery encounters immunologic barriers, including preexisting NAbs that exclude a substantial proportion of patients and postdosing immune reactions that may necessitate high-dose immunosuppression. Because AAV cannot be safely redosed, the initial treatment window is particularly critical, as therapeutic expression diminishes relative to body mass as children grow.

Additional challenges include incomplete cardiac transduction, leaving patients vulnerable to cardiomyopathy—one of the leading causes of mortality in DMD—and reduced efficacy in advanced disease due to fibrosis and fat infiltration that limit vector penetration. CRISPR-based editing strategies add unique risks, such as variable editing efficiency across muscles, mosaic dystrophin restoration, potential off-target activity, and stochastic DNA repair outcomes in postmitotic myofibers. Manufacturing constraints further restrict access, as DMD requires exceptionally high vector doses, driving cost, production complexity, and global supply limitations. Importantly, neither microdystrophin replacement nor genome editing addresses secondary drivers of disease progression—including inflammation, calcium dysregulation, mitochondrial dysfunction, fibrosis, and cardiac remodeling—meaning that restored dystrophin alone cannot reverse established pathology.

Together, these safety considerations and biological constraints underscore the need to develop lower-dose or repeat-dosable delivery systems, improved cardiac and whole-body targeting methods, next-generation genome-editing tools with high precision and efficiency, and combination therapeutic strategies capable of modifying both primary and downstream disease mechanisms in DMD.

11 Conclusion

Gene therapy has fundamentally reshaped the therapeutic landscape of DMD by shifting the focus from symptomatic management toward molecular correction of the primary genetic defect. Among current strategies, AAV-mediated microdystrophin replacement has emerged as the most clinically advanced approach, offering systemic dystrophin restoration and measurable functional benefits in early-stage disease. However, limitations related to vector capacity, immune barriers, incomplete cardiac transduction, and the imperfect correlation between dystrophin expression and clinical outcomes underscore that microdystrophin functions as a surrogate rather than a full biological replacement.

Future progress in DMD gene therapy will depend on next-generation capsid engineering, immune-evasive or readministrable delivery platforms, improved cardiac targeting, and maturation of genome-editing technologies capable of durable correction. Importantly, combination strategies integrating dystrophin-restoring gene therapy with adjunctive approaches targeting inflammation, fibrosis, and muscle regeneration are likely required to achieve maximal and sustained clinical benefit. Collectively, these advances position gene therapy not as a single solution, but as the central pillar of a multimodal therapeutic paradigm for DMD.

AI disclosure

During the preparation of this work, the author(s) used ChatGPT to improve the readability and grammar of the text. All the contents were verified, and the author(s) take full responsibility for all content in the text. After using this tool/service, the author(s) reviewed and edited the content as needed and take(s) full responsibility for the content of the publication.

References

't Hoen, P.A.C., de Meijer, E.J., Boer, J.M., Vossen, R.H.A.M., Rolf, T., Maatman, R.G.H.J., Davies, K.E., van Ommen, G.-J.B., van Deutekom, J.C.T., Dunnen, J.T., 2008. Generation and characterization of transgenic mice with the full-length human DMD gene. J. Biol. Chem. 00219258. 283 (9), 5899–5907. doi:10.1074/jbc.m709410200.

Aartsma-Rus, A., Corey, D.R., 2020. The 10th oligonucleotide therapy approved: golodirsen for duchenne muscular dystrophy. Nucleic Acid Ther. 21593345. 30 (2), 67–70. Available from: http://www.liebertonline.com/oli. doi:10.1089/nat.2020.0845.

Aartsma-Rus, A., Goemans, N., 2019. A sequel to the eteplirsen saga: eteplirsen is approved in the United States but was not approved in Europe. Nucleic Acid Ther. 21593345. 29 (1), 13–15. Available from: http://www.liebertonline.com/oli. doi:10.1089/nat.2018.0756.

Aartsma-Rus, A., Krieg, A.M., 2017. FDA Approves eteplirsen for Duchenne muscular dystrophy: the next chapter in the eteplirsen saga. Nucleic Acid Ther. 21593345. 27 (1), 1–3. Available from: http://www.liebertonline.com/oli. doi:10.1089/nat.2016.0657.

Aartsma-Rus, A., Van Deutekom, J.C.T., Fokkema, I.F., Van Ommen, G.B., Den Dunnen, J.T., 2006. Entries in the Leiden Duchenne muscular dystrophy mutation database: an overview of mutation types and paradoxical cases that confirm the reading-frame rule. Muscle Nerve. 0148-639X. 34 (2), 135–144. doi:10.1002/mus.20586.

Agency, E.M. EU Clinical Trials Register entry: eudraCT 2020-002093-27. 2026 8 February 2026]; Available from: https://www.clinicaltrialsregister.eu/ctr-search/search?query=2020-002093-27.

Ahn, A.H., Kunkel, L.M., 1993. The structural and functional diversity of dystrophin. Nat. Genet. 15461718. 3 (4), 283–291. doi:10.1038/ng0493-283.

Akshaya, R., Mohan, S., Vellapandian, C., 2024. A voyage on the role of nuclear factor kappa B (NF-kB) signaling pathway in duchenne muscular dystrophy: an inherited muscle disorder. Cureus. 2168-8184. 16 (8). doi:10.7759/cureus.67901.

Allen, D.G., Whitehead, N.P., Froehner, S.C., 2015. Absence of dystrophin disrupts skeletal muscle signaling: roles of Ca2+, reactive oxygen species, and nitric oxide in the development of muscular dystrophy. Physiol. Rev. 15221210. 96 (1), 253–305. Available from: http://physrev.physiology.org/content/96/1/253.full.pdf. doi:10.1152/physrev.00007.2015.

Amann, K.J., Renley, B.A., Ervasti, J.M., 1998. A cluster of basic repeats in the dystrophin rod domain binds F-actin through an electrostatic interaction. J. Biol. Chem. 00219258. 273 (43), 28419–28423. doi:10.1074/jbc.273.43.28419.

Amoasii, L., Hildyard, J.C.W., Li, H., Sanchez-Ortiz, E., Mireault, A., Caballero, D., Harron, R., Stathopoulou, T.-R., Massey, C., Shelton, J.M., Bassel-Duby, R., Piercy, R.J., Olson, E.N., 2018. Gene editing restores dystrophin expression in a canine model of Duchenne muscular dystrophy. Science. 0036-8075. 362 (6410), 86–91. doi:10.1126/science.aau1549.

Amoasii, L., Long, C., Hui, L., Mireault, A.A., Shelton, J.M., Sanchez-Ortiz, E., McAnally, J.R., Bhattacharyya, S., Schmidt, F., Grimm, D., Hauschka, S.D., Bassel-Duby, R., Olson, E.N., 2017. Single-cut genome editing restores dystrophin expression in a new mouse model of muscular dystrophy. Sci. Transl. Med. 1946-6234. 9 (418). doi:10.1126/scitranslmed.aan8081.

Anthony, K., Feng, L., Arechavala-Gomeza, V., Guglieri, M., Straub, V., Bushby, K., Cirak, S., Morgan, J., Muntoni, F., 2012. Exon skipping quantification by quantitative reverse-transcription polymerase chain reaction in Duchenne muscular dystrophy patients treated with the antisense oligomer eteplirsen. Hum. Gene Ther. Methods, 23 (5), 336–345. doi:10.1089/hgtb.2012.117.

Araki, E., Nakamura, K., Nakao, K., Kameya, S., Kobayashi, O., Nonaka, I., Kobayashi, T., Katsuki, M., 1997. Targeted disruption of Exon 52 in the mouse dystrophin gene induced muscle degeneration similar to that observed in duchenne muscular dystrophy. Biochem. Biophys. Res. Commun. 0006291X. 238 (2), 492–497. doi:10.1006/bbrc.1997.7328.

Arpke, R.W., Darabi, R., Mader, T.L., Zhang, Y., Toyama, A., Lonetree, C.L., Nash, N., Lowe, D.A., Perlingeiro, R.C.R., Kyba, M., 2013. A new immuno-, dystrophin-deficient model, the NSG-mdx4Cv mouse, provides evidence for functional improvement following allogeneic satellite cell transplantation. Stem Cells. 10665099. 31 (8), 1611–1620. Available from: http://onlinelibrary.wiley.com/doi/10.1002/stem.1402/pdf. doi:10.1002/stem.1402.

Asher, D.R., Thapa, K., Dharia, S.D., Khan, N., Potter, R.A., Rodino-Klapac, L.R., Mendell, J.R., 2020. Clinical development on the frontier: gene therapy for duchenne muscular dystrophy. Expert Opin. Biol. Ther. 17447682. 20 (3), 263–274. Available from: https://www.tandfonline.com/loi/iebt20. doi:10.1080/14712598.2020.1725469.

Atencia-Fernandez, S., Shiel, R.E., Mooney, C.T., Nolan, C.M., 2015. Muscular dystrophy in the Japanese Spitz: an inversion disrupts the DMD and RPGR genes. Anim. Genet. 02689146. 46 (2), 175–184. doi:10.1111/age.12266.

Audhya, I., Nacson, A.B., Gooch, K., Basnyat, B., Slota, C., Martin, S., Murphy, A., Lansdall, C.J., Ciobanu, T., Nascimento, A., Veerapandiyan, A., 2026. Caregiver-reported patient experiences with duchenne muscular dystrophy: qualitative in-trial interviews 1 year after delandistrogene moxeparvovec in the pivotal EMBARK trial. Neurol. Ther. 2193-8253. 15 (1), 41–60. doi:10.1007/s40120-025-00842-7.

Ayyar Gupta, V., Pitchforth, J.M., Domingos, J., Ridout, D., Iodice, M., Rye, C., Chesshyre, M., Wolfe, A., Selby, V., Mayhew, A., Mazzone, E.S., Ricotti, V., Hogrel, J.-Y., Niks, E.H., de Groot, I., Servais, L., Straub, V., Mercuri, E., Manzur, A.Y., Muntoni, F., Alway, S.E., 2023. Determining minimal clinically important differences in the North Star Ambulatory Assessment (NSAA) for patients with Duchenne muscular dystrophy. PLoS One. 1932-6203. 18 (4), e0283669. doi:10.1371/journal.pone.0283669.

Bai, R., Guo, W., Zhang, T., Ren, S., Liu, J., Xiao, P., Zhang, J., Sun, W., Yang, J., Yue, M., Liu, S., Zhou, C., Shangang, L., Wang, H., Zhang, S., Weizhi, J., Shiwen, W., Chen, Y., 2025. Single-cut gene therapy in a one-step generated rhesus monkey model of Duchenne muscular dystrophy. Cell Rep. Med. 26663791. 6 (4), 102037. doi:10.1016/j.xcrm.2025.102037.

Baine, S., Wier, C., Lemmerman, L., Cooper-Olson, G., Kempton, A., Haile, A., Endres, J., Fedoce, A., Nesbit, E., Rodino-Klapac, L.R., Potter, R.A., 2024. Long-term survival and myocardial function following systemic delivery of delandistrogene moxeparvovec in DMD MDX Rats. Hum. Gene Ther. 1043-0342. 35 (23-24), 978–988. doi:10.1089/hum.2024.013.

Banks, G.B., Judge, L.M., Allen, J.M., Chamberlain, J.S., Cox, G.A., 2010. The polyproline site in hinge 2 influences the functional capacity of truncated dystrophins. PLoS Genet. 1553-7404. 6 (5), e1000958. doi:10.1371/journal.pgen.1000958.

Bello, L., Hoffman, E.P., Pegoraro, E., 2023. Is it time for genetic modifiers to predict prognosis in Duchenne muscular dystrophy? Nat. Rev. Neurol. 17594766. 19 (7), 410–423. Available from: https://www.nature.com/nrneurol/. doi:10.1038/s41582-023-00823-0.

Bian, X., Snow, Z.K., Zinn, C.J., Gowan, C.C., Conley, S.M., Bratulin, A.L., Elhusseiny, K.M., Miller, J., Tchkonia, T., Kirkland, J.L., Lerman, L.O., Hickson, L.J., 2025. Activin A antagonism with follistatin reduces kidney fibrosis, injury, and cellular senescence-associated inflammation in murine diabetic kidney disease. Kidney360. 2641-7650. 6 (8), 1278–1291. doi:10.34067/kid.0000000776.

Binks, M., Early,, 2019. Initial Data from C3391001, a First-In-Human Safety Study of PF-06939926, a Mini-Dystrophin Gene Therapy for the Potential Treatment of DMD, Parent Project Muscular Dystrophy.

Birnkrant, D.J., Bushby, K., Bann, C.M., Apkon, S.D., Blackwell, A., Brumbaugh, D., Case, L.E., Clemens, P.R., Hadjiyannakis, S., Pandya, S., Street, N., Tomezsko, J., Wagner, K.R., Ward, L.M., Weber, D.R., 2018. Diagnosis and management of Duchenne muscular dystrophy, part 1: diagnosis, and neuromuscular, rehabilitation, endocrine, and gastrointestinal and nutritional management. Lancet Neurol. 14744465. 17 (3), 251–267. Available from: http://www.journals.elsevier.com/the-lancet-neurology/. doi:10.1016/S1474-4422(18)30024-3.

Bladen, C.L., et al., 2015. The TREAT-NMD DMD global database: analysis of more than 7,000 Duchenne muscular dystrophy mutations. Hum. Mutat 36 (4), 395–402. doi:10.1002/humu.22758.

Bladen, C.L., Salgado, D., Monges, S., Foncuberta, M.E., Kekou, K., Kosma, K., Dawkins, H., Lamont, L., Roy, A.J., Chamova, T., Guergueltcheva, V., Chan, S., Korngut, L., Campbell, C., Dai, Y., Wang, J., Barišić, N., Brabec, P., Lahdetie, J., Walter, M.C., Schreiber-Katz, O., Karcagi, V., Garami, M., Viswanathan, V., Bayat, F., Buccella, F., Kimura, E., Koeks, Z., van den Bergen, J.C., Rodrigues, M., Roxburgh, R., Lusakowska, A., Kostera-Pruszczyk, A., Zimowski, J., Santos, R., Neagu, E., Artemieva, S., Rasic, V.M., Vojinovic, D., Posada, M., Bloetzer, C., Jeannet, P.Y., Joncourt, F., Díaz-Manera, J., Gallardo, E., Karaduman, A.A., Topaloğlu, H., Sherif, R.E., Stringer, A., Shatillo, A.V., Martin, A.S., Peay, H.L., Bellgard, M.I., Kirschner, J., Flanigan, K.M., Straub, V., Bushby, K., Verschuuren, J., Aartsma-Rus, A., Béroud, C., Lochmüller, H., 2015. The TREAT-NMD DMD global database: analysis of more than 7,000 duchenne muscular dystrophy mutations. Hum. Mutat. 10981004. 36 (4), 395–402. Available from: http://onlinelibrary.wiley.com/journal/10.1002/(ISSN)1098-1004. doi:10.1002/humu.22758.

Boehler, J.F., Brown, K.J., Ricotti, V., Morris, C.A., 2024. N-terminal titin fragment: a non-invasive, pharmacodynamic biomarker for microdystrophin efficacy. Skelet. Muscle. 20445040. 14 (1). Available from: https://skeletalmusclejournal.biomedcentral.com/. doi:10.1186/s13395-023-00334-y.

Boland, B.J., Silbert, P.L., Groover, R.V., Wollan, P.C., Silverstein, M.D., 1996. Skeletal, cardiac, and smooth muscle failure in Duchenne muscular dystrophy. Pediatr. Neurol. 08878994. 14 (1), 7–12. Available from: https://www.elsevier.com/locate/pedneu. doi:10.1016/0887-8994(95)00251-0.

Bonilla, E., Samitt, C.E., Miranda, A.F., Hays, A.P., Salviati, G., DiMauro, S., Kunkel, L.M., Hoffman, E.P., Rowland, L.P., 1988. Duchenne muscular dystrophy: deficiency of dystrophin at the muscle cell surface. Cell. 00928674. 54 (4), 447–452. doi:10.1016/0092-8674(88)90065-7.

Bresolin, N., Castelli, E., Comi, G.P., Felisari, G., Bardoni, A., Perani, D., Grassi, F., Turconi, A., Mazzucchelli, F., Gallotti, D., Moggio, M., Prelle, A., Ausenda, C., Fazio, G., Scarlato, G., 1994. Cognitive impairment in Duchenne muscular dystrophy. Neuromuscul. Disord. 09608966. 4 (4), 359–369. doi:10.1016/0960-8966(94)90072-8.

Brooke, M.H., Fenichel, G.M., Griggs, R.C., Mendell, J.R., Moxley, R., Florence, J., King, W.M., Pandya, S., Robison, J., Schierbecker, J., Signore, L., Miller, J.P., Gilder, B.F., Kaiser, K.K., Mandel, S., Arfken, C., 1989. Duchenne muscular dystrophy: patterns of clinical progression and effects of supportive therapy. Neurology. 0028-3878. 39 (4), 475. doi:10.1212/WNL.39.4.475.

Brusa, R., Magri, F., Bresolin, N., Comi, G.P., Corti, S., 2020. Noncoding RNAs in Duchenne and Becker muscular dystrophies: role in pathogenesis and future prognostic and therapeutic perspectives. Cell. Mol. Life Sci. 1420-682X. 77 (21), 4299–4313. doi:10.1007/s00018-020-03537-4.

Budzinska, M., Zimna, A., Kurpisz, M., 2021. The role of mitochondria in duchenne muscular dystrophy. J. Physiol. Pharmacol. 18991505. 72 (2), 1–10. Available from: http://www.jpp.krakow.pl/journal/archive/04_21/pdf/10.26402/jpp.2021.2.01.pdf. doi:10.26402/jpp.2021.2.01.

Bulfield, G., Siller, W.G., Wight, P.A., Moore, K.J., 1984. X chromosome-linked muscular dystrophy (mdx) in the mouse. Proc. Natl. Acad. Sci. USA. 0027-8424. 81 (4), 1189–1192. doi:10.1073/pnas.81.4.1189.

Bushby, K., Connor, E., 2011. Clinical outcome measures for trials in Duchenne muscular dystrophy: report from International Working Group meetings. Clin. Investig. 2041-6792. 1 (9), 1217–1235. doi:10.4155/cli.11.113.

Butterfield, R.J., Shieh, P.B., Li, H., Binks, M., McDonnell, T.G., Ryan, K.A., Delnomdedieu, M., Belluscio, B.A., Neelakantan, S., Levy, D.I., Schwartz, P.F., Smith, E.C., 2025. AAV mini-dystrophin gene therapy for Duchenne muscular dystrophy: a phase 1b trial. Nat. Med. 1546170X. 31 (8), 2712–2721. Available from: https://www.nature.com/nm/. doi:10.1038/s41591-025-03750-3.

Byrne, B.J., Butterfield, R.J., Shieh, P.B., Smith, E.C., Licht, C., Binks, M., Casinghino, S., Delnomdedieu, M., Ravindra, K.C., McDonnell, T., Ryan, K., Schulz, M., Shen, Q., Shi, H., Sirivelu, M.P., Vaidya, V.S., Whiteley, L., Levy, D.I., 2025a. Complement activation in a phase Ib study of fordadistrogene movaparvovec for Duchenne muscular dystrophy. Mol. Ther. 15250024. 33 (9), 4226–4238. Available from: https://www.sciencedirect.com/science/journal/15250016. doi:10.1016/j.ymthe.2025.06.032.

Byrne, B.J., Flanigan, K.M., Matesanz, S.E., Finkel, R.S., Waldrop, M.A., D'Ambrosio, E.S., Johnson, N.E., Smith, B.K., Bönnemann, C., Carrig, S., Rossano, J.W., Greenberg, B., Lalaguna, L., Lara-Pezzi, E., Subramony, S., Corti, M., Mercado-Rodriguez, C., Leon-Astudillo, C., Ahrens-Nicklas, R., Bharucha-Goebel, D., Gao, G., Gessler, D.J., Hwu, W.L., Chien, Y.H., Lee, N.C., Boye, S.L., Boye, S.E., George, L.A., 2025b. Current clinical applications of AAV-mediated gene therapy. Mol. Ther. 15250024. 33 (6), 2479–2516. Available from: https://www.sciencedirect.com/science/journal/15250016. doi:10.1016/j.ymthe.2025.04.045.

Chamberlain, J.R., Chamberlain, J.S., 2017. Progress toward gene therapy for duchenne muscular dystrophy. Mol. Ther. 15250024. 25 (5), 1125–1131. Available from: https://www.journals.elsevier.com/molecular-therapy. doi:10.1016/j.ymthe.2017.02.019.

Chamberlain, J.S., Metzger, J., Reyes, M., Townsend, D.W., Faulkner, J.A., 2007. Dystrophin-deficient mdx mice display a reduced life span and are susceptible to spontaneous rhabdomyosarcoma. FASEB J. 08926638. 21 (9), 2195–2204. Available from: http://www.fasebj.org/cgi/reprint/21/9/2195. doi:10.1096/fj.06-7353com.

Chapman, V.M., Miller, D.R., Armstrong, D., Caskey, C.T., 1989. Recovery of induced mutations for X chromosome-linked muscular dystrophy in mice. Proc. Natl. Acad. Sci. USA. 0027-8424. 86 (4), 1292–1296. doi:10.1073/pnas.86.4.1292.

Cheeran, D., Khan, S., Khera, R., Bhatt, A., Garg, S., Grodin, J.L., Morlend, R., Araj, F.G., Amin, A.A., Thibodeau, J.T., Das, S., Drazner, M.H., Mammen, P.P.A., 2017. Predictors of death in adults with duchenne muscular dystrophy–associated cardiomyopathy. J. Am. Heart Assoc. 2047-9980. 6 (10). doi:10.1161/jaha.117.006340.

Chemello, F., Bassel-Duby, R., Olson, E.N., 2020. Correction of muscular dystrophies by CRISPR gene editing. J. Clin. Invest. 15588238. 130 (6), 2766–2776. Available from: https://www.jci.org/articles/view/136873/pdf. doi:10.1172/JCI136873.

Chemello, F., Olson, E.N., Bassel-Duby, R., 2023. CRISPR-Editing therapy for Duchenne muscular dystrophy. Hum. Gene Ther. 15577422. 34 (9-10), 379–387. Available from: https://www.liebertonline.com/hum. doi:10.1089/hum.2023.053.

Chen, Y., Zheng, Y., Kang, Y., Yang, W., Niu, Y., Guo, X., Zhuchi, T., Chenyang, S., Wang, H., Xing, R., Xiuqiong, P., Yang, S.-H., Shihua, L., Weizhi, J., Xiao-Jiang, L., 2015. Functional disruption of the dystrophin gene in rhesus monkey using CRISPR/Cas9. Hum. Mol. Genet. 1460-2083. 24 (13), 3764–3774. doi:10.1093/hmg/ddv120.

Chicoine, L.G., Montgomery, C.L., Bremer, W.G., Shontz, K.M., Griffin, D.A., Heller, K.N., Lewis, S., Malik, V., Grose, W.E., Shilling, C.J., Campbell, K.J., Preston, T.J., Coley, B.D., Martin, P.T., Walker, C.M., Clark, K.R., Sahenk, Z., Mendell, J.R., Rodino-Klapac, L.R., 2014a. Plasmapheresis eliminates the negative impact of AAV antibodies on microdystrophin gene expression following vascular delivery. Mol. Ther. 15250016. 22 (2), 338–347. doi:10.1038/mt.2013.244.

Chicoine, L.G., Rodino-Klapac, L.R., Shao, G., Xu, R., Bremer, W.G., Camboni, M., Golden, B., Montgomery, C.L., Shontz, K., Heller, K.N., Griffin, D.A., Lewis, S., Coley, B.D., Walker, C.M., Clark, K.R., Sahenk, Z., Mendell, J.R., Martin, P.T., 2014b. Vascular delivery of rAAVrh74.MCK. GALGT2 to the gastrocnemius muscle of the rhesus macaque stimulates the expression of dystrophin and laminin α2 surrogates. Mol. Ther. 15250024. 22 (4), 713–724. Available from: https://www.journals.elsevier.com/molecular-therapy. doi:10.1038/mt.2013.246.

Chulanova, Y., Breier, D., Peer, D., 2025. Delivery of genetic medicines for muscular dystrophies. Cell Rep. Med. 26663791. 6 (1). Available from: https://www.sciencedirect.com/science/journal/26663791. doi:10.1016/j.xcrm.2024.101885.

Chwalenia, K., Feng, V.Y., Hemmer, N., Friedrichsen, H.J., Vorobieva, I., Wood, M.J.A., Roberts, T.C., 2025. AAV microdystrophin gene replacement therapy for Duchenne muscular dystrophy: progress and prospects. Gene Ther. 14765462. 32 (5), 447–461. Available from: https://www.nature.com/gt/. doi:10.1038/s41434-025-00561-6.

Clemens, P.R., Rao, V.K., Connolly, A.M., Harper, A.D., Mah, J.K., McDonald, C.M., Smith, E.C., Zaidman, C.M., Nakagawa, T., Hoffman, E.P., 2023. Efficacy and safety of viltolarsen in boys with Duchenne muscular dystrophy: results from the phase 2, open-label, 4-year extension study. J. Neuromuscul. Dis. 22143602. 10 (3), 439–447. doi:10.3233/JND-221656.

Clemens, P.R., Rao, V.K., Connolly, A.M., Harper, A.D., Mah, J.K., Smith, E.C., McDonald, C.M., Zaidman, C.M., Morgenroth, L.P., Osaki, H., Satou, Y., Yamashita, T., Hoffman, E.P., 2020. Safety, tolerability, and efficacy of viltolarsen in boys with duchenne muscular dystrophy amenable to exon 53 skipping: a phase 2 randomized clinical trial. JAMA Neurol. 21686157. 77 (8), 982–991. Available from: http://archneur.jamanetwork.com/issues.aspx. doi:10.1001/jamaneurol.2020.1264.

ClinicalTrials.gov. Clinical Trial Registration: NCT05429372. 2026a. Available from: https://clinicaltrials.gov/study/NCT05429372. (10 February 2026).

ClinicalTrials.gov. Clinical Trial Registration: NCT05683379. 2026b. Available from: https://clinicaltrials.gov/study/NCT05683379. (10 February 2026).

ClinicalTrials.gov. Clinical Trial Registration: NCT05689164. 2026c. Available from: https://clinicaltrials.gov/study/NCT05689164. (10 February 2026).

ClinicalTrials.gov. Clinical Trial Registration: NCT05693142. 2026d. Available from: https://clinicaltrials.gov/study/NCT05693142. (10 February 2026).

ClinicalTrials.gov. Clinical Trial Registration: NCT05881408. 2026e. Available from: https://clinicaltrials.gov/study/NCT05881408. (10 February 2026).

ClinicalTrials.gov. Clinical Trial Registration: NCT06128564. 2026f. Available from: https://clinicaltrials.gov/study/NCT06128564. (10 February 2026).

ClinicalTrials.gov. Clinical Trial Registration: NCT06138639. 2026g. Available from: https://clinicaltrials.gov/study/NCT06138639. (10 February 2026).

ClinicalTrials.gov. Clinical Trial Registration: NCT06241950. 2026h; Available from: https://clinicaltrials.gov/study/NCT06241950 (10 February 2026).

ClinicalTrials.gov. Clinical Trial Registration: NCT06491927. 2026i. Available from: https://clinicaltrials.gov/study/NCT06491927. (10 February 2026).

ClinicalTrials.gov. Clinical Trial Registration: NCT06597656. 2026j. Available from: https://clinicaltrials.gov/study/NCT06597656. (10 February 2026).

Counterman, K.J., Furlong, P., Wang, R.T., Martin, A.S., 2020. Delays in diagnosis of Duchenne muscular dystrophy: an evaluation of genotypic and sociodemographic factors. Muscle Nerve. 10974598. 61 (1), 36–43. Available from: http://onlinelibrary.wiley.com/journal/10.1002/(ISSN)1097-4598. doi:10.1002/mus.26720.

Crisafulli, S., Sultana, J., Fontana, A., Salvo, F., Messina, S., Trifirò, G., 2020. Global epidemiology of Duchenne muscular dystrophy: an updated systematic review and meta-analysis. Orphanet J. Rare Dis. 1750-1172. 15 (1). doi:10.1186/s13023-020-01430-8.

Cully, T.R., Rodney, G.G., 2020. Nox4—RyR1—Nox2: regulators of micro-domain signaling in skeletal muscle. Redox Biol. 2213-2317. 36, 101557. doi:10.1016/j.redox.2020.101557.

D'Ambrosio, E.S., Mendell, J.R., 2023. Evolving therapeutic options for the treatment of Duchenne muscular dystrophy. Neurotherapeutics. 18787479. 20 (6), 1669–1681. Available from: https://www.springer.com/journal/13311. doi:10.1007/s13311-023-01423-y.

Danko, I., Chapman, V., Wolff, J.A., 1992. The frequency of revertants in mdx mouse genetic models for duchenne muscular dystrophy. Pediatr. Res. 15300447. 32 (1), 128–131. doi:10.1203/00006450-199207000-00025.

Darmahkasih, A.J., Rybalsky, I., Tian, C., Shellenbarger, K.C., Horn, P.S., Lambert, J.T., Wong, B.L., 2020. Neurodevelopmental, behavioral, and emotional symptoms common in Duchenne muscular dystrophy. Muscle Nerve. 10974598. 61 (4), 466–474. Available from:: http://onlinelibrary.wiley.com/journal/10.1002/(ISSN)1097-4598. doi:10.1002/mus.26803.

Davidson, Z.E., Vidmar, S., Griffiths, A., Howard, M.E., Berlowitz, D.J., Jones, E.F., Scully, T., Treanor, D., Singh, B., Prior, D., Frost, M.G., Forbes, R., Billich, N., Adams, J., Carroll, K., Lawati, T.A., Bourne, H., Kornberg, A.J., Ryan, M.M., Woodcock, I.R., Yiu, E.M., Cheung, M.M.H., 2025. Survival in Duchenne muscular dystrophy in Australia: a 50 year retrospective cohort study. Lancet Reg. Health West. Pac. 26666065. 58. Available from: https://www.sciencedirect.com/science/journal/26666065. doi:10.1016/j.lanwpc.2025.101568.

Deconinck, A.E., Rafael, J.A., Skinner, J.A., Brown, S.C., Potter, A.C., Metzinger, L., Watt, D.J., Dickson, J.G., Tinsley, J.M., Davies, K.E., 1997. Utrophin-dystrophin-deficient mice as a model for Duchenne muscular dystrophy. Cell. 00928674. 90 (4), 717–727. Available from: https://www.sciencedirect.com/journal/cell. doi:10.1016/S0092-8674(00)80532-2.

Duan, D., 2018a. Micro-dystrophin gene therapy goes systemic in Duchenne muscular dystrophy patients. Hum. Gene Ther. 1043-0342. 29 (7), 733–736. doi:10.1089/hum.2018.012.

Duan, D., 2018b. Systemic AAV micro-dystrophin gene therapy for Duchenne muscular dystrophy. Mol. Ther. 15250016. 26 (10), 2337–2356. doi:10.1016/j.ymthe.2018.07.011.

Duan, D., et al., 2021. Duchenne muscular dystrophy. Nat. Rev. Dis. Primers. 7 (1), 13.

Duan, D., 2024. Full-length dystrophin gene therapy for Duchenne muscular dystrophy. Mol. Ther. 15250016. 32 (9), 2817–2818. doi:10.1016/j.ymthe.2024.07.026.

Ducret, T., Vandebrouck, C., Cao, M.L., Lebacq, J., Gailly, P., 2006. Functional role of store-operated and stretch-activated channels in murine adult skeletal muscle fibres. J. Physiol. 0022-3751. 575 (3), 913–924. doi:10.1113/jphysiol.2006.115154.

Dystrophy, P.P.M. Pfizer discontinues development of investigational mini-dystrophin gene therapy. 2024. Available from: https://www.parentprojectmd.org/pfizer-discontinues-development-of-investigational-mini-dystrophin-gene-therapy/. (18 May 2025).

Eagle, M., Baudouin, S.V., Chandler, C., Giddings, D.R., Bullock, R., Bushby, K., 2002. Survival in Duchenne muscular dystrophy: improvements in life expectancy since 1967 and the impact of home nocturnal ventilation. Neuromuscul. Disord. 09608966. 12 (10), 926–929. doi:10.1016/s0960-8966(02)00140-2.

Echigoya, Y., Trieu, N., Duddy, W., Moulton, H.M., Yin, H., Partridge, T.A., Hoffman, E.P., Kornegay, J.N., Rohret, F.A., Rogers, C.S., Yokota, T., 2021. A dystrophin exon-52 deleted miniature pig model of duchenne muscular dystrophy and evaluation of exon skipping. Int. J. Mol. Sci. 1422-0067. 22 (23), 13065. doi:10.3390/ijms222313065.

Efthymiou, S., Manole, A., Houlden, H., 2016. Next-generation sequencing in neuromuscular diseases. Curr. Opin. Neurol. 14736551. 29 (5), 527–536. Available from: http://journals.lww.com/co-neurology/pages/default.aspx. doi:10.1097/WCO.0000000000000374.

Egorova, T.V., Zotova, E.D., Reshetov, D.A., Polikarpova, A.V., Vassilieva, S.G., Vlodavets, D.V., Gavrilov, A.A., Ulianov, S.V., Buchman, V.L., Deykin, A.V., 2019. CRISPR/Cas9-generated mouse model of Duchenne muscular dystrophy recapitulating a newly identified large 430 kb deletion in the human DMD gene. Dis. Model. Mech. 1754-8411. 12 (4). doi:10.1242/dmm.037655.

Ehrhardt, A., Xu, H., Kay, M.A., 2003. Episomal persistence of recombinant adenoviral vector genomes during the cell cycle in vivo. J. Virol. 77 (13), 7689–7695. doi:10.1128/JVI.77.13.7689-7695.2003.

Emery, A.E.H., 1991. Population frequencies of inherited neuromuscular diseases—a world survey. Neuromuscul. Disord. 09608966. 1 (1), 19–29. doi:10.1016/0960-8966(91)90039-u.

England, S.B., Nicholson, L.V.B., Johnson, M.A., Forrest, S.M., Love, D.R., Zubrzycka-Gaarn, E.E., Bulman, D.E., Harris, J.B., Davies, K.E., 1990. Very mild muscular dystrophy associated with the deletion of 46% of dystrophin. Nature. 0028-0836. 343 (6254), 180–182. doi:10.1038/343180a0.

Erkut, E., Yokota, T., 2022. CRISPR Therapeutics for Duchenne muscular dystrophy. Int. J. Mol. Sci. 14220067. 23 (3). Available from: https://www.mdpi.com/1422-0067/23/3/1832/pdf. doi:10.3390/ijms23031832.

Ervasti, J.M., 2007. Dystrophin, its interactions with other proteins, and implications for muscular dystrophy. Biochim. Biophys. Acta Mol. Basis Dis. 09254439. 1772 (2), 108–117. doi:10.1016/j.bbadis.2006.05.010.

Escobar-Huertas, J.F., Vaca-González, J.J., María Guevara, J., Ramirez-Martinez, A.M., Trabelsi, O., Garzón-Alvarado, D.A., 2024. Duchenne and Becker muscular dystrophy: cellular mechanisms, image analysis, and computational models: a review. Cytoskeleton. 1949-3584. 81 (6-7), 269–286. doi:10.1002/cm.21826.

Farini, A., Meregalli, M., Belicchi, M., Battistelli, M., Parolini, D., D'Antona, G., Gavina, M., Ottoboni, L., Constantin, G., Bottinelli, R., Torrente, Y., 2007. T and B lymphocyte depletion has a marked effect on the fibrosis of dystrophic skeletal muscles in the scid/mdx mouse. J. Pathol. 0022-3417. 213 (2), 229–238. doi:10.1002/path.2213.

Ferizovic, N., Summers, J., de Zárate, I.B.O., Werner, C., Jiang, J., Landfeldt, E., Buesch, K., Coelho-Filho, O.R., 2022. Prognostic indicators of disease progression in Duchenne muscular dystrophy: a literature review and evidence synthesis. PLoS One. 1932-6203. 17 (3), e0265879. doi:10.1371/journal.pone.0265879.

Flanigan, K.M., Dunn, D.M., von Niederhausern, A., Howard, M.T., Mendell, J., Connolly, A., Saunders, C., Modrcin, A., Dasouki, M., Comi, G.P., Bo, R.D., Pickart, A., Jacobson, R., Finkel, R., Medne, L., Weiss, R.B., 2009a. DMD Trp3X nonsense mutation associated with a founder effect in North American families with mild Becker muscular dystrophy. Neuromuscul. Disord. 09608966. 19 (11), 743–748. doi:10.1016/j.nmd.2009.08.010.

Flanigan, K.M., Dunn, D.M., Von Niederhausern, A., Soltanzadeh, P., Gappmaier, E., Howard, M.T., Sampson, J.B., Mendell, J.R., Wall, C., King, W.M., Pestronk, A., Florence, J.M., Connolly, A.M., Mathews, K.D., Stephan, C.M., Laubenthal, K.S., Wong, B.L., Morehart, P.J., Meyer, A., Finkel, R.S., Bonnemann, C.G., Medne, L., Day, J.W., Dalton, J.C., Margolis, M.K., Hinton, V.J., Weiss, R.B., Bromberg, M., Swoboda, K., Kerr, L., Gurvich, O., Tuohy, T., Taylor, L., Zhao, L., Hart, K., Moural, C., Hak, K., Duval, B., Hamil, C., Mahmoud, M., Aoyagi, A., Viollet, L., Gailey, S., Lopate, G., Golumbek, P., Schierbecker, J., Malkus, B., Siener, C., Baldwin, K., Glanzman, A.M., Flickinger, J., Naughton, C.E., 2009b. Mutational spectrum of DMD mutations in dystrophinopathy patients: application of modern diagnostic techniques to a large cohort. Hum. Mutat. 10597794. 30 (12), 1657–1666. Available from: http://www3.interscience.wiley.com/cgi-bin/fulltext/122580737/PDFSTART. doi:10.1002/humu.21114.

Flanigan, K.M., Vetter, T.A., Simmons, T.R., Iammarino, M., Frair, E.C., Rinaldi, F., Chicoine, L.G., Harris, J., Cheatham, J.P., Cheatham, S.L., Boe, B., Waldrop, M.A., Zygmunt, D.A., Packer, D., Martin, P.T., 2022. A first-in-human phase I/IIa gene transfer clinical trial for Duchenne muscular dystrophy using rAAVrh74.MCK.GALGT2. Mol. Ther. Methods Clin. Dev. 23290501. 27, 47–60. Available from: https://www.journals.elsevier.com/molecular-therapy-methods-and-clinical-development/. doi:10.1016/j.omtm.2022.08.009.

Fortunato, F., Ferlini, A., 2023. Biomarkers in Duchenne muscular dystrophy: current status and future directions. J. Neuromuscul. Dis. 22143602. 10 (6), 987–1002. Available from: http://www.iospress.nl/journal/journal-of-neuromuscular-diseases/. doi:10.3233/JND-221666.

Frank, D.E., Schnell, F.J., Akana, C., El-Husayni, S.H., Desjardins, C.A., Morgan, J., Charleston, J.S., Sardone, V., Domingos, J., Dickson, G., Straub, V., Guglieri, M., Mercuri, E., Servais, L., Muntoni, F., 2020. Increased dystrophin production with golodirsen in patients with Duchenne muscular dystrophy. Neurology. 1526632X. 94 (21), E2270. Available from: http://www.neurology.org. doi:10.1212/WNL.0000000000009233.

Gailly, P., Boland, B., Himpens, B., Casteels, R., Gillis, J.M., 1993. Critical evaluation of cytosolic calcium determination in resting muscle fibres from normal and dystrophic (mdx) mice. Cell Calcium. 01434160. 14 (6), 473–483. doi:10.1016/0143-4160(93)90006-r.

Gailly, P., De Backer, F., Van Schoor, M., Gillis, J.M., 2007. In situ measurements of calpain activity in isolated muscle fibres from normal and dystrophin-lacking mdx mice. J. Physiol. 0022-3751. 582 (3), 1261–1275. doi:10.1113/jphysiol.2007.132191.

Gallagher, T., 2025. European Society of Gene & Cell Therapy Spring School 2025. Hum. Gene Ther. 1043-0342. 36 (11-12), 856–857. doi:10.1089/hum.2025.088.

Gao, Q.Q., McNally, E.M., 2015. The dystrophin complex: structure, function, and implications for therapy. Compr. Physiol. 20404603. 5 (3), 1223–1239. Available from: http://onlinelibrary.wiley.com/store/10.1002/cphy.c140048/asset/c140048.pdf?v=1&t=imvu9z5o&s=2119f13d1c07e8a84880f0aacf0c7895c54d6811. doi:10.1002/cphy.c140048.

Gatto, F., Benemei, S., Piluso, G., Bello, L., 2024. The complex landscape of DMD mutations: moving towards personalized medicine. Front. Genet. 1664-8021. 15. doi:10.3389/fgene.2024.1360224.

Gillman, A., Ciobanu, T., Barrett, L., Davies, E.W., Murphy, A.P., Johnson, A., Mills, J., Heinrich, P., Przydzial, K., Ewens, B., Vandenberg, G., Cano, S., Mayhew, A., 2025. A patient-centered qualitative evaluation of meaningful change on the NSAA and PUL in Duchenne Muscular Dystrophy. Front. Neurol. 1664-2295. 16. doi:10.3389/fneur.2025.1509174.

Grady, R.M., Teng, H., Nichol, M.C., Cunningham, J.C., Wilkinson, R.S., Sanest, J.R., 1997. Skeletal and cardiac myopathies in mice lacking utrophin and dystrophin: a model for Duchenne muscular dystrophy. Cell. 00928674. 90 (4), 729–738. Available from: https://www.sciencedirect.com/journal/cell. doi:10.1016/S0092-8674(00)80533-4.

Griggs, R.C., Miller, J.P., Greenberg, C.R., Fehlings, D.L., Pestronk, A., Mendell, J.R., Moxley, R.T., King, W., Kissel, J.T., Cwik, V., Vanasse, M., Florence, J.M., Pandya, S., Dubow, J.S., Meyer, J.M., 2016. Efficacy and safety of deflazacort vs prednisone and placebo for Duchenne muscular dystrophy. Neurology. 1526632X. 87 (20), 2123–2131. Available from: http://www.neurology.org. doi:10.1212/WNL.0000000000003217.

Gurvich, O.L., Maiti, B., Weiss, R.B., Aggarwal, G., Howard, M.T., Flanigan, K.M., 2009. DMD exon 1 truncating point mutations: amelioration of phenotype by alternative translation initiation in exon 6. Hum. Mutat. 10597794. 30 (4), 633–640. Available from: http://www3.interscience.wiley.com/cgi-bin/fulltext/122193359/PDFSTART. doi:10.1002/humu.20913.

Hammers, D.W., Hart, C.C., Matheny, M.K., Wright, L.A., Armellini, M., Barton, E.R., Sweeney, H.L., 2020. The D2.mdx mouse as a preclinical model of the skeletal muscle pathology associated with Duchenne muscular dystrophy. Sci. Rep. 20452322. 10 (1). Available from: www.nature.com/srep/index.html. doi:10.1038/s41598-020-70987-y.

Harper, S.Q., Hauser, M.A., DelloRusso, C., Duan, D., Crawford, R.W., Phelps, S.F., Harper, H.A., Robinson, A.S., Engelhardt, J.F., Brooks, S.V., Chamberlain, J.S., 2002. Modular flexibility of dystrophin: implications for gene therapy of Duchenne muscular dystrophy. Nat. Med. 10788956. 8 (3), 253–261. doi:10.1038/nm0302-253.

Hibma, J.E., Jayachandran, P., Neelakantan, S., Harnisch, L.O., 2023. Disease progression modeling of the North Star Ambulatory Assessment for Duchenne Muscular Dystrophy. CPT Pharmacometrics Syst. Pharmacol. 21638306. 12 (3), 375–386. Available from: http://onlinelibrary.wiley.com/journal/10.1002/(ISSN)2163-8306. doi:10.1002/psp4.12921.

High, K.A., Roncarolo, M.G., 2019. Gene therapy. N. Engl. J. Med. 15334406. 381 (5), 455–464. Available from: http://www.nejm.org/medical-index. doi:10.1056/NEJMra1706910.

Hinderer, C., Katz, N., Buza, E.L., Dyer, C., Goode, T., Bell, P., Richman, L.K., Wilson, J.M., 2018. Severe toxicity in nonhuman primates and piglets following high-dose intravenous administration of an adeno-associated virus vector expressing human SMN. Hum. Gene Ther. 1043-0342. 29 (3), 285–298. doi:10.1089/hum.2018.015.

Hoffman, E.P., Brown, R.H., Kunkel, L.M., 1987. Dystrophin: the protein product of the duchenne muscular dystrophy locus. Cell. 00928674. 51 (6), 919–928. doi:10.1016/0092-8674(87)90579-4.

Hoffman, E.P., Kunkel, L.M., 1989. Dystrophin abnormalities in Duchenne/Becker muscular dystrophy. Neuron. 08966273. 2 (1), 1019–1029. doi:10.1016/0896-6273(89)90226-2.

Hordeaux, J., Lamontagne, R.J., Song, C., Buchlis, G., Dyer, C., Buza, E.L., Ramezani, A., Wielechowski, E., Greig, J.A., Chichester, J.A., Bell, P., Wilson, J.M., 2024. High-dose systemic adeno-associated virus vector administration causes liver and sinusoidal endothelial cell injury. Mol. Ther. 15250016. 32 (4), 952–968. doi:10.1016/j.ymthe.2024.02.002.

Howard, M.T., Shirts, B.H., Petros, L.M., Flanigan, K.M., Gesteland, R.F., Atkins, J.F., 2000. Sequence specificity of aminoglycoside-induced stop codon readthrough: potential implications for treatment of Duchenne muscular dystrophy. Ann. Neurol. 03645134. 48 (2), 164–169. doi:10.1002/1531-8249(200008)48:2<164::AID-ANA5>3.0.CO;2-B.

Hwa, H., Chang, Y., Chen, C., Kao, Y., Jong, Y., Chao, M., Ko, T., 2007. Multiplex ligation-dependent probe amplification identification of deletions and duplications of the Duchenne muscular dystrophy gene in taiwanese subjects. J. Formos. Med. Assoc. 106 (5), 339–346. doi:10.1016/S0929-6646(09)60318-1.

Im, W.B., Phelps, S.F., Copen, E.H., Adams, E.G., Slightom, J.L., Chamberlain, J.S., 1996. Differential expression of dystrophin isoforms in strains of mdx mice with different mutations. Hum. Mol. Genet. 09646906. 5 (8), 1149–1153. Available from: http://hmg.oxfordjournals.org/. doi:10.1093/hmg/5.8.1149.

Iyer, S.R., Xu, S., Shah, S.B., Lovering, R.M., 2020. Muscle phenotype of a rat model of Duchenne muscular dystrophy. Muscle Nerve. 10974598. 62 (6), 757–761. Available from: http://onlinelibrary.wiley.com/journal/10.1002/(ISSN)1097-4598. doi:10.1002/mus.27061.

Kariyawasam, D., D'Silva, A., Mowat, D., Russell, J., Sampaio, H., Jones, K., Taylor, P., Farrar, M., 2022. Incidence of Duchenne muscular dystrophy in the modern era; an Australian study. Eur. J. Hum. Genet. 1018-4813. 30 (12), 1398–1404. doi:10.1038/s41431-022-01138-2.

Keam S.J., 2024. Vamorolone: first approval. Drugs 84 (1), 111–117. doi:10.1007/s40265-023-01986-2.

Kenjo, E., Hozumi, H., Makita, Y., Iwabuchi, K.A., Fujimoto, N., Matsumoto, S., Kimura, M., Amano, Y., Ifuku, M., Naoe, Y., Inukai, N., Hotta, A., 2021. Low immunogenicity of LNP allows repeated administrations of CRISPR-Cas9 mRNA into skeletal muscle in mice. Nat. Commun. 2041-1723. 12 (1). doi:10.1038/s41467-021-26714-w.

Khan, G., Hussain, M.S., 2026. Expanding the potential of gene therapy for duchenne muscular dystrophy. Curr. Pharm. Des. 18734286. 32 (5), 330–337. Available from: http://www.eurekaselect.com/606/journal/current-pharmaceutical-design. doi:10.2174/0113816128386290250507101412.

Kim, K., Ryu, S.-M., Kim, S.-T., Baek, G., Kim, D., Lim, K., Chung, E., Kim, S., Kim, J.-S., 2017. Highly efficient RNA-guided base editing in mouse embryos. Nat. Biotechnol. 1087-0156. 35 (5), 435–437. doi:10.1038/nbt.3816.

Klymiuk, N., Blutke, A., Graf, A., Krause, S., Burkhardt, K., Wuensch, A., Krebs, S., Kessler, B., Zakhartchenko, V., Kurome, M., Kemter, E., Nagashima, H., Schoser, B., Herbach, N., Blum, H., Wanke, R., Aartsma-Rus, A., Thirion, C., Lochmüller, H., Walter, M.C., Wolf, E., 2013. Dystrophin-deficient pigs provide new insights into the hierarchy of physiological derangements of dystrophic muscle. Hum. Mol. Genet. 1460-2083. 22 (21), 4368–4382. doi:10.1093/hmg/ddt287.

Koenig, M., Hoffman, E.P., Bertelson, C.J., Monaco, A.P., Feener, C., Kunkel, L.M., 1987. Complete cloning of the duchenne muscular dystrophy (DMD) cDNA and preliminary genomic organization of the DMD gene in normal and affected individuals. Cell. 00928674. 50 (3), 509–517. doi:10.1016/0092-8674(87)90504-6.

Koo, T., Lu-Nguyen, N.B., Malerba, A., Kim, E., Kim, D., Cappellari, O., Cho, H.-Y., Dickson, G., Popplewell, L., Kim, J.-S., 2018. Functional rescue of dystrophin deficiency in mice caused by frameshift mutations using campylobacter jejuni Cas9. Mol. Ther. 15250016. 26 (6), 1529–1538. doi:10.1016/j.ymthe.2018.03.018.

Kornegay, J.N., Bogan, J.R., Bogan, D.J., Childers, M.K., Li, J., Nghiem, P., Detwiler, D.A., Larsen, C.A., Grange, R.W., Bhavaraju-Sanka, R.K., Tou, S., Keene, B.P., Howard, J.F., Wang, J., Fan, Z., Schatzberg, S.J., Styner, M.A., Flanigan, K.M., Xiao, X., Hoffman, E.P., 2012. Canine models of Duchenne muscular dystrophy and their use in therapeutic strategies. Mamm. Genome. 14321777. 23 (1-2), 85–108. doi:10.1007/s00335-011-9382-y.

Kota, J., Handy, C.R., Haidet, A.M., Montgomery, C.L., Eagle, A., Rodino-Klapac, L.R., Tucker, D., Shilling, C.J., Therlfall, W.R., Walker, C.M., Weisbrode, S.E., Janssen, P.M.L., Clark, K.R., Sahenk, Z., Mendell, J.R., Kaspar, B.K., 2009. Follistatin gene delivery enhances muscle growth and strength in nonhuman primates. Sci. Transl. Med. 1946-6234. 1 (6). doi:10.1126/scitranslmed.3000112.

Krawczak, M., Ball, E.V., Cooper, D.N., 1998. Neighboring-nucleotide effects on the rates of germ-line single-base- pair substitution in human genes. Am. J. Hum. Genet. 00029297. 63 (2), 474–488. Available from: http://www.elsevier.com/wps/find/journaldescription.cws_home/713561/description#description. doi:10.1086/301965.

Kunkel,, 2005. William Allan Award address. Cloning of the DMD gene. Am. J. Hum. Genet. (2005). 76 (2), 205–214.

Lai, Y., Thomas, G.D., Yue, Y., Yang, H.T., Li, D., Long, C., Judge, L., Bostick, B., Chamberlain, J.S., Terjung, R.L., Duan, D., 2009. Dystrophins carrying spectrin-like repeats 16 and 17 anchor nNOS to the sarcolemma and enhance exercise performance in a mouse model of muscular dystrophy. J. Clin. Invest. 0021-9738. 119 (3), 624–635. doi:10.1172/jci36612.

Larcher, T., Lafoux, A., Tesson, L., Remy, S., Thepenier, V., François, V., Guiner, C.L., Goubin, H., Dutilleul, M., Guigand, L., Toumaniantz, G., De Cian, A., Boix, C., Renaud, J.-B., Cherel, Y., Giovannangeli, C., Concordet, J.-P., Anegon, I., Huchet, C., Asakura, A., 2014. Characterization of dystrophin deficient rats: a new model for Duchenne muscular dystrophy. PLoS One. 1932-6203. 9 (10), e110371. doi:10.1371/journal.pone.0110371.

Le Guiner, C., Servais, L., Montus, M., Larcher, T., Fraysse, B., Moullec, S., Allais, M., François, V., Dutilleul, M., Malerba, A., Koo, T., Thibaut, J.-L., Matot, B., Devaux, M., Duff, J.L., Deschamps, J.-Y., Barthelemy, I., Blot, S., Testault, I., Wahbi, K., Ederhy, S., Martin, S., Veron, P., Georger, C., Athanasopoulos, T., Masurier, C., Mingozzi, F., Carlier, P., Gjata, B., Hogrel, J.-Y., Adjali, O., Mavilio, F., Voit, T., Moullier, P., Dickson, G., 2017. Long-term microdystrophin gene therapy is effective in a canine model of Duchenne muscular dystrophy. Nat. Commun. 2041-1723. 8 (1). doi:10.1038/ncomms16105.

Le, S., Yu, M., Hovan, L., Zhao, Z., Ervasti, J., Yan, J., 2018. Dystrophin as a molecular shock absorber. ACS Nano. 1936-0851. 12 (12), 12140–12148. doi:10.1021/acsnano.8b05721.

Leckie, J., Zia, A., Yokota, T., 2024. An updated analysis of exon-skipping applicability for Duchenne muscular dystrophy using the UMD-DMD database. Genes. 20734425. 15 (11). Available from: http://www.mdpi.com/journal/genes/. doi:10.3390/genes15111489.

Lee, I., Turnage, C., Sutyla, R., Mitchell, P., Lindahl, H., Jesus, A., Scharf, R.J., 2022. The hidden disease: delayed diagnosis in Duchenne muscular dystrophy and co-occurring conditions. J. Dev. Behav. Pediatr. 0196-206X. 43 (8), e541. doi:10.1097/dbp.0000000000001105.

Lee, S.H., Lee, J.H., Lee, K.-A., Choi, Y.-C., 2015. Clinical and genetic characterization of female dystrophinopathy. J. Clin. Neurol. 1738-6586. 11 (3), 248. doi:10.3988/jcn.2015.11.3.248.

Lek, A., Wong, B., Keeler, A., Blackwood, M., Kaiyue, M., Huang, S., Sylvia, K., Rita Batista, A., Artinian, R., Kokoski, D., Parajuli, S., Putra, J., Carreon, C.K., Lidov, H., Woodman, K., Pajusalu, S., Spinazzola, J.M., Gallagher, T., LaRovere, J., Balderson, D., Black, L., Sutton, K., Horgan, R., Lek, M., Flotte, T., 2023. Death after high-dose rAAV9 gene therapy in a patient with Duchenne's muscular dystrophy. N. Engl. J. Med. 0028-4793. 389 (13), 1203–1210. doi:10.1056/nejmoa2307798.

Li, B., Lu, Z., Zhou, T., Mou, H., Zou, Z., Yao, K., Yu, Z., Ma, W., Wang, T., Miao, H., Ma, L., 2025. Loop engineering improves prime editing efficiency. Mol. Ther. Nucleic Acids. 21622531. 36 (4), 102764. doi:10.1016/j.omtn.2025.102764.

Liao, H., Wu, J., VanDusen, N.J., Li, Y., Zheng, Y., 2024. CRISPR-Cas9-mediated homology-directed repair for precise gene editing. Mol. Ther. Nucleic Acids. 21622531. 35 (4), 102344. doi:10.1016/j.omtn.2024.102344.

Long, C., McAnally, J.R., Shelton, J.M., Mireault, A.A., Bassel-Duby, R., Olson, E.N., 2014. Prevention of muscular dystrophy in mice by CRISPR/Cas9–mediated editing of germline DNA. Science. 0036-8075. 345 (6201), 1184–1188. doi:10.1126/science.1254445.

Magrath, P., Maforo, N., Renella, P., Nelson, S.F., Halnon, N., Ennis, D.B., 2018. Cardiac MRI biomarkers for Duchenne muscular dystrophy. Biomark. Med. 1752-0363. 12 (11), 1271–1289. doi:10.2217/bmm-2018-0125.

Maguire, K., Suzuki, T., DiMatteo, D., Parekh-Olmedo, H., Kmiec, E., 2009. Genetic correction of splice site mutation in purified and enriched myoblasts isolated from mdx5cv mice. BMC Mol. Biol. 1471-2199. 10 (1). doi:10.1186/1471-2199-10-15.

Mah, J., Selby, K., Campbell, C., Nadeau, A., Tarnopolsky, M., McCormick, A., Dooley, J., Kolski, H., Skalsky, A., Smith, R., Buckley, D., Ray, P., Yoon, G., 2011. A population-based study of dystrophin mutations in Canada. Can. J. Neurol. Sci. 03171671. 38 (3), 465–474. Available from: http://journals.cambridge.org/action/displayBackIssues?jid=CJN. doi:10.1017/S0317167100011896.

Mali, P., Esvelt, K.M., Church, G.M., 2013. Cas9 as a versatile tool for engineering biology. Nat. Methods. 15487105. 10 (10), 957–963. doi:10.1038/nmeth.2649.

Manzur, A.Y., Kuntzer, T., Pike, M., Swan, A., 2008. Glucocorticoid corticosteroids for Duchenne muscular dystrophy. Cochrane Database Syst. Rev. 1469493X. (1). Available from http://www.mrw.interscience.wiley.com/cochrane/clsysrev/articles/CD003725/pdf_fs.html. doi:10.1002/14651858.CD003725.pub3.

Mareedu, S., Million, E.D., Duan, D., Babu, G.J., 2021. Abnormal calcium handling in duchenne muscular dystrophy: mechanisms and potential therapies. Front. Physiol. 1664-042X. 12. doi:10.3389/fphys.2021.647010.

Masson, J.D., Taglietti, V., Ruby, F., Ono, H., Mouri, N., Jorge, A., Guillaud, L., Tiret, L., Relaix, F., 2025. Extensive striated muscle damage in a rat model of Duchenne muscular dystrophy with Dmd exons 10–17 duplication. Skelet. Muscle. 20445040. 15 (1). Available from: https://skeletalmusclejournal.biomedcentral.com/. doi:10.1186/s13395-025-00386-2.

Mbakam, C.H., Lamothe, G., Tremblay, J.P., 2022. Therapeutic strategies for dystrophin replacement in Duchenne muscular dystrophy. Front. Med. 2296858X. 9. Available from: https://journal.frontiersin.org/journal/medicine. doi:10.3389/fmed.2022.859930.

McDonald, C.M., et al, 1995. Profiles of neuromuscular diseases: duchenne muscular dystrophy. Am. J. Phys. Med. Rehabil. 74 (5), S93.

McDonald, C.M., Elkins, J.S., Dharmarajan, S., Gooch, K., Ciobanu, T., Lansdall, C.J., Murphy, A.P., McDougall, F., Mercuri, E.M., Audhya, I., 2025. Caregiver global impression observations from EMBARK: a phase 3 study evaluating delandistrogene moxeparvovec in ambulatory patients with Duchenne muscular dystrophy. Neurol. Ther. 21936536. 14 (1), 211–225. Available from: https://www.springer.com/journal/40120. doi:10.1007/s40120-024-00685-8.

Mcdonald, C.M., Henricson, E.K., Abresch, R.T., Florence, J., Eagle, M., Gappmaier, E., Glanzman, A.M., Spiegel, R., Barth, J., Elfring, G., Reha, A., Peltz, S.W., 2013. The 6-minute walk test and other clinical endpoints in duchenne muscular dystrophy: reliability, concurrent validity, and minimal clinically important differences from a multicenter study. Muscle Nerve. 10974598. 48 (3), 357–368. doi:10.1002/mus.23905.

McGreevy, J.W., Hakim, C.H., McIntosh, M.A., Duan, D., 2015. Animal models of Duchenne muscular dystrophy: from basic mechanisms to gene therapy. Dis. Model. Mech. 1754-8411. 8 (3), 195–213. doi:10.1242/dmm.018424.

McLaughlin, J.E., Foka, I.C.K., Lawton, M.A., Di, R., 2025. CRISPR activation: identifying and using novel genes for plant disease resistance breeding. Front. Genome Ed. 26733439. 7. Available from: https://www.frontiersin.org/journals/genome-editing/about. doi:10.3389/fgeed.2025.1596600.

Mendell, J.R., et al., 2025. AAV gene therapy for Duchenne muscular dystrophy: the EMBARK phase 3 randomized trial. Nat. Med. 31 (1), 332–341. doi:10.1038/s41591-024-03304-z.

Mendell, J.R., Chicoine, L.G., Al-Zaidy, S.A., Sahenk, Z., Lehman, K., Lowes, L., Miller, N., Alfano, L., Galliers, B., Lewis, S., Murrey, D., Peterson, E., Griffin, D.A., Church, K., Cheatham, S., Cheatham, J., Hogan, M.J., Rodino-Klapac, L.R., 2019. Gene delivery for Limb-Girdle Muscular dystrophy type 2D by isolated limb infusion. Hum. Gene Ther. 15577422. 30 (7), 794–801. Available from: www.liebertonline.com/hum. doi:10.1089/hum.2019.006.

Mendell, J.R., Muntoni, F., McDonald, C.M., Mercuri, E.M., Ciafaloni, E., Komaki, H., Leon-Astudillo, C., Nascimento, A., Proud, C., Schara-Schmidt, U., Veerapandiyan, A., Zaidman, C.M., Furgerson, M., Ding, K., Singh, P., Potter, R., Asher, D.R., Murphy, A.P., Reid, C., Hooper, G., Torre, C.O., Manfrini, M., Rodino-Klapac, L.R., 2026. Two-year outcomes following delandistrogene moxeparvovec treatment in ambulatory patients with duchenne muscular dystrophy: phase 3 EMBARK trial. Neurol. Ther. 21936536. Available from: https://www.springer.com/journal/40120. doi:10.1007/s40120-025-00879-8.

Mendell, J.R., Muntoni, F., McDonald, C.M., Mercuri, E.M., Ciafaloni, E., Komaki, H., Leon-Astudillo, C., Nascimento, A., Proud, C., Schara-Schmidt, U., Veerapandiyan, A., Zaidman, C.M., Guridi, M., Murphy, A.P., Reid, C., Wandel, C., Asher, D.R., Darton, E., Mason, S., Potter, R.A., Singh, T., Zhang, W., Fontoura, P., Elkins, J.S., Rodino-Klapac, L.R., 2025. AAV gene therapy for Duchenne muscular dystrophy: the EMBARK phase 3 randomized trial. Nat. Med. 1546170X. 31 (1), 332–341. Available from: https://www.nature.com/nm/. doi:10.1038/s41591-024-03304-z.

Mendell, J.R., Rodino-Klapac, L., Sahenk, Z., Malik, V., Kaspar, B.K., Walker, C.M., Clark, K.R., 2012. Gene therapy for muscular dystrophy: lessons learned and path forward. Neurosci. Lett. 18727972. 527 (2), 90–99. doi:10.1016/j.neulet.2012.04.078.

Mendell, J.R., Sahenk, Z., Lehman, K.J., Lowes, L.P., Reash, N.F., Iammarino, M.A., Alfano, L.N., Lewis, S., Church, K., Shell, R., Potter, R.A., Griffin, D.A., Hogan, M., Wang, S., Mason, S., Darton, E., Rodino-Klapac, L.R., 2024. Long-term safety and functional outcomes of delandistrogene moxeparvovec gene therapy in patients with Duchenne muscular dystrophy: a phase 1/2a nonrandomized trial. Muscle Nerve. 10974598. 69 (1), 93–98. Available from: http://onlinelibrary.wiley.com/journal/10.1002/(ISSN)1097-4598. doi:10.1002/mus.27955.

Mendell, J.R., Sahenk, Z., Lehman, K., Nease, C., Lowes, L.P., Miller, N.F., Iammarino, M.A., Alfano, L.N., Nicholl, A., Al-Zaidy, S., Lewis, S., Church, K., Shell, R., Cripe, L.H., Potter, R.A., Griffin, D.A., Pozsgai, E., Dugar, A., Hogan, M., Rodino-Klapac, L.R., 2020. Assessment of systemic delivery of rAAVrh74.MHCK7.micro-dystrophin in children with Duchenne muscular dystrophy: a nonrandomized controlled trial. JAMA Neurol. 21686157. 77 (9), 1122–1131. Available from: http://archneur.jamanetwork.com/issues.aspx. doi:10.1001/jamaneurol.2020.1484.

Mendell, J.R., Sahenk, Z., Malik, V., Gomez, A.M., Flanigan, K.M., Lowes, L.P., Alfano, L.N., Berry, K., Meadows, E., Lewis, S., Braun, L., Shontz, K., Rouhana, M., Clark, K.R., Rosales, X.Q., Al-Zaidy, S., Govoni, A., Rodino-Klapac, L.R., Hogan, M.J., Kaspar, B.K., 2015. A phase 1/2a follistatin gene therapy trial for becker muscular dystrophy. Mol. Ther. 15250024. 23 (1), 192–201. Available from: https://www.journals.elsevier.com/molecular-therapy. doi:10.1038/mt.2014.200.

Mendell, J.R., Shieh, P.B., McDonald, C.M., Sahenk, Z., Lehman, K.J., Lowes, L.P., Reash, N.F., Iammarino, M.A., Alfano, L.N., Sabo, B., Woods, J.D., Skura, C.L., Mao, H.C., Staudt, L.A., Griffin, D.A., Lewis, S., Wang, S., Potter, R.A., Singh, T., Rodino-Klapac, L.R., 2023. Expression of SRP-9001 dystrophin and stabilization of motor function up to 2 years post-treatment with delandistrogene moxeparvovec gene therapy in individuals with Duchenne muscular dystrophy. Front. Cell Dev. Biol. 2296634X. 11. Available from: https://www.frontiersin.org/journals/cell-and-developmental-biology#. doi:10.3389/fcell.2023.1167762.

Mercuri, E., Nascimento Osorio, A., Muntoni, F., Buccella, F., Desguerre, I., Kirschner, J., Tulinius, M., de Resende, M.B.D., Morgenroth, L.P., Gordish-Dressman, H., Johnson, S., Kristensen, A., Werner, C., Trifillis, P., Henricson, E.K., McDonald, C.M., 2023. Safety and effectiveness of ataluren in patients with nonsense mutation DMD in the STRIDE Registry compared with the CINRG Duchenne Natural History Study (2015–2022): 2022 interim analysis. J. Neurol. 0340-5354. 270 (8), 3896–3913. doi:10.1007/s00415-023-11687-1.

Mesa, L.E., Dubrovsky, A.L., Corderi, J., Marco, P., Flores, D., 1991. Steroids in duchenne muscular dystrophy—deflazacort trial. Neuromuscul. Disord. 09608966. 1 (4), 261–266. doi:10.1016/0960-8966(91)90099-e.

Messina, S., Vita, G.L., 2018. Clinical management of Duchenne muscular dystrophy: the state of the art. Neurol. Sci. 15903478. 39 (11), 1837–1845. Available from: https://link.springer.com/journal/10072. doi:10.1007/s10072-018-3555-3.

Miller, T.M., Kim, S.H., Yamanaka, K., Hester, M., Umapathi, P., Arnson, H., Rizo, L., Mendell, J.R., Gage, F.H., Cleveland, D.W., Kaspar, B.K., 2006. Gene transfer demonstrates that muscle is not a primary target for non-cell-autonomous toxicity in familial amyotrophic lateral sclerosis. Proc. Natl. Acad. Sci. USA. 00278424. 103 (51), 19546–19551. doi:10.1073/pnas.0609411103.

Min, Y.L., Chemello, F., Li, H., Rodriguez-Caycedo, C., Sanchez-Ortiz, E., Mireault, A.A., McAnally, J.R., Shelton, J.M., Zhang, Y., Bassel-Duby, R., Olson, E.N., 2020. Correction of three prominent mutations in mouse and human models of Duchenne muscular dystrophy by single-cut genome editing. Mol. Ther. 15250024. 28 (9), 2044–2055. Available from: https://www.journals.elsevier.com/molecular-therapy. doi:10.1016/j.ymthe.2020.05.024.

Min, Y.L., Li, H., Rodriguez-Caycedo, C., Mireault, A.A., Huang, J., Shelton, J.M., McAnally, J.R., Amoasii, L., Mammen, P.P.A., Bassel-Duby, R., Olson, E.N., 2019. CRISPR-Cas9 corrects Duchenne muscular dystrophy exon 44 deletion mutations in mice and human cells. Sci. Adv. 23752548. 5 (3). Available from: http://advances.sciencemag.org/content/5/3/eaav4324/tab-pdf. doi:10.1126/sciadv.aav4324.

Monaco, A.P., Bertelson, C.J., Liechti-Gallati, S., Moser, H., Kunkel, L.M., 1988. An explanation for the phenotypic differences between patients bearing partial deletions of the DMD locus. Genomics. 10898646. 2 (1), 90–95. doi:10.1016/0888-7543(88)90113-9.

Monaco, A.P., Neve, R.L., Colletti-Feener, C., Bertelson, C.J., Kurnit, D.M., Kunkel, L.M., 1986. Isolation of candidate cDNAs for portions of the Duchenne muscular dystrophy gene. Nature. 00280836. 323 (6089), 646–650. doi:10.1038/323646a0.

Monceau, A., Moutachi, D., Lemaitre, M., Garcia, L., Trollet, C., Furling, D., Klein, A., Ferry, A., 2022. Dystrophin restoration after adeno-associated virus U7–mediated Dmd exon skipping is modulated by muscular exercise in the severe D2-Mdx Duchenne muscular dystrophy murine model. Am. J. Pathol. 00029440. 192 (11), 1604–1618. doi:10.1016/j.ajpath.2022.07.016.

Mourkioti, F., Kustan, J., Kraft, P., Day, J.W., Zhao, -M.-M., Kost-Alimova, M., Protopopov, A., DePinho, R.A., Bernstein, D., Meeker, A.K., Blau, H.M., 2013. Role of telomere dysfunction in cardiac failure in Duchenne muscular dystrophy. Nat. Cell Biol. 1465-7392. 15 (8), 895–904. doi:10.1038/ncb2790.

Mullard, A., 2023. FDA Approves first gene therapy for Duchenne muscular dystrophy, despite internal objections. Nat. Rev. Drug Discov. 14741784. 22 (8), 610. Available from: https://www.nature.com/nrd/. doi:10.1038/d41573-023-00103-y.

Muraine, L., Bensalah, M., Dhiab, J., Cordova, G., Arandel, L., Marhic, A., Chapart, M., Vasseur, S., Benkhelifa-Ziyyat, S., Bigot, A., Butler-Browne, G., Mouly, V., Negroni, E., Trollet, C., 2020. Transduction efficiency of adeno-associated virus serotypes after local injection in mouse and human skeletal muscle. Hum. Gene Ther. 1043-0342. 31 (3-4), 233–240. doi:10.1089/hum.2019.173.

Naarding, K.J., Reyngoudt, H., Van Zwet, E.W., Hooijmans, M.T., Tian, C., Rybalsky, I., Shellenbarger, K.C., Louër, J.L., Wong, B.L., Carlier, P.G., Kan, H.E., Niks, E.H., 2020. MRI Vastus lateralis fat fraction predicts loss of ambulation in Duchenne muscular dystrophy. Neurology. 1526632X. 94 (13), E1386. Available from: http://www.neurology.org. doi:10.1212/WNL.0000000000008939.

Nakamura, K., Fujii, W., Tsuboi, M., Tanihata, J., Teramoto, N., Takeuchi, S., Naito, K., Yamanouchi, K., Nishihara, M., 2014. Generation of muscular dystrophy model rats with a CRISPR/Cas system. Sci. Rep. 2045-2322. 4 (1). doi:10.1038/srep05635.

Nakamura, K., Kanou, M., Fujii, W., Kouzaki, K., Jimbo, T., Yamanouchi, K., Nakazato, K., Ueda, H., Hirata, J., Yamana, K., 2025. New selective androgen receptor modulator TEI-SARM2 improves muscle function in a Duchenne muscular dystrophy rat model. Hum. Mol. Genet. 0964-6906. 34 (10), 852–863. doi:10.1093/hmg/ddaf028.

NCT06270719, C.g.T.R. Clinical Trial Registration: NCT06270719. 2026. Available from: https://clinicaltrials.gov/study/NCT06270719. (10 February 2026).

Nelson, J.W., Randolph, P.B., Shen, S.P., Everette, K.A., Chen, P.J., Anzalone, A.V., An, M., Newby, G.A., Chen, J.C., Hsu, A., Liu, D.R., 2022. Engineered pegRNAs improve prime editing efficiency. Nat. Biotechnol. 15461696. 40 (3), 402–410. Available from: http://www.nature.com/nbt/index.html. doi:10.1038/s41587-021-01039-7.

Nonneman, D.J., Brown-Brandl, T., Jones, S.A., Wiedmann, R.T., Rohrer, G.A., 2012. A defect in dystrophin causes a novel porcine stress syndrome. BMC Genomics. 1471-2164. 13 (1), 233. doi:10.1186/1471-2164-13-233.

O'Brien, K.F., Kunkel, L.M., 2001. Dystrophin and muscular dystrophy: past, present, and future. Mol. Genet. Metab. 10967192. 74 (1-2), 75–88. Available from: http://www.elsevier.com/inca/publications/store/6/2/2/9/2/0/index.htt. doi:10.1006/mgme.2001.3220.

Oh, H.J., Chung, E., Kim, J., Kim, M.J., Kim, G.A., Lee, S.H., Ra, K., Eom, K., Park, S., Chae, J.H., Kim, J.S., Lee, B.C., 2022. Generation of a dystrophin mutant in dog by nuclear transfer using CRISPR/Cas9-mediated somatic cells: a preliminary study. Int. J. Mol. Sci. 14220067. 23 (5). Available from: https://www.mdpi.com/1422-0067/23/5/2898/pdf. doi:10.3390/ijms23052898.

Osorio, A.N., Cantillo, J.M., Salas, A.C., Garrido, M.M., Padilla, J.J.V., 2019. Consensus on the diagnosis, treatment and follow-up of patients with Duchenne muscular dystrophy. Neurología (English Edition. 21735808. 34 (7), 469–481. doi:10.1016/j.nrleng.2018.01.001.

Padmaswari, M.H., Agrawal, S., Jia, M.S., Ivy, A., Maxenberger, D.A., Burcham, L.A., Nelson, C.E., 2023. Delivery challenges for CRISPR-Cas9 genome editing for Duchenne muscular dystrophy. Biophys. Rev. 4 (1), 011307. doi:10.1063/5.0131452.

Palma, M., Lejeune, F., 2021. Deciphering the molecular mechanism of stop codon readthrough. Biol. Rev. 1469185X. 96 (1), 310–329. Available from: http://onlinelibrary.wiley.com/journal/10.1111/(ISSN)1469-185X. doi:10.1111/brv.12657.

Pan, C., Gen, L., Malzahn, A.A., Cheng, Y., Leyson, B., Sretenovic, S., Gurel, F., Coleman, G.D., Qi, Y., 2022. Boosting plant genome editing with a versatile CRISPR-Combo system. Nat. Plants. 2055-0278. 8 (5), 513–525. doi:10.1038/s41477-022-01151-9.

Pane, M., Mazzone, E.S., Sivo, S., Sormani, M.P., Messina, S., D'Amico, A., Carlesi, A., Vita, G., Fanelli, L., Berardinelli, A., Torrente, Y., Lanzillotta, V., Viggiano, E., D'Ambrosio, P., Cavallaro, F., Frosini, S., Barp, A., Bonfiglio, S., Scalise, R., De Sanctis, R., Rolle, E., Graziano, A., Magri, F., Palermo, C., Rossi, F., Donati, M.A., Sacchini, M., Arnoldi, M.T., Baranello, G., Mongini, T., Pini, A., Battini, R., Pegoraro, E., Previtali, S., Bruno, C., Politano, L., Comi, G.P., Bertini, E., Mercuri, E., Cohn, R., 2014. Long term natural history data in ambulant boys with Duchenne muscular dystrophy: 36-month changes. PLoS One. 1932-6203. 9 (10), e108205. doi:10.1371/journal.pone.0108205.

Philippidis, A., 2022. Food and drug administration lifts clinical hold on pfizer duchenne muscular dystrophy gene therapy linked to patient death. Hum. Gene Ther. 1043-0342. 33 (11-12), 573–576. doi:10.1089/hum.2022.29211.bfs.

Philippidis, A., 2024. Pfizer Weighs Next Steps after DMD Therapy Linked to Boy's death fails phase III Trial. Hum. Gene Ther. 1043-0342. 35 (13-14), 413–415. doi:10.1089/hum.2024.38567.bfs.

Phillips, M.F., Quinlivan, R.C.M., Edwards, R.H.T., Calverley, P.M.A., 2001. Changes in spirometry over time as a prognostic marker in patients with Duchenne muscular dystrophy. Am. J. Respir. Crit. Care Med. 1073-449X. 164 (12), 2191–2194. doi:10.1164/ajrccm.164.12.2103052.

Pichavant, C., Aartsma-Rus, A., Clemens, P.R., Davies, K.E., Dickson, G., Takeda, S., Wilton, S.D., Wolff, J.A., Wooddell, C.I., Xiao, X., Tremblay, J.P., 2011. Current status of pharmaceutical and genetic therapeutic approaches to treat DMD. Mol. Ther. 15250016. 19 (5), 830–840. doi:10.1038/mt.2011.59.

Pickar-Oliver, A., Gough, V., Bohning, J.D., Liu, S., Robinson-Hamm, J.N., Daniels, H., Majoros, W.H., Devlin, G., Asokan, A., Gersbach, C.A., 2021. Full-length dystrophin restoration via targeted exon integration by AAV-CRISPR in a humanized mouse model of Duchenne muscular dystrophy. Mol. Ther. 15250016. 29 (11), 3243–3257. doi:10.1016/j.ymthe.2021.09.003.

Potter, R.A., Moeller, I.H., Khan, S., Haegel, H., Hollenstein, A., Steiner, G., Wandel, C., Murphy, A.P., Asher, D.R., Palatinsky, E., Griffin, D.A., Mason, S., Iannaccone, S.T., Zaidman, C.M., Rodino-Klapac, L.R., 2025. Immunologic investigations into transgene directed immune-mediated myositis following delandistrogene moxeparvovec gene therapy. Sci. Rep. 20452322. 15 (1). Available from: https://www.nature.com/srep/. doi:10.1038/s41598-024-84077-w.

Ramos, J.N., Hollinger, K., Bengtsson, N.E., Allen, J.M., Hauschka, S.D., Chamberlain, J.S., 2019. Development of novel micro-dystrophins with enhanced functionality. Mol. Ther. 15250024. 27 (3), 623–635. Available from: https://www.journals.elsevier.com/molecular-therapy. doi:10.1016/j.ymthe.2019.01.002.

Rees, H.A., Liu, D.R., 2018. Base editing: precision chemistry on the genome and transcriptome of living cells. Nat. Rev. Genet. 14710064. 19 (12), 770–788. Available from: http://www.nature.com/reviews/genetics. doi:10.1038/s41576-018-0059-1.

Ricotti, V., Ridout, D.A., Pane, M., Main, M., Mayhew, A., Mercuri, E., Manzur, A.Y., Muntoni, F., Robb, S., Quinlivan, R., Sarkozy, A., Butler, J., Bushby, K., Straub, V., Guglieri, M., Eagle, M., Roper, H., McMurchie, H., Childs, A., Pysden, K., Pallant, L., Spinty, S., Peachey, G., Shillington, A., Wraige, E., Jungbluth, H., Sheehan, J., Spahr, R., Hughes, I., Bateman, E., Cammiss, C., Willis, T., Groves, L., Emery, N., Baxter, P., Senior, M., Scott, E., Hartley, L., Parsons, B., Majumdar, A., Jenkins, L., Toms, B., Naismith, K., Keddie, A., Horrocks, I., Di Marco, M., Chow, G., Miah, A., De Goede, C., Thomas, N., Geary, M., Palmer, J., White, C., Greenfield, K., Wilson, I., 2016. The NorthStar Ambulatory Assessment in Duchenne muscular dystrophy: considerations for the design of clinical trials. Journal of Neurology, Neurosurgery and Psychiatry. 1468330X. 87 (2), 149–155. Available from: http://jnnp.bmj.com/content/87/2/149.full.pdf. doi:10.1136/jnnp-2014-309405.

Rodino-Klapac, L., et al, 2019. Systemic gene transfer with rAAVrh74. MHCK7. SGCB Increased β-sarcoglycan expression in patients with Limb Girdle muscular dystrophy type 2E. Neuropediatrics 50 (S 02), GNP–PO59.

Rodino-Klapac, L.R., Haidet, A.M., Kota, J., Handy, C., Kaspar, B.K., Mendell, J.R., 2009. Inhibition of myostatin with emphasis on follistatin as a therapy for muscle disease. Muscle Nerve. 10974598. 39 (3), 283–296. Available from: http://www3.interscience.wiley.com/cgi-bin/fulltext/122194851/PDFSTART. doi:10.1002/mus.21244.

Rooney, J.E., Welser, J.V., Dechert, M.A., Flintoff-Dye, N.L., Kaufman, S.J., Burkin, D.J., 2006. Severe muscular dystrophy in mice that lack dystrophin and α7 integrin. J. Cell Sci. 00219533. 119 (11), 2185–2195. doi:10.1242/jcs.02952.

Sacco, A., Mourkioti, F., Tran, R., Choi, J., Llewellyn, M., Kraft, P., Shkreli, M., Delp, S., Pomerantz, J.H., Artandi, S.E., Blau, H.M., 2010. Short telomeres and stem cell exhaustion model Duchenne muscular dystrophy in mdx/mTR mice. Cell. 00928674. 143 (7), 1059–1071. doi:10.1016/j.cell.2010.11.039.

Salari, N., Fatahi, B., Valipour, E., Kazeminia, M., Fatahian, R., Kiaei, A., Shohaimi, S., Mohammadi, M., 2022. Global prevalence of Duchenne and Becker muscular dystrophy: a systematic review and meta-analysis. J. Orthop. Surg. Res. 1749-799X. 17 (1). doi:10.1186/s13018-022-02996-8.

Salari, N., Fatahi, B., Valipour, E., Kazeminia, M., Fatahian, R., Kiaei, A., Shohaimi, S., Mohammadi, M., 2022. Global prevalence of Duchenne and Becker muscular dystrophy: a systematic review and meta-analysis. J. Orthop. Surg. Res. 17 (1). doi:10.1186/s13018-022-02996-8.

Schwotzer, N., Sissy, C.E., Desguerre, I., Frémeaux-Bacchi, V., Servais, L., Fakhouri, F., 2024. Thrombotic microangiopathy as an emerging complication of viral vector–based gene therapy. Kidney Int. Rep. 24680249. 9 (7), 1995–2005. doi:10.1016/j.ekir.2024.04.024.

Servais, L., Mercuri, E., Straub, V., Guglieri, M., Seferian, A.M., Scoto, M., Leone, D., Koenig, E., Khan, N., Dugar, A., Wang, X., Han, B., Wang, D., Muntoni, F., Brusa, C., Antonaci, L., Brogna, C., Merli, L., Monforte, M., Norcia, G., Pane, M., Ferrantini, G., Dickson, G., Morgan, J., Sardone, V., Akana, C., Charleston, J.S., Desjardins, C.A., El-Husayni, S.H., Frank, D.E., Schnell, F.J., 2022. Long-term safety and efficacy data of golodirsen in ambulatory patients with Duchenne muscular dystrophy amenable to exon 53 skipping: a first-in-human, multicenter, two-part, open-label, phase 1/2 trial. Nucleic Acid Ther. 2159-3337. 32 (1), 29–39. doi:10.1089/nat.2021.0043.

Sharp, N.J.H., Kornegay, J.N., Van Camp, S.D., Herbstreith, M.H., Secore, S.L., Kettle, S., Hung, W.-Y., Constantinou, C.D., Dykstra, M.J., Roses, A.D., Bartlett, R.J., 1992. An error in dystrophin mRNA processing in golden retriever muscular dystrophy, an animal homologue of Duchenne muscular dystrophy. Genomics. 08887543. 13 (1), 115–121. doi:10.1016/0888-7543(92)90210-j.

Sheehan, D.W., Birnkrant, D.J., Benditt, J.O., Eagle, M., Finder, J.D., Kissel, J., Kravitz, R.M., Sawnani, H., Shell, R., Sussman, M.D., Wolfe, L.F., 2018. Respiratory management of the patient with Duchenne muscular dystrophy. Pediatrics. 10984275. 142, S62. Available from: http://pediatrics.aappublications.org/content/pediatrics/142/Supplement_2/S53.full.pdf?download=true. doi:10.1542/peds.2018-0333H.

Sherlock, S.P., Levy, D.I., McIntosh, A., Shieh, P.B., Smith, E.C., McDonnell, T.G., Ryan, K.A., Delnomdedieu, M., Binks, M., Lal, A.K., Butterfield, R.J., 2025. Cardiac safety of fordadistrogene movaparvovec gene therapy in Duchenne muscular dystrophy: initial observations from a phase 1b trial. Mol. Ther. 15250024. 33 (9), 4216–4225. Available from: https://www.sciencedirect.com/science/journal/15250016. doi:10.1016/j.ymthe.2025.06.031.

Shimatsu, Y., Katagiri, K., Furuta, T., Nakura, M., Tanioka, Y., Yuasa, K., Tomohiro, M., Kornegay, J.N., Nonaka, I., Takeda, S., 2003. Canine X-Linked Muscular Dystrophy in Japan (CXMDJ). Exp. Anim. 1341-1357. 52 (2), 93–97. doi:10.1538/expanim.52.93.

Shimizu-Motohashi, Y., 2025. Gene therapy for Duchenne muscular dystrophy. Brain Dev. 03877604. 47 (5), 104424. doi:10.1016/j.braindev.2025.104424.

Shirley, M., 2021. Casimersen: first approval. Drugs. 0012-6667. 81 (7), 875–879. doi:10.1007/s40265-021-01512-2.

Sicinski, P., Geng, Y., Ryder-Cook, A.S., Barnard, E.A., Darlison, M.G., Barnard, P.J., 1989. The molecular basis of muscular Dystrophy in the mdx mouse: a point mutation. Science. 0036-8075. 244 (4912), 1578–1580. doi:10.1126/science.2662404.

Smith, B.F., Kornegay, J.N., Duan, D., 2007. Independent canine models of Duchenne muscular dystrophy due to intronic insertions of repetitive DNA. Mol. Ther. 15 (Suppl 1), S51.

Song, B., Yang, S., Hwang, G.-H., Jihyeon, Y., Bae, S., 2021. Analysis of NHEJ-based DNA repair after CRISPR-mediated DNA cleavage. Int. J. Mol. Sci. 1422-0067. 22 (12), 6397. doi:10.3390/ijms22126397.

Stark, A.E., 2015. Determinants of the incidence of Duchenne muscular dystrophy. Ann. Transl. Med. 23055847. 3 (19). Available from: http://atm.amegroups.com/. doi:10.3978/j.issn.2305-5839.2015.10.45.

Stephenson, A.A., Flanigan, K.M., 2021. Gene editing and modulation for Duchenne muscular dystrophy. Prog. Mol. Biol. Transl. Sci. 182, 225–255. Available from: http://www.elsevier.com/books/book-series/progress-in-molecular-biology-and-translational-science#. doi:10.1016/bs.pmbts.2021.01.029.

Sui, T., Lau, Y.S., Liu, D., Liu, T., Xu, L., Gao, Y., Lai, L., Zhanjun, L., Han, R., 2018. A novel rabbit model of Duchenne muscular dystrophy generated by CRISPR/Cas9. Dis. Model. Mech. 1754-8411. 11 (6). doi:10.1242/dmm.032201.

Suthar, R., Sankhyan, N., 2018. Duchenne Muscular dystrophy: a practice update. Indian J. Pediatr. 09737693. 85 (4), 276–281. Available from: http://www.springerlink.com/content/0019-5456. doi:10.1007/s12098-017-2397-y.

Syed, Y.Y., 2016. Eteplirsen: first global approval. Drugs. 11791950. 76 (17), 1699–1704. Available from: http://rd.springer.com/journal/40265. doi:10.1007/s40265-016-0657-1.

Taglietti, V., Kefi, K., Bronisz-Budzyńska, I., Mirciloglu, B., Rodrigues, M., Cardone, N., Coulpier, F., Periou, B., Gentil, C., Goddard, M., Authier, F.-J., Pietri-Rouxel, F., Malfatti, E., Lafuste, P., Tiret, L., Relaix, F., 2022. Duchenne muscular dystrophy trajectory in R-DMDdel52 preclinical rat model identifies COMP as biomarker of fibrosis. Acta Neuropathol. Commun. 2051-5960. 10 (1). doi:10.1186/s40478-022-01355-2.

Tasfaout, H., Halbert, C.L., McMillen, T.S., Allen, J.M., Reyes, T.R., Flint, G.V., Grimm, D., Hauschka, S.D., Regnier, M., Chamberlain, J.S., 2024. Split intein-mediated protein trans-splicing to express large dystrophins. Nature. 0028-0836. 632 (8023), 192–200. doi:10.1038/s41586-024-07710-8.

Taylor, M., Jefferies, J., Byrne, B., Lima, J., Ambale-Venkatesh, B., Ostovaneh, M.R., Makkar, R., Goldstein, B., Smith, R.R., Fudge, J., Malliaras, K., Fedor, B., Rudy, J., Pogoda, J.M., Marbán, L., Ascheim, D.D., Marbán, E., Victor, R.G., 2019. Cardiac and skeletal muscle effects in the randomized HOPE-Duchenne trial. Neurology. 1526632X. 92 (8), E866. Available from: http://www.neurology.org. doi:10.1212/WNL.0000000000006950.

Teramoto, N., Sugihara, H., Yamanouchi, K., Nakamura, K., Kimura, K., Okano, T., Shiga, T., Shirakawa, T., Matsuo, M., Nagata, T., Daimon, M., Matsuwaki, T., Nishihara, M., 2020. Pathological evaluation of rats carrying in-frame mutations in the dystrophin gene: a new model of Becker muscular dystrophy. Dis. Model. Mech. 1754-8411. 13 (9). doi:10.1242/dmm.044701.

Theadom, A., Rodrigues, M., Roxburgh, R., Balalla, S., Higgins, C., Bhattacharjee, R., Jones, K., Krishnamurthi, R., Feigin, V., 2015. Prevalence of muscular dystrophies: a systematic literature review. Neuroepidemiology. 0251-5350. 43 (3-4), 259–268. doi:10.1159/000369343.

Therapeutics, H. HuidaGene therapeutics initiates M.U.S.C.L.E. clinical trial of HG302 for Duchenne muscular dystrophy and completes first patient dosed. 2024. Available from: https://www.huidagene.com/new/news/72. (18 May 2025).

Todd, M., Kang, S., Shunwen, W., Adhin, D., Yoon, D.Y., Willcocks, R., Kim, S., 2025. Use of imaging biomarkers and ambulatory functional endpoints in Duchenne muscular dystrophy clinical trials: systematic review and machine learning-driven trend analysis. J. Neuromuscul. Dis. 2214–3599. doi:10.1177/22143602251360664.

Tuffery-Giraud, S., Béroud, C., Leturcq, F., Yaou, R.B., Hamroun, D., Michel-Calemard, L., Moizard, M.-P., Bernard, R., Cossée, M., Boisseau, P., Blayau, M., Creveaux, I., Guiochon-Mantel, A., de Martinville, B., Philippe, C., Monnier, N., Bieth, E., Van Kien, P.K., Desmet, F.-O., Humbertclaude, V., Kaplan, J.-C., Chelly, J., Claustres, M., 2009. Genotype-phenotype analysis in 2,405 patients with a dystrophinopathy using the UMD-DMD database: a model of nationwide knowledgebase. Hum. Mutat. 10597794. 30 (6), 934–945. doi:10.1002/humu.20976.

U.S. Food and Drug Administration, 2025. FDA Approves New Safety Warning and Revised Indication that Limits Use for Elevidys Following Reports of Fatal Liver Injury. U.S. Food and Drug Administration, Silver Spring, MD. https://www.fda.gov/news-events/press-announcements/fda-approves-new-safety-warning-and-revised-indication-limits-use-elevidys-following-reports-fatal. (Accessed 27 July 2026).

Vaillend, C., Aoki, Y., Mercuri, E., Hendriksen, J., Tetorou, K., Goyenvalle, A., Muntoni, F., 2025. Duchenne muscular dystrophy: recent insights in brain related comorbidities. Nat. Commun. 2041-1723. 16 (1). doi:10.1038/s41467-025-56644-w.

Valentine, B.A., Cooper, B.J., de Lahunta, A., O'Quinn, R., Blue, J.T., 1988. Canine X-linked muscular dystrophy. An animal model of Duchenne muscular dystrophy: clinical studies. J. Neurol. Sci. 0022510X. 88 (1-3), 69–81. doi:10.1016/0022-510X(88)90206-7.

van Putten, M., Kumar, D., Hulsker, M., Hoogaars, W.M.H., Plomp, J.J., van Opstal, A., van Iterson, M., Admiraal, P., van Ommen, G.-J.B., Hoen, P.A.C., Aartsma-Rus, A., 2012. Comparison of skeletal muscle pathology and motor function of dystrophin and utrophin deficient mouse strains. Neuromuscul. Disord. 09608966. 22 (5), 406–417. doi:10.1016/j.nmd.2011.10.011.

Vandebrouck, C., Martin, D., Schoor, M.C.-V., Debaix, H., Gailly, P., 2002. Involvement of TRPC in the abnormal calcium influx observed in dystrophic (mdx) mouse skeletal muscle fibers. J. Cell Biol. 1540-8140. 158 (6), 1089–1096. doi:10.1083/jcb.200203091.

Vandenborne, K., Walter, G.A., Straub, V., Willcocks, R.J., Forbes, S.C., Mercuri, E.M., Muntoni, F., Ding, K., Ennamuri, S., Reid, C., Murphy, A.P., Manfrini, M., Mendell, J.R., Elkins, J.S., Rodino-Klapac, L.R., 2025. Quantitative muscle magnetic resonance outcomes in patients With Duchenne muscular dystrophy an exploratory analysis from the EMBARK randomized clinical trial. JAMA Neurol. 21686157. 82 (7), 734–744. Available from: http://archneur.jamanetwork.com/issues.aspx. doi:10.1001/jamaneurol.2025.0992.

Venugopal, V., Pavlakis, S., 2023. Duchenne Muscular Dystrophy, in StatPearls. Internet. StatPearls Publishing.

Verma, S., Nwosu, S.N., Razdan, R., Upadhyayula, S.R., Phan, H.C., Koroma, A.A., Leguizamo, I., Correa, N.S., Kuipa, M., Lee, D., Vanderford, T.H., Gardner, M.R., 2023. Seroprevalence of adeno-associated virus neutralizing antibodies in males with Duchenne muscular dystrophy. Hum. Gene Ther. 1043-0342. 34 (9-10), 430–438. doi:10.1089/hum.2022.081.

Vulin, A., Barthélémy, I., Goyenvalle, A., Thibaud, J.-L., Beley, C., Griffith, G., Benchaouir, R., le Hir, M., Unterfinger, Y., Lorain, S., Dreyfus, P., Voit, T., Carlier, P., Blot, S., Garcia, L., 2012. Muscle function recovery in golden retriever muscular dystrophy after AAV1-U7 exon skipping. Mol. Ther. 15250016. 20 (11), 2120–2133. doi:10.1038/mt.2012.181.

Vulin, A., Wein, N., Simmons, T.R., Rutherford, A.M., Findlay, A.R., Yurkoski, J.A., Kaminoh, Y., Flanigan, K.M., 2015. The first exon duplication mouse model of Duchenne muscular dystrophy: a tool for therapeutic development. Neuromuscul. Disord. 09608966. 25 (11), 827–834. doi:10.1016/j.nmd.2015.08.005.

Walmsley, G.L., Arechavala-Gomeza, V., Fernandez-Fuente, M., Burke, M.M., Nagel, N., Holder, A., Stanley, R., Chandler, K., Marks, S.L., Muntoni, F., Shelton, G.D., Piercy, R.J., Feany, M.B., 2010. A Duchenne muscular dystrophy gene hot spot mutation in dystrophin-deficient cavalier king charles spaniels is amenable to Exon 51 skipping. PLoS One. 1932-6203. 5 (1), e8647. doi:10.1371/journal.pone.0008647.

Walsh, J., Palandra, J., Duriga, N., Beidler, D., McIntosh, A., Binks, M., Neubert, H., 2025. Dystrophin/mini-dystrophin expression analysis by immunoaffinity liquid chromatography–tandem mass spectrometry after gene therapy for DMD. Gene Ther. 0969-7128. 32 (6), 573–580. doi:10.1038/s41434-025-00554-5.

Walsh, S., Nygren, J., Pontén, A., Jovinge, S., Rojas, M., 2011. Myogenic reprogramming of bone marrow derived cells in a W41Dmdmdx deficient mouse model. PLoS One. 1932-6203. 6 (11), e27500. doi:10.1371/journal.pone.0027500.

Wang, B., Li, J., Xiao, X., 2000. Adeno-associated virus vector carrying human minidystrophin genes effectively ameliorates muscular dystrophy in mdx mouse model. Proc. Natl. Acad. Sci. USA. 00278424. 97 (25), 13714–13719. doi:10.1073/pnas.240335297.

Wang, T., Daoud, C., Dubois, A., Corre, G., Bellec, J., Bovolenta, M., Philidet, L., Dorval, A., Bourg, N., Roudaut, C., Albini, S., Warthi, G., Jaber, A., Richard, I., 2026. A new dystrophin-deficient rat model mirroring exon skipping in patients with DMD exon 45 deletions. Dis. Model. Mech. 1754-8403. 19 (1). doi:10.1242/dmm.052578.

Weber, F.J., Latshang, T.D., Blum, M.R., Kohler, M., Wertli, M.M., 2022. Prognostic factors, disease course, and treatment efficacy in Duchenne muscular dystrophy: a systematic review and meta-analysis. Muscle Nerve. 10974598. 66 (4), 462–470. Available from: http://onlinelibrary.wiley.com/journal/10.1002/(ISSN)1097-4598. doi:10.1002/mus.27682.

Wein, N., Vulin, A., Falzarano, M.S., Szigyarto, C.A.-K., Maiti, B., Findlay, A., Heller, K.N., Uhlén, M., Bakthavachalu, B., Messina, S., Vita, G., Passarelli, C., Passarelli, S., Bovolenta, M., Neri, M., Gualandi, F., Wilton, S.D., Rodino-Klapac, L.R., Yang, L., Dunn, D.M., Schoenberg, D.R., Weiss, R.B., Howard, M.T., Ferlini, A., Flanigan, K.M., 2014. Translation from a DMD exon 5 IRES results in a functional dystrophin isoform that attenuates dystrophinopathy in humans and mice. Nat. Med. 1078-8956. 20 (9), 992–1000. doi:10.1038/nm.3628.

Willmann, R., De Luca, A., Benatar, M., Grounds, M., Dubach, J., Raymackers, J.-M., Nagaraju, K., 2012. Enhancing translation: guidelines for standard pre-clinical experiments in mdx mice. Neuromuscul. Disord. 09608966. 22 (1), 43–49. doi:10.1016/j.nmd.2011.04.012.

Wilson, D.G.S., Tinker, A., Iskratsch, T., 2022. The role of the dystrophin glycoprotein complex in muscle cell mechanotransduction. Commun. Biol. 23993642. 5 (1). Available from: https://www.springer.com/journal/11852. doi:10.1038/s42003-022-03980-y.

Xie, J., Weikai, G., Nan, L., Liu, Q., Chen, F., Yang, X., Huang, X., Ouyang, Z., Zhang, Q., Zhao, Y., Liu, Z., Gou, S., Han, W., Lai, C., Fan, N., Jin, Q., Shi, H., Liang, Y., Lan, T., Quan, L., Xiaoping, L., Wang, K., Lai, L., 2019. Efficient base editing for multiple genes and loci in pigs using base editors. Nat. Commun. 2041-1723. 10 (1). doi:10.1038/s41467-019-10421-8.

Xu, R., DeVries, S., Camboni, M., Martin, P.T., 2009. Overexpression of Galgt2 reduces dystrophic pathology in the skeletal muscles of alpha sarcoglycan-deficient mice. Am. J. Pathol. 00029440. 175 (1), 235–247. doi:10.2353/ajpath.2009.080967.

Yao, T., Yuan, G., Haiwei, L., Liu, Y., Zhang, J., Tuskan, G.A., Muchero, W., Chen, J.-G., Yang, X., 2023. CRISPR/Cas9-based gene activation and base editing in Populus. Hortic. Res. 2052-7276. 10 (6). doi:10.1093/hr/uhad085.

Young, C.S., Mokhonova, E., Quinonez, M., Pyle, A.D., Spencer, M.J., 2017. Creation of a novel humanized dystrophic mouse model of Duchenne muscular dystrophy and application of a CRISPR/Cas9 gene editing therapy. J. Neuromuscul. Dis. 22143602. 4 (2), 139–145. Available from: http://www.iospress.nl/journal/journal-of-neuromuscular-diseases/. doi:10.3233/JND-170218.

Yuan, J., Yunqing, M., Huang, T., Chen, Y., Peng, Y., Bing, L., Jia, L., Zhang, Y., Song, B., Sun, X., Ding, Q., Song, Y., Chang, X., 2018. Genetic modulation of RNA splicing with a CRISPR-guided cytidine deaminase. Mol. Cell. 10972765. 72 (2), 380. doi:10.1016/j.molcel.2018.09.002.

Yucel, N., Chang, A.C., Day, J.W., Rosenthal, N., Blau, H.M., 2018. Humanizing the mdx mouse model of DMD: the long and the short of it. Npj Regen. Med. 2057-3995. 3 (1). doi:10.1038/s41536-018-0045-4.

Zabłocka, B., Górecki, D.C., Zabłocki, K., 2021. Disrupted calcium homeostasis in duchenne muscular dystrophy: a common mechanism behind diverse consequences. Int. J. Mol. Sci. 14220067. 22 (20). Available from: https://www.mdpi.com/1422-0067/22/20/11040/pdf. doi:10.3390/ijms222011040.

Zaidman, C.M., et al, 2023. Delandistrogene moxeparvovec gene therapy in ambulatory patients (aged ≥4 to ESCAPED_CHARACTER_LESS_THAN8 Years) with Duchenne muscular dystrophy: 1-Year interim results from study SRP-9001-103 (ENDEAVOR. Ann. Neurol. 94 (5), 955–968.

Zanoteli, E., França, M.C., Marques, W., 2024. Gene-based therapies for neuromuscular disorders. Arquivos de Neuro-Psiquiatria. 16784227. 82 (6). Available from: https://www.scielo.br/j/anp/a/HDCLFstBv4VWnPhGT5bPNWw/?lang=en#. doi:10.1055/s-0043-1777755.

Zhao, J., Kodippili, K., Yue, Y., Hakim, C.H., Wasala, L., Pan, X., Zhang, K., Yang, N.N., Duan, D., Lai, Y., 2016. Dystrophin contains multiple independent membrane-binding domains. Hum. Mol. Genet. 14602083. 25 (17), 3647–3653. Available from: http://hmg.oxfordjournals.org/. doi:10.1093/hmg/ddw210.

Zhou, Y., Zhang, C., Xiao, W., Herzog, R.W., Han, R., 2024. Systemic delivery of full-length dystrophin in Duchenne muscular dystrophy mice. Nat. Commun. 2041-1723. 15 (1). doi:10.1038/s41467-024-50569-6.

Zincarelli, C., Soltys, S., Rengo, G., Rabinowitz, J.E., 2008. Analysis of AAV serotypes 1–9 mediated gene expression and tropism in mice after systemic injection. Mol. Ther. 15250016. 16 (6), 1073–1080. doi:10.1038/mt.2008.76.

Zygmunt, D.A., Crowe, K.E., Flanigan, K.M., Martin, P.T., 2017. Comparison of serum rAAV serotype-specific antibodies in patients with Duchenne muscular dystrophy, becker muscular dystrophy, inclusion body myositis, or GNE myopathy. Hum. Gene Ther. 15577422. 28 (9), 737–746. Available from: https://www.liebertonline.com/hum. doi:10.1089/hum.2016.141.

Chapter 7

Gene therapy in limb-girdle muscular dystrophy (LGMD)

Zahra Oushyani Roudsari and Parnaz Niazi
Department of Medical Biotechnology, Zanjan University of Medical Sciences, Zanjan, Iran

1 Introduction

Limb-girdle muscular dystrophy (LGMD) refers to a spectrum of neuromuscular conditions characterized by the progressive weakening of the proximal girdle muscles, specifically those of the shoulders and pelvis (Anwar and Yokota, 2024; Barton et al., 2020; Johnson and Statland, 2022). A defining characteristic of these diseases is their extensive genetic heterogeneity (Nallamilli et al., 2018). Recent advancements in molecular genetics, imaging, and natural history research have led to a revised understanding of the definition, classification, and epidemiology of LGMDs, which were historically known as "muscular dystrophies affecting the limb girdles" (Johnson and Statland, 2022; Liewluck, 2025). This progress has yielded greater knowledge of the global prevalence, clinical presentation, and prognosis of LGMD (Stokholm et al., 2022; Georganopoulou et al., 2021).

Although each LGMD subtype is rare, collectively they represent a significant challenge among primary myopathies (Georganopoulou et al., 2021; D'Este et al., 2025; Alonso-Pérez et al., 2022). Metaanalytic data estimate the global incidence at approximately 0.7 per 100,000 and the prevalence at about 3 per 100,000 for the entire group (Doody et al., 2024; Bouchard and Tremblay, 2023). However, significant regional epidemiological variance exists; for instance, CAPN3 mutations are more frequent in Eurasian populations, while FKRP variants are more prevalent in Northern Europe (Lin et al., 2023; Jensen et al., 2024; Mroczek et al., 2022). Population-based statistics are constrained by the rarity of each genetic subtype, and prevalence may be underestimated due to studies from referral centers potentially missing milder cases; founder effects in specific populations can also skew subtype frequencies (Lin et al., 2023; Zambon et al., 2024).

LGMD is defined by a gradual decline in muscular function and a slow breakdown of muscle fibers, leading to impaired movement, wasting, and weakness (Johnson and Statland, 2022; Costa et al., 2022). Despite genetic diversity, a similar pattern of muscle involvement and clinical progression allows for diagnosis, as the pathophysiological outcome is consistent across subtypes (Johnson and Statland, 2022; Taghizadeh et al., 2019). The core diagnostic criteria for LGMD are as follows: primary skeletal muscle involvement, pronounced weakness in the limb-girdle regions, and genetic confirmation in unrelated families (Jensen et al., 2023). This classification encompasses a wide clinical spectrum, from severe early-onset to mild late-onset cases, with the range of acknowledged subtypes continually broadening due to advances in genetic testing (Li et al., 2022).

Clinically, the onset is often insidious, with initial signs, including difficulty running, climbing stairs, or rising from a chair due to weakness in pelvic and thigh muscles (LGMD, 2020). This can manifest as a waddling Trendelenburg gait, caused by weakened hip abductors and pelvic tilting (Gandbhir et al., 2024; Fujita et al., 2025). Children with recessive forms may experience frequent falls, while those with dominant forms may initially notice exercise-induced fatigue or asymmetry (D'Este et al., 2025; Doody et al., 2024). Shoulder girdle involvement leads to difficulties with overhead activities and scapular winging (Johnson and Statland, 2022; Alonso-Pérez et al., 2022; Park et al., 2025). Patients experience persistent fatigue and reduced capacity, but sensory and cognitive functions remain intact (Angelini and Rodríguez, 2024; Vrist et al., 2023). Over time, muscles show atrophy, pseudohypertrophy, and fatty replacement on imaging (Morishima and Schoser, 2023). Systemic complications are significant and include cardiac issues (cardiomyopathy, arrhythmias) requiring

Gene Therapy in Neurodegenerative Disorders. https://doi.org/10.1016/B978-0-443-40467-2.00013-0

lifelong monitoring, and respiratory insufficiency due to muscle weakness, potentially leading to restrictive lung disease and ventilator dependence (Libell et al., 2020; Muni-Lofra et al., 2023). Later stages involve contractures and skeletal problems like scoliosis and Achilles tendon shortening (Chu and Moran, 2018).

Diagnostically, elevated serum creatine kinase (CK) levels (5–50 times normal) are a common early sign (Howard et al., 2024; Tanboon and Nishino, 2024; Kim et al., 2021). Electromyography shows a myopathic pattern, and muscle MRI is essential for detecting early fatty replacement (Morishima and Schoser, 2023; Kim et al., 2021). However, genetic confirmation remains the gold standard, enabling subtype classification and prognostication for extramuscular complications (Sarıkaya Uzan et al., 2024; Luglio et al., 2024). As the disease advances, patients often lose ambulation and require assistive devices, with weakness in trunk, neck, and oropharyngeal muscles, leading to postural problems, dysphagia, and dysarthria (Alonso-Pérez et al., 2022; Bouchard and Tremblay, 2023; Angelini and Rodríguez, 2024; Allen et al., 2024). Disease progression is highly variable, even among individuals with the same mutation (D'Este et al., 2025).

The prognosis of LGMD is highly variable and depends on the genetic subtype, age of onset, and systemic involvement (Johnson and Statland, 2022; Pegoraro and Hoffman, 2012). Early-onset forms often progress faster than adult-onset variants, which may have a milder course with prolonged ambulation (Sun et al., 2025). Loss of independent ambulation is a key milestone, with timing varying significantly between subtypes (Audhya et al., 2022; Kale et al., 2025; Møller et al., 2022). Rapidly progressive forms like sarcoglycanopathies can lead to loss of walking in adolescence (LGMD, 2020; Vainzof et al., 2021). Postural problems from trunk weakness and contractures further impair mobility (D'Este et al., 2025). Critically, cardiac and respiratory complications are major determinants of prognosis (Angelini and Pegoraro, 2020). Cardiac manifestations include cardiomyopathy and arrhythmias, requiring continuous monitoring and medication (Chowdhury et al., 2022; Nikhanj, 2022). Respiratory insufficiency develops from progressive weakness of the diaphragm and intercostal muscles (Muni-Lofra et al., 2023; Mauro and Aliverti, 2016). Early noninvasive ventilation (NIV) can defer morbidity and improve survival, and overall, a multidisciplinary care approach has significantly extended patient life expectancy (D'Este et al., 2025; Nikhanj, 2022; Nandanwar et al., 2024). Since LGMDs are monogenic disorders with accessible target tissue, they are particularly amenable to gene therapy approaches.

2 Disease pathogenesis and genetic pattern

LGMD is a highly heterogeneous condition, and its classification is based on the underlying genetic defect and mode of inheritance, categorized into autosomal dominant (type 1, now also referred to as LGMD1) and autosomal recessive (type 2, now also referred to as LGMD2) forms. The various forms of this disease, along with the associated genes and proteins, are outlined in Table 7.1 (Bouchard and Tremblay, 2023).

As mentioned above, the pathogenesis of LGMD subtypes involves a range of diverse molecular mechanisms. The disease arises from defects in key cellular components and pathways, including cellular structures, protein processing, mitochondrial dysfunction, intracellular trafficking, as well as various signal transduction pathways and nuclear functions. Furthermore, these signaling pathways are frequently interconnected, and the impairment of multiple pathways concurrently contributes to the pathology in certain LGMD types (Fig. 7.1) (Barton et al., 2020 Barton et al., 2020;Taghizadeh et al., 2019).

2.1 Cellular structures

Among the key cellular components that are modified in LGMD are the dystrophin–glycoprotein complex and the sarcomere complex (Taghizadeh et al., 2019). The dystrophin–glycoprotein complex establishes a critical link between the cell's internal cytoskeleton and the extracellular matrix, thereby preserving the muscle fiber's physical structure along with the stability of the neuromuscular junctions. This complex is composed of several proteins, including dystrophin/utrophin, dystroglycans (α and β), sarcoglycans (SGs) (α, β, γ, and δ), sarcospan, syntrophins (α1, β1, β2, γ1, and γ2), and α-dystrobrevin (Belhasan and Akaaboune, 2020). Mutations that cause structural changes in these proteins are associated with the development of LGMD. For example, mutations in dystrophin and SGs are related to the development of LGMD2C–F and P, and a deficiency in any single component of the SGs usually leads to the absence of the other components from the sarcolemma (Taghizadeh et al., 2019; Laval and Bushby, 2004).

Muscle fibers are made of repeating units called sarcomeres. These contractile units are precise, organized structures built from a complex of proteins such as titin, nebulin, actin, myosin, troponin, and tropomyosin (Rui et al., 2010). A mutation in protein components of sarcomeres leads to various types of LGMD, comprising LGMD1A (mutations in the myotilin gene, *MYOT*), LGMD1E and LGMD2Q (mutations in the desmin gene, *DES*), LGMD2A (mutations in the calpain 3 gene, *CAPN3*), LGMD2G (mutations in the titin cap gene, *TCAP*), LGMD2J (mutations in the massive titin gene), and LGMD2Q which is associated with mutations in the plectin gene, *PLEC1*(Taghizadeh et al., 2019).

TABLE 7.1 Classification of LGMDs.

LGMD type (old/new nomenclature)	Gene	Protein	Key pathological and clinical features
Autosomal dominant (LGMDD)			
LGMD 1A	*MYOT*	Myotilin	Late-onset distal myopathy (ankles, feet, calves); respiratory/cardiac failure reported. Cross-links actin filaments.
LGMD 1B	*LMNA*	Lamin A/C	Causes nuclear envelope alterations. Associated with cardiac arrhythmia.
LGMD 1C	*CAV3*	Caveolin-3	Impaired mitochondrial function. Phenotype variable: hyperCKemia, ankle contracture, calf hypertrophy, myalgia, muscle rippling.
LGMD 1D/D1	*DNAJB6*	DNAJB6 (cochaperone)	Fat infiltration on MRI; rimmed vacuoles in biopsies. Affects proximal and distal muscles (lower/proximal limbs most severely).
LGMD 1E	*DES*	Desmin	Desmin aggregate formation; distal weakness; cardiomyopathy. Fatty replacement in the semitendinosus, gastrocnemius, soleus.
LGMD 1F/D2	*TNPO3*	Transportin 3	Abnormal sarcomeric assembly. Pelvic weakness first and then axial/shoulder girdle atrophy. Rimmed vacuoles on biopsy.
LGMD 1G/D3	*HNRNPDL*	hnRNPDL	Regulates transcription/splicing. Muscles show rimmed vacuoles.
LGMD 1H	*Unknown (Chr 3p)*	Unknown	Unknown
LGMD 1I/D4	*CAPN3*	Calpain 3	Dominant mutations. Fat infiltration in thigh adductors, semimembranosus, calf muscles on MRI.
Bethlem Myopathy/D5	*COL6A1, A2, A3*	Collagen VI	Higher fat fraction in muscles (e.g., psoas major). "Target sign" in the rectus femoris, "sandwich sign" in the vastus lateralis on MRI.
Autosomal recessive (LGMDR)			
LGMD 2A/R1	*CAPN3*	Calpain 3	Recessive mutations. Regulates the sarcomere.
LGMD 2B/R2	*DYSF*	Dysferlin	Impairs muscle repair. Posterior thigh/leg muscles most affected. Diamond-shaped quadriceps bulge when active.
LGMD 2C/R5	*SGCG*	γ-Sarcoglycan	Part of DAP complex. Phenotype severity varies; severe if ambulation lost before age 13.
LGMD 2D/R3	*SGCA*	α-Sarcoglycan	Pelvic girdle weakness. Fatty replacement in the vastus intermedius, semimembranosus, biceps femoris.
LGMD 2E/R4	*SGCB*	β-Sarcoglycan	Weakness and fatty replacement in the latissimus dorsi, spine extensors, glutei, adductors, thigh muscles.
LGMD 2F/R6	*SGCD*	δ-Sarcoglycan	Progressive proximal atrophy; high CK; some patients have cardiomyopathy.
LGMD 2G/R7	*TCAP*	Telethonin	Role in myofibrillogenesis. Thigh weakness first and then calf muscles. Fatty infiltration in glutei, hip, thigh.
LGMD 2H/R8	*TRIM32*	TRIM32	Regulates Ca^{2+} movement. Calf hypertrophy; muscle fibers of varying sizes, small vacuoles, internal nuclei.
LGMD 2I/R9	*FKRP*	Fukutin Related Protein	Glycosylation of α-dystroglycan. Muscle biopsies show hypertrophic, split, necrotic fibers.
LGMD 2J/R10	*TTN*	Titin	Responsible for muscle tension. Weakened thighs; rimmed vacuoles with amyloid in biopsies.
LGMD 2K/R11	*POMT1*	POMT1	Necessary for O-mannosylation of α-dystroglycan. Proximal limb weakness and cognitive impairment.
LGMD 2L/R12	*ANO5*	Anoctamin 5	Calcium-activated chloride channel. Prominent asymmetrical quadriceps femoris and biceps brachii atrophy.
LGMD 2M/R13	*FKTN*	Fukutin	Glycosylation of α-dystroglycan. Proximal weakness in childhood, hypotonia, hypertrophy; possible ocular, cognitive, cardiac issues.

Continued

TABLE 7.1 Classification of LGMDs.—Cont'd

LGMD type (old/new nomenclature)	Gene	Protein	Key pathological and clinical features
LGMD 2N/R14	*POMT2*	POMT2	Necessary for O-mannosylation. Walking difficulties, exercise pain. Weakness in hamstrings, gluteal, paraspinal muscles; cognitive impairment.
LGMD 2O/R15	*POMGnT1*	POMGnT1	Catalyzes α-dystroglycan modification. Proximal weakness around puberty in calves, quadriceps, hamstrings, deltoids; myopia.
LGMD 2P/R16	*DAG1*	Dystroglycan	Part of the dystroglycan complex. Proximal weakness and cognitive impairment.
LGMD 2Q/R17	*PLEC1*	Plectin	Roles in cell survival, growth, actin organization. Weakness in the deltoids, erector spinae, glutei, biceps femoris, adductor magnus.
LGMD 2R	*DES*	Desmin	Connects sarcomeres. Induces distal weakness.
LGMD 2S/R18	*TRAPPC11*	TRAPPC11	Transport between ER and Golgi. School-age onset, progressive proximal weakness (hips > shoulders); possible respiratory issues.
LGMD 2T/R19	*GMPPB*	GMPPB	Produces GDP-mannose for α-dystroglycan glycosylation. Affects paraspinal/hamstrings; hypotonia, epilepsy, cognitive issues, cataracts, cardiomyopathy.
LGMD 2U/R20	*ISPD*	Isoprenoid synthase	Glycosylation of α-dystroglycan. Sural hypertrophy; moderate proximal weakness.
LGMD 2V	*GAA*	α-1,4-Glucosidase	Important for lysozyme function. Progressive proximal weakness; respiratory/cardiac involvement.
LGMD 2W	*LIMS2*	LIMS2	Involved in cell spreading/migration. Early-onset proximal weakness, calf hypertrophy, severe quadriparesis, dilated cardiomyopathy.
LGMD 2X/R25	*BVES*	POPDC1	Structure/function of muscle cells. Exercise-induced myalgia, cardiac arrhythmia.
LGMD 2Y	*TOR1AIP1*	Torsin 1AIP1	Link between the nuclear membrane and lamina. Skeletal muscle weakness and cardiac involvement.
LGMD 2Z/R21	*POGLUT1*	POGLUT1	Involved in protein transport. Limb-girdle weakness, decreased α-dystroglycan glycosylation, fatty replacement.
LGMDR22	*COL6A1, A2, A3*	Collagen VI	Recessive form of Bethlem myopathy. Progressive proximal weakness, ankle contractures, interphalangeal joint contractures.
LGMDR23	*LAMA2*	Laminin α2	Myotube stability and apoptosis. Progressive proximal weakness; 36% of patients have seizures.
LGMDR24	*POMGNT2*	POMGNT2	Catalyzes α-dystroglycan modification. Progressive proximal lower limb weakness; possible ocular and cognitive impairments.

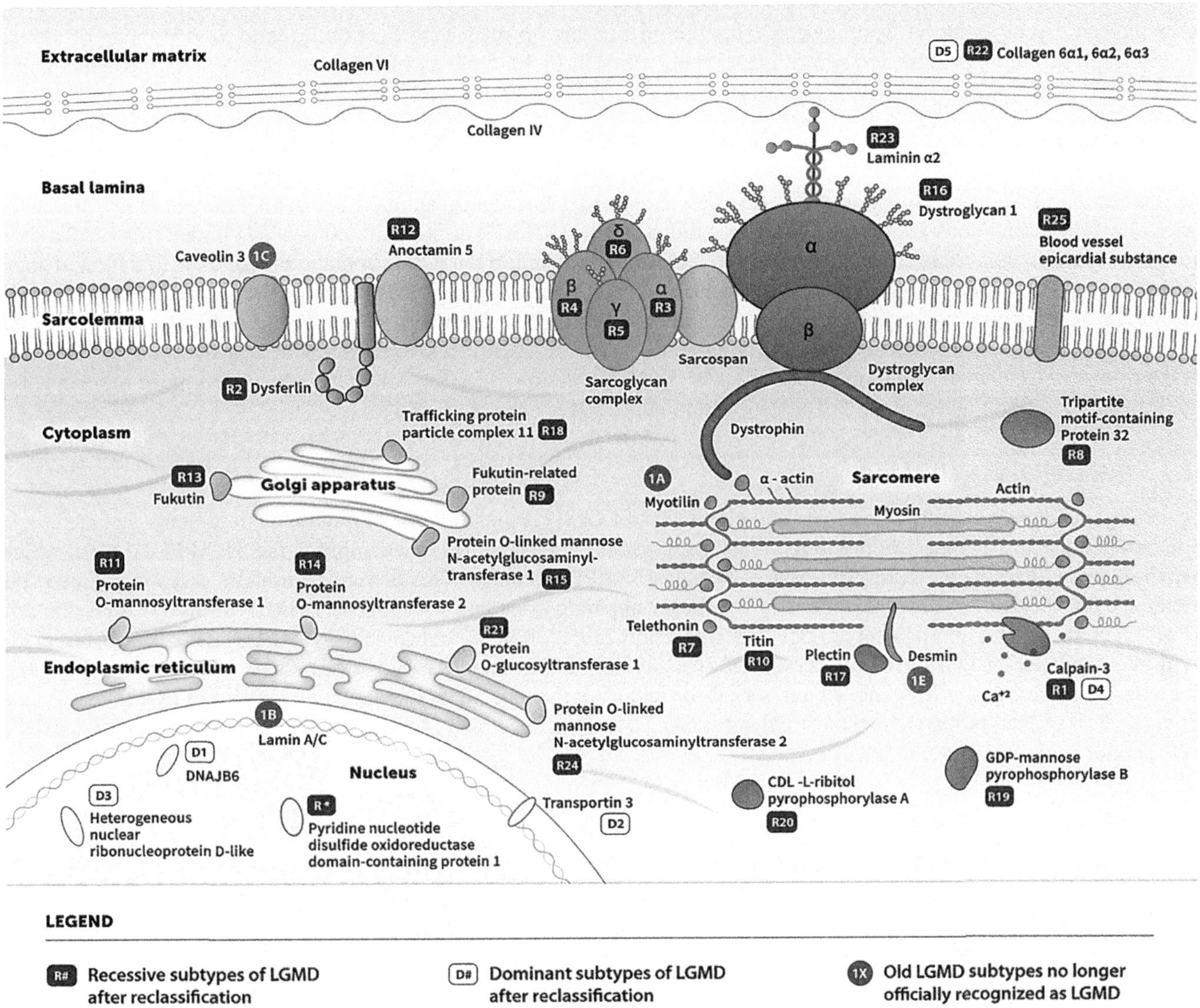

FIG. 7.1 This schematic provides a spatial representation of the subcellular distribution of proteins associated with limb-girdle muscular dystrophy (LGMD) subtypes in myocytes. The figure integrates current (R#, D#) and select decommissioned dominant (1#) subtypes, mapping individual proteins to specific compartments—including the extracellular matrix, sarcolemma, and organelles such as the nucleus and endoplasmic reticulum—to illustrate their pathogenic loci. *(From Georganopoulou, D.G., Moisiadis, V.G., Malik, F.A., Mohajer, A., Dashevsky, T.M., Wuu, S.T., Hu, C.K., 2021. A journey with LGMD: from protein abnormalities to patient impact. Protein J. 40 (4), 466–488. https://doi.org/10.1007/s10930-021-10006-9.)*

LGMD pathologies also extend to proteins of the sarcolemma and cytosol. Specific genetic defects in these cellular components are linked to distinct subtypes. LGMD1C is caused by mutations in the sarcolemmal protein caveolin-3, disrupting its role in cellular signaling. LGMD2B arises from an insufficiency of the sarcolemmal protein Dysferlin. LGMD2H is associated with mutations in TRIM32, a cytosolic E3-ubiquitin ligase involved in the ubiquitination process (Laval and Bushby, 2004).

The nuclear lamina is a protein-based meshwork that gives the nucleus its structural shape and supports the inner nuclear membrane and pores. Lamins A and C, which are key components of this structure, are both produced from the *LMNA* gene (Dubik and Mai, 2020). Mutations in this single gene are linked to a variety of muscular dystrophy diseases, including LGMD1B (Donnaloja et al., 2020).

2.2 Protein processing

A critical mechanism in several LGMDs involves defective glycosylation of dystroglycan, a core component of the muscle cell membrane's dystrophin–glycoprotein complex. Specifically, mutations in eight genes (*FKRP, POMT1, FKTN, POMT2, POMGnT1, ISPD, POMGNT2,* and *GMPPB*), encoding sarcoplasmic reticulum and Golgi apparatus enzymes, disrupt the

O-mannosyl glycosylation of dystroglycan. This failure in proper protein modification leads to a wide range of clinical severities, from mild conditions to severe LGMD subtypes (2I, K, M–O, T, and U) (Barton et al., 2020; Georganopoulou et al., 2021; Taghizadeh et al., 2019).

2.3 Mitochondrial dysfunction

Variations in mitochondrial function are increasingly recognized as a significant modifier of LGMD pathology, and divergent clinical presentations are observed among patients with identical causative mutations. The underlying mechanisms may involve mitochondrial deficits in energy (ATP) production, calcium ion (Ca^{2+}) homeostasis, and the initiation of apoptosis. Studies have documented mitochondrial impairment in six LGMD subtypes, with a direct correlation between the regulation of mitochondria-mediated cell death and disease severity in genetically identical individuals. An example of this phenomenon has been observed in people dealing with LGMD2C. A patient's condition with LGMD2C can be worsened if the primary *SGCG* gene mutation is accompanied by dysregulation of BCL2, which leads to variations in mitochondria-mediated cell death (Vincent et al., 2016; Hadj Salem et al., 2012; Saha et al., 2018).

2.4 Intracellular trafficking

An additional element contributing to the development of LGMD is the presence of mutations in proteins that are crucial for intracellular trafficking, leading to dysfunction in vesicle transport. A key example is the TRAPPC11 gene, which, when mutated, disrupts the Transport Protein Particle (TRAPP) complex. This disruption impairs essential cellular trafficking from the endoplasmic reticulum to the Golgi apparatus, leading to multiple LGMD forms (1C, 2F, 2S, 2X) (Frattini et al., 2017; Domingos et al., 2017). A frameshift mutation in the transportin 3 (*TNPO3*) gene is also associated with the incidence of LGMD1F and LGMD D2 (Costa et al., 2022; Torella et al., 2013). Other involved mutated genes include the dysferlin (*DYSF*) gene, which is vital for repairing the muscle membrane and is linked to LGMD2B (Poudel et al., 2024), and the blood vessel epicardial substance (*BVES*) gene, which is considered the leading cause of LGMD2X (De Ridder et al., 2019).

2.5 Signaling pathways

Skeletal muscle adapts to workload through shared molecular pathways. In muscular dystrophies, such as LGMD, these signaling pathways are disrupted during rest, activity, and passive movement. A central player in this "mechanosensing" is the SG complex. Evidence suggests the SG complex is a critical signaling hub, and its absence or mutation (e.g., in LGMD2C) leads to aberrant signaling, such as hyperactivity in ERK1/2. Phosphorylation of specific residues on SGs appears crucial for their function, and mutations in these regions cause severe disease. This implies that proteins regulating SG phosphorylation could themselves be disease-causing genes or modifiers of symptom severity (Barton et al., 2020). In this case, the role of p38 is more complicated. Reduced p38 activity is seen in conditions like LGMD2A (calpain 3 deficiency), leading to blunted adaptation and poor regeneration (Kramerova et al., 2012; Kramerova et al., 2018). Elevated p38 activity is also found in other LGMD subtypes (e.g., LGMD2B, 2C, 2F, DMD) (Oliva et al., 2019; Wissing et al., 2014). Together, these findings suggest that both hypo- and hyperactivation of p38 can contribute to muscle degeneration in different LGMD subtypes, complicating simple therapeutic targeting of this pathway. Studies also demonstrated that the pathogenesis of *LMNA*-related muscular dystrophy involves a molecular pathway where *LMNA* gene mutations activate the NF-κB-mediated inflammatory response, leading to cellular apoptosis (Fan et al., 2020).

3 Currently approved therapeutic approaches for LGMD

The heterogeneity of LGMD disease presents a significant challenge. With over 35 distinct conditions, a uniform therapeutic strategy is not feasible. The diversity in inheritance patterns, clinical presentation, genetic mutations, and disease mechanisms directly explains the limited development of targeted treatments. So far, drug-based treatments for LGMDs, including corticosteroids, have shown limited and inconsistent benefits, and their efficacy remains unproven in controlled trials. Other approaches, like growth hormone, myoblast transfer, and myostatin inhibition, have also proven ineffective (Pozsgai et al., 2021). Since there is no definitive treatment for all subtypes of this disease, supportive and symptom-relieving treatments are considered the first line of LGMD management. Treatments focusing on symptoms encompass physical therapy and rehabilitation, which seek to preserve mobility, joint flexibility, and muscle strength. Key symptoms such as weakness and fatigue reduce patients' ability to walk and perform everyday activities, frequently resulting in decreased physical activity and weight gain. As a result, rehabilitation efforts are essential for reducing muscle strength decline and maintaining the highest level of independence for patients (D'Este et al., 2025; Gianola et al., 2013). Recent findings suggest that exercise can be advantageous for individuals with LGMD; however, the specific type and intensity of exercise should be customized to each person and their particular LGMD subtype to prevent any further complications (D'Este et al., 2025). Aerobic exercise, such as riding a stationary bike, is both safe and beneficial, as it enhances cardiovascular health and improves certain aspects of muscle strength without leading to greater muscle damage (Nandanwar et al., 2024; Sveen et al., 2007; Sheikh and Vissing, 2019). These exercises are beneficial, particularly in the early stages of the disease when a considerable amount of muscle mass is still intact (D'Este et al., 2025). Apart from the physical health of the patient, these exercises can also positively impact their mental condition by enhancing their quality of life and alleviating depression related to the illness (Granero-Jiménez et al., 2022; O'Dowd et al., 2022). However, there are also considerations in this regard, such as ensuring the exercise intensity aligns with the disease stage and the patient's physical condition (Brun et al., 2017; Preisler and Orngreen, 2018). In addition, it is important to consider the specific subtype of the disease, as in certain cases, physical activities may even lead to adverse effects (Kilmer et al., 1994; Milner-Brown and Miller, 1988).

Early respiratory intervention is also essential, since respiratory muscle weakness is a frequent and severe issue in LGMD. The deterioration of lung function, typically assessed using Forced Vital Capacity (FVC), varies depending on the subtype and can result in sleep hypoventilation, which may be indicated by initial symptoms such as snoring, excessive daytime sleepiness, and headaches (Muni-Lofra et al., 2023; Mori-Yoshimura et al., 2017). To address this, guidelines suggest routine pulmonary function evaluations, and the use of NIV often becomes vital, especially when nocturnal FVC falls below 50% (Topaloglu et al., 2015). The approaches to management are diverse, incorporating NIV to assist with breathing, aerobic exercise to enhance respiratory capacity indirectly, and targeted physical therapy like deep breathing exercises to strengthen respiratory muscles directly. The main objectives include enhancing respiratory function, averting complications, and significantly boosting the patient's quality of life and longevity (Angelini and Pegoraro, 2020; Nandanwar et al., 2024; Sveen et al., 2007).

Although less frequent than in other muscular dystrophies, cardiac involvement is a recognized feature of certain LGMD subtypes (LGMD1B and LGMD2A-I). This complication can manifest as cardiomyopathy, impairing the heart's pumping function, or as conduction abnormalities leading to arrhythmias. Standard management involves routine cardiac surveillance, including ECG, echocardiogram, and/or cardiac MRI, to guide appropriate intervention. Therapeutic strategies range from pharmacotherapy and implantable devices like pacemakers to heart transplantation for advanced heart failure (Narayanaswami et al., 2014).

In LGMD, dysphagia and arm weakness can contribute to nutritional deficits. To address this, individuals with swallowing problems, weight loss, or risk of aspiration should undergo a swallowing evaluation and may require interventions like dietary texture modification or a feeding tube (Narayanaswami et al., 2014).

TABLE 7.2 Limb-girdle muscular dystrophy (LGMD) management overview.

Management	Current approaches and therapies
Rehabilitation and physical therapy	**Customized exercise:** Type and intensity must be tailored to the individual and their specific LGMD subtype. **Aerobic exercise:** e.g., stationary biking. Safe and beneficial for cardiovascular health and some muscle strength.
Respiratory management	**Routine monitoring:** Regular pulmonary function tests. **Intervention:** Noninvasive ventilation (NIV), especially when nocturnal FVC is below 50%. **Supportive therapies:** Aerobic exercise and deep breathing exercises.
Cardiac management	**Routine surveillance:** ECG, echocardiogram, cardiac MRI. **Interventions:** Pharmacotherapy, pacemakers, or heart transplantation for advanced cases.
Nutritional management	**Evaluation:** Swallowing assessments. **Interventions:** Dietary texture modification, feeding tubes.
Emerging/investigational therapies	• **Stem cell therapy:** • **Autologous transfer:** Using the patient's own genetically modified cells. • **Allogenic transfer:** Using cells from a healthy donor • **Myoblast transfer:** Injecting donor muscle cells directly into the muscle. • **Gene therapy:** Other proposed treatments are under research.

One approach that has received attention in recent decades is the use of stem cells in the treatment of genetic disorders. This area has also been researched in the context of LGMD. The goal of stem cell therapy is to introduce cells capable of producing a needed protein. This can be achieved either by modifying the patient's own cells (autologous transfer) or by utilizing cells from a donor (allogeneic transfer). Early pilot studies suggest potential benefits in some outcome measures, but data remain limited, and stem cell therapy is not yet part of standard care. A similar approach, myoblast transfer, involves directly injecting a donor's muscle cells into the patient's muscle tissue. The results of the studies indicate the desirable and satisfactory effects of stem cell therapy on LGMD patients (Sharma et al., 2023; Sharma et al., 2013; Skuk and Tremblay, 2017) (Table 7.2). Other remaining proposed treatments for LGMD are based on gene therapy methods (Bouchard and Tremblay, 2023).

4 Gene therapy in LGMD

The limitations of pharmacological interventions have shifted the therapeutic focus toward molecular strategies. Translational research has identified exon skipping and gene therapy as promising avenues, with the latter being particularly ideal for targeting the clear genetic origins of these monogenic, recessive diseases (Chamakioti et al., 2022). Exon skipping utilizes synthetic DNA segments known as antisense oligonucleotides (ASOs) to modify gene expression. The initial ASOs were unstable; however, chemical enhancements, such as substituting parts of their backbone with sulfur or morpholino rings, have greatly improved their durability and lifespan within the body (Rinaldi and Wood, 2018). These enhanced ASOs operate via two main mechanisms: some attract enzymes that degrade the target RNA, while others serve as a "steric block" to obstruct RNA translation or splicing physically. Various delivery methods, including conjugation with targeting molecules like GalNAc and encapsulation in lipid nanoparticles, are employed to deliver ASOs effectively (Roberts et al., 2020). Clinical trials addressing conditions such as Duchenne muscular dystrophy (DMD) have shown encouraging results, with specific ASOs successfully reinstating dystrophin protein in muscle fibers postinjection or intravenous administration (Kinali et al., 2009; Cirak et al., 2011). These experiences in DMD form the conceptual and regulatory basis for exon-skipping approaches being explored in LGMD subtypes. ASO techniques have demonstrated effectiveness in exon skipping for LGMDs in laboratory experiments. An example is a study conducted by Wyatt et al. on LGMD 2C. For many cases of limb-girdle muscular dystrophy type 2C (LGMD2C), researchers have pinpointed a prospective treatment involving the skipping of multiple exons. LGMD2C, caused by mutations in the autosomal recessive SGCG gene, frequently leads to a frameshift that results in the loss of γ-sarcoglycan expression. Studies indicate that skipping four specific exons can rectify this frameshift, facilitating the production of a shortened but functional protein referred to as "Mini-Gamma." Experimental application of vivo-phosphorodiamidate morpholino oligomers has successfully promoted this exon skipping in human myogenic cells harboring various SGCG mutations. This illustrates the practicality of a universal skipping approach suitable for a wide range of LGMD2C patients (Wyatt et al., 2018). Further research by Demonbreun et al. involved engineering a murine model of LGMD2C via CRISPR/Cas9 to introduce the Sgcg 521ΔT point mutation. Subsequent intramuscular delivery of a multiplex ASO cocktail designed for mice demonstrated effective correction of the mutant reading frame 5 (Demonbreun et al., 2020). In another study, Lee et al. addressed dysferlinopathy, a muscle-wasting disease caused by mutations in the dysferlin (DYSF) gene, which is essential for repairing muscle cell membranes. While exon-skipping therapy for exon 32 has been previously identified, this study successfully discovered two new multiexon skipping targets—exons 26–27 and 28–29—that effectively restore the protein's membrane-repair function in patient cells (Lee et al., 2018). Researchers have also explored suppressing dominant mutations by degrading the relevant pathogenic alleles. For example, Liu et al. applied RNA interference to knock down the expression of a mutant myotilin gene that causes LGMD1A (allele-specific silencing). This strategy proved effective in mice, eliminating the mutant protein and reversing the disease symptoms, as seen by restored muscle strength and growth (Liu et al., 2012). Allele-specific silencing may be advantageous for treating LGMD, as certain LGMD subtypes like LGMD1C, LGMD1D, and LGMD1E are severe and progressive in homozygous forms, while heterozygous animal models show normal conditions (Liu and Harper, 2012).

Gene delivery represents an important area of research aimed at addressing protein deficiencies in patients with LGMD. This strategy involves transferring healthy genes to produce proteins that are currently lacking (Pozsgai et al., 2021). Various types of vectors, such as viral and nonviral vectors, can be utilized for this purpose. Among viral vectors, the predominant approach involves using recombinant adeno-associated virus (rAAV) for gene transport. Achievements include the successful delivery of genes encoding calpain and SGs in both animal studies and human trials (e.g., SRP-9003), leading to sustained protein expression and improved muscle strength. For larger genes such as dysferlin, a dual-AAV system has been created to effectively deliver the genetic information. To mitigate potential immune responses to AAVs, some research groups opt for plasmid DNA injections directly into muscle, followed by electroporation. This approach has shown effectiveness in restoring proteins like calpain and dysferlin in mice and presents a promising alternative to viral vector techniques (Bouchard and Tremblay, 2023).

Alongside gene transfer, strategies utilizing gene editing are also being implemented to address genetic disorders. The area of gene editing has progressed through various technological generations. It started with engineered nucleases such as ZFNs and TALENs, which utilize protein-based systems to induce specific DNA double-strand breaks (DSBs) and promote repair mechanisms (Durai et al., 2005; Sun and Zhao, 2013). A significant advancement occurred with the introduction of the CRISPR–Cas9 system, which employs a guide RNA to target a specific DNA sequence, simplifying and broadening the process (Ramachandran, 2020). Recently, precision in gene editing has improved through innovations like "base editing," which chemically transforms one nucleotide into another without breaking the DNA, and "prime editing," which allows direct rewriting of genetic information using a reverse transcriptase template. To mitigate the possible inaccuracies associated with DSBs, these new methods have been developed. Base editors operate by using a catalytically inactive Cas protein attached to a deaminase enzyme to convert one base pair into another without severing the DNA backbone. Prime editors provide even greater precision by pairing a Cas9 nickase with a reverse transcriptase and a specialized guide RNA (pegRNA) to "write" new genetic sequences directly into a chosen site (Ramachandran, 2020; Gaudelli et al., 2017). These cutting-edge tools have already been utilized in laboratory studies to correct mutations linked to LGMDs, such as LGMD2A and LGMD2B (Müthel et al., 2023; Turan et al., 2016).

5 Recent clinical trials of gene therapy for LGMD

The landscape of treatment for LGMD is rapidly evolving, with gene therapy emerging as a promising therapeutic strategy. Recent clinical trials have primarily focused on gene therapy for LGMD disease. Here, we describe examples of gene therapy methods that are in clinical trials for LGMD (Tables 7.3 and 7.4).

The small size of the SG genes makes them ideal candidates for adeno-associated virus (AAV)-mediated gene replacement, a strategy that has now entered clinical trials for several subtypes. One of the pioneering gene therapy trials in LGMD was a Phase 1 study by Herson et al. for LGMD2C. This study involved nine nonambulatory patients with a homozygous del525T mutation and absent γ-sarcoglycan protein. Researchers used intramuscular injections of an AAV1 vector expressing the human *γ-sarcoglycan* gene into a forearm muscle at three escalating doses. The trial demonstrated excellent safety with no serious adverse events over 6 months. Critically, immunohistochemical analysis of injected-muscle biopsy specimens confirmed new γ-sarcoglycan expression in all three patients who received the highest dose, with mRNA and protein also detected via RT-PCR and Western blot in some patients. This study provided the first proof-of-concept that AAV-mediated gene transfer could successfully induce SG protein expression in LGMD2C patients (Herson et al., 2012).

As the most common sarcoglycanopathy, LGMD2D has been extensively studied. Groundbreaking work by Mendell's group established the feasibility of this approach. An initial double-blind, placebo-controlled trial involved intramuscular injection of "rAAV1.tMCK.hSGCA" (a vector containing the human α-sarcoglycan gene under a muscle-specific promoter) into a foot muscle, and the results were highly encouraging. Persistent α-sarcoglycan gene expression was observed for 6 months in two of the three subjects. The therapy also led to improved histopathological markers, including a reduction in major histocompatibility complex I expression, an increase in muscle fiber size, and, most importantly, restoration of the full SG complex (Mendell et al., 2009; Mendell et al., 2010).

Building on this foundational work, a subsequent Phase 1 trial was completed to further assess safety and determine the effective dose, though its results are not yet published. Importantly, the field is now moving toward systemic delivery to target a larger muscle mass. An ongoing Phase I/IIa dose escalation study is investigating the safety and maximum tolerated dose of a novel vector, "scAAVrh74.tMCK.hSGCA," administered via isolated limb perfusion to the legs of LGMD2D patients (Chu and Moran, 2018).

Following the precedent set by LGMD2C and 2D, gene replacement therapy (GRT) for LGMD2E is now underway. A recent clinical trial (NCT05876780) has been initiated to evaluate "SRP-9003," an AAV vector delivering the gene for β-sarcoglycan. This trial aims to assess the safety of this treatment and its ability to restore β-sarcoglycan expression systemically (Bouchard and Tremblay, 2023).

LGMD2I is a severe, progressive muscular dystrophy caused by mutations in the *FKRP* gene, which is essential for muscle stability. Two GRTs are now in clinical trials to address this. LION-101 (Asklepios BioPharmaceutical) therapy uses a recombinant AAV vector to deliver a functional FKRP gene. It is currently in a Phase ½ dose escalation study (NCT05230459) to evaluate its safety. GNT0006 (Atamyo Therapeutics) is another AAV-based *FKRP* gene therapy. Its ongoing Phase ½ trial (NCT05224505) is assessing the treatment's safety, efficacy, and how patients' immune systems respond to it (Chamakioti et al., 2022).

TABLE 7.3 Published human AAV gene therapy trials in LGMD.

Study period (start year)	LGMD subtype (old/new)	Agent/vector	ClinicalTrials.gov ID	Phase and design	Route	Sample size	Follow-up	Assessments	Main findings and publication
2008–2011	LGMD2D/LGMDR3 (α-sarcoglycanopathy)	rAAV1.tMCK.hSGCA	NCT00494195	Phase I, randomized, double-blind, intramuscular proof-of-concept	Intramuscular into EDB (treated vs. saline control side)	3 treated subjects (low-dose cohort of planned N = 6)	6 weeks (2 pts) and 3 months (1 pt)	Safety; α-SG expression by IHC and qPCR; restoration of the full sarcoglycan complex; muscle fiber size; serum AAV1 neutralizing antibodies; ELISpot T-cell responses.	Local α-SG and sarcoglycan complex restoration with sustained expression up to 3 months; no serious adverse events. Ann Neurol. 2009;66:290–297 (Mendell et al.).
2006–2011	LGMD2C/LGMDR5 (γ-sarcoglycanopathy)	AAV1.des.hSGCG	NCT01344798	Phase I, open-label, dose escalation (three dose cohorts)	Intramuscular into extensor carpi radialis	9 nonambulant patients (3 per cohort)	6 months	Safety; γ-SG expression by IHC and Western; γ-SG mRNA (RT-PCR); vector genomes; immune monitoring (AAV1 serology, CTL response); clinical status.	Detectable γ-SG expression in all high-dose patients (~5%–10% positive fibers) and low-level expression in some lower-dose patients; no serious adverse events; one transient CTL response to AAV1 capsid. Brain. 2012;135:483–492 (Herson et al.).
2018–ongoing	LGMD2E/LGMDR4 (β-sarcoglycanopathy)	SRP-9003 (bidridistrogene xeboparvovec; AAVrh74.MHCK7.SGCB)	NCT03652259	Phase I/II, open-label, single IV dose with long-term follow-up	Intravenous infusion	6 children (4–13 years)	2-year primary analysis; safety follow-up reported at 4–5 years	Safety and β-SG expression; NSAA; timed function tests (10 m walk/run, time-to-rise); 6MWD; CK; cardiac function; skeletal muscle MRI fat fraction.	Robust β-SG expression, improved or stabilized NSAA and other functional endpoints versus natural history; acceptable safety profile. Nat Med. 2024;30:199–206 (Mendell et al.).

TABLE 7.4 Ongoing/planned LGMD gene therapy clinical trials.

Agent	Target subtype	Vector/ promoter	Registry ID	Phase and design	Route	Planned sample size	Follow-up and key assessments	Status/ sponsor
SRP-9003 (bidridistrogene xeboparvovec)	LGMD2E/LGMDR4	AAVrh74. MHCK7. SGCB	NCT03652259 (+ Phase 3 Sarepta program)	Phase ½ completed ($n = 6$); Phase 3 randomized efficacy/safety trial underway	Intravenous single infusion	Small Phase ½ (Stokholm et al., 2022); Phase 3 larger pediatric/ young adult cohort (screening)	Long-term safety; NSAA and timed tests; respiratory and cardiac function; MRI; QoL scales	Ongoing program; Sarepta Therapeutics
AB-1003 (LION-CS101)	LGMD2I/LGMDR9 (FKRP-related)	AAV9-FKRP	NCT05230459	Phase ½, randomized, placebo-controlled, dose escalation in adults	Intravenous single infusion	≈10 adults in Part 1 (with potential expansion)	Safety/tolerability; FKRP expression and α-DG glycosylation (biopsy); 6MWT; strength; cardiac and respiratory function	Ongoing; AskBio/Bayer
ATA-100 (formerly GNT0006)	LGMD2I/LGMDR9	AAV vector carrying human FKRP	NCT05224505 (Europe)	Phase 1b/2b, multicenter, dose-finding with extension	Intravenous single infusion	Small dose escalation cohorts (exact N per cohort not yet fully public)	Safety; pharmacodynamics (matriglycan/α-DG glycosylation); muscle MRI; strength (MRC); 6MWT	Ongoing; Atamyo Therapeutics/ Genethon
ATA-200	LGMD2C/LGMDR5 (γ-sarcoglycanopathy)	AAV8 expressing human SGCG	NCT05973630	Phase ½, pediatric dose escalation (two dose cohorts)	Intravenous single infusion	Planned ≈6 pediatric patients (ages 6–13 years)	Safety/tolerability; pharmacodynamic biomarkers; functional tests; cardiac and respiratory assessments	Recruiting; Atamyo Therapeutics
ATA-300/ GNT0008 (candidate)	LGMD2A/LGMDR1 (calpainopathy)	AAV vector delivering CAPN3 (muscle-specific promoter)	Preclinical (no clinical ID yet)	Planned first-in-human gene replacement trial	Planned intravenous systemic delivery	Not yet defined	Expected: safety; CAPN3 expression; muscle strength/ function; cardiac and respiratory monitoring	Advanced preclinical; Genethon/ Atamyo

6 Current advances in preclinical gene therapy for LGMD

The translation of gene therapy from the laboratory to the clinic for LGMD is underpinned by extensive preclinical research. These studies in animal models are crucial for establishing proof-of-concept, optimizing delivery methods, and ensuring the safety and efficacy of therapeutic strategies. Significant progress has been made for several LGMD subtypes, moving them closer to human clinical trials (Table 7.5).

The SGs genes are ideally sized for AAV delivery, making LGMD2D (α-sarcoglycan deficiency) and LGMD2E (β-sarcoglycan deficiency) strong candidates for GRT. Regarding LGMD2D (*SGCA*), studies in sgca$^{-/-}$ mice have demonstrated the efficacy of systemic gene delivery. A single intravenous injection of a self-complementary AAVrh74 vector carrying the human *α-sarcoglycan* gene ("scAAVrh74.tMCK.hSGCA") led to robust sarcolemmal protein expression across all doses tested. This was accompanied by significant functional improvements, including reduced fibrosis, increased specific force generation in the diaphragm and limb muscles, protection against eccentric contraction-induced damage, and enhanced locomotor activity. The treatment was well-tolerated, showing no vector-related toxicity and supporting the move toward clinical trials using systemic administration (Griffin et al., 2021). Perhaps one of the most compelling preclinical demonstrations comes from LGMD2E research. Systemic delivery of "scAAV.MHCK7.hSGCB" in sgcb$^{-/-}$ mice resulted in near-universal transgene expression (98.1%) across all skeletal muscles. The treatment yielded dramatic, multisystemic benefits, including improved muscle pathology and an 85.5% reduction in serum CK, a 94.4% increase in diaphragm force production, a 48% reduction in kyphoscoliosis, and dramatic increases in overall ambulation (57%) and vertical rearing

TABLE 7.5 Preclinical gene therapy advances for LGMD subtypes.

Year	Subtype/gene	Animal model	Vector and delivery	Follow-up	Assessments	Major findings	Publication
2021	LGMD2D/LGMDR3 (SGCA)	Sgca−/− mice	Systemic AAVrh74.tMCK.hSGCA (IV)	Up to ~6 months	Muscle histology; α-SG and DGC expression; CK; force; biodistribution and toxicology	Dose-dependent widespread α-SG expression and functional rescue; supported IV dosing for SRP-9004.	(Griffin et al., 2021).
2017	LGMD2E/LGMDR4 (SGCB)	Sgcb−/− mice	Systemic AAVrh74-SGCB	Several months	β-SG levels; fibrosis; MRI-T2; contractility; cardiac function	Near-complete β-SG rescue with marked improvement in skeletal and cardiac pathology.	(Pozsgai et al., 2017)
2014	LGMD2I/LGMDR9 (FKRP)	FKRP L276I knock-in mice	Systemic AAV9-FKRP (neonatal/adult dosing)	3–9 months after treatment	α-DG glycosylation; skeletal and cardiac histology; CK; treadmill; echocardiography	Restored α-DG glycosylation and improved skeletal/cardiac pathology and function.	(Qiao et al., 2014)
2018	LGMD2I/LGMDR9 (FKRP)	FKRP P448L mice	Systemic AAV9-FKRP at multiple doses	Up to 12 months	Fibrosis; diaphragm and limb muscle function; survival	Dose-dependent functional rescue and increased longevity; defined therapeutic window.	(Vannoy et al., 2018)
2023	LGMD2I/LGMDR9 (FKRP)	FKRP P448L mice	AAV-FKRP ± ribitol supplementation	Several months	Matriglycan levels; grip strength; treadmill; histology	Ribitol plus AAV-FKRP further enhanced matriglycan and functional outcomes vs monotherapy.	(Cataldi et al., 2023)
2023	LGMDR25 (BVES)	Bves-knockout mice	Systemic AAV9-BVES	Several months	Histology; cardiac MRI; ex vivo muscle force; locomotor activity; body mass	AAV9-BVES restored BVES expression and markedly improved skeletal and cardiac phenotypes.	(Li et al., 2023)

(132%). These results, achieved with no adverse effects, provide a powerful rationale for clinical trials aimed at treating both skeletal and cardiac muscle deficits in LGMD2E (Pozsgai et al., 2017).

The primary challenge in treating LGMD2B (dysferlin deficiency) is the enormous size of the DYSF gene, which exceeds the packaging capacity of a single AAV vector. To address this, researchers developed a novel dual-vector approach ("AAV.DYSF.DV"). This system splits the full-length dysferlin cDNA into two separate AAVrh74 vectors. Once inside the muscle cell, the two parts recombine via homologous recombination to produce a functional, full-length dysferlin protein. This strategy has proven safe and effective in both dysferlin-deficient mice and nonhuman primates (rhesus macaques), restoring dysferlin expression and its critical membrane repair function after both intramuscular and intravascular delivery. Long-term studies showed sustained expression for over 12 months, laying a solid foundation for preclinical safety studies and future clinical applications (Potter et al., 2018; Sondergaard et al., 2015).

For LGMD2I/R9, caused by mutations in the *FKRP* gene, research has explored both direct gene replacement and innovative compensatory approaches. In the gene replacement approach, systemic delivery of an AAV9 vector carrying the human FKRP gene in a knock-in mouse model (L276I) successfully restored α-dystroglycan glycosylation—a key biochemical defect—in both skeletal and cardiac muscle. This correction led to the amelioration of dystrophic pathology and cardiomyopathy, restoring muscle and heart contractile function even when treatment was administered to older, symptomatic animals. A critical finding, however, was that very high doses of the vector could cause toxicity, underscoring the need for precise control of gene expression levels (Gicquel et al., 2017). In the compensatory gene therapy (LARGE and GALGT2) approach, since *FKRP* operates in the α-dystroglycan glycosylation pathway, researchers investigated whether overexpressing the *LARGE* gene could compensate for its deficiency. AAV-mediated *LARGE* delivery successfully induced functional hyperglycosylation of α-dystroglycan and improved dystrophic pathology in *FKRP*-deficient mice, presenting a viable alternative therapeutic strategy (Vannoy et al., 2014). Similarly, therapy with "rAAVrh74.MCK.GALGT2" significantly reduced muscle pathology in an FKRP mutant model. Interestingly, the mechanism appeared distinct from its action in other dystrophies, primarily promoting muscle regeneration rather than inducing surrogate complex expression, highlighting a novel pathway for therapeutic benefit (Thomas et al., 2016).

Early attempts at calpain 3 ("*CAPN3*") gene replacement for LGMD2A were hampered by cardiotoxicity when the transgene was expressed in the heart. This hurdle was overcome by engineering an AAV vector ("rAAVrh74.tMCK.CAPN3") that restricts expression to skeletal muscle. This was achieved by using a muscle-specific promoter and incorporating a target sequence for a cardiac-specific microRNA, which actively suppresses the transgene in heart tissue. Systemic delivery of this optimized vector in **CAPN3**-deficient mice restored proteolytic activity, corrected histology, and rescued contractile force deficits without any evidence of cardiotoxicity. Treated mice showed significantly improved performance in exhaustive treadmill tests, correlated with a beneficial remodeling of muscle fibers toward a more fatigue-resistant type (Roudaut et al., 2013).

7 Current challenges and limitations in gene therapy for LGMD

The preclinical and clinical landscape for LGMD gene therapy has yielded several promising directions, though significant hurdles remain. Researchers have not only demonstrated the efficacy of direct gene replacement for suitably sized genes but have also engineered sophisticated solutions to overcome major obstacles, such as large gene size and tissue-specific toxicity. The robust functional recovery and excellent safety profiles observed in animal models provide a strong and compelling foundation for clinical trials, opening horizons for transformative treatments for these debilitating diseases. Currently, key successes in clinical trials include demonstrated safety, sustained transgene expression, and histopathological evidence of disease reversal. However, critical challenges remain, including preexisting immunity to AAV vectors, the need for efficient whole-body muscle targeting via systemic delivery, and the development of therapies for dominant and more complex forms of LGMD. As trials progress from local to systemic administration and new vectors are engineered, the prospect of effective, widely applicable gene therapies for LGMD continues to grow.

Despite the general safety of AAV in the central nervous system, gene therapy faces significant hurdles. Key limitations include toxicity from unpredictable transgene expression (leading to potential neurotoxicity) and off-target immune reactions, as well as limited efficacy in patients with preexisting immunity to AAV capsids. The goal of gene therapy is to deliver a therapeutic gene that produces a protein at a level that is both safe and effective, a "therapeutic window." However, achieving this precise control is immensely challenging. Unlike a natural gene, the therapeutic transgene delivered by AAV is often under the control of a strong, constitutively active promoter. This can lead to the production of the protein at levels far higher than what is physiological. This overexpression can itself be toxic to cells. For example, in the case of LGMD, overexpressing an SG protein might disrupt the delicate stoichiometry of the entire dystrophin–glycoprotein complex, potentially leading to cellular stress or malfunction. In other disorders, overexpression has been linked to neurotoxicity, where excessive protein production overwhelms neuronal cells, leading to inflammation, seizures, or even neuronal death. This was a concern in some early trials for disorders like giant axonal neuropathy and has been observed in animal models for various CNS diseases.

While promoters are designed to be tissue-specific (e.g., muscle-specific), they are rarely perfectly so. Low-level "leaky" expression in off-target tissues like the liver or the central nervous system can trigger unexpected immune responses or toxicity in those organs.

The relationship between the number of AAV vectors administered (the dose) and the resulting level of protein expression is not always linear or predictable. Finding the "Goldilocks" dose—one that is high enough to be therapeutic but low enough to avoid toxicity—is a central challenge in clinical trial design.

The human immune system is adept at recognizing and attacking foreign invaders, and AAV vectors, despite being engineered, can be perceived as such. If the body presents fragments of the AAV capsid or the *newly synthesized* transgene product as foreign, cytotoxic T-lymphocytes (CTLs) can be activated. These CTLs can then identify and destroy the very cells that have successfully taken up the therapeutic gene, nullifying the treatment's benefit. This was a significant issue in early hemophilia and DMD trials. Antibodies can also bind to the AAV capsid and prevent the virus from entering its target cells, rendering the therapy ineffective.

Preexisting immunity to AAV capsids is perhaps one of the most significant practical barriers to treatment. AAVs are common in the environment, and a large percentage of the human population (estimates range from 30% to 70%, depending on the serotype) has already been exposed to wild-type AAVs and has developed neutralizing antibodies. A patient with preexisting NAbs is typically excluded from receiving AAV gene therapy because the treatment is almost guaranteed to fail. The antibodies will clear the vector from the bloodstream before it can reach its target tissue. This automatically disqualifies a huge portion of the potential patient population. Furthermore, a single administration of an AAV therapy will itself trigger a strong, long-lasting antibody response. This makes redosing impossible if the first treatment's effect wanes over time, which is a concern for pediatric patients who may outgrow the initial therapy (Chandler et al., 2017). Given that many LGMD patients are children or young adults, the impossibility of redosing is particularly relevant as their muscle mass and functional demands change over time. To overcome these challenges, future strategies are focusing on using regulated promoters for precise gene control, developing new chemical modifications to improve delivery and stability, and advancing genome-editing techniques like CRISPR/Cas9 (Pozsgai et al., 2021; Chamakioti et al., 2022). Although applying these to the genetically diverse LGMD population is complex, ongoing research and clinical trials are actively working to unlock the full potential of gene therapy for this condition (Turan et al., 2016).

8 Conclusion

LGMD is a complex and diverse group of diseases unified by a common pattern of muscle weakness but distinguished by extensive genetic and pathophysiological heterogeneity. The journey from a clinical diagnosis to a precise genetic classification is crucial, as it informs prognosis and management. Currently, patient care is centered on a multidisciplinary supportive approach aimed at maintaining mobility, managing respiratory and cardiac complications, and optimizing quality of life. However, the horizon of LGMD treatment is being reshaped by groundbreaking advances in molecular medicine. Gene therapy has emerged as the most promising avenue for developing targeted, potentially curative interventions. Preclinical studies have successfully demonstrated the efficacy of innovative strategies, including dual-vector systems for large genes such as dysferlin and cardioprotective vectors for calpain-3, thereby overcoming previous technical barriers. This strong foundation has catalyzed the transition to human clinical trials, with several GRTs for recessive forms like LGMD2D, 2E, and 2I now underway. These early trials have provided critical proof-of-concept, established safety, and demonstrated successful transgene expression and histological improvement.

Nevertheless, significant challenges remain. The clinical application of gene therapy is constrained by issues of preexisting immunity to AAV vectors, the need for more efficient systemic delivery to all muscles, and the difficulty of treating dominant-negative forms of the disease. Furthermore, the extreme genetic diversity of LGMD demands the development of subtype-specific therapies, making a universal cure unlikely in the near future. Future efforts must focus on refining vector design, exploring novel delivery platforms, and advancing gene-editing technologies like CRISPR/Cas9 to address a broader spectrum of mutations. In conclusion, while the path forward is complex, the remarkable progress in gene therapy research offers unprecedented hope for altering the natural history of LGMD and providing effective treatments for patients affected by these challenging diseases.

AI disclosure

During the preparation of this work, the author(s) used Deepseek and Grammarly to paraphrase. After using these tools/services, the author(s) reviewed and edited the content as needed and take(s) full responsibility for the content of the publication.

References

Allen, J., Stone-Ghariani, A., Quezada, G., Banks, D., Rose, F., Knight, W., Newman, J., Newman, W., Anderson, P., Smith, C., 2024. Living with Dysphagia: a Survey Exploring the Experiences of Adults Living with Neuromuscular Disease and their Caregivers in the United Kingdom. J. Neuromuscul. Dis. 2214-3599. 11 (2), 389–410. doi:10.3233/jnd-230002.

Alonso-Pérez, J., González-Quereda, L., Bruno, C., Panicucci, C., Alavi, A., Nafissi, S., Nilipour, Y., Zanoteli, E., de Augusto Isihi, L.M., Melegh, B., Hadzsiev, K., Muelas, N., Vílchez, J.J., Dourado, M.E., Kadem, N., Kutluk, G., Umair, M., Younus, M., Pegorano, E., Bello, L., Crawford, T.O., Suárez-Calvet, X., Töpf, A., Guglieri, M., Marini-Bettolo, C., Gallano, P., Straub, V., Díaz-Manera, J., 2022. Clinical and genetic spectrum of a large cohort of patients with δ-sarcoglycan muscular dystrophy. Brain. 0006-8950. 145 (2), 596–606. doi:10.1093/brain/awab301.

Angelini, C., Pegoraro, V., 2020. Assessing diagnosis and managing respiratory and cardiac complications of sarcoglycanopathy. Expert Opin. Orphan Drugs. 2167-8707. 8 (12), 515–523. doi:10.1080/21678707.2020.1865916.

Angelini, C., Rodríguez, A.A., 2024. Assessment of the quality of life in patients with LGMD. The case of transportinopathy. Acta Myologica. 25321900. 43 (1), 16–20. doi:10.36185/2532-1900-397.

Anwar, S., Yokota, T., 2024. The Dysferlinopathies Conundrum: clinical Spectra, Disease Mechanism and Genetic Approaches for Treatments. Biomolecules. 2218-273X. 14 (3), 256. doi:10.3390/biom14030256.

Audhya, I.F., Cheung, A., Szabo, S.M., Flint, E., Weihl, C.C., Gooch, K.L., 2022. Progression to Loss of Ambulation Among Patients with Autosomal Recessive Limb-girdle Muscular Dystrophy: a Systematic Review. J. Neuromuscul. Dis. 22143602. 9 (4), 477–492. . doi:10.3233/JND-210771.

Barton, E.R., Pacak, C.A., Stoppel, W.L., Kang, P.B., 2020. The ties that bind: functional clusters in limb-girdle muscular dystrophy. Skelet. Muscle. 20445040. 10 (1). doi:10.1186/s13395-020-00240-7.

Belhasan, D.C., Akaaboune, M., 2020. The role of the dystrophin glycoprotein complex on the neuromuscular system. Neurosci. Lett. 03043940. 722, 134833. doi:10.1016/j.neulet.2020.134833.

Bouchard, C., Tremblay, J.P., 2023. Limb–Girdle Muscular Dystrophies Classification and Therapies. J. Clin. Med. 2077-0383. 12 (14), 4769. doi:10.3390/jcm12144769.

Brun, B.N., Mockler, S.R.H., Laubscher, K.M., Stephan, C.M., Collison, J.A., Zimmerman, M.B., Mathews, K.D., 2017. Childhood Activity on Progression in Limb Girdle Muscular Dystrophy 2I. J. Child Neurol. 17088283. 32 (2), 204–209. doi:10.1177/0883073816677680.

Cataldi, M.P., Vannoy, C.H., Blaeser, A., Tucker, J.D., Leroy, V., Rawls, R., Killilee, J., Holbrook, M.C., Lu, Q.L., 2023. Improved efficacy of FKRP AAV gene therapy by combination with ribitol treatment for LGMD2I. Mol. Ther. 31 (12), 3478–3489. doi:10.1016/j.ymthe.2023.10.022.

Chamakioti, M., Karantzelis, N., Taraviras, S., 2022. Advanced Gene-Targeting Therapies for Motor Neuron Diseases and Muscular Dystrophies. Int. J. Mol. Sci. 1422-0067. 23 (9), 4824. doi:10.3390/ijms23094824.

Chandler, R.J., Sands, M.S., Venditti, C.P., 2017. Recombinant Adeno-Associated Viral Integration and Genotoxicity: insights from Animal Models. Hum. Gene Ther. 15577422. 28 (4), 314–322. doi:10.1089/hum.2017.009.

Chowdhury, A.W., Uddin, N., Khalequzzaman, M., Sarkar, M., Sina, H., Haque, A.T.M.M., Haque, A., 2022. A Patient of Limb-girdle Muscular Dystrophy Type 1B Presenting with Heart Failure and Cardiac Conduction Defects: a Case Report. Bangladesh Med. Res. Counc. Bull. 03779238. 48 (1), 76–80. doi:10.3329/bmrcb.v48i1.60664.

Chu, M.L., Moran, E., 2018. The Limb–Girdle Muscular Dystrophies: is Treatment on the Horizon? Neurotherapeutics. 18787479. 15 (4), 849–862. doi:10.1007/s13311-018-0648-x.

Cirak, S., Arechavala-Gomeza, V., Guglieri, M., Feng, L., Torelli, S., Anthony, K., Abbs, S., Garralda, M.E., Bourke, J., Wells, D.J., Dickson, G., Wood, M.J.A., Wilton, S.D., Straub, V., Kole, R., Shrewsbury, S.B., Sewry, C., Morgan, J.E., Bushby, K., Muntoni, F., 2011. Exon skipping and dystrophin restoration in patients with Duchenne muscular dystrophy after systemic phosphorodiamidate morpholino oligomer treatment: an open-label, phase 2, dose-escalation study. Lancet. 01406736. 378 (9791), 595–605. doi:10.1016/s0140-6736(11)60756-3.

Costa, R., Rodia, M.T., Pacilio, S., Angelini, C., Cenacchi, G., 2022. LGMD D2 TNPO3-Related: from Clinical Spectrum to Pathogenetic Mechanism. Front. Neurol. 16642295. 13. doi:10.3389/fneur.2022.840683.

D'Este, G., Spagna, M., Federico, S., Cacciante, L., Cieślik, B., Kiper, P., Barresi, R., 2025. Limb-girdle muscular dystrophies: a scoping review and overview of currently available rehabilitation strategies. Muscle Nerve. 0148-639X. 71 (2), 138–146. doi:10.1002/mus.28284.

De Ridder, W., Nelson, I., Asselbergh, B., De Paepe, B., Beuvin, M., Yaou, R.B., Masson, C., Boland, A., Deleuze, J.-F., Maisonobe, T., Eymard, B., Symoens, S., Schindler, R., Brand, T., Johnson, K., Töpf, A., Straub, V., De Jonghe, P., De Bleecker, J.L., Bonne, G., Baets, J., 2019. Muscular dystrophy with arrhythmia caused by loss-of-function mutations in BVES. Neurol. Genet. 2376-7839. 5 (2). doi:10.1212/nxg.0000000000000321.

Demonbreun, A.R., Wyatt, E.J., Fallon, K.S., Oosterbaan, C.C., Page, P.G., Hadhazy, M., Quattrocelli, M., Barefield, D.Y., McNally, E.M., 2020. A gene-edited mouse model of limb-girdle muscular dystrophy 2C for testing exon skipping. Dis. Model. Mech. 1754-8411. 13 (2). doi:10.1242/dmm.040832.

Domingos, J., Sarkozy, A., Scoto, M., Muntoni, F., 2017. Dystrophinopathies and Limb-Girdle Muscular Dystrophies. Neuropediatrics. 0174-304X. 48 (4), 262–272. doi:10.1055/s-0037-1601860.

Donnaloja, F., Carnevali, F., Jacchetti, E., Raimondi, M.T., 2020. Lamin A/C Mechanotransduction in Laminopathies. Cells. 2073-4409. 9 (5), 1306. doi:10.3390/cells9051306.

Doody, A., Alfano, L., Diaz-Manera, J., Lowes, L., Mozaffar, T., Mathews, K.D., Weihl, C.C., Wicklund, M., Hung, M., Statland, J., Johnson, N.E., 2024. Defining clinical endpoints in limb girdle muscular dystrophy: a GRASP-LGMD study. BMC Neurol. 14712377. 24 (1). doi:10.1186/s12883-024-03588-1.

Dubik, N., Mai, S., 2020. Lamin A/C: function in Normal and Tumor Cells. Cancers. 2072-6694. 12 (12), 3688. doi:10.3390/cancers12123688.

Durai, S., Mani, M., Kandavelou, K., Wu, J., Porteus, M.H., Chandrasegaran, S., 2005. Zinc finger nucleases: custom-designed molecular scissors for genome engineering of plant and mammalian cells. Nucleic Acids Res. 13624962. 33 (18), 5978–5990. doi:10.1093/nar/gki912.

Fan, Y., Tan, D., Zhang, X., Song, D., Chang, X., Wang, S., Yan, H., Lin, G., Yang, H., Bönnemann, C., Liu, J., Wang, S., Xiru, W., Zhang, H., Xiong, H., 2020. Nuclear Factor-κB Pathway Mediates the Molecular Pathogenesis of LMNA-Related Muscular Dystrophies. Biochem. Genet. 0006-2928. 58 (6), 966–980. doi:10.1007/s10528-020-09989-4.

Frattini, P., Villa, C., De Santis, F., Meregalli, M., Belicchi, M., Erratico, S., Bella, P., Raimondi, M.T., Lu, Q., Torrente, Y., 2017. Autologous intramuscular transplantation of engineered satellite cells induces exosome-mediated systemic expression of Fukutin-related protein and rescues disease phenotype in a murine model of limb-girdle muscular dystrophy type 2I. Hum. Mol. Genet. 0964-6906. 26 (19), 3682–3698. doi:10.1093/hmg/ddx252.

Fujita, T., Hamai, S., Hara, D., Kawahara, S., Yamaguchi, R., Ikemura, S., Motomura, G., Kawaguchi, K., Nakashima, Y., 2025. Trendelenburg gait after total hip arthroplasty due to reduced muscle contraction of the hip abductors and extensors. J. Orthop. 0972978X. 59, 57–63. doi:10.1016/j.jor.2024.07.020.

Gandbhir, V.N., Lam, J.C., Lui, F., Rayi, A., 2024. Trendelenburg Gait. [Internet]. StatPearls Publishing.

Gaudelli, N.M., Komor, A.C., Rees, H.A., Packer, M.S., Badran, A.H., Bryson, D.I., Liu, D.R., 2017. Programmable base editing of T to G C in genomic DNA without DNA cleavage. Nature. 14764687. 551 (7681), 464–471. doi:10.1038/nature24644.

Georganopoulou, D.G., Moisiadis, V.G., Malik, F.A., Mohajer, A., Dashevsky, T.M., Wuu, S.T., Hu, C.K., 2021. A Journey with LGMD: from Protein Abnormalities to Patient Impact. Protein J. 15734943. 40 (4), 466–488. doi:10.1007/s10930-021-10006-9.

Gianola, S., Pecoraro, V., Lambiase, S., Gatti, R., Banfi, G., Moja, L., Cohn, R., 2013. Efficacy of Muscle Exercise in Patients with Muscular Dystrophy: a Systematic Review Showing a Missed Opportunity to Improve Outcomes. PLoS One. 1932-6203. 8 (6), e65414. doi:10.1371/journal.pone.0065414.

Gicquel, E., Maizonnier, N., Foltz, S.J., Martin, W.J., Bourg, N., Svinartchouk, F., Charton, K., Beedle, A.M., Richard, I., 2017. AAV-mediated transfer of FKRP shows therapeutic efficacy in a murine model but requires control of gene expression. Hum. Mol. Genet. 14602083. 26 (10), 1952–1965. doi:10.1093/hmg/ddx066.

Granero-Jiménez, J., Mar López-Rodríguez, M., Dobarrio-Sanz, I., 2022. Alda Elena Cortés-Rodríguez, Influence of Physical Exercise on Psychological Well-Being of Young Adults: a Quantitative Study. Int. J. Environ. Res. Public Health. 1660-4601. 19 (7), 4282. doi:10.3390/ijerph19074282.

Griffin, D.A., Pozsgai, E.R., Heller, K.N., Potter, R.A., Peterson, E.L., Rodino-Klapac, L.R., 2021. Preclinical Systemic Delivery of Adeno-Associated a-Sarcoglycan Gene Transfer for Limb-Girdle Muscular Dystrophy. Hum. Gene Ther. 15577422. 32 (7-8), 390–404. doi:10.1089/hum.2019.199.

Herson, S., Hentati, F., Rigolet, A., Behin, A., Romero, N.B., Leturcq, F., Laforêt, P., Maisonobe, T., Amouri, R., Haddad, H., Audit, M., Montus, M., Masurier, C., Gjata, B., Georger, C., Cheraï, M., Carlier, P., Hogrel, J.Y., Herson, A., Allenbach, Y., Lemoine, F.M., Klatzmann, D., Sweeney, H.L., Mulligan, R.C., Eymard, B., Caizergues, D., Voït, T., Benveniste, O., 2012. A phase i trial of adeno-associated virus serotype 1-γ-sarcoglycan gene therapy for limb girdle muscular dystrophy type 2C. Brain. 14602156. 135 (2), 483–492. doi:10.1093/brain/awr342.

Howard, R., Kullmann, D., Werring, D., Zandi, M., 2024. Disorders of Muscle Neurology: A Queen Square Textbook. Wiley, pp. 575–602. doi:10.1002/9781119715672.ch18.

Jensen, S.M., Friborg, O., Mellgren, S.I., Müller, K.I., Bergvik, S., Arntzen, K.A., 2024. Health-Related Quality of Life in FKRP-Related Limb-Girdle Muscular Dystrophy R9. J. Neuromuscul. Dis. 22143602. 11 (1), 59–74. doi:10.3233/JND-221629.

Jensen, S.M., Müller, K.I., Mellgren, S.I., Bindoff, L.A., Rasmussen, M., Ørstavik, K., Jonsrud, C., Tveten, K., Nilssen, Ø., Van Ghelue, M., Arntzen, K.A., 2023. Epidemiology and natural history in 101 subjects with FKRP-related limb-girdle muscular dystrophy R9. The Norwegian LGMDR9 cohort study (2020). Neuromuscul. Disord. 18732364. 33 (2), 119–132. doi:10.1016/j.nmd.2022.11.005.

Johnson, N.E., Statland, J.M., 2022. The Limb-Girdle Muscular Dystrophies. Continuum. 1080-2371. 28 (6), 1698–1714. doi:10.1212/con.0000000000001178.

Kale, M.Y., Korkmaz, H.A., Ozyilmaz, B., Taşkın, A.Ç., Boluk, E., 2025. Demographic, clinical, and genetic characteristics of patients with Limb-Girdle Muscular Dystrophies (LGMD): a single tertiary-center experience. Eur. J. Paediatr. Neurol. 15322130. 58, 64–73. doi:10.1016/j.ejpn.2025.07.009.

Kilmer, D.D., McCrory, M.A., Wright, N.C., Aitkens, S.G., Bernauer, E.M., 1994. The effect of a high resistance exercise program in slowly progressive neuromuscular disease. Arch. Phys. Med. Rehabil. 00039993. 75 (5), 560–563. doi:10.1016/s0003-9993(21)01620-8.

Kim, M.J., Oh, Y.J., Hong, Y.H., Park, S.H., Oh, J.S., Kim, M.J., Chae, J.H., Shin, K., 2021. A Patient with Late-onset Limb-girdle Muscular Dystrophy Type 2B Mimicking Dermatomyositis: a Case Report and Review. J. Rheum. Dis. 22334718. 28 (2), 101–106. doi:10.4078/jrd.2021.28.2.101.

Kinali, M., Arechavala-Gomeza, V., Feng, L., Cirak, S., Hunt, D., Adkin, C., Guglieri, M., Ashton, E., Abbs, S., Nihoyannopoulos, P., Garralda, M.E., Rutherford, M., Mcculley, C., Popplewell, L., Graham, I.R., Dickson, G., Wood, M.J.A., Wells, D.J., Wilton, S.D., Kole, R., Straub, V., Bushby, K., Sewry, C., Morgan, J.E., Muntoni, F., 2009. Local restoration of dystrophin expression with the morpholino oligomer AVI-4658 in Duchenne muscular dystrophy: a single-blind, placebo-controlled, dose-escalation, proof-of-concept study. Lancet Neurol. 14744422. 8 (10), 918–928. doi:10.1016/s1474-4422(09)70211-x.

Kramerova, I., Kudryashova, E., Ermolova, N., Saenz, A., Jaka, O., de Munain, A.L., Spencer, M.J., 2012. Impaired calcium calmodulin kinase signaling and muscle adaptation response in the absence of calpain 3. Hum. Mol. Genet. 1460-2083. 21 (14), 3193–3204. doi:10.1093/hmg/dds144.

Kramerova, I., Torres, J.A., Eskin, A., Nelson, S.F., Spencer, M.J., 2018. Calpain 3 and CaMKIIβ signaling are required to induce HSP70 necessary for adaptive muscle growth after atrophy. Hum. Mol. Genet. 14602083. 27 (9), 1642–1653. doi:10.1093/hmg/ddy071.

Laval, S.H., Bushby, K.M.D., 2004. Limb-girdle muscular dystrophies – from genetics to molecular pathology. Neuropathol. Appl. Neurobiol. 0305-1846. 30 (2), 91–105. doi:10.1111/j.1365-2990.2004.00555.x.

Lee, J.J.A., Maruyama, R., Duddy, W., Sakurai, H., Yokota, T., 2018. Identification of Novel Antisense-Mediated Exon Skipping Targets in DYSF for Therapeutic Treatment of Dysferlinopathy. Mol. Ther. Nucleic Acids. 21622531. 13, 596–604. doi:10.1016/j.omtn.2018.10.004.

LGMD, A.C., 2020. Identification, description and classification. Acta Myologica. 39 (4), 207.

Li, C., Haller, G., Weihl, C.C., 2022. Current and Future Approaches to Classify VUSs in LGMD-Related Genes. Genes. 2073-4425. 13 (2), 382. doi:10.3390/genes13020382.

Li, H., Wang, P., Hsu, E., Pinckard, K.M., Stanford, K.I., Han, R., 2023. Systemic AAV9.BVES delivery ameliorates muscular dystrophy in a mouse model of LGMDR25. Mol. Ther. 31 (2), 398–408. doi:10.1016/j.ymthe.2022.11.012.

Libell, E.M., Richardson, J.A., Lutz, K.L., Ng, B.Y., Mockler, S.R.H., Laubscher, K.M., Stephan, C.M., Zimmerman, B.M., Edens, E.R., Reinking, B.E., Mathews, K.D., 2020. Cardiomyopathy in limb girdle muscular dystrophy R9, FKRP related. Muscle Nerve. 10974598. 62 (5), 626–632. doi:10.1002/mus.27052.

Liewluck, T., 2025. Limb-Girdle Muscular Dystrophies. Continuum. 1080-2371. 31 (5), 1344–1371. doi:10.1212/cont.0000000000001615.

Lin, F., Yang, K., Lin, X., Jin, M., Chen, L., Zheng, F.Z., Qiu, L.L., Ye, Z.X., Chen, H.Z., Lin, M.T., Wang, N., Wang, Z.Q., 2023. Clinical features, imaging findings and molecular data of limb-girdle muscular dystrophies in a cohort of Chinese patients. Orphanet J. Rare Dis. 17501172. 18 (1). doi:10.1186/s13023-023-02897-x.

Liu, J., Harper, S., 2012. RNAi-based Gene Therapy for Dominant Limb Girdle Muscular Dystrophies. Curr. Gene Ther. 15665232. 12 (4), 307–314. doi:10.2174/156652312802083585.

Liu, J., Wallace, L., Garwick-Coppens, S., Hauser, M., Mendell, J., Harper, S., 2012. TO 1 RNAi therapy for LGMD1A. Neuromuscul. Disord. 22 (9), 906.

Luglio, A., Maggi, E., Riviello, F.N., Conforti, A., Sorrentino, U., Zuccarello, D., 2024. Hereditary Neuromuscular Disorders in Reproductive Medicine. Genes. 2073-4425. 15 (11), 1409. doi:10.3390/genes15111409.

Mauro, A.L., Aliverti, A., 2016. Physiology of respiratory disturbances in muscular dystrophies. Breathe. 20734735. 12 (4), 318–327. doi:10.1183/20734735.012716.

Mendell, J.R., Rodino-Klapac, L.R., Rosales-Quintero, X., Kota, J., Coley, B.D., Galloway, G., Craenen, J.M., Lewis, S., Malik, V., Shilling, C., Byrne, B.J., Conlon, T., Campbell, K.J., Bremer, W.G., Viollet, L., Walker, C.M., Sahenk, Z., Clark, K.R., 2009. Limb-girdle muscular dystrophy type 2D gene therapy restores α-sarcoglycan and associated proteins. Ann. Neurol. 15318249. 66 (3), 290–297. doi:10.1002/ana.21732.

Mendell, J.R., Rodino-Klapac, L.R., Rosales, X.Q., Coley, B.D., Galloway, G., Lewis, S., Malik, V., Shilling, C., Byrne, B.J., Conlon, T., Campbell, K.J., Bremer, W.G., Taylor, L.E., Flanigan, K.M., Gastier-Foster, J.M., Astbury, C., Kota, J., Sahenk, Z., Walker, C.M., Clark, K.R., 2010. Sustained alpha-sarcoglycan gene expression after gene transfer in limb-girdle muscular dystrophy, type 2D. Ann. Neurol. 15318249. 68 (5), 629–638. doi:10.1002/ana.22251.

Milner-Brown, H.S., Miller, R.G., 1988. Muscle strengthening through high-resistance weight training in patients with neuromuscular disorders. Arch. Phys. Med. Rehabil. 69 (1), 14–19.

Møller, L.A., Martinsen, B., Werlauff, U., Dreyer, P., 2022. Ageing with neuromuscular disease: getting lost in transitions. Disabil. Rehabil. 14645165. 44 (16), 4311–4318. doi:10.1080/09638288.2021.1902577.

Mori-Yoshimura, M., Segawa, K., Minami, N., Oya, Y., Komaki, H., Nonaka, I., Nishino, I., Murata, M., 2017. Cardiopulmonary dysfunction in patients with limb-girdle muscular dystrophy 2A. Muscle Nerve. 0148-639X. 55 (4), 465–469. doi:10.1002/mus.25369.

Morishima, R., Schoser, B., 2023. A Straightforward Approach to Analyze Skeletal Muscle MRI in Limb-Girdle Muscular Dystrophy for Differential Diagnosis: a Systematic Review. Muscles. 2813-0413. 2 (4), 374–388. doi:10.3390/muscles2040029.

Mroczek, M., Inashkina, I., Stavusis, J., Zayakin, P., Khrunin, A., Micule, I., Kenina, V., Zdanovica, A., Zídková, J., Fajkusová, L., Limborska, S., van der Kooi, A.J., Brusse, E., Leonardis, L., Maver, A., Pajusalu, S., Õunap, K., Puusepp, S., Dobosz, P., Sypniewski, M., Burnyte, B., Lace, B., 2022. CAPN3 c.1746-20C>G variant is hypomorphic for LGMD R1 calpain 3-related. Hum. Mutat. 1059-7794. 43 (10), 1347–1353. doi:10.1002/humu.24421.

Muni-Lofra, R., Juanola-Mayos, E., Schiava, M., Moat, D., Elseed, M., Michel-Sodhi, J., Harris, E., McCallum, M., Moore, U., Richardson, M., Trainor, C., Wong, K., Malinova, M., Bolano-Diaz, C., Keogh, M.J., Ghimenton, E., Verdu-Diaz, J., Mayhew, A., Guglieri, M., Straub, V., James, M.K., Marini-Bettolo, C., Diaz-Manera, J., 2023 Longitudinal Analysis of Respiratory Function of Different Types of Limb Girdle Muscular Dystrophies Reveals Independent Trajectories . Neurol. Genet. 2376-7839. 9 (4). doi:10.1212/nxg.0000000000200084.

Müthel, S., Marg, A., Ignak, B., Kieshauer, J., Escobar, H., Stadelmann, C., Spuler, S., 2023. Cas9-induced single cut enables highly efficient and template-free repair of a muscular dystrophy causing founder mutation. Mol. Ther. Nucleic Acids. 21622531. 31, 494–511. doi:10.1016/j.omtn.2023.02.005.

Nallamilli, B.R.R., Chakravorty, S., Kesari, A., Tanner, A., Ankala, A., Schneider, T., da Silva, C., Beadling, R., Alexander, J.J., Askree, S.H., Whitt, Z., Bean, L., Collins, C., Khadilkar, S., Gaitonde, P., Dastur, R., Wicklund, M., Mozaffar, T., Harms, M., Rufibach, L., Mittal, P., Hegde, M., 2018. Genetic landscape and novel disease mechanisms from a large LGMD cohort of 4656 patients. Ann. Clin. Transl. Neurol. 23289503. 5 (12), 1574–1587. doi:10.1002/acn3.649.

Nandanwar, S.P., Udhoji, S.P., Raghuveer, R., 2024. Management of a 25-Year-Old Female Patient With Limb-Girdle Muscular Dystrophy With Physiotherapy: a Case Report. Cureus. 2168-8184. 16 (1). doi:10.7759/cureus.51428.

Narayanaswami, P., Weiss, M., Selcen, D., David, W., Raynor, E., Carter, G., Wicklund, M., Barohn, R.J., Ensrud, E., Griggs, R.C., Gronseth, G., Amato, A.A., 2014. Evidence-based guideline summary: diagnosis and treatment of limb-girdle and distal dystrophies. Neurology. 0028-3878. 83 (16), 1453–1463. doi:10.1212/WNL.0000000000000892.

Nikhanj, A., 2022. Understanding heart disease in patients with muscular dystrophy.

O'Dowd, D.N., Bostock, E.L., Smith, D., Morse, C.I., Orme, P., Payton, C.J., 2022. The effects of 12 weeks' resistance training on psychological parameters and quality of life in adults with Facioscapulohumeral, Becker, and Limb–girdle dystrophies. Disabil. Rehabil. 14645165. 44 (20), 5950–5956. doi:10.1080/09638288.2021.1955306.

Oliva, J., Galasinski, S., Richey, A., Campbell, A.E., Meyers, M.J., Modi, N., Zhong, J.W., Tawil, R., Tapscott, S.J., Sverdrup, F.M., 2019. Clinically advanced p38 inhibitors suppress DUX4 expression in cellular and animal models of facioscapulohumeral muscular dystrophys. J. Pharmacol. Exp. Ther. 15210103. 370 (2), 219–230. doi:10.1124/jpet.119.259663.

Park, S.B., Weishuhn, L.J., Taqi, M., 2025. Winging of the Scapula. StatPearls Publishing.

Pegoraro, E., Hoffman, E.P., 2012. Limb-girdle muscular dystrophy overview–retired chapter, for historical reference only.

Potter, R.A., Griffin, D.A., Sondergaard, P.C., Johnson, R.W., Pozsgai, E.R., Heller, K.N., Peterson, E.L., Lehtimäki, K.K., Windish, H.P., Mittal, P.J., Albrecht, D.E., Mendell, J.R., Rodino-Klapac, L.R., 2018. Systemic Delivery of Dysferlin Overlap Vectors Provides Long-Term Gene Expression and Functional Improvement for Dysferlinopathy. Hum. Gene Ther. 15577422. 29 (7), 749–762. doi:10.1089/hum.2017.062.

Poudel, B.H., Fletcher, S., Wilton, S.D., Aung-Htut, M., 2024. Limb Girdle Muscular Dystrophy Type 2B (LGMD2B): diagnosis and Therapeutic Possibilities. Int. J. Mol. Sci. 1422-0067. 25 (11), 5572. doi:10.3390/ijms25115572.

Pozsgai, E., Griffin, D., Potter, R., Sahenk, Z., Lehman, K., Rodino-Klapac, L.R., Mendell, J.R., 2021. Unmet needs and evolving treatment for limb girdle muscular dystrophies. Neurodegener. Dis. Manag. 17582032. 11 (5), 411–429. doi:10.2217/nmt-2020-0066.

Pozsgai, E.R., Griffin, D.A., Heller, K.N., Mendell, J.R., Rodino-Klapac, L.R., 2017. Systemic AAV-Mediated β-Sarcoglycan Delivery Targeting Cardiac and Skeletal Muscle Ameliorates Histological and Functional Deficits in LGMD2E Mice. Mol. Ther. 15250024. 25 (4), 855–869. doi:10.1016/j.ymthe.2017.02.013.

Preisler, N., Orngreen, M.C., 2018. Exercise in muscle disorders: what is our current state? Curr. Opin. Neurol. 14736551. 31 (5), 610–617. doi:10.1097/WCO.0000000000000597.

Qiao, C., Wang, C.H., Zhao, C., Lu, P., Awano, H., Xiao, B., Li, J., Yuan, Z., Dai, Y., Martin, C.B., 2014. Muscle and heart function restoration in a limb girdle muscular dystrophy 2I (LGMD2I) mouse model by systemic FKRP gene delivery. Mol. Ther. 22 (11), 1890–1899. doi:10.1038/mt.2014.141.

Ramachandran, R., 2020. CRISPR/Cas9 System: 2020 Nobel Prize in Chemistry. Resonance. 0973712X. 25 (12), 1669–1680. doi:10.1007/s12045-020-1088-6.

Rinaldi, C., Wood, M.J.A., 2018. Antisense oligonucleotides: the next frontier for treatment of neurological disorders. Nat. Rev. Neurol. 1759-4758. 14 (1), 9–21. doi:10.1038/nrneurol.2017.148.

Roberts, T.C., Langer, R., Wood, M.J.A., 2020. Advances in oligonucleotide drug delivery. Nat. Rev. Drug Discov. 14741784. 19 (10), 673–694. doi:10.1038/s41573-020-0075-7.

Roudaut, C., Roy, F.L., Suel, L., Poupiot, J., Charton, K., Bartoli, M., Richard, I., 2013. Restriction of Calpain3 Expression to the Skeletal Muscle Prevents Cardiac Toxicity and Corrects Pathology in a Murine Model of Limb-Girdle Muscular Dystrophy. Circulation. 0009-7322. 128 (10), 1094–1104. doi:10.1161/circulationaha.113.001340.

Rui, Y., Bai, J., Perrimon, N., Barsh, G.S., 2010. Sarcomere Formation Occurs by the Assembly of Multiple Latent Protein Complexes. PLoS Genet. 1553-7404. 6 (11), e1001208. doi:10.1371/journal.pgen.1001208.

Saha, M., Reddy, H.M., Salih, M.A., Estrella, E., Jones, M.D., Mitsuhashi, S., Cho, K.A., Suzuki-Hatano, S., Rizzo, S.A., Hamad, M.H., Mukhtar, M.M., Hamed, A.A., Elseed, M.A., Lek, M., Valkanas, E., Macarthur, D.G., Kunkel, L.M., Pacak, C.A., Draper, I., Kang, P.B., 2018. Impact of PYROXD1 deficiency on cellular respiration and correlations with genetic analyses of limb-girdle muscular dystrophy in Saudi Arabia and Sudan. Physiol. Genomics. 15312267. 50 (11), 929–939. doi:10.1152/physiolgenomics.00036.2018.

Salem, I.H., Kamoun, F., Louhichi, N., Trigui, M., Triki, C., Fakhfakh, F., 2012. Impact of single-nucleotide polymorphisms at the TP53-binding and responsive promoter region of BCL2 gene in modulating the phenotypic variability of LGMD2C patients. Mol. Biol. Rep. 0301-4851. 39 (7), 7479–7486. doi:10.1007/s11033-012-1581-4.

Sarıkaya Uzan, G., Yılmaz Uzman, C., Çinleti, T., Günay, Ç., Ülgenalp, A., Kurul, S.H., Yiş, U., 2024. Molecular Diagnosis of Limb-Girdle Muscular Dystrophy Using Next-Generation Sequencing Panels. Mol. Syndromol. 1661-8769. 15 (1), 14–21. doi:10.1159/000533976.

Sharma, A., Sane, H., Badhe, P., Gokulchandran, N., Kulkarni, P., Lohiya, M., Hema Biju, V.C.J., 2013. A Clinical Study Shows Safety and Efficacy of Autologous Bone Marrow Mononuclear Cell Therapy to Improve Quality of Life in Muscular Dystrophy Patients. Cell Transplant. 0963-6897. 22 (1_suppl), 127–138. doi:10.3727/096368913x672136.

Sharma, A., Sane, H., Gokulchandran, N., Paranjape, A., Shaikh, Z., KM, A., et al., 2023. Cell therapy for muscular dystrophy. In: Potential Therapeutic Strategies for Muscular Dystrophy. IntechOpen.

Sheikh, A.M., Vissing, J., 2019. Exercise therapy for muscle and lower motor neuron diseases. Acta Myologica. 25321900. 38 (4), 215–232. Available from https://www.actamyologica.it/wp-content/uploads/2019/12/04_Art_Original_Vissing-1.pdf.

Skuk, D., Tremblay, J.P., 2017. The Process of Engraftment of Myogenic Cells in Skeletal Muscles of Primates: understanding Clinical Observations and Setting Directions in Cell Transplantation Research. Cell Transplant. 15553892. 26 (11), 1763–1779. doi:10.1177/0963689717724798.

Sondergaard, P.C., Griffin, D.A., Pozsgai, E.R., Johnson, R.W., Grose, W.E., Heller, K.N., Shontz, K.M., Montgomery, C.L., Liu, J., Clark, K.R., Sahenk, Z., Mendell, J.R., Rodino-Klapac, L.R., 2015. AAV. Dysferlin overlap vectors restore function in dysferlinopathy animal models. Ann. Clin. Transl. Neurol. 23289503. 2 (3), 256–270. doi:10.1002/acn3.172.

Stokholm, R.N., Handberg, C., Knudsen, L.F., 2022. Prevalence of chronic pain in a national cohort of patients with limb-girdle muscular dystrophy: a cross-sectional study. Disabil. Rehabil. 14645165. 44 (25), 7802–7810. doi:10.1080/09638288.2021.1998669.

Sun, C.C., Xiao, J.L., Zhao, Z., Liu, H.Y., Tang, C.F., 2025. Recent insights into limb-girdle muscular dystrophy: impacts, therapy, and challenges. Histol. Histopathol. 16995848. 40 (12), 1927–1939. doi:10.14670/HH-18-929.

Sun, N., Zhao, H., 2013. Transcription activator-like effector nucleases (TALENs): a highly efficient and versatile tool for genome editing. Biotechnol. Bioeng. 0006-3592. 110 (7), 1811–1821. doi:10.1002/bit.24890.

Sveen, M.L., Jeppesen, T.D., Hauerslev, S., Krag, T.O., Vissing, J., 2007. Endurance training: an effective and safe treatment for patients with LGMD2I. Neurology. 00283878. 68 (1), 59–61. doi:10.1212/01.wnl.0000250358.32199.24.

Taghizadeh, E., Rezaee, M., Barreto, G.E., Sahebkar, A., 2019. Prevalence, pathological mechanisms, and genetic basis of limb-girdle muscular dystrophies: a review. J. Cell. Physiol. 10974652. 234 (6), 7874–7884. doi:10.1002/jcp.27907.

Tanboon, J., Nishino, I., 2024. Autosomal recessive limb-girdle muscular dystrophies. In: Principles and Practice of the Muscular Dystrophies. Springer, pp. 93–121.

Thomas, P.J., Xu, R., Martin, P.T., 2016. B4GALNT2 (GALGT2) Gene Therapy Reduces Skeletal Muscle Pathology in the FKRP P448L Mouse Model of Limb Girdle Muscular Dystrophy 2I. Am. J. Pathol. 00029440. 186 (9), 2429–2448. doi:10.1016/j.ajpath.2016.05.021.

Topaloglu, H., Narayanaswami, P., Carter, G., David, W., Weiss, M., Amato, A.A., 2015. Evidence-based guideline summary: diagnosis and treatment of limb-girdle and distal dystrophies: report of the Guideline Development Subcommittee of the American Academy of Neurology and the Practice Issues Review Panel of the American Association of Neuromuscular & Electrodiagnostic Medicine. Neurology. 0028-3878. 84 (16), 1720–1721. doi:10.1212/WNL.0000000000001508.

Torella, A., Fanin, M., Mutarelli, M., Peterle, E., Blanco, F.D.V., Rispoli, R., Savarese, M., Garofalo, A., Piluso, G., Morandi, L., Ricci, G., Siciliano, G., Angelini, C., Nigro, V., McNeil, P., 2013. Next-Generation Sequencing Identifies Transportin 3 as the Causative Gene for LGMD1F. PLoS One. 1932-6203. 8 (5), e63536. doi:10.1371/journal.pone.0063536.

Turan, S., Farruggio, A.P., Srifa, W., Day, J.W., Calos, M.P., 2016. Precise correction of disease mutations in induced pluripotent stem cells derived from patients with limb girdle muscular dystrophy. Mol. Ther. 15250024. 24 (4), 685–696. doi:10.1038/mt.2016.40.

Vainzof, M., Souza, L.S., Gurgel-Giannetti, J., Zatz, M., 2021. Sarcoglycanopathies: an update. Neuromuscul. Disord. 18732364. 31 (10), 1021–1027. doi:10.1016/j.nmd.2021.07.014.

Vannoy, C.H., Leroy, V., Lu, Q.L., 2018. Dose-dependent effects of FKRP gene-replacement therapy on functional rescue and longevity in dystrophic mice. Molecular Therapy – Methods & Clinical Development 11, 106–120. doi:10.1016/j.omtm.2018.10.004.

Vannoy, C.H., Xu, L., Keramaris, E., Lu, P., Xiao, X., Lu, Q.L., 2014. Adeno-associated virus-mediated overexpression of LARGE rescues α-dystroglycan function in dystrophic mice with mutations in the fukutin-related protein. Hum. Gene Ther. Methods. 19466544. 25 (3), 187–196. doi:10.1089/hgtb.2013.151.

Vincent, A.E., Rosa, H.S., Alston, C.L., Grady, J.P., Rygiel, K.A., Rocha, M.C., Barresi, R., Taylor, R.W., Turnbull, D.M., 2016. Dysferlin mutations and mitochondrial dysfunction. Neuromuscul. Disord. 18732364. 26 (11), 782–788. doi:10.1016/j.nmd.2016.08.008.

Vrist, L.T.H., Knudsen, L.F., Handberg, C., 2023. 'It becomes the new everyday life'–experiences of chronic pain in everyday life of people with limb-girdle muscular dystrophy. Disabil. Rehabil. 14645165. 45 (23), 3875–3882. doi:10.1080/09638288.2022.2142679.

Wissing, E.R., Boyer, J.G., Kwong, J.Q., Sargent, M.A., Karch, J., McNally, E.M., Otsu, K., Molkentin, J.D., 2014. P38α MAPK underlies muscular dystrophy and myofiber death through a Bax-dependent mechanism. Hum. Mol. Genet. 14602083. 23 (20), 5452–5463. doi:10.1093/hmg/ddu270.

Wyatt, E.J., Demonbreun, A.R., Kim, E.Y., Puckelwartz, M.J., Vo, A.H., Dellefave-Castillo, L.M., Gao, Q.Q., Vainzof, M., Pavanello, R.C.M., Zatz, M., McNally, E.M., 2018. Efficient exon skipping of SGCG mutations mediated by phosphorodiamidate morpholino oligomers. JCI Insight. 23793708. 3 (9). doi:10.1172/jci.insight.99357.

Zambon, A.A., Falzone, Y.M., Bolino, A., Previtali, S.C., 2024. Molecular mechanisms and therapeutic strategies for neuromuscular diseases. Cell. Mol. Life Sci. 1420-682X. 81 (1). doi:10.1007/s00018-024-05229-9.

Section II

Lysosomal storage disorders

Chapter 8

Gene therapy in leukodystrophies

Asma Vafadar[a,b], Zahra Heidari[c,d] and Amir Savardashtaki[b,e]
[a]*Student Research Committee, Shiraz University of Medical Sciences, Shiraz, Iran*
[b]*Department of Medical Biotechnology, School of Advanced Medical Sciences and Technologies, Shiraz University of Medical Sciences, Shiraz, Iran*
[c]*Shahid Chamran Hospital, Shiraz University of Medical Sciences, Shiraz, Iran*
[d]*Maternal-Fetal Medicine Research Center, Shiraz University of Medical Sciences, Shiraz, Iran*
[e]*Infertility Research Center, Shiraz University of Medical Sciences, Shiraz, Iran*

1 Introduction

Leukodystrophies are a heterogeneous group of inherited, mostly progressive encephalopathies characterized by selective involvement of the white matter in the central nervous system (CNS). Traditionally defined by pathological demyelination, recent genomic and imaging advances have expanded and refined this classification, distinguishing primary leukodystrophies from secondary leukoencephalopathies and enabling molecular diagnosis across >30 causative genes. The occurrence of leukodystrophies is approximately one in every 7500–8000 live births worldwide (Heim et al., 1997; Vanderver et al., 2012). Although leukodystrophies are rare diseases, they profoundly affect children, leading to significant health challenges. These conditions are associated with a high rate of early death, with an average age of around 8 years. They reduce neurological function and place considerable emotional, physical, and economic strain on families. Annual medical expenses are approximately US$22,500 per patient per year due to frequent hospitalizations and long-term care, while complications like epilepsy and feeding difficulties further increase healthcare demands and caretaker stress (Bonkowsky et al., 2010; Zampini et al., 2025).

Underlying genetic architecture, historical founder effects, and patterns of consanguinity shape symptom variability and disease frequency across populations. These population-specific genetic factors not only influence the frequency of certain leukodystrophies but also contribute to differences in clinical presentation, including age of onset, disease severity, and the rate of neurological decline. For example, X-linked adrenoleukodystrophy (X-ALD) is the most common peroxisomal disorder worldwide, with an occurrence rate of approximately 1 in 17,000 births, predominantly affecting males, reflecting its X-linked inheritance pattern (Engelen et al., 2012; Matos et al., 2023). In contrast, metachromatic leukodystrophy (MLD) shows a higher prevalence in parts of Europe and the Middle East, where specific founder mutations are associated with earlier onset and more aggressive disease courses (Chang et al., 2024). Other leukodystrophies, such as globoid cell leukodystrophy (GLD; Krabbe disease [KD]) and Pelizaeus–Merzbacher disease (PMD), remain rare globally, but their phenotypic spectrum can vary by region depending on the distribution of pathogenic variants within local genetic backgrounds (Elitt and Tesar, 2024; Noureldeen et al., 2025).

Clinical manifestations of leukodystrophies depend on the underlying gene defect, the affected metabolic pathway, and the age of onset. Most forms present in infancy or early childhood and exhibit progressive neurological decline with motor dysfunction, spasticity, ataxia, seizures, cognitive impairment, behavioral changes, and sensory deficits (Bonkowsky and Keller, 2021). In contrast, specific variants emerge in adolescence or adulthood, generally with a milder and more slowly progressive course. Early-onset forms are typically severe and can lead to rapid neurodevelopmental regression (Muthusamy et al., 2023).

Overall prognosis for pediatric inherited leukodystrophies remains poor, with an estimated mortality rate of 30%–35% and an average age of death around 8 years, emphasizing the severity in early-onset cases (Bonkowsky et al., 2010; Muthusamy et al., 2023). Diagnostic delays are common due to extensive clinical and genetic heterogeneity, leading to irreversible neurological damage before effective interventions can be implemented. Conventional treatments, including enzyme replacement therapy (ERT), dietary modifications (e.g., Lorenzo's oil), or hematopoietic stem cell transplantation

Gene Therapy in Neurodegenerative Disorders. https://doi.org/10.1016/B978-0-443-40467-2.00010-5

(HSCT), provide limited efficacy, mainly due to poor blood–brain barrier (BBB) penetration and the inability to reverse existing white matter injury (Vaia et al., 2024; Audouard et al., 2025; Lin et al., 2025).

Due to the challenges presented by leukodystrophies, there is increasing interest in therapeutic strategies that target their genetic basis. This chapter provides a comprehensive overview of current gene-based therapies, emphasizing clinical advancements, ongoing trials, and the preclinical models that have assisted in their development.

2 Leukodystrophy: Pathogenesis and genetic patterns

2.1 Historical and molecular overview

Over a century ago, Pelizaeus and Merzbacher independently described a hereditary, slowly progressive white matter disorder known as "diffuse sclerosis." Unlike multiple sclerosis, this condition is characterized by profound myelin loss and a rigid, sclerotic white matter structure (Merzbacher, 1910; Nowacki et al., 2022). The term *leukodystrophy,* introduced in 1928, derives from the Greek *leuko* ("white") and *dystrophia* ("degeneration" or "disordered nourishment"), referring to the pathological hallmark of these disorders: degeneration of the cerebral white matter (van der Knaap and Bugiani, 2017).

To date, more than 50 genetically distinct leukodystrophies have been identified, and new subtypes continue to be discovered through advances in genomic and biochemical analysis. Key groups of leukodystrophies include Alexander disease (AxD), Canavan disease (CD), KD (also known as GLD), MLD, PMD, X-ALD, vanishing white matter disease (VWM), and adult-onset leukoencephalopathy with axonal spheroids and pigmented glia (ALSP).

Among these, MLD, GLD, X-ALD, CD, and AxD have been most extensively studied and serve as prototypical models for understanding leukodystrophy pathogenesis and guiding therapeutic innovation (van der Knaap and Bugiani, 2017; Metovic et al., 2024).

2.2 Classification and core pathogenic mechanisms

Leukodystrophies were historically classified based on radiographic and histopathological characteristics, including demyelination, hypomyelination, spongiform changes, or cystic degeneration (Perlman and Mar, 2012; Waldman, 2018). Although these categories provided early diagnostic insight, they lacked pathobiological specificity. With the advent of next-generation sequencing (NGS), advanced magnetic resonance imaging (MRI), and molecular biology, a more mechanistic classification has emerged, one that relates genotype, cellular pathology, and therapeutic potential (van der Knaap et al., 2019; Nowacki et al., 2022; Muthusamy et al., 2023). In 2017 van der Knaap and Bugiani introduced a new classification system for leukodystrophies, organizing these conditions into five categories based on their pathological changes and underlying mechanisms. These categories include myelin disorders (myelinopathies), astrocytopathies, leukoaxonopathies, microgliopathies, and leukovasculopathies. Recognizing these classes is vital for conducting gene therapy strategies. For example, microgliopathies can be targeted using hematopoietic stem cell (HSC)-based gene therapies, while astrocytopathies may be managed through antisense oligonucleotide (ASO) interventions. An overview of leukodystrophies is provided in Table 8.1 (van der Knaap and Bugiani, 2017).

Myelinopathies are conditions that affect oligodendrocytes or the myelin sheath. They can be categorized into three main types: hypomyelination, demyelination, and vacuolated myelin conditions. When these disorders also impact neurons or their axons, they are referred to as leukoaxonopathies. On the contrary, astrocytopathies and microgliopathies arise from dysfunction in astrocytes or microglia, which can disrupt inflammatory and repair processes in the brain. Leukovasculopathies, meanwhile, are caused by issues in small cerebral blood vessels. It is important to note that microglia can be replenished from precursors derived from HSCs, which hold significant therapeutic potential (van der Knaap and Bugiani, 2017).

Mutations that cause leukodystrophy disrupt critical cell functions. These defects specifically affect multiple essential biological processes. Some mutations impair cellular recycling mechanisms managed by peroxisomes and lysosomes, with peroxisomal β-oxidation disrupted in X-ALD and lysosomal dysfunction predominant in MLD and GLD. Others compromise mitochondrial energy production or structural stability, affecting components such as myelin, the cytoskeleton, the extracellular matrix, and the BBB. In addition, genetic changes can interfere with protein synthesis, particularly during translation and assembly of proteins from messenger RNA (mRNA) (Nowacki et al., 2022; Metovic et al., 2024). Understanding these pathogenic mechanisms (Fig. 8.1) is essential for developing effective gene therapies.

2.3 Genetic patterns and therapeutic relevance

Although the etiology of some leukodystrophies remains unclear, most have a well-established genetic origin with diverse patterns of inheritance. According to extensive cohort studies, autosomal recessive inheritance is the most frequently observed, whereas autosomal dominant (including de novo) and X-linked forms are relatively less common (Vanderver

TABLE 8.1 Overview of major leukodystrophies based on their principal pathogenic mechanism.

Mechanistic category	Representative disorders	Principal gene(s)	Prevalence	Primary affected cell	Inheritance	Key pathology/ mechanism	Typical onset	Ref.
Myelin disorders (Demyelination)	Cerebral X-ALD	*ABCD1*	1 per 14,000–17,000 in males	Oligodendrocytes, neurons	X-linked recessive	Impaired peroxisomal β-oxidation leads to VLCFA accumulation and inflammation	Childhood	(Mosser et al., 1993, Dubois-Dalcq et al., 1999, Moser et al., 2007, Cartier et al., 2012, Waldman, 2018, Gupta et al., 2022, Aerts-Kaya and van Til, 2023)
	MLD	*ARSA*, *PSAP*	1 per 40,000–160,000	Oligodendrocytes, microglia	Autosomal recessive	Lack of *ARSA* leads to sulfatide accumulation (both galactosyl and lactosyl sulfatide)	Late-infantile/ **juvenile/** adult	(DeLuca et al., 1979, Barth et al., 1994, Gieselmann, 2008, Cesani et al., 2016, Shaimardanova et al., 2020, Aerts-Kaya and van Til, 2023)
	KD (GLD)	*GALC*	1–5 per 100,000	Oligodendrocytes	Autosomal recessive	Psychosine buildup leads to oligodendrocyte apoptosis	Early-infantile	(Kass, 1953, Hagberg et al., 1963, Hagberg, 1971, Chen et al., 1993, Compston, 2013, Bradbury et al., 2021, Aerts-Kaya and van Til, 2023)
Myelin disorders (Hypomyelination)	PMD	*PLP1*	1 per 100,000	Oligodendrocytes	X-linked recessive	Defective proteolipid protein 1 leads to impaired myelin compaction	Childhood–adult	(van der Knaap and Valk, 1989, Hudson, 2003, Osorio and Goldman, 2018, Bahrambeigi et al., 2019, Inoue, 2019, Matsumoto et al., 2019, Aerts-Kaya and van Til, 2023)
	PMLD	*GJC2*	Unknown/ rare	Oligodendrocytes, astrocytes	Autosomal recessive	Impaired formation of gap junctions disrupts cell-to-cell communication	Childhood	(Charzewska et al., 2016, Aerts-Kaya and van Til, 2023)
Myelin disorders (Myelin vacuolization)	CD	*ASPA*	1 per 100,000	Oligodendrocytes	Autosomal recessive	N-acetylaspartate accumulation leads to spongy white matter degeneration	Infantile	(Kumar et al., 2009, Hoshino and Kubota, 2014, A et al., 2017, de Rosa et al., 2019, Nowacki et al., 2022, Aerts-Kaya and van Til, 2023, Bokhari et al., 2026)

TABLE 8.1 Overview of major leukodystrophies based on their principal pathogenic mechanism—cont'd

Mechanistic category	Representative disorders	Principal gene(s)	Prevalence	Primary affected cell	Inheritance	Key pathology/ mechanism	Typical onset	Ref.
Astrocytopathies	AxD	*GFAP*	1 per 1,000,000	Astrocytes	Autosomal dominant (often de novo)	Toxic gain-of-function in glial fibrillary acidic protein leads to astrocyte stress	Infantile–adult	(Li et al., 2002, Messing et al., 2012, Sosunov et al., 2018, Aerts-Kaya and van Til, 2023)
	VWM	*EIF2B1–5*	1–4 per 1,000,000	Astrocytes, oligodendrocytes	Autosomal recessive	Mutations in the eIF2B complex disrupt translation regulation and stress response, leading to the progressive loss of white matter	Infancy to adolescence	(van der Knaap et al., 2002, van der Knaap et al., 2006, Aerts-Kaya and van Til, 2023)
Leukoaxonopathies	POLR3-related leukodystrophy or 4H leukodystrophy	*POLR3A, POLR3B, POLR1C, POLR3K*	*Unknown/ rare*	Neurons, oligodendrocytes	Autosomal recessive	Defective assembly and role of RNA polymerase III	Childhood	(Daoud et al., 2013, Thiffault et al., 2015, Dorboz et al., 2018, Coulombe et al., 2021, Aerts-Kaya and van Til, 2023, Metovic et al., 2024, Kok et al., 2025)
Microgliopathies	ALSP	*CSF1R*	*5 per 100,000*	*Microglia*	Autosomal dominant	Microglial depletion leads to axonal spheroids and pigmented glia	Adult	(Sassi et al., 2018, Han et al., 2020, Papapetropoulos et al., 2021, Aerts-Kaya and van Til, 2023)
Vasculopathies	CADASIL	*NOTCH3*	Population-dependent	Vascular smooth muscle cells	Autosomal dominant	NOTCH3 extracellular domain accumulation is associated with small-vessel arteriopathy, subcortical infarcts, and leukoencephalopathy	Adulthood	(Young et al., 2021, Hack et al., 2022, Aerts-Kaya and van Til, 2023, Zhao et al., 2025)
	CARASIL	*HTRA1*	Very rare	Vascular smooth muscle cells	Autosomal recessive	HTRA1 deficiency is related to small-vessel arteriopathy with leukoencephalopathy and early ischemic strokes	Early adulthood	(Fukutake, 2011, Aerts-Kaya and van Til, 2023, Haddad et al., 2024)

Abbreviations: X-ALD: X-linked adrenoleukodystrophy; ***ABCD1:*** ATP-binding cassette sub-family D member 1; **VLCFA:** Very-long-chain fatty acid; **MLD:** Metachromatic leukodystrophy; ***ARSA:*** Arylsulfatase A; ***PSAP:*** Prosaposin; **KD (GLD):** Krabbe disease (Globoid cell leukodystrophy); ***GALC:*** Galactosylceramidase; **PMD:** Pelizaeus-Merzbacher disease; ***PLP1:*** Proteolipid protein 1; **PMLD:** Pelizaeus-Merzbacher-like disease; ***GJC2:*** Gap junction protein gamma 2; **CD:** Canavan disease; **ASPA:** Aspartoacylase; **AxD:** Alexander disease; **GFAP:** Glial fibrillary acidic protein; **VWM:** Vanishing white matter disease; **EIF2B1–5:** Eukaryotic translation initiation factor 2B subunits 1–5; **POLR3A:** RNA polymerase III subunit A; **POLR3B:** RNA polymerase III subunit B; **ALSP:** Adult-onset leukoencephalopathy with axonal spheroids and pigmented glia; **CSF1R:** Colony-stimulating factor 1 receptor; **CADASIL:** Cerebral autosomal dominant arteriopathy with subcortical infarcts and leukoencephalopathy; **NOTCH3**: Notch receptor 3; **CARASIL:** Cerebral autosomal recessive arteriopathy with subcortical infarcts and leukoencephalopathy; **HTRA1:** HtrA serine peptidase 1.

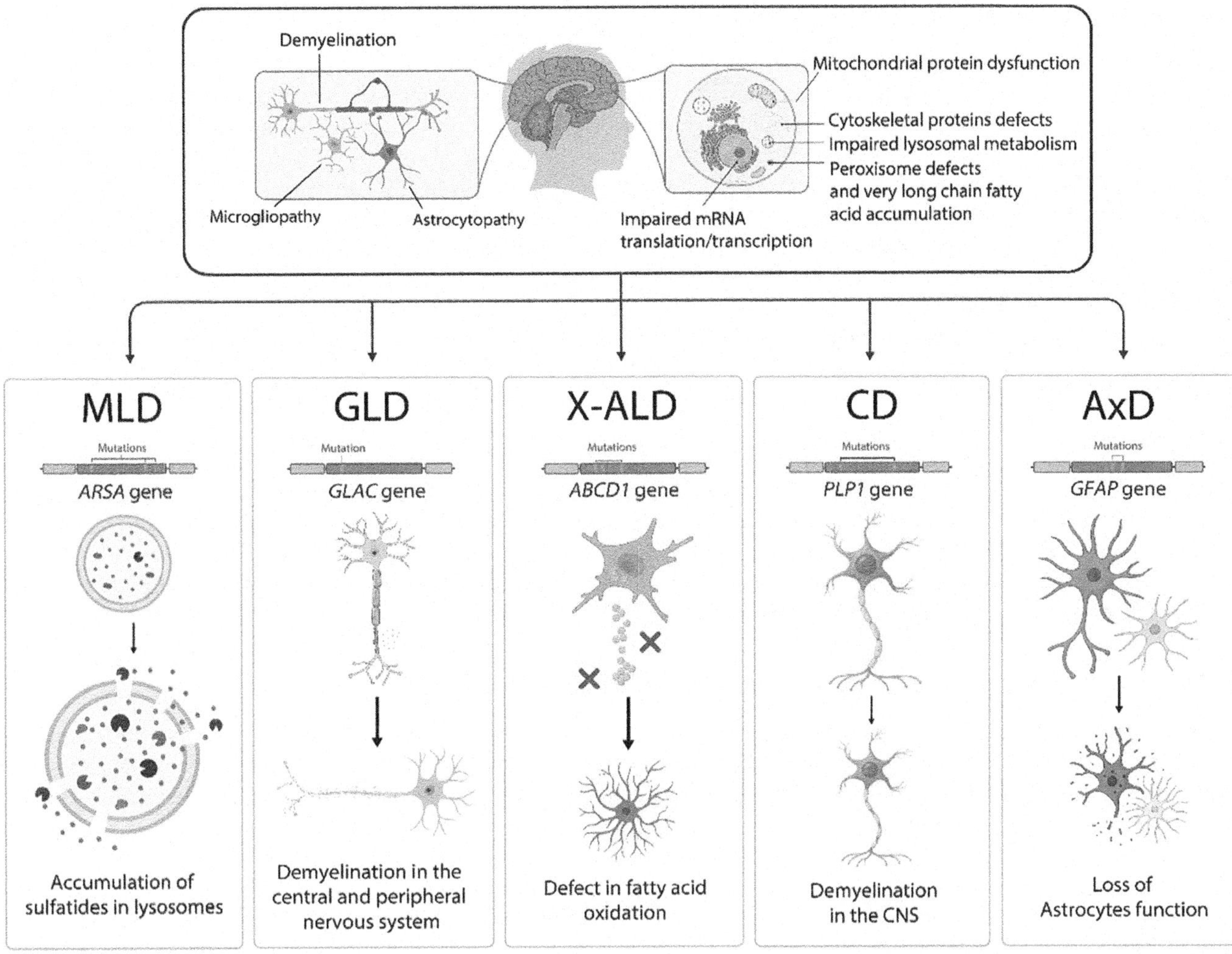

FIG. 8.1 Pathogenic mechanisms of the five most common types of leukodystrophies.

et al., 2015; Sarret, 2020). Understanding these patterns of inheritance is crucial for effective genetic counseling and the development of targeted therapies, as they influence whether a mutation leads to a loss or gain of function at the molecular level.

In autosomal recessive leukodystrophies, pathogenic variants often result in the loss of enzymatic or structural protein function. For example, mutations in *ARSA* in MLD lead to the loss of enzyme function. These conditions are generally amenable to restorative strategies such as enzyme replacement, gene supplementation, or stem cell-based therapies (Chanoumidou et al., 2020). Conversely, autosomal dominant leukodystrophies typically arise from toxic gain-of-function or dominant-negative effects, such as *GFAP* mutations in AxD. These disorders require gene-silencing interventions, including ASOs, RNA interference (RNAi), or small-molecule inhibitors, to counteract the toxic protein effect (van der Knaap and Bugiani, 2017).

From a therapeutic perspective, treatment strategies are directly linked to mechanistic classes: autosomal recessive, enzyme-deficient disorders such as *ARSA* (MLD), *GALC* (Krabbe/GLD), and *ABCD1* (X-ALD) are primary candidates for gene addition or HSC-based gene therapy, whereas autosomal dominant gain-of-function disorders like *GFAP* (AxD), *PLP1* (PMD), and *LMNB1* (adult-onset leukodystrophy) are better suited for ASO- or RNAi-based approaches, illustrating the importance of understanding disorder mechanisms for therapy development (Lin et al., 2025; Wolf et al., 2025).

Comprehending these inheritance-driven molecular mechanisms enhances diagnostic precision and informs personalized treatment design, reinforcing the mechanistic rationale for the disease-specific gene therapy strategies discussed in subsequent sections.

2.4 Translational and diagnostic implications

Exploring the mechanisms and genetics of leukodystrophies has translational implications for both diagnostic precision and therapeutic innovation. The mechanistic categorization of leukodystrophies into oligodendrocytopathic, astrocytopathic, microgliopathic, axonopathic, or metabolic forms has significantly influenced both diagnosis and therapy design (Vanderver et al., 2015; van der Knaap and Bugiani, 2017).

Accurate diagnosis requires an integrated approach that combines advanced MRI analysis, biochemical profiling, and NGS methods. Although MRI can indicate disease mechanisms, its specific diagnostic utility is limited due to overlap among leukodystrophies. Thus, proper classification depends on interpreting MRI findings in conjunction with molecular, biochemical, and clinical evidence (Parikh et al., 2015; Ashrafi et al., 2020). In addition, genotype-based newborn screening programs, such as those targeting *ABCD1* (X-ALD), *ARSA* (MLD), and *GALC* (GLD), demonstrate how early molecular detection can identify candidates for HSC or gene-based therapies before irreversible neurodegeneration occurs. Importantly, newborn screening is vital because HSCT and HSC-GT provide the most significant clinical benefit when given during the presymptomatic or minimally symptomatic stages, before rapid neurological decline starts (Vanderver et al., 2015).

From a translational perspective, this mechanistic–genetic classification facilitates the design of disease-specific cellular and animal models, which are indispensable for evaluating candidate therapies and biomarkers. It also enhances patient classification in clinical trials, ensuring that therapeutic interventions target relevant molecular pathways rather than broad phenotypic categories (Vanderver et al., 2015).

Collectively, these advances connect molecular insights with practical strategies, including the basis of precision medicine in leukodystrophies and supporting current and future therapeutic options.

3 Current therapeutic approaches for leukodystrophies

Treatment options for leukodystrophies remain limited due to their genetic heterogeneity and complexity. While advances in molecular diagnostics and disease modeling have enhanced early detection and understanding of pathophysiology, no universally curative therapy is currently available. Historically, management has focused on supportive care to alleviate symptoms and slow the progression of neurological decline (Gordon-Lipkin and Fatemi, 2018).

For many years, the only intervention capable of altering disease progression has been HSCT. More recently, autologous HSC gene therapy has emerged as an alternative disease-modifying approach, particularly for disorders mediated by loss-of-function mechanisms. HSCT is most commonly performed as an allogeneic procedure (allo-HSCT), in which healthy donor-derived HSCs are infused to replace the patient's defective hematopoietic system. Allo-HSCT has been used in leukodystrophies since the 1980s, and although comprehensive outcome data are limited, eligibility criteria and optimal timing for transplantation have become better defined (Sevin and Mochel, 2024).

Following engraftment, donor-derived myeloid cells migrate into the CNS, differentiate into microglia-like cells, and secrete the deficient enzymes required for metabolic homeostasis. This enzymatic cross-correction reduces neuroinflammation, mitigates demyelination, and supports neuronal survival. However, the extent of cross-correction differs among leukodystrophies. For instance, in MLD, remyelination and clinical stabilization can occur even without complete biochemical normalization. This indicates specific differences in microglial replacement and enzyme kinetics related to each disease (Nieder et al., 2011; Weinhard et al., 2020; Wolf et al., 2020).

Clinical evidence suggests that HSCT can be particularly effective in disorders such as X-ALD, MLD, KD, and also ALSP (Escolar et al., 2005; van Egmond et al., 2013; Kuhl et al., 2017; Dulski et al., 2022). For example, a Japanese registry cohort study on cerebral X-ALD showed that HSCT can halt disease progression and improve 10-year overall survival. The procedure is most effective when done early, usually before the Loes MRI score—an imaging scale for cerebral white matter involvement that ranges from 0 (normal) to 34 (severe)—reaches 9. This score indicates the shift from moderate to more advanced neurodegeneration (Matsukawa et al., 2020).

Landmark studies have reported 5-year survival rates of up to 76% in patients with cerebral X-linked adrenoleukodystrophy (cALD) undergoing HSCT. However, these benefits are primarily restricted to patients treated before advanced white matter involvement (ClinicalTrials.gov ID: NCT00176904, NCT00668564, and NCT00383448) (Miller et al., 2011). Similar positive outcomes have been observed in MLD and GLD, where HSCT restores partial enzymatic activity and delays neurological decline (Groeschel et al., 2016; Allewelt et al., 2018; Riedel et al., 2024).

Furthermore, outside of transplantation, ERT and metabolic modulation have been studied, but to date, no ERT has been approved for primary leukodystrophies. Trials of recombinant ERT in MLD (*ARSA*; NCT03771898) (C et al., 2020) and GLD (*GALC*, including small-molecule chaperones) (Spratley and Deane, 2016) have shown partial biochemical improvements; however, clinical benefits remain limited due to inadequate BBB penetration and insufficient CNS enzyme delivery. Metabolic strategies, such as Lorenzo's oil in X-ALD, can normalize very-long-chain fatty acids (VLCFA) levels, but they do not reliably prevent neurological deterioration (Berger et al., 2010).

Although allo-HSCT is the standard treatment for several leukodystrophies, years of clinical experience have highlighted significant limitations. The effectiveness of this therapy depends heavily on performing the transplantation early, before substantial white matter damage occurs, emphasizing the need for early diagnosis. These challenges have led to the development of next-generation strategies, such as autologous gene-corrected HSCT and adeno-associated virus (AAV)-based gene therapies. The following section focuses on these gene-based approaches as the evolutionary progression of HSCT, designed to achieve more consistent and durable metabolic correction while eliminating donor-related constraints.

4 Gene therapy in leukodystrophies

Gene therapy has become a transformative approach for leukodystrophies because most of these disorders are monogenic and mainly impact the CNS (Wolf et al., 2020; Lin et al., 2025). Over the past decade, gene therapy has progressed from experimental work to approved clinical treatments, marked by the authorization of two autologous HSC gene therapies.

4.1 Mechanisms and approaches

Gene therapy strategies for leukodystrophies are classified into three main categories: gene replacement, gene editing, and RNA modulation (Fig. 8.2).

The most established strategy is gene replacement, in which a functional transgene is introduced using viral vectors, delivered either ex vivo or in vivo (Aerts-Kaya and van Til, 2023; Zhou et al., 2024). In the ex vivo autologous

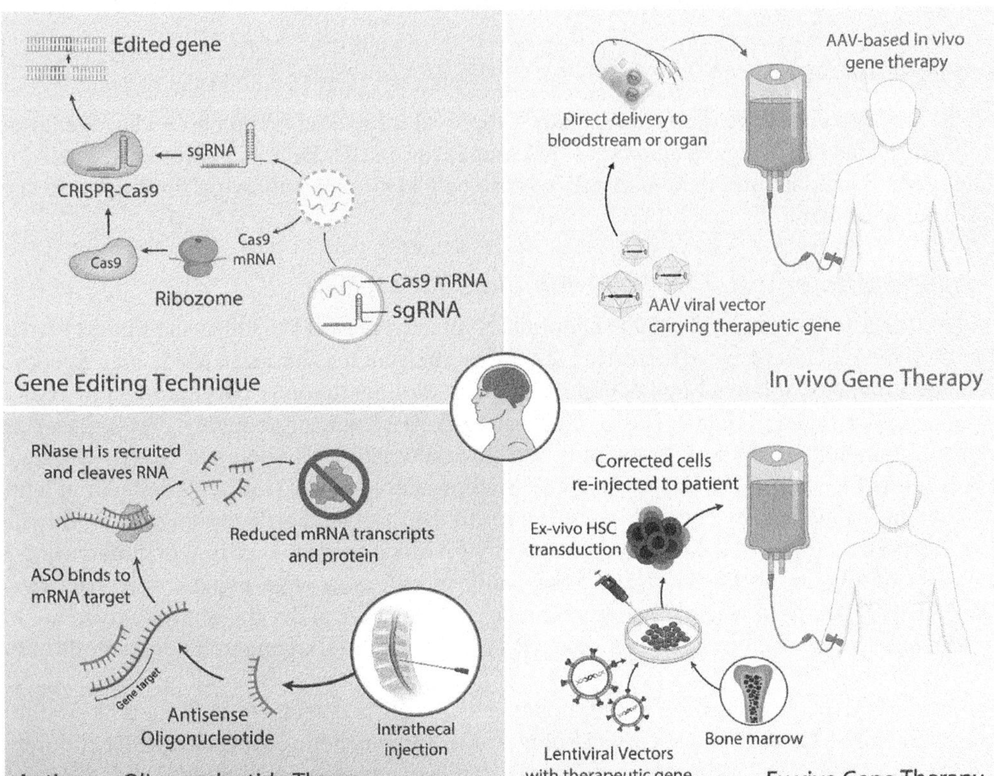

FIG. 8.2 Gene therapy approaches for leukodystrophies. Antisense oligonucleotides (ASOs) bind to complementary mRNA sequences and recruit RNase H to degrade the target transcript, thereby reducing the expression of mutant RNA. Genome editing using CRISPR–Cas9 employs a single-guide RNA (sgRNA) to direct Cas9 to disease-specific genomic sequences; efficient targeting requires a short protospacer-adjacent motif (PAM). Ex vivo gene therapy modifies autologous hematopoietic stem cells using lentiviral vectors before reinfusion, generating enzyme-secreting microglia and myeloid cells in the CNS. In vivo, AAV-based gene therapy delivers therapeutic genes directly to the CNS or systemically, transducing neurons, astrocytes, or oligodendrocytes depending on the vector serotype and route.

hematopoietic stem cell gene therapy (HSC-GT) approach, $CD34^+$ cells are collected from the patient, transduced with a lentiviral vector carrying the therapeutic gene, and reinfused after conditioning (Biffi et al., 2013; Lin et al., 2025). This approach enables long-term enzyme delivery by corrected myeloid lineage cells and forms the basis of approved therapies such as atidarsagene autotemcel for MLD (Sessa et al., 2016; Fumagalli et al., 2022) and elivaldogene autotemcel for X-ALD (Eichler et al., 2017; Duncan et al., 2024).

The second strategy is in vivo viral vector delivery, which can be administered directly into the CNS via intrathecal or intracerebral injection, or systemically through intravenous (IV) injection. In contrast to HSC-GT, which targets microglia-lineage cells, in vivo AAV-mediated delivery enables direct transduction of CNS parenchyma, with the predominant cellular targets, neurons, astrocytes, or oligodendrocytes, determined by the AAV serotype and route of administration. This approach aims to achieve long-term transgene expression but remains largely experimental (Metovic et al., 2024; Ramachandran et al., 2025a).

Gene editing is a precision tool that utilizes genome-editing techniques such as CRISPR–Cas9 to correct disease-causing mutations at their original locations within the genome (Klinkovskij et al., 2023). Proof-of-concept studies have shown that these methods can restore *ARSA* activity in MLD and correct *GALC* deficiency in GLD models. However, these approaches are still in the preclinical evaluation phase due to safety concerns, including off-target effects and limited treatment delivery efficiency (Cabre-Romans and Cuella-Martin, 2025).

RNA modulation includes the use of ASOs and RNAi platforms to target abnormal transcripts or suppress harmful gain-of-function mutations (Chery, 2016). For example, ASOs targeting mutant GFAP mRNA in AxD significantly reduce aberrant protein accumulation and astrocytic stress, supporting the first human clinical trial of ASO therapy for AxD (NCT04849741). ASOs exert their effects by binding to complementary RNA sequences and recruiting RNase H to degrade the target transcript, thereby reducing the expression of the mutant RNA.

In all strategies, therapeutic efficacy depends on efficient vector delivery, stable gene expression, and the ability of the transduced or donor-derived cells to provide enzyme cross-correction via diffusion or exosome-mediated transport across affected CNS regions.

4.2 Approved gene therapies

Two autologous HSC-GT products have received regulatory approval for leukodystrophies: atidarsagene autotemcel (OTL-200/Lenmeldy) for MLD and elivaldogene autotemcel (Skysona) for cALD. Both therapies utilize self-inactivating lentiviral vectors to deliver a functional transgene into patient-derived $CD34^+$ cells, allowing for long-term enzyme or protein expression following engraftment.

4.2.1 Atidarsagene autotemcel (OTL-200/Lenmeldy) for MLD

Atidarsagene autotemcel, also known as OTL-200 and formerly referred to as OTL-100, is the first ex vivo autologous HSC gene therapy approved for MLD. It received marketing authorization from the European Medicines Agency (EMA) in 2020 under the brand name Libmeldy, followed by US Food and Drug Administration (FDA) approval in 2024 as Lenmeldy. It is indicated for presymptomatic late-infantile (PSLI, children aged 30 months or younger), presymptomatic early-juvenile (PSEJ, children aged 30 months to under 7 years), and early-symptomatic early-juvenile (ESEJ, children under 7 years) MLD. Treatment is limited to specialized centers that can perform autologous $CD34^+$ cell collection, lentiviral modification, myeloablative conditioning, and extended follow-up care. To date, only a few dozen children worldwide have received Lenmeldy, mainly due to the rarity of MLD and the complexity involved in individualized manufacturing processes.

Lenmeldy consists of autologous $CD34^+$ HSCs genetically modified ex vivo using a self-inactivating, replication-incompetent HIV-1-derived lentiviral vector encoding human *ARSA* cDNA under the phosphoglycerate kinase promoter. The transduced cells incorporate one or more functional *ARSA* copies and, following infusion, engraft in the bone marrow to reconstitute hematopoiesis. This generates circulating and tissue-resident cells expressing functional *ARSA* (Biffi et al., 2013). *ARSA* enzymatic activity in peripheral blood mononuclear cells is detectable by 3 months postinfusion, reaching supraphysiologic levels in PSLI and PSEJ patients and normal-to-supraphysiologic levels in ESEJ children, with sustained activity observed for up to 12 years. This biochemical restoration is mechanistically linked to reduced sulfatide accumulation, providing strong evidence of disease modification beyond symptomatic improvement.

The therapy requires careful preparation of autologous HSCs combined with myeloablative conditioning. HSCs are obtained from bone marrow or mobilized peripheral blood, the latter achieved with granulocyte-colony-stimulating factor (G-CSF) and, when needed, plerixafor, a CXCR4 antagonist that enhances stem cell release. Collected $CD34^+$ cells undergo lentiviral transduction, cryopreservation, and batch release. Before infusion, patients receive myeloablative conditioning with busulfan, a chemotherapeutic agent that temporarily suppresses bone marrow to allow engraftment of genetically modified HSCs. Busulfan dosing is pharmacokinetically adjusted to achieve exposure within standard myeloablation ranges. Lenmeldy is administered as a single IV infusion consisting of one to eight thawed bags (10–20 mL each) containing approximately 2–11.8 × 10^6 $CD34^+$ cells/mL. Median infused doses associated with clinical benefit were 14.2 × 10^6 $CD34^+$ cells/kg for PSLI patients and 9–10 × 10^6 $CD34^+$ cells/kg for PSEJ and ESEJ populations. Standard posttransplant supportive care includes the use of irradiated blood products during the first 3 months and long-term precautions, such as avoiding blood, tissue, or organ donation.

The clinical efficacy of Lenmeldy has been evaluated in 39 treated children across two open-label clinical studies and a European expanded-access program, with continuous follow-up extending up to 12.2 years (median: 6.8 years) (Fahim et al., 2024). Outcomes were benchmarked against an external untreated natural history cohort. In children with PSLI MLD, Lenmeldy demonstrated remarkable preservation of neurodevelopmental milestones. This achievement, observed over 5 years, stands in clear contrast to the rapid and universal deterioration seen in untreated children. The 6-year survival rate among treated PSLI participants was 100%, compared with 58% in the natural history cohort. Most treated children maintained independent ambulation beyond age 5, and severe cognitive decline, typically observed in untreated disease, was rare or absent (Kehrer et al., 2014).

Outcomes in PSEJ MLD participants were more variable, reflecting the heterogeneity of disease progression in this subgroup. Despite this variability, treatment conferred substantial benefit compared with untreated siblings or natural history comparators. Several participants maintained a normal developmental progress and stable cognitive function into subsequent childhood (up to 11–12 years), although one death due to a cerebral thrombotic event occurred, highlighting the need for ongoing monitoring and cautious clinical follow-up. In ESEJ MLD, Lenmeldy mitigated but did not entirely prevent progressive neurologic deterioration. However, most surviving participants preserved normal cognitive function, approximately half lost independent ambulation over time, consistent with the narrower therapeutic window once symptoms have emerged. Two deaths attributable to MLD progression occurred within this cohort. Across all treated groups, gallbladder pathology, a recognized manifestation of MLD, remained incompletely responsive to therapy. Among 28 children with preexisting gallbladder disease, half exhibited persistent abnormalities after treatment, and five additional cases were diagnosed during follow-up. These findings indicate that visceral involvement may evolve independently of neurologic outcomes (Kehrer et al., 2014; Fumagalli et al., 2021).

Radiologic and neuropathologic data are limited, but sustained supraphysiologic *ARSA* activity strongly supports long-term biochemical and mechanistic disease modification.

The safety profile reflects anticipated toxicities from busulfan conditioning and autologous HSC transplantation. Common early reactions include cytopenias, febrile neutropenia, stomatitis, infections, hepatomegaly, rash, and elevated liver enzymes (Fumagalli et al., 2022). Serious risks include thrombotic events, encephalitis, venoocclusive disease, and delayed platelet engraftment (~10%). No cases of insertional oncogenesis or engraftment failure were observed, though lifelong surveillance is required due to lentiviral integration. Other considerations include theoretical dimethyl sulfoxide (DMSO) hypersensitivity, false-positive HIV polymerase chain reaction (PCR) results, and interference from antiretroviral medications. No absolute contraindications exist, but the ability to tolerate myeloablation is essential. The therapy has not been studied in children with significant viral infections or hepatic/renal impairment. Pregnancy must be excluded; breastfeeding is not recommended. Busulfan-related infertility is an essential long-term risk, with ovarian failure occurring in ~50% of females.

Lenmeldy is manufactured and marketed by Orchard Therapeutics (Kyowa Kirin). Its individualized ex vivo process contributes to its high cost, with a US price of approximately USD 4.25 million, one of the highest among approved gene therapies. Despite the substantial upfront cost, its ability to prevent severe neurodegeneration and lifelong disability supports significant clinical and economic value in appropriately selected patients. Long-term postmarketing surveillance is mandated for at least 15 years to monitor for late hematologic malignancies, assess integration-site safety, and evaluate sustained neurologic benefit. Ongoing studies include long-term analysis of cryopreserved lots, expanded outcome collection, and exploratory programs investigating adjunctive therapeutic strategies for MLD (Gangji et al., 2025).

4.2.2 Elivaldogene autotemcel (Skysona) for cerebral X-ALD

Elivaldogene autotemcel (Skysona) is an autologous HSC gene therapy designed to slow neurologic progression in boys with early, active cerebral X-ALD. It received conditional approval from the European Medicines Agency in 2021 and FDA Accelerated Approval in 2022, supported by multiple expedited designations such as Orphan Drug, Breakthrough Therapy, Priority Review, and Rare Pediatric Disease status (Scott et al., 2023). Access is restricted to specialized gene therapy centers capable of autologous stem cell mobilization, lentiviral modification, busulfan conditioning, infusion, and long-term monitoring, and more than 60 patients worldwide have been treated through clinical trials, expanded-access programs, and commercial use.

The therapy uses autologous $CD34^+$ HSCs transduced ex vivo with a self-inactivating lentiviral vector expressing human ABCD1 cDNA under the myelin basic protein promoter. The restored ALD protein (ALDP) enables improved peroxisomal metabolism of VLCFAs and attenuates the inflammatory demyelination characteristic of cALD. After infusion, genetically corrected cells engraft in the bone marrow and differentiate into monocytes and microglia capable of long-term ALDP expression, detectable within weeks and sustained for several years in patients with stable engraftment.

Treatment requires HSC mobilization with G-CSF and, when necessary, plerixafor, followed by leukapheresis, $CD34^+$ purification, and lentiviral transduction. Before infusion, patients receive pharmacokinetically adjusted myeloablative busulfan to permit durable engraftment of the gene-corrected cells. Skysona is administered as a single IV infusion of one or more thawed bags (4–20 mL each) containing approximately 5×10^6 to 20×10^6 $CD34^+$ cells/kg, followed by standard posttransplant supportive care, including irradiated blood products, infection prophylaxis, and routine surveillance.

The therapeutic potential of Skysona has been investigated in two open-label clinical trials (ALD-102 and ALD-104) that enrolled 67 boys with early, active cALD. A landmark study by Cartier et al. demonstrated that autologous transplantation of $CD34^+$ cells engineered with a lentiviral vector led to functional expression of the ALDP in two patients, resulting in stabilization of cerebral demyelination 14–16 months after treatment and maintenance of clinical stability. These initial results supported further enrollment in the STARBEAM trial (ALD-102) (Cartier et al., 2009).

Interim analysis reported by Eichler et al. for 17 boys treated with Skysona showed that the majority (15 of 17) remained alive without major functional disability, with brain MRI indicating arrested lesion progression. Twelve patients preserved a cALD-specific neurologic function score of 0 over 18–36 months of follow-up. Across both trials, overall survival at 2 years posttreatment was approximately 90%. Compared with natural history cohorts, treated patients exhibited slower progression to significant functional disability or death, emphasizing the importance of early intervention (Eichler et al., 2017). Long-term observations suggest that sustained engraftment of genetically corrected HSCs can confer durable neurologic stabilization. However, individual outcomes may vary, particularly in patients with incomplete engraftment or ongoing MRI activity, underscoring the need for continued monitoring.

Skysona's radiologic benefit primarily involves slowing, not reversing, lesion progression, and long-term neurologic stabilization varies with initial disease severity. Human data on pathology remain limited, but animal studies confirm the restoration of peroxisomal function after engraftment of corrected microglial precursors.

Safety risks reflect both lentiviral integration and myeloablation. Skysona carries an FDA boxed warning for hematologic malignancies, including myelodysplastic syndrome and acute myeloid leukemia, attributed to insertional oncogenesis, with some cases requiring allogeneic HSCT and fatal outcomes reported. Additional serious complications include prolonged cytopenias, severe infections, hepatic venoocclusive disease, delayed platelet recovery, and mucositis. Common adverse events include febrile neutropenia, elevated hepatic enzymes, and viral or bacterial infections. False-positive HIV PCR tests may occur due to vector sequences, and antiretroviral therapy must be avoided before cell collection.

Although there are no formal contraindications, the therapy is not suitable for patients who cannot tolerate myeloablative conditioning. It has not been studied in individuals with HIV, HTLV, hepatitis B or C, or significant liver or kidney problems, and it is contraindicated during pregnancy and breastfeeding due to conditioning toxicity. Busulfan-related infertility is a known risk, especially in boys treated before puberty.

Skysona is developed and marketed by Bluebird Bio, with a US list price of approximately USD 3 million per infusion. Despite the high upfront cost, modeling suggests potential cost-effectiveness due to the prevention of rapid neurologic decline, permanent disability, and premature mortality in early-treated cALD. Administration occurs only in accredited centers with GMP-compliant cell processing and long-term follow-up capabilities.

Given the integrating lentiviral vector, patients are enrolled in a mandatory 15-year long-term follow-up program that includes hematologic surveillance, integration-site analysis, and MRI and neurologic assessments. Postmarketing research continues to evaluate the durability of microglial replacement, refinements to conditioning regimens, long-term adrenal and neurologic outcomes, and the potential applicability of the platform to other peroxisomal or demyelinating disorders (Scott et al., 2023).

5 Recent and Ongoing Clinical Trials of Gene Therapy for Leukodystrophies

Clinical development of gene therapy for leukodystrophies has expanded rapidly over the past decade. Multiple first-in-human, dose-escalation, and long-term follow-up studies are currently evaluating AAV-, lentiviral-, and ASO-based therapeutic platforms. These clinical programs primarily target MLD, GLD, CD, AxD, and several rare adult-onset leukodystrophies. These trials are evaluating safety, biodistribution, pharmacodynamic responses, biomarker correction, and preliminary evidence of clinical efficacy. Collectively, these studies provide essential translational evidence supporting the success of approved therapies such as Lenmeldy and Skysona.

5.1 Metachromatic leukodystrophy

Several clinical gene therapy programs are currently evaluating for MLD, highlighting the increasing interest in both in vivo and ex vivo strategies. These initiatives complement the approved autologous HSC gene therapy, OTL-200.

The most advanced ongoing trial is led by Orchard Therapeutics (NCT04283227). This Phase III study, initiated in 2022, is designed for individuals with late-juvenile MLD. It evaluates the efficacy and safety of a single infusion of OTL-200, with follow-up monitoring extending up to 8 years. Qualified participants include symptomatic patients aged 7–17 years, as well as presymptomatic individuals under 17 years with an affected sibling who has the same subtype. All participants must have two pathogenic *ARSA* variants and demonstrate reduced *ARSA* enzymatic activity.

The primary endpoints of the trial include assessing changes in CSF *ARSA* activity and the NAA/creatine ratio at 24 months, with extended pharmacodynamic evaluations planned through year 8. This long-term monitoring aims to determine whether OTL-200, already approved for early-onset MLD, can provide durable stabilization in late-onset cases, which typically progress more slowly and are more heterogeneous.

To complement these efforts, the Shenzhen Geno-Immune Medical Institute launched two parallel first-in-human studies evaluating an optimized lentiviral vector, TYF-ARSA. The first trial (NCT03725670), initiated in May 2025, investigates direct in vivo delivery through a combination of intrathecal and IV administration. The study registers patients aged 1 month to 50 years with genetically confirmed *ARSA* deficiency and MRI-based evidence of the disease. Individuals with active infections, HIV, malignancies, or significant organ dysfunction are excluded. TYF-ARSA is administered at a dosage of 1–2 × 10^9 TU/mL, with 1 year of safety monitoring (using the Common Terminology Criteria for Adverse Events v4.0) and a 3-year evaluation of biochemical and imaging biomarkers. This program represents one of the earliest clinical attempts to deliver lentiviral vectors directly to both the CNS and peripheral tissues.

The second trial (NCT07046338), which was launched in June 2025 by the same sponsor, focuses on an ex vivo approach, using the same TYF-ARSA vector to transduce autologous $CD34^+$ HSC. In this trial, patients receive 1–10 × 10^6 gene-modified cells per kilogram following cell culture transduction. The inclusion and exclusion criteria are similar to those of the in vivo study, allowing for cross-comparison between direct CNS delivery and HSC-based systemic approaches. The primary objectives of the ex vivo trial include safety assessment and vector integration profiling. At the same time, long-term follow-up will evaluate transgene durability and disease stabilization through MRI over 3 years.

These ongoing trials emphasize the rapidly evolving clinical landscape for MLD. They include various approaches, including in vivo lentiviral administration, ex vivo modification of HSCs, and long-term evaluation of an already licensed gene therapy in new patient subgroups. The goal of these efforts is toexpand treatment eligibility across age groups, address the unmet needs of late-onset variants, and clarify the comparative advantages of direct CNS delivery over traditional autologous HSC platforms.

5.2 Cerebral X-ALD

Clinical development of gene therapy for cALD has progressed substantially over the past decade. Several completed and ongoing studies are evaluating the use of autologous HSC gene therapy with lentiviral vectors. Genetix Biotherapeutics has established a major clinical program that includes three interconnected trials: ALD-102, ALD-104, and the long-term follow-up study LTF-304.

The pivotal ALD-102 trial (NCT01896102) is a Phase II/III study that began in 2013. It evaluated the safety and efficacy of $CD34^+$ HSCs transduced with the Lenti-D vector encoding the *ABCD1* gene following myeloablative conditioning. Male patients aged 17 years or younger with elevated VLCFAs, MRI-confirmed cerebral disease, and Loes scores of 0.5–9 with gadolinium enhancement were enrolled in the study. The results demonstrated a clinical benefit, leading to conditional approval from the EMA in 2021 and accelerated approval from the FDA for elivaldogene autotemcel (eli-cel/Skysona) in 2022.

The subsequent ALD-104 trial (NCT03852498) used the same Lenti-D product and conditioning regimen to further evaluate safety and engraftment kinetics. Both trials aimed to achieve primary endpoints, including survival without major functional disabilities (MFDs) at 24 months, the absence of graft failure, and successful neutrophil engraftment. Secondary measures included assessments of neurological function scores, radiographic progression, survival rates, and vector copy numbers in peripheral blood.

Participants from these trials are currently being monitored in the ongoing LTF-304 long-term study (NCT02698579), which is designed to follow safety and the durability of benefits for up to 15 years. The study's outcomes include long-term MFD-free survival, the incidence of GVHD and immune-related adverse events, hematologic stability, the need for additional transplantation, and the emergence of new neurological deficits.

Together, these studies represent the most comprehensive gene therapy program for cALD to date and provide the clinical foundation for the regulatory approval of *Skysona* as a disease-modifying therapy.

5.3 Krabbe disease

Gene therapy efforts for GLD have intensified in recent years, with several clinical investigations focused on systemic or CNS-targeted AAV-mediated *GALC* delivery, often evaluated alongside or in comparison with HSCT. The first clinical gene therapy program was the RESKUE trial (NCT04693598), sponsored by Forge Biologics. This trial began in November 2021 as a first-in-human, open-label Phase I/II dose escalation study. It evaluated FBX-101, an AAVrh10 vector encoding human *GALC*, administered intravenously to infants undergoing umbilical cord blood transplantation (UCBT).

Suitable participants included infants aged 1 day to 12 months with biallelic *GALC* mutations representative of the infantile phenotype, along with biochemical and neurophysiological findings supporting a diagnosis of GLD. Key exclusion criteria included prior exposure to gene therapy, congenital anomalies affecting neurodevelopment, and unrelated neurological injuries. The trial was designed to include two dosing cohorts: a lower-dose cohort and a higher-dose cohort, each with three participants.

FBX-101 was administered as a single IV infusion through a peripheral venous catheter after UCBT. The goal was to deliver the *GALC* systemically during the early stages of immune reconstitution. The first two infants who received treatment were given a dose of 3×10^{13} vector genomes per kilogram (vg/kg) at 25 and 29 days post-UCBT, respectively. The treatment was well tolerated, with no serious adverse events reported. There were no increases in liver enzyme levels, and no anti-AAVrh10 antibodies were detected.

Both treated infants demonstrated significant increases in *GALC* activity in plasma and CSF, reductions in psychosine levels, improvements in gross motor skills, and MRI results indicating preservation of white matter integrity comparable to that of age-matched controls. Remarkably, one infant achieved entirely typical developmental motor milestones, while the other exceeded the outcomes typically observed in transplanted GLD patients (Escolar et al., 2023).

Based on these findings, Forge Biologics initiated a second trial in February 2023, identified as NCT05739643. This trial aimed to expand treatment options for both early- and late-infantile GLD patients up to age 18. The study utilized FBX-101-REKLAIM, the trial name for the next-generation version of FBX-101. The clinical trial maintained the UCBT followed by gene therapy, while incorporating updated eligibility criteria. One significant change was the exclusion of patients with anti-AAVrh10 antibody titers greater than 1:100. In the first cohort, five subjects received a low dose of 1.6×10^{13} gc/kg, administered at least 21 days after UCBT.

The therapy demonstrated excellent tolerability, with no treatment-related serious adverse events observed through 18 months. Among the two myeloablated subjects, there were significant increases in GALC levels in plasma and CSF, along with a reduction in psychosine to undetectable levels. These subjects also demonstrated normal maturation of white-matter tracts, as confirmed by diffusion tensor imaging. The three immunosuppressed subjects experienced no toxicity, and one child, who was followed for 6 months, showed notable improvement in gross motor development (Escolar et al., 2024).

In August 2024, Forge Biologics launched a long-term follow-up study (NCT06308718) called FBX-101-LTFU. This multicenter, noninterventional observational program is designed to monitor participants who previously received FBX-101 in earlier interventional trials. The study includes children, adults, and older adults who completed their participation in an FBX-101 trial before the company discontinued the development of the interventional FBX-101 program. The goal of this LTFU study is to evaluate long-term vector persistence, safety, neurological outcomes, and *GALC*-related biomarkers.

Another AAV-based approach, known as PBKR03, was developed by Passage Bio/Gemma Biotherapeutics in February 2022 and evaluated in the clinical trial NCT04771416. This treatment was designed to deliver the *GALC* gene using the oligodendrocyte-tropic AAVHu68 vector, which was injected into the cisterna magna. The multicenter, dose escalation Phase I/II trial targeted early infantile GLD. Although the preliminary results showed encouraging safety and pharmacodynamic trends, such as reduced psychosine levels and early signs of myelin preservation, the trial has been suspended due to strategic changes within the company, which have paused further development of PBKR03.

Overall, clinical experience with GLD emphasizes two key points: (1) systemic delivery of AAV-mediated *GALC* can result in strong, long-lasting biomarker correction and (2) combining gene therapy with early myeloablative HSCT might offer a synergistic advantage, especially for rapidly progressing infantile cases.

5.4 Canavan disease

In CD, ongoing clinical research is focusing on innovative AAV-based platforms to restore *ASPA* function. Current trials are investigating both oligodendrocyte-targeted vectors and systemic delivery strategies using rAAV9. The most advanced approach involves rAAV-Olig001, a recombinant AAV vector engineered to target oligodendrocytes selectively. This vector is being evaluated in an open-label Phase I/II clinical trial (NCT04833907). Olig001 is a chimeric construct derived from various AAV serotypes (AAV1, 2, 6, 8, and 9) and shows approximately ninefold greater binding affinity to oligodendrocytes compared to AAV8, representing a substantial improvement over previously available AAV serotypes (Powell et al., 2016).

The trial began in April 2021 and enrolled children aged 3–60 months. It aims to assess the safety, pharmacodynamic profile, and preliminary efficacy of a single intracerebroventricular (ICV) dose of rAAV-Olig001-ASPA (MYR-101). A neurosurgical procedure was performed to deliver the treatment using two predefined ICV injection sites, resulting in a total intracranial dose of 3.7×10^{13} vg. Safety outcomes indicated that 87.5% of participants experienced at least one serious adverse event; however, none were linked to MYR-101. All children encountered at least one adverse effect, including procedure-related seizures and CSF leakage, but all problems resolved without any lasting complications. No deaths occurred during the study. In addition, one participant experienced two mild-to-moderate adverse events (fever and rash), which were considered possibly related to MYR-101 but fully resolved. Preliminary efficacy signals were promising. CSF NAA levels decreased, and MRI showed increased myelination. In addition, improvements in developmental landmarks were observed using the Mullen Scales of Early Learning (MSEL) (Leone et al., 2025). Overall, these preliminary findings suggest that MYR-101 is generally well-tolerated and provides initial evidence of both biochemical and clinical benefits.

A second gene therapy program for CD is currently being investigated in a Phase I/II open-label clinical trial (NCT04998396). This trial is evaluating BBP-812, a recombinant AAV9 vector that delivers the *ASPA* gene using a ubiquitous promoter. This systemic approach aims to restore ASPA expression in both neuronal and nonneuronal populations. Suitable participants include children aged 30 months or younger who have confirmed biochemical, genetic, and clinical diagnoses of CD. This includes those with elevated urinary NAA levels and biallelic *ASPA* mutations. Participants with positive anti-AAV9 antibodies or those who have previously received AAV-based therapeutics or vaccines are excluded from the study.

Cases in the trial will receive a single IV infusion of either a low or high dose of the treatment. The follow-up period for the study will last 52 weeks. Key endpoints include the frequency of adverse events, changes in urine and CNS NAA levels (measured by magnetic resonance spectroscopy), and overall tolerability of the treatment. Recruitment for the trial is currently ongoing.

Taken together, the development of CD gene therapy has made rapid progress, with both CNS-targeted and systemic AAV platforms demonstrating early biochemical improvements and acceptable safety profiles. This positions CD as one of the leukodystrophies most likely to achieve therapeutic advancements.

5.5 Alexander disease

Therapeutic development for AxD has primarily focused on ASO-based inhibition of mutant *GFAP*. An ongoing clinical trial is currently evaluating intrathecal ASO administration across a broad age range.

A Phase I–III, double-blind, randomized, placebo-controlled trial (NCT04849741) is assessing ION373 (zilganersen), an ASO administered intrathecally and specifically designed to reduce *GFAP* expression. Sponsored by Ionis Pharmaceuticals, this trial started in June 2021 and is expected to conclude in late 2025. Although no results have been published yet, the study design is comprehensive and rigorous, with approximately 73 patients aged 2–65 years registered, representing pediatric, adult, and older adult populations with AxD.

Inclusion in the study requires a genetically confirmed *GFAP* mutation. Exclusion criteria include prior exposure to siRNA within specified timeframes, a history of gene therapy or cell transplantation, and any previous experimental neurosurgical procedures. All participants will follow a multiple-ascending-dose regimen, receiving intrathecal bolus injections every 12 weeks during the initial 60-week treatment period. A 120-week extension will follow to assess long-term safety, pharmacokinetic stability, and sustained suppression of *GFAP* levels. Zilganersen has demonstrated significant *GFAP* reduction and reversal of pathology in preclinical models of AxD, making it one of the most promising ASO strategies currently advancing toward potential regulatory evaluation.

Overall, AxD represents a strong candidate for mutation-directed ASO therapy, and the ongoing ION373 program is the most advanced clinical effort aimed at modifying the natural progression of this otherwise progressive and fatal disorder.

5.6 Autosomal dominant leukodystrophy

As part of new precision medicine approaches for ultra-rare adult-onset leukodystrophies, a personalized ASO strategy has recently been explored for ADLD. This disorder is caused by a duplication of the *LMNB1* gene, leading to overexpression of lamin B1 and progressive myelin degeneration. Unlike inflammatory leukodystrophies, such as cALD, ADLD represents a distinct category characterized by autonomic dysfunction, gait disturbances, and slowly progressive corticospinal tract involvement.

An interventional N-of-1 study (NCT06816498), sponsored by the n-Lorem Foundation, was launched in March 2025 to evaluate the safety and potential clinical benefits of a fully personalized ASO (nL-LMNB1-001) in a single, 51-year-old male with genetically confirmed ADLD. The study aims to determine whether targeted suppression of *LMNB1* overexpression can alter the clinical course of the disease. The primary outcomes include: (1) longitudinal changes in disease progression rate over 24 months following ASO administration, (2) changes in neurological function, including memory, visual processing, and language, assessed through standardized neurocognitive tests, and (3) radiologic changes in brain atrophy quantified by MRI from baseline to 24 months after treatment.

Although the study design outlines the dosing schedule and follow-up intervals, there is no publicly available information regarding the exact route of administration, dosing concentration, or dose escalation strategy. This lack of disclosure is consistent with confidentiality practices typically applied to individualized ASO therapeutics. This trial highlights the feasibility of developing personalized, mutation-specific ASOs for ultra-rare leukodystrophies. However, scalability remains a challenge, and broader applicability will depend on advancements in manufacturing, regulatory pathways, and long-term safety assessments.

The key molecular targets and therapeutic mechanisms being investigated in gene therapy trials for leukodystrophies are outlined in Tables 8.2 and 8.3, which include the vector platforms, delivery routes, and therapeutic goals.

6 Current advances in preclinical gene therapy for leukodystrophies

Leukodystrophies are a group of over 50 genetically distinct disorders characterized by problems with the development or maintenance of myelin, the protective covering of nerves. These conditions present a wide range of clinical symptoms due to the variety of underlying molecular defects, including deficiencies in lysosomal enzymes, metabolic imbalances, and abnormalities in structural proteins. Therefore, it is crucial to have vital preclinical programs that can clarify the mechanisms of these diseases and guide the development of gene-based therapies. Experimental studies have focused on improving vector engineering, optimizing delivery routes, and developing animal models that accurately mimic human pathology. The following subsections provide an overview of significant preclinical advances in key leukodystrophies.

6.1 Metachromatic leukodystrophy

Preclinical research in MLD has concentrated on AAV capsid engineering, CNS delivery optimization, and long-term enzyme restoration. One of the most extensively studied vectors is HSC15/ARSA, a recombinant version of the naturally occurring rAAV-15 serotype, engineered to express human *ARSA* cDNA. IV administration of HSC15/ARSA in *ARSA* knockout mice significantly increased *ARSA* enzyme activity, transcript levels, and vector genomes in targeted tissues. Importantly, transgene expression persisted for up to 12weeks in neonatal mice and 52weeks in adult mice, correlating with improved motor performance. These findings highlight the ability of HSC15/ARSA to efficiently cross the blood–nerve barrier, blood–spinal cord barrier, and BBB, restoring near-physiological ARSA activity and preventing motor decline in adult MLD mice (Martin et al., 2023).

AAVPHP.eBh-ARSAHA is an innovative platform that represents a next-generation AAV variant, known for its exceptional ability to penetrate the BBB after IV administration. In a study by Audouard et al., treating MLD mice at 6 or 9 months of age resulted in extensive CNS transduction, complete normalization of sulfatide accumulation, and significant suppression of neuroinflammation. This was achieved consistently at low doses and in later-stage animals. These findings provide strong preclinical support for advancing AAVPHP.eBh-ARSAHA into early-phase (I/II) clinical trials (Audouard et al., 2024).

A complementary safety assessment conducted by Rosenberg et al. examined the delivery of AAVrh.10hARSA through direct intraparenchymal infusion at 12 white-matter sites in nonhuman primates (NHPs). The study found that low doses of AAVrh.10hARSA were well tolerated; however, high doses led to localized MRI abnormalities and immune cell infiltrates at the catheter sites. Although clinical endpoints were not evaluated in this assessment, the findings help define safe dosing thresholds for intraparenchymal AAVrh.10 vectors in CNS applications (Rosenberg et al., 2021).

Further studies have also considered the delivery of AAV9/ARSA intrathecally to both presymptomatic and symptomatic 1-year-old mice with MLD. The highest expression of *ARSA* was observed in the cerebellum and olfactory bulbs, and there was a significant reduction in sulfatide accumulation in late-stage mice. However, motor performance did not improve in those already exhibiting symptoms. In contrast, mice that were treated early at 6 weeks of age showed widespread correction of sulfatide storage throughout the brain (Miyake et al., 2021).

TABLE 8.2 Clinical trials with outcomes reported in leukodystrophies.

Type of leukodys-trophy	Vector, targeted gene	Phase/NCT number/ sponsor/study start and status	Agent name/ dose and administration route	Targeted patient population/ participants number	Follow-up and monitoring	Complications/ management	Result	Refs.
MLD	AAVrh.10, *ARSA*	NCT01801709, Phase I/II, Open-label/ Single-arm/National de la Santé et de la Recherche Médicale, France/2014/Completed	AAVrh.10cuARSA/ 1–4 × 10^{12} vg/ IC Administration	Boys or girls with an early-onset form of MLD/between 6 months and 5 years/4 MLD children	Over a 2-year follow-up/*ARSA* level	SAEs, including intracranial suffusion, seizures/resolved spontaneously	Increased ARSA activity in CSF to 20%–70% of the control value	(Seyrantepe et al., 2018)
	Lentiviral vector, *ARSA*	NCT01560182/ NCT03392987 Phase I/ II, open-label/Orchard Therapeutics/2010/ Active, not recruiting	OTL-200	Pre or earlysymptomatic MLD patients with the early and late juvenile and infantile variants/ up to 7 years of age/39 treated patients and 49 untreated	Over a 6.76-year follow-up/*ASPA* level	Febrile neutropenia	Reduced the risk of severe motor impairment. Preserved cognitive function and motor development in the majority of patients	(Fumagalli et al., 2025)
CD	rAAV-Olig001, *ASPA*	NCT04833907, Phase I/ II, open-label/Myrtelle Incorporated/2021/ Enrolling by invitation	MYR-101/3.7 × 10^{13} vg/ ICV Administration	Definitive diagnosis of typical CD by a board-certified neurologist/3–60 months, Child/24 patients	Over a 1-year follow-up/NAA concentration	All participants experienced ≥ 1 adverse event and resolved	Diminished NAA concentrations in CSF, elevated myelination, enhanced MSEL developmental outcomes	(Leone et al., 2025)
	AAV serotype 9, *ASPA*	NCT04998396, Phase I/II, open label/*ASPA* Therapeutics/2021/ Recruiting	AAV9 BBP-812	Biallelic mutation of the ASPA gene, determined at screening or documented in the participant's medical history/children up to 30 months/26 patients	Over a 1-year follow-up/NAA concentration	Not applicable	Reduced NAA levels found in urine, CSF, and brain	(Eichler et al., 2023)
Infantile GLD/GLD	AAVrh10 vector, *GALC*	NCT05739643, Phase I/II, Forge Biologics, Incorporated, 2023, Active, not recruiting	FBX-101-REKLAIM/1.6 × 10^{13} gc/kg/ IV Administration	Two GALC mutations predictive of an infantile-onset phenotype/ up to 18 years (child, adult)	Over a 1.5-year follow-up/GALC level	No serious adverse events	Restored GALC activity and psychosine levels to normal ranges	(Escolar, Vander Lugt et al. 2024)
	AAVrh10 vector, *GALC*	NCT04693598, Phase I/II, Forge Biologics, Incorporated, 2021, Active, not recruiting	FBX-101/3 × 10^{13} vg/kg IV Administration	Two GALC mutations predictive of an infantile-onset phenotype/1 day to 12 months (child)/two patients	Over a 1 to 2-year follow-up/GALC level	Non-AEs	Enhanced plasma CSF GALC enzyme activity, improvement in gross motor skills	(Escolar, Vander Lugt et al. 2023)
cALD	Lentiviral vector, *ABCD1*	NCT01896102, Phase II/ III/Genetix Biotherapeutics Incorporated/2013/ Completed	ALD-102/5 × 10^{6} cells/kg/ IV Administration	Gadolinium enhancement on MRI of demyelinating lesions/ males aged 17 years and younger/32 patients	Over a 6-year follow-up/ MRI	Not applicable	The majority of patients exhibited no significant functional disabilities	(Eichler et al., 2024)

Abbreviation: **AAVrh.10**: Adeno-associated virus serotype rh.10; ***ARSA***: Arylsulfatase A; **NAA**: N-acetylaspartate; **MLD**: Metachromatic leukodystrophy; **GALC**: Galactosylceramidase; **IV**: Intravenous; **ASPA**: Aspartoacylase; **MSEL**: Mullen Scales of Early Learning; **CD**: Canavan disease; **CSF**: Cerebrospinal fluid; **cALD:** Cerebral X-linked adrenoleukodystrophy; **KD**: Krabbe disease; **AE**: Adverse event; **SAE**: Serious adverse event; **VLCFA**: Very-long-chain fatty acids; **ICV**: Intracerebroventricular; **IC**: Intracerebral.

TABLE 8.3 Ongoing clinical trials in leukodystrophies.

Type of leukodystrophy	Vector, targeted gene	Phase/NCT number/sponsor/ study start/status	Product name/dose and administration route	Targeted patient population/participant number	Follow-up and monitoring
MLD	Lentiviral vector, *ARSA*	NCT03725670 Phase I/II/Shenzhen Geno-Immune Medical Institute/ Recruiting/2025	GIMI-IRB-18005/1–2 × 10^9 cell/kg/ IT and IV Administration	ARSA mutations /1 month to 50 years (child, adult)/10 patients	Up to 3-year follow-up
	Lentiviral vector, *ARSA*	NCT07046338 Phase I/II/Shenzhen Geno-Immune Medical Institute/ Recruiting/2025	GIMI-IRB-25001/1–10 × 10^6 cell/kg/ Lentiviral TYF-ARSA correction of the patient's autologous HSCs.	ARSA mutations/1 month to 50 years (child, adult)/10 patients	Up to 3-year follow-up/MRI
	Lentiviral vector, *ARSA*	NCT02559830 Phase I/II/ Shenzhen Second People's Hospital/ Recruiting/2015	ChiCTR-OPC-15005802/2–20 × 10^6 cell/kg Transduced CD34+ hematopoietic stem cell	Low ARSA activity (below 20% of the normal level)/1–16 years (child)/50 patients	Up to 8-year follow-up/MRI, ARSA activity.
	Lentiviral vector, *ARSA*	NCT04283227 Phase IIIA/Orchard Therapeutics/ Active, not recruiting/2022	OTL-200	Low ARSA activity and mutation in ARSA/child, adult, older adult	Up to 8-year follow-up/ARSA activity/NAA concentration/MRI
cALD	Lentiviral vector, ABCD*1*	NCT02698579/ Genetix Biotherapeutics Incorporated, active, not recruiting/2016	LTF-304	Individuals who have been administered the Lenti-D drug product in a parent clinical study/64 patients	Up to 13-year follow-up/MRI, behavioral testing
	Lentiviral vector, ABCD*1*	NCT02559830/ Shenzhen Second People's Hospital/ Unknown status/2015	ChiCTR-OPC-15005802/2–20 × 10^6 cell/kg Transduced CD34 + hematopoietic stem cell	Abnormally high level of VLCFA/1–16 years (child)/50 patients	Up to 8-year follow-up/MRI, VLCFA level
	Lentiviral vector, ABCD*1*	NCT03727555/ Shenzhen Geno-Immune Medical Institute/ Recruiting/2025	GIMI-IRB-18006/1–2 × 10^9/ IT and IV Administration	Abnormally high levels of VLCFA/1–60 years (child, adult)/30 patients	Up to 3-year follow-up/MRI, VLCFA level (biochemical and neurological analysis)
GLD	AAVrh.10-*hGALC*	NCT06308718/ Forge Biologics, Incorporated/ Terminated/2024	FBX-101/ IV Administration	Child, adult, older adult/2 patients	36-month follow-up
	AAV, HU 68 Serotype, *GALC*	NCT04771416 Phase I/II/Gemma Biotherapeutics/ Suspended/2022	PBKR03/1.5–5 × 10^8 ICM Administration	Presymptomatic or symptomatic, with the first symptoms of GLD, children aged 1–9 years old/24 patients	Up to 8-year follow-up
CD	AAV, *ASPA*	NCT05317780/ University of Florida/Not applicable	rAAV9-CB6-AspA/N/A/ IV and ICV Administration	Elevated brain NAA levels and/or 18 months to 24 months (child)	Not applicable

Type of leukodys-trophy	Vector, targeted gene	Phase/NCT number/sponsor/study start/status	Product name/dose and administration route	Targeted patient population/participant number	Follow-up and monitoring
TUBB4A-related Leukodys-trophy	ASO-*TUBB4A*	NCT06369974 Phase I/II/ Massachusetts General Hospital/Not yet recruiting/2024	ASO (2024P000386)	Mutation in TUBB4A/4 years and older (child, adult, older adult)/1 patient	1-year follow-up
	ASO-*TUBB4A*	NCT07222371 Phase I/II/University of California, San Diego/Active, not recruiting/2025	nL-TUBB4-001	Mutation in *TUBB4A*/child, adult, older adult /1 patient	1–2 follow-ups
AxD	ASO-*GFAP*	NCT04849741 Phase III/Ionis Pharmaceuticals Incorporated/Active not recruiting/2021	Zilganersen/IT Administration	Mutation in the *GFAP* gene/2–65 years (child, adult, older adult) /54 patients	61 weeks of follow-up

Abbreviation: AAV: Adeno-associated virus serotype rh.10; **ARSA**: Arylsulfatase A; **MLD**: Metachromatic leukodystrophy; **GALC**: Galactosylceramidase; **CD**: Canavan disease; **ASO**: Antisense oligonucleotide; **KD**: Krabbe disease; **IT**: Intrathecal; **IV**: Intravenous; **ICV**: Intracerebroventricular; **NAA**: N-acetylaspartate; VLCFA: Very-long-chain fatty acid; **GLD**: Globoid cell leukodystrophy; **cALD**: Cerebral adrenoleukodystrophy; **AxD**: Alexander disease; **MRI**: Magnetic resonance imaging; **ICM**: Intracisterna magna.

Recently, Ramachandran et al. introduced an engineered capsid called AAV.GMU01, which demonstrated improved biodistribution in the CNS and robust transgene expression in NHPs following minimally invasive intracerebral micro-injection. Unlike previous approaches using the AAVrh.10 vector, such as in the clinical trial NCT01801709 for MLD, AAV.GMU01 demonstrated near-regular *ARSA* activity without requiring extensive intracerebral delivery. These findings highlight the potential of capsid engineering to overcome previous translational challenges (Ramachandran et al., 2025a).

Preclinical findings show advances in AAV capsid engineering, CNS delivery, and *ARSA* restoration, forming a foundation for early human trials. Effective BBB penetration, sustained correction, and early intervention are key for gene therapy success in MLD.

6.2 Globoid cell leukodystrophy

Research on GLD has primarily focused on large-animal models, particularly the naturally occurring canine GLD model, which closely recapitulates infantile GLD in humans. A significant study by Bradbury et al. involved administering AAV9-GALC intrathecally, successfully preventing neurological deterioration and extending the lifespan of treated subjects to over 2.5 years, >7 times that of untreated controls. The study observed dose-dependent improvements, including increased *GALC* activity, normalized psychosine levels, enhanced myelination, and reduced inflammation throughout both the CNS and peripheral nervous system (Bradbury et al., 2020).

A subsequent study performed in 2024 by the same research group evaluated the effects of AAVrh10-GALC, both as an independent treatment and in combination with HSCT. The results indicated that HSCT alone did not significantly extend the lifespan of untreated GLD dogs, which is typically around 16 weeks. However, when HSCT was combined with AAVrh10 treatment, disease progression was delayed, and the dogs' survival was extended beyond 1 year. Furthermore, a fivefold increase in the AAVrh10 dosage maintained near-normal neurological function in dogs up to 2 years of age. Both peripheral nerve conduction velocity and morphology, as well as *GALC* enzyme activity and psychosine levels, were normalized in dogs receiving AAVrh10 either alone or in combination with HSCT (Bradbury et al., 2024).

In investigations using mouse models, Herdt et al. administered AAV1-GALC directly into the brains of newborn twitcher mice via ventricular injection. This treatment led to improved weight gain, increased survival rates, and complete normalization of psychosine levels in the forebrain. These findings emphasize the therapeutic importance of early CNS restoration of *GALC* in slowing down disease progression (Herdt et al., 2023).

Preclinical studies in GLD show that restoring *GALC* activity via AAV9, AAVrh10, or ICV delivery can normalize psychosine toxicity, preserve myelination, and extend survival in mice and large animals. These findings suggest that combining AAV-based CNS delivery with HSCT could overcome rapid neurodegeneration in infantile GLD.

6.3 Canavan disease

Preclinical development for CD has increasingly concentrated on optimizing AAV vectors that specifically target oligodendrocytes and can restore ASPA function across the CNS. A significant area of investigation has focused on AAV/Olig001, a specifically engineered capsid variant known for its strong affinity for oligodendrocytes.

In a nur7 CD mouse model, which features a nonsense mutation in the murine *ASPA* gene that leads to spongiform degeneration, elevated NAA levels, and early motor decline, researchers systematically compared various administration routes: intraparenchymal, intrathecal, ICV, and intracisterna magna, using an AAV/Olig001-GFP reporter. Among these methods, ICV delivery yielded the most extensive distribution throughout the CNS and achieved robust oligodendrocyte transduction, with oligotropism exceeding 70% in nearly all examined brain regions. Dose escalation testing (ranging from 1×10^{10} to 1×10^{11} viral genomes) was conducted in symptomatic 6-week-old nur7 mice. The results indicated a restoration of *ASPA* activity, significant reductions in CNS vacuolation, and substantial improvement in motor performance. These findings highlight the therapeutic potential of early oligodendrocyte-targeted gene replacement for treating CD (Francis et al., 2021).

A complementary study was developed based on these findings, evaluating the metabolic and myelination effects of AAV-Olig001 delivering the *ASPA* transgene (AAV-Olig001-ASPA). In this study, homozygous nur7 animals were genotyped using single nucleotide polymorphism (SNP)-based qPCR. The mice received four intraparenchymal injections, each containing 4×10^9 viral genomes. Quantification of brain metabolites showed a twofold reduction in NAA levels compared with GFP-treated and naïve mutant controls, confirming functional recovery of the *ASPA* enzyme. These metabolic improvements were associated with increased availability of acetyl-CoA, restoration of energetic homeostasis, and significant enhancements in daily activity and overall neurological function (Francis et al., 2016).

Additional preclinical research has utilized a unique murine model, the *ASPA*-deficient lacZ-knock-in mice (AKO) This model expresses bacterial β-galactosidase under the regulatory elements of *ASPA* and mimics key pathological features of CD, including progressive myelin loss, elevated NAA levels, astrogliosis, and motor deficits. In this model, intracranial administration of 3×10^9 vg of AAV2 encoding a codon-optimized human *ASPA* sequence (enhanced with a Kozak motif) resulted in significant and lasting phenotypic improvement over 12 weeks. The treated animals showed correction of metabolic abnormalities, as evidenced by proton magnetic resonance spectroscopy, reduced vacuolation observed via MRI, and improved motor performance on rotarod testing. Importantly, this study provided substantial evidence that late-stage intervention can reverse established pathology, including body weight loss, demyelination, and neural inflammation, thus highlighting a therapeutic window for *ASPA* restoration strategies (Fröhlich et al., 2022).

Overall, these preclinical investigations establish a compelling rationale for oligodendrocyte-targeted AAV gene replacement as a key therapeutic approach in CD. The consistent recovery in metabolic, histopathological, and functional measures across the nur7 and AKO models emphasizes the translational potential of these vectors and supports ongoing efforts toward clinical application.

6.4 X-linked adrenoleukodystrophy

Understanding the molecular basis of X-ALD and evaluating therapeutic approaches require robust cellular and animal models that accurately replicate the disease-specific metabolic and pathological features.

A key preclinical study conducted by Benhamida et al. showed that lentiviral transduction of hematopoietic stem and progenitor cells (HSPCs) from X-ALD patients can achieve stable, long-term expression of the ALDP after transplantation into immunodeficient mice. $CD34^+$ cells were mobilized from X-ALD patients using granulocyte colony-stimulating factor and then transduced with a lentiviral vector encoding the human ALDP at a density of 1×10^6 cells/mL in a serum-free medium. The expression of ALDP in the transduced hematopoietic cells was confirmed through immunocytochemistry using an anti-human ALD polyclonal antibody that does not cross-react with the murine protein.

A total of 1.5×10^6 transduced $CD34^+$ cells were then transplanted intravenously into sublethally irradiated NOD-LtSz-scid/scid (NOD/SCID) mice. VLCFA levels were measured in hematopoietic lineages using gas/liquid chromatography combined with isotope-dilution mass spectrometry. The study demonstrated sustained expression of ALDP in monocytes and macrophages derived from engrafted, patient-derived HSPCs, along with measurable biochemical correction. These findings provided early proof-of-concept for HSC-based gene therapy in X-ALD and laid the groundwork for later clinical translation (Benhamida et al., 2003).

6.5 Alexander disease

AxD is caused by autosomal dominant gain-of-function mutations in the *GFAP* gene. This leads to pathological overexpression and accumulation of GFAP protein, resulting in the formation of Rosenthal fibers, astrocyte dysfunction, white matter abnormalities, and progressive neurological decline. Since AxD is driven by toxic gain-of-function rather than a loss of protein activity, gene-silencing strategies, particularly ASOs, have emerged as a promising preclinical therapeutic approach.

A key study conducted by Hagemann et al. assessed a series of ASOs specifically designed to target and suppress the murine GFAP transcript in various AxD mouse models. Following ICV administration, ASO treatment led to near-complete and sustained depletion of GFAP protein throughout the brain and spinal cord. This substantial suppression also resulted in the removal of *GFAP* from the CSF, supporting the potential use of CSF *GFAP* as a sensitive pharmacodynamic biomarker for future clinical applications.

At the cellular level, the ASO-mediated knockdown of *GFAP* prevented the formation of Rosenthal fibers, decreased astrocytic stress responses, and normalized inflammatory markers. Notably, the treatment produced functional improvements, including enhanced overall body condition and restoration of hippocampal neurogenesis, suggesting a reversal of downstream pathogenic processes rather than merely reducing *GFAP* levels (Hagemann et al., 2018).

These findings confirm ASO-mediated *GFAP* suppression as one of the most advanced and well-validated preclinical strategies for treating AxD. In addition, this research has paved the way for the first human ASO program (ION373/zilganersen), which is currently in clinical trials. This highlights the promising potential of gene-targeted therapies for this astrocytopathy. These studies collectively emphasize the potential of gene therapy strategies in treating leukodystrophies, as summarized in Table 8.4.

7 Current challenges and limitations in gene therapy for leukodystrophies

Gene therapy for leukodystrophies faces several biological, technical, and translational challenges that complicate the development of safe and effective treatments. These disorders primarily affect the cerebral white matter and have multiple genetic causes, resulting in significant heterogeneity. This diversity makes it difficult to employ a single therapeutic strategy for all leukodystrophy subtypes (Aubourg, 2016).

One of the most well-known challenges is the efficient delivery of therapeutic constructs into the CNS. Physical and physiological barriers, such as the skull, the BBB, and the cerebrospinal fluid-brain barrier, limit the ability of large biomolecules and viral vectors to penetrate these areas (Aerts-Kaya and van Til, 2023). The BBB tightly regulates the endothelial interface, which is crucial for maintaining neural homeostasis, and also restricts the effectiveness of systemically administered vectors. This presents a significant obstacle to widespread gene transfer within the CNS (Piguet et al., 2021).

Limitations associated with viral vectors also hinder therapeutic advancements. Current viral vectors differ significantly in their ability to target specific cells (tropism), their distribution throughout the body (biodistribution), and their potential to provoke immune responses (immunogenicity). These differences can result in inconsistent therapeutic outcomes, challenges in targeting relevant cell populations, and inadequate transgene expression. Furthermore, the toxicities that depend on the dose, including inflammatory reactions and localized neurological changes, raise ongoing safety concerns, especially since higher doses of vectors are often required to ensure adequate coverage of the CNS (Rosenberg et al., 2021; Hordeaux et al., 2022).

In addition to biological limitations, the economics of gene therapy present significant challenges to its implementation. While single-dose gene therapies have the potential to terminate years of progressive neurological deterioration, their high upfront costs raise concerns regarding repayment and impartial access for patients (Salzman et al., 2018). Without strong financial support mechanisms in place, the transformative benefits of gene therapy are unlikely to reach all suitable patients.

Ultimately, differences between animal models and human neurobiology can undermine the reliability of predictions made in preclinical studies. Variations in CNS size, cellular composition, vector tropism, and immune responses complicate the process of applying safety and efficacy results to human patients. These translational gaps highlight the persistent delay involved in progressing from preclinical studies to clinical trials.

8 Conclusion

Over the past two decades, gene therapy for leukodystrophies has evolved from an experimental idea to a potential treatment. Advances in vector design have driven this progress, improved strategies for delivering treatments to the CNS, and enabled the development of more effective preclinical models. As a result of these developments, two autologous HSC-based gene therapy products have been approved: atidarsagene autotemcel for MLD and elivaldogene autotemcel for cerebral X-ALD. These approvals demonstrate the feasibility of correcting specific subtypes of leukodystrophy through gene therapy.

Despite these advances, important biological, technical, and economic challenges remain, including efficient delivery across the blood–brain barrier, sustained and cell-specific transgene expression, long-term safety, and the high cost of individualized therapies. Continued progress in capsid engineering, genome editing, precision targeting, and immune-evasive

TABLE 8.4 Preclinical studies in leukodystrophies.

Type of diseases	Animal model	Targeted genes and vectors	Administration root/ dose	Safety/toxicity and efficacy	Follow-up and monitoring	Results	Refs.
MLD	MLD Mice (C57BLy6J)/ NHPs	AAV9, AAVHSC15 *ARSA*	IV/IT Administration (.5 to 6E + 13 vg/kg)	Efficacy: *ARSA* activity levels increase from 1 week after dosing and remain elevated through 52 weeks postdose. Methods: A colorimetric enzyme assay and mass spectrometry for quantifying ARSA protein	1, 4, 12, and 52 weeks/*ARSA* level/ behavioral testing	Enhanced *ARSA* activity ($p \leq 0.05$ at low dose), prevention of motor impairment ($p < 0.01$), and normalization of disease biomarkers.	(Martin et al., 2023)
	Early- and late-treated MLD mice (6 or 9 months old) Fertilized oocytes from Bl6 wild-type mice	AAVPHP.Eb, *ARSA*	IV Administration (2.5×10^{11} to 5×10^{11} vg)	Efficacy: Persistent ARSA activity in the cerebral cortex and brain Methods: spectrometry for ARSA protein quantitation	3 or 6 months after injection/*ARSA* level	Enhancement in neuroinflammation at minimal dosage ($p < 0.01$) and delayed intervention. Suggesting a Phase I/II clinical trial in patients with MLD.	(Audouard et al., 2024)
	NHP, adult Chlorocebus aethiops sabaeus monkey (2- to 4-year-old males and females)	AAVrh.10, *ARSA*	IP Administration (2.85×10^{10} to 1.5×10^{12} gc/site)	Safe in a low dose of AAVrh.10 h ARSA and localized CNS abnormality at a high dose	3 months after injection/ behavioral testing and MRI	No clinical consequences.	(Rosenberg et al., 2021)
	MLD mice (6 weeks and 1 year old C57BL/6J)	AAV9, *ARSA*	IT Administration (4.0×10^{11} vg)	Safe and efficient	3 or 13.5 months after the injection/ARSA level and behavioral testing	Enhance *ARSA* expression in the cerebellum and spinal cord. Decrease in sulfatide storage and sulfatide/*GALC* ratios within the CNS ($p < 0.01$). Notable enhancement in a behavioral assessment ($p < 0.03$).	(Miyake et al., 2021)
GLD	Twitcher mouse model	AAV1, *GALC*	ICV Administration (6.0×10^{10} particles/ dose)	Safe and efficient Increased median lifespan	78-day follow-up/survival rate and psychosine levels	Mitigate the accumulation of psychosine ($p < 0.0001$).	(Herdt et al., 2023)
	GLD dog (missense mutation in *GALC*/4 weeks to 2 years old)	AAVrh10, *GALC*	IV Administration	Safe and efficient/translation to clinical trials (NCT04693598, NCT05739643)	2-year follow-up/survival rate, *GALC* level, and psychosine levels	Increase average survival, neurological function, GALC enzyme activity, and reduce psychosine levels.	(Bradbury et al., 2024)
	Twitcher mice (C57BL/6 W339X mutation in the *GALC* gene)	AAVrh10, *mGALC*	IV Administration (4×10^{13} gc/kg)	Enhancing *GALC* activity through a fourfold increase in viral vector administration	500-day follow-up/survival rate and *GALC* level	Enhance the survival rate ($P < 0.0001$) and increased *GALC* enzyme activity.	(Rafi et al., 2020)
	GLD dog (missense mutation in the *GALC* gene)	AAV9, *cGALC*	IT Administration (High dose: 1×10^{14} vector genomes, low dose: 2×10^{13})		2.5-year follow-upMRI, GALC level, concentrations of CSF psychosine, galactosylceramide, and total protein	Enhance *GALC* enzyme activity ($P = 0.02$), Restore psychosine levels ($P < 0.05$), facilitate myelination, and reduce inflammation in the CNS and PNS.	(Bradbury et al., 2020)

CD	Mice (C57BL/6J, nur7 mouse/ nonsense point mutation in the *ASPA* gene/male and female)	AAV Olig001, *ASPA*	ICV Administration (1×10^{10}, 1×10^{11}, and 1×10^{10} total vector genomes)	The administration of AAV/ Olig001-GFP achieved over 70% oligotropism across all targeted regions/dose-dependent of ASPA activity	4-month follow-up/*ASPA* level and behavioral testing	Dose-dependent restoration of *ASPA* activity, improvement in motor function.	(Francis et al., 2021)
	Mice (homozygous nur7 mouse)	AAV-Olig001, *ASPA*	IP Administration (4 injections at a dose of 4×10^9 vg per animal)	Safe and efficient	1-week follow-up/ behavioral testing, NAA, acetyl-CoA concentration, and daily activity	Notable two-fold decrease in NAA ($p < 0.001$). Improvement in daily activity ($p = 0.00021$).	(Francis et al., 2016)
	Mice (*ASPA* deficiency, lacZ knock-in)	AAV2, *ASPA*	Intracranial Administration (3×10^9 vg)	Safe and efficient	12-week follow-up/ MRI, behavioral testing, H-MR spectroscopy, and N-acetyl-L-aspartate level	Reduction in brain vacuoles, increased levels of N-acetyl-L-aspartate, astrogliosis, and demyelination.	(Fröhlich et al., 2022)
AxD	Mice (FVB/N point mutations at the *GFAP* locus)	ASO targeting *GFAP*	ICV Administration (10 μL of 100 mg/mL)	Efficacy: Luciferase assay/Western analysis An extended and extensive decrease in protein levels was observed 16 weeks following ICV	16-week follow-up/*GFAP* level	Suppress *GFAP* expression, enhance physical condition, and restore hippocampal neurogenesis.	(Hagemann et al., 2018)

Abbreviation: **MLD**: Metachromatic leukodystrophy; ***ARSA***: Arylsulfatase A; **HSGT**: Hematopoietic stem cell gene therapy; **FIM**: Functional Independence Measure; **MRI**: Magnetic resonance imaging; **NHP**: nonhuman primates; **GLD**: Globoid cell leukodystrophy; ***GALC***: Galactocerebrosidase; **CNS**: Central nervous system; **IV**: Intravenous; **IT**: Intrathecal; **PNS**: Peripheral nervous system; **ICV**: Intracerebroventricular; **IP**: Intraparenchymal; ***ASPA***: Aspartoacylase; ***GFAP***: Glial fibrillary acidic protein.

vector platforms is expected to overcome many of these limitations and expand treatment to a broader spectrum of leukodystrophies.

Collectively, these advances position gene therapy as one of the most promising disease-modifying strategies for inherited white matter disorders and are likely to redefine their clinical management in the coming years.

References

A, I.A., Hadera, M.G., Kotter, M., Sonnewald, U., 2017. Oligodendrocytes Do Not Export NAA-Derived Aspartate In Vitro. Neurochem. Res. 42 (3), 827–837.

Aerts-Kaya, F., van Til, N.P., 2023. Gene and Cellular Therapies for Leukodystrophies. Pharmaceutics 15 (11), 2522.

Allewelt, H., Taskindoust, M., Troy, J., Page, K., Wood, S., Parikh, S., Prasad, V.K., Kurtzberg, J., 2018. Long-Term Functional Outcomes after Hematopoietic Stem Cell Transplant for Early Infantile Krabbe Disease. Biol. Blood Marrow Transplant. 24 (11), 2233–2238.

Ashrafi, M.R., Amanat, M., Garshasbi, M., Kameli, R., Nilipour, Y., Heidari, M., Rezaei, Z., Tavasoli, A.R., 2020. An update on clinical, pathological, diagnostic, and therapeutic perspectives of childhood leukodystrophies. Expert Rev. Neurother. 20 (1), 65–84.

Aubourg, P., 2016. Gene therapy for leukodystrophy: progress, challenges and opportunities. Expert Opin. Orphan Drugs 4 (4), 359–367.

Audouard, E., Chereau, F., Lupiet, C., Mansat, C., Roger, L., Libert, P., Banchi, E.G., Khefif, N., Farabos, D., Lamaziere, A., Colle, M.A., Folcher, M., Sevin, C., Piguet, F., 2025. Encapsulated cells as an enzyme replacement therapy for metachromatic leukodystrophy. J. Control. Release 388 (Pt 1), 114307.

Audouard, E., Khefif, N., Mansat, C., Nelcha, O., Banchi, E.G., Lupiet, C., Farabos, D., Lamaziere, A., Sevin, C., Piguet, F., 2024. Dose-response evaluation of intravenous gene therapy in a symptomatic mouse model of metachromatic leukodystrophy. Mol. Ther. Methods Clin. Dev. 32 (2), 101248.

Bahrambeigi, V., Song, X., Sperle, K., Beck, C.R., Hijazi, H., Grochowski, C.M., Gu, S., Seeman, P., Woodward, K.J., Carvalho, C.M.B., Hobson, G.M., Lupski, J.R., 2019. Distinct patterns of complex rearrangements and a mutational signature of microhomeology are frequently observed in PLP1 copy number gain structural variants. Genome Med. 11 (1), 80.

Barth, M.L., Fensom, A., Harris, A., 1994. The arylsulphatase A gene and molecular genetics of metachromatic leucodystrophy. J. Med. Genet. 31 (9), 663–666.

Benhamida, S., Pflumio, F., Dubart-Kupperschmitt, A., Zhao-Emonet, J.C., Cavazzana-Calvo, M., Rocchiccioli, F., Fichelson, S., Aubourg, P., Charneau, P., Cartier, N., 2003. Transduced CD34+ cells from adrenoleukodystrophy patients with HIV-derived vector mediate long-term engraftment of NOD/SCID mice. Mol. Ther. 7 (3), 317–324.

Berger, J., Pujol, A., Aubourg, P., Forss-Petter, S., 2010. Current and future pharmacological treatment strategies in X-linked adrenoleukodystrophy. Brain Pathol. 20 (4), 845–856.

Biffi, A., Montini, E., Lorioli, L., Cesani, M., Fumagalli, F., Plati, T., Baldoli, C., Martino, S., Calabria, A., Canale, S., Benedicenti, F., Vallanti, G., Biasco, L., Leo, S., Kabbara, N., Zanetti, G., Rizzo, W.B., Mehta, N.A., Cicalese, M.P., Casiraghi, M., Boelens, J.J., Carro, U.D., Dow, D.J., Schmidt, M., Assanelli, A., Neduva, V., Di Serio, C., Stupka, E., Gardner, J., von Kalle, C., Bordignon, C., Ciceri, F., Rovelli, A., Roncarolo, M.G., Aiuti, A., Sessa, M., Naldini, L., 2013. Lentiviral hematopoietic stem cell gene therapy benefits metachromatic leukodystrophy. Science 341 (6148), 1233158.

Bokhari, M.R., Samanta, D., Bokhari, S.R.A., 2026. Canavan Disease. StatPearls. Treasure Island (FL) Ineligible Companies. Disclosure: Debopam Samanta Declares No Relevant Financial Relationships with Ineligible Companies. Disclosure: Syed Rizwan Bokhari Declares No Relevant Financial Relationships with Ineligible Companies., StatPearls Publishing. StatPearls Publishing LLC. Copyright © 2025.

Bonkowsky, J.L., Keller, S.,; AAP Section on Neurology, Council on Genetics (2021), 2021. Leukodystrophies in Children: diagnosis, Care, and Treatment. Pediatrics 148 (3), e2021053126. doi:10.1542/peds.2021-053126.

Bonkowsky, J.L., Nelson, C., Kingston, J.L., Filloux, F.M., Mundorff, M.B., Srivastava, R., 2010. The burden of inherited leukodystrophies in children. Neurology 75 (8), 718–725.

Bradbury, A.M., Bagel, J.H., Nguyen, D., Lykken, E.A., Salvador, J.P., Jiang, X., Swain, G.P., Assenmacher, C.A., Hendricks, I.J., Miyadera, K., Hess, R.S., Ostrager, A., P, O.D., Sands, M.S., Ory, D.S., Shelton, G.D., Bongarzone, E.R., Gray, S.J., Vite, C.H., 2020. Krabbe disease successfully treated via monotherapy of intrathecal gene therapy. J. Clin. Invest. 130 (9), 4906–4920.

Bradbury, A.M., Bagel, J., Swain, G., Miyadera, K., Pesayco, J.P., Assenmacher, C.A., Brisson, B., Hendricks, I., Wang, X.H., Herbst, Z., Pyne, N., Odonnell, P., Shelton, G.D., Gelb, M., Hackett, N., Szabolcs, P., Vite, C.H., Escolar, M., 2024. Combination HSCT and intravenous AAV-mediated gene therapy in a canine model proves pivotal for translation of Krabbe disease therapy. Mol. Ther. 32 (1), 44–58.

Bradbury, A.M., Bongarzone, E.R., Sands, M.S., 2021. Krabbe disease: new hope for an old disease. Neurosci. Lett. 752, 135841.

C, I.D., Sevin, C., Krageloh-Mann, I., Giugliani, R., Sakai, N., Wu, J., Wasilewski, M., 2020. Safety of intrathecal delivery of recombinant human arylsulfatase A in children with metachromatic leukodystrophy: results from a phase 1/2 clinical trial. Mol. Genet. Metab. 131 (1-2), 235–244.

Cabre-Romans, J.J., Cuella-Martin, R., 2025. CRISPR-dependent base editing as a therapeutic strategy for rare monogenic disorders. Front. Genome Ed. 7, 1553590.

Cartier, N., Hacein-Bey-Abina, S., Bartholomae, C.C., Bougneres, P., Schmidt, M., Kalle, C.V., Fischer, A., Cavazzana-Calvo, M., Aubourg, P., 2012. Lentiviral hematopoietic cell gene therapy for X-linked adrenoleukodystrophy. Methods Enzymol. 507, 187–198.

Cartier, N., Hacein-Bey-Abina, S., Bartholomae, C.C., Veres, G., Schmidt, M., Kutschera, I., Vidaud, M., Abel, U., Dal-Cortivo, L., Caccavelli, L., Mahlaoui, N., Kiermer, V., Mittelstaedt, D., Bellesme, C., Lahlou, N., Lefrere, F., Blanche, S., Audit, M., Payen, E., Leboulch, P., l'Homme, B., Bougneres, P., Kalle, C.V., Fischer, A., Cavazzana-Calvo, M., Aubourg, P., 2009. Hematopoietic stem cell gene therapy with a lentiviral vector in X-linked adrenoleukodystrophy. Science 326 (5954), 818–823.

Cesani, M., Lorioli, L., Grossi, S., Amico, G., Fumagalli, F., Spiga, I., Filocamo, M., Biffi, A., 2016. Mutation Update of ARSA and PSAP Genes Causing Metachromatic Leukodystrophy. Hum. Mutat. 37 (1), 16–27.

Chang, S.C., Bergamasco, A., Bonnin, M., Bisono, T.A., Moride, Y., 2024. A systematic review on the birth prevalence of metachromatic leukodystrophy. Orphanet J. Rare Dis. 19 (1), 80.

Chanoumidou, K., Mozafari, S., Baron-Van Evercooren, A., Kuhlmann, T., 2020. Stem cell derived oligodendrocytes to study myelin diseases. Glia 68 (4), 705–720.

Charzewska, A., Wierzba, J., Izycka-Swieszewska, E., Bekiesinska-Figatowska, M., Jurek, M., Gintowt, A., Klosowska, A., Bal, J., Hoffman-Zacharska, D., 2016. Hypomyelinating leukodystrophies - a molecular insight into the white matter pathology. Clin. Genet. 90 (4), 293–304.

Chen, Y.Q., Rafi, M.A., de Gala, G., Wenger, D.A., 1993. Cloning and expression of cDNA encoding human galactocerebrosidase, the enzyme deficient in globoid cell leukodystrophy. Hum. Mol. Genet. 2 (11), 1841–1845.

Chery, J., 2016. RNA therapeutics: rNAi and antisense mechanisms and clinical applications. Postdoc J. 4 (7), 35–50.

Compston, A., 2013. A new familial infantile form of diffuse brain-sclerosis. Brain 136 (Pt 9), 2649–2651.

Coulombe, B., Derksen, A., Piana, R.L., Brais, B., Gauthier, M.S., Bernard, G., 2021. POLR3-related leukodystrophy: how do mutations affecting RNA polymerase III subunits cause hypomyelination? Fac. Rev. 10, 12.

Daoud, H., Tetreault, M., Gibson, W., Guerrero, K., Cohen, A., Gburek-Augustat, J., Synofzik, M., Brais, B., Stevens, C.A., Sanchez-Carpintero, R., Goizet, C., Naidu, S., Vanderver, A., Bernard, G., 2013. Mutations in POLR3A and POLR3B are a major cause of hypomyelinating leukodystrophies with or without dental abnormalities and/or hypogonadotropic hypogonadism. J. Med. Genet. 50 (3), 194–197.

de Rosa, V., Secondo, A., Pannaccione, A., Ciccone, R., Formisano, L., Guida, N., Crispino, R., Fico, A., Polishchuk, R., D'Aniello, A., Annunziato, L., Boscia, F., 2019. D-Aspartate treatment attenuates myelin damage and stimulates myelin repair. EMBO Mol. Med. 11 (1), e9278.

DeLuca, C., Brown, J.A., Shows, T.B., 1979. Lysosomal arylsulfatase deficiencies in humans: chromosome assignments for arylsulfatase A and B. Proc. Natl. Acad. Sci. USA 76 (4), 1957–1961.

Dorboz, I., Dumay-Odelot, H., Boussaid, K., Bouyacoub, Y., Barreau, P., Samaan, S., Jmel, H., Eymard-Pierre, E., Cances, C., Bar, C., Poulat, A.L., Rousselle, C., Renaldo, F., Elmaleh-Berges, M., Teichmann, M., Boespflug-Tanguy, O., 2018. Mutation in POLR3K causes hypomyelinating leukodystrophy and abnormal ribosomal RNA regulation. Neurol. Genet. 4 (6), e289.

Dubois-Dalcq, M., Feigenbaum, V., Aubourg, P., 1999. The neurobiology of X-linked adrenoleukodystrophy, a demyelinating peroxisomal disorder. Trends Neurosci. 22 (1), 4–12.

Dulski, J., Heckman, M.G., White, L.J., Zur-Wyrozumska, K., Lund, T.C., Wszolek, Z.K., 2022. Hematopoietic Stem Cell Transplantation in CSF1R-Related Leukoencephalopathy: retrospective Study on Predictors of Outcomes. Pharmaceutics 14 (12), 2778.

Duncan, C.N., Bledsoe, J.R., Grzywacz, B., Beckman, A., Bonner, M., Eichler, F.S., Kuhl, J.S., Harris, M.H., Slauson, S., Colvin, R.A., Prasad, V.K., Downey, G.F., Pierciey, F.J., Kinney, M.A., Foos, M., Lodaya, A., Floro, N., Parsons, G., Dietz, A.C., Gupta, A.O., Orchard, P.J., Thakar, H.L., Williams, D.A., 2024. Hematologic Cancer after Gene Therapy for Cerebral Adrenoleukodystrophy. N. Engl. J. Med. 391 (14), 1287–1301.

Eichler, F., Duncan, C., Musolino, P.L., Orchard, P.J., De Oliveira, S., Thrasher, A.J., Armant, M., Dansereau, C., Lund, T.C., Miller, W.P., Raymond, G.V., Sankar, R., Shah, A.J., Sevin, C., Gaspar, H.B., Gissen, P., Amartino, H., Bratkovic, D., Smith, N.J.C., Paker, A.M., Shamir, E., O'Meara, T., Davidson, D., Aubourg, P., Williams, D.A., 2017. Hematopoietic Stem-Cell Gene Therapy for Cerebral Adrenoleukodystrophy. N. Engl. J. Med. 377 (17), 1630–1638.

Eichler, F., Duncan, C.N., Musolino, P.L., Lund, T.C., Gupta, A.O., De Oliveira, S., Thrasher, A.J., Aubourg, P., Kuhl, J.S., Loes, D.J., Amartino, H., Smith, N., Fernandes, J.F., Sevin, C., Sankar, R., Hussain, S.A., Gissen, P., Dalle, J.H., Platzbecker, U., Downey, G.F., McNeil, E., Demopoulos, L., Dietz, A.C., Thakar, H.L., Orchard, P.J., Williams, D.A., 2024. Lentiviral Gene Therapy for Cerebral Adrenoleukodystrophy. N. Engl. J. Med. 391 (14), 1302–1312.

Eichler, F., Nagy, A., Laforet, G., Burton, C., Kirby, K.M., Kinane, B., Townsend, E., Kiefer, M., Leiro, B., Williams, R., 2023. Initial Biomarker and Clinical Findings from the CANaspire Canavan Disease Gene Therapy Trial: Exploration of Connections between NAA and Disease Severity. 26th Annual Meeting of the American Society of Gene and Cell Therapy (ASGCT) May 2023. Los Angeles, CA, USA.

Elitt, M.S., Tesar, P.J., 2024. Pelizaeus-Merzbacher disease: on the cusp of myelin medicine. Trends Mol. Med. 30 (5), 459–470.

Engelen, M., Kemp, S., de Visser, M., van Geel, B.M., Wanders, R.J., Aubourg, P., Poll-The, B.T., 2012. X-linked adrenoleukodystrophy (X-ALD): clinical presentation and guidelines for diagnosis, follow-up and management. Orphanet J. Rare Dis. 7, 51.

Escolar, M.L., Poe, M.D., Provenzale, J.M., Richards, K.C., Allison, J., Wood, S., Wenger, D.A., Pietryga, D., Wall, D., Champagne, M., Morse, R., Krivit, W., Kurtzberg, J., 2005. Transplantation of umbilical-cord blood in babies with infantile Krabbe's disease. N. Engl. J. Med. 352 (20), 2069–2081.

Escolar, M., Lugt, M.V., Poe, M., Greco, M., Werling, K., De Silva, E., Ruiz, J., Szabolcs, P., 2023. P014: first-in-human phase 1/2 trial of intravenous FBX-101 with cord blood transplantation increases GALC, brain and motor development in infantile Krabbe. Genet. Med. Open 1 (1),100363.

Escolar, M., Lugt, M.V., Wang, R., Ruiz, J., Poe, M., Szabolcs, P., 2024. O30: REKLAIM: a firstin-human phase Ib clinical trial of FBX-101 (AAVrh10. GALC) intravenously administered after UCBT for infantile Krabbe disease. Genet. Med. Open 2, 100875.

Fahim, S.M., Lin, G., Suh, K., Carlson, J.J., Richardson, M., Herce-Hagiwara, B., Dickerson, R., Pearson, S.D., Rind, D.M., Agboola, F., 2024. Atidarsagene autotemcel for metachromatic leukodystrophy. J. Manag. Care Spec. Pharm. 30 (2), 201–205.

Francis, J.S., Markov, V., Wojtas, I.D., Gray, S., McCown, T., Samulski, R.J., Figueroa, M., Leone, P., 2021. Preclinical biodistribution, tropism, and efficacy of oligotropic AAV/Olig001 in a mouse model of congenital white matter disease. Mol. Ther. Methods Clin. Dev. 20, 520–534.

Francis, J.S., Wojtas, I., Markov, V., Gray, S.J., McCown, T.J., Samulski, R.J., Bilaniuk, L.T., Wang, D.J., De Vivo, D.C., Janson, C.G., Leone, P., 2016. N-acetylaspartate supports the energetic demands of developmental myelination via oligodendroglial aspartoacylase. Neurobiol. Dis. 96, 323–334.

Fröhlich, D., Kalotay, E., von Jonquieres, G., Bongers, A., Lee, B., Suchowerska, A.K., Housley, G.D., Klugmann, M., 2022. Dual-function AAV gene therapy reverses late-stage Canavan disease pathology in mice. Front. Mol. Neurosci. 15, 1061257.

Fukutake, T., 2011. Cerebral autosomal recessive arteriopathy with subcortical infarcts and leukoencephalopathy (CARASIL): from discovery to gene identification. J. Stroke Cerebrovasc. Dis. 20 (2), 85–93.

Fumagalli, F., Calbi, V., Gallo, V., Zambon, A.A., Recupero, S., Ciotti, F., Sarzana, M., Fraschini, M., Scarparo, S., De Mattia, F., Miglietta, S., Pierini, C., Soncini, M., Morena, F., Montini, E., Barzaghi, F., Consiglieri, G., Ferrua, F., Migliavacca, M., Tucci, F., Fratini, E.S., Ippolito, A., Silvani, P., Calvi, M.R., Clerici, A., Corti, A., Facchini, M., Locatelli, S., Sangalli, M., Zancan, S., Miotto, F., Sora, M.G.N., Baldoli, C., Martino, S., Cordoba-Claros, A., Moro, S.L., Gollop, N.D., Abate, J., Yarzi, M.N., Nutkins, P., Shenker, A., Calissano, M., Brooks, J., Richardson, A., Campbell, L., Filippi, M., Naldini, L., Cicalese, M.P., Ciceri, F., Bernardo, M.E., Aiuti, A., 2025. Long-Term Effects of Atidarsagene Autotemcel for Metachromatic Leukodystrophy. N. Engl. J. Med. 392 (16), 1609–1620.

Fumagalli, F., Calbi, V., Sora, M.G.N., Sessa, M., Baldoli, C., Rancoita, P.M.V., Ciotti, F., Sarzana, M., Fraschini, M., Zambon, A.A., Acquati, S., Redaelli, D., Attanasio, V., Miglietta, S., De Mattia, F., Barzaghi, F., Ferrua, F., Migliavacca, M., Tucci, F., Gallo, V., Carro, U.D., Canale, S., Spiga, I., Lorioli, L., Recupero, S., Fratini, E.S., Morena, F., Silvani, P., Calvi, M.R., Facchini, M., Locatelli, S., Corti, A., Zancan, S., Antonioli, G., Farinelli, G., Gabaldo, M., Garcia-Segovia, J., Schwab, L.C., Downey, G.F., Filippi, M., Cicalese, M.P., Martino, S., Di Serio, C., Ciceri, F., Bernardo, M.E., Naldini, L., Biffi, A., Aiuti, A., 2022. Lentiviral haematopoietic stem-cell gene therapy for early-onset metachromatic leukodystrophy: long-term results from a non-randomised, open-label, phase 1/2 trial and expanded access. Lancet 399 (10322), 372–383.

Fumagalli, F., Zambon, A.A., Rancoita, P.M.V., Baldoli, C., Canale, S., Spiga, I., Medaglini, S., Penati, R., Facchini, M., Ciotti, F., Sarzana, M., Lorioli, L., Cesani, M., Sora, M.G.N., Carro, U.D., Cugnata, F., Antonioli, G., Recupero, S., Calbi, V., Di Serio, C., Aiuti, A., Biffi, A., Sessa, M., 2021. Metachromatic leukodystrophy: a single-center longitudinal study of 45 patients. J. Inherit. Metab. Dis. 44 (5), 1151–1164.

Gangji, R.N., Ali, Q.A., Dhamija, R., Corado, A.M., Kharbanda, S., A. T. C. *documents@acmg.net, 2025. Lenmeldy (atidarsagene autotemcel) for individuals with early metachromatic leukodystrophy (MLD): a therapeutics bulletin of the American College of Medical Genetics and Genomics (ACMG). Genet. Med. Open 3, 103432.

Gieselmann, V., 2008. Metachromatic leukodystrophy: genetics, pathogenesis and therapeutic options. Acta Paediatr. 97 (457), 15–21.

Gordon-Lipkin, E., Fatemi, A., 2018. Current Therapeutic Approaches in Leukodystrophies: a Review. J. Child Neurol. 33 (13), 861–868.

Groeschel, S., Kuhl, J.S., Bley, A.E., Kehrer, C., Weschke, B., Doring, M., Bohringer, J., Schrum, J., Santer, R., Kohlschutter, A., Krageloh-Mann, I., Muller, I., 2016. Long-term Outcome of Allogeneic Hematopoietic Stem Cell Transplantation in Patients With Juvenile Metachromatic Leukodystrophy Compared With Nontransplanted Control Patients. JAMA Neurol. 73 (9), 1133–1140.

Gupta, A.O., Raymond, G., Pierpont, E.I., Kemp, S., McIvor, R.S., Rayannavar, A., Miller, B., Lund, T.C., Orchard, P.J., 2022. Treatment of cerebral adrenoleukodystrophy: allogeneic transplantation and lentiviral gene therapy. Expert Opin. Biol. Ther. 22 (9), 1151–1162.

Hack, R.J., Gravesteijn, G., Cerfontaine, M.N., Hegeman, I.M., Mulder, A.A., Oberstein, S.A.J.L., Rutten, J.W., 2022. Cerebral Autosomal Dominant Arteriopathy With Subcortical Infarcts and Leukoencephalopathy Family Members With a Pathogenic NOTCH3 Variant Can Have a Normal Brain Magnetic Resonance Imaging and Skin Biopsy Beyond Age 50 Years. Stroke 53 (6), 1964–1974.

Haddad, E.N., Kumar, P., Shearn-Nance, G., Kharal, G.A., Dhawan, A., 2024. Clinical Approach to Genetic Cerebral Arteriopathy in the Adult Patient With Ischemic Stroke. Neurol. Genet. 10 (5), e200182.

Hagberg, B., 1971. Clinical aspects of globoid cell and metachromatic leukodystrophies. Birth Defects Orig. Artic. Ser. 7 (1), 103–112.

Hagberg, B., Sourander, P., Svennerholm, L., 1963. Diagnosis of Krabbe's infantile leucodystrophy. J. Neurol. Neurosurg. Psychiatry 26 (3), 195–198.

Hagemann, T.L., Powers, B., Mazur, C., Kim, A., Wheeler, S., Hung, G., Swayze, E., Messing, A., 2018. Antisense suppression of glial fibrillary acidic protein as a treatment for Alexander disease. Ann. Neurol. 83 (1), 27–39.

Han, J., Sarlus, H., Wszolek, Z.K., Karrenbauer, V.D., Harris, R.A., 2020. Microglial replacement therapy: a potential therapeutic strategy for incurable CSF1R-related leukoencephalopathy. Acta Neuropathol. Commun. 8 (1), 217.

Heim, P., Claussen, M., Hoffmann, B., Conzelmann, E., Gartner, J., Harzer, K., Hunneman, D.H., Kohler, W., Kurlemann, G., Kohlschutter, A., 1997. Leukodystrophy incidence in Germany. Am. J. Med. Genet. 71 (4), 475–478.

Herdt, A.R., Peng, H., Dickson, D.W., Golde, T.E., Eckman, E.A., Lee, C.W., 2023. Brain Targeted AAV1-GALC Gene Therapy Reduces Psychosine and Extends Lifespan in a Mouse Model of Krabbe Disease. Genes 14 (8), 1517.

Hordeaux, J., Jeffrey, B.A., Jian, J., Choudhury, G.R., Michalson, K., Mitchell, T.W., Buza, E.L., Chichester, J., Dyer, C., Bagel, J., Vite, C.H., Bradbury, A.M., Wilson, J.M., 2022. Efficacy and Safety of a Krabbe Disease Gene Therapy. Hum. Gene Ther. 33 (9-10), 499–517.

Hoshino, H., Kubota, M., 2014. Canavan disease: clinical features and recent advances in research. Pediatr. Int. 56 (4), 477–483.

Hudson, L.D., 2003. Pelizaeus-Merzbacher disease and spastic paraplegia type 2: two faces of myelin loss from mutations in the same gene. J. Child Neurol. 18 (9), 616–624.

Inoue, K., 2019. Pelizaeus-Merzbacher Disease: molecular and Cellular Pathologies and Associated Phenotypes. Adv. Exp. Med. Biol. 1190, 201–216.

Kass, A., 1953. Acute diffuse infantile sclerosis of the brain (Krabbe's disease); a report of two cases in sibs. Acta Paediatr. 42 (1), 70–76.

Kehrer, C., Groeschel, S., Kustermann-Kuhn, B., Burger, F., Kohler, W., Kohlschutter, A., Bley, A., Steinfeld, R., Gieselmann, V., Krageloh-Mann, I., German, L., 2014. Language and cognition in children with metachromatic leukodystrophy: onset and natural course in a nationwide cohort. Orphanet J. Rare Dis. 9, 18.

Klinkovskij, A., Shepelev, M., Isaakyan, Y., Aniskin, D., Ulasov, I., 2023. Advances of Genome Editing with CRISPR/Cas9 in Neurodegeneration: the Right Path towards Therapy. Biomedicines 11 (12), 3333.

Kok, L.M.L., van Zyl, H., Gotte, F., Lichtendahl, E., Breeuwsma, N., Wolf, N.I., Heine, V.M., 2025. POLR3 gene and protein expression dynamics in 4H leukodystrophy using iPSC-derived neuronal lineages. Stem Cell Res. 88, 103805.

Kuhl, J.S., Suarez, F., Gillett, G.T., Hemmati, P.G., Snowden, J.A., Stadler, M., Vuong, G.L., Aubourg, P., Kohler, W., Arnold, R., 2017. Long-term outcomes of allogeneic haematopoietic stem cell transplantation for adult cerebral X-linked adrenoleukodystrophy. Brain 140 (4), 953–966.

Kumar, S., Biancotti, J.C., Matalon, R., de Vellis, J., 2009. Lack of aspartoacylase activity disrupts survival and differentiation of neural progenitors and oligodendrocytes in a mouse model of Canavan disease. J. Neurosci. Res. 87 (15), 3415–3427.

Leone, P., Lober, R.M., Francis, J., Flamini, O., Cecil, K.M., Shera, D., Janson, C.G., 2025. Oligodendrocyte-targeted adeno-associated virus gene therapy for Canavan disease in children: a phase 1/2 trial. Nat. Med. 31 (11), 3772–3779.

Li, R., Messing, A., Goldman, J.E., Brenner, M., 2002. GFAP mutations in Alexander disease. Int. J. Dev. Neurosci. 20 (3-5), 259–268.

Lin, W., Zhang, M., Zheng, S., Lian, Q., 2025. Stem cell and gene therapies for leukodystrophies. Mol. Ther. Methods Clin. Dev. 33 (3), 101527.

Martin, T.S., Seabrook, T.A., Gall, K., Newman, J., Avila, N., Hayes, A., Kivaa, M., Lotterhand, J., Mercaldi, M., Patel, K., 2023. Single systemic administration of a gene therapy leading to disease treatment in metachromatic leukodystrophy arsa knock-out mice. J. Neurosci. 43 (19), 3567–3581.

Matos, T., Costa, C., Araujo, A.N., Do Vale, S., 2023. Clinical course and endocrine dysfunction in X-linked adrenoleukodystrophy: a case series. Endocrinol. Diabetes Nutr. (Engl. Ed 70 (6), 421–428.

Matsukawa, T., Yamamoto, T., Honda, A., Toya, T., Ishiura, H., Mitsui, J., Tanaka, M., Hao, A., Shinohara, A., Ogura, M., Kataoka, K., Seo, S., Kumano, K., Hosoi, M., Narukawa, K., Yasunaga, M., Maki, H., Ichikawa, M., Nannya, Y., Imai, Y., Takahashi, T., Takahashi, Y., Nagasako, Y., Yasaka, K., Mano, K.K., Matsukawa, M.K., Miyagawa, T., Hamada, M., Sakuishi, K., Hayashi, T., Iwata, A., Terao, Y., Shimizu, J., Goto, J., Mori, H., Kunimatsu, A., Aoki, S., Hayashi, S., Nakamura, F., Arai, S., Momma, K., Ogata, K., Yoshida, T., Abe, O., Inazawa, J., Toda, T., Kurokawa, M., Tsuji, S., 2020. Clinical efficacy of haematopoietic stem cell transplantation for adult adrenoleukodystrophy. Brain Commun. 2 (1), fcz048.

Matsumoto, N., Watanabe, N., Iibe, N., Tatsumi, Y., Hattori, K., Takeuchi, Y., Oizumi, H., Ohbuchi, K., Torii, T., Miyamoto, Y., Yamauchi, J., 2019. Hypomyelinating leukodystrophy-associated mutation of RARS leads it to the lysosome, inhibiting oligodendroglial morphological differentiation. Biochem. Biophys. Rep. 20, 100705.

Merzbacher, L., 1910. Eine eigenartige familiär-hereditäre erkrankungsform (Aplasia axialis extracorticalis congenita). Zeitschrift Für Die Gesamte Neurologie Und Psychiatrie 3 (1), 1–138.

Messing, A., Brenner, M., Feany, M.B., Nedergaard, M., Goldman, J.E., 2012. Alexander disease. J. Neurosci. 32 (15), 5017–5023.

Metovic, J., Li, Y., Gong, Y., Eichler, F., 2024. Gene therapy for the leukodystrophies: from preclinical animal studies to clinical trials. Neurotherapeutics 21 (4), e00443.

Miller, W.P., Rothman, S.M., Nascene, D., Kivisto, T., DeFor, T.E., Ziegler, R.S., Eisengart, J., Leiser, K., Raymond, G., Lund, T.C., Tolar, J., Orchard, P.J., 2011. Outcomes after allogeneic hematopoietic cell transplantation for childhood cerebral adrenoleukodystrophy: the largest single-institution cohort report. Blood 118 (7), 1971–1978.

Miyake, N., Miyake, K., Sakai, A., Yamamoto, M., Suzuki, H., Shimada, T., 2021. Treatment of adult metachromatic leukodystrophy model mice using intrathecal administration of type 9 AAV vector encoding arylsulfatase A. Sci. Rep. 11 (1), 20513.

Moser, H.W., Mahmood, A., Raymond, G.V., 2007. X-linked adrenoleukodystrophy. Nat. Clin. Pract. Neurol. 3 (3), 140–151.

Mosser, J., Douar, A.M., Sarde, C.O., Kioschis, P., Feil, R., Moser, H., Poustka, A.M., Mandel, J.L., Aubourg, P., 1993. Putative X-linked adrenoleukodystrophy gene shares unexpected homology with ABC transporters. Nature 361 (6414), 726–730.

Muthusamy, K., Sivadasan, A., Dixon, L., Sudhakar, S., Thomas, M., Danda, S., Wszolek, Z.K., Wierenga, K., Dhamija, R., Gavrilova, R., 2023. Adult-onset leukodystrophies: a practical guide, recent treatment updates, and future directions. Front. Neurol. 14, 1219324.

Nieder, M.L., McDonald, G.B., Kida, A., Hingorani, S., Armenian, S.H., Cooke, K.R., Pulsipher, M.A., Baker, K.S., 2011. National Cancer Institute-National Heart, Lung and Blood Institute/pediatric Blood and Marrow Transplant Consortium First International Consensus Conference on late effects after pediatric hematopoietic cell transplantation: long-term organ damage and dysfunction. Biol. Blood Marrow Transplant. 17 (11), 1573–1584.

Noureldeen, M.M., Zaki, M.S., Rafat, K., Abdel-Hamid, M.S., Salem, A.M.S., 2025. The landscape of pediatric genetic white matter disorders at a tertiary referral hospital in Upper Egypt and the report of 31 novel variants. Ital. J. Pediatr. 51 (1), 187.

Nowacki, J.C., Fields, A.M., Fu, M.M., 2022. Emerging cellular themes in leukodystrophies. Front. Cell Dev. Biol. 10, 902261.

Osorio, M.J., Goldman, S.A., 2018. Neurogenetics of Pelizaeus-Merzbacher disease. Handb. Clin. Neurol. 148, 701–722.

Papapetropoulos, S., Pontius, A., Finger, E., Karrenbauer, V., Lynch, D.S., Brennan, M., Zappia, S., Koehler, W., Schoels, L., Hayer, S.N., Konno, T., Ikeuchi, T., Lund, T., Orthmann-Murphy, J., Eichler, F., Wszolek, Z.K., 2021. Adult-Onset Leukoencephalopathy With Axonal Spheroids and Pigmented Glia: review of Clinical Manifestations as Foundations for Therapeutic Development. Front. Neurol. 12, 788168.

Parikh, S., Bernard, G., Leventer, R.J., van der Knaap, M.S., van Hove, J., Pizzino, A., McNeill, N.H., Helman, G., Simons, C., Schmidt, J.L., Rizzo, W.B., Patterson, M.C., Taft, R.J., Vanderver, A., Consortium, G., 2015. A clinical approach to the diagnosis of patients with leukodystrophies and genetic leukoencephelopathies. Mol. Genet. Metab. 114 (4), 501–515.

Perlman, S.J., Mar, S., 2012. Leukodystrophies. Adv. Exp. Med. Biol. 724, 154–171.

Piguet, F., de Saint Denis, T., Audouard, E., Beccaria, K., Andre, A., Wurtz, G., Schatz, R., Alves, S., Sevin, C., Zerah, M., Cartier, N., 2021. The Challenge of Gene Therapy for Neurological Diseases: strategies and Tools to Achieve Efficient Delivery to the Central Nervous System. Hum. Gene Ther. 32 (7-8), 349–374.

Powell, S.K., Khan, N., Parker, C.L., Samulski, R.J., Matsushima, G., Gray, S.J., McCown, T.J., 2016. Characterization of a novel adeno-associated viral vector with preferential oligodendrocyte tropism. Gene Ther 23 (11), 807–814.

Rafi, M.A., Luzi, P., Wenger, D.A., 2020. Conditions for combining gene therapy with bone marrow transplantation in murine Krabbe disease. Bioimpacts 10 (2), 105–115.

Ramachandran, S., Ardinger, J., Bu, J., Ramos, M., Guo, L., Ghosh, D., Hossain, M., Chou, S.-C., Chen, Y., Wischhof, E., Ayloo, S., Trullo, R., Luo, Y., Hogestyn, J., DuBreuil, D., Crosier, E., Adams, J., Richards, A., Tsabar, M., Gaglia, G., Nass, S., Nambiar, B., Woodcock, D., O'Riordan, C., Tang, Q., Elmer, B., Zhang, B., Goulet, M., Mueller, C. (2025a). "AAV-mediated ARSA replacement for the treatment of Metachromatic Leukodystrophy." bioRxiv: 2025.2003.2012.642609.

Riedel, A., Faul, C., Reuss, K., Schroder, J.C., Lang, P.J., Lengerke, C., Weissert, N., Hengel, H., Groschel, S., Schoels, L., Bethge, W.A., 2024. Allogeneic hematopoietic cell transplantation for adult metachromatic leukodystrophy: a case series. Blood Adv. 8 (6), 1504–1508.

Rosenberg, J.B., Chen, A., De, B.P., Dyke, J.P., Ballon, D.J., Monette, S., Arbona, R.J.R., Kaminsky, S.M., Crystal, R.G., Sondhi, D., 2021. Safety of Direct Intraparenchymal AAVrh.10-Mediated Central Nervous System Gene Therapy for Metachromatic Leukodystrophy. Hum. Gene Ther. 32 (11-12), 563–580.

Salzman, R., Cook, F., Hunt, T., Malech, H.L., Reilly, P., Foss-Campbell, B., Barrett, D., 2018. Addressing the Value of Gene Therapy and Enhancing Patient Access to Transformative Treatments. Mol. Ther. 26 (12), 2717–2726.

Sarret, C., 2020. Leukodystrophies and genetic leukoencephalopathies in children. Rev. Neurol. (Paris 176 (1-2), 10–19.

Sassi, C., Nalls, M.A., Ridge, P.G., Gibbs, J.R., Lupton, M.K., Troakes, C., Lunnon, K., Al-Sarraj, S., Brown, K.S., Medway, C., Lord, J., Turton, J., Bras, J., Consortium, A., Blumenau, S., Thielke, M., Josties, C., Freyer, D., Dietrich, A., Hammer, M., Baier, M., Dirnagl, U., Morgan, K., Powell, J.F., Kauwe, J.S., Cruchaga, C., Goate, A.M., Singleton, A.B., Guerreiro, R., Hodges, A., Hardy, J., 2018. Mendelian adult-onset leukodystrophy genes in Alzheimer's disease: critical influence of CSF1R and NOTCH3. Neurobiol. Aging 66, 179e117–179e129.

Scott, A.I., Mallhi, K.K., Ganesh, J., Chen, W.L.;CMG Therapeutics Committee, 2023. Elivaldogene autotemcel approved for treatment of cerebral adrenoleukodystrophy (CALD) in males: A therapeutics bulletin of the American College of Medical Genetics and Genomics (ACMG). Genet. Med. Open 1 (1), 100835. doi:10.1016/j.gimo.2023.100835.

Sessa, M., Lorioli, L., Fumagalli, F., Acquati, S., Redaelli, D., Baldoli, C., Canale, S., Lopez, I.D., Morena, F., Calabria, A., Fiori, R., Silvani, P., Rancoita, P.M., Gabaldo, M., Benedicenti, F., Antonioli, G., Assanelli, A., Cicalese, M.P., Carro, U.D., Sora, M.G., Martino, S., Quattrini, A., Montini, E., Di Serio, C., Ciceri, F., Roncarolo, M.G., Aiuti, A., Naldini, L., Biffi, A., 2016. Lentiviral haemopoietic stem-cell gene therapy in early-onset metachromatic leukodystrophy: an ad-hoc analysis of a non-randomised, open-label, phase 1/2 trial. Lancet 388 (10043), 476–487.

Sevin, C., Mochel, F., 2024. Hematopoietic stem cell transplantation in leukodystrophies. Handb. Clin. Neurol. 204, 355–366.

Seyrantepe, V., Ates, N., Can, M., Sengul, T., Demir, S.A., 2018. Intracerebral gene therapy in children with metachromatic leukodystrophy: results of a phase I/II trial. Abstracts/Molecular Genetics and Metabolism 123, S15–S153.

Shaimardanova, A.A., Chulpanova, D.S., Solovyeva, V.V., Mullagulova, A.I., Kitaeva, K.V., Allegrucci, C., Rizvanov, A.A., 2020. Metachromatic Leukodystrophy: diagnosis, Modeling, and Treatment Approaches. Front. Med. (Lausanne) 7, 576221.

Sosunov, A., Olabarria, M., Goldman, J.E., 2018. Alexander disease: an astrocytopathy that produces a leukodystrophy. Brain Pathol. 28 (3), 388–398.

Spratley, S.J., Deane, J.E., 2016. New therapeutic approaches for Krabbe disease: the potential of pharmacological chaperones. J. Neurosci. Res. 94 (11), 1203–1219.

Thiffault, I., Wolf, N.I., Forget, D., Guerrero, K., Tran, L.T., Choquet, K., Lavallee-Adam, M., Poitras, C., Brais, B., Yoon, G., Sztriha, L., Webster, R.I., Timmann, D., van de Warrenburg, B.P., Seeger, J., Zimmermann, A., Mate, A., Goizet, C., Fung, E., van der Knaap, M.S., Fribourg, S., Vanderver, A., Simons, C., Taft, R.J., Yates 3rd, J.R., Coulombe, B., Bernard, G., 2015. Recessive mutations in POLR1C cause a leukodystrophy by impairing biogenesis of RNA polymerase III. Nat. Commun. 6, 7623.

Vaia, Y., Bruschi, F., Tagi, V.M., Tosi, M., Montanari, C., Zuccotti, G., Tonduti, D., Verduci, E., 2024. Microbiota gut-brain axis: implications for pediatric-onset leukodystrophies. Front. Nutr. 11, 1417981.

van der Knaap, M.S., Bugiani, M., 2017. Leukodystrophies: a proposed classification system based on pathological changes and pathogenetic mechanisms. Acta Neuropathol. 134 (3), 351–382.

van der Knaap, M.S., Leegwater, P.A., Konst, A.A., Visser, A., Naidu, S., Oudejans, C.B., Schutgens, R.B., Pronk, J.C., 2002. Mutations in each of the five subunits of translation initiation factor eIF2B can cause leukoencephalopathy with vanishing white matter. Ann. Neurol. 51 (2), 264–270.

van der Knaap, M.S., Pronk, J.C., Scheper, G.C., 2006. Vanishing white matter disease. Lancet Neurol. 5 (5), 413–423.

van der Knaap, M.S., Schiffmann, R., Mochel, F., Wolf, N.I., 2019. Diagnosis, prognosis, and treatment of leukodystrophies. Lancet Neurol. 18 (10), 962–972.

van der Knaap, M.S., Valk, J., 1989. The reflection of histology in MR imaging of Pelizaeus-Merzbacher disease. AJNR Am. J. Neuroradiol. 10 (1), 99–103.

van Egmond, M.E., Pouwels, P.J., Boelens, J.J., Lindemans, C.A., Barkhof, F., Steenwijk, M.D., van Hasselt, P.M., van der Knaap, M.S., Wolf, N.I., 2013. Improvement of white matter changes on neuroimaging modalities after stem cell transplant in metachromatic leukodystrophy. JAMA Neurol. 70 (6), 779–782.

Vanderver, A., Hussey, H., Schmidt, J.L., Pastor, W., Hoffman, H.J., 2012. Relative incidence of inherited white matter disorders in childhood to acquired pediatric demyelinating disorders. Semin. Pediatr. Neurol. 19 (4), 219–223.

Vanderver, A., Prust, M., Tonduti, D., Mochel, F., Hussey, H.M., Helman, G., Garbern, J., Eichler, F., Labauge, P., Aubourg, P., Rodriguez, D., Patterson, M.C., Van Hove, J.L., Schmidt, J., Wolf, N.I., Boespflug-Tanguy, O., Schiffmann, R., van der Knaap, M.S., Consortium, G., 2015. Case definition and classification of leukodystrophies and leukoencephalopathies. Mol. Genet. Metab. 114 (4), 494–500.

Waldman, A.T., 2018. Leukodystrophies. Continuum (Minneap Minn 24 (1, Child Neurology), 130–149.

Weinhard, S., Wiedemann, A., Leheup, B., Dalle, J.H., Labich, B.L., Pochon, C., 2020. Pubertal outcomes of children transplanted with allogeneic stem cells after myeloablative total body irradiation or busulfan: influence of age and sex is confirmed, while a role of chronic graft-versus-host disease in delayed puberty onset is revealed. Pediatr. Transplant. 24 (6), e13773.

Wolf, N.I., Breur, M., Plug, B., Beerepoot, S., Westerveld, A.S.R., van Rappard, D.F., de Vries, S.I., Kole, M.H.P., Vanderver, A., van der Knaap, M.S., Lindemans, C.A., van Hasselt, P.M., Boelens, J.J., Matzner, U., Gieselmann, V., Bugiani, M., 2020. Metachromatic leukodystrophy and transplantation: remyelination, no cross-correction. Ann. Clin. Transl. Neurol. 7 (2), 169–180.

Wolf, N.I., van der Knaap, M.S., Engelen, M., 2025. Treatment of leukodystrophies: advances and challenges. Eur. J. Paediatr. Neurol. 56, 46–50.

Young, K.Z., Xu, G., Keep, S.G., Borjigin, J., Wang, M.M., 2021. Overlapping Protein Accumulation Profiles of CADASIL and CAA: is There a Common Mechanism Driving Cerebral Small-Vessel Disease? Am. J. Pathol. 191 (11), 1871–1887.

Zampini, L., Cordolcini, L., Draghi, L., Zanchi, P., Vaia, Y., Bonaventura, E., Tonduti, D., 2025. The impact of leukodystrophies on parents' lives. J. Pediatr. Psychol. 50 (12), 1091–1098.

Zhao, Y., Lu, Y., Wang, F., Wang, Y., Li, Y., Sun, R., Shang, J., Jiang, C., Zhang, J., 2025. Mechanistic advances in factors influencing phenotypic variability in cerebral autosomal dominant arteriopathy with subcortical infarcts and leukoencephalopathy: a review. Front. Neurol. 16, 1573052.

Zhou, L., Wang, Y., Xu, Y., Zhang, Y., Zhu, C., 2024. Advances in AAV-mediated gene replacement therapy for pediatric monogenic neurological disorders. Mol. Ther. Methods Clin. Dev. 32 (4), 101357.

Chapter 9

Gene therapy in neuronal ceroid lipofuscinosis (NCL)

Mojdeh Salehi Namini[a,b], Sayed Ali Reza Mousavi[c], and Asad Kazemi[c]

[a]*Stem Cell and Regenerative Medicine Research Center, Iran University of Medical Sciences, Tehran, Iran*

[b]*Department of Tissue Engineering, School of Advanced Technologies in Medicine, Tehran University of Medical Sciences, Tehran, Iran*

[c]*Department of Medical Biotechnology, School of Advanced Technologies in Medicine, Tehran University of Medical Sciences, Tehran, Iran*

1 Introduction

Neuronal ceroid lipofuscinoses (NCLs) are a group of rare inherited neurodegenerative disorders that primarily affect children and young adults (Nita et al., 2016). These neurodegenerative disorders are also known as a type of lysosomal storage disease, in which abnormal materials accumulate within cells due to the body's inability to degrade them properly (Wang et al., 2011).

Over time, this causes the brain and nervous system to stop functioning as they should, gradually. As a result, patients with NCLs experience progressive neurological decline, with typical symptoms such as epilepsy, loss of vision, cognitive impairment, struggle with movement, and ultimately, this condition usually leads to an early death (Schulz et al., 2013; Hasibuan and Arthy, 2023). Therefore because there is no specific treatment for most types of NCLs, developing and applying new treatment strategies is incredibly valuable. NCL disorders have several forms, all of which share similar symptoms and effects on the brain. These forms are classified (Table 9.1). Each type of NCLs is caused by a different genetic mutation and has its own unique underlying mechanisms (Williams and Mole, 2012). NCL disorders are inherited in an autosomal recessive (AR) manner, except CLN4 disease, which is inherited in an autosomal dominant (AD) manner (Kohlschütter et al., 2019). The childhood forms are clinically marked by progressive vision loss, cognitive and motor decline, epileptic seizures, and ultimately, premature death (Hasibuan and Arthy, 2023). In contrast, the rarer adult forms are primarily characterized by the development of dementia (Huang et al., 2024).

Although the clinical symptoms of different forms of NCLs vary, they all share the same underlying pathological features. In these cases, granules accumulate inside the cytoplasm of nerve cells. These granules appear under specific lighting, react to histochemical stains (such as Periodic acid–Schiff (PAS) and Sudan Black B), and do not break down with fat-dissolving chemicals (Martin et al., 1999).

As a consequence, NCL disorders lead to a gradual and selective loss of nerve cells, often accompanied by the growth and increased activity of supporting brain cells (astrocytes) and the presence of immune cells like macrophages (Takahashi et al., 2022). Furthermore, in the most severe early-onset forms, such as the congenital and infantile types, nerve cells in the brain and retina can be almost entirely lost, causing significant atrophy of these tissues. Hence, variations in the ultrastructure of accumulating storage cytosomes across different NCL subtypes have historically played a key role in their classification, alongside the age at which clinical symptoms first appear (Haltia, 2003). In this regard, researchers have identified 14 types of NCLs (CLN1–CLN8 and CLN10–CLN14), each of which is caused by changes in a specific gene (Radke et al., 2015).

2 Overview, classification, and genetics of NCLs

As of 2015, researchers have identified 14 types of NCL: CLN1/PPT1, CLN2/TPP1, CLN3, CLN4/DNAJC5, CLN5, CLN6, CLN7/MFSD8, CLN8, CLN10/CTSD, CLN11/GRN, CLN12/ATP13A2, CLN13/CTSF, and CLN14/KCTD7.

Gene Therapy in Neurodegenerative Disorders. https://doi.org/10.1016/B978-0-443-40467-2.00002-6

TABLE 9.1 Various forms of NCL with specific characteristics.

Form of NCLs	Genotype–phenotype correlation	Onset age (respectively)	Type of deficiency	Gene	Encoded protein	Typical clinical manifestations
CLN1/Haltia–Santavuori disease	Infantile, late infantile, juvenile, adult	6–12 months, 1.5–4 years, 5–7 years, above 16 years	Lysosomal enzyme deficiency	*CLN1/PPT1*	PPT1[a]	Progressive blindness, cognitive decline, motor deficits, and seizures
CLN2/Jansky-Bielschowsky disease	Infantile, late infantile, juvenile	Under 18 months, 2–4 years, above 5 years		*CLN2/TPP1*	TPP1[b]	Motor decline, speech delay, and seizures
CLN3/Spielmeyer-Vogt-Sjögren-Battendisease	Juvenile	4–7 years	Nonenzymatic protein deficiency	*CLN3*	Transmembrane protein	Seizures, visual deficits, motor impairments, and cognitive decline
CLN4/Parry disease	Adult autosomal dominant	Above 20 years		*CLN4/DNAJC5*	Soluble cysteine string protein α	Progressive myoclonus epilepsy
CLN5	Late infantile, juvenile, adult	4–5 years, 5–7 years, above 16–50 years		*CLN5*	Soluble lysosomal enzyme	Motor clumsiness, progressive visual failure resulting in blindness, motor and mental deterioration, seizures
CLN6/Kufs disease type A	Late infantile, teenage-adult	Above 18 months, above 7 years		*CLN6*	ER[c] membrane protein	Blindness, seizures, and cognitive decline
CLN7	Late infantile, juvenile, adult	1.5–6 years; protracted: above 7 years; above 29–65 years		*CLN7/MFSD8*	Lysosomal membrane protein	Seizures, developmental regression, loss of vision, and premature death
CLN8	Late infantile	2–7 years, protracted: 5–10 years		*CLN8*	ER membrane protein	Seizures, progressive vision loss, and developmental regression
CLN9	Infantile, juvenile	Around 4 years	Not identified	*CLN9*	Not identified	Vision loss, seizures, ataxia, and myoclonus
CLN10	Congenital, late infantile, juvenile, adult	Prenatal and perinatal, 4 years, 8–15 years, after 20 years	Lysosomal enzyme	*CTSD/CLN10*	Cathepsin D	Cognitive decline, loss of speech, visual dysfunction, and loss of motor functions
CLN11	Adult	Above 20 years, 50–70 years	Lysosomal enzyme chaperone	*CLN11/GRN*	Progranulin	Retinal dystrophy, seizures, myoclonus, cerebellar ataxia, and atrophy
CLN12	Juvenile	13–16 years	Nonenzymatic protein	*ATP13A2*	Ion pump	Motor clumsiness, myoclonus, behavioral changes, spinocerebellar ataxia, bulbar syndrome, extrapyramidal and pyramidal symptoms, cognitive decline
CLN13/Kufs disease type B	Adult	Above 20 years	Lysosomal enzyme	*CLN13/CTSF*	Cysteine protease cathepsin F	Tremor, ataxia, seizures
CLN14/progressive myoclonic epilepsy type 3	Infantile	8–24 months	Membrane protein	*KCTD7*	Potassium channel tetramerization domain containing 7	Developmental regression, visual impairment

[a]*PPT1 (palmitoyl-protein thioesterase 1).*
[b]*TPP1 (tripeptidyl-peptidase 1).*
[c]*ER (endoplasmic reticulum).*

Although NCL disorders have been widely studied, the exact roles of the affected proteins and how they contribute to disease development are still not fully understood. According to the NCL mutation database (https://www.ucl.ac.uk/ncl-disease), the most prevalent types of NCLs are CLN1 disease (infantile NCL [INCL]), CLN2 disease (late infantile), CLN3 disease (juvenile NCL), and CLN6 disease (late infantile NCL [LINCL]) (Mole et al., 2019). A comparative overview of these types, such as key clinical features, associated proteins, subcellular localizations, and common genotypes, is provided in the following and summarized in Table 9.1 and Fig. 9.1.

2.1 CLN1

CLN1 disease, a form of INCL, was previously known as Santavuori–Haltia disease. This form results from mutations in the CLN1 gene. The *CLN1* gene encodes the enzyme palmitoyl-protein thioesterase 1 (PPT1), which is responsible for breaking down lipid-modified proteins. CLN1 is located on chromosome 1p34.2 and follows an AR pattern of inheritance (Kim et al., 2008). Clinical symptoms in these patients usually appear by 18 months of age, beginning with developmental regression and rapidly progressing to hypotonia, ataxia, myoclonus, epileptic seizures, and gradual loss of vision. In these

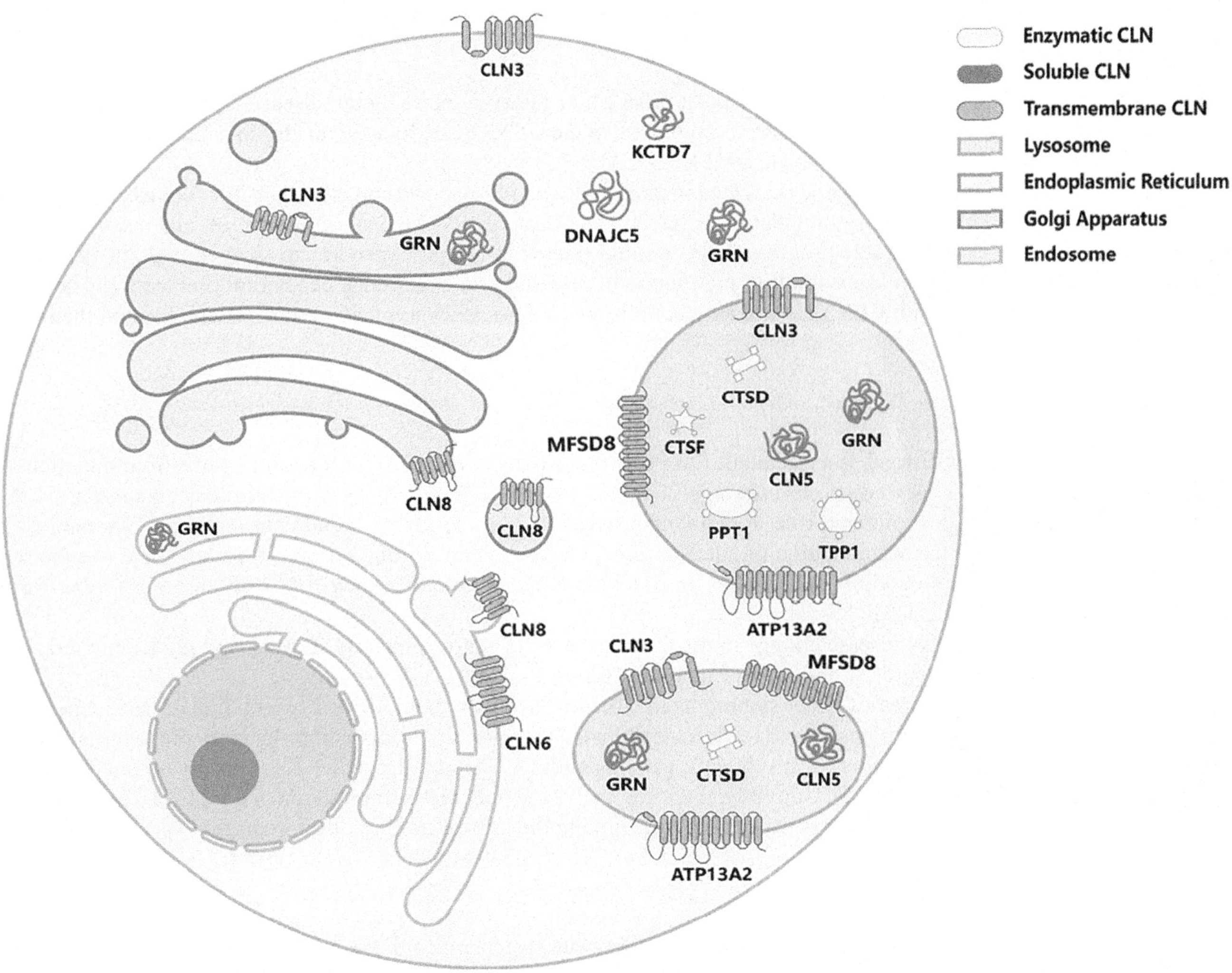

FIG. 9.1 The locations of NCL proteins. CLN1 (PPT1), CLN2 (TPP1), CLN10 (CTSD), and CLN13 (CTSF) encode soluble NCL proteins with known enzymatic activity; CLN4 (DNAJC5), CLN5, CLN11 (GRN), and CLN14 (KCTD7) encode soluble proteins, with one protein (encoded by GRN) localized to compartments in the secretory pathway. CLN3, CLN6, CLN7 (MFSD8), CLN8, and CLN12 (ATP13A2) encode transmembrane proteins with different subcellular locations: CLN3 mainly localizes to the plasma membrane, Golgi, endosome, and lysosome; CLN6 localizes to the ER; MFSD8 mainly localizes to the lysosome and endosome; CLN8 shuttles between the ER and the ERGIC compartment; ATP13A2 mainly localizes to the membrane of lysosomes and late endosomes. *From Zhang, Y., Bingying, D., Zou, M., Peng, B., Rao, Y., 2025. Neuronal ceroid lipofuscinosis—concepts, classification, and avenues for therapy. CNS Neurosci. Ther. 1755–5930. 31 (2). https://doi.org/10.1111/cns.70261.*

cases, ongoing neurodegeneration often leads to microcephaly, and most of these children eventually require feeding tubes. However, most affected children do not survive beyond early childhood (Nita et al., 2016; Mink et al., 2013). Other later-onset forms of CLN1 disease have been associated with particular CLN1 mutations; these cases also have a significantly limited lifespan (Jeung et al., 2015).

2.2 CLN2

CLN2 disease, a form of LINCL, results from mutations in the *CLN2* gene, which is located on chromosome 11p15.4. This gene encodes the lysosomal enzyme tripeptidyl peptidase 1 (TPP1), which plays a critical role in normal lysosomal function and the degradation of proteins (Kohan et al., 2013; Priyanka et al., 2024). Clinical symptoms typically appear between 2 and 4 years of age, beginning with ataxia (Nickel et al., 2018). As CLN2 disease progresses, affected children usually experience a gradual loss of speech and motor skills, accompanied by myoclonus, progressive visual impairment, and cognitive deterioration (Bryzik et al., 2025).

CLN2 disease often progresses quickly, and most affected individuals typically do not survive beyond 8–12 years of age (Bryzik et al., 2025). Several studies have demonstrated the clinical progression and neurological decline associated with this severe phenotype (Nickel et al., 2018; Gao et al., 2011)

2.3 CLN3

CLN3 disease, juvenile neuronal ceroid lipofuscinosis (JNCL), or juvenile-onset Batten disease typically occurs between the ages of 4 and 10 years. This form results from mutations in the *CLN3* gene, located on chromosome 16p12.1, and follows an AR inheritance pattern (Grisolia et al., 2016).

The protein encoded by the *CLN3* gene is located in endosomes and lysosomes and is responsible for vesicular trafficking, lysosomal pH regulation, and osmoregulation (Cotman and Staropoli, 2012). One of the earliest and most significant symptoms of CLN3 disease is rapidly progressive vision loss related to retinal degeneration (Kuper et al., 2019). As the disease progresses, patients often lose cognitive and motor abilities and also experience behavioral problems and seizures. Consequently, most patients with CLN3 disease have a limited life expectancy, typically not surviving beyond their early twenties (Mole et al., 2025).

2.4 CLN4

CLN4, also known as Parry disease, is a rare adult form of NCL (Zhang et al., 2025). This form results from mutations in the *DNAJC5* gene, which is located on chromosome 20q13.33 (Singh, 2025). This gene encodes cysteine string protein alpha (CSPα), which plays a significant role in maintaining synaptic vesicle function and neuronal protein homeostasis. These mutations can cause the accumulation of autofluorescent lipopigments within neurons, a pathological symptom of NCLs (Borjini et al., 2023). In addition, CLN4 has an AD pattern and is unique among NCLs because most other forms have an AR pattern (Masola, 2025).

Clinical symptoms of CLN4 disease emerge in the thirties. These symptoms include seizures and myoclonic jerks and progress to a decline in both cognitive and motor functions (Vanier et al., 2022).

Also, when the disease progresses, other symptoms like dementia, ataxia, tremors, and speech impairments emerge. In these cases, neurodegeneration progresses and leads to death, which commonly occurs within 15 years of the initial presentation of symptoms. Unlike most other NCL subtypes, patients with CLN4, similar to CLN13, typically do not experience visual deterioration (Zhang et al., 2025). Currently, there are no disease-specific therapies known for CLN4. Care is limited to managing symptoms and supporting quality of life, highlighting the critical need to develop disease-specific therapies.

2.5 CLN5

CLN5 disease, the Finnish variant late-infantile NCL, is a rare neurodegenerative disease that results from mutations in the *CLN5* gene. This gene is located on 13q22.3 and encodes a soluble lysosomal glycoprotein; however, the precise biological function of CLN5 remains unclear (Breitenbach et al., 2013).

This form has an AR inheritance pattern, and symptoms typically appear between the ages of 4 and 7. The main symptoms are motor impairments or psychomotor regression (Basak et al., 2021). During CLN5 disease progression, other symptoms such as clumsiness, seizures, vision impairment, ataxia, and myoclonus may occur (Madhavi et al., 2024). Life expectancy varies with disease progression, but death usually occurs between 10 and 30 years of age (Basak et al., 2021).

2.6 CLN6

CLN6 disease is a variant late-infantile or early juvenile form of NCL. This form results from mutations in the *CLN6* gene. This gene is located on chromosome 15q23, which encodes an endoplasmic reticulum (ER) membrane protein (Butz et al., 2020). CLN6 disease follows an AR pattern of inheritance (Rus et al., 2022). Clinical symptoms are variable and typically appear between 18 months and 8 years of age. Common symptoms include vision loss, motor decline, seizures, dysarthria, and ataxia (Nicolaou et al., 2021). This form of the disease typically progresses quickly, and survival is markedly reduced, often into late childhood or adolescence (Rus et al., 2022).

2.7 CLN7

CLN7 disease is caused by mutations in the *MFSD8* gene, which is located on chromosome 4q28.2 (Qiao et al., 2022). This gene encodes a lysosomal membrane protein, but the precise function of this protein is unknown (Kousi et al., 2009). CLN7 disease symptoms typically occur between 2 and 7 years of age and include epilepsy, developmental regression, myoclonus, ataxia, vision loss, and speech impairment (Kayani et al., 2024). The progression of CLN7 disease is characterized by a decline in motor and cognitive abilities, ultimately leading to death during the teenage years (Mohammed, 2020). To date, only one patient with CLN7 disease has received a personalized antisense oligonucleotide treatment. Initial outcomes indicate that this approach helped stop the disease's progression and reduced both the frequency and severity of seizures (Kim et al., 2019).

2.8 CLN8

CLN8 disease has two primary forms: the Turkish variant, a late-infantile NCL, and Northern epilepsy. Both forms are caused by mutations in the *CLN8* gene, located on chromosome 8p23.3, which encodes a transmembrane protein of the ER involved in lysosome biogenesis (Baltar et al., 2024). The Turkish variant is the more severe type, with symptoms usually starting between 2 and 7 years of age (Mitchell et al., 2001). Children with this form often experience myoclonic epilepsy, ataxia, and developmental regression, losing the ability to walk and speak. Survival of these patients rarely extends beyond late childhood (Badura-Stronka et al., 2021). The Northern epilepsy form generally emerges between 5 and 10 years of age and usually presents with recurrent seizures. As CLN8 disease progresses, patients experience ataxia and other motor impairments, gradual cognitive decline, and vision problems that worsen over time. Consequently, life expectancy is reduced (Ranta and Lehesjoki, 2000).

2.9 CLN9

CLN9 disease was first identified in 2004, and only a few cases have been reported. Therefore the genetic cause is still unknown. Symptoms typically begin in early childhood, around the age of 4. These symptoms include vision loss, seizures, ataxia, and myoclonus. Similar to other forms of NCL, patients with CLN9 also exhibit cerebral atrophy and developmental regression. Most reported cases demonstrated a lifespan of 15–20 years (Schulz et al., 2004). It remains unclear whether CLN9 is a separate NCL subtype, as some cases overlap with the CLN5 region but show no confirmed disease-causing mutations (Specchio et al., 2021).

2.10 CLN10

CLN10 disease occurs due to mutations in the *CTSD* gene, located at chromosome 11p15.5, which encodes the lysosomal aspartic protease cathepsin D (Jalanko and Braulke, 2009). CLN10 is one of the rarest and most severe forms of NCL. The disease symptoms usually emerge very early, sometimes even before birth, with seizures. Most affected infants experience problems such as heart failure. As a consequence, these infants typically survive only a few days. In addition, neurological symptoms such as microencephaly, absent reflexes, epileptic seizures, and respiratory difficulties have been detected in this form. Notably, the symptoms of this form are varied and depend on disease progression, the specific mutation, and the activity of cathepsin D. CLN10 disease leads to death quickly; however, these patients may initially appear normal (Specchio et al., 2021; Doccini et al., 2016). Over time, symptoms such as problems with coordination, learning, and eventually loss of vision, speech, and movement slowly appear (Specchio et al., 2021).

2.11 CLN11

CLN11 is an uncommon form of NCL, and limited case numbers have been reported. CLN11 disease occurs due to mutations in the *GRN* gene, located on chromosome 17q21.31). Symptoms typically appear around 25 years of age, indicating a late form of NCL, similar in timing to CLN4 but with different clinical characteristics. CLN11 is characterized by rapidly progressive retinal dystrophy, along with seizures, myoclonus, cerebellar ataxia, and cerebellar atrophy. This condition is caused by homozygous mutations in the *GRN* gene, which encodes the granulin precursor protein (Huin et al., 2020; Maximiano-Alves et al., 2025).

2.12 CLN12

CLN12 disease, also known as juvenile parkinsonism–NCL, typically appears between 13 and 16 years of age. It is caused by mutations in the *ATP13A2* gene, which is located on chromosome 1p36.13 (Park et al., 2011). This gene encodes an enzyme involved in regulating ion balance (Park et al., 2011). Initially reported in a canine model, CLN12 was later identified in four related individuals from a family in Belgium (Farias et al., 2011). These patients exhibited learning difficulties around the age of 8, followed by symptoms such as unsteady walking, myoclonus, and mood changes during their preteen years. As the disease progresses, affected patients experience loss of speech and mobility problems (Rosenberg et al., 2019). The family members also displayed a combination of progressive spinocerebellar ataxia, bulbar syndrome, extrapyramidal and pyramidal symptoms, along with cognitive decline (Rosenberg et al., 2019; Bras et al., 2012).

2.13 CLN13

CLN13 disease, also known as Kufs disease (CLN13) type B, typically appears around the age of 30. It is caused by mutations in the CTSF gene, which is located on chromosome 11q13.2. This gene encodes the lysosomal cysteine protease cathepsin F (Magrinelli et al., 2017). Early clinical symptoms include memory impairment and behavioral changes. As CLN13 disease progresses, motor symptoms such as tremors, bradykinesia, and rigidity with extrapyramidal features are reported (Zhang et al., 2025). In contrast to most other forms of NCLs, CLN13 is not associated with vision loss. Currently, no approved treatment has been identified for this disorder.

2.14 CLN14

CLN14 or progressive myoclonic epilepsy type 3 (EPM3) is a rare neurodegenerative disorder that follows an AR pattern of inheritance. CLN14 results from mutations in the *KCTD7* gene, which is located on chromosome 7q11.21 (Staropoli et al., 2012). The protein encoded by this gene is potassium channel tetramerization domain-containing protein 7. The initial clinical symptoms are myoclonic seizures, which are typically detected in infants between 8 and 24 months of age. As the disease progresses, affected children exhibit developmental regression and significant visual impairment. Characteristic pathological findings demonstrate the accumulation of autofluorescent lipopigment storage material within fibroblasts, neurons, and eccrine secretory epithelial cells (Metz et al., 2018). Currently, no therapeutic strategies have been reported for CLN14.

3 Current approved therapeutic approaches for the disease

For decades, researchers have worked hard to develop effective treatments for the various forms of NCLs to lessen the range of symptoms. Several approved and experimental treatment options look promising for managing the disease and reducing symptoms, although a definitive cure remains out of reach. The following sections elaborate on these strategies (Fig. 9.2).

3.1 Enzyme replacement

The most well-known treatment for CLN2 disease is enzyme replacement therapy (ERT), which has been approved by both the Food and Drug Administration (FDA) and European Medicines Agency (EMA) (Mole et al., 2019). This therapy involves delivering recombinant TPP1, known as cerliponase alfa (Brineura), into the cerebrospinal fluid every 2 weeks over a period of more than 4 hours per infusion. The drug is administered through a specialized reservoir that is implanted under the skin of the skull. After infusion, the TPP1 enzyme enters neuronal cells through a process mediated by the mannose-6-phosphate receptor. Once inside, it moves to the lysosome, where it helps reduce the buildup of storage material. This treatment raises levels of TPP1 in the brain, slowing disease progression and leading to measurable improvements in motor and language function (Schulz et al., 2018; Markham, 2017). This drug has been approved in the United States for patients under 3 years old and for all ages in the EU for CLN2 patients. It is administered as 300 mg once every

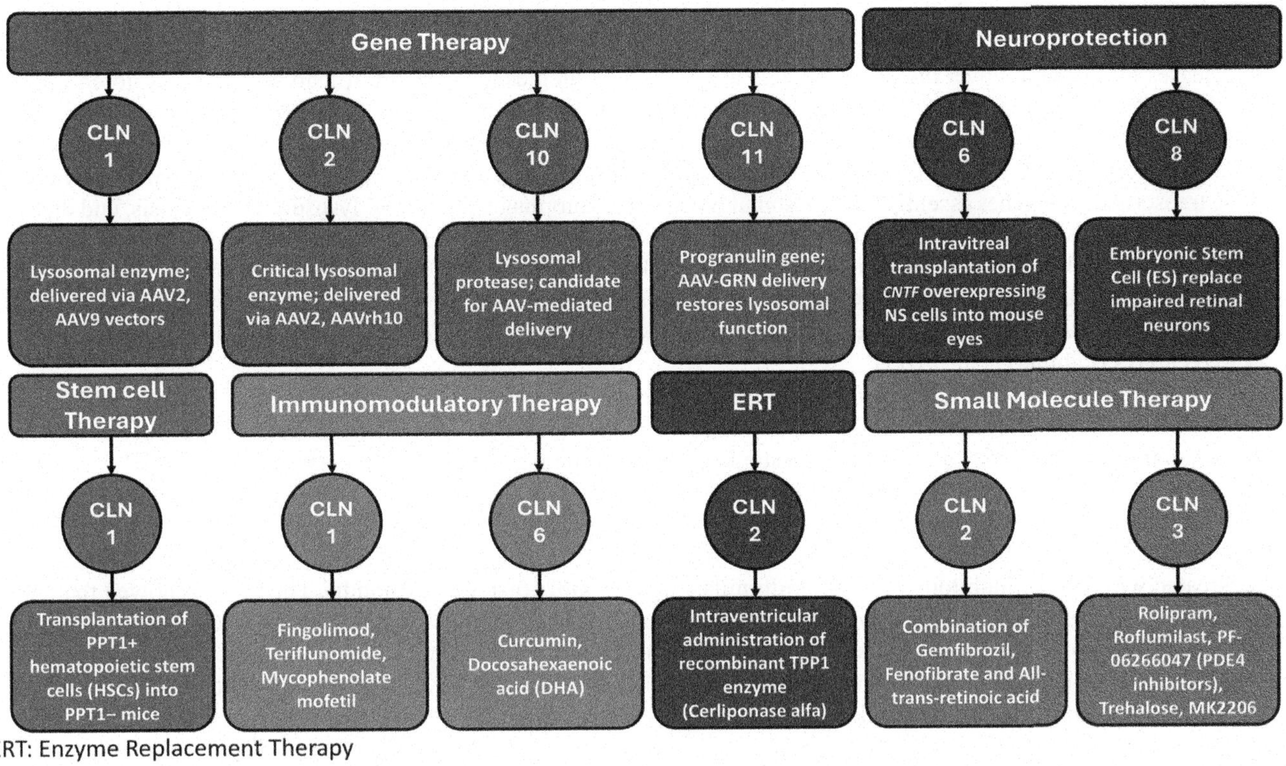

FIG. 9.2 Schematic overview of therapeutic strategies being explored for neuronal ceroid lipofuscinoses (CLN disorders), organized by modality. Approaches include gene replacement using AAV vectors for key disease genes/enzymes (e.g., CLN1/PPT1 via AAV2/AAV9; CLN2/TPP1 via AAV2/AAVrh10; CLN10; CLN11/GRN), neuroprotective retinal interventions (e.g., intravitreal transplantation of CNTF-overexpressing neural stem cells), stem-cell-based strategies (embryonic stem cell replacement of impaired retinal neurons; transplantation of PPT1 + hematopoietic stem cells in PPT1 − models), immunomodulatory agents (fingolimod, teriflunomide, mycophenolate mofetil), enzyme replacement therapy for CLN2 (intraventricular recombinant TPP1; cerliponase alfa), and small-molecule candidates (curcumin, DHA; gemfibrozil/fenofibrate/all-trans-retinoic acid; PDE4 inhibitors, including rolipram/roflumilast/PF-06266047; trehalose; MK2206). *ERT, enzyme replacement therapy.*

other week via intracerebroventricular (ICV) injection through a reservoir implanted under the calvaria. Each injection comprises 10 mL of cerliponase alfa with 2 mL of intraventricular electrolytes (U.S. Food and Drug Administration, 2024). The duration of treatment varies based on the case, ranging from 48 to 240 weeks (Markham, 2017). As with most medical treatments, ERT does have side effects, which are generally mild and can include fever, vomiting, allergic reactions, and rare infections related to the device (Kohlschütter et al., 2019). Although the immune system has developed antibodies against this enzyme, clinical trials have reported no significant downsides or any interference in the treatment process (Cherukuri et al., 2018). Furthermore, MRI studies support the benefits of this therapy, showing that treated patients lose brain volume at a rate of around 3%, compared to over 16% in untreated historical controls (Gaur et al., 2024). However, risks also exist; for instance, implanting the ICV device carries a risk of infection (Schulz et al., 2018). These procedures are complex and require careful surgical technique and postoperative care. Due to these challenges, researchers are also exploring alternative strategies, such as gene therapy, which may offer safer, longer-term solutions. Other potential treatments under investigation include small molecules, stem cell-based approaches, and immunomodulatory/neuroprotective strategies (Kohlschütter et al., 2019) that will be elaborated.

3.2 Small molecules

Treatment with gemfibrozil and fenofibrate, in combination with all-trans retinoic acid, has been shown to enhance TPP1 expression by activating the Transcription Factor EB (TFEB). This activation results from PPARα binding to its response element within the TPP1 promoter region, following the displacement of HSP-90. This approach can increase TFEB expression and TPP1 mRNA, which leads to a reduction in the amount of autofluorescent material in lysosomes. After administering gemfibrozil and fenofibrate to mice with CLN2, the fibrate drugs were replaced with HSP-90 in a competitive reaction, leading to PPARα activation. This trend resulted in PPARα binding to the PPAR-binding site (RXR) in the *CLN2* gene and regulation of TPP1, resulting in increased TPP1 expression (Ghosh et al., 2012). Given the complexity of neurodegenerative diseases, developing novel therapies requires a variety of strategies. As a result, researchers have explored interventions that target specific pathways to find more effective and improved treatments. One promising direction focuses on manipulating transcription factors involved in CLN progression, with the hope that this could lead to reliable therapeutic results (Palmieri et al., 2011; Sardiello et al., 2009; Settembre et al., 2011). TFEB orchestrates cellular clearance by governing the transcription of genes essential for lysosomal biogenesis, autophagic processes, lysosomal exocytosis, and the maintenance of proteostatic capacity within the lysosomal compartment. However, TFEB activity is suppressed by the serine/threonine kinase (Akt). Researchers believe that blocking Akt would boost TFEB-driven gene regulation. Hence, scientists used α-D-glucopyranosyl α-D-glucopyranoside (trehalose) and MK2206 to suppress Akt activity in CLN3 mouse model. This treatment successfully reduced disease symptoms and significantly increased the maximum lifespan of the mice from 544 to 699 days (Palmieri et al., 2017).

Aside from transcriptional regulation, signaling pathways are also crucial in CLN development. The second messenger cyclic adenosine monophosphate (cAMP) is a key regulator of various signaling events (Kandel, 2012), but cAMP levels are regulated by other compounds. Phosphodiesterases (PDEs) are responsible for the degradation of cAMP, and, in contrast, inhibitors for a specific form of PDE, PDE4, can elevate cAMP levels (Francis et al., 2011). It is found that cAMP has much lower levels in brains affected by CLN3. Because cAMP is broken down by PDEs, specifically the PDE4 subtype, researchers hypothesized that inhibiting PDE4 could restore cAMP levels and improve symptoms. To test this, CLN3 mice were treated with PDE4 inhibitors, rolipram, roflumilast, and PF-06266047, for 6 months. The results showed that raising cAMP levels through this method led to significant improvements in symptoms (Aldrich et al., 2016).

Other small molecules under investigation include phosphocysteamine and N-acetylcysteine (Zhang et al., 2001; Roberts et al., 2012), cysteamine bitartrate (Levin et al., 2014), N-(tert-butyl) hydroxylamine (Sarkar et al., 2013), α-amino-3-hydroxy-5-methyl-4-isoxazolepropionic acid-type glutamate receptor antagonist (Kovács et al., 2011), and memantine (Kovács et al., 2012).

3.3 Stem cell therapy

Stem cell treatments are receiving considerable attention in research. In a proof-of-concept study for CLN1, hematopoietic stem cells (HSCs) and bone marrow from PPT1 + mice were transplanted into mice with PPT1 deficiency. The results showed significant improvements in all disease markers related to PPT1 deficiency, suggesting that this method can deliver functional enzymes to the affected nervous system (Peviani et al., 2023). However, studies related to this approach remain preliminary.

3.4 Immunomodulatory and neuroprotection

The immune system plays a leading role in causing detrimental effects in patients with CLN. In fact, the immune response to the lack of the PPT1 and TPP1 enzymes is the underlying cause of CLN1 and CLN2, respectively (Itagaki et al., 2018). These primary enzyme deficiencies in CLN1 and CLN2 trigger harmful neuroinflammation. This involves activating microglia and allowing CD8+ T lymphocytes to enter the central nervous system, which exacerbates neuronal loss and accelerates symptoms such as vision loss and motor decline. Preclinical studies support this connection; for example, CLN1 mice that lack an adaptive immune system showed significantly less neurodegeneration. As a result, immunomodulation has become a promising treatment option. Compounds such as fingolimod, teriflunomide, and mycophenolate mofetil, an immunosuppressant, have demonstrated effectiveness in animal models by reducing neuroinflammation and improving key clinical symptoms. Natural substances, such as curcumin, have also shown potential in lowering this damaging immune response (Groh et al., 2013; Seehafer et al., 2011; Jung et al., 2006; Mirza et al., 2013). Along with immunomodulation, neuroprotection serves as a complementary strategy aimed at preserving neuronal health and function. This approach seeks to protect nerve cells from damage, often employing techniques such as stem cell therapy. For example, transplanting embryonic stem cells into the eyes of CLN8 mice delayed the crucial loss of photoreceptor cells. In a more advanced approach, researchers engineered neural stem cells to overexpress ciliary neurotrophic factor (CNTF), a protein that is known to protect retinal neurons. When these modified cells were transplanted, they significantly slowed visual degeneration in a CLN6 model, showing the potential of combining cell-based delivery with neuroprotective factors (Khalilpour et al., 2017; Meyer et al., 2006; Jankowiak et al., 2015). Among all the possible treatments mentioned, only enzyme replacement has gained FDA and EMA approval. However, other approaches seem promising; further experiments and clinical trials are needed to discover the safest and most optimal treatment.

4 Gene therapy in NCL

4.1 Targets for gene therapy and its mechanism of action in the disease

Gene therapy is an intriguing approach to discovering novel treatments for diseases that are difficult to cure. Gene therapy encompasses two fundamental methods. First, DNA can be transported into the patient's system via a vector. Second, patients' cells are harvested, genetically modified, and then infused back into the patient's body (Mali, 2013). The first method is quite popular, and many studies are conducted based on gene delivery via a vector. Adeno-associated virus (AAV) is one of the main vectors used for gene therapy. AAV is nonintegrating, forms episomes, is long-lasting in non-dividing cells, and has a packaging limit of ~4.7–5 kb. To function in host cells, AAV requires another virus, primarily adenoviruses, to be present in the host cell, and this attribute is called coinfection (Samulski and Muzyczka, 2014). The recombinant adeno-associated virus (rAAV) is an essential vector for gene therapy. An interesting feature of rAAV is that it can provide a long-lasting treatment by being a temporary element inside the cell. This is different from the wild-type AAV. The wild-type AAV uses its Rep protein to insert its genome into a specific place on human chromosome 19. This creates a permanent genetic change. However, the rAAV vector is engineered differently. All the original viral genes, including the Rep gene, are removed. This removal is done to add a therapeutic gene, which is also a key safety improvement. Without the Rep protein, the vector cannot integrate its DNA into the host cell's chromosomes. Instead, the rAAV genome follows a different path. Inside the nucleus, its single-stranded DNA becomes double-stranded. The ends of this DNA then connect to form circular molecules called episomes. These episomes stay in the nucleus as separate structures from the cell's own chromosomes. The cell can read the therapeutic gene from these episomes and make the needed protein to treat the disease. The key point is that the viral DNA and the host DNA remain separate. This separation gives the rAAV vector its main safety benefit and its main limitation. Because the DNA does not integrate, it avoids a risk called insertional mutagenesis. However, how long the treatment lasts depends on the cell type. In cells that do not divide, like neurons or muscle cells, the episome is stable and can work for a long time. But in tissues where cells are actively dividing, the rAAV episome is not copied when the cell divides. It gets diluted and lost over time, making the therapeutic effect temporary. Therefore rAAV is designed as a safe vector that provides a long-lasting, but not always permanent, genetic correction (Wang et al., 2019; Zwi-Dantsis et al., 2025; Wang et al., 2024). In the gene therapy process via rAAV, the target gene is placed between the two ITRs. However, since the length of the AAV genome is approximately 4.8 kilobases, delivering genes longer than 5 kilobases via AAV is a challenge. To address this issue, scientists use a dual-vector approach in which the target gene, which is most often longer than 5 kilobases, is split into segments of <5 kilobases and packaged into multiple AAV vectors for delivery purposes (Chamberlain et al., 2016). rAAV vectors are classified into different types based on factors such as serotype, genome structure, and payload design. These vectors encompass the following types: AAV1, AAV2, AAV5, AAV6, AAV8, and AAV9 (Wang et al., 2024). Scientists have tested various types of AAV vectors engineered with genes to find a better approach for curing NCL.

Targets of gene therapy for NCLs are broad. One of the primary targets is the PPT1 gene, which is responsible for producing a lysosomal enzyme that prevents lysosomal storage material aggregation. rAAVs have been frequently used to deliver this gene into mammalian cells. The TPP1 gene, with a similar function to PPT1, is another target of gene therapy (Griffey et al., 2004; Macauley et al., 2012; Shyng et al., 2017; Katz et al., 2017, Katz et al. (2015)). Progranulin haploinsufficiency can cause CLN11. In fact, mutations in the progranulin gene (Grn) are directly related to the development of a lysosomal storage disorder. Hence, the infusion of Grn into an AAV vector can deliver the normal gene to cells (Arrant et al., 2018). Furthermore, cathepsin D (CtsD) is a type of protease that is active in the lysosome, and its deficiency causes lysosomal storage material aggregation. Therefore CtsD can be considered a candidate for gene delivery via AAV vectors (Shevtsova et al., 2010). In addition, carrying the functional CLN3, CLN5, and CLN6 genes via AAV vectors is another potential target (Bosch et al., 2016; Sondhi et al., 2014; Kleine Holthaus et al., 2018; Mitchell et al., 2018).

4.2 Current role of gene therapy in the treatment of the disease

Gene therapy has emerged as one of the most compelling therapeutic approaches for disease-modifying treatment in NCLs. However, none of the gene therapy strategies for curing NCLs have gained FDA approval to date. The primary strategy in this approach involves using recombinant AAV vectors to mediate the delivery of functional genes to affected tissues, mainly the Central Nervous System (CNS) and the retina (Kleine Holthaus et al., 2020), and preclinical experiments on murine and large animal models have proven the favorable efficacy of this strategy, which led to several clinical trials for various forms of NCLs (Rosenberg et al., 2019). A critical consideration in this therapeutic strategy is the distinct pathophysiology of CNS and ocular manifestations. Once CNS-directed gene therapy often does not entirely prevent retinal degeneration, dedicated ocular approaches are needed (Kleine Holthaus et al., 2018). Consequently, the parallel development of ocular gene therapy is crucial. Research indicates that therapeutic efficacy is contingent upon targeting the specific retinal cell populations affected by the disease pathology. For instance, in CLN3 disease, which primarily involves the inner retina, AAV-mediated transgene delivery targeted to bipolar cells has demonstrated significant cytoprotection and functional rescue in murine models (Bartsch and Storch, 2022). Similar success in attenuating retinal degeneration has been observed following ocular AAV administration in models of CLN1, CLN5, CLN6, CLN10, and CLN11 diseases (Mirza et al., 2013). In addition, the defective protein further determines the therapeutic strategy. For deficiencies in lysosomal enzymes such as PPT1 and TPP1, gene therapy can leverage the process by which the secreted gene product from transduced cells is endocytosed by nontransduced cells, achieving a broader therapeutic impact. Conversely, for mutations in nonsecreted transmembrane proteins (e.g., CLN3, CLN6, CLN7), a direct transduction of a high proportion of target cells is likely required, as these proteins are not amenable to cross-correction (Zhang et al., 2025). Although there are no widely approved, commercially available gene therapies for NCLs at present, the FDA has granted fast-track designation to gene therapies for different forms of NCLs.

4.3 Current advances in preclinical gene therapy for the disease

Preclinical studies in gene therapy are essential for understanding the complex biological processes underlying genetic disorders and for exploring new treatment approaches. To date, several preclinical gene therapies have been designed and reported; this approach has significant therapeutic potential for NCLs (Table 9.2).

In this regard, Griffey et al. designed a study in which neonatal Ppt1 knockout mice received multiple intracranial administrations of an AAV2 vector carrying the human *PPT1* gene. The results demonstrated statistically significant therapeutic effects of this approach both in vitro and in vivo. In vitro, PPT1-deficient fibroblasts treated with the vector showed a nearly threefold increase in enzyme activity (13.2 vs. 4.8 nmol/mg/h; $P < 0.01$), confirming that the recombinant enzyme was functionally produced. In addition, intracranial administration of rAAV–PPT1 to newborn PPT1–/– mice restored approximately 15% of normal PPT1 activity near the injection sites and significantly reduced secondary β-glucuronidase elevation and cortical storage material ($P < 0.05$). Treated mice also exhibited enhanced brain weights (0.38 ± 0.015 g vs. 0.31 ± 0.019 g; $P < 0.01$) and thicker cortices across several regions ($P < 0.01$–0.042), with clear histological evidence of preserved neurons and reduced lysosomal accumulation. Thus the outcomes from the study by Griffey et al. demonstrated that this therapeutic approach led to localized restoration of PPT1 enzymatic activity at the injection sites. This restoration was associated with a reduction in lysosomal storage accumulation, mitigation of neurodegenerative changes, and significant enhancement in behavior compared to untreated mutant controls. However, the intervention did not significantly alter seizure frequency or extend the lifespan in the treated animals (Griffey et al., 2004).

In another study, Macauley et al. demonstrated that gene therapy targeting PPT1 in a murine model of NCL caused significant improvements in neurobehavioral function and led to increased lifespan in the treated animals. In the Macauley et al. study, the therapeutic effects on motor function and lifespan were significantly enhanced when gene therapy was combined with bone marrow transplantation, whereas bone marrow transplantation alone showed no significant therapeutic impact. Indeed, the results of this study demonstrated that brain tissue from combination-treated mice showed thicker cortical regions, less storage material, and a significant reduction in neuroinflammation, with decreased levels of activated microglia and astrocytes ($P < 0.001$). In addition, biochemical investigations revealed a sixfold increase in PPT1 enzyme activity compared to wild-type levels, which demonstrated enhanced gene expression. Therefore these findings indicate that AAV2/5 gene therapy with BMT not only extended lifespan but also significantly preserved brain structure and function (Macauley et al., 2012). Furthermore, Shyng et al. applied AAV9-PPT1 into the CNS of Ppt1 knockout mice.

The results of this study demonstrated that treated mice lived nearly 19 months, compared to about 8 months for untreated animals ($p < 0.0001$), and showed enhanced motor coordination and strength ($p < 0.001$). Moreover, pathological analysis revealed reduced inflammation in both the brain and spinal cord ($p < 0.01$), with preservation of motor neurons ($p < 0.001$).

As a consequence, this study revealed that AAV9-PPT1 caused less neuropathological injury in the targeted areas of the CNS, improved motor skills, and increased survival. In addition, this study illustrated that combined delivery to both the brain and spinal cord led to significantly superior therapeutic outcomes compared to treatment of either region alone. Therefore applying AAV9-PPT1 in the brain and spinal cord at the same time is important for effective gene therapy in NCLs disease (Shyng et al., 2017).

Moreover, in another study, Katz et al. demonstrated that intraventricular administration of the AAV2-TPP1 vector in a canine model of CLN2 disease caused the induction of high TPP1 expression in ependymal cells and extensive enzyme distribution across the brain.

TABLE 9.2 Preclinical gene therapies for various forms of NCLs.

Form of NCLs	Treatment	Target	Subject model(s)	Dosage	Injection route/site(s)	Outcome	Toxicity	Follow-up and monitoring	Complications	Suggested and used complication management strategies
CLN1	AAV2-PPT1	Restoring PPT1 function	*Ppt1* knockout mice	0.5 μL ($8.5*10^5$ IU) per injection, 4 injections	Intracranial	Reduction in lysosomal storage materials, mitigating neurodegeneration, improves behavioral features	No significant toxicity	Monitoring until 7 months of age**Methods:** Measuring PPT1 and β-glucuronidase levels in the brain, qPCR* to track rAAV-PPT1, and electron microscopy to monitor AFSM* changes	Minor efficacy, 15% restoration of PPT1 activity, brain mass, and cortical thickness did not reach normal levels	**Suggestions:** use of alternative AAV serotypes, use of protein transduction domains for broader enzyme distribution, and direct injections into the cerebellum
	AAV2/5-PPT1 BMT*	Restoring PPT1 function		Vector: 2 μL (1×10^{11} vg/mL) injected into 6 sites of the brain BMT: 100 μL	Intracranial, temporal vein injection	Increasing lifespan, improving motor function	7 out of 9 mice treated with AAV2/5 + BMT developed liver tumors	Monitoring up to 18.5 months **Methods:** Kaplan–Meier analysis for survival tracking, assessing motor function using a constant-speed rotarod test, measuring PPT1 activity, flow cytometry for bone marrow engraftment levels, qPCR for measuring AAV-PPT1 levels, histological analysis for assessing AFSM and neuroinflammation	Minor efficacy, sole BMT was harmful and aggravated motor disturbances	**Suggestions:** targeting the peripheral nervous system and visceral organs, intrathecal and systemic delivery of gene therapy, in addition to the CNS*-directed approach, neonatal period should be considered as a risk factor in treatment
	AAV9-PPT1	Restoring PPT1 function		1×10^{12} vg/mL per injection, one injection	Intracranial, intrathecal	Improving neuropathological injury, motor function, and lifespan	No significant toxicity	Monitoring for up to 22 months **Methods:** Kaplan-Meier analysis for survival tracking, assessing motor function using a constant-speed rotarod test, measuring PPT1 and β-glucuronidase levels in the brain and spinal cord, histological analysis for cortical thickness and brain weight, AFSM, and neuroinflammation, and measuring levels of inflammatory markers in the brain	Minor efficacy	**Implemented:** Combination of intracranial and intrathecal injections **Suggestion(s):** Directly targeting the thalamus and exploring additional modalities to address peripheral and cardiac manifestations

continued

TABLE 9.2 Preclinical gene therapies for various forms of NCLs—cont'd

Form of NCLs	Treatment	Target	Subject model(s)	Dosage	Injection route/site(s)	Outcome	Toxicity	Follow-up and monitoring	Complications	Suggested and used complication management strategies
CLN2	AAV2-TPP1	Restoring TPP1 function	Canine (TPP1-deficient Dachshunds)	2×10^{12} vg/mL per injection, one injection	Intraventricular	Extending lifespan, improving cognitive function, delaying neurological symptoms, cerebellar ataxia, and visual deficits	No significant toxicity	Monitoring up to ~ 21 months **Methods:** Clinical neurological examinations to track the onset of specific symptoms, assessing cognitive function via the T-maze task, measuring PPT1 activity, performing a neutralization assay for measuring anti-TPP1 NAbs*, conducting qPCR for measuring AAV2-TPP1 distribution, utilizing immunohistochemistry for TPP1 distribution, and staining for neuroinflammation	Minor efficacy, less effective for the retina	**Implemented:** Complication of NAbs was managed with immunosuppressive drugs **Suggestion(s):** Intraocular injection of AAV-TPP1 for retinal correction
	AAVrh.10 hCLN2	Restoring TPP1 function	Rats (Sprague–Dawley rats) NHPs* (African green monkeys)	Rats: 5.0 μL containing 5×10^{10} GC* per injection, 2 injections-Nonhuman Primates: 15 μL containing 1.5×10^{11} GC per injection, 12 injections	Rats: Intraparenchymal into the striatum-Nonhuman primates: intraparenchymal at 12 CNS locations	An increase in TPP1 expression and widespread distribution of TPP1	Mild-to-moderate localized histopathological changes at the injection site, including perivascular cuffing (lymphocytes/plasma cells), gliosis, and swollen microvesiculated neurons in rats; in NHPs, focal gliosis and hemosiderin-laden macrophages at injection sites, which progressed to mild-to-moderate white matter edema, inflammation, and gliosis by 90 days	90-days monitoring **Methods:** Daily health checks, videotaped sessions for behavioral assessment, hematology and serum chemistry, measurement of anti-AAVrh.10 neutralizing antibodies, and comprehensive postmortem examinations	Localized inflammatory and histopathological responses at the injection sites, anticapsid neutralizing antibodies in NHPs	No specific management strategies were employed or deemed necessary; localized brain inflammation was considered an acceptable risk, and the immune response in NHPs might be more pronounced than in humans

CLN3	scAAV9/MeCP2-hCLN3sc AAV9/β-actin-hCLN3	Restoring CLN3 function	Cln3ex7/8 mice	2×10^{12} GC per injection, one injection	Intravenous	Improving motor function, attenuating microglial and astrocyte activation, and reducing lysosomal pathology	No significant toxicity	Monitoring up to 13 months**Methods:** Weekly body weight measurements, blood chemistry analysis, accelerating rotarod test for motor function, postmortem brain analysis, Western blot for expression levels of control vectors, and qPCR for expression levels of *CLN3*	Minor efficacy, the MeCP2 promoter led to expression primarily in neurons, while the β-actin promoter led to expression primarily in astrocytes	**Implemented:** Using only MeCP2 promoter
CLN3	AAV9.Mecp2.hCLN3	Delivering the human CLN3 gene to the CNS		4 μL (2.2×10^{10} vg per animal), one injection	Intracerebroventricular	Highly reduced lysosomal storage material, attenuating astrocytosis and microgliosis, improving motor function	No significant toxicity	Monitoring up to 24 months of age**Methods:** RT-qPCR for assessing transgene expression, immunohistochemistry for assessing hCLN3 protein, biomarker analysis, rotarod, pole climb, force plate actimeter, neurological score, optokinetic tracking, home cage running wheel activity, and neonatal reflex tests for behavioral analysis	Rescue of lysosomal storage and neuroinflammation was not 100% complete; male mice showed a more pronounced reduction in SubC accumulation at 12 months and in microglial activation (CD68) at 18 months, treatment did not rescue all behavioral deficits	**Suggestion:** Testing in more severe animal models or by developing strategies for even earlier administration (e.g., in utero) or broader transduction
CLN5	sAAV9-MND-ovineCLN5 & LV-MND-ovineCLN5	Restoring CLN5 function	Borderdale sheep with a naturally occurring CLN5 deficiency	ssAAV9: 3.1×10^{12} vg per animal, multiple injections Lentivirus: 3.6×10^{9} TU* per animal, multiple injections	Intracerebroventricular, intraparenchymal	Extending lifespan, improving neurological function, attenuating brain atrophy, and delaying disease progression	No significant toxicity	Monitoring up to 27 months **Methods:** Monthly assessments using OBDRS*, closed-field maze tests to assess vision and cognition, menace response tests, funduscopic eye examinations, and ERG* for visual function; longitudinal CT* scans to measure intracranial volume and monitor brain atrophy; MRI* to confirm brain structure; and immunohistochemistry on brain tissue to confirm CLN5 protein expression (postmortem)	Vision loss has not been entirely prevented	**Suggestion:** Adding an ocular-specific injection

continued

TABLE 9.2 Preclinical gene therapies for various forms of NCLs—cont'd

Form of NCLs	Treatment	Target	Subject model(s)	Dosage	Injection route/site(s)	Outcome	Toxicity	Follow-up and monitoring	Complications	Suggested and used complication management strategies
CLN8	AAV9. pT-MecP2. hCLN8	Delivering the human CLN8 gene to the CNS	Cln8mnd mice	4 μL (5×10^{10} vg per animal), one injection	Intracerebro-ventricular	High expression of hCLN8, significant reduction in lysosomal storage material, attenuation of astrocytosis and microgliosis, improvement in motor function, prevention of tremors, partial improvement in MWM*, and an increase in lifespan	No significant toxicity	Monitoring up to 24 months of age **Methods:** RNAscope and qRT-PCR for assessing transgene expression, biomarker analysis, rotarod, vertical pole climb, force plate actimeter (for tremors), a three-component neurological score for motor function analysis, and Morris water maze for visual analysis	Incomplete late-stage rescue limited cognitive/visual recovery	No complication management strategies were implemented or suggested
CLN11	AAV2-Grn	Delivering the progranulin gene to neurons in the brain	Progranulin knockout mice	1 μL (7.36×10^{11} GC/mL) per hemisphere, one injection in each cerebral hemisphere	Medial prefrontal cortex	Increase in progranulin levels, reduction in microgliosis, and correction of abnormal cathepsin D activity	Local inflammatory response at the injection site; strong elevation of MHC II	Monitoring 8–10 weeks postinj ection**Methods:**Immunostaining and autofluorescence imaging for assessing microgliosis, ELISA* for assessing progranulin protein, qRT-PCR for assessing progranulin RNA, Western blot for lysosomal proteins, and enzymatic activity assays for assessing cathepsin D	Localized injection site inflammation, reduced transgene expression in knockout mice, and generation of antiprogranulin antibodies	No complication management strategies were implemented or suggested

In this study, dogs treated with rAAV.caTPP1 exhibited enhanced TPP1 levels, approximately 30-fold above baseline levels ($p < 0.01$). This result confirmed efficient gene expression; however, enzyme activity initially declined due to immune responses.

Also, treated dogs demonstrated significant delays (4–6 months, $p < 0.05$) in the onset of neurological symptoms such as ataxia, nystagmus, and proprioceptive deficits compared to untreated animals. Furthermore, the survival of treated dogs was extended from an average of 10.4 months in untreated dogs to 17.5 months ($p < 0.05$). In addition, histological assays revealed two–tenfold increases in TPP1 enzyme activity across multiple brain regions, reduced glial activation and storage burden, and promoted neuronal preservation. Notably, this study showed increased TPP1 levels in the heart and spleen ($p < 0.05$). Therefore this study suggested using AAV to deliver the gene to brain cells to provide long-term TPP1 protein production, wide distribution in the central nervous system, and a significant therapeutic effect on TPP1 deficiency (Katz et al., 2015).

Sondhi et al. conducted a preclinical study to demonstrate the safety of intracerebral delivery of AAVrh10-TPP1 in rats and African green monkeys. Previous clinical reports have illustrated that direct delivery of the human CLN2 cDNA to the central nervous system, using an AAV2 vector, is safe in pediatric patients. In this regard, Sondhi et al. demonstrated in previous gene therapy investigations that the rhesus macaque-derived vector AAVrh.10hCLN2 enabled widespread distribution of TPP1 throughout the central nervous system. Therefore based on these previous outcomes, this study was conducted to investigate the safety and effectiveness of CNS-delivered AAVrh.10hCLN2. The outcomes of this study showed that bilateral striatal administration of 1×10^{11} genome copies of AAVrh.10hCLN2 in rat models caused no significant systemic or neurological toxicity. Similarly, in African green monkeys that were administered 1.8×10^{12} genome copies, no significant clinical or biochemical abnormalities were observed.

Furthermore, quantitative enzyme assays showed TPP-I activity more than two standard deviations above baseline in 31.7% ± 8.1% of brain tissue at 90 days, which confirmed efficient transgene expression. These results revealed appropriate AAVrh.10hCLN2 distribution with an acceptable safety profile in the CNS. In addition, TPP-I activity exceeded normal levels, demonstrating the potential of AAVrh.10hCLN2 as a therapeutic option and its suitability for use in clinical trial studies (Sondhi et al., 2012).

Furthermore, Bosch et al. studied a gene therapy approach for JNCL, which is caused by mutations in the CLN3 gene. In this study, the researchers investigated two self-complementary AAV9 (scAAV9) vectors designed to deliver the human CLN3 gene at different expression levels, using either the MeCP2 or β-actin promoters.

After one injection in young NCL mice, both designed vectors increased CLN3 levels. However, only the MeCP2-based vector significantly improved motor coordination to near wild-type levels ($p < 0.05$). In addition, scAAV9/MeCP2-hCLN3 attenuated brain inflammation by reducing microglial activation, astrocytosis, and lysosomal markers (LAMP-1 and SCMAS) to near-normal levels. This effect occurred because MeCP2 targeted neurons, while β-actin primarily targeted astrocytes. This study demonstrated that systemic delivery of scAAV9 can restore CLN3 in the brain, and it highlights the importance of choosing the right promoter for effective treatment (Bosch et al., 2016).

In addition, Mitchell et al. conducted a study to investigate the therapeutic effect of gene therapy for CLN5 disease in a sheep disease model. This study utilized either lentiviral or AAV9 vectors to deliver the CLN5 gene through ICV and intraparenchymal injections at 2–3 months of age, prior to the appearance of disease symptoms.

The results revealed that this intervention was well tolerated, without complicated immune reactions or behavioral abnormalities during 2 years of follow-up. The treatment was well tolerated, without complications, inflammation, or behavioral changes over 2 years. Treated sheep stayed healthy far longer than untreated animals, which normally live up to 22 months. However, treated animals demonstrated delayed loss of vision around 21–25 months. Brain scans confirmed that treated animals maintained normal brain size and structure. Notably, one AAV9-treated sheep is still alive and active at 57 months. These results show that early CLN5 gene therapy causes slow disease progression and decreased symptoms like brain atrophy and visual deterioration and also significantly extends survival (Mitchell et al., 2018).

In this regard, Arrant et al. investigated a gene therapy strategy for CLN11 caused by GRN mutations. In this study, AAV was applied to deliver the progranulin gene into a mouse model. This AAV-based therapy reduced lipofuscinosis, decreased microgliosis, and restored lysosomal function in brain cells.

Indeed, AAV-Grn significantly promoted progranulin levels in the brain ($F(1,10) = 49.78$, $p < 0.001$). In addition, this approach significantly decreased lipofuscin accumulation in the CA3 and VPM/VPL thalamus ($F(1,29) = 10.09$, $p = 0.0035$) and enhanced microglial activation in the motor cortex and CA3 ($F(3,26) = 3.03$, $p = 0.0473$). Furthermore, behavioral tests performed on treated mice were similar to those of wild-type controls (Mann–Whitney test, $p < 0.01$). Notably, this gene therapy demonstrated effectiveness even after symptoms had begun. Moreover, this gene therapy not only targeted neurons but also microglia, leading to an improvement in neuron health. These outcomes demonstrated that AAV gene therapy delivering progranulin could be a promising treatment for CLN11 disease (Arrant et al., 2018).

Also, Johnson et al. investigated the safety and efficacy of an AAV9-based gene therapy in CLN3 disease. In this study, AAV9 was delivered directly into the cerebrospinal fluid and led to widespread expression of the human CLN3 gene for 2 years. This intervention significantly attenuated pathological symptoms, such as a 70–85% decrease in lysosomal auto fluorescent storage material and a 65% reduction in mitochondrial ATP synthase ($p < 0.01$). Furthermore, neuroinflammation decreased due to reduced astrocyte and microglial activation ($p < 0.05$). In addition, treated mice demonstrated significant progression in function and activity patterns ($p < 0.05$).

So, the behavioral and pathological investigations revealed that the gene therapy approach was effective and ameliorated disease symptoms.

Notably, this approach was safe and well-tolerated (Johnson et al., 2023). Furthermore, Johnson et al. conducted another gene therapy study for CLN8 disease by applying a self-complementary AAV9 vector to deliver the human CLN8 gene via ICV injection in neonatal mice. This gene encodes an ER protein involved in lysosomal enzyme transport. The primary purpose of this study was to demonstrate the safety and efficacy of this approach. The results revealed the safety and long-term expression of the therapeutic gene across the central nervous system. In addition, this approach significantly reduced pathological and behavioral symptoms of the disease and extended survival (Johnson et al., 2023).

4.4 Recent clinical trials or ongoing clinical trials on gene therapy for the disease

Recently, gene therapy has gained attention as a potential treatment for NCLs disease. However, gene therapy presents several challenges, such as accurately targeting affected CNS regions, ensuring safety, and demonstrating effectiveness, all of which must be carefully considered (Mole et al., 2019). Among the various types of viral vectors, AAVs are the most commonly used vectors in NCLs (Ziółkowska et al., 2025). Based on preclinical outcomes, AAV vectors have been proposed due to several key factors, such as their capacity to transport appropriately sized genes, effectively target the CNS, ensure safety, and support long-term gene expression (Byrne et al., 2012; Weinberg et al., 2013). Moreover, AAV vectors can also be modified to improve their target specificity (Weinberg et al., 2013). On the contrary, for gene therapy in NCL disease patients, eligibility must include specific inclusion and exclusion criteria to ensure the safety and efficacy of clinical gene therapy. Typically, the patient must have a confirmed genetic diagnosis of a specific NCLs type and demonstrate symptoms of early to mid-stage disease with enough healthy brain cells remaining to benefit from the application of gene therapy. Furthermore, patients must fall within a certain age range, as progressed NCLs may reduce the efficacy of gene delivery (Donsante and Boulis, 2018; Schulz et al., 2024). Exclusion criteria for NCLs gene therapy include severe cerebral atrophy on neuroimaging, uncontrolled epilepsy, or serious systemic illnesses that could affect the outcomes of the gene therapy approach (Donsante and Boulis, 2018; Schulz et al., 2024).

Contraindications include hypersensitivity to any vector component, significant coagulopathies, or anatomical issues (Aylward et al., 2020). Ethical considerations are also required through informed consent and careful review by a multidisciplinary team (Kohlschütter, 2021). The mentioned criteria are essential for clinical trials to help with patient selection and treatment outcomes. Up to now, some gene therapy studies have been recorded for the treatment of NCLs, although no results have been reported at this time.

For instance, a completed clinical trial study (NCT00151216) investigated the use of intracranial gene therapy for treating CLN2 disease (Table 9.3). As mentioned, CLN2 disease is caused by TPP1 deficiency.

In this trial, five children with severe disease (disability score 0–4) and six children with moderate disease (disability score 5–6) participated. In this gene therapy study, AAV2 was applied to deliver a functional copy of the CLN2 gene directly into multiple sites within the brain region of the central nervous system, with a total dose of approximately 3×10^{12} viral particles, to restore TPP-I activity and slow disease progression and symptoms by preventing further neuronal loss. The follow-up period of this study was approximately 42–48 months.

The neurological outcomes demonstrated a slower rate of disease progression, such as declines in motor function, seizure frequency, and communication abilities in several treated cases. In addition, neuroimaging results verified these clinical findings, showing reduced brain atrophy and sustained levels of N-acetyl-aspartate in brain regions. This suggests that the gene therapy is safe and effective for CLN2 disease. This trial study was an important step in developing other gene therapies for CLN disorders.

Notably, this study continued and led to the initiation of another clinical trial (NCT01161576) for CLN2 disease. In this trial, patients must have at least one of the five most common CLN2 mutations: C3670T (nonsense mutation resulting in Arg208 stop), G3556C (intron seven splice site mutation), G5271C (missense mutation Gln422His), T4396G (aberrant splicing in intron 8), or G4655A (missense mutation Cys365Tyr).

In this clinical trial study, AAV2 was also applied to deliver a functional copy of the CLN2 gene directly into the brain through stereotactic injections at multiple sites. In this trial, patients were divided into two groups based on the total amount

TABLE 9.3 Overview of completed and ongoing clinical trials for NCLs.

ID	Status	CLN type	Title	Type of treatment	Phase	Name of the product	Dosage	Eligibility criteria	Number of patients	Targeted genes and vectors	Adminis-tration route	Follow-up and monitoring	Brief result/ safety and efficacy	Statistical interpre-tation
NCT 01161576	Com-pleted	CLN2	Safety Study of a Gene Transfer Vector (Rh.10) for Children with Late Infantile Neuronal Ceroid Lipofuscinosis (LINCL)	Gene therapy	Phase 1	AAVrh. 10CUhCLN2	Group A: 9×10^{11} GC[a]group GC group B: 2.85×10^{11}GC	2–18 years, having a Weill-Cornell LINCL scale[b] score between 4 and 12, and have not participated in gene or stem cell therapy	16	**Gene:** TPP1 **Vector:** AAVrh.10	Intracranial injections	Visit at the month 18	No result has been reported	No statistical outcomes have been reported
NCT 01414985	Com-pleted	CLN2	Study of CNS Administration of a Replication-Dependent AAV Vector for CLN2 Disease	Gene therapy	Phase ½	AAVrh. 10CUCLN2	Group A: 9×10^{11} GCGroup GC Group B: 2.85×10^{11}GC	3–18 years, definitive diagnosis of LINCL[c], Weill-Cornell LINCL scale score between 1 and <4, or an uncommon genotype (not one of the five most common CLN2 mutations), and have not participated in gene or stem cell therapy	8	**Gene:** TPP1 **Vector:** AAVrh.10	Intracranial injections	Visit at months 1, 6, 12, and 18	No result has been reported	No statistical outcomes have been reported
NCT 05791864	Recruit-ing	CLN2	Study to Evaluate the Safety and Tolerability of Gene Therapy with TTX-381 for Ocular Manifes-tations in CLN2 Disease	Gene therapy	Phase ½	TTX-381	Cohort 1:$2 \times 10^{1\circ}$ GCCohort 2:$6 \times 10^{1\circ}$GC	12–84 months, having biallelic CLN2 mutations and decreased TPP1 enzyme activity, exhibiting clinical symptoms of CLN2 disease or having an older sibling with a confirmed CLN2 diagnosis, meet specific CRT[d] criteria measured by SD-OCT[e]; CRT must be between 140 μm and 210 μm in both eyes	16	**Gene:** TPP1 **Vector:** AAV9	Subretinal injections	5 years of follow-up	No result has been reported	No statistical outcomes have been reported

continued

TABLE 9.3 Overview of completed and ongoing clinical trials for NCLs—cont'd

ID	Status	CLN type	Title	Type of treatment	Phase	Name of the product	Dosage	Eligibility criteria	Number of patients	Targeted genes and vectors	Adminis-tration route	Follow-up and monitoring	Brief result/ safety and efficacy	Statistical interpre-tation
NCT 03770572	Active, not recruit-ing	CLN3	A Study of AT-GTX-502 in Children with CLN3 Type Batten Disease	Gene therapy	Phase ½	CLN-301	Cohort 1:6 × 10^{13}vg[f] Cohort 2:1.2 × 10^{14}vg	3–10 years, UBDRS[g] physical impairment score of ≤ 7, able to walk independently at least 50 feet	7	**Gene:** CLN3 **Vector:** AAV9	Intrathecal injections	Frequent visits on days 7, 14, 21, and 30. Year 1: Assessments every 3 months. Years 2–5: Assessments every 6 months	No result has been reported	No statistical outcomes have been reported
NCT 05228145	Active, not recruit-ing	CLN5	Gene Therapy Study for Children with CLN5 Batten Disease	Gene therapy	Phase ½	NGN-101	No detail has been provided	3–9 years, having impaired motor function, language function, and/or visual acuity, must be able to walk with or without assistance, and must be able to comply with all protocol-required assessments	6	**Gene:** CLN5 **Vector:** AAV9	Intracere-broven-tricular and intravitreal	Multiple visits in 5 years	No result has been reported	No statistical outcomes have been reported
NCT 04273243	Active, not recruit-ing	CLN7	Gene Therapy for CLN7 Batten Disease	Gene therapy	Phase ½	AT-GTX-501	No detail has been provided	3–18 years, must have previously received the specific gene therapy AT-GTX-501 (either completed or discontinued)	10	**Gene**: CLN7 **Vector**: AAV9	Intrathecal	Multiple visits in 5 years	No result has been reported	No statistical outcomes have been reported

[a] *GC (genome copy)*

[b] *Weill-Cornell LINCL Scale (Weill-Cornell Late Infantile Neuronal Ceroid Lipofuscinosis Scale)*

[c] *LINCL (late infantile neuronal ceroid lipofuscinoses)*

[d] *CRT (central retinal thickness)*

[e] *SD-OCT (spectral-domain optical coherence tomography)*

[f] *vg (vector genome)*

[g] *UBDRS (Unified Batten Disease Rating Scale)*

of the AAVrh.10.CUhCLN2 vector administered. Group A included 6 participants who received a total dose of 9.0×10^{11} genome copies, and Group B included 10 participants who received 2.85×10^{11} genome copies. The follow-up period of this study was approximately 18 months.

The outcomes of this study were not reported. In this regard, a similar clinical trial study (NCT01414985) Phase ½ was recorded, which evaluated the safety and effectiveness of a gene therapy using the AAVrh.10 viral vectors to deliver a healthy copy of the CLN2 gene to patients. Unlike previous studies that focused on patients with the five most common CLN2 mutations and early-stage disease, this clinical trial study expands eligibility criteria to include patients with any genotype who also have moderate to severe neurological symptoms.

In this study, participants were divided into two dose groups. Group A included two participants that received a single dose of 9×10^{11} AAVrh.10.CUhCLN2 vector. Group B included six participants that received a single dose of 2.85×10^{11} AAVrh.10.CUhCLN2 vector. These patients were followed-up for 18 months to monitor the safety and efficacy of the applied vector. This clinical trial registry has not reported results. Another clinical trial study (NCT02725580) evaluated the safety and effectiveness of AT-GTX-501, a gene therapy for children with NCL6 disorder. In the AT-GTX-501 approach, a self-complementary AAV serotype 9 (scAAV9) was used to deliver a functional copy of the CLN6 gene directly into the CNS.

In this study, 13 children received a single intrathecal injection of AT-GTX-501, which contained 1.5×10^{13} scAAV9 vectors. The therapy was administered into the lumbar spinal cord region to restore some gene function in the brain and spinal cord to help slow the progression of the disease.

Participants were followed for 24 months, with regular evaluations to assess safety and functional outcomes, including motor, language, cognitive, and overall neurological health. Based on the outcomes of this Phase ½ clinical trial, AT-GTX-501 was shown to be safe, well-tolerated, and some children experienced slower disease progression and improvements in symptoms. However, the results of this study varied, not from child to child. In this regard, another long-term follow-up study (NCT04273243) has been recorded to evaluate the safety and efficacy of AT-GTX-501 in individuals with CLN6 Batten disease.

So, in this Phase I/IIa study of AT-GTX-501, 10 out of 13 participants continued into a long-term follow-up study to evaluate the safety and efficacy of this approach for an additional three years.

The main goal is to monitor long-term safety, focusing on any treatment-related side effects. A secondary goal is to assess how well the treatment works over time, especially in maintaining motor and language abilities, using the Hamburg Scale. These kinds of extended studies are invaluable for determining the long-term safety and efficacy of gene therapy in currently untreatable neurodegenerative disorders. Another Phase ½ clinical trial study (NCT03770572) delivered a functional copy of the *CLN3* gene via a self-complementary AAV9 vector into the lower spinal cord. The goal of this study is to investigate the safety and potential effectiveness of the AAV9 vector for gene therapy in CLN3 disease. This study is ongoing and has not yet been completed. Furthermore, another active Phase ½ clinical trial study (NCT05228145) is being conducted to evaluate the safety and efficacy of a gene therapy-based approach for CLN5 disease. In this study, seven children participated and received ICV and intravitreal (IVT) injections of AAV9 to deliver a functional copy of the human *CLN5* gene.

So, in this approach, AAV9 is designed to target both the central nervous system and the retina to improve disease symptoms. These participants were divided into two groups. One group received a low dose of 6×10^{13} AAV9 in ≤ 5 mL, and the second group received a high dose of 1.2×10^{14} AAV9 in ≤ 10 mL. The long-term safety of participants, tolerability, and therapeutic efficacy, including changes in motor, language, cognitive, and visual functions, will be assessed for 5 years. Thus based on these clinical trial studies, AAVs are the recommended and commonly used vectors for delivering functional genes, with no major safety concerns reported to date. In addition, early outcomes suggest that the gene therapy approach is effective and offers potential benefits for various forms of CLN disorders.

5 Current challenges and limitations in NCL gene therapy

NCLs are a rare group of diseases caused by mutations in different genes that lead to a wide range of symptoms, ages of onset, and disease mechanisms. However, most types share common symptoms, such as seizures, vision loss, and a decline in movement and mental abilities. On the contrary, treatment options remain limited. Recently, gene therapy has gained attention as a potential treatment for severe, untreatable genetic disorders like NCLs. The primary goal of the gene therapy approach for NCL disorders is to deliver a functional copy of the defective gene to restore enzyme activity or modify protein dysfunction in the CNS. Generally, AAV vectors, due to their established safety profile and ability to transduce neurons, are the most recommended vectors for this approach. However, achieving effective outcomes remains one of the main challenges for NCL gene therapy and requires further investigation. For delivering vectors to the CNS, a major obstacle is the blood–brain barrier, which prevents vectors from spreading widely within the CNS (Kimura and Harashima, 2020). To address this, invasive delivery routes, such as ICV or intrathecal administration, are applied for NCL gene therapy (Jensen et al., 2021). However, vector biodistribution is limited, resulting in decreased efficacy in specific brain regions. Furthermore, immune responses against the viral vectors can hinder long-term expression (Mingozzi and High, 2013). On the contrary, recent studies have explored AAV serotypes with advances in synthetic capsid engineering, such as AAV9 and AAVrh10. These serotypes have demonstrated enhanced CNS distribution and allow for less invasive delivery routes (Sharma et al., 2025). However, these serotypes have also raised new concerns, such as overexpression of the therapeutic protein, which can lead to toxicity (Tanguy et al., 2015). Moreover, applying novel and effective treatments requires a deeper understanding of the molecular mechanisms underlying NCLs. For some subtypes, such CLN3 and CLN6, these mechanisms remain poorly defined. Without detailed molecular data, the application of gene therapy or other treatments will not be effective. Therefore understanding the mechanisms of NCLs in the body is essential for improving and personalizing future therapies. It is also important to note that NCLs affect multiple organs, including the retina, brain, heart, and liver, which necessitates more systemic AAV administration. These additional administrations raise further concerns related to immune responses and vector-associated toxicity. Finally, combining gene therapy with other approved treatments, such as antiinflammatory or lysosome-targeting drugs, is recommended to manage symptoms and achieve better outcomes.

6 Conclusion

NCL is one of the complex neurodegenerative disorders, with limited effective treatment strategies. Recently, gene therapy has emerged as an attractive therapeutic strategy for NCLs. However, significant challenges, such as efficient delivery methods and the risk of immune responses or toxicity, still remain. Therefore more preclinical and clinical gene therapy investigations, as well as combination treatments, are required to develop long-term, effective gene therapies. These advances in gene therapy have the potential to improve the quality of life for patients with NCLs in the coming years.

Therefore NCLs serve as a key example of neurodegenerative disorders in which gene therapy has rapidly advanced from preclinical proof-of-concept to the early stages of clinical translation.

AI disclosure

During the preparation of this work, the author(s) used Grammarly for grammar editing. After using this tool/service, the author(s) reviewed and edited the content as needed and take(s) full responsibility for the content of the publication.

References

Aldrich, A., Bosch, M.E., Fallet, R., Odvody, J., Burkovetskaya, M., Rao, K.V.R., Cooper, J.D., Drack, A.V., Kielian, T., 2016. Efficacy of phosphodiesterase-4 inhibitors in juvenile Batten disease (CLN3). Ann. Neurol. 15318249. 80 (6), 909–923. doi:10.1002/ana.24815.

Arrant, A.E., Onyilo, V.C., Unger, D.E., Roberson, E.D., 2018. Progranulin gene therapy improves lysosomal dysfunction and microglial pathology associated with frontotemporal dementia and neuronal ceroid lipofuscinosis. J. Neurosci. 15292401. 38 (9), 2341–2358. doi:10.1523/JNEUROSCI.3081-17.2018.

Aylward, S.C., Pindrik, J., Abreu, N.J., Cherny, W.B., O'Neal, M., de Los Reyes, E., 2020. Cerliponase alfa for CLN2 disease, a promising therapy. Expert Opin. Orphan Drugs. 21678707. 8 (11), 445–454. doi:10.1080/21678707.2020.1856654.

Badura-Stronka, M., Winczewska-Wiktor, A., Pietrzak, A., Hirschfeld, A.S., Zemojtel, T., Wołyńska, K., Bednarek-Rajewska, K., Seget-Dubaniewicz, M., Matheisel, A., Latos-Bielenska, A., Steinborn, B., 2021. CLN8 Mutations presenting with a phenotypic continuum of neuronal ceroid lipofuscinosis—literature review and case report. Genes. 2073-4425. 12 (7), 956. doi:10.3390/genes12070956.

Baltar, F., Simoes, C., Garagorry, F., Graña, M., Rodríguez, S., Haydée Aunchayna, M., Tapié, A., Cerisola, A., González, G., Naya, H., Spangenberg, L., Raggio, V., 2024. Two compound heterozygous variants in the CLN8 gene are responsible for neuronal cereidolipofuscinoses disorder in a child: a case report. Front. Pediatr. 2296-2360. 12. doi:10.3389/fped.2024.1379254.

Bartsch, U., Storch, S., 2022. Experimental therapeutic approaches for the treatment of retinal pathology in neuronal ceroid lipofuscinoses. Front. Neurol. 16642295. 13. doi:10.3389/fneur.2022.866983.

Basak, I., Wicky, H.E., McDonald, K.O., Xu, J.B., Palmer, J.E., Best, H.L., Lefrancois, S., Lee, S.Y., Schoderboeck, L., Hughes, S.M., 2021. A lysosomal enigma CLN5 and its significance in understanding neuronal ceroid lipofuscinosis. Cell. Mol. Life Sci. 1420-682X. 78 (10), 4735–4763. doi:10.1007/s00018-021-03813-x.

Borjini, N., López-Begines, S., Fernández-Chacón, R., 2023. Deciphering the role of microglia in kufs disease/CLN4: insights from transgenic mouse models expressing mutant forms of CSPA/DNAJC5. IBRO Neurosci. Rep. 26672421. 15, S216. doi:10.1016/j.ibneur.2023.08.346.

Bosch, M.E., Aldrich, A., Fallet, R., Odvody, J., Burkovetskaya, M., Schuberth, K., Fitzgerald, J.A., Foust, K.D., Kielian, T., 2016. Self-complementary AAV9 gene delivery partially corrects pathology associated with juvenile neuronal ceroid lipofuscinosis (CLN3). J. Neurosci. 15292401. 36 (37), 9669–9682. doi:10.1523/JNEUROSCI.1635-16.2016.

Bras, J., Verloes, A., Schneider, S.A., Mole, S.E., Guerreiro, R.J., 2012. Mutation of the parkinsonism gene ATP13A2 causes neuronal ceroid-lipofuscinosis. Hum. Mol. Genet. 0964-6906. 21 (12), 2646–2650. doi:10.1093/hmg/dds089.

Breitenbach, M., Ralser, M., Perrone, G.G., Iglseder, B., Rinnerthaler, M., Dawes, I.W., 2013. Oxidative stress and neurodegeneration: the yeast model system. Front. Biosci. 27686698. 18 (3), 1174–1193. doi:10.2741/4171.

Bryzik, A., Larysz, D., Larysz, P., Karpierz, J.I., Paprocka, J., 2025. Case report: the window that closed too soon: lessons from a late CLN2 diagnosis and death of a 9-year-old boy. Front. Genet. 1664-8021. 16. doi:10.3389/fgene.2025.1622185.

Butz, E.S., Chandrachud, U., Mole, S.E., Cotman, S.L., 2020. Moving towards a new era of genomics in the neuronal ceroid lipofuscinoses. Biochim. Biophys. Acta Mol. Basis Dis. 09254439. 1866 (9), 165571. doi:10.1016/j.bbadis.2019.165571.

Byrne, B.J., Falk, D.J., Clément, N., Mah, C.S., 2012. Gene therapy approaches for lysosomal storage disease: next-generation treatment. Hum. Gene Ther. 15577422. 23 (8), 808–815. doi:10.1089/hum.2012.140.

Chamberlain, K., Riyad, J.M., Weber, T., 2016. Expressing transgenes that exceed the packaging capacity of adeno-associated virus capsids. Hum. Gene Ther. Methods. 19466544. 27 (1), 1–12. doi:10.1089/hgtb.2015.140.

Cherukuri, A., Cahan, H., de Hart, G., Van Tuyl, A., Slasor, P., Bray, L., Henshaw, J., Ajayi, T., Jacoby, D., O'Neill, C.A., Schweighardt, B., 2018. Immunogenicity to cerliponase alfa intracerebroventricular enzyme replacement therapy for CLN2 disease: results from a Phase 1/2 study. Clin. Immunol. 15216616. 197, 68–76. doi:10.1016/j.clim.2018.09.003.

Cotman, S.L., Staropoli, J.F., 2012. The juvenile Batten disease protein, CLN3, and its role in regulating anterograde and retrograde post-Golgi trafficking. Clin. Lipidol. 17584302. 7 (1), 79–91. doi:10.2217/clp.11.70.

Doccini, S., Sartori, S., Maeser, S., Pezzini, F., Rossato, S., Moro, F., Toldo, I., Przybylski, M., Santorelli, F.M., Simonati, A., 2016. Early infantile neuronal ceroid lipofuscinosis (CLN10 disease) associated with a novel mutation in CTSD. J. Neurol. 0340-5354. 263 (5), 1029–1032. doi:10.1007/s00415-016-8111-6.

Donsante, A., Boulis, N.M., 2018. Progress in gene and cell therapies for the neuronal ceroid lipofuscinoses. Expert Opin. Biol. Ther. 17447682. 18 (7), 755–764. doi:10.1080/14712598.2018.1492544.

Farias, F.H.G., Zeng, R., Johnson, G.S., Wininger, F.A., Taylor, J.F., Schnabel, R.D., McKay, S.D., Sanders, D.N., Lohi, H., Seppälä, E.H., Wade, C.M., Lindblad-Toh, K., O'Brien, D.P., Katz, M.L., 2011. A truncating mutation in ATP13A2 is responsible for adult-onset neuronal ceroid lipofuscinosis in Tibetan terriers. Neurobiol. Dis. 09699961. 42 (3), 468–474. doi:10.1016/j.nbd.2011.02.009.

Francis, S.H., Blount, M.A., Corbin, J.D., 2011. Mammalian cyclic nucleotide phosphodiesterases: molecular mechanisms and physiological functions. Physiol. Rev. 15221210. 91 (2), 651–690. doi: 10.1152/physrev.00030.2010.

Gao, Y., Yang, K., Shujiang, X., Wang, C., Liu, J., Zhang, Z., Yuan, M., Luo, X., Liu, M., Wang, Q.K., Liu, J.Y., 2011. Identification of compound heterozygous mutations in GNPTG in three siblings of a Chinese family with mucolipidosis type III gamma. Mol. Genet. Metab. 10967192. 102 (1), 107–109. doi:10.1016/j.ymgme.2010.09.007.

Gaur, P., Gissen, P., Biswas, A., Mankad, K., Sudhakar, S., D'Arco, F., Schulz, A., Fiehler, J., Sedlacik, J., Löbel, U., 2024. Enzyme replacement therapy for CLN2 disease: MRI volumetry shows significantly slower volume loss compared with a natural history cohort. Am. J. Neuroradiol. 0195-6108. 45 (11), 1791–1797. doi:10.3174/ajnr.a8408.

Ghosh, A., Corbett, G.T., Gonzalez, F.J., Pahan, K., 2012. Gemfibrozil and fenofibrate, food and drug administration-approved lipid-lowering drugs, up-regulate tripeptidyl-peptidase 1 in brain cells via peroxisome proliferator-activated receptor α: implications for late infantile batten disease therapy. J. Biol. Chem. 1083351X. 287 (46), 38922–38935. doi:10.1074/jbc.M112.365148.

Griffey, M., Bible, E., Vogler, C., Levy, B., Gupta, P., Cooper, J., Sands, M.S., 2004. Adeno-associated virus 2-mediated gene therapy decreases autofluorescent storage material and increases brain mass in a murine model of infantile neuronal ceroid lipofuscinosis. Neurobiol. Dis. 09699961. 16 (2), 360–369. doi:10.1016/j.nbd.2004.03.005.

Grisolia, M., Sestito, S., Ceravolo, F., Invernizzi, F., Salpietro, V., Polizzi, A., Ruggieri, M., Garavaglia, B., Concolino, D., 2016. The neuronal ceroid lipofuscinoses: a case-based overview. J. Pediatr. Biochem. 1879-5390. 6 (1), 060–065. doi:10.1055/s-0036-1582222.

Groh, J., Kühl, T.G., Ip, C.W., Nelvagal, H.R., Sri, S., Duckett, S., Mirza, M., Langmann, T., Cooper, J.D., Martini, R., 2013. Immune cells perturb axons and impair neuronal survival in a mouse model of infantile neuronal ceroid lipofuscinosis. Brain. 1460-2156. 136 (4), 1083–1101. doi:10.1093/brain/awt020.

Haltia, M., 2003. The neuronal ceroid-lipofuscinoses. J. Neuropathol. Exp. Neurol. 0022-3069. 62 (1), 1–13. doi:10.1093/jnen/62.1.1.

Hasibuan, A.M., Arthy, C.C. NCL Disorders: common causes of dementia in children: a case report. In Sixth International Conference of Neuroscience, Neurology and Psychiatry Universitas Sumatera Utara (ICONAP 2023). 2023. Atlantis Press.

Huang, H., Liao, Y., Yanni, Y., Qin, H., Wei, Y.Z., Cao, L., 2024. Adult-onset neuronal ceroid lipofuscinosis misdiagnosed as autoimmune encephalitis and normal-pressure hydrocephalus: a 10-year case report and case-based review. Medicine. 1536-5964. 103 (43), e40248. doi:10.1097/md.0000000000040248.

Huin, V., Barbier, M., Bottani, A., Lobrinus, J.A., Clot, F., Lamari, F., Chat, L., Rucheton, B., Fluchère, F., Auvin, S., Myers, P., Gelot, A., Camuzat, A., Caillaud, C., Jornéa, L., Forlani, S., Saracino, D., Duyckaerts, C., Brice, A., Durr, A., Ber, I.L., 2020. Homozygous GRN mutations: new phenotypes and new insights into pathological and molecular mechanisms. Brain. 0006-8950. 143 (1), 303–319. doi:10.1093/brain/awz377.

Itagaki, R., Endo, M., Yanagisawa, H., Hossain, M.A., Akiyama, K., Yaginuma, K., Miyajima, T., Chen, W., Iwamoto, T., Igarashi, J., Kobayashi, Y., Tohyama, J., Iwama, K., Matsumoto, N., Shintaku, H., Eto, Y., 2018. Characteristics of PPT1 and TPP1 enzymes in neuronal ceroid lipofuscinosis (NCL) 1 and 2 by dried blood spots (DBS) and leukocytes and their application to newborn screening. Mol. Genet. Metab. 10967192. 124 (1), 64–70. doi:10.1016/j.ymgme.2018.03.007.

Jalanko, A., Braulke, T., 2009. Neuronal ceroid lipofuscinoses. Biochim. Biophys. Acta Mol. Cell Res. 01674889. 1793 (4), 697–709. doi:10.1016/j.bbamcr.2008.11.004.

Jankowiak, W., Kruszewski, K., Flachsbarth, K., Skevas, C., Richard, G., Rüther, K., Braulke, T., Bartsch, U., Langmann, T., 2015. Sustained neural stem cell-based intraocular delivery of CNTF attenuates photoreceptor loss in the nclf mouse model of neuronal ceroid lipofuscinosis. PLOS ONE. 1932-6203. 10 (5), e0127204. doi:10.1371/journal.pone.0127204.

Jensen, T.L., Gøtzsche, C.R., Woldbye, D.P.D., 2021. Current and future prospects for gene therapy for rare genetic diseases affecting the brain and spinal cord. Front. Mol. Neurosci. 16625099. 14. doi:10.3389/fnmol.2021.695937.

Jeung, H., Thomann, P.A., Wolf, R.C., 2015. Novel gene variations in early-onset frontotemporal dementia with positive family history of neural ceroid lipofuscinosis-1. Neurol. Clin. Pract. 21630933. 5 (6), 484–487. doi:10.1212/CPJ.0000000000000134.

Johnson, T.B., Brudvig, J.J., Likhite, S., Pratt, M.A., White, K.A., Cain, J.T., Booth, C.D., Timm, D.J., Davis, S.S., Meyerink, B., Pineda, R., Dennys-Rivers, C., Kaspar, B.K., Meyer, K., Weimer, J.M., 2023. Early postnatal administration of an AAV9 gene therapy is safe and efficacious in CLN3 disease. Front. Genet. 16648021. 14. doi:10.3389/fgene.2023.1118649.

Jung, K.K., Lee, H.S., Cho, J.Y., Shin, W.C., Rhee, M.H., Kim, T.G., Kang, J.H., Kim, S.H., Hong, S., Kang, S.Y., 2006. Inhibitory effect of curcumin on nitric oxide production from lipopolysaccharide-activated primary microglia. Life Sci. 00243205. 79 (21), 2022–2031. doi:10.1016/j.lfs.2006.06.048.

Kandel, E.R., 2012. The molecular biology of memory: CAMP, PKA, CRE, CREB-1, CREB-2, and CPEB. Mol. Brain. 17566606. 5 (1). doi:10.1186/1756-6606-5-14.

Katz, M.L., Johnson, G.C., Leach, S.B., Williamson, B.G., Coates, J.R., Whiting, R.E.H., Vansteenkiste, D.P., Whitney, M.S., 2017. Extraneuronal pathology in a canine model of CLN2 neuronal ceroid lipofuscinosis after intracerebroventricular gene therapy that delays neurological disease progression. Gene Ther. 0969-7128. 24 (4), 215–223. doi:10.1038/gt.2017.4.

Katz, M.L., Tecedor, L., Chen, Y., Williamson, B.G., Lysenko, E., Wininger, F.A., Young, W.M., Johnson, G.C., Whiting, R.E.H., Coates, J.R., Davidson, B.L., 2015. AAV gene transfer delays disease onset in a TPP1-deficient canine model of the late infantile form of Batten disease. Sci. Transl. Med. 19466242. 7 (313). doi:10.1126/scitranslmed.aac6191.

Kayani, S., BordesEdgar, V., Lowden, A., Nettesheim, E.R., Dahshi, H., Messahel, S., Minassian, B.A., Greenberg, B.M., 2024. Neuronal ceroid lipofuscinoses type 7 (CLN7): a case series reporting cross sectional and retrospective clinical data to evaluate validity of standardized tools to assess disease progression, quality of life, and adaptive skills. Orphanet J. Rare Dis. 1750-1172. 19 (1). doi:10.1186/s13023-024-03448-8.

Khalilpour, S., Latifi, S., Behnammanesh, G., Majid, A.M.S.A., Majid, A.S.A., Tamayol, A., 2017. Ischemic optic neuropathy as a model of neurodegenerative disorder: a review of pathogenic mechanism of axonal degeneration and the role of neuroprotection. J. Neurol. Sci. 0022510X. 375, 430–441. doi:10.1016/j.jns.2016.12.044.

Kim, J., Chunguang, H., El Achkar, C.M., Black, L.E., Douville, J., Larson, A., Pendergast, M.K., Goldkind, S.F., Lee, E.A., Kuniholm, A., Soucy, A., Vaze, J., Belur, N.R., Fredriksen, K., Stojkovska, I., Tsytsykova, A., Armant, M., DiDonato, R.L., Choi, J., Cornelissen, L., Pereira, L.M., Augustine, E.F., Genetti, C.A., Dies, K., Barton, B., Williams, L., Goodlett, B.D., Riley, B.L., Pasternak, A., Berry, E.R., Pflock, K.A., Chu, S., Reed, C., Tyndall, K., Agrawal, P.B., Beggs, A.H., Grant, P.E., Urion, D.K., Snyder, R.O., Waisbren, S.E., Poduri, A., Park, P.J., Patterson, A., Biffi, A., Mazzulli, J.R., Bodamer, O., Berde, C.B., Yu, T.W., 2019. Patient-customized oligonucleotide therapy for a rare genetic disease. N. Engl. J. Med. 0028-4793. 381 (17), 1644–1652. doi:10.1056/nejmoa1813279.

Kim, S.J., Zhang, Z., Sarkar, C., Tsai, P.C., Lee, Y.C., Dye, L., Mukherjee, A.B., 2008. Palmitoyl protein thioesterase-1 deficiency impairs synaptic vesicle recycling at nerve terminals, contributing to neuropathology in humans and mice. J. Clin. Invest. 15588238. 118 (9), 3075–3086. doi: 10.1172/JCI33482.

Kimura, S., Harashima, H., 2020. Current status and challenges associated with CNS-targeted gene delivery across the BBB. Pharmaceutics. 19994923. 12 (12), 1–33. doi:10.3390/pharmaceutics12121216.

Kleine Holthaus, S.M., Aristorena, M., Maswood, R., Semenyuk, O., Hoke, J., Hare, A., Smith, A.J., Mole, S.E., Ali, R.R., 2020. Gene therapy targeting the inner retina rescues the retinal phenotype in a mouse model of CLN3 batten disease. Hum. Gene Ther. 15577422. 31 (13-14), 709–718. doi:10.1089/hum.2020.038.

Kleine Holthaus, S.M., Ribeiro, J., Abelleira-Hervas, L., Pearson, R.A., Duran, Y., Georgiadis, A., Sampson, R.D., Rizzi, M., Hoke, J., Maswood, R., Azam, S., Luhmann, U.F.O., Smith, A.J., Mole, S.E., Ali, R.R., 2018. Prevention of photoreceptor cell loss in a Cln6nclf mouse model of batten disease requires CLN6 gene transfer to bipolar cells. Mol. Ther. 1525-0024. 26 (5), 1343–1353. doi:10.1016/j.ymthe.2018.02.027.

Kohan, R., Noelia Carabelos, M., Xin, W., Sims, K., Guelbert, N., Adriana Cismondi, I., Pons, P., Alonso, G.I., Troncoso, M., Witting, S., Pearce, D.A., de Kremer, R.D., María Oller-Ramírez, A., de Halac, I.N., 2013. Neuronal ceroid lipofuscinosis type CLN2: a new rationale for the construction of phenotypic subgroups based on a survey of 25 cases in South America. Gene. 03781119. 516 (1), 114–121. doi:10.1016/j.gene.2012.12.058.

Kohlschütter, A., 2021. Ethical issues in care and treatment of neuronal ceroid lipofuscinoses (NCL)–a personal view. Front. Neurol. 1664-2295. 12. doi:10.3389/fneur.2021.692527.

Kohlschütter, A., Schulz, A., Bartsch, U., Storch, S., 2019. Current and emerging treatment strategies for neuronal ceroid lipofuscinoses. CNS Drugs. 1172-7047. 33 (4), 315–325. doi:10.1007/s40263-019-00620-8.

Kousi, M., Siintola, E., Dvorakova, L., Vlaskova, H., Turnbull, J., Topcu, M., Yuksel, D., Gokben, S., Minassian, B.A., Elleder, M., Mole, S.E., Lehesjoki, A.-E., 2009. Mutations in CLN7/MFSD8 are a common cause of variant late-infantile neuronal ceroid lipofuscinosis. Brain. 1460-2156. 132 (3), 810–819. doi:10.1093/brain/awn366.

Kovács, A.D., Saje, A., Wong, A., Ramji, S., Cooper, J.D., Pearce, D.A., 2012. Age-dependent therapeutic effect of memantine in a mouse model of juvenile Batten disease. Neuropharmacology. 18737064. 63 (5), 769–775. doi:10.1016/j.neuropharm.2012.05.040.

Kovács, A.D., Saje, A., Wong, A., Szénási, G., Kiricsi, P., Szabó, E., Cooper, J.D., Pearce, D.A., 2011. Temporary inhibition of AMPA receptors induces a prolonged improvement of motor performance in a mouse model of juvenile Batten disease. Neuropharmacology. 00283908. 60 (2-3), 405–409. doi:10.1016/j.neuropharm.2010.10.010.

Kuper, W.F.E., Van Alfen, C., Van Eck, L., Huijgen, B.C.H., Nieuwenhuis, E.E.S., Van Brussel, M., Van Hasselt, P.M., 2019. Motor function impairment is an early sign of CLN3 disease. Neurology. 1526632X. 93 (3), E293. doi:10.1212/WNL.0000000000007773.

Levin, S.W., Baker, E.H., Zein, W.M., Zhang, Z., Quezado, Z.M.N., Miao, N., Gropman, A., Griffin, K.J., Bianconi, S., Chandra, G., Khan, O.I., Caruso, R.C., Liu, A., Mukherjee, A.B., 2014. Oral cysteamine bitartrate and N-acetylcysteine for patients with infantile neuronal ceroid lipofuscinosis: a pilot study. Lancet Neurol. 14744465. 13 (8), 777–787. doi:10.1016/S1474-4422(14)70142-5.

Macauley, S.L., Roberts, M.S., Wong, A.M., McSloy, F., Reddy, A.S., Cooper, J.D., Sands, M.S., 2012. Synergistic effects of central nervous system-directed gene therapy and bone marrow transplantation in the murine model of infantile neuronal ceroid lipofuscinosis. Ann. Neurol. 15318249. 71 (6), 797–804. doi:10.1002/ana.23545.

Madhavi, K., Kandadai, R.M., Kola, S., Borgohain, R., Alugolu, R., Prasad, V., Nandeesh, B.N., Govindaraj, P., 2024. Adult-onset neuronal ceroid lipofuscinosis: CLN5 variant presenting as focal dystonia. Tremor Other Hyperkinet. Mov. 2160-8288. 14 (1). doi:10.5334/tohm.941.

Magrinelli, F., Pezzini, F., Moro, F., Santorelli, F.M., Simonati, A., 2017. Diagnostic methods and emerging treatments for adult neuronal ceroid lipofuscinoses (Kufs disease). Expert Opin. Orphan Drugs. 2167-8707. 5 (6), 487–501. doi:10.1080/21678707.2017.1325359.

Mali, S., 2013. Delivery systems for gene therapy. Indian J. Hum. Genet. 0971-6866. 19 (1), 3. doi:10.4103/0971-6866.112870.

Markham, A., 2017. Cerliponase alfa: first global approval. Drugs. 0012-6667. 77 (11), 1247–1249. doi:10.1007/s40265-017-0771-8.

Martin, J., et al, 1999. Adult NCL. HH, Mole SE, Lake BD the Neuronal Ceroid Lipofuscinoses (Batten Disease) IOS, Amsterdam. pp. 77–90.

Masola, S.N., 2025. Neuronal ceroid lipofuscinosis across species: a fatal hereditary neurodegenerative disease. An Overv. Dis. Health Res. 1, 28–50.

Maximiano-Alves, G., Caravelas, R.D.A.M., Gonçalves, T.A.P., Corniani, K.F., Nather, J.C., Geraldi-Tomaselli, C.V., Frezatti, R.S.S., Fernandes, R.M.F., Carlos Dos Santos, A., Marques, W., Tomaselli, P.J., 2025. Neuronal ceroid lipofuscinosis 11 (CLN11) presenting with early-onset cone-rod dystrophy and learning difficulties. Neurogenetics. 1364-6753. 26 (1). doi:10.1007/s10048-025-00800-3.

Metz, K.A., Teng, X., Coppens, I., Lamb, H.M., Wagner, B.E., Rosenfeld, J.A., Chen, X., Zhang, Y., Kim, H.J., Meadow, M.E., Wang, T.S., Haberlandt, E.D., Anderson, G.W., Leshinsky-Silver, E., Bi, W., Markello, T.C., Pratt, M., Makhseed, N., Garnica, A., Danylchuk, N.R., Burrow, T.A., Jayakar, P., McKnight, D., Agadi, S., Gbedawo, H., Stanley, C., Alber, M., Prehl, I., Peariso, K., Ong, M.T., Mordekar, S.R., Parker, M.J., Crooks, D., Agrawal, P.B., Berry, G.T., Loddenkemper, T., Yang, Y., Maegawa, G.H.B., Aouacheria, A., Markle, J.G., Wohlschlegel, J.A., Hartman, A.L., Hardwick, J.M., 2018. KCTD7 deficiency defines a distinct neurodegenerative disorder with a conserved autophagy-lysosome defect. Ann. Neurol. 15318249. 84 (5), 766–780. doi:10.1002/ana.25351.

Meyer, J.S., Katz, M.L., Maruniak, J.A., Kirk, M.D., 2006. Embryonic stem cell-derived neural progenitors incorporate into degenerating retina and enhance survival of host photoreceptors. Stem Cells. 10665099. 24 (2), 274–283. doi: 10.1634/stemcells.2005-0059.

Mingozzi, F., High, K.A., 2013. Immune responses to AAV vectors: overcoming barriers to successful gene therapy. Blood. 15280020. 122 (1), 23–36. doi:10.1182/blood-2013-01-306647.

Mink, J.W., Augustine, E.F., Adams, H.R., Marshall, F.J., Kwon, J.M., 2013. Classification and natural history of the neuronal ceroid lipofuscinoses. J. Child Neurol. 17088283. 28 (9), 1101–1105. doi:10.1177/0883073813494268.

Mirza, M., Volz, C., Karlstetter, M., Langiu, M., Somogyi, A., Ruonala, M.O., Tamm, E.R., Jägle, H., Langmann, T., Swaroop, A., 2013. Progressive retinal degeneration and glial activation in the CLN6nclf mouse model of neuronal ceroid lipofuscinosis: a beneficial effect of DHA and curcumin supplementation. PLoS ONE. 1932-6203. 8 (10), e75963. doi:10.1371/journal.pone.0075963.

Mitchell, N.L., Russell, K.N., Wellby, M.P., Wicky, H.E., Schoderboeck, L., Barrell, G.K., Melzer, T.R., Gray, S.J., Hughes, S.M., Palmer, D.N., 2018. Longitudinal in vivo monitoring of the CNS demonstrates the efficacy of gene therapy in a sheep model of CLN5 Batten disease. Mol. Ther. 15250024. 26 (10), 2366–2378. doi:10.1016/j.ymthe.2018.07.015.

Mitchell, W.A., Wheeler, R.B., Sharp, J.D., Bate, S.L., Gardiner, R.M., Ranta, U.S., Lonka, L., Williams, R.E., Lehesjoki, A.E., Mole, S.E., 2001. Turkish variant late infantile neuronal ceroid lipofuscinosis (CLN7) may be allelic to CLN8. Eur. J. Paediatr. Neurol. 10903798. 5, 21–27. doi:10.1053/ejpn.2000.0429.

Mohammed, A., 2020. Roles of CLN7 in Neurodevelopment, University of Birmingham.

Mole, S.E., Anderson, G., Band, H.A., Berkovic, S.F., Cooper, J.D., Holthaus, S.M.K., McKay, T.R., Medina, D.L., Rahim, A.A., Schulz, A., Smith, A.J., 2019. Clinical challenges and future therapeutic approaches for neuronal ceroid lipofuscinosis. Lancet Neurol. 14744465. 18 (1), 107–116. doi:10.1016/S1474-4422(18)30368-5.

Mole, S.E., Gissen, P., Nordstrom, S., Wait, S., Allen, L., Antonini, M., Brownnutt, L., Brown, R., Cole, B., Gibbon, F., Henderson, R.H., Kenrick, S., Sisic, Z., Thompson, B., Nightingale, J., 2025. Evidence of the impact of CLN2 and CLN3 Batten disease on families in the United Kingdom. Orphanet J. Rare Dis. 17501172. 20 (1). doi:10.1186/s13023-025-03747-8.

Nickel, M., Simonati, A., Jacoby, D., Lezius, S., Kilian, D., Van de Graaf, B., Pagovich, O.E., Kosofsky, B., Yohay, K., Downs, M., Slasor, P., Ajayi, T., Crystal, R.G., Kohlschütter, A., Sondhi, D., Schulz, A., 2018. Disease characteristics and progression in patients with late-infantile neuronal ceroid lipofuscinosis type 2 (CLN2) disease: an observational cohort study. Lancet Child Adolesc. Health. 23524642. 2 (8), 582–590. doi:10.1016/s2352-4642(18)30179-2.

Nicolaou, P., Tanteles, G.A., Votsi, C., Zamba-Papanicolaou, E., Papacostas, S.S., Christodoulou, K., Christou, Y.-P., 2021. A novel CLN6 variant associated with juvenile neuronal ceroid lipofuscinosis in patients with absence of visual loss as a presenting feature. Front. Genet. 1664-8021. 12. doi:10.3389/fgene.2021.746101.

Nita, D.A., Mole, S.E., Minassian, B.A., 2016. Neuronal ceroid lipofuscinoses. Epileptic Disord. 19506945. 18, S73. doi:10.1684/epd.2016.0844Canada.

Palmieri, M., Impey, S., Kang, H., Di Ronza, A., Pelz, C., Sardiello, M., Ballabio, A., 2011. Characterization of the CLEAR network reveals an integrated control of cellular clearance pathways. Hum. Mol. Genet. 14602083. 20 (19), 3852–3866. doi:10.1093/hmg/ddr306.

Palmieri, M., Pal, R., Nelvagal, H.R., Lotfi, P., Stinnett, G.R., Seymour, M.L., Chaudhury, A., Bajaj, L., Bondar, V.V., Bremner, L., Saleem, U., Tse, D.Y., Sanagasetti, D., Wu, S.M., Neilson, J.R., Pereira, F.A., Pautler, R.G., Rodney, G.G., Cooper, J.D., Sardiello, M., 2017. MTORC1-independent TFEB activation via Akt inhibition promotes cellular clearance in neurodegenerative storage diseases. Nat. Commun. 20411723. 8. doi:10.1038/ncomms14338.

Park, J.S., Mehta, P., Cooper, A.A., Veivers, D., Heimbach, A., Stiller, B., Kubisch, C., Fung, V.S., Krainc, D., Mackay-Sim, A., Sue, C.M., 2011. Pathogenic effects of novel mutations in the P-type ATPase ATP13A2 (PARK9) causing Kufor-Rakeb syndrome, a form of early-onset parkinsonism. Hum. Mutat. 10981004. 32 (8), 956–964. doi:10.1002/humu.21527.

Peviani, M., Das, S., Patel, J., Jno-Charles, O., Kumar, R., Zguro, A., Mathews, T.D., Cabras, P., Milazzo, R., Cavalca, E., Poletti, V., Biffi, A., 2023. An innovative hematopoietic stem cell gene therapy approach benefits CLN1 disease in the mouse model. EMBO Mol. Med. 1757-4676. 15 (4). doi:10.15252/emmm.202215968.

Priyanka, K., Madhana, P.N., Eswaramoorthy, R., Ramasamy, M., 2024. A computational approach to analyzing the functional and structural impacts of Tripeptidyl-Peptidase 1 missense mutations in neuronal ceroid lipofuscinosis. Metab. Brain Dis. 15737365. 39 (4), 545–558. doi:10.1007/s11011-024-01341-8.

Qiao, Y., Yang, G., Cheng, Y., Su, Y., Nan, L., Shang, Q., Xing, Q., 2022. Case Report: novel MFSD8 Variants in a Chinese Family With Neuronal Ceroid Lipofuscinoses 7. Front. Genet. 1664-8021. 13. doi:10.3389/fgene.2022.807515.

Radke, J., Stenzel, W., Goebel, H.H., 2015. Human NCL neuropathology. Biochim. Biophys. Acta Mol. Basis Dis. 1879260X. 1852 (10), 2262–2266. doi:10.1016/j.bbadis.2015.05.007.

Ranta, S., Lehesjoki, A.-E., 2000. Northern epilepsy, a new member of the NCL family. Neurol. Sci. 1590-1874. 21 (0), S43. doi:10.1007/s100720070039.

Roberts, M.S., Macauley, S.L., Wong, A.M., Yilmas, D., Hohm, S., Cooper, J.D., Sands, M.S., 2012. Combination small molecule PPT1 mimetic and CNS-directed gene therapy as a treatment for infantile neuronal ceroid lipofuscinosis. J. Inherit. Metab. Dis. 15732665. 35 (5), 847–857. doi:10.1007/s10545-011-9446-x.

Rosenberg, J.B., Chen, A., Kaminsky, S.M., Crystal, R.G., Sondhi, D., 2019. Advances in the treatment of neuronal ceroid lipofuscinosis. Expert Opin. Orphan Drugs. 21678707. 7 (11), 473–500. doi:10.1080/21678707.2019.1684258.

Rus, C.M., Weissensteiner, T., Pereira, C., Susnea, I., Danquah, B.D., Torres, G.M., Rocha, M.E., Cozma, C., Saravanakumar, D., Mannepalli, S., Kandaswamy, K.K., Di Bucchianico, S., Zimmermann, R., Rolfs, A., Bauer, P., Beetz, C., 2022. Clinical and genetic characterization of a cohort of 97 CLN6 patients tested at a single center. Orphanet J. Rare Dis. 17501172. 17 (1). doi:10.1186/s13023-022-02288-8.

Samulski, R.J., Muzyczka, N., 2014. AAV-mediated gene therapy for research and therapeutic purposes. Annu. Rev. Virol. 23270578. 1 (1), 427–451. doi:10.1146/annurev-virology-031413-085355.

Sardiello, M., Palmieri, M., Ronza, A.D., Medina, D.L., Valenza, M., Gennarino, V.A., Di Malta, C., Donaudy, F., Embrione, V., Polishchuk, R.S., Banfi, S., Parenti, G., Cattaneo, E., Ballabio, A., 2009. A gene network regulating lysosomal biogenesis and function. Science. 10959203. 325 (5939), 473–477. doi:10.1126/science.1174447.

Sarkar, C., Chandra, G., Peng, S., Zhang, Z., Liu, A., Mukherjee, A.B., 2013. Neuroprotection and lifespan extension in Ppt1-/- mice by NtBuHA: therapeutic implications for INCL. Nat. Neurosci. 15461726. 16 (11), 1608–1617. doi:10.1038/nn.3526.

Schulz, A., Ajayi, T., Specchio, N., de Los Reyes, E., Gissen, P., Ballon, D., Dyke, J.P., Cahan, H., Slasor, P., Jacoby, D., Kohlschütter, A., 2018. Study of Intraventricular Cerliponase Alfa for CLN2 disease. N. Engl. J. Med. 0028-4793. 378 (20), 1898–1907. doi:10.1056/nejmoa1712649.

Schulz, A., Dhar, S., Rylova, S., Dbaibo, G., Alroy, J., Hagel, C., Artacho, I., Kohlschütter, A., Lin, S., Boustany, R., 2004. Impaired cell adhesion and apoptosis in a novel CLN9 Batten disease variant. Ann. Neurol. 0364-5134. 56 (3), 342–350. doi:10.1002/ana.20187.

Schulz, A., Kohlschütter, A., Mink, J., Simonati, A., Williams, R., 2013. NCL diseases—clinical perspectives. Biochim. Biophys. Acta Mol. Basis Dis. 09254439. 1832 (11), 1801–1806. doi:10.1016/j.bbadis.2013.04.008.

Schulz, A., Specchio, N., de Los Reyes, E., Gissen, P., Nickel, M., Trivisano, M., Aylward, S.C., Chakrapani, A., Schwering, C., Wibbeler, E., Westermann, L.M., Ballon, D.J., Dyke, J.P., Cherukuri, A., Bondade, S., Slasor, P., Pfeffer, J.C., 2024. Safety and efficacy of cerliponase alfa in children with neuronal ceroid lipofuscinosis type 2 (CLN2 disease): an open-label extension study. Lancet Neurol. 14744422. 23 (1), 60–70. doi:10.1016/s1474-4422(23)00384-8.

Seehafer, S.S., Ramirez-Montealegre, D., Wong, A.M.S., Chan, C.H., Castaneda, J., Horak, M., Ahmadi, S.M., Lim, M.J., Cooper, J.D., Pearce, D.A., 2011. Immunosuppression alters disease severity in juvenile Batten disease mice. J. Neuroimmunol. 01655728. 230 (1-2), 169–172. doi:10.1016/j.jneuroim.2010.08.024.

Settembre, C., Di Malta, C., Polito, V.A., Arencibia, M.G., Vetrini, F., Erdin, S., Erdin, S.U., Huynh, T., Medina, D., Colella, P., Sardiello, M., Rubinsztein, D.C., Ballabio, A., 2011. TFEB links autophagy to lysosomal biogenesis. Science. 10959203. 332 (6036), 1429–1433. doi:10.1126/science.1204592Italy.

Sharma, S., Joshi, V., Kumar, V., 2025. Implications of AAV serotypes in neurological disorders: current clinical applications and challenges. Clin. Transl. Neurosci. 2514183X. 9 (3). doi:10.3390/ctn9030032.

Shevtsova, Z., Garrido, M., Weishaupt, J., Saftig, P., Bähr, M., Lühder, F., Kügler, S., 2010. CNS-Expressed cathepsin D prevents lymphopenia in a murine model of congenital neuronal ceroid lipofuscinosis. Am. J. Pathol. 00029440. 177 (1), 271–279. doi:10.2353/ajpath.2010.091267.

Shyng, C., Nelvagal, H.R., Dearborn, J.T., Tyynelä, J., Schmidt, R.E., Sands, M.S., Cooper, J.D., 2017. Synergistic effects of treating the spinal cord and brain in CLN1 disease. Proc. Natl. Acad. Sci. U. S. A. 0027-8424. 114 (29). doi:10.1073/pnas.1701832114.

Singh, J., 2025. The Localization and Functional Characterization of a DNAJC5-Like Protein in Dictyostelium Discoideum, Trent University (Canada).

Sondhi, D., Johnson, L., Purpura, K., Monette, S., Souweidane, M.M., Kaplitt, M.G., Kosofsky, B., Yohay, K., Ballon, D., Dyke, J., Kaminksy, S.M., Hackett, N.R., Crystal, R.G., 2012. Long-term expression and safety of administration of AAVrh.10hCLN2 to the brain of rats and nonhuman primates for the treatment of late infantile neuronal ceroid lipofuscinosis. Hum. Gene Ther. Methods. 1946-6536. 23 (5), 324–335. doi:10.1089/hgtb.2012.120.

Sondhi, D., Scott, E.C., Chen, A., Hackett, N.R., Wong, A.M.S., Kubiak, A., Nelvagal, H.R., Pearse, Y., Cotman, S.L., Cooper, J.D., Crystal, R.G., 2014. Partial correction of the CNS lysosomal storage defect in a mouse model of juvenile neuronal ceroid lipofuscinosis by neonatal CNS administration of an adeno-associated virus serotype rh.10 vector expressing the human CLN3 Gene. Hum. Gene Ther. 1043-0342. 25 (3), 223–239. doi:10.1089/hum.2012.253.

Specchio, N., Ferretti, A., Trivisano, M., Pietrafusa, N., Pepi, C., Calabrese, C., Livadiotti, S., Simonetti, A., Rossi, P., Curatolo, P., Vigevano, F., 2021. Neuronal ceroid lipofuscinosis: potential for targeted therapy. Drugs. 0012-6667. 81 (1), 101–123. doi:10.1007/s40265-020-01440-7.

Staropoli, J.F., Karaa, A., Lim, E.T., Kirby, A., Elbalalesy, N., Romansky, S.G., Leydiker, K.B., Coppel, S.H., Barone, R., Xin, W., MacDonald, M.E., Abdenur, J.E., Daly, M.J., Sims, K.B., Cotman, S.L., 2012. A homozygous mutation in KCTD7 links neuronal ceroid lipofuscinosis to the ubiquitin-proteasome system. Am. J. Hum. Genet. 15376605. 91 (1), 202–208. doi:10.1016/j.ajhg.2012.05.023.

Takahashi, K., Nelvagal, H.R., Lange, J., Cooper, J.D., 2022. Glial dysfunction and its contribution to the pathogenesis of the neuronal ceroid lipofuscinoses. Front. Neurol. 1664-2295. 13. doi:10.3389/fneur.2022.886567.

Tanguy, Y., Biferi, M.G., Besse, A., Astord, S., Cohen-Tannoudji, M., Marais, T., Barkats, M., 2015. Systemic AAVrh10 provides higher transgene expression than AAV9 in the brain and the spinal cord of neonatal mice. Front. Mol. Neurosci. 1662-5099. 8 (July). doi:10.3389/fnmol.2015.00036.

U.S. Food and Drug Administration, 2024. Highlights of Prescribing Information: Cerliponase Alfa. Silver Spring, M.F.c.O.A.f.h. Internet.

Vanier, M.T., Caillaud, C., Levade, T., 2022. Disorders of sphingolipid synthesis, sphingolipidoses, niemann-pick disease type c and neuronal ceroid lipofuscinoses. In: Inborn Metabolic Diseases: Diagnosis and Treatment. Springer Berlin Heidelberg, France, 735–764. doi:10.1007/978-3-662-63123-2_40.

Wang, D., Tai, P.W.L., Gao, G., 2019. Adeno-associated virus vector as a platform for gene therapy delivery. Nat. Rev. Drug Discov. 14741784. 18 (5), 358–378. doi:10.1038/s41573-019-0012-9.

Wang, J.H., Gessler, D.J., Zhan, W., Gallagher, T.L., Gao, G., 2024. Adeno-associated virus as a delivery vector for gene therapy of human diseases. Signal Transduct. Target. Ther. 20593635. 9 (1). doi:10.1038/s41392-024-01780-w.

Wang, R.Y., Bodamer, O.A., Watson, M.S., Wilcox, W.R., 2011. Lysosomal storage diseases: diagnostic confirmation and management of presymptomatic individuals. Genet. Med. 10983600. 13 (5), 457–484. doi:10.1097/GIM.0b013e318211a7e1.

Weinberg, M.S., Samulski, R.J., McCown, T.J., 2013. Adeno-associated virus (AAV) gene therapy for neurological disease. Neuropharmacology. 18737064. 69, 82–88. doi:10.1016/j.neuropharm.2012.03.004.

Williams, R.E., Mole, S.E., 2012. New nomenclature and classification scheme for the neuronal ceroid lipofuscinoses. Neurology. 0028-3878. 79 (2), 183–191. doi:10.1212/WNL.0b013e31825f0547.

Zhang, Y., Bingying, D., Zou, M., Peng, B., Rao, Y., 2025. Neuronal ceroid lipofuscinosis—concepts, classification, and avenues for therapy. CNS Neurosci. Ther. 1755-5930. 31 (2). doi:10.1111/cns.70261.

Zhang, Z., Butler, J.D.B., Levin, S.W., Wisniewski, K.E., Brooks, S.S., Mukherjee, A.B., 2001. Lysosomal ceroid depletion by drugs: therapeutic implications for a hereditary neurodegenerative disease of childhood. Nat. Med. 10788956. 7 (4), 478–484. doi:10.1038/86554.

Ziółkowska, E.A., Takahashi, K., Dickson, P.I., Sardiello, M., Sands, M.S., Cooper, J.D., 2025. Neuronal ceroid lipofuscinosis: underlying mechanisms and emerging therapeutic targets. Nat. Rev. Neurol. 17594766. 21 (11), 606–622. doi:10.1038/s41582-025-01132-4.

Zwi-Dantsis, L., Mohamed, S., Massaro, G., Moeendarbary, E., 2025. Adeno-associated virus vectors: principles, practices, and prospects in gene therapy. Viruses. 1999-4915. 17 (2), 239. doi:10.3390/v17020239.

Section III

Neurodegenerative disorders

Chapter 10

Gene therapy in Alzheimer's disease (AD)

Maryam Bemanalizadeh[a,*], Mitra Ashrafi[b,*], Mahdieh Eslami[c], Shiva Moein[d], Asad Kazemi[e], and Vahid Mansouri[a]

[a]*Gene Therapy Research Center, Shariati Hospital, Tehran University of Medical Sciences, Tehran, Iran*

[b]*School of Medicine, Zanjan University of Medical Sciences, Zanjan, Iran*

[c]*School of Medicine, Isfahan University of Medical Sciences, Isfahan, Iran*

[d]*Department of Pathology, Brigham and Women's Hospital, Boston, MA, United States*

[e]*Department of Medical Biotechnology, School of Advanced Technologies in Medicine, Tehran University of Medical Sciences, Tehran, Iran*

1 Introduction

Alzheimer's disease (AD), a neurodegenerative disorder characterized by progressive cognitive decline, memory impairment, and functional deterioration, remains one of the most significant global health challenges. Despite decades of research, current therapeutic strategies provide only symptomatic relief and have limited efficacy in altering the disease trajectory. The multifaceted pathophysiology of AD, involving amyloid-beta (Aβ) plaques, tau protein aggregation, neuroinflammation, synaptic dysfunction, and neuronal loss, underscores the complexity of the disorder and the need for innovative therapeutic approaches (Sakakibara et al., 2018).

Gene therapy has emerged as a promising modality in the treatment of neurodegenerative diseases, including AD. By leveraging the ability to modulate gene expression, deliver therapeutic genes, or silence pathological ones, gene therapy offers the potential to address the underlying molecular and cellular dysfunctions driving AD pathology. Advances in vector technology, particularly the development of adeno-associated viruses (AAVs) and lentiviral vectors, have enhanced the precision, efficiency, and safety of gene delivery to the central nervous system (CNS). These breakthroughs, combined with a growing understanding of the genetic and molecular contributors to AD, have spurred interest in translating preclinical findings into clinical applications (Arabi et al., 2022; Ling et al., 2023).

Recent preclinical studies have demonstrated the potential of gene therapy to target key mechanisms of AD, such as reducing Aβ production, enhancing Aβ clearance, stabilizing tau protein, and modulating neuroinflammatory pathways. These findings have laid the foundation for clinical trials exploring novel gene therapy approaches aimed at modifying disease progression. However, significant challenges remain, including optimizing delivery systems for widespread CNS distribution, minimizing immunogenicity, ensuring sustained therapeutic effects, and addressing the heterogeneity of AD pathology.

This review provides a comprehensive overview of the current state of knowledge in AD gene therapy, synthesizing insights from preclinical models and early clinical trials. It highlights key achievements, identifies critical limitations, and discusses future directions for advancing this therapeutic approach. By bridging the gap between preclinical research and clinical application, this manuscript aims to elucidate the potential of gene therapy to transform the landscape of AD treatment and address the unmet needs of patients and caregivers worldwide.

1.1 AD pathogenesis

The pathogenesis of AD is a complex interplay of molecular mechanisms and cellular dysfunctions that drive progressive neurodegeneration. The hallmark features of AD include Aβ plaque accumulation, tau protein hyperphosphorylation, synaptic dysfunction, and neuroinflammation, all of which contribute to cognitive decline and neuronal loss (John and Reddy, 2021) (Fig. 10.1).

* Maryam Bemanalizadeh and Mitra Ashrafi made an equal contribution as the first authors.

Gene Therapy in Neurodegenerative Disorders. https://doi.org/10.1016/B978-0-443-40467-2.00014-2

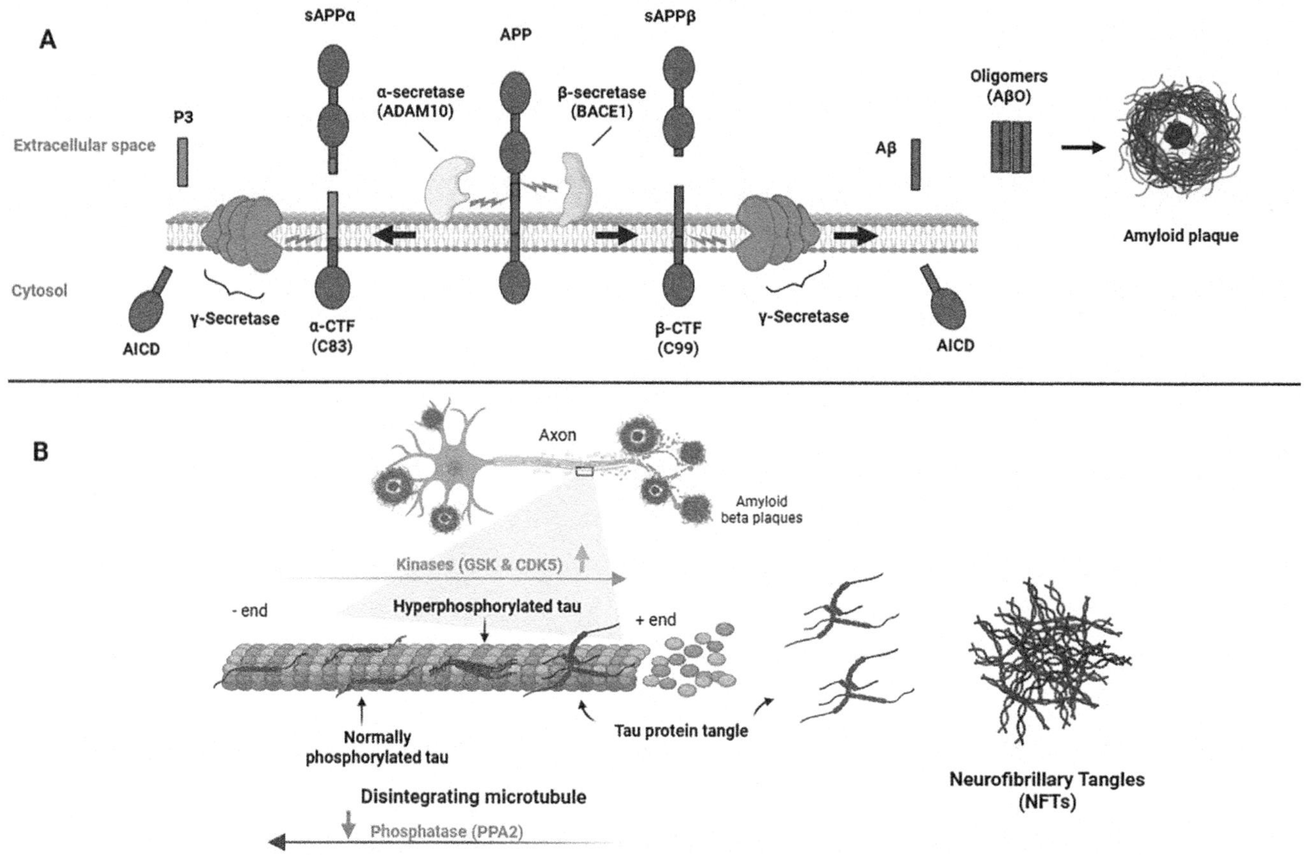

FIG. 10.1 (A) Proteolytic processing pathways of the amyloid precursor protein (APP). APP is a type I transmembrane protein featuring a large N-terminal extracellular domain, a single transmembrane helix, and a short C-terminal cytoplasmic tail. Its processing occurs via two competing pathways. The nonamyloidogenic pathway involves cleavage by α-secretase (ADAM10) within the Aβ domain, producing the soluble fragment sAPPα and the membrane-bound C-terminal fragment α-CTF (C83). Subsequent γ-secretase-mediated cleavage of α-CTF releases the P3 peptide and the APP intracellular domain (AICD). Conversely, the amyloidogenic pathway is initiated by β-secretase (BACE1) cleavage at the N-terminus of the Aβ domain, generating sAPPβ and β-CTF (C99). Further processing of β-CTF by γ-secretase produces the Aβ peptide, which aggregates into oligomers (AβO), and AICD. The juxtamembrane region contains the α- and β-cleavage sites, while the γ-cleavage site resides within the transmembrane domain. (B) Pathological hyperphosphorylation of the microtubule-associated Tau protein, driven by an imbalance between kinase and phosphatase activities, disrupts its normal function, leading to its dissociation from axonal microtubules. Mislocalized Tau subsequently aggregates into intracellular neurofibrillary tangles (NFTs). The consequent loss of microtubule integrity disrupts essential axonal transport, resulting in chronic axonal degeneration and, ultimately, neuronal death. *(Created in https://BioRender.com).*

One of the central events in AD pathogenesis is the accumulation of extracellular Aβ plaques, which disrupt neuronal function, induce oxidative stress, and trigger neuroinflammation. Aβ is generated through sequential cleavage of the amyloid precursor protein (APP) by β-secretase (BACE1) and γ-secretase. An imbalance between Aβ production and clearance leads to the formation of toxic Aβ oligomers and fibrils, which aggregate into plaques. Several factors contribute to impaired Aβ clearance, including the apolipoprotein E (APOE) ε4 allele, which reduces clearance efficiency and promotes deposition. In addition, dysfunction of Aβ-degrading enzymes, such as neprilysin (NEP), endothelin-converting enzyme (ECE), and insulin-degrading enzyme (IDE), further accelerates plaque accumulation. Moreover, vascular dysfunction and blood–brain barrier (BBB) impairment hinder Aβ transport across the cerebrovascular system, exacerbating its accumulation in the brain (Shi et al., 2024).

Another major pathological hallmark of AD is the intracellular aggregation of hyperphosphorylated tau proteins, forming neurofibrillary tangles (NFTs) that disrupt neuronal function. Normally, tau stabilizes microtubules and supports intracellular transport. However, in AD, tau becomes hyperphosphorylated, detaches from microtubules, and aggregates into toxic filaments. This process leads to cytoskeletal destabilization, impaired axonal transport, and synaptic dysfunction, ultimately contributing to cognitive decline. In addition, misfolded tau proteins spread from cell to cell, facilitating the propagation of tau pathology and accelerating neurodegeneration throughout the brain (Zhang et al., 2021).

Neurotrophic factors, such as nerve growth factor (NGF) and brain-derived neurotrophic factor (BDNF), are critical for neuronal survival, synaptic plasticity, and cognitive function. In AD, NGF and BDNF signaling pathways are disrupted, particularly affecting cholinergic neurons in the basal forebrain and hippocampus. Reduced NGF signaling contributes to cholinergic neuronal loss, impairing learning and memory, while decreased BDNF levels weaken synaptic plasticity, leading to accelerated cognitive decline (Rafii et al., 2018; Girotra et al., 2022).

Growing evidence suggests that metabolic dysfunction and insulin resistance are key contributors to AD pathogenesis. Impaired insulin signaling in the brain affects synaptic function and neuroprotection, drawing comparisons between AD and "type 3 diabetes." In addition, dysregulation of cholesterol metabolism, particularly in APOE ε4 carriers, exacerbates Aβ accumulation and neurodegeneration. Oxidative stress and mitochondrial dysfunction further compromise neuronal energy metabolism, increasing vulnerability to apoptosis (Kshirsagar et al., 2021).

Chronic neuroinflammation also plays a pivotal role in AD progression, largely mediated by activated microglia and astrocytes responding to Aβ and tau pathology. Overactivated microglia fail to clear Aβ efficiently, leading to a sustained inflammatory response that exacerbates neuronal damage. In addition, proinflammatory cytokines such as TNF-α, IL-1β, and IL-6 further amplify neurotoxicity and synaptic loss. Dysfunction of the renin–angiotensin system has also been implicated in neuroinflammation and vascular dysregulation, contributing to progressive cognitive decline. Moreover, BBB dysfunction compromises nutrient delivery, impairs Aβ clearance, and allows infiltration of peripheral immune cells, exacerbating neurodegeneration (Leng and Edison, 2021).

Multiple intracellular signaling pathways are disrupted in AD, leading to synaptic failure and neuronal apoptosis. Dysregulation of pathways such as phospholipase C-gamma, Ras/Raf/MAPK, and PI3K leads to impaired neurotrophic signaling and reduced neuronal survival. In addition, disruptions in calcium homeostasis increase excitotoxicity, further contributing to synaptic loss (Palasz et al., 2023).

The complex pathophysiology of AD involves amyloid and tau pathology, neuroinflammation, metabolic dysfunction, and vascular impairment. Understanding these interconnected mechanisms is essential for developing targeted therapeutic strategies aimed at enhancing neurotrophic support, modulating Aβ clearance, and restoring metabolic balance. Given the multifaceted nature of AD, future treatments must address these diverse pathological processes to halt or slow disease progression.

1.2 Genetics of AD

AD is a genetically complex disorder influenced by multiple genes, epigenetic modifications, and environmental factors (Fenoglio et al., 2018). Rather than being caused by a single gene, AD susceptibility arises from the interplay of numerous genetic variants that can either increase or decrease disease risk (Fenoglio et al., 2018; Bellenguez et al., 2022). Understanding these genetic contributors is crucial for developing new therapeutic strategies, as well as for advancing early diagnosis and prevention approaches. Among the most well-established genetic determinants of AD are the APP, presenilin 1 (PSEN1), presenilin 2 (PSEN2), and APOE genes (Lanoiselée et al., 2017; Raulin et al., 2022).

2 Early-onset familial Alzheimer's disease and causative mutations

A small proportion of AD cases, known as early-onset familial Alzheimer's disease (EOFAD), follow an autosomal dominant inheritance pattern and typically manifest before the age of 65. These cases are primarily caused by mutations in three genes: APP, PSEN1, and PSEN2.

The APP gene, located on chromosome 21, encodes the APP, which undergoes proteolytic cleavage by β-secretase and γ-secretase to generate Aβ peptides (Thinakaran and Koo, 2008). Pathogenic mutations in APP, particularly those near the β- and γ-secretase cleavage sites, lead to excessive production of Aβ42, the most aggregation-prone isoform, which contributes to amyloid plaque formation and neurotoxicity (Thinakaran and Koo, 2008; Janelidze et al., 2016; Bekris et al., 2010). To date, 49 APP mutations have been linked to AD, with most exhibiting dominant inheritance, although two recessive variants (A673V and E693Δ) have also been identified (Cruts et al., 2012; Guerreiro et al., 2012). Notably, individuals with Down syndrome (trisomy 21), who have an extra copy of APP, exhibit an increased risk of developing AD, further supporting APP's role in disease pathogenesis.

Mutations in PSEN1 (chromosome 14) and PSEN2 (chromosome 1), which encode essential components of the γ-secretase complex, are the most common genetic causes of EOFAD. These mutations alter γ-secretase activity, increasing the Aβ42/Aβ40 ratio, leading to enhanced amyloid aggregation and plaque deposition (Cruts et al., 2012; Tanzi and Bertram, 2005; Esler and Wolfe, 2001; Steiner, 2004). PSEN1 mutations account for approximately 50% of EOFAD cases, and carriers often exhibit an aggressive disease course, with earlier symptom onset and faster cognitive decline. In contrast, PSEN2 mutations are rarer and are associated with greater phenotypic variability (Wakabayashi and De Strooper, 2008).

3 Late-onset Alzheimer's disease and APOE genotype

The vast majority of AD cases fall under late-onset Alzheimer's disease (LOAD), which develops after the age of 65. Unlike EOFAD, which is driven by deterministic mutations, LOAD is influenced by multiple genetic risk factors, the most significant being the APOE gene on chromosome 19q13.2 (Corder et al., 1993). APOE encodes APOE, a key lipid transporter in the brain that modulates Aβ clearance, aggregation, and deposition (Long and Holtzman, 2019; Silva et al., 2019).

Three common APOE alleles, ε2, ε3, and ε4, are known to affect AD risk. The APOE ε4 allele is the strongest genetic risk factor for sporadic AD, increasing susceptibility in a dose-dependent manner (Saunders, 2000; Roses, 1996). Individuals carrying one copy of APOE ε4 have a threefold increased risk, while those with two copies (homozygous carriers) face a 12-fold greater risk of developing AD compared to APOE ε3 homozygotes (Mahley et al., 2009). Conversely, the APOE ε2 allele appears protective, reducing AD risk. Studies suggest that APOE ε4 contributes to AD pathogenesis by impairing Aβ clearance, promoting tau-mediated neurodegeneration, and influencing microglial function and neuroinflammation (Yamazaki et al., 2019; Kok et al., 2009; Polvikoski et al., 1995). APOE ε4 carriers also show faster Aβ deposition in the brain, even in preclinical stages of AD. In addition, recent findings indicate that APOE ε4 may exert Aβ-independent effects, disrupting synaptic plasticity, cholesterol metabolism, and vascular integrity, further contributing to neurodegeneration (Liu et al., 2013).

4 Epigenetic mechanisms in AD

In addition to genetic variations, epigenetic modifications play a critical role in AD development (Roubroeks et al., 2017; Cavalli and Heard, 2019). Epigenetics refers to heritable changes in gene expression that occur without altering the DNA sequence, including DNA methylation, histone modifications, and noncoding RNAs (ncRNAs) (Kiselev et al., 2021).

DNA methylation abnormalities have been extensively studied in AD, particularly in genes associated with disease pathogenesis. Altered methylation patterns in APP, PSEN1, and APOE suggest a link between epigenetic regulation and disease progression (Yokoyama et al., 2017). Studies have shown that hypomethylation of the APP promoter may lead to increased Aβ production, while hypermethylation of PSEN1 could impair γ-secretase function (Iwata et al., 2014). In addition, differential DNA methylation at loci such as SORL1, ABCA7, HLA-DRB5, and BIN1 has been associated with pathological AD changes, further underscoring the importance of epigenetic regulation in disease onset and progression (Yu et al., 2015).

Histone modifications, which regulate chromatin accessibility and gene transcription, are also implicated in AD (Santana et al., 2023). Dysregulation of histone acetylation and methylation has been linked to synaptic dysfunction, cognitive decline, and neuroinflammation. Targeting these modifications with histone deacetylase inhibitors is currently being explored as a potential therapeutic strategy to restore gene expression homeostasis in AD (Nativio et al., 2018).

Another emerging area of research is the role of ncRNAs, including microRNAs (miRNAs) and long ncRNAs, which regulate gene expression posttranscriptionally (Cai et al., 2022). Recent studies suggest that specific miRNAs can modulate Aβ aggregation, tau phosphorylation, and neuroinflammation, making them promising therapeutic targets (Cai et al., 2022; Swanson et al., 2021; Gámez-Valero et al., 2020; Li et al., 2021; Jiang and Xu, 2023).

4.1 Current therapeutic approaches for AD: From conventional therapies to novel approaches

Recently, it has been acknowledged that the multifaceted and partially known pathophysiological changes leading to AD may start decades earlier than the onset of clinical presentations, by which time considerable and likely permanent damage has occurred broadly on multiple levels, from molecular to cellular networks. Despite rapid improvement in our knowledge regarding the pathophysiology of AD, no cure has been developed for AD (Atri, 2019).

Current US Food and Drug Administration (FDA)-approved symptomatic treatments for AD include cholinesterase inhibitors (donepezil, rivastigmine, and galantamine) and the N-methyl-D-aspartate (NMDA)-receptor antagonist memantine; a fixed-dose combination of memantine ER and donepezil (Namzaric) is also approved for moderate-to-severe AD. These agents do not cure AD and generally provide modest, temporary symptomatic benefits. In addition, brexpiprazole (Rexulti) is FDA-approved for agitation associated with dementia due to AD, providing a symptomatic option for a major behavioral complication; however, it does not modify disease progression and carries clinically important safety warnings typical of antipsychotic-class agents in older adults (Atri, 2019; Mangialasche et al., 2010). After a prolonged period without an FDA-approved disease-modifying therapy, antiamyloid monoclonal antibodies were introduced: aducanumab received accelerated approval in 2021 (later discontinued by the manufacturer), lecanemab (Leqembi) received accelerated approval in January 2023 and traditional approval in July 2023, and donanemab (Kisunla) was approved in July 2024. These agents are indicated for initiation in early symptomatic AD (typically MCI due to AD or mild AD dementia) with confirmed amyloid-β pathology, reflecting the populations studied in pivotal trials (Table 10.1).

TABLE 10.1 Classification and characteristics of FDA-approved AD medications.

Drug	Indication	Dose and description	Common side effects
Cognitive symptoms treatments			
Donepezil	Mild-to-severe AD	Starting dose: 5 mg/d; can be increased to 10 mg/d after 4–6 wk. Before starting donepezil 23 mg/d, patients should be on donepezil 10 mg/d for at least 3 mo.	Nausea, vomiting, loss of appetite, muscle cramps, and increased frequency of bowel movements
Rivastigmine	Mild-to-moderate AD	Oral: starting dose: 1.5 mg twice daily. If well-tolerated, the dose may be increased to 3 mg twice daily after 2 wk. Subsequent increases to 4.5 and 6 mg twice daily should be attempted after a minimum of 2 wk at the previous dose. Maximum dose: 6 mg twice daily. Oral rivastigmine can be difficult to tolerate.Patch: starting dose: one 4.6 mg patch once daily for a period of 24 h. Maintenance dose: one 9.5 mg or 13.3 mg patch once daily for a period of 24 h. Before initiating the maintenance dose, patients should undergo a minimum of 4 wk of treatment at the initial dose (or at the lower patch dose of 9.5 mg) with good tolerability.	Nausea, vomiting, loss of appetite, and increased frequency of bowel movements
Galantamine	Mild-to-moderate AD	Extended-release: start at 8 mg once daily for 4 wk; increase to 16 mg once daily for 4 wk; increase to 24 mg once daily.Generic: start at 4 mg twice daily for 4 wk; increase to 8 mg twice daily for 4 wk; increase to 12 mg twice daily.	Nausea, vomiting, loss of appetite, and increased frequency of bowel movements
Memantine	Moderate-to-severe AD	Immediate-release: starting dose: 5 mg once daily; increase the dose in 5 mg increments to a maximum of 20 mg daily (divided doses taken twice daily) with a minimum of 1 wk between dose increases. In earlier stages, a 10 mg-daily dose may be considered. The maximum recommended dose for severe renal impairment is 5 mg twice daily. Extended-release (XR): for patients new to memantine, the recommended starting dose of memantine XR is 7 mg once daily, and the recommended target dose is 28 mg once daily. The dose should be increased in 7 mg increments every seventh day. The minimum recommended interval between dose increases is 1 week, and only if the previous dose has been well-tolerated. The maximum recommended dose for severe renal impairment is 14 mg once daily.	Headache, constipation, confusion, and dizziness
Donepezil and memantine	Moderate-to-severe AD	A combination capsule consisting of 7–28 mg memantine/10 mg donepezil is given orally once daily. It can be started in patients already on a background of stable donepezil 10 mg daily (with memantine dose titration) or in patients already on combination treatment with each agent. The maximum recommended dose in severe renal impairment is 14 mg memantine XR/10 mg donepezil once daily.	Nausea, vomiting, loss of appetite, increased frequency of bowel movements, headache, constipation, confusion, and dizziness
Disease-modifying treatment			
Aducanumab	MCI or mild AD	Once every 4 weeks	ARIA, headache, and falls
Lecanemab	MCI or mild AD	Biweekly	ARIA, infusion-related reactions, and brain atrophy
Donanemab	MCI or mild AD	Once every 4 weeks	ARIA and headache
Noncognitive symptom treatments (psychological and behavioral)			
Suvorexant	Insomnia in mild-to-moderate AD	10 mg once per night.Maximum recommended dose: 20 mg once per night	Impaired alertness and motor coordination, worsening of depression or suicidal thinking, complex sleep behaviors, sleep paralysis, and compromised respiratory function

Abbreviations: AD: Alzheimer's disease; MCI: mild cognitive impairment; ARIA: amyloid-related imaging abnormalities; wk: week; d: day; mo: month; FDA: Food and Drug Administration.

Efficacy of antiamyloid antibodies is described as slowing (not reversing) clinical decline, with average benefits that many clinicians consider modest and variably "clinically meaningful" across patients. Their use also requires intensive safety management because they carry boxed warnings for amyloid-related imaging abnormalities (ARIA), including edema and micro/macrohemorrhage, with higher ARIA risk in APOE ε4 homozygotes and potential for serious intracerebral hemorrhage, particularly in higher-risk patients and in the setting of antithrombotic exposure; therefore baseline and follow-up MRI monitoring and careful risk stratification are integral to prescribing (Swanson et al., 2021; Reish et al., 2023). From a health-economics perspective, the cost-effectiveness of antiamyloid monoclonal antibodies remains debated because overall "real-world" costs extend well beyond the drug itself to include amyloid confirmation testing, infusion or injection delivery infrastructure, repeated MRI monitoring, and management of ARIA and other adverse events. Independent value assessments have suggested that current pricing may exceed commonly cited cost-effectiveness thresholds in some scenarios, although estimates vary by healthcare system, patient selection, treatment duration, and negotiated discounts. Recent moves toward more convenient dosing options (e.g., subcutaneous maintenance after IV initiation) may reduce some logistical burden, but they do not eliminate the need for ongoing safety monitoring and associated costs.

4.2 Gene therapy in AD

With the failure of conventional and disease-modifying therapies to halt or reverse AD progression, gene therapy has emerged as a promising alternative, offering a targeted approach to modulate disease-related genes, enhance neurotrophic support, and promote Aβ clearance (Long and Holtzman, 2019; Tuszynski et al., 2015). Gene therapy aims to correct genetic abnormalities, modify gene expression, or introduce therapeutic genes using advanced molecular tools and delivery systems. Over the past decade, preclinical and early-phase clinical studies have investigated gene therapy's potential in AD, focusing on key targets such as neurotrophic factors, Aβ-degrading enzymes, and APOE modulation.

4.3 Targets for gene therapy

The molecular targets of AD gene therapy investigated so far can be categorized into three groups: (1) neurotrophins (NTs), including NGF and BDNF; (2) enzymes directly implicated in the degradation of Aβ, such as NEP, ECE, and cathepsin B (CTSB); and (3) factors associated with Aβ burden, notably APOE.

4.3.1 Nerve growth factor

The decline of cholinergic neurons in the basal forebrain correlates with cognitive decline in AD (Whitehouse et al., 1982; Schmitz and Nathan Spreng, 2016). NGF, an intrinsic NT, plays a pivotal role in governing the growth and survival of cholinergic neurons by functionally activating the TrkA receptor (Mobley et al., 1986; Holtzman et al., 1992). Early and substantial cholinergic neurodegeneration significantly contributes to cell loss and cognitive decline in AD, with a decrease in NGF levels observed in the basal forebrain region during the disease. Animal models of AD have demonstrated that NGF stimulates cholinergic neurons, crucial for maintaining cognitive function and preventing neuronal death (Counts et al., 2004). In both rats and monkeys, exogenous NGF has been shown to delay the degeneration of axotomized basal forebrain cholinergic neurons (Tuszynski et al., 2015; Koliatsos et al., 1990). Furthermore, it exhibits neuroprotective properties and enhances memory in aged rats (Fischer et al., 1987; Bishop et al., 2008), while also reversing the loss of cholinergic neurons in elderly monkeys (Nagahara et al., 2009). These findings solidify NGF's potential as a neuroprotective treatment for AD.

4.3.2 Brain-derived neurotrophic factor

Another noteworthy NT is BDNF, which possesses a significant capacity to decelerate neuronal aging and enhance neuronal activity in cell systems susceptible to deterioration in AD. Analogous to NGF, BDNF plays a role in promoting target innervation, axonal extension, and neuronal survival during brain development. In contrast to NGF, which primarily influences cholinergic neurons in the basal forebrain, BDNF directly impacts the survival and functionality of cortical and hippocampal neurons (Conner et al., 1997; Yan et al., 1997). The cortex and hippocampal neurons, both expressing the BDNF receptor trkB, are pivotal for the manifestation of the characteristic symptoms of AD (Conner et al., 1997; Yan et al., 1997). Theoretically, BDNF gene therapy holds the potential to more effectively or enduringly prevent or delay the degeneration of critical hippocampal and cortical circuitry observed in AD.

4.3.3 Apolipoprotein E

As previously mentioned, APOE competitively binds to Aβ receptors, hindering Aβ uptake and modifying its clearance (Long and Holtzman, 2019; Silva et al., 2019). Intriguing and promising results emerged from the initial investigation involving gene delivery in mice, wherein various human APOE isoforms were introduced using a lentiviral vector. In APP transgenic mice, PDGF promoter driven amyloid precursor protein (PDAPP) which lacks murine APOE, the introduction of human APOE ε4 led to elevated Aβ levels and increased Aβ burden after 5 weeks. Conversely, the expression of APOE ε2 and APOE ε3 had no impact on Aβ levels, affirming the detrimental nature of APOE ε4. Moreover, when compared to APOE ε3 and APOE ε4, the lentivirus (LV)-mediated production of APOE ε2 significantly reduced both Aβ42 concentration and Aβ load in PDAPP mice expressing endogenous APOE (Dodart et al., 2005). It is noteworthy that since APOE is widely distributed throughout the CNS, delivering the APOE gene is expected to exert a broad range of effects on the brain and facilitate Aβ clearance across the BBB. Despite recent evidence indicating a reduction in Aβ clearance across the BBB when the APOE complex forms, the clearance occurred more rapidly when Aβ was associated with APOE ε2/3 as opposed to APOE ε4 (Bell et al., 2007b; Deane et al., 2008). Consequently, gene delivery of APOE ε2 holds promise as a potential treatment for AD, although it may also carry unintended side effects, necessitating precise control of APOE ε2 expression.

4.3.4 Neprilysin

NEP, a glycoprotein located on the plasma membrane and classified as a neutral zinc metalloendopeptidase, serves as a pivotal enzyme responsible for breaking down Aβ (Kanemitsu et al., 2003). During the aging process and the early stages of AD, there is observed inactivation and downregulation of NEP (Grimm et al., 2013). These findings suggest that the diminished NEP activity during aging contributes to the accumulation of Aβ, thereby playing a role in the progression of AD (Grimm et al., 2013). Notably, increasing evidence indicates that augmenting NEP levels through neuronal overexpression promotes the degradation of Aβ, preventing the formation of Aβ plaques and the associated cytopathology (Poirier et al., 2006). As a strategy for AD therapy, gene therapy has recently been employed to enhance NEP levels or activity (Nayerossadat et al., 2012). In addition to reducing oxidative stress and inflammation, NEP gene therapy administered in the early stages of AD demonstrated a 50% improvement in memory impairment (El-Amouri et al., 2008). These conclusive findings underscore the potential value of NEP gene delivery as a therapeutic approach for reducing Aβ levels in AD.

4.3.5 Endothelin-converting enzyme

ECEs belong to the M13 zinc metallopeptidase subclass, which is a family of type II integral membrane-bound proteases. Their name stems from their ability to convert the inert precursor "big endothelin" into the potent vasoactive peptide endothelin-1 (ET-1). Furthermore, ECEs have been recognized for their capability to hydrolyze various physiologically active peptides by cleaving hydrophobic residues on the amino side (Hoang and Turner, 1997; Johnson et al., 1999), and studies have confirmed their involvement in degrading Aβ (Eckman et al., 2001). Both in vitro and in vivo, ECE demonstrates the capacity to break down Aβ, and the deletion of ECE1 or ECE2 genes in mice leads to a 1.3-fold increase in Aβ levels in the brain (Eckman et al., 2001, 2003). Microarray research in late-onset AD reveals a significant reduction in ECE2 levels in the inferior parietal lobe (Weeraratna et al., 2007). A study by Carty et al. administering AAV5-ECE-1 to APP/PS1 transgenic mice results in substantial ECE1 expression near the injection sites, leading to a decrease in Aβ and plaques in the anterior cortex and hippocampus (Carty et al., 2008). These findings underscore ECE's role in Aβ degradation, making it a potential target for gene therapy.

4.3.6 Cathepsin B

CTSB is a lysosomal cysteine protease with conflicting, context-dependent roles in AD-related Aβ biology. Recent work reports that CTSB activity can exacerbate Aβ proteotoxicity (and CTSB levels are elevated in AD patient brains), and in a *Caenorhabditis elegans* Aβ model, knockdown of the CTSB homolog (cpr-6) alleviated Aβ toxicity while reducing swsn-3 expression and increasing SMK-1 protein levels, implicating aging regulatory pathways. Some pharmacologic studies show that CTSB inhibitors (e.g., CA074Me, E64d) can improve memory and reduce Aβ40/Aβ42 and CTFβ in transgenic mice expressing APP with the wild-type β-secretase site. At the same time, evidence also supports a protective clearance role in certain models: in APP transgenic mice (J20), CTSB deficiency increased plaque load and the Aβ42/40 ratio, whereas lentiviral CTSB delivery reduced anti-Aβ antibody, and thioflavin S-positive plaques, consistent with Aβ degradation activity (Hook et al., 2008; Mueller-Steiner et al., 2006; Hook et al., 2009; Nilsson et al., 2010). Overall, CTSB likely exhibits dual proteolytic effects (intersecting with both APP processing and Aβ degradation/proteotoxicity pathways), so its net impact depends strongly on the model, APP context, and compartment, warranting careful interpretation when considering it as a therapeutic or gene therapy target.

4.3.7 Other targets

Angiotensin-converting enzyme (ACE) has drawn attention as an enzyme involved in Aβ degradation. ACE typically participates in converting angiotensin I to angiotensin II to regulate vasoconstriction and subsequently control blood pressure. An indel polymorphism in intron 16 of the ACE gene appears to be associated with AD, where the deletion variant is linked to a reduced risk of AD (Lehmann et al., 2005). The administration of ACE inhibitors to various animal models of AD has shown some positive effects on cognition, yet the impact on Aβ levels and the molecular mechanisms underlying the cognitive improvement remain unclear. Moreover, studies with ACE knockout (KO) mice indicate that ACE does not directly affect Aβ levels, suggesting an indirect role in cognition enhancement that does not involve Aβ degradation (Saido, 2013). The effects of ACE gene delivery are not immediately evident and may potentially influence blood pressure management, which warrants careful consideration. Further investigation is necessary to elucidate the specific role of ACE in AD and better understand its implications for therapeutic interventions.

Implicated in the degradation of Aβ, IDE, another metalloendopeptidase, has drawn attention in research. Studies indicate that mice with a knocked-out IDE gene exhibit slightly elevated Aβ levels (Miller et al., 2003; Farris et al., 2003). IDE demonstrates the ability to degrade Aβ, along with insulin and amylin peptides associated with type 2 diabetes. Notably, both hyperinsulinemia and genetic variants of IDE are correlated with an increased risk of AD, hinting at IDE as a potential common factor (Qiu and Folstein, 2006). In the context of gene therapy, the delivery of the IDE gene holds promise for enhancing Aβ degradation and potentially altering the pathogenesis of AD. However, careful investigation into the potential consequences on insulin levels is imperative.

4.4 Animal models of AD

To assess new therapeutic approaches, several animal models have been created to study specific aspects of AD. Animal models are designed to reveal the primary symptoms of AD, such as amyloid and tau pathology or neuron degeneration. Furthermore, the development of AD pathology should follow a predictable pattern in specific regions of the brain, mirroring the progression of the disease in human AD patients. Only a portion of the clinical signs of AD in humans can be exhibited by these animal models (Zhong et al., 2024).

Aging is the greatest risk factor for developing AD, and aging-related animal models aim to mimic age-related cognitive decline and neuropathology. Aged rodents, such as aged rats and mice, were commonly used to study the natural progression of AD-like pathology with age. Aging models are valuable for understanding the impact of aging on AD pathogenesis and evaluating interventions to slow down cognitive decline in elderly individuals. On the contrary, transgenic and KO/knock-in (KI) mouse models have been instrumental in advancing our understanding of AD, particularly in comparison to naturally aging models that develop AD-like pathology. Transgenic models, typically overexpressing human APP (Bellenguez et al., 2022), presenilin (PSEN), or tau mutations associated with familial Alzheimer's disease (FAD), are engineered to recapitulate specific genetic mutations known to drive early-onset AD. These models often exhibit accelerated development of amyloid plaques, NFTs, and cognitive deficits, providing a valuable framework for studying disease progression and testing therapeutic strategies. In contrast, KO/KI models allow for the precise manipulation of endogenous genes, enabling the study of the functional consequences of specific genetic risk factors, such as the APOE ε4 allele or tau phosphorylation. Aging models, on the other hand, involve wild-type mice that develop AD-like features over time, mimicking late-onset sporadic AD, which is the most common form of the disease. While these aging models reflect the multifactorial nature of AD more closely, they lack the rapid disease onset and the clear genetic triggers seen in transgenic and KO/KI models. As such, each model offers unique insights into the molecular mechanisms underlying AD, and combining these approaches can provide a more comprehensive understanding of both early- and late-onset forms of the disease (Jankowsky and Zheng, 2017).

The first section of Table 10.2 provides a comprehensive summary of eight commonly used transgenic mouse models in AD research. These models, including Tg2576, TgCRND8, and 3xTg-AD, highlight various genetic mutations associated with AD, such as the Swedish and Indiana mutations in *APP* and *MAPT* mutations for tau pathology. Each model demonstrates distinct phenotypes, such as amyloid plaque formation, neurodegeneration, and tau hyperphosphorylation, which are key pathological features of AD. Notably, these models also exhibit behavioral impairments, like memory deficits and spatial learning difficulties, along with sex-specific differences in disease progression. For example, *Tg2576* females display more pronounced cognitive decline and disrupted sleep patterns compared to males, reflecting the importance of sex as a variable in AD studies (Table 10.2).

The second section of Table 10.2 focuses on KO and KI models, summarizing six critical models like APP KI, Tau KI, and TREM2 KO, which explore AD-associated mutations and their effects on neuroinflammation, amyloidosis, and synaptic dysfunction. The models highlight the contributions of various mutations to Aβ deposition, tau pathology, and microglial activation, crucial for understanding the immune response in AD. For instance, APP KI models, such as NL-G-F, show cortical amyloid plaques by 2 months of age, with increased microgliosis and astrocytosis, while APOE KI mice exhibit significant tau phosphorylation and BBB deterioration. These models are integral for investigating the complex interaction between genetic mutations and disease mechanisms in AD (Table 10.2).

Beyond rodent systems, invertebrate models such as *Drosophila melanogaster* and *C. elegans* are also widely used to interrogate discrete AD mechanisms. In *Drosophila*, expression of human Aβ and/or tau enables rapid, low-cost genetic

TABLE 10.2 The transgenic and knock-in/knockout mouse models of Alzheimer's disease.

	Transgenic mouse models				
Name	**Genetic modification**	**Pathological features and sex differences**	**Behavioral characteristics**	**Strengths and limitations**	**References**
Tg2576	*hAPP695* with the Swedish mutation	Dense amyloid plaques are visible by 7–8 months, with significant deposition by 11–13 months in the brain parenchyma and vasculature; females show more plaque accumulation and cognitive decline with delayed sleep onset, while males exhibit increased aggression with disrupted sleep EEG rhythms by 22 months.	Cognitive impairments in spatial and working memory from 9 to 12 months; increased seizure susceptibility observed within 12–14 months.	Lack of tau pathology, shortened lifespan in certain genetic backgrounds.	(Jankowsky and Zheng, 2017; Götz et al., 2018; Hsiao et al., 1996; Takeuchi et al., 2000; Kobayashi and Chen, 2005; Deacon et al., 2009; King et al., 1999; Wisor et al., 2005; Toljan and Homolak, 2021; Chan et al., 2015; Silva et al., 2022; Westmark et al., 2010; Jacobsen et al., 2006; Apelt et al., 2002; Klingner et al., 2003; Shirvan et al., 2009; Woodhouse et al., 2009; Lilja et al., 2013; Chishti et al., 2001; Shekari and Fahnestock, 2021; Ashe and Zahs, 2010)
PS19	*hMAPT (P301S)*, mixed background	Development of tau seeding at 1.5 months, and neurodegeneration begins by 9 months in the hippocampus and entorhinal cortex; females have a higher survival rate compared to males.	Memory and learning impairments, limb weakness, and hyperactivity are observed at 3 months, while paralysis develops at 7 months.	Lack of amyloid pathology, shortened lifespan.	(Ashe and Zahs, 2010; Holmes et al., 2014; Sun et al., 2020; Takeuchi et al., 2011)
	hMAPT (P301S), congenic background	NFTs are observed at 6 months; the median lifespan ranges from 11 to 15 months.	Hyperactivity, altered pain perception, and startle response noted from 3 months.	More stable in the pathogenesis in comparison with the mixed background.	(Takeuchi et al., 2011; Maruyama et al., 2013; Lasagna-Reeves et al., 2016; Zhang et al., 2012)
TgCRND8	Double mutant: *hAPP 695 (KM670/671 NL* and *V717F)*	Stabilized Aβ40 levels and increased Aβ42 within 4–10 weeks, amyloid deposition noted in the cerebral cortex by 2–3 months, dense-cored plaque formation by 4–5 months in the hippocampus, midbrain, brainstem, and cerebellum; females demonstrate earlier cognitive deficits, while males show altered sleep patterns.	Sleep cycle disturbance (reduced REM and NREM sleep), metabolic dysregulation due to reduced NAA levels in the hippocampus and cortex, and dysregulation of myoinositol levels throughout mice aging.	Aggressive behavior, lack of tau pathology, and shortened lifespan.	(Chishti et al., 2001; Kosel et al., 2020; Granger et al., 2016; Janus et al., 2000; Götz and Ittner, 2008; Colby-Milley et al., 2015; Salek et al., 2010; Clark, 1998; Moffett et al., 2007; Ross et al., 1998; Meyerhoff et al., 1993; Rothman et al., 2003)
rTg4510	*hMAPTP301L* crossed with tTA allele	Pretangles form by 2.5 months; significant tau hyperphosphorylation and neuronal loss are noted by 5.5 months; females exhibit more severe tau pathology despite similar transgene expression levels between sexes, with worsened deterioration of spatial learning and memory.	Decline in spatial memory function observed from 4 months, increased anxiety and hyperactivity at later ages.	Restricted pathology to the cortex and hippocampus, robust tau expression, and neurodegeneration.	(Ramsden et al., 2005; Santacruz et al., 2005; Yue et al., 2011; Spires et al., 2006; Ludvigson et al., 2011; Sahara et al., 2014; Menkes-Caspi et al., 2015; Cook et al., 2014)

Continued

TABLE 10.2 The transgenic and knock-in/knockout mouse models of Alzheimer's disease—Cont'd

	Transgenic mouse models				
APP/PS1	Double mutant: *hAPP695* (*KM670/671 NL*) and *PS1* (*delEx9*)	Aβ40 and Aβ42 deposits, microglial and astrocytic activation, and an accumulation of endosomes and lysosomes by 4 months; modest neuronal loss adjacent to amyloid plaques in the hippocampus and cortex by 8–10 months; no significant differences reported between sexes regarding cognitive deficits.	Memory deficits detected at 6 months; impaired spatial navigation and learning noted by 12 months; nest-building and burrowing within 8–14 months.	Lack of tau pathology, hippocampal neuronal circuit dysfunction, LTP impairment (8–10 mo), but no PPF deficits (8–9 mo).	(Jankowsky et al., 2004; Jackson et al., 2016; Joshi et al., 2015; Ma, 2013; Nixon et al., 2005; Nixon and Yang, 2011, Nixon and Yang, 2012; Kuhla et al., 2017; van der Spek et al., 2021; Gelman et al., 2018; Kilgore et al., 2010; Lalonde et al., 2005; Volianskis et al., 2010; Janus et al., 2015; Hulshof et al., 2022; Zhou et al., 2020)
P301L	*Human 4R/2N* introduced to *P301L* mutation	NFTs without axonal dilations in the brainstem and spinal cord appear at 6 months, with a lifespan of approximately 8–12 months.	Impairments in passive avoidance tests at 5 months, motor deficits develop by 7 months.	Younger P301L mice may have better cognitive abilities compared to WT controls.	(Terwel et al., 2005; Dutschmann et al., 2010; Maurin et al., 2014)
5xFAD	Combination of Swedish, London, and Florida *hAPP* mutations with *M146L* and *L286V* in *PS1* mutations: Tg6799, Tg7092, Tg7031	Early amyloid aggregates and plaques were observed from 1.5 to 2 months, and neuroinflammation phenotypes (astrogliosis and microgliosis) are evident from as early as 2 months, with progressive neurodegeneration within 6 months; females show more pronounced amyloid pathology than males, while males show more significantly reduced PPF and decreases in HDL levels.	Spatial memory impairments evident from around 4 months, with motor dysfunction, increased hyperactivity, and reduced anxiety noted at later ages.	Lack of tau pathology, alignment of molecular signatures with human AD brains, LTP deficits (12 mo).	(Oakley et al., 2006; Maarouf et al., 2013; Sadleir et al., 2015; Devi et al., 2010; Forner et al., 2021; Oblak et al., 2021; Xiao et al., 2015; Kimura and Ohno, 2009; O'Leary et al., 2018)
	5xFAD (Bred on C57BL6 background)	Amyloid plaques develop aggressively in the subiculum and layer V pyramidal neurons within 16 days of birth, with a 40% loss by 1 year, intraneuronal plaque formation within 6 weeks, which progresses in the cortex, hippocampus, and thalamus in 2 months, and the spinal cord in 3 months; thinner myelin sheathes and shorter axon calibers develop in 1 and 2–3 months, respectively.	Impaired memory in cross-maze tests with reduced anxiety in the elevated plus maze at 3–6 months.	Lack of tau pathology, aggressive onset of amyloid pathology.	(Richard et al., 2015; Jawhar et al., 2012; Gu et al., 2018)
	5xFAD (AD-BXD variant)	Varying levels of amyloid pathology and cognitive impairment without correlation between them; females display more motor impairments compared to males.	Cognitive function varies widely due to age and transgene presence.	Used to model the genetic variation in humans and to identify transcriptional networks protective against AD-related cognitive decline.	(Neuner et al., 2019)
3xTg-AD	Triple mutant: *hPS1* (*M146V*), *hAPP* (*KM670/671 NL*), *MAPT P301L*	Extracellular amyloid deposits in the frontal cortex at 6 months, significant tau pathology in the hippocampus developing by the age of 1 year; females show earlier development of plaques and tangles associated with cognitive decline compared to males, while males exhibit earlier and increased morbidity/mortality rates.	Spatial learning deficits noted from as early as 6 months, worsening with age.	Lack of neuronal loss, intraneuronal Aβ immunoreactivity within 3–4 months, genetic drift observed within this model.	(Oddo et al., 2003; Belfiore et al., 2019; Billings et al., 2005; Carroll and Pike, 2008; Perez et al., 2011; Giménez-Llort et al., 2008)

	Knock-in and knockout mouse models				
APP KI	APP NL	No tau pathology observed; increased anxiety-like behavior noted at 15 months.	No decline in spatial learning or memory detected.	Lack of tau pathology and no decline in spatial learning and memory.	(Sakakibara et al., 2018; Saito et al., 2014; Xia et al., 2022; Watamura et al., 2022; Hirota et al., 2022)
	APP GF	Initial Aβ deposition observed at 4 months; significant deposition across larger brain areas *than in APP NLF or APP NLGF mice* by 12 months.	N/A	N/A	
	APP NLGF	Cortical Aβ amyloidosis starts at 2 months, with saturation by 7 months, consistent subcortical amyloidosis by 4 months, and greater microgliosis and astrocytosis compared to APP NL-F.	Spatial learning decline noted by 8 months, but memory retention remains intact, exhibiting anxiolytic-like behavior at 15 months and hyperreactivity to pain stimuli by 15–18 months.	N/A	
	APP SAA	Amyloid deposition detectable from 4 months, with a peak burden in the cortex and hippocampus by 8 months; females show more pronounced hyperactivity at 8 months.	Significant hyperactivity observed at 18 months.	N/A	
	APP NL-F	High levels of Aβ42 with a notable Aβ42/Aβ40 ratio; initial Aβ deposition occurs by 6 months, leading to cortical amyloidosis by 24 months. Neuroinflammation (microglia accumulation and astrocyte activation) is observed near Aβ plaques.	Memory impairment detected at 18 months.	Provides insights into upstream factors affecting Aβ amyloidosis compared to other mutations.	
Tau KI	m*MAPT* exons 1–14 replaced with *hMAPT* gene	Normal axonal localization of tau observed; no significant differences compared to WT mice.	Little variation in phenotype compared to wild type; often crossed with other models for further study.	MAPT P290S KI-developed murine tau aggregates.	(Saito et al., 2019; Huang et al., 2022; Barendrecht et al., 2023)
	MAPT KI x APP NLGF	Accelerated spread of pathological tau observed within 19 months; tau humanization does not affect Aβ levels or neuroinflammation.	Tau humanization did not affect memory.	More tau accumulation in the presence of amyloidosis. APP NLGF x MAPT P290S dKI mice demonstrated more tau inclusions than age-matched MAPT P290S KI mice. dKI mice also demonstrated tau-seeding abilities.	
	5xFAD x MAPT KI	Tau humanization appears protective against AD-related impairments and offsets LTP deficits compared to WT.	Reduced anxiogenic-like behavior and improved spatial learning compared to 5xFAD mice.	Enrichment in lysosome, phagocytosis, and oxidative phosphorylation by GSEA, compared to 5xFAD and human coexpression modules.	

Continued

TABLE 10.2 The transgenic and knock-in/knockout mouse models of Alzheimer's disease.

	Transgenic mouse models				
APOE KI	Targeted replacement of the APOE gene (APOE KI)	E4FAD variant leads to increased Aβ42 accumulation, tau hyperphosphorylation, neuronal loss, deterioration of the BBB, and reduced cerebral blood flow compared to E3FAD mice; female E3FAD and E4FAD mice have significantly higher Aβ42 and Aβ40 levels than their male counterparts. Female E4FAD mice exhibit significant deficits in learning and memory tasks, along with other deficits, including phospholipid and cholesterol dysregulation, microglial dysfunction, neuroinflammation, and tauopathy-related neurodegeneration.	E4FAD mice developed hippocampalassociated memory deficits and had a substantial drop in nest construction scores compared to E3FAD mice.	NA	(Barendrecht et al., 2023; Davignon et al., 1988; Sullivan et al., 2004; Knouff et al., 1999; Liraz et al., 2013; Montagne et al., 2021; Shi et al., 2017; Andrews-Zwilling et al., 2010; Salomon-Zimri et al., 2014; TCW et al., 2022; Zhu et al., 2015; Blanchard et al., 2022; Machlovi et al., 2022; Lynch et al., 2003; Colton et al., 2002; Balu et al., 2019; Tai et al., 2017; Guo et al., 2023; McLean et al., 2022; Koutsodendris et al., 2023a; Koutsodendris et al., 2023b; Huynh et al., 2019; Bien-Ly et al., 2012; Rodriguez et al., 2014)
	Floxed APOE KI	Presence of APOE protein in astrocytes without reactive Iba-1 positive microglial involvement.	Overall, cerebral accumulation of amyloid plaques not significantly affected when crossed with APP/PS1 mice.	Typically used to cross with other transgenic models, such as APP/PS1 mice or PS19 mice. Overall, cerebral accumulation of amyloid plaques in APOE4 KI mice crossed with APP/PS1 mice was not affected. PS19-E4 crossed mice demonstrated a higher degree of neurodegeneration.	
	APOE KI: JAX	Female APOE4 KI JAX mice had lower plasma Aβ42 and a decreased Aβ42/40 ratio. However, there were no differences between APOE4 and APOE3 KImice. Aβ40 levels did not differ regardless of APOE genotype or sex. It is suggested that there are higher levels of aggregate-prone Aβ42 in the brain in female APOE4 KI mice compared to their male counterparts.	Similar locomotor activity, motor coordination, and working memory tested by open field, rotarod, and Y-maze tests, respectively, between APOE4 KI and control mice, with an age-dependent decline observed from 2 to 12 months.	NA	
TREM2 KO	Deletion of exons 3 and 4	Reduced microglial numbers and impaired myelin repair; prolonged microgliosis and impaired cholesterol transport.	Decreased performance on motor coordination tests observed at 12 months when fed a specific diet.	No apparent neurological phenotypes, except for impaired immune response and altered transcriptome. TREM2 deficiency increased early-stage plaque growth but did not affect overall plaque deposition in an APP/PS1 dE9 mouse model with human APOE3 or APOE4.	(Ulland et al., 2017; Poliani et al., 2015; Otero et al., 2012; Wang et al., 2015; Turnbull et al., 2006; Cantoni et al., 2015; Nugent et al., 2020; Andreone et al., 2020; Fitz et al., 2020; Leyns et al., 2017)

	TREM2 KO x 5xFAD; TREM2 R47H x5xFAD	High levels of amyloid in the hippocampus with reduced microglial viability near plaques observed at 8 months.	Decreased inflammatory markers suggest that severe microglial responses may contribute to neurodegeneration.	Microglia are less viable than TREM2 +/ + 5xFAD mice, with reduced levels of CSF1 and decreased TREM2 shedding, accompanied by impaired downstream signaling in TREM2 R47H x5xFAD mice.	
	TREM2 KO x PS19	Less neurodegeneration and microgliosis compared to PS19 mice. No differences in p-tau levels and tau solubility.	NA	Decreased inflammatory markers suggest that a severe microglia response can contribute to neurodegeneration.	
hTREM2 KI	TREM2 CV and TREM2 R47H	Impaired lipid sensing and DAM responses to amyloid observed, impaired soluble TREM2 cell-surface interactions with decreased TREM2 shedding noted on neurons.	NA	Mice developed less brain atrophy and synaptic loss, with diminished microglial reactivity and phagocytosis, when compared to PS19-TREM2(CV) mice.	
	TREM2 R47H and APPPS21-TREM2 + / R47H	Reduced microglial response to amyloid with reduced amounts of dense-cored plaques; TREM2 R47H mice with cuprizone-induced neuroinflammation demonstrated age-dependent impairments in microglial interaction with plaques by 4 months, as well as LTP and synaptic loss within 12 months.	NA	Attenuated microglial response to amyloid with increased neurite dystrophy.	
	LOAD1, LOAD2, and others by JAX	No amyloid plaques or other AD hallmark changes observed in LOAD1 mouse models, even at 24 months of age. However, after HFD treatment, LOAD2 mice demonstrated neuronal loss and elevated brain Aβ42 levels within 16 months.	No cognitive decline observed in LOAD1 mouse models, even at 24 months of age; however, LOAD2 mice on HFD exhibited behavioral impairment.	Reduction in brain TREM2 protein levels and changes in circulating cytokine levels. Regional changes in glycolysis and vascular perfusion. Female LOAD1 mice showed increased risks of mortality, and glycolysis was significantly altered (4 mo–12 mo).	
hTREM2 Tg	BAC TREM2 Tg	Reduced amyloid plaques with associated gene signature changes, including dampened damage-associated microglial gene expression and upregulated neuronal gene expression.	NA	Cognitive performance improved in BAC hTREM2 Tg x 5xFAD mice compared to 5xFAD mice, with increased phagocytic microglia and reduced neurite dystrophy observed.	

Abbreviations: HFD: high-fat diet; NAA: N-acetyl aspartate; DAM: Disease-associated microglia; LTP: Long-term potentiation; WT: wild type; PFF: paired-pulse facilitation; HDL: high-density lipoprotein.

modifier screens and in vivo therapeutic screening. *C. elegans* models similarly support efficient mechanistic studies and high-throughput screening using transgenic strains expressing human Aβ and/or tau, although both invertebrate systems lack mammalian brain architecture and thus complement (rather than replace) mouse models for translational questions (Fernandez-Funez et al., 2015; Alvarez et al., 2022).

5 Current advances in preclinical gene therapy for AD

Preclinical research into gene therapy for AD has accelerated, with numerous studies employing various animal models, including transgenic mice expressing human genes associated with AD pathology. These models typically focus on Aβ and tau protein pathologies. For instance, many studies utilize APP/PSEN1 transgenic mice, which display similar disease mechanisms to human AD. These models allow for an in-depth study of gene therapies aimed at modulating molecular pathways implicated in AD progression.

Several studies have targeted genes that either reduce Aβ production or promote its clearance. These efforts include the use of CRISPR–Cas9 gene-editing technology, such as in the study by Duan et al. (2022), where AAV9-Cas9-SW1 was used to reduce Aβ deposition and neuroinflammation. Other studies, like those focusing on APOE genes, leverage the well-known risk-modifying effects of APOE isoforms. For instance, APOE4-to-APOE2 corrections via CRISPR technologies have been shown to mitigate disease phenotypes in transgenic mice.

Moreover, innovative gene delivery vectors, such as AAV (Rafii et al., 2018) and LV, have facilitated the transfection of target genes into the CNS. These approaches have shown promise in improving cognitive deficits in preclinical models, suggesting potential for future clinical applications in AD gene therapy (Table 10.3).

6 Evaluating the safety, toxicity, and efficacy of gene therapies

The safety and efficacy of gene therapy in AD models have been key considerations in preclinical studies. The efficacy of gene therapy has been primarily evaluated through behavioral and cognitive testing, such as the Morris water maze and T-maze tasks. Results consistently demonstrate significant improvements in memory, spatial learning, and cognitive performance. For instance, Wang et al. demonstrated that ultrasound-mediated delivery of BDNF via retroviral vectors enhanced memory and learning in a spatial task, a promising indicator of efficacy (Wang et al., 2021).

Regarding safety, studies have focused on monitoring the potential toxicity of viral vectors and genetic modifications. AAV and LV vectors have proven to be relatively safe, with minimal adverse effects reported in most preclinical models. However, long-term safety data are still limited, and potential risks, such as off-target effects or immune responses to viral vectors, need further investigation. Moreover, CRISPR–Cas9 technology, while potent, raises concerns about potential off-target genetic modifications, which could lead to unintended consequences. Notably, accumulating clinical experience across systemically administered AAV gene therapy programs has raised awareness that very high vector doses can, in rare cases, be associated with serious toxicity, particularly liver injury, and have prompted regulatory pauses/clinical holds following severe adverse events, including fatalities, especially in individuals with preexisting hepatic vulnerability (Agarwal, 2020).

7 Target genes and delivery vectors in AD gene therapy

Target genes for AD gene therapy are diverse, reflecting the multifaceted pathology of the disease. Key targets include APP, tau, APOE, and neurotrophic factors such as BDNF and NGF. These genes are implicated in Aβ production, tau hyperphosphorylation, synaptic plasticity, and neuroprotection. For instance, Richard et al. (2008) targeted TLR2 to reduce neuroinflammation and amyloid deposition, while studies like that of He et al. (2017) focused on the CIP gene, with positive effects on cognitive and anxiety-related behaviors.

Vectors used for gene delivery are predominantly viral, with AAV and LV being the most commonly employed. AAV vectors, such as those used in the study by Zhao et al. (2016), show high transfection efficiency and lower immunogenicity, making them a preferred option. Lentiviral vectors, as used by Wang et al. (2021), offer sustained gene expression and are effective in targeting both dividing and nondividing cells. Emerging technologies, such as CRISPR–Cas9 and siRNA, also show promise in gene silencing or editing strategies, providing more precise control over gene expression.

8 Administration routes and tools for gene delivery

Gene delivery in preclinical AD studies has been achieved through various routes, including direct intracranial injections, intravenous delivery, and stereotactic administration. The choice of route often depends on the specific gene therapy and its intended target within the brain. For instance, hippocampal injections are commonly used when targeting genes associated with memory and learning.

TABLE 10.3 Gene therapy preclinical trials in Alzheimer's disease.

Year	First author	Country	Animal model	Target gene	Agent/ vector	Gene therapy strategy	Sample size	Time to follow-up	Cognitive assessment	Main findings
1995	(Chen and Gage, 1995)	United States	Aged behaviorally impaired and young adult nonimpaired rats	NGF	Retrovirus	Intraparenchymal grafts (bilateral, dorsal to the nucleus basalis) of autologous fibroblasts genetically modified to produce and secrete human NGF	N = 20	6 weeks	Morris water maze test	Improved performance
1998	(Rosenberg et al., 1988)	United States	Rats	NGF	Retrovirus		N = 16	2 weeks	Not assessed	Not assessed
1998	(Tuszynski et al., 1998)	United States	Aged and nonaged Macaca mulatta (rhesus) monkeys	NGF	Retrovirus	Intraparenchymal grafts of autologous fibroblasts, genetically modified to produce and secrete human NGF	N = 16	3 months	Not assessed	Not assessed
1999	(Smith et al., 1999)	United States	Aged and nonaged rhesus monkeys	NGF	Retrovirus	Intraparenchymal grafts of autologous fibroblasts, genetically modified to produce and secrete human NGF	N = 15	8 months	Not assessed	Not assessed
2000	(Conner et al., 2001)	United States	Aged and nonaged Macaca mulatta (rhesus) monkeys	NGF	Retrovirus	Intraparenchymal grafts of autologous fibroblasts, genetically modified to produce and secrete human NGF	N = 18	3 months	Not assessed	Not assessed
2000	(Klein et al., 2000)	United States	Male Sprague–Dawley rats	NGF	AAV	Intraseptal injections	N = 15–16 per group	12 months	Morris water maze test	Maintenance of memory and learning
2004	(Wu et al., 2004)	United States	Aged rats	NGF	AAV	Intracranial injection	N = 3–5 per group	5 months	Morris water maze test	No significant differences between the aged and adult rats

Continued

TABLE 10.3 Gene therapy preclinical trials in Alzheimer's disease—Cont'd

Year	First author	Country	Animal model	Target gene	Agent/ vector	Gene therapy strategy	Sample size	Time to follow-up	Cognitive assessment	Main findings
2004	(Dodart et al., 2005)	United States	PDAPP mice	Apo E	lentivirus	Bilateral hippocampal injection of lentiviral vectors expressing apoE2, apoE3, or apoE4	N = 17	5 weeks	Not assessed	Not assessed
2004	(Iwata et al., 2004)	United States	Tg2576 mice	Neprilysin	AAV	dentate gyrus injection	N = 5–7 per group (71)	3 months	Not assessed	Not assessed
2007	(Hemming et al., 2007)	United States	APP $_{Swe/Ind}$ heterozygous transgenic mice	Ab-degrading protease neprilysin	Lentivirus	Unilateral hippocampal injection of fibroblasts expressing NEP	NM	4 weeks	Not assessed	Not assessed
2008	(Richard et al., 2008)	Canada	Triple-transgenic mice (APP + PSEN1 + TLR2-/-)	TLR2	Lentivirus	Hippocampal injection of TLR2 gene delivered by a lentiviral vector	N = 22	3–3.5 months	T-water maze task	Improved memory and learning
2009	(Guan et al., 2009)	United States	3 × Tg-AD mice	NEP gene	lentivirus	Bone marrow transplantation	N = 22	6 months	Morris water maze test	No significant behavioral impairment
2009	(Kanninen et al., 2009)	Finland	APP/PSEN1 transgenic mice	Nrf2	Lentivirus	Hippocampal injection of the Nrf2 gene delivered by a lentiviral vector	N = 11	6 months	Morris water maze test	Improved memory and learning
2009	Rose (Rose et al., 2009)	United States	NEP and APP transgenic mice	Neprilysin	Lentivirus	Intracerebral injection	N = 6 per group		Not assessed	Not assessed
2010	(Ryan et al., 2010)	United States	AD triple-transgenic (3xTg-AD) mice	scFv	Adenovirus	Hippocampal injection of the scFv gene delivered by an adenoviral vector	N = 36 (10–12 per group)	5 and 9 months	Morris water maze test	Improved memory and learning
2010	(Wang et al., 2010)	China	APP/PSEN1 transgenic mice	p75ECD	AAV	Intramuscular injection of the p75ECD gene delivered by a viral vector and IgG Fc fragment (AAV-ECD-Fc)	N = 32	9 months	Morris water maze test, ANY-maze software	Improved memory and learning

2012	(Hafez et al., 2012)	United States	APP (TASD41) and APP/PSEN1 transgenic mice (TASTPM)	F-spondin	Lentivirus	Injection of the F-spondin gene delivered by a lentiviral vector	N = 36	2 months	Morris water maze test, NORT	Improved memory, exploratory behavior, and a decrease in anxiety
2012	(Iwasaki et al., 2012)	Japan	APP transgenic mice (Tg2576)	BDNF	Sendai virus vector	Hippocampal injection of the BDNF gene delivered by the SeV viral vector	N = 18	14 days	Morris water maze test	Improved memory and reduced latency time
2012	(Kiyota et al., 2012)	United States	APP/PSEN1 transgenic mice	IL-10	Lentivirus	Hippocampal injection of the IL-10 gene delivered by a lentiviral vector	N = 9	5 months	Radial arm water maze	Improved memory and learning
2013	(Zhang et al., 2013)	China	APP/PSEN1 transgenic mice	sphK1	Adenovirus	Silencing of the sphK1 gene through a hippocampal injection containing siRNA loaded via an adenoviral vector	N = 30	4 weeks	Morris water maze test	Less learning and memory capacity. Worsening of motor disabilities
2014	(Pérez-González et al., 2014)	Spain	APP/PSEN1 transgenic mice	LEP gene	HIV-1	Intraventricular injection of a lentivirus vector expressing leptin protein in a self-inactivating HIV-1 vector (HIV-leptin)	N = 28	3 months	T-maze, NORT	Improved memory. Reduced choice time
2014	(Wen et al., 2014)	China	SAMP8 transgenic mice	ROCK II-siRNA	siRNA	Injection intracranial of ROCK II-siRNA delivered by PEG-PEI vector	N = 20	4 weeks	Morris water maze test	Improved memory and spatial learning
2015	(Kemppainen et al., 2015)	Finland	APP/PSEN1 transgenic mice	CDNF	AAV	Hippocampal microinjections with the CDNF protein gene via AAV2 viral vector and injections with AAV2 + CDNF protein	N = 116	2 weeks	Morris water maze test, NORT, spontaneous exploratory activity	Improved memory and exploratory behavior
2015	(Kiyota et al., 2015)	United States	APP transgenic mice	CD74	AAV	Hippocampal injection of CD74 gene delivered by AAV2 viral vector	N = NR	4 months	Morris water maze test	Improved memory and learning
2016	(Chen et al., 2016)	China	Modified mice (injection Aβ1-42)	ABN: Ascl1, Brn2, Ngn2	Lentivirus	Hippocampal injection of three transcription factors delivered by a lentiviral vector	N = 40	7 days	Morris water maze test	Improved memory and spatial learning

Continued

TABLE 10.3 Gene therapy preclinical trials in Alzheimer's disease—Cont'd

Year	First author	Country	Animal model	Target gene	Agent/ vector	Gene therapy strategy	Sample size	Time to follow-up	Cognitive assessment	Main findings
2016	(Spilman et al., 2016)	United States	PDAβPPSwe/ Ind transgenic mice	Netrin-1	AAV	Hippocampal injection of the netrin-1 gene delivered by a viral vector	N = 65	2 and 4 weeks	NORT	Improved working memory
2016	(Wang et al., 2016)	China	APP/PSEN1 transgenic mice	scFv	Lentivirus	Intramuscular injection of scFv gene delivered by a lentiviral vector	N = 30	6 months	Morris water maze test	Improved memory and spatial learning
2016	(Zhao et al., 2016)	China	APP/PSEN1 transgenic mice	APOE	AAV-APOE2	Intracerebral AAV delivery of APOE2	N = 15	6 months	Morris water maze	Significant reduction in amyloid plaques and cognitive improvement
2017	(Bustos et al., 2017)	Chile	APP/PSEN1 transgenic mice	Dlg4/PSD95	Lentivirus	Hippocampal injection of Dlg4/ PSD95 gene delivery via a viral vector	N = NR	1 month	NORT, OLMT	Improved memory
2017	(Embury et al., 2017)	United States	APP/PSEN1 transgenic mice	CatB	Adenovirus	Hippocampal injection of the CatB gene delivered by an adenoviral vector	N = 37	3-4 months	Radial arm water maze	Improved memory and learning
2017	(He et al., 2017)	China	APP/PSEN1 transgenic mice	CIP	Adenovirus	Intracranial injection of the CIP gene delivered by an adenoviral vector	N = 48	3 months	Morris water maze test, TSB	Improved memory and learning. Reduction in anxiety levels
2017	(Li et al., 2017)	China	Modified rats (injection Aβ1-42)	ABCA7	Lentivirus	Injection of a gene delivered by a lentiviral vector	N = 40	NR	Morris water maze test	Decreased exhaust latency and longer distance traveled on a platform
2017	(Mukham-edyarov et al., 2018)	Russia	APP/PSEN1 transgenic mice	NGF	Adenovirus	Intravenous transplantation of human UCBMCs overexpressing NGF	N = 78	16 days	T-maze, Plus- Maze	Improved memory, learning, and spatial memory. Possible reduction in anxiety levels
2017	(Zheng et al., 2017)	China	Transgenic mice APP/ ApoEKO and APP/GFAP-ApoEKO	TGF-β	AAV	Intracranial injection of TGF-β delivered by AAV vector	N = 10	12 months	Morris water maze test, Y-Maze, Radial Arm Maze	Improved memory and learning
2018	(Martinez-Losa et al., 2018)	United States	APP transgenic mice	SCN1A (Nav1.1)	–	Brain transplantation of interneurons that overexpress Nav1.1	N = 105	5–6 days	Morris water maze test, fear conditioning tests, Y-maze	Improved spatial memory

2018	(Tan et al., 2018)	New Zealand	APP/PSEN1 transgenic rats	APP- α	Lentivirus	Hippocampal injection of viral vectors	N = 69	8–10 months	Morris water maze test	Probable increased anxiety. Improved memory, as well as spatial memory
2019	(Chavoshine-zhad et al., 2019)	Iran	Swedish (K670N/M671L) and Indiana (V717F) mutations (LV-hAPPSw/Ind) rats	IFNβ	Lentivirus	LV-APP + IFNβ intracerebral injection	N = 10 per group	49–50 days	Passive avoidance test	Improved memory impairments
2019	(Guo et al., 2019)	China	APPswe + PSEN1dE9 double-transgenic mice	BACE 1 siRNA	siRNA	Intravenous injection	N = 9	2 days	Morris water maze test	Improved spatial learning and memory
2019	(Petukhova et al., 2019)	Russia	Transgenic mice (genotype B6C3: Tg(APP695) 85DboTg (PSEN1) 85Dbo)	GDNF	Adenovirus	Transplant of UCBMCs transduced with adenoviral vectors expressing GDNF	N = 9	7 days	T-maze	Improved spatial memory
2019	(Zeng et al., 2019)	China	APPswe/ PSEN1dE9 transgenic mice	KL	Lentivirus	Intracerebroventricular injection of LV-KL	N = 15 per group	1, 2, and 6 days	Morris water maze test, Y-maze, passive avoidance test, space exploration test	Improved neurobehavioral performance
2020	(Barros-Viegas et al., 2020)	Portugal	AD triple-transgenic (3xTg-AD) mice	pri-miR-31	Lentivirus	Injection with lentiviral mediated primiR-31	N = 9	4 months	T-maze, NORT	Reduction of anxiety, improved cognition, and long-term memory
2020	(Li et al., 2020)	China	APP/PSEN1 transgenic mice	hNEP	Plasmid	hNEP plasmid injections in the muscles surgically exposed	N = 20	1 month	Morris water maze test	Improved memory performance
2020	(Selles et al., 2020)	Brazil	APPswe/ PSEN1dE9 transgenic mice	IGF1	Adenovirus	RAd-IGF1 injections	N = 17-22 per mice group	7 days	NORT	Prevention of memory impairments
2021	(Brookhouser et al., 2021)	United States	hiPSC-based model	APOE	CRISPR–Cas9	CRISPR correction of APOE4 to APOE2	N = 20	8 weeks	Cellular assays	Mitigation of disease-related phenotypes and reduction in amyloid-beta accumulation

Continued

TABLE 10.3 Gene therapy preclinical trials in Alzheimer's disease—Cont'd

Year	First author	Country	Animal model	Target gene	Agent/ vector	Gene therapy strategy	Sample size	Time to follow-up	Cognitive assessment	Main findings
2021	(Duan et al., 2022)	China	5XFAD transgenic mice	Mutated APP	AAV	Intrahippocampal administration of AAV9-Cas9-SW1; intravenous injection of AAV-PHP.eB:Cas9-SW1	N = 3-12 per group	3-9 months	Open-field test, Y-maze spontaneous alternation test, elevated plus maze test.	Reduced deposition of amyloid-beta, as well as microgliosis, neurite dystrophy, and the impairment of cognitive performance
2021	(Patel et al., 2021)	United States	Aged female squirrel monkeys	TLR9 agonist CpG oligodeoxynucleotides (ODNs)	Oligodeoxy-nucleotides	Subcutaneous injection of CpG ODN into the interclavicular space	N = 15	2 years	Inhibitory control of behavior test, delayed nonmatching-to-sample T-maze test	Improving learning and spatial working memory evaluated, increased immunoreactivity against amyloid plaques
2021	(Wan et al., 2021)	China	AAV-hTau-induced C57 AD model	STAT3	AAV	AAV-STAT3 injected into the hippocampus	N = 19	1 month	Morris water maze test	Ameliorated hTau-induced cognitive deficits
2021	(Wang et al., 2021)	China	SD rats	BDNF	Retrovirus	Retroviral ultrasound microbubbles loading BDNF (MpLXSN-BDNF).	N = 32	2 months	Morris water maze, a spatial learning task	Ultrasound combined with viral microbubbles carrying BDNF can increase the transfection efficiency of brain neurons
2022	(Arora et al., 2022)	United States	Transgenic APP/PS1 AD mice	BDNF	Plasmid	Enhancing BDNF expression	N = 24	6 days	Nesting behavior	Reduced amyloid beta plaques and improved synapse functions without significant adverse effects
2022	(Wang et al., 2022)	China	APP/PS1 mice	TREM2-encoding plasmid	Plasmid	Microglia-targeted gene delivery systems (PHSA@PF/pTREM2)	Not accessible	Not accessible	Not accessible	Remodeling the inflammatory microenvironment and enhancing Aβ clearance, leading to an improvement in cognitive performance
2023	(Colettis et al., 2023)	Argentina	McGill-R-Thy1-APP (Tg +/–) heterozygous transgenic	NUsc1	AAV	Injection of AAV-NUsc1 into lateral ventricles	N = 31	2 months	Short-term exploratory behavior, habituation to an open field, object discrimination, and long-term memory for objects.	Recovery of LT discrimination capacity and LTM formation

2023	(Litvinchuk et al., 2024)	United States	APP/PSEN1 transgenic mice	APOE4	LXR agonist	Pharmacological activation of APOE4	N = 12	4 months	Morris water maze	Improvement in lipid metabolism and reduction of glial lipid accumulation
2024	(Lee et al., 2024)	South Korea	Fibril amyloid-beta (fAβ)-induced AD mice and okadaic acid (OA)-induced AD mice	AAV-aMTD-Parkin	AAV	AAV-aMTD-Parkin injected into the right ST	N = 6	4-8 weeks	Y-maze test	Improved cognitive function and recover
2024	(Li et al., 2024)	China	Aβ1-42-induced AD mouse model	NDI1	AAV	AAV9-NDI1 was injected stereotaxically into the hippocampal area	N = 3–4 per group	2 weeks	Novel object recognition test, Morris water maze test	Improve short-term cognitive memory ability, long-term spatial memory ability, tau histopathology, and neural tissue inflammation
2024	(Luo et al., 2024)	Japan	APP/PS1 mice	P2Y1R shRNA	AAV	P2Y1R gene knockdown by using shRNAs for P2Y1R in an adeno-associated virus vector	N = 12–15 per group	1–11 days	Morris water maze, Y-maze, and contextual fear conditioning tests	Reduced amyloid-beta accumulation, glial neuroinflammation, blood–brain barrier dysfunction, and cognitive impairment

Advanced delivery methods have been explored to increase transfection efficiency and minimize invasiveness. Ultrasound-mediated delivery, as shown by Wang et al. (2021), uses microbubbles to enhance viral vector penetration into brain tissues. Similarly, the use of nanoparticles and polymers, such as PEG-PEI vectors for siRNA delivery, provides alternative nonviral approaches that may reduce potential immune responses or toxicity associated with viral vectors.

In addition, the development of gene-editing tools like CRISPR–Cas9 has revolutionized the administration process. Techniques for delivering CRISPR components directly into the brain through viral vectors or nanoparticle systems are currently being optimized for efficient gene editing with minimal off-target effects.

9 Outcome measures and assessment techniques in preclinical studies

Behavioral and cognitive assessments form the core of outcome measures in AD gene therapy research. Common tests include the Morris water maze, Y-maze, and Novel Object Recognition tasks, which evaluate spatial memory, working memory, and recognition memory, respectively. These tests have consistently shown that gene therapy can mitigate cognitive decline in AD models.

Biological outcome measures also include reductions in Aβ plaques, tau phosphorylation, and neuroinflammation, as well as improvements in synaptic plasticity and neuronal survival. Positron emission tomography (PET) imaging and histopathological examinations were also used to assess these changes in the brain.

10 Follow-up strategies and long-term monitoring in gene therapy experiments

Follow-up durations in preclinical studies vary, with most ranging from a few weeks to several months, depending on the gene therapy's expected timeframe for functional changes. For example, Duan et al. (2022) employed a follow-up of up to 9 months to assess the long-term effects of CRISPR–Cas9 interventions on cognitive performance and neuroinflammation. Monitoring throughout the study period typically includes repeated cognitive testing and imaging techniques, such as MRI or PET scans, to evaluate brain pathology changes over time.

In some studies, longitudinal monitoring has revealed that cognitive improvements often persist for months after gene therapy administration. For instance, Richard et al. (2008)) observed sustained memory improvements in AD mice up to 3.5 months postgene therapy. Such findings are encouraging, suggesting that gene therapy can produce long-lasting benefits. However, the need for extended follow-up in future studies is critical to understanding the durability of therapeutic effects and potential late-onset side effects.

11 Managing complications and mitigating risks in gene therapy for AD

Complications in gene therapy studies have generally been minimal in preclinical models, but risks do exist, particularly concerning the immune response to viral vectors. AAV and lentiviral vectors are generally well-tolerated, but immune reactions or toxicity could pose significant challenges, especially in translating these therapies to human trials. In the study by Guo et al. (2019), siRNA delivery was associated with transient inflammation, though no severe long-term adverse effects were observed.

Management of complications involves careful vector design to minimize immune responses and the use of immunosuppressive drugs where necessary. Advanced strategies, such as using modified viral capsids that evade immune detection, are being developed. Furthermore, CRISPR-based therapies must address the issue of off-target gene editing, which can be mitigated through improved guide RNA design and delivery systems.

In conclusion, while preclinical studies demonstrate that gene therapy holds promise for treating AD, the translation from animal models to clinical applications will require careful attention to safety, efficacy, and long-term monitoring.

11.1 Recent clinical trials on gene therapy in AD

Overall, by December 2025, eight clinical trials have been submitted to ClinicalTrials.gov, and the results of three of them have already been published (Rafii et al., 2018, 2014; Tuszynski et al., 2005). An overview of two Phase 1 and one Phase 2 clinical trials is shown in Table 10.4. Moreover, two studies discussed the postmortem analysis of the patients who underwent gene therapy intervention (Tuszynski et al., 2015; Castle et al., 2020) (Table 10.4).

Tuszynski et al. conducted the first Phase 1 clinical trial of ex vivo NGF (Rafii et al., 2018) gene transfer in eight patients with mild AD to determine whether the extensive cholinergic neuronal loss in the disease can be halted. In this study, participants underwent implantation of autologous fibroblasts genetically modified by a Moloney leukemia virus

TABLE 10.4 Completed gene therapy clinical trials in Alzheimer's disease.

First author, publication year, registration	Study phase, sample (F/M), mean age, country	Population	Target gene	Gene therapy administration, mean follow-up	Assessments
Phase 1 and 2 clinical trials					
Tuszynski (2005), NCT00017940	Phase I; N = 8 (5/3) (initial), N = 6 (final); 67.2 ± 2.6; United States	AD based on NINCDS-ADRDA criteria, 50–80 years of age, with a baseline MMSE range of 20–28	NGF	Ex vivo gene therapy (grafts of autologous cells); 22 months	Primary outcome: safety Secondary outcome: MMSE, ADAS-Cog, PET scans
Rafii (2014), NCT00087789	Phase I; N = 10 (6/4); 68.1 ± 10.6 years; United States	Mild-to-moderate AD, based on NINCDS-ADRDA criteria, with a baseline MMSE range of 16–28 and a modified HIS of ≤ 4 points	NGF	In vivo gene therapy (AAV2-NGF vector: CERE-110); 24 months	Primary outcome: safety Secondary outcome: MMSE, ADAS-Cog, ADCS-CGIC, PET scans
Rafii et al. (2018), NCT00876863	Phase II; N = 49 (21/28), 68 ± 6.4 years, United States	Mild-to-moderate AD, 55–80 years of age, with a baseline MMSE range of 17–26 and a modified HIS of ≤ 4 points	NGF	In vivo gene therapy (AAV2-NGF vectors); stereotactically guided intracerebral injections; 24 months	ADAS-Cog
Postmortem analysis					
Tuszynski et al. (2015); NCT00017940–NCT00087789	Phase I; N = 10 (6/4); 69.3 ± 2.2 years; United States	AD based on NINDS criteria, 50–80 years of age, with a baseline MMSE range of 20–30 for the ex vivo trial and 16–26 for the in vivo trial	NGF	8 ex vivo gene therapy (grafts of autologous cells) and 2 in vivo gene therapy (AAV2-NGF vectors); 1–10 years	Anatomicopathological investigation (postmortem analysis)
Castle et al. (2020), NCT00087789	Phase I; N = 3 (2/1); United States	AD based on NINDS criteria, 50–80 years of age, with a baseline MMSE range of 16–26	NGF	In vivo gene therapy (AAV2-NGF vectors); stereotactically guided intracerebral injections; 10 months–7.8 years	Histopathological investigation (postmortem analysis)

vector to express human NGF into the nucleus basalis of Meynert (nbM) using stereotaxic neurosurgery. The first two injections were administered solely to the right brain of patients, while in the subsequent six patients, injections were administered bilaterally. In this study, the trial began with injections performed while subjects were conscious but sedated, which led to sudden movement while the injection needle was still inside the brain, resulting in subcortical hemorrhage in two patients. Therefore to prevent unexpected movements, all subsequent surgeries were performed under general anesthesia, and the final analysis was conducted on six patients. After 18–24 months of follow-up, no long-term adverse events attributable to the delivery of NGF were reported. Assessment of both the Mini-Mental State Examination (MMSE) and the Alzheimer Disease Assessment Scale-Cognitive subscale (ADAS-Cog) indicated a decrease in the rate of cognitive decline (22-month postoperative mean MMSE change of − 3.0 ± 1.0 points per year vs. pretreatment mean MMSE change of − 6.1 ± 2.7 points per year, and 22-month postoperative mean ADAS-Cog change of − 6.2 ± 1.9 points per year vs. mean ADAS-Cog decline of 3–10 points per year in moderately severe AD patients). Overall, the cognitive status of five out of six patients improved or stabilized. Moreover, PET scans in four bilaterally treated patients revealed a significant enhancement in 18-fluorodeoxyglucose uptake on the second scan in comparison with the first in the majority of cortical regions receiving cholinergic input from the nucleus basalis and also the cerebellum ($P < 0.05$) (Tuszynski et al., 2005).

Another Phase 1 clinical trial was conducted by Rafii et al., in which ten mild-to-moderate patients with AD were included. These patients underwent genetically engineered adeno-associated virus serotype 2 vector delivering NGF

(AAV2-NGF [CERE-110]) combined with stereotactic surgery into the nbM in three cohorts with escalating doses (cohort A: 1.2×10^{10} vector genomes [vg], cohort B: 5.8×10^{10} vg, and cohort C: 1.2×10^{11} vg). Over 24 months of follow-up, AAV2-NGF was found to be safe and well tolerated, and imaging/neuropsychological testing showed no evidence of accelerated clinical decline relative to expectations for this disease stage. On the ADAS-Cog, participants deteriorated by a mean ± SE of 12.7 ± 4.3, 21.7 ± 9.3, and 23.1 ± 9.7 points at 24 months compared to baseline for the highest dose to lowest dose, respectively. Furthermore, the change from baseline at month 24 on the MMSE was − 7.3 ± 2.9, − 12.3 ± 0.3, and − 10.3 ± 5.2 points for the highest dose to lowest dose, respectively. There was also minimal deterioration on the Alzheimer's Disease Cooperative Study Clinical Global Impression of Change (ADCS-CGIC) in all three dose groups (e.g., mean ADCS-CGIC at month 24 was 5.3 ± 0.25, 5.3 ± 0.33, and 5.7 ± 0.33, points for the highest dose to lowest dose, respectively. Furthermore, FDG PET scans did not reveal the predicted metabolic declines in any of the dose groups (Rafii et al., 2014).

The only completed Phase 2 gene therapy randomized clinical trial in AD patients was conducted by Rafii et al. in a multicenter setting, in which 49 participants with mild to moderate AD were randomly assigned in a 1:1 ratio to receive stereotactically guided bilateral intracerebral injections of AAV2-NGF (with a total dose of 2.0×10^{11} vg) or sham surgery. AAV2-NGF was safe and well-tolerated through 24-month follow-up. No significant differences were noted between the treatment group and placebo in terms of various neuropsychological tests (the ADAS-Cog mean change of 14.52 ± 4.66 vs. 9.11 ± 4.65, $p = 0.17$; the Clinical Dementia Rating—Sum of Boxes mean change of 4.75 ± 1.55 vs. 2.81 ± 1.47, $p = 0.09$; modified Clinical Global Impression of Change mean change of 5.59 ± 0.33 vs. 5.33 ± 0.27, $p = 0.21$; MMSE mean change of − 6.18 ± 2.18 vs. − 4.17 ± 2.67, $p = 0.16$; Neuropsychiatric Inventory (Swanson et al., 2021) mean change of 6.61 ± 4.76 vs. 9.18 ± 9.89, $p = 0.95$; Alzheimer Disease Cooperative Study—Activities of Daily Living mean change of − 17.65 ± 6.84 vs. − 12.94 ± 9.19, $p = 0.61$ in the treatment group vs. placebo, respectively, in all reports). There was also no statistical difference between the two groups at 24 months regarding the standardized uptake value ratio change in bilateral posterior cingulate gyri (−0.152 ± 0.024 vs. − 0.142 ± 0.018, $p = 0.47$). In addition, no significant hippocampal and ventricular volume changes were reported between the two groups (mean hippocampal volume change of − 0.41 ± 0.16 vs. − 0.43 ± 0.07 cubic cm, $p = 0.95$, and mean ventricular volume change of 52.16 ± 5.01 vs. 44.84 ± 9.94 cubic cm, $p = 0.64$ in the treatment group vs. placebo, respectively, in both reports) (Rafii et al., 2018).

In addition to the main findings reported in the aforementioned studies, two distinct studies detailed the postmortem analysis of patients who underwent gene therapy for AD (Tuszynski et al., 2015; Castle et al., 2020). In one study, ten patients with early AD who received NGF gene therapy using ex vivo or in vivo gene transfer (8 and 2 patients, respectively), in an extended survey of two Phase 1 clinical trials, were examined for in vivo gene expression, cholinergic neuronal responses to NGF, and activation of NGF-related cell signaling. Among all ten patients, a trophic response in the form of axonal sprouting toward the NGF source in the AD brains was observed. Moreover, cholinergic neuronal hypertrophy was found on the NGF-treated side in comparison with the nontreated side of the brain in three patients who underwent unilateral gene transfer ($p < 0.05$). In addition, two patients who underwent AAV2–NGF gene transfer exhibited activation of cellular signaling and functional markers. Sprouting induced by NGF also persisted for 10 years following gene transfer (Tuszynski et al., 2015). In another study, the brains of three deceased patients from the Phase 1 clinical trial of AAV2-NGF gene therapy for mild-to-moderate AD, who received a cumulative dose of 2.0×10^{11} vg of AAV2-NGF by stereotactic injection into the nbM, were analyzed. The vast majority of injections failed to reach the nbM directly as expected. Although NGF gene expression continued for at least 7 years at the injection sites, some injections were mistargeted, and most precisely targeted injections revealed limited dissemination of AAV2-NGF.

All completed clinical trials targeted NGF and used stereotactic injections for gene delivery (Rafii et al., 2018, 2014; Tuszynski et al., 2005). Although NGF gene delivery appears feasible, safe, and well-tolerated over extended periods (up to 10 years), its efficacy cannot be accurately measured. Even though the neuropsychological and radiological outcomes were not as intended, we cannot conclude that NGF gene delivery was ineffective. The small number of participants who underwent gene therapy (44 in total) (Rafii et al., 2018, 2014; Tuszynski et al., 2005) and the lack of prospective control subjects in Phase 1 clinical trials (Rafii et al., 2014; Tuszynski et al., 2005) precluded interpretation regarding potential efficacy. Moreover, some injections were mistargeted, and almost all targeted injections showed the limited spread of AAV2-NGF (Rafii et al., 2014; Tuszynski et al., 2005; Castle et al., 2020); we thus cannot reliably conclude that AAV2-NGF failed in terms of efficacy because effective target engagement is essential for gene therapy in humans. New methods, such as MR-guided AAV infusion, intrathecal AAV infusion with Trendelenburg positioning, and the evolution of AAV capsids, have the potential to significantly enhance targeting and gene delivery in humans (Castle et al., 2020). It is remarkable to note that all neuropsychological assessment tools used in clinical studies have been validated and are broadly utilized in clinical practice. All three studies have chosen to use the MMSE, which includes specific assessments for orientation, attention, memory, language, and visual–spatial skills (Folstein et al., 1975) as well as ADAS-Cog, which is considered the gold standard for assessing the efficacy of antidementia treatments by evaluating several cognitive domains, including memory, language, and praxis (Rosen et al., 1984). The advantage of MMSE includes its rapid administration and high levels of acceptance as a diagnostic instrument among health professionals and researchers; nonetheless, it may not be sensitive enough to detect discrete cognitive changes. Therefore we recommend adding more sensitive cognition tools to detect these changes.

The study design aspects utilized in these trials can guide upcoming studies of other targeted genes, such as APOE2, BDNF, and hTERT, which are being conducted for AD gene therapy. Ongoing gene delivery clinical trials are summarized in Table 10.5. Gene therapy plays a key role in the fight against AD due to its ability to focus on specific targets and deliver treatment precisely to affected brain regions. In addition, it has the potential to shed light on the beneficial effects of modulating pathways, which will help in better understanding the physiopathology of the disease and halting its progression. As with most clinical conditions, the success of new drug development strongly relies on critical factors, including detailed knowledge of the natural history of the disease; its clinical, radiological, and biological endpoints; the selection of the patient population with genetic background considerations; and the identification of valuable outcome predictors and biomarkers of disease progression. This is particularly fundamental for novel individualized treatments, in which the number of patients included in Phase 1 and Phase 2 trials will be low (Tables 10.4 and 10.5).

TABLE 10.5 Ongoing gene therapy clinical trials in Alzheimer's disease.

PI, study start date, status, registration	The study, sample, country	Population	Target	Gene therapy administration, follow-up	Assessments
Mark Tuszynski; 2022; recruiting; NCT05040217	Phase I; N = 12; United States	MCI and mild AD based on NIA-AA criteria, 50–80 years of age, with a baseline MMSE range of 24–29 and a modified HIS of ≤ 4 points	BDNF	AAV2-BDNF; stereotaxically administered into the brain; 24 months	Primary outcome: safety and tolerability; secondary outcomes: RAVLT (Rey Auditory Verbal Learning Task), BCFDM, MMSE, ADAS-Cog, PET scans, MRI scans, CSF analysis
Jorge Ulloa; 2019; unknown; NCT04133454	Phase I; N = 5; Colombia	AD, ≥ 45 years of age	hTERT	AAV-hTERT; IV or IT treatment; 52 months	Primary outcome: safety and tolerability; secondary outcome: telomere length
Lexeo Clinical Trials: Lexeo Therapeutics; 2022; Not yet recruiting; NCT05400330	Phase I; N = 15; United States	APOE4 homozygote, MCI or AD, ≥ 50 years of age	APOE2	AAVrh.10hAPOE2; IT treatment; 260 weeks	Primary outcome: long-term safety; secondary outcomes: CSF APOE isoform biomarkers, amyloid PET scan, CSF markers (including Aβ42, Aβ42/Aβ40 ratio, T–tau, and P-tau), quantitative MRI, cognitive and clinical assessment
Trials: Lexeo Therapeutics; 2019; recruiting; NCT03634007	Phase 1 and 2; N = 15; United States	APOE4 homozygote, MCI or AD, ≥ 50 years of age	APOE2	AAVrh.10hAPOE2; IT treatment; 52 weeks	Primary outcome: safety
Maria E. Jönhagen; 2008; unknown; NCT01163825	Phase 1; N = 6; Sweden	AD based on NINCDS-ADRDA criteria, 50–80 years of age, with a baseline MMSE range of 15–24	NGF	Encapsulated cell biodelivery of NGF: multiple implantable devices housing NGF-secreting human cells in the brain; 12 months	Primary outcome: safety; secondary outcome: ADAS-Cog, neuropsychological test battery

Abbreviations: NIA-AA: National Institute on Aging—Alzheimer's Association; NINCDS-ADRDA: National Institute of Neurological and Communicative Disorders and Stroke—Alzheimer's Disease and Related Disorders Association; HIS: Hachinski Ischemia Scale; MMSE: Mini-Mental State Examination; BCFDM: Benson Complex Figure Drawing and Memory.

11.2 Current challenges and obstacles to overcome in future studies

Gene therapy holds immense promise for AD treatment, but it faces several significant challenges. One of the primary obstacles is delivering therapeutic genes to the appropriate brain regions efficiently and safely. The BBB restricts the passage of large molecules, including gene therapy vectors, into the brain. Overcoming this barrier without causing undue harm to the delicate brain tissue remains a critical challenge. Researchers are exploring various strategies, such as viral vectors engineered to penetrate the BBB or nonviral delivery systems, to improve gene delivery to the brain. In addition, ensuring the long-term stability and specificity of gene expression in targeted brain cells presents another hurdle in the development of effective gene therapy for AD.

Another challenge in gene therapy for AD is selecting the most suitable gene targets and designing therapeutic constructs that effectively modulate disease progression. AD is a multifactorial disorder with complex genetic and molecular mechanisms. Identifying the key genes and pathways involved in disease pathology is crucial for developing targeted gene therapy approaches. Moreover, gene therapy strategies must balance the delicate equilibrium between reducing Aβ accumulation, enhancing neuronal function, and mitigating neuroinflammation without causing adverse effects. Achieving precise and selective modulation of gene expression in the brain while minimizing off-target effects remains a formidable task.

Furthermore, safety concerns surrounding gene therapy for AD pose significant barriers to its clinical translation. Off-target effects, immune responses to viral vectors, and potential unintended consequences of gene manipulation raise ethical and safety considerations. Long-term monitoring of treated individuals is essential to assess the therapy's safety and efficacy over time. In addition, regulatory approval processes for gene therapy in AD require stringent evaluation of potential risks and benefits. Collaborative efforts among researchers, clinicians, regulatory agencies, and industry partners are necessary to address safety concerns and advance gene therapy as a viable treatment option for AD. Despite these challenges, the rapid progress in gene-editing technologies and our growing understanding of AD pathophysiology offer hope for overcoming these obstacles and realizing the potential of gene therapy in combating this devastating neurodegenerative disorder.

12 Conclusion

Gene therapy represents a transformative frontier in AD treatment, offering a targeted approach to modify key pathological processes beyond the scope of conventional therapeutics. With accumulating evidence from preclinical models and early clinical trials, therapeutic gene delivery, particularly of neurotrophic factors and Aβ-degrading enzymes, has shown potential to restore cognitive function, reduce neuropathology, and slow disease progression. However, challenges such as precise brain targeting, sustained transgene expression, immune response mitigation, and safe vector delivery remain critical barriers to clinical translation. The modest efficacy observed in NGF-related human trials underscores the importance of optimizing delivery systems, improving targeting accuracy, and expanding the scope to other promising genes like APOE2 and BDNF. Moreover, integrating advanced gene-editing technologies and developing more sensitive outcome measures will be essential for evaluating clinical benefits. The future of AD gene therapy lies in interdisciplinary collaboration, improved vector engineering, and the design of precision-medicine strategies tailored to genetic and pathological heterogeneity. With continued innovation and rigorously designed clinical trials, gene therapy may ultimately reshape the therapeutic landscape of AD.

AI disclosure

During the preparation of this work, the author(s) used ChatGPT for content organization and improving language and readability. After using this tool/service, the author(s) reviewed and edited the content as needed and take(s) full responsibility for the content of the publication.

References

Agarwal, S.J.N.B., 2020. High-dose AAV gene therapy deaths. Nat. Biotechnol. 1546-1696. 38 (8), 910. doi:10.1038/s41587-020-0642-9.

Alvarez, J., Alvarez-Illera, P., Santo-Domingo, J., Fonteriz, R.I., Montero, M., 2022. Modeling Alzheimer's Disease in Caenorhabditis elegans. Biomedicines. 2227-9059. 10 (2), 288. doi:10.3390/biomedicines10020288.

Andreone, B.J., Przybyla, L., Llapashtica, C., Rana, A., Davis, S.S., van Lengerich, B., Lin, K., Shi, J., Mei, Y., Astarita, G., Di Paolo, G., Sandmann, T., Monroe, K.M., Lewcock, J.W., 2020. Alzheimer's-associated PLCγ2 is a signaling node required for both TREM2 function and the inflammatory response in human microglia. Nat. Neurosci. 15461726. 23 (8), 927–938. doi:10.1038/s41593-020-0650-6.

Andrews-Zwilling, Y., Bien-Ly, N., Qin, X., Gang, L., Bernardo, A., Yoon, S.Y., Zwilling, D., Yan, T.X., Chen, L., Huang, Y., 2010. ApolipoproteinE4 Causes Age- and Tau-Dependent Impairment of GABAergic Interneurons, Leading to Learning and Memory Deficits in Mice. J. Neurosci. 0270-6474. 30 (41), 13707–13717. doi:10.1523/jneurosci.4040-10.2010.

Apelt, J., Kumar, A., Schliebs, R., 2002. Impairment of cholinergic neurotransmission in adult and aged transgenic Tg2576 mouse brain expressing the Swedish mutation of human β-amyloid precursor protein. Brain Res. 00068993. 953 (1-2), 17–30. doi:10.1016/s0006-8993(02)03262-6.

Arabi, F., Mansouri, V., Ahmadbeigi, N., 2022. Gene therapy clinical trials, where do we go? An overview. Biomed. Pharmacother. 07533322. 153, 113324. doi:10.1016/j.biopha.2022.113324.

Arora, S., Kanekiyo, T., Singh, J., 2022. Functionalized nanoparticles for brain targeted BDNF gene therapy to rescue Alzheimer's disease pathology in transgenic mouse model. Int. J. Biol. Macromol. 01418130. 208, 901–911. doi:10.1016/j.ijbiomac.2022.03.203.

Ashe, K.H., Zahs, K.R., 2010. Probing the Biology of Alzheimer's Disease in Mice. Neuron. 08966273. 66 (5), 631–645. doi:10.1016/j.neuron.2010.04.031.

Atri, A., 2019. Current and Future Treatments in Alzheimer's Disease. Semin. Neurol. 10989021. 39 (2), 227–240. doi:10.1055/s-0039-1678581.

Balu, D., Karstens, A.J., Loukenas, E., Weng, J.M., York, J.M., Valencia-Olvera, A.C., LaDu, M.J., 2019. The role of APOE in transgenic mouse models of AD. Neurosci. Lett. 18727972. 707. doi:10.1016/j.neulet.2019.134285.

Barendrecht, S., Schreurs, A., Geissler, S., Sabanov, V., Ilse, V., Rieckmann, V., Eichentopf, R., Künemund, A., Hietel, B., Wussow, S., Hoffmann, K., Körber-Ferl, K., Pandey, R., Carter, G.W., Demuth, H.-U., Holzer, M., Roßner, S., Schilling, S., Preuss, C., Balschun, D., Cynis, H., 2023. A novel human tau knock-in mouse model reveals interaction of Abeta and human tau under progressing cerebral amyloidosis in 5xFAD mice. Alzheimers Res. Ther. 1758-9193. 15 (1). doi:10.1186/s13195-022-01144-y.

Barros-Viegas, A.T., Carmona, V., Ferreiro, E., Guedes, J., Cardoso, A.M., Cunha, P., de Almeida, L.P., de de Oliveira, C.R., de Magalhães, J.P., Peça, J., Cardoso, A.L., 2020. miRNA-31 Improves Cognition and Abolishes Amyloid-β Pathology by Targeting APP and BACE1 in an Animal Model of Alzheimer's Disease. Mol. Ther. Nucleic Acids. 21622531. 19, 1219–1236. doi:10.1016/j.omtn.2020.01.010.

Bekris, L.M., Yu, C.E., Bird, T.D., Tsuang, D.W., 2010. Genetics of Alzheimer disease. J. Geriatr. Psychiatry Neurol. 23 (4), 213–227. doi:https://doi.org/10.1177/0891988710383571.

Belfiore, R., Rodin, A., Ferreira, E., Velazquez, R., Branca, C., Caccamo, A., Oddo, S., 2019. Temporal and regional progression of Alzheimer's disease-like pathology in 3xTg-AD mice. Aging Cell. 1474-9718. 18 (1). doi:10.1111/acel.12873.

Bell, R.D., Sagare, A.P., Friedman, A.E., Bedi, G.S., Holtzman, D.M., Deane, R., Zlokovic, B.V., 2007b. Transport pathways for clearance of human Alzheimer's amyloid β-peptide and apolipoproteins E and J in the mouse central nervous system. J. Cereb. Blood Flow Metab. 0271678X. 27 (5), 909–918. doi:10.1038/sj.jcbfm.9600419.

Bellenguez C., Kucukali F., Jansen I.E., Kleineidam L., Moreno-Grau S., Amin N., et al. 2022. New insights into the genetic etiology of Alzheimer's disease and related dementias. Nat. Genet. 54 (4), 412–36. doi:10.1038/s41588-022-01024-z.

Bien-Ly, N., Gillespie, A.K., Walker, D., Yoon, S.Y., Huang, Y., 2012. Reducing human Apolipoprotein E levels attenuates age-dependent Aβ accumulation in mutant human amyloid precursor protein transgenic mice. J. Neurosci. 15292401. 32 (14), 4803–4811. doi:10.1523/JNEUROSCI.0033-12.2012.

Billings, L.M., Oddo, S., Green, K.N., McGaugh, J.L., LaFerla, F.M., 2005. Intraneuronal Aβ causes the onset of early Alzheimer's disease-related cognitive deficits in transgenic mice. Neuron. 08966273. 45 (5), 675–688. doi:10.1016/j.neuron.2005.01.040.

Bishop, K.M., Hofer, E.K., Mehta, A., Ramirez, A., Sun, L., Tuszynski, M., Bartus, R.T., 2008. Therapeutic potential of CERE-110 (AAV2-NGF): targeted, stable, and sustained NGF delivery and trophic activity on rodent basal forebrain cholinergic neurons. Exp. Neurol. 10902430. 211 (2), 574–584. doi:10.1016/j.expneurol.2008.03.004.

Blanchard, J.W., Akay, L.A., Davila-Velderrain, J., von Maydell, D., Mathys, H., Davidson, S.M., Effenberger, A., Chen, C.Y., Maner-Smith, K., Hajjar, I., Ortlund, E.A., Bula, M., Agbas, E., Ng, A., Jiang, X., Kahn, M., Blanco-Duque, C., Lavoie, N., Liu, L., Reyes, R., Lin, Y.T., Ko, T., R'Bibo, L., Ralvenius, W.T., Bennett, D.A., Cam, H.P., Kellis, M., Tsai, L.H., 2022. APOE4 impairs myelination via cholesterol dysregulation in oligodendrocytes. Nature. 14764687. 611 (7937), 769–779. doi:10.1038/s41586-022-05439-w.

Brookhouser, N., Raman, S., Frisch, C., Srinivasan, G., Brafman, D.A., 2021. APOE2 mitigates disease-related phenotypes in an isogenic hiPSC-based model of Alzheimer's disease. Mol. Psychiatry. 14765578. 26 (10), 5715–5732. doi:10.1038/s41380-021-01076-3.

Bustos, F.J., Ampuero, E., Jury, N., Aguilar, R., Falahi, F., Toledo, J., Ahumada, J., Lata, J., Cubillos, P., Henríquez, B., Guerra, M.V., Stehberg, J., Neve, R.L., Inestrosa, N.C., Wyneken, U., Fuenzalida, M., Härtel, S., Sena-Esteves, M., Varela-Nallar, L., Rots, M.G., Montecino, M., Van Zundert, B., 2017. Epigenetic editing of the Dlg4/PSD95 gene improves cognition in aged and Alzheimer's disease mice. Brain. 14602156. 140 (12), 3252–3268. doi:10.1093/brain/awx272.

Cai, Z., Zhao, K., Zeng, L., Liu, M., Sun, T., Zhuorong, L., Liu, R., 2022. The Relationship between the Aberrant Long Non-Coding RNA-Mediated Competitive Endogenous RNA Network and Alzheimer's Disease Pathogenesis. Int. J. Mol. Sci. 1422-0067. 23 (15), 8497. doi:10.3390/ijms23158497.

Cantoni, C., Bollman, B., Licastro, D., Xie, M., Mikesell, R., Schmidt, R., Yuede, C.M., Galimberti, D., Olivecrona, G., Klein, R.S., Cross, A.H., Otero, K., Piccio, L., 2015. TREM2 regulates microglial cell activation in response to demyelination in vivo. Acta Neuropathol. 0001-6322. 129 (3), 429–447. doi:10.1007/s00401-015-1388-1.

Carroll, J.C., Pike, C.J., 2008. Selective estrogen receptor modulators differentially regulate Alzheimer-like changes in female 3xTg-AD mice. Endocrinology. 00137227. 149 (5), 2607–2611. doi:10.1210/en.2007-1346.

Carty, N.C., Nash, K., Lee, D., Mercer, M., Gottschall, P.E., Meyers, C., Muzyczka, N., Gordon, M.N., Morgan, D., 2008. Adeno-associated viral (AAV) serotype 5 vector mediated gene delivery of endothelin-converting enzyme reduces Aβ deposits in APP + PS1 transgenic mice. Mol. Ther. 15250024. 16 (9), 1580–1586. doi:10.1038/mt.2008.148.

Castle, M.J., Baltanás, F.C., Kovacs, I., Nagahara, A.H., Barba, D., Tuszynski, M.H., 2020. Postmortem analysis in a clinical trial of AAV2-NGF gene therapy for Alzheimer's disease identifies a need for improved vector delivery. Hum. Gene Ther. 15577422. 31 (7-8), 415–422. doi:10.1089/hum.2019.367.

Cavalli, G., Heard, E., 2019. Advances in epigenetics link genetics to the environment and disease. Nature. 0028-0836. 571 (7766), 489–499. doi:10.1038/s41586-019-1411-0.

Chan, J., Jones, N.C., Bush, A.I., O'Brien, T.J., Kwan, P., 2015. A mouse model of Alzheimer's disease displays increased susceptibility to kindling and seizure-associated death. Epilepsia. 15281167. 56 (6), e73. doi:10.1111/epi.12993Australia.

Chavoshinezhad, S., Kouchesfahani, H.M., Salehi, M.S., Pandamooz, S., Ahmadiani, A., Dargahi, L., 2019. Intranasal interferon beta improves memory and modulates inflammatory responses in a mutant APP-overexpressing rat model of Alzheimer's disease. Brain Res. Bull. 18732747. 150, 297–306. doi:10.1016/j.brainresbull.2019.06.015.

Chen, K.S., Gage, F.H., 1995. Somatic gene transfer of NGF to the aged brain: behavioral and morphological amelioration. J. Neurosci. 0270-6474. 15 (4), 2819–2825. doi:10.1523/jneurosci.15-04-02819.1995.

Chen, P., Yan, Q., Wang, S., Wang, C., Zhao, P., 2016. Transfer of three transcription factors via a lentiviral vector ameliorates spatial learning and memory impairment in a mouse model of Alzheimer's disease. Gene. 03781119. 587 (1), 59–63. doi:10.1016/j.gene.2016.04.032.

Chishti, M.A., Yang, D.-S., Janus, C., Phinney, A.L., Horne, P., Pearson, J., Strome, R., Zuker, N., Loukides, J., French, J., Turner, S., Lozza, G., Grilli, M., Kunicki, S., Morissette, C., Paquette, J., Gervais, F., Bergeron, C., Fraser, P.E., Carlson, G.A., Peter, S.G.-H., Westaway, D., 2001. Early-onset Amyloid Deposition and Cognitive Deficits in Transgenic Mice Expressing a Double Mutant Form of Amyloid Precursor Protein 695. J. Biol. Chem. 00219258. 276 (24), 21562–21570. doi:10.1074/jbc.m100710200.

Clark, J.B., 1998. N-Acetyl Aspartate: a Marker for Neuronal Loss or Mitochondrial Dysfunction. Dev. Neurosci. 0378-5866. 20 (4-5), 271–276. doi:10.1159/000017321.

Colby-Milley, J., Cavanagh, C., Jego, S., Breitner, J.C.S., Quirion, R., Adamantidis, A., Vyazovskiy, V., 2015. Sleep-Wake Cycle Dysfunction in the TgCRND8 Mouse Model of Alzheimer's Disease: from Early to Advanced Pathological Stages. PLOS ONE. 1932-6203. 10 (6), e0130177. doi:10.1371/journal.pone.0130177.

Colettis, N.-C., Oberholzer, M.-V., Cercato, M., Habif, M., Selles, M.-C., Salas, D., Sebollela, A., Klein, W.-L., Epstein, A.-L., Salvetti, A., Ferreira, S.-T., Jerusalinsky, D., 2023. Memory recovery through gene therapy with a single chain antibody fragment selective for Aβ oligomers in a model of Alzheimer's disease in rats. J. Appl. Cogn. Neurosci. 2745-0031. 4 (1), e00344688. doi:10.17981/jacn.4.1.2023.2.

Colton, C.A., Brown, C.M., Cook, D., Needham, L.K., Xu, Q., Czapiga, M., Saunders, A.M., Schmechel, D.E., Rasheed, K., Vitek, M.P., 2002. APOE and the regulation of microglial nitric oxide production: a link between genetic risk and oxidative stress. Neurobiol. Aging. 01974580. 23 (5), 777–785. doi:10.1016/S0197-4580(02)00016-7.

Conner, J.M., Darracq, M.A., Roberts, J., Tuszynski, M.H., 2001. Nontropic actions of neurotrophins: subcortical nerve growth factor gene delivery reverses age-related degeneration of primate cortical cholinergic innervation. Proc. Natl. Acad. Sci. U. S. A. 00278424. 98 (4), 1941–1946. doi:10.1073/pnas.98.4.1941.

Conner, J.M., Lauterborn, J.C., Yan, Q., Gall, C.M., Varon, S., 1997. Distribution of brain-derived neurotrophic factor (BDNF) protein and mRNA in the normal adult rat CNS: evidence for anterograde axonal transport. J. Neurosci. 02706474. 17 (7), 2295–2313. doi:10.1523/jneurosci.17-07-02295.1997.

Cook, C., Dunmore, J.H., Murray, M.E., Scheffel, K., Shukoor, N., Tong, J., Castanedes-Casey, M., Phillips, V., Rousseau, L., Penuliar, M.S., Kurti, A., Dickson, D.W., Petrucelli, L., Fryer, J.D., 2014. Severe amygdala dysfunction in a MAPT transgenic mouse model of frontotemporal dementia. Neurobiol. Aging. 15581497. 35 (7), 1769–1777. doi:10.1016/j.neurobiolaging.2013.12.023.

Corder, E.H., Saunders, A.M., Strittmatter, W.J., Schmechel, D.E., Gaskell, P.C., Small, G.W., Roses, A.D., Haines, J.L., Pericak-Vance, M.A., 1993. Gene dose of apolipoprotein E type 4 allele and the risk of Alzheimer's disease in late onset families. Science 00368075. 261 (5123), 921–923. Available from www.sciencemag.org.

Counts, S.E., Nadeem, M., Wuu, J., Ginsberg, S.D., Saragovi, H.U., Mufson, E.J., 2004. Reduction of cortical TrkA but not p75 NTR protein in early-stage Alzheimer's disease. Ann. Neurol. 03645134. 56 (4), 520–531. doi:10.1002/ana.20233.

Cruts, M., Theuns, J., Van Broeckhoven, C., 2012. Locus-specific mutation databases for neurodegenerative brain diseases. Hum. Mutat. 1059-7794. 33 (9), 1340–1344. doi:10.1002/humu.22117.

Davignon, J., Gregg, R.E., Sing, C.F., 1988. Apolipoprotein E polymorphism and atherosclerosis. Arteriosclerosis. 02765047. 8 (1), 1–21. doi:10.1161/01.atv.8.1.1.

Deacon, R.M.J., Koros, E., Bornemann, K.D., Rawlins, J.N.P., 2009. Aged Tg2576 mice are impaired on social memory and open field habituation tests. Behav. Brain Res. 01664328. 197 (2), 466–468. doi:10.1016/j.bbr.2008.09.042.

Deane, R., Sagare, A., Hamm, K., Parisi, M., Lane, S., Finn, M.B., Holtzman, D.M., Zlokovic, B.V., 2008. apoE isoform-specific disruption of amyloid β peptide clearance from mouse brain. J. Clin. Invest. 15588238. 118 (12), 4002–4013. doi:10.1172/JCI36663.

Devi, L., Ohno, M., Combs, C., 2010. Phospho-eIF2α Level Is Important for Determining Abilities of BACE1 Reduction to Rescue Cholinergic Neurodegeneration and Memory Defects in 5XFAD Mice. PLoS ONE. 1932-6203. 5 (9), e12974. doi:10.1371/journal.pone.0012974.

Dodart, J.C., Marr, R.A., Koistinaho, M., Gregersen, B.M., Malkani, S., Verma, I.M., Paul, S.M., 2005. Gene delivery of human apolipoprotein E alters brain Aβ burden in a mouse model of Alzheimer's disease. Proc. Natl. Acad. Sci. U. S. A. 00278424. 102 (4), 1211–1216. doi:10.1073/pnas.0409072102.

Duan, Y., Ye, T., Qu, Z., Chen, Y., Miranda, A., Zhou, X., Lok, K.-C., Chen, Y., Fu, A.K.Y., Gradinaru, V., Ip, N.Y., 2022. Brain-wide Cas9-mediated cleavage of a gene causing familial Alzheimer's disease alleviates amyloid-related pathologies in mice. Nat. Biomed. Eng. 2157-846X. 6 (2), 168–180. doi:10.1038/s41551-021-00759-0.

Duan, Y., Ye, T., Qu, Z., Chen, Y., Miranda, A., Zhou, X., Lok, K.C., Chen, Y., Fu, A.K.Y., Gradinaru, V., Ip, N.Y., 2022. Brain-wide Cas9-mediated cleavage of a gene causing familial Alzheimer's disease alleviates amyloid-related pathologies in mice. Nat. Biomed. Eng.. 6 (2), 168–180. doi: 10.1038/s41551-021-00759-0.

Dutschmann, M., Menuet, C., Stettner, G.M., Gestreau, C., Borghgraef, P., Devijver, H., Gielis, L., Hilaire, G., Van Leuven, F., 2010. Upper airway dysfunction of tau-P301L mice correlates with tauopathy in midbrain and ponto-medullary brainstem nuclei. J. Neurosci. 15292401. 30 (5), 1810–1821. doi:10.1523/JNEUROSCI.5261-09.2010.

Eckman, E.A., Reed, D.K., Eckman, C.B., 2001. Degradation of the Alzheimer's Amyloid β Peptide by Endothelin-converting Enzyme. J. Biol. Chem. 276 (27), 24540–24548. doi:10.1074/jbc.M007579200.

Eckman, E.A., Watson, M., Marlow, L., Sambamurti, K., Eckman, C.B., 2003. Alzheimer's disease β-amyloid peptide is increased in mice deficient in endothelin-converting enzyme. J. Biol. Chem. 278 (4), 2081–2084. doi:10.1074/jbc.C200642200.

El-Amouri, S.S., Zhu, H., Yu, J., Marr, R., Verma, I.M., Kindy, M.S., 2008. Neprilysin: an enzyme candidate to slow the progression of Alzheimer's disease. Am. J. Pathol. 00029440. 172 (5), 1342–1354. doi:10.2353/ajpath.2008.070620.

Embury, C.M., Dyavarshetty, B., Lu, Y., Wiederin, J.L., Ciborowski, P., Gendelman, H.E., Kiyota, T., 2017. Cathepsin B Improves ß-Amyloidosis and Learning and Memory in Models of Alzheimer's Disease. J. Neuroimmune Pharmacol. 15571904. 12 (2), 340–352. doi:10.1007/s11481-016-9721-6.

Esler, W.P., Wolfe, M.S., 2001. A portrait of Alzheimer secretases - New features and familiar faces. Science 293 (5534), 1449–1454. doi:10.1126/science.1064638.

Farris, W., Mansourian, S., Chang, Y., Lindsley, L., Eckman, E.A., Frosch, M.P., Eckman, C.B., Tanzi, R.E., Selkoe, D.J., Guénette, S., 2003. Insulin-degrading enzyme regulates the levels of insulin, amyloid β-protein, and the β-amyloid precursor protein intracellular domain in vivo. Proc. Natl. Acad. Sci. U. S. A. 100 (7), 4162–4167. doi:10.1073/pnas.0230450100.

Fenoglio, C., Scarpini, E., Serpente, M., Galimberti, D., 2018. Role of Genetics and Epigenetics in the Pathogenesis of Alzheimer's Disease and Frontotemporal Dementia. J. Alzheimers Dis. 1387-2877. 62 (3), 913–932. doi:10.3233/jad-170702.

Fernandez-Funez, P., de Mena, L., Rincon-Limas, D.E., 2015. Modeling the complex pathology of Alzheimer's disease in Drosophila. Exp. Neurol. 10902430. 274, 58–71. doi:10.1016/j.expneurol.2015.05.013.

Fischer, W., Wictorin, K., Björklund, A., Williams, L.R., Varon, S., Gage, F.H., 1987. Amelioration of cholinergic neuron atrophy and spatial memory impairment in aged rats by nerve growth factor. Nature. 00280836. 329 (6134), 65–68. doi:10.1038/329065a0.

Fitz, N.F., Wolfe, C.M., Playso, B.E., Biedrzycki, R.J., Lu, Y., Nam, K.N., Lefterov, I., Koldamova, R., 2020. Trem2 deficiency differentially affects phenotype and transcriptome of human APOE3 and APOE4 mice. Mol. Neurodegener. 17501326. 15 (1). doi:10.1186/s13024-020-00394-4.

Folstein, M.F., Folstein, S.E., McHugh, P.R., 1975. "Mini-mental state". A practical method for grading the cognitive state of patients for the clinician. J. Psychiatr. Res. 00223956. 12 (3), 189–198. doi:10.1016/0022-3956(75)90026-6.

Forner, S., Kawauchi, S., Balderrama-Gutierrez, G., Kramár, E.A., Matheos, D.P., Phan, J., Javonillo, D.I., Tran, K.M., Hingco, E., da Cunha, C., Rezaie, N., Alcantara, J.A., Baglietto-Vargas, D., Jansen, C., Neumann, J., Wood, M.A., MacGregor, G.R., Mortazavi, A., Tenner, A.J., LaFerla, F.M., Green, K.N., 2021. Systematic phenotyping and characterization of the 5xFAD mouse model of Alzheimer's disease. Sci. Data. 2052-4463. 8 (1). doi:10.1038/s41597-021-01054-y.

Gámez-Valero, A., Guisado-Corcoll, A., Herrero-Lorenzo, M., Solaguren-Beascoa, M., Martí, E., 2020. Non-Coding RNAs as Sensors of Oxidative Stress in Neurodegenerative Diseases. Antioxidants. 2076-3921. 9 (11), 1095. doi:10.3390/antiox9111095.

Gelman, S., Palma, J., Tombaugh, G., Ghavami, A., 2018. Differences in Synaptic Dysfunction between rTg4510 and APP/PS1 Mouse Models of Alzheimer's Disease. J. Alzheimers Dis. 18758908. 61 (1), 195–208. doi:10.3233/JAD-170457.

Giménez-Llort, L., Arranz, L., Maté, I., De la Fuente, M., 2008. Gender-Specific Neuroimmunoendocrine Aging in a Triple-Transgenic 3×Tg-AD Mouse Model for Alzheimer's Disease and Its Relation with Longevity. Neuroimmunomodulation. 1021-7401. 15 (4-6), 331–343. doi:10.1159/000156475.

Girotra, P., Behl, T., Sehgal, A., Singh, S., Bungau, S., 2022. Investigation of the Molecular Role of Brain-Derived Neurotrophic Factor in Alzheimer's Disease. J. Mol. Neurosci. 0895-8696. 72 (2), 173–186. doi:10.1007/s12031-021-01824-8.

Götz, J., Bodea, L.G., Goedert, M., 2018. Rodent models for Alzheimer disease. Nat. Rev. Neurosci. 14710048. 19 (10), 583–598. doi:10.1038/s41583-018-0054-8.

Götz, J., Ittner, L.M., 2008. Animal models of Alzheimer's disease and frontotemporal dementia. Nat. Rev. Neurosci. 14710048. 9 (7), 532–544. doi:10.1038/nrn2420.

Granger, M.W., Franko, B., Taylor, M.W., Messier, C., George-Hyslop, P.S., Bennett, S.A.L., de la Torre, J., 2016. A TgCRND8 Mouse Model of Alzheimer's Disease Exhibits Sexual Dimorphisms in Behavioral Indices of Cognitive Reserve. J. Alzheimers Dis. 13872877. 51 (3), 757–773. doi:10.3233/jad-150587.

Grimm, M.O.W., Mett, J., Stahlmann, C.P., Haupenthal, V.J., Zimmer, V.C., Hartmann, T., 2013. Neprilysin and Aβ clearance: impact of the APP intracellular domain in NEP regulation and implications in Alzheimer's disease. Front. Aging Neurosci. 16634365. 5. doi:10.3389/fnagi.2013.00098.

Gu, L., Wu, D., Tang, X., Xinyang, Q., Xiaoli, L., Bai, F., Chen, X., Ren, Q., Zhang, Z., 2018. Myelin changes at the early stage of 5XFAD mice. Brain Res. Bull. 03619230. 137, 285–293. doi:10.1016/j.brainresbull.2017.12.013.

Guan, H., Liu, Y., Daily, A., Police, S., Kim, M.H., Oddo, S., LaFerla, F.M., Pauly, J.R., Murphy, M.P., Hersh, L.B., 2009. Peripherally expressed neprilysin reduces brain amyloid burden: a novel approach for treating Alzheimer's disease. J. Neurosci. Res. 10974547. 87 (6), 1462–1473. doi:10.1002/jnr.21944.

Guerreiro, R.J., Gustafson, D.R., Hardy, J., 2012. The genetic architecture of Alzheimer's disease: beyond APP, PSENS and APOE. Neurobiol. Aging. 15581497. 33 (3), 437–456. doi:10.1016/j.neurobiolaging.2010.03.025.

Guo, L., Cao, J., Hou, J., Li, Y., Huang, M., Zhu, L., Zhang, L., Lee, Y., Duarte, M.L., Zhou, X., Wang, M., Liu, C.-C., Martens, Y., Chao, M., Goate, A., Guojun, B., Haroutunian, V., Cai, D., Zhang, B., 2023. Sex specific molecular networks and key drivers of Alzheimer's disease. Mol. Neurodegener. 1750-1326. 18 (1). doi:10.1186/s13024-023-00624-5.

Guo, Q., Zheng, X., Yang, P., Pang, X., Qian, K., Wang, P., Shuting, X., Sheng, D., Wang, L., Cao, J., Wei, L., Zhang, Q., Jiang, X., 2019. Small interfering RNA delivery to the neurons near the amyloid plaques for improved treatment of Alzheimer's disease. Acta Pharm. Sin. B. 22113835. 9 (3), 590–603. doi:10.1016/j.apsb.2018.12.010.

Hafez, D.M., Huang, J.Y., Richardson, J.C., Masliah, E., Peterson, D.A., Marr, R.A., 2012. F-spondin gene transfer improves memory performance and reduces amyloid-β levels in mice. Neuroscience. 03064522. 223, 465–472. doi:10.1016/j.neuroscience.2012.07.038.

Hemming, M.L., Patterson, M., Reske-Nielsen, C., Lin, L., Isacson, O., Selkoe, D.J., Gandy, S., 2007. Reducing Amyloid Plaque Burden via Ex Vivo Gene Delivery of an Aβ-Degrading Protease: a Novel Therapeutic Approach to Alzheimer Disease. PLoS Med. 1549-1676. 4 (8), e262. doi:10.1371/journal.pmed.0040262.

He, Pan, Xu, He, Huang, Song, et al., 2017. Adeno-associated virus 9-mediated Cdk5 inhibitory peptide reverses pathologic changes and behavioral deficits in the Alzheimer's disease mouse model. FASEB J. 31 (8), 3383–3392. doi:10.1096/fj.201700064R.

Hirota, Y., Sakakibara, Y., Ibaraki, K., Takei, K., Iijima, K.M., Sekiya, M., 2022. Distinct brain pathologies associated with Alzheimer's disease biomarker-related phospho-tau 181 and phospho-tau 217 in App knock-in mouse models of amyloid-β amyloidosis. Brain Commun. 26321297. 4 (6). doi:10.1093/braincomms/fcac286.

Hoang, M.V., Turner, A.J., 1997. Novel activity of endothelin-converting enzyme: hydrolysis of bradykinin. Biochem. J. 02646021. 327 (1), 23–26. doi:10.1042/bj3270023.

Holmes, B.B., Furman, J.L., Mahan, T.E., Yamasaki, T.R., Mirbaha, H., Eades, W.C., Belaygorod, L., Cairns, N.J., Holtzman, D.M., Diamond, M.I., 2014. Proteopathic tau seeding predicts tauopathy in vivo. Proc. Natl. Acad. Sci. U. S. A. 10916490. 111 (41), E4376. doi:10.1073/pnas.1411649111.

Holtzman, D.M., Li, Y., Parada, L.F., Kinsman, S., Chen, C.K., Valletta, J.S., Zhou, J., Long, J.B., Mobley, W.C., 1992. p140trk mRNA marks NGF-responsive forebrain neurons: evidence that trk gene expression is induced by NGF. Neuron. 08966273. 9 (3), 465–478. doi:10.1016/0896-6273(92)90184-F.

Hook, V.Y.H., Kindy, M., Hook, G., 2008. Inhibitors of Cathepsin B Improve Memory and Reduce β-Amyloid in Transgenic Alzheimer Disease Mice Expressing the Wild-type, but Not the Swedish Mutant, β-Secretase Site of the Amyloid Precursor Protein. J. Biol. Chem. 00219258. 283 (12), 7745–7753. doi:10.1074/jbc.m708362200.

Hook, V.Y.H., Kindy, M., Reinheckel, T., Peters, C., Hook, G., 2009. Genetic cathepsin B deficiency reduces β-amyloid in transgenic mice expressing human wild-type amyloid precursor protein. Biochem. Biophys. Res. Commun. 0006291X. 386 (2), 284–288. doi:10.1016/j.bbrc.2009.05.131.

Hsiao, K., Chapman, P., Nilsen, S., Eckman, C., Harigaya, Y., Younkin, S., Yang, F., Cole, G., 1996. Correlative memory deficits, Aβ elevation, and amyloid plaques in transgenic mice. Science. 00368075. 274 (5284), 99–102. doi:10.1126/science.274.5284.99.

Huang, M., Macdonald, J., Lavenir, I., Chen, R., Craxton, M., Slavik-Smith, E., Davies, S.W., Goedert, M., 2022. Increase in Tau Pathology in P290S Mapt Knock-In Mice Crossed with App NL-G-F Mice. eneuro. 2373-2822. 9 (6), ENEURO.0247-22.2022. doi:10.1523/eneuro.0247-22.2022.

Hulshof, L.A., Frajmund, L.A., van Nuijs, D., van der Heijden, D.C.N., Middeldorp, J., Hol, E.M., 2022. Both male and female APPswe/PSEN1dE9 mice are impaired in spatial memory and cognitive flexibility at 9 months of age. Neurobiol. Aging. 15581497. 113, 28–38. doi:10.1016/j.neurobiolaging.2021.12.009.

Huynh, T.P.V., Wang, C., Tran, A.C., Tabor, G.T., Mahan, T.E., Francis, C.M., Finn, M.B., Spellman, R., Manis, M., Tanzi, R.E., Ulrich, J.D., Holtzman, D.M., 2019. Lack of hepatic apoE does not influence early Aβ deposition: observations from a new APOE knock-in model. Mol. Neurodegener. 17501326. 14 (1). doi:10.1186/s13024-019-0337-1.

Iwasaki, Y., Negishi, T., Inoue, M., Tashiro, T., Tabira, T., Kimura, N., 2012. Sendai virus vector-mediated brain-derived neurotrophic factor expression ameliorates memory deficits and synaptic degeneration in a transgenic mouse model of Alzheimer's disease. J. Neurosci. Res. 0360-4012. 90 (5), 981–989. doi:10.1002/jnr.22830.

Iwata, A., Nagata, K., Hatsuta, H., Takuma, H., Bundo, M., Iwamoto, K., Tamaoka, A., Murayama, S., Saido, T., Tsuji, S., 2014. Altered CpG methylation in sporadic Alzheimer's disease is associated with APP and MAPT dysregulation. Hum. Mol. Genet. 1460-2083. 23 (3), 648–656. doi:10.1093/hmg/ddt451.

Iwata, N., Mizukami, H., Shirotani, K., Takaki, Y., Muramatsu, S.-I., Bao, L., Gerard, N.P., Gerard, C., Ozawa, K., Saido, T.C., 2004. Presynaptic Localization of Neprilysin Contributes to Efficient Clearance of Amyloid-β Peptide in Mouse Brain. J. Neurosci. 0270-6474. 24 (4), 991–998. doi:10.1523/jneurosci.4792-03.2004.

Jackson, R.J., Rudinskiy, N., Herrmann, A.G., Croft, S., Kim, J.S.M., Petrova, V., Ramos-Rodriguez, J.J., Pitstick, R., Wegmann, S., Garcia-Alloza, M., Carlson, G.A., Hyman, B.T., Spires-Jones, T.L., 2016. Human tau increases amyloid β plaque size but not amyloid β-mediated synapse loss in a novel mouse model of Alzheimer's disease. Eur. J. Neurosci. 14609568. 44 (12), 3056–3066. doi:10.1111/ejn.13442.

Jacobsen, J.S., Wu, C.C., Redwine, J.M., Comery, T.A., Arias, R., Bowlby, M., Martone, R., Morrison, J.H., Pangalos, M.M., Reinhart, P.H., Bloom, F.E., 2006. Early-onset behavioral and synaptic deficits in a mouse model of Alzheimer's disease. Proc. Natl. Acad. Sci. U. S. A. 00278424. 103 (13), 5161–5166. doi:10.1073/pnas.0600948103.

Janelidze, S., Zetterberg, H., Mattsson, N., Palmqvist, S., Vanderstichele, H., Lindberg, O., et al., 2016. CSF Abeta42/Abeta40 and Abeta42/Abeta38 ratios: better diagnostic markers of Alzheimer disease. Ann. Clin. Transl. Neurol. 3 (3), 154–165. doi:https://doi.org/10.1002/acn3.274.

Jankowsky, J.L., Slunt, H.H., Gonzales, V., Jenkins, N.A., Copeland, N.G., Borchelt, D.R., 2004. APP processing and amyloid deposition in mice haplo-insufficient for presenilin 1. Neurobiol. Aging 25 (7), 885–892. doi:10.1016/j.neurobiolaging.2003.09.008.

Jankowsky, J.L., Zheng, H., 2017. Practical considerations for choosing a mouse model of Alzheimer's disease. Mol. Neurodegener. 17501326. 12 (1). doi:10.1186/s13024-017-0231-7.

Janus, C., Flores, A.Y., Xu, G., Borchelt, D.R., 2015. Behavioral abnormalities in APPSwe/PS1dE9 mouse model of AD-like pathology: comparative analysis across multiple behavioral domains. Neurobiol. Aging. 15581497. 36 (9), 2519–2532. doi:10.1016/j.neurobiolaging.2015.05.010.

Janus, C., Pearson, J., McLaurin, J.A., Mathews, P.M., Jiang, Y., Schmidt, S.D., Chishti, M.A., Horne, P., Heslin, D., French, J., Mount, H.T.J., Nixon, R.A., Mercken, M., Bergeron, C., Fraser, P.E., George-Hyslop, P.S., Westaway, D., 2000. Aβ peptide immunization reduces behavioural impairment and plaques in a model of Alzheimer's disease. Nature. 00280836. 408 (6815), 979–982. doi:10.1038/35050110.

Jawhar, S., Trawicka, A., Jenneckens, C., Bayer, T.A., Wirths, O., 2012. Motor deficits, neuron loss, and reduced anxiety coinciding with axonal degeneration and intraneuronal Aβ aggregation in the 5XFAD mouse model of Alzheimer's disease. Neurobiol. Aging. 01974580. 33 (1), 196.e29. doi:10.1016/j.neurobiolaging.2010.05.027.

Jiang, Y., Xu, N., 2023. The Emerging Role of Autophagy-Associated lncRNAs in the Pathogenesis of Neurodegenerative Diseases. Int. J. Mol. Sci. 1422-0067. 24 (11), 9686. doi:10.3390/ijms24119686.

John, A., Reddy, P.H., 2021. Synaptic basis of Alzheimer's disease: focus on synaptic amyloid beta, P-tau and mitochondria. Ageing Res. Rev. 15681637. 65, 101208. doi:10.1016/j.arr.2020.101208.

Johnson, G.D., Stevenson, T., Ahn, K., 1999. Hydrolysis of peptide hormones by endothelin-converting enzyme-1: a comparison with neprilysin. J. Biol. Chem. 00219258. 274 (7), 4053–4058. doi:10.1074/jbc.274.7.4053.

Joshi, G., Gan, K.A., Johnson, D.A., Johnson, J.A., 2015. Increased Alzheimer's disease-like pathology in the APP/ PS1δE9mouse model lacking Nrf2 through modulation of autophagy. Neurobiol. Aging. 15581497. 36 (2), 664–679. doi:10.1016/j.neurobiolaging.2014.09.004.

Kanemitsu, H., Tomiyama, T., Mori, H., 2003. Human neprilysin is capable of degrading amyloid β peptide not only in the monomeric form but also the pathological oligomeric form. Neurosci. Lett. 03043940. 350 (2), 113–116. doi:10.1016/s0304-3940(03)00898-x.

Kanninen, K., Heikkinen, R., Malm, T., Rolova, T., Kuhmonen, S., Leinonen, H., Ylä-Herttuala, S., Tanila, H., Levonen, A.-L., Koistinaho, M., Koistinaho, J., 2009. Intrahippocampal injection of a lentiviral vector expressing Nrf2 improves spatial learning in a mouse model of Alzheimer's disease. Proc. Natl. Acad. Sci. U. S. A. 0027-8424. 106 (38), 16505–16510. doi:10.1073/pnas.0908397106.

Kemppainen, S., Lindholm, P., Galli, E., Lahtinen, H.M., Koivisto, H., Hämäläinen, E., Saarma, M., Tanila, H., 2015. Cerebral dopamine neurotrophic factor improves long-term memory in APP/PS1 transgenic mice modeling Alzheimer's disease as well as in wild-type mice. Behav. Brain Res. 18727549. 291, 1–11. doi:10.1016/j.bbr.2015.05.002.

Kilgore, M., Miller, C.A., Fass, D.M., Hennig, K.M., Haggarty, S.J., Sweatt, J.D., Rumbaugh, G., 2010. Inhibitors of class 1 histone deacetylases reverse contextual memory deficits in a mouse model of Alzheimer's disease. Neuropsychopharmacology. 0893133X. 35 (4), 870–880. doi:10.1038/npp.2009.197.

Kimura, R., Ohno, M., 2009. Impairments in remote memory stabilization precede hippocampal synaptic and cognitive failures in 5XFAD Alzheimer mouse model. Neurobiol. Dis. 09699961. 33 (2), 229–235. doi:10.1016/j.nbd.2008.10.006.

King, D.L., Arendash, G.W., Crawford, F., Sterk, T., Menendez, J., Mullan, M.J., 1999. Progressive and gender-dependent cognitive impairment in the APP(SW) transgenic mouse model for Alzheimer's disease. Behav. Brain Res. 01664328. 103 (2), 145–162. doi:10.1016/S0166-4328(99)00037-6.

Kiselev, I.S., Kulakova, O.G., Boyko, A.N., Favorova, O.O., 2021. DNA Methylation As an Epigenetic Mechanism in the Development of Multiple Sclerosis. Acta Naturae. 20758251. 13 (2), 45–57. doi:10.32607/actanaturae.11043.

Kiyota, T., Ingraham, K.L., Swan, R.J., Jacobsen, M.T., Andrews, S.J., Ikezu, T., 2012. AAV serotype 2/1-mediated gene delivery of anti-inflammatory interleukin-10 enhances neurogenesis and cognitive function in APP+PS1 mice. Gene Ther. 0969-7128. 19 (7), 724–733. doi:10.1038/gt.2011.126.

Kiyota, T., Zhang, G., Morrison, C.M., Bosch, M.E., Weir, R.A., Lu, Y., Dong, W., Gendelman, H.E., 2015. AAV2/1 CD74 gene transfer reduces β-amyloidosis and improves learning and memory in a mouse model of Alzheimer's disease. Mol. Ther. 15250024. 23 (11), 1712–1721. doi:10.1038/mt.2015.142.

Klein, R.L., Hirko, A.C., Meyers, C.A., Grimes, J.R., Muzyczka, N., Meyer, E.M., 2000. NGF gene transfer to intrinsic basal forebrain neurons increases cholinergic cell size and protects from age-related, spatial memory deficits in middle-aged rats. Brain Res. 875 (1-2), 144–151. doi:10.1016/S0006-8993(00)02634-2.

Klingner, M., Apelt, J., Kumar, A., Sorger, D., Sabri, O., Steinbach, J., Scheunemann, M., Schliebs, R., 2003. Alterations in cholinergic and non-cholinergic neurotransmitter receptor densities in transgenic Tg2576 mouse brain with β-amyloid plaque pathology. Int. J. Dev. Neurosci. 0736-5748. 21 (7), 357–369. doi:10.1016/j.ijdevneu.2003.08.001.

Knouff, C., Hinsdale, M.E., Mezdour, H., Altenburg, M.K., Watanabe, M., Quarfordt, S.H., Sullivan, P.M., Maeda, N., 1999. Apo E structure determines VLDL clearance and atherosclerosis risk in mice. J. Clin. Invest. 00219738. 103 (11), 1579–1586. doi:10.1172/JCI6172.

Kobayashi, D.T., Chen, K.S., 2005. Behavioral phenotypes of amyloid-based genetically modified mouse models of Alzheimer's disease. Genes, Brain and Behavior. 1601-1848. 4 (3), 173–196. doi:10.1111/j.1601-183x.2005.00124.x.

Kok, E., Haikonen, S., Luoto, T., Huhtala, H., Goebeler, S., Haapasalo, H., Karhunen, P.J., 2009. Apolipoprotein E–dependent accumulation of Alzheimer disease–related lesions begins in middle age. Ann. Neurol. 0364-5134. 65 (6), 650–657. doi:10.1002/ana.21696.

Koliatsos, V.E., Nauta, H.J., Clatterbuck, R.E., Holtzman, D.M., Mobley, W.C., Price, D.L., 1990. Mouse nerve growth factor prevents degeneration of axotomized basal forebrain cholinergic neurons in the monkey. J. Neurosci. 0270-6474. 10 (12), 3801–3813. doi:10.1523/jneurosci.10-12-03801.1990.

Kosel, F., Pelley, J.M.S., Franklin, T.B., 2020. Behavioural and psychological symptoms of dementia in mouse models of Alzheimer's disease-related pathology. Neurosci. Biobehav. Rev. 18737528. 112, 634–647. doi:10.1016/j.neubiorev.2020.02.012.

Koutsodendris, N., Blumenfeld, J., Agrawal, A., Traglia, M., Grone, B., Zilberter, M., Yip, O., Rao, A., Nelson, M.R., Hao, Y., Thomas, R., Yoon, S.Y., Arriola, P., Huang, Y., 2023b. Neuronal APOE4 removal protects against tau-mediated gliosis, neurodegeneration and myelin deficits. Nat. Aging. 2662-8465. 3 (3), 275–296. doi:10.1038/s43587-023-00368-3.

Koutsodendris, N., Blumenfeld, J., Agrawal, A., Traglia, M., Yip, O., Rao, A., Kim, M.J., Nelson, M.R., Wang, Y.-H., Grone, B., Hao, Y., Thomas, R., Zilberter, M., Yoon, S.Y., Arriola, P., Huang, Y., 2023a. APOE4-promoted gliosis and degeneration in tauopathy are ameliorated by pharmacological inhibition of HMGB1 release. Cell Rep. 22111247. 42 (10), 113252. doi:10.1016/j.celrep.2023.113252.

Kshirsagar, V., Thingore, C., Juvekar, A., 2021. Insulin resistance: a connecting link between Alzheimer's disease and metabolic disorder. Metab. Brain Dis. 0885-7490. 36 (1), 67–83. doi:10.1007/s11011-020-00622-2.

Kuhla, A., Rühlmann, C., Lindner, T., Polei, S., Hadlich, S., Krause, B.J., Vollmar, B., Teipel, S.J., 2017. APPswe/PS1dE9 mice with cortical amyloid pathology show a reduced NAA/Cr ratio without apparent brain atrophy: a MRS and MRI study. Neuroimage Clin. 22131582. 15, 581–586. doi:10.1016/j.nicl.2017.06.009.

Lalonde, R., Kim, H.D., Maxwell, J.A., Fukuchi, K., 2005. Exploratory activity and spatial learning in 12-month-old APP695SWE/co+PS1/ΔE9mice with amyloid plaques. Neurosci. Lett. 03043940. 390 (2), 87–92. doi:10.1016/j.neulet.2005.08.028.

Lanoiselée, H.-M., Nicolas, G., Wallon, D., Rovelet-Lecrux, A., Lacour, M., Rousseau, S., Richard, A.-C., Pasquier, F., Rollin-Sillaire, A., Martinaud, O., Quillard-Muraine, M., de la Sayette, V., Boutoleau-Bretonniere, C., Etcharry-Bouyx, F., Chauviré, V., Sarazin, M., le le Ber, I., Epelbaum, S., Jonveaux, T., Rouaud, O., Ceccaldi, M., Félician, O., Godefroy, O., Formaglio, M., Croisile, B., Auriacombe, S., Chamard, L., Vincent, J.-L., Sauvée, M., Marelli-Tosi, C., Gabelle, A., Ozsancak, C., Pariente, J., Paquet, C., Hannequin, D., Campion, D., Miller, B.L., 2017. APP, PSEN1, and PSEN2 mutations in early-onset Alzheimer disease: a genetic screening study of familial and sporadic cases. PLoS Med. 1549-1676. 14 (3), e1002270. doi:10.1371/journal.pmed.1002270.

Lasagna-Reeves, C.A., de Haro, M., Hao, S., Park, J., Rousseaux, M.W.C., Al-Ramahi, I., Jafar-Nejad, P., Vilanova-Velez, L., See, L., De Maio, A., Nitschke, L., Wu, Z., Troncoso, J.C., Westbrook, T.F., Tang, J., Botas, J., Zoghbi, H.Y., 2016. Reduction of Nuak1 Decreases Tau and Reverses Phenotypes in a Tauopathy Mouse Model. Neuron. 10974199. 92 (2), 407–418. doi:10.1016/j.neuron.2016.09.022.

Lee, S.H., Bae, E.J., Perez-Acuna, D., Jung, M.K., Han, J.W., Mook-Jung, I., Lee, S.J., 2024. Amyloid-β-activated microglia can induce compound proteinopathies. Brain. 14602156. 147 (12), 4105–4120. doi:10.1093/brain/awae221.

Lehmann, D.J., Cortina-Borja, M., Warden, D.R., Smith, A.D., Sleegers, K., Prince, J.A., Van Duijn, C.M., Kehoe, P.G., 2005. Large meta-analysis establishes the ACE insertion-deletion polymorphism as a marker of Alzheimer's disease. Am. J. Epidemiol. 162 (4), 305–317. doi:10.1093/aje/kwi202.

Leng, F., Edison, P., 2021. Neuroinflammation and microglial activation in Alzheimer disease: where do we go from here? Nat. Rev. Neurol. 1759-4758. 17 (3), 157–172. doi:10.1038/s41582-020-00435-y.

Leyns, C.E.G., Ulrich, J.D., Finn, M.B., Stewart, F.R., Koscal, L.J., Serrano, J.R., Robinson, G.O., Anderson, E., Colonna, M., Holtzman, D.M., 2017. TREM2 deficiency attenuates neuroinflammation and protects against neurodegeneration in a mouse model of tauopathy. Proc. Natl. Acad. Sci. U. S. A. 10916490. 114 (43), 11524–11529. doi:10.1073/pnas.1710311114.

Li, C., Ni, Y.Q., Xu, H., Xiang, Q.Y., Zhao, Y., Zhan, J.K., He, J.Y., Li, S., Liu, Y.S., 2021. Roles and mechanisms of exosomal non-coding RNAs in human health and diseases. Signal Transduct. Target. Ther. 20593635. 6 (1). doi:10.1038/s41392-021-00779-x.

Li, H., Chen, Z., Shen, Y., Xiong, T., Chen, A., Chen, L., Yifan, Y., Jiang, Q., Zhang, Y., Sun, J., Shen, L., 2024. Gene therapy in Aβ-induced cell and mouse models of Alzheimer's disease through compensating defective mitochondrial complex I function. J. Transl. Med. 1479-5876. 22 (1). doi:10.1186/s12967-024-05571-3.

Li, M., Yuan, Y., Hu, B., Wu, L., 2017. Study on Lentivirus-Mediated ABCA7 Improves Neurocognitive Function and Related Mechanisms in the C57BL/6 Mouse Model of Alzheimer's Disease. J. Mol. Neurosci. 0895-8696. 61 (4), 489–497. doi:10.1007/s12031-017-0889-x.

Li, Y., Wang, Y., Wang, J., Chong, K.Y., Xu, J., Liu, Z., Shan, C., 2020. Expression of Neprilysin in Skeletal Muscle by Ultrasound-Mediated Gene Transfer (Sonoporation) Reduces Amyloid Burden for AD. Mol. Ther. Methods Clin. Dev. 23290501. 17, 300–308. doi:10.1016/j.omtm.2019.12.012.

Lilja, A.M., Röjdner, J., Mustafiz, T., Thomé, C.M., Storelli, E., Gonzalez, D., Unger-Lithner, C., Greig, N.H., Nordberg, A., Marutle, A., Reddy, H., 2013. Age-Dependent Neuroplasticity Mechanisms in Alzheimer Tg2576 Mice Following Modulation of Brain Amyloid-β Levels. PLoS ONE. 1932-6203. 8 (3), e58752. doi:10.1371/journal.pone.0058752.

Ling, Q., Herstine, J.A., Bradbury, A., Gray, S.J., 2023. AAV-based in vivo gene therapy for neurological disorders. Nat. Rev. Drug Discov. 14741784. 22 (10), 789–806. doi:10.1038/s41573-023-00766-7.

Liraz, O., Boehm-Cagan, A., Michaelson, D.M., 2013. ApoE4 induces Aβ42, tau, and neuronal pathology in the hippocampus of young targeted replacement apoE4 mice. Mol. Neurodegener. 17501326. 8 (1). doi:10.1186/1750-1326-8-16.

Litvinchuk, A., Suh, J.H., Guo, J.L., Lin, K., Davis, S.S., Bien-Ly, N., Tycksen, E., Tabor, G.T., Serrano, J.R., Manis, M., Bao, X., Lee, C., Bosch, M., Perez, E.J., Yuede, C.M., Cashikar, A.G., Ulrich, J.D., Di Paolo, G., Holtzman, D.M., 2024. Amelioration of Tau and ApoE4-linked glial lipid accumulation and neurodegeneration with an LXR agonist. Neuron. 10974199. 112 (3), 384. doi:10.1016/j.neuron.2023.10.023.

Liu, C.C., Kanekiyo, T., Xu, H., Bu, G., 2013. Apolipoprotein e and Alzheimer disease: risk, mechanisms and therapy. Nat. Rev. Neurol. 17594766. 9 (2), 106–118. doi:10.1038/nrneurol.2012.263.

Long, J.M., Holtzman, D.M., 2019. Alzheimer Disease: an Update on Pathobiology and Treatment Strategies. Cell. 10974172. 179 (2), 312–339. doi:10.1016/j.cell.2019.09.001.

Ludvigson, A.E., Luebke, J.I., Lewis, J., Peters, A., 2011. Structural abnormalities in the cortex of the rTg4510 mouse model of tauopathy: a light and electron microscopy study. Brain Struct. Funct. 1863-2653. 216 (1), 31–42. doi:10.1007/s00429-010-0295-4.

Luo, S., Tamada, A., Saikawa, Y., Wang, Y., Yu, Q., Hisatsune, T., 2024. P2Y1R silencing in Astrocytes Protected Neuroinflammation and Cognitive Decline in a Mouse Model of Alzheimer's Disease. Aging Dis. 21525250. 15 (4), 1969–1988. doi:10.14336/AD.2023.1006.

Lynch, J.R., Tang, W., Wang, H., Vitek, M.P., Bennett, E.R., Sullivan, P.M., Warner, D.S., Laskowitz, D.T., 2003. APOE genotype and an ApoE-mimetic peptide modify the systemic and central nervous system inflammatory response. J. Biol. Chem. 278 (49), 48529–48533. doi:10.1074/jbc.M306923200.

Ma, Q., 2013. Role of Nrf2 in Oxidative Stress and Toxicity. Annu. Rev. Pharmacol. Toxicol. 0362-1642. 53 (1), 401–426. doi:10.1146/annurev-pharmtox-011112-140320.

Maarouf, C.L., Kokjohn, T.A., Whiteside, C.M., Macias, M.P., Kalback, W.M., Sabbagh, M.N., Beach, T.G., Vassar, R., Roher, A.E., 2013. Molecular Differences and Similarities between Alzheimer's Disease and the 5XFAD Transgenic Mouse Model of Amyloidosis. Biochem. Insights. 1178-6264. 6, BCI.S13025. doi:10.4137/bci.s13025.

Machlovi, S.I., Neuner, S.M., Hemmer, B.M., Khan, R., Liu, Y., Huang, M., Zhu, J.D., Castellano, J.M., Cai, D., Marcora, E., Goate, A.M., 2022. APOE4 confers transcriptomic and functional alterations to primary mouse microglia. Neurobiol. Dis. 09699961. 164, 105615. doi:10.1016/j.nbd.2022.105615.

Mahley, R.W., Weisgraber, K.H., Huang, Y., 2009. Apolipoprotein E: structure determines function, from atherosclerosis to Alzheimer's disease to AIDS. J. Lipid Res. 15397262. 50, S183. doi:10.1194/jlr.R800069-JLR200.

Mangialasche, F., Solomon, A., Winblad, B., Mecocci, P., Kivipelto, M., 2010. Alzheimer's disease: clinical trials and drug development. Lancet Neurol. 14744422. 9 (7), 702–716. doi:10.1016/s1474-4422(10)70119-8.

Martinez-Losa, M., Tracy, T.E., Ma, K., Verret, L., Clemente-Perez, A., Khan, A.S., Cobos, I., Ho, K., Gan, L., Mucke, L., Alvarez-Dolado, M., Palop, J.J., 2018. Nav1.1-Overexpressing Interneuron Transplants Restore Brain Rhythms and Cognition in a Mouse Model of Alzheimer's Disease. Neuron. 10974199. 98 (1), 75. doi:10.1016/j.neuron.2018.02.029.

Maruyama, M., Shimada, H., Suhara, T., Shinotoh, H., Ji, B., Maeda, J., Zhang, M.-R., Trojanowski, J.Q., Lee, V.M.-Y., Ono, M., Masamoto, K., Takano, H., Sahara, N., Iwata, N., Okamura, N., Furumoto, S., Kudo, Y., Chang, Q., Saido, T.C., Takashima, A., Lewis, J., Jang, M.-K., Aoki, I., Ito, H., Higuchi, M., 2013. Imaging of tau pathology in a tauopathy mouse model and in Alzheimer patients compared to normal control. Neuron. 08966273 79 (6), 1094–1108. doi: 10.1016/j.neuron.2013.07.037.

Maurin, H., Chong, S.-A., Kraev, I., Davies, H., Kremer, A., Seymour, C.M., Lechat, B., Jaworski, T., Borghgraef, P., Devijver, H., Callewaert, G., Stewart, M.G., Leuven, F.V., Vitorica, J., 2014. Early Structural and Functional Defects in Synapses and Myelinated Axons in Stratum Lacunosum Moleculare in Two Preclinical Models for Tauopathy. PLoS ONE. 1932-6203. 9 (2), e87605. doi:10.1371/journal.pone.0087605.

McLean, J.W., Bhattrai, A., Vitali, F., Raikes, A.C., Wiegand, J.P.L., Brinton, R.D., 2022. Contributions of sex and genotype to exploratory behavior differences in an aged humanized APOE mouse model of late-onset Alzheimer's disease. Learn. Mem. 15495485. 29 (9), 321–331. doi:10.1101/lm.053588.122.

Menkes-Caspi, N., Yamin, H.G., Kellner, V., Spires-Jones, T.L., Cohen, D., Stern, E.A., 2015. Pathological tau disrupts ongoing network activity. Neuron. 10974199. 85 (5), 959–966. doi:10.1016/j.neuron.2015.01.025.

Meyerhoff, D.J., MacKay, S., Bachman, L., Poole, N., Dillon, W.P., Weiner, M.W., Fein, G., 1993. Reduced brain N-acetylaspartate suggests neuronal loss in cognitively impaired human immunodeficiency virus–seropositive individuals. Neurology. 1526632X. 43 (3), 509–515. doi:10.1212/wnl.43.3_part_1.509.

Miller, B.C., Eckman, E.A., Sambamurti, K., Dobbs, N., Martin Chow, K., Eckman, C.B., Hersh, L.B., Thiele, D.L., 2003. Amyloid-β peptide levels in brain are inversely correlated with insulysin activity levels in vivo. Proc. Natl. Acad. Sci. U. S. A. 0027-8424. 100 (10), 6221–6226. doi:10.1073/pnas.1031520100.

Mobley, W.C., Rutkowski, J.L., Tennekoon, G.I., Gemski, J., Buchanan, K., Johnston, M.V., 1986. Nerve growth factor increases choline acetyltransferase activity in developing basal forebrain neurons. Mol. Brain Res. 0169328X. 1 (1), 53–62. doi:10.1016/0169-328X(86)90020-3.

Moffett, J.R., Ross, B., Arun, P., Madhavarao, C.N., Namboodiri, A.M.A., 2007. N-Acetylaspartate in the CNS: from neurodiagnostics to neurobiology. Prog. Neurobiol. 03010082. 81 (2), 89–131. doi:10.1016/j.pneurobio.2006.12.003.

Montagne, A., Nikolakopoulou, A.M., Huuskonen, M.T., Sagare, A.P., Lawson, E.J., Lazic, D., Rege, S.V., Grond, A., Zuniga, E., Barnes, S.R., Prince, J., Sagare, M., Hsu, C.J., LaDu, M.J., Jacobs, R.E., Zlokovic, B.V., 2021. APOE4 accelerates advanced-stage vascular and neurodegenerative disorder in old Alzheimer's mice via cyclophilin A independently of amyloid-β. Nat. Aging. 26628465. 1 (6), 506–520. doi:10.1038/s43587-021-00073-z.

Mueller-Steiner, S., Zhou, Y., Arai, H., Roberson, E.D., Sun, B., Chen, J., Wang, X., Yu, G., Esposito, L., Mucke, L., Gan, L., 2006. Antiamyloidogenic and Neuroprotective Functions of Cathepsin B: implications for Alzheimer's Disease. Neuron. 08966273. 51 (6), 703–714. doi:10.1016/j.neuron.2006.07.027.

Mukhamedyarov, M.A., Leushina, A.V., Tikhonova, A.E., Petukhova, E.O., Garanina, E.E., Taleb, R.B., Kaligin, M.S., Mukhamedshina, Y.O., Rizvanov, A.A., Zefirov, A.L., Islamov, R.R., 2018. Intravenous Transplantation of Human Umbilical Cord Blood Mononuclear Cells Overexpressing Nerve Growth Factor Improves Spatial Memory in APP/PS1 Transgenic Mice with a Model of Alzheimer's Disease. BioNanoScience. 21911649. 8 (1), 473–480. doi:10.1007/s12668-017-0497-9.

Nagahara, A.H., Bernot, T., Moseanko, R., Brignolo, L., Blesch, A., Conner, J.M., Ramirez, A., Gasmi, M., Tuszynski, M.H., 2009. Long-term reversal of cholinergic neuronal decline in aged non-human primates by lentiviral NGF gene delivery. Exp. Neurol. 10902430. 215 (1), 153–159. doi:10.1016/j.expneurol.2008.10.004.

Nativio, R., Donahue, G., Berson, A., Lan, Y., Amlie-Wolf, A., Tuzer, F., Toledo, J.B., Gosai, S.J., Gregory, B.D., Torres, C., Trojanowski, J.Q., Wang, L.-S., Johnson, F.B., Bonini, N.M., Berger, S.L., 2018. Dysregulation of the epigenetic landscape of normal aging in Alzheimer's disease. Nat. Neurosci. 1097-6256. 21 (4), 497–505. doi:10.1038/s41593-018-0101-9.

Nayerossadat, N., Maedeh, T., Ali, P., 2012. Viral and nonviral delivery systems for gene delivery. Adv. Biomed. Res. 2277-9175. 1 (1), 27. doi:10.4103/2277-9175.98152.

Neuner, S.M., Heuer, S.E., Huentelman, M.J., O'Connell, K.M.S., Kaczorowski, C.C., 2019. Harnessing Genetic Complexity to Enhance Translatability of Alzheimer's Disease Mouse Models: a Path toward Precision Medicine. Neuron. 10974199. 101 (3), 399. doi:10.1016/j.neuron.2018.11.040.

Nilsson, P., Iwata, N., Muramatsu, S.I., Tjernberg, L.O., Winblad, B., Saido, T.C., 2010. Gene therapy in Alzheimer's disease - potential for disease modification. J. Cell. Mol. Med. 15821838. 14 (4), 741–757. doi:10.1111/j.1582-4934.2010.01038.x.

Nixon, R.A., Wegiel, J., Kumar, A., Yu, W.H., Peterhoff, C., Cataldo, A., Cuervo, A.M., 2005. Extensive involvement of autophagy in Alzheimer disease: an immuno-electron microscopy study. J. Neuropathol. Exp. Neurol. 00223069. 64 (2), 113–122. doi:10.1093/jnen/64.2.113.

Nixon, R.A., Yang, D.S., 2011. Autophagy failure in Alzheimer's disease-locating the primary defect. Neurobiol. Dis. 1095953X. 43 (1), 38–45. doi:10.1016/j.nbd.2011.01.021.

Nixon, R.A., Yang, D.S., 2012. Autophagy and neuronal cell death in neurological disorders. Cold Spring Harb. Perspect. Biol. 19430264. 4 (10). doi:10.1101/cshperspect.a008839.

Nugent, A.A., Lin, K., van Lengerich, B., Lianoglou, S., Przybyla, L., Davis, S.S., Llapashtica, C., Wang, J., Kim, D.J., Xia, D., Lucas, A., Baskaran, S., Haddick, P.C.G., Lenser, M., Earr, T.K., Shi, J., Dugas, J.C., Andreone, B.J., Logan, T., Solanoy, H.O., Chen, H., Srivastava, A., Poda, S.B., Sanchez, P.E., Watts, R.J., Sandmann, T., Astarita, G., Lewcock, J.W., Monroe, K.M., Di Paolo, G., 2020. TREM2 Regulates Microglial Cholesterol Metabolism upon Chronic Phagocytic Challenge. Neuron. 10974199. 105 (5), 837. doi:10.1016/j.neuron.2019.12.007.

O'Leary, T.P., Robertson, A., Chipman, P.H., Rafuse, V.F., Brown, R.E., 2018. Motor function deficits in the 12 month-old female 5xFAD mouse model of Alzheimer's disease. Behav. Brain Res. 01664328. 337, 256–263. doi:10.1016/j.bbr.2017.09.009.

Oakley, H., Cole, S.L., Logan, S., Maus, E., Shao, P., Craft, J., Guillozet-Bongaarts, A., Ohno, M., Disterhoft, J., Van Eldik, L., Berry, R., Vassar, R., 2006. Intraneuronal β-Amyloid Aggregates, Neurodegeneration, and Neuron Loss in Transgenic Mice with Five Familial Alzheimer's Disease Mutations: potential Factors in Amyloid Plaque Formation. J. Neurosci. 0270-6474. 26 (40), 10129–10140. doi:10.1523/jneurosci.1202-06.2006.

Oblak, A.L., Lin, P.B., Kotredes, K.P., Pandey, R.S., Garceau, D., Williams, H.M., Uyar, A., O'Rourke, R., O'Rourke, S., Ingraham, C., Bednarczyk, D., Belanger, M., Cope, Z.A., Little, G.J., Williams, S.P.G., Ash, C., Bleckert, A., Ragan, T., Logsdon, B.A., Mangravite, L.M., Rizzo, S.J.S., Territo, P.R., Carter, G.W., Howell, G.R., Sasner, M., Lamb, B.T., 2021. Comprehensive Evaluation of the 5XFAD Mouse Model for Preclinical Testing Applications: a MODEL-AD Study. Front. Aging Neurosci. 16634365. 13. doi:10.3389/fnagi.2021.713726.

Oddo, S., Caccamo, A., Shepherd, J.D., Murphy, M.P., Golde, T.E., Kayed, R., Metherate, R., Mattson, M.P., Akbari, Y., LaFerla, F.M., 2003. Triple-transgenic model of Alzheimer's Disease with plaques and tangles: intracellular Aβ and synaptic dysfunction. Neuron. 08966273. 39 (3), 409–421. doi:10.1016/S0896-6273(03)00434-3.

Otero, K., Shinohara, M., Zhao, H., Cella, M., Gilfillan, S., Colucci, A., Faccio, R., Ross, F.P., Teitelbaum, S.L., Takayanagi, H., Colonna, M., 2012. TREM2 and β-catenin regulate bone homeostasis by controlling the rate of osteoclastogenesis. J. Immunol. 15506606. 188 (6), 2612–2621. doi:10.4049/jimmunol.1102836.

Palasz, E., Wilkaniec, A., Stanaszek, L., Andrzejewska, A., Adamczyk, A., 2023. Glia-Neurotrophic Factor Relationships: possible Role in Pathobiology of Neuroinflammation-Related Brain Disorders. Int. J. Mol. Sci. 1422-0067. 24 (7), 6321. doi:10.3390/ijms24076321.

Patel, A.G., Nehete, P.N., Krivoshik, S.R., Pei, X., Cho, E.L., Nehete, B.P., Ramani, M.D., Shao, Y., Williams, L.E., Wisniewski, T., Scholtzova, H., 2021. Innate immunity stimulation via CpG oligodeoxynucleotides ameliorates Alzheimer's disease pathology in aged squirrel monkeys. Brain. 14602156. 144 (7), 2146–2165. doi:10.1093/brain/awab129.

Pérez-González, R., Alvira-Botero, M.X., Robayo, O., Antequera, D., Garzón, M., Martín-Moreno, A.M., Brera, B., de Ceballos, M.L., Carro, E., 2014. Leptin gene therapy attenuates neuronal damages evoked by amyloid-β and rescues memory deficits in APP/PS1 mice. Gene Ther. 0969-7128. 21 (3), 298–308. doi:10.1038/gt.2013.85.

Perez, S.E., He, B., Muhammad, N., Oh, K.J., Fahnestock, M., Ikonomovic, M.D., Mufson, E.J., 2011. Cholinotrophic basal forebrain system alterations in 3xTg-AD transgenic mice. Neurobiol. Dis. 09699961. 41 (2), 338–352. doi:10.1016/j.nbd.2010.10.002.

Petukhova, E.O., Mukhamedshina, Y.O., Salafutdinov, I.I., Garanina, E.E., Kaligin, M.S., Leushina, A.V., Rizvanov, A.A., Reis, H.J., Palotás, A., Zefirov, A.L., Mukhamedyarov, M.A., 2019. Effects of Transplanted Umbilical Cord Blood Mononuclear Cells Overexpressing GDNF on Spatial Memory and Hippocampal Synaptic Proteins in a Mouse Model of Alzheimer's Disease. J. Alzheimers Dis. 18758908. 69 (2), 443–453. doi:10.3233/JAD-190150.

Poirier, R., Wolfer, D.P., Welzl, H., Tracy, J., Galsworthy, M.J., Nitsch, R.M., Mohajeri, M.H., 2006. Neuronal neprilysin overexpression is associated with attenuation of Aβ-related spatial memory deficit. Neurobiol. Dis. 09699961. 24 (3), 475–483. doi:10.1016/j.nbd.2006.08.003.

Poliani, P.L., Wang, Y., Fontana, E., Robinette, M.L., Yamanishi, Y., Gilfillan, S., Colonna, M., 2015. TREM2 sustains microglial expansion during aging and response to demyelination. J. Clin. Invest. 15588238. 125 (5), 2161–2170. doi:10.1172/JCI77983.

Polvikoski, T., Sulkava, R., Haltia, M., Kainulainen, K., Vuorio, A., Verkkoniemi, A., Niinistö, L., Halonen, P., Kontula, K., Apolipoprotein, E., 1995. Dementia, and Cortical Deposition of β-Amyloid Protein. N. Engl. J. Med. 0028-4793. 333 (19), 1242–1248. doi:10.1056/nejm199511093331902.

Qiu, W., Folstein, M., 2006. Insulin, insulin-degrading enzyme and amyloid-β peptide in Alzheimer's disease: review and hypothesis. Neurobiol. Aging. 01974580. 27 (2), 190–198. doi:10.1016/j.neurobiolaging.2005.01.004.

Rafii, M.S., Baumann, T.L., Bakay, R.A.E., Ostrove, J.M., Siffert, J., Fleisher, A.S., Herzog, C.D., Barba, D., Pay, M., Salmon, D.P., Chu, Y., Kordower, J.H., Bishop, K., Keator, D., Potkin, S., Bartus, R.T., 2014. A phase1 study of stereotactic gene delivery of AAV2-NGF for Alzheimer's disease. Alzheimers Dement. 1552-5260. 10 (5), 571–581. doi:10.1016/j.jalz.2013.09.004.

Rafii, M.S., Tuszynski, M.H., Thomas, R.G., Barba, D., Brewer, J.B., Rissman, R.A., Siffert, J., Aisen, P.S., Mintzer, J., Lerner, A., Levey, A., Burke, J., Sano, M., Turner, S., Zamrini, E., Grill, J., Marson, D., 2018. Adeno-associated viral vector (serotype 2)-nerve growth factor for patients with Alzheimer disease a randomized clinical trial. JAMA Neurol. 21686149. 75 (7), 834–841. doi:10.1001/jamaneurol.2018.0233.

Ramsden, M., Kotilinek, L., Forster, C., Paulson, J., McGowan, E., SantaCruz, K., Guimaraes, A., Yue, M., Lewis, J., Carlson, G., Hutton, M., Ashe, K.H., 2005. Age-Dependent Neurofibrillary Tangle Formation, Neuron Loss, and Memory Impairment in a Mouse Model of Human Tauopathy (P301L). J. Neurosci. 0270-6474. 25 (46), 10637–10647. doi:10.1523/jneurosci.3279-05.2005.

Raulin, A.C., Doss, S.V., Trottier, Z.A., Ikezu, T.C., Bu, G., Liu, C.C., 2022. ApoE in Alzheimer's disease: pathophysiology and therapeutic strategies. Mol. Neurodegener. 17501326. 17 (1). doi:10.1186/s13024-022-00574-4.

Reish, N.J., Jamshidi, P., Stamm, B., Flanagan, M.E., Sugg, E., Tang, M., Donohue, K.L., Mccord, M., Krumpelman, C., Mesulam, M.M., Castellani, R., Chou, S.H.Y., 2023. Multiple Cerebral Hemorrhages in a Patient Receiving Lecanemab and Treated with t-PA for Stroke. N. Engl. J. Med. 15334406. 388 (5), 478–479. doi:10.1056/NEJMc2215148.

Richard, B.C., Kurdakova, A., Baches, S., Bayer, T.A., Weggen, S., Wirths, O., 2015. Gene dosage dependent aggravation of the neurological phenotype in the 5XFAD mouse model of Alzheimer's disease. J. Alzheimers Dis. 18758908. 45 (4), 1223–1236. doi:10.3233/JAD-143120.

Richard, K.L., Filali, M., Préfontaine, P., Rivest, S., 2008. Toll-like receptor 2 acts as a natural innate immune receptor to clear amyloid β1-42 and delay the cognitive decline in a mouse model of Alzheimer's disease. J. Neurosci. 02706474. 28 (22), 5784–5793. doi:10.1523/JNEUROSCI.1146-08.2008Canada.

Rodriguez, G.A., Tai, L.M., LaDu, M.J., Rebeck, G.W., 2014. Human APOE4 increases microglia reactivity at Aβ plaques in a mouse model of Aβ deposition. J. Neuroinflammation. 17422094. 11. doi:10.1186/1742-2094-11-111.

Rose, J.B., Crews, L., Rockenstein, E., Adame, A., Mante, M., Hersh, L.B., Gage, F.H., Spencer, B., Potkar, R., Marr, R.A., Masliah, E., 2009. Neuropeptide Y fragments derived from neprilysin processing are neuroprotective in a transgenic model of Alzheimer's disease. J. Neurosci. 02706474. 29 (4), 1115–1125. doi:10.1523/JNEUROSCI.4220-08.2009.

Rosen, W.G., Mohs, R.C., Davis, K.L., 1984. A new rating scale for Alzheimer's disease. . The American Journal of Psychiatry 141 (11), 1356–1364. doi:10.1176/ajp.141.11.1356.

Rosenberg, M.B., Friedmann, T., Robertson, R.C., Tuszynski, M.A.K., Wolff, J.A., Breakefield, X.O., Gage, F.H., 1988. Grafting genetically modified cells to the damaged brain: restorative effects of ngf expression. Science. 00368075. 242 (4885), 1575–1578. doi:10.1126/science.3201248.

Roses, A.D., 1996. Apolipoprotein E alleles as risk factors in Alzheimer's disease. Annu. Rev. Med. 00664219. 47, 387–400. doi:10.1146/annurev.med.47.1.387.

Ross, B.D., Ernst, T., Kreis, R., Haseler, L.J., Bayer, S., Danielsen, E., Blüml, S., Shonk, T., Mandigo, J.C., Caton, W., Clark, C., Jensen, S.W., Lehman, N.L., Arcinue, E., Pudenz, R., Shelden, C.H., 1998. 1H MRS in acute traumatic brain injury. J. Magn. Reson. Imaging. 10531807. 8 (4), 829–840. doi:10.1002/jmri.1880080412.

Rothman, D.L., Behar, K.L., Hyder, F., Shulman, R.G., 2003. In vivo NMR Studies of the Glutamate Neurotransmitter Flux and Neuroenergetics: implications for Brain Function. Annu. Rev. Physiol. 00664278. 65, 401–427. doi:10.1146/annurev.physiol.65.092101.142131.

Roubroeks, J.A.Y., Smith, R.G., van den Hove, D.L.A., Lunnon, K., 2017. Epigenetics and DNA methylomic profiling in Alzheimer's disease and other neurodegenerative diseases. J. Neurochem. 14714159. 143 (2), 158–170. doi:10.1111/jnc.14148.

Ryan, D.A., Mastrangelo, M.A., Narrow, W.C., Sullivan, M.A., Federoff, H.J., Bowers, W.J., 2010. AB-directed single-chain antibody delivery via a serotype-1 AAV vector improves learning behavior and pathology in Alzheimer's disease mice. Mol. Ther. 15250024. 18 (8), 1471–1481. doi:10.1038/mt.2010.111.

Sadleir, K.R., Eimer, W.A., Cole, S.L., Vassar, R., 2015. Aβ reduction in BACE1 heterozygous null 5XFAD mice is associated with transgenic APP level. Mol. Neurodegener. 17501326. 10 (1). doi:10.1186/1750-1326-10-1.

Sahara, N., Perez, P.D., Lin, W.L., Dickson, D.W., Ren, Y., Zeng, H., Lewis, J., Febo, M., 2014. Age-related decline in white matter integrity in a mouse model of tauopathy: an invivo diffusion tensor magnetic resonance imaging study. Neurobiol. Aging. 15581497. 35 (6), 1364–1374. doi:10.1016/j.neurobiolaging.2013.12.009.

Saido, T.C., 2013. Metabolism of amyloid β peptide and pathogenesis of Alzheimer's disease. Proceedings of the Japan Academy, Series B. 0386-2208. 89 (7), 321–339. doi:10.2183/pjab.89.321.

Saito, T., Matsuba, Y., Mihira, N., Takano, J., Nilsson, P., Itohara, S., Iwata, N., Saido, T.C., 2014. Single App knock-in mouse models of Alzheimer's disease. Nat. Neurosci. 1097-6256. 17 (5), 661–663. doi:10.1038/nn.3697.

Saito, T., Mihira, N., Matsuba, Y., Sasaguri, H., Hashimoto, S., Narasimhan, S., Zhang, B., Murayama, S., Higuchi, M., Lee, V.M.Y., Trojanowski, J.Q., Saido, T.C., 2019. Humanization of the entire murine Mapt gene provides a murine model of pathological human tau propagation. J. Biol. Chem. 00219258. 294 (34), 12754–12765. doi:10.1074/jbc.ra119.009487.

Sakakibara, Y., Sekiya, M., Saito, T., Saido, T.C., Iijima, K.M., 2018. Cognitive and emotional alterations in App knock-in mouse models of Aβ amyloidosis. BMC Neurosci. 14712202. 19 (1). doi:10.1186/s12868-018-0446-8.

Salek, R.M., Xia, J., Innes, A., Sweatman, B.C., Adalbert, R., Randle, S., McGowan, E., Emson, P.C., Griffin, J.L., 2010. A metabolomic study of the CRND8 transgenic mouse model of Alzheimer's disease. Neurochem. Int. 01970186. 56 (8), 937–947. doi:10.1016/j.neuint.2010.04.001.

Salomon-Zimri, S., Boehm-Cagan, A., Liraz, O., Michaelson, D.M., 2014. Hippocampus-related cognitive impairments in young apoE4 targeted replacement mice. Neurodegener. Dis. 16602862. 13 (2-3), 86–92. doi:10.1159/000354777.

Santacruz, K., Lewis, J., Spires, T., Paulson, J., Kotilinek, L., Ingelsson, M., Guimaraes, A., DeTure, M., Ramsden, M., McCowan, E., Forster, C., Yue, M., Orne, J., Janus, C., Mariash, A., Kuskowski, M., Hyman, B., Hutton, M., Ashe, K.H., 2005. Medicine: tau suppression in a neurodegenerative mouse model improves memory function. Science 309 (5733), 476–481. doi:10.1126/science.1113694.

Santana, D.A., Smith, M.D.A.C., Chen, E.S., 2023. Histone Modifications in Alzheimer's Disease. Genes. 20734425. 14 (2). doi:10.3390/genes14020347.

Saunders, A.M., 2000. Apolipoprotein E and Alzheimer disease: an update on genetic and functional analyses. J. Neuropathol. Exp. Neurol. 00223069. 59 (9), 751–758. doi:10.1093/jnen/59.9.751.

Schmitz, T.W., Nathan Spreng, R., 2016. Alzheimer's Disease Neuroimaging I. Basal forebrain degeneration precedes and predicts the cortical spread of Alzheimer's pathology. Nat. Commun. 7 (2041-1723 (Electronic)), 13249. doi:https://doi.org/10.1038/ncomms13249.

Selles, M.C., Fortuna, J.T.S., Zappa-Villar, M.F., de Faria, Y.P.R., Souza, A.S., Suemoto, C.K., Leite, R.E.P., Rodriguez, R.D., Grinberg, L.T., Reggiani, P.C., Ferreira, S.T., 2020. Adenovirus-Mediated Transduction of Insulin-Like Growth Factor 1 Protects Hippocampal Neurons from the Toxicity of Aβ Oligomers and Prevents Memory Loss in an Alzheimer Mouse Model. Mol. Neurobiol. 15591182. 57 (3), 1473–1483. doi:10.1007/s12035-019-01827-y.

Shekari, A., Fahnestock, M., 2021. Cholinergic neurodegeneration in Alzheimer disease mouse models. Handb. Clin. Neurol. 182, 191–209. doi:10.1016/B978-0-12-819973-2.00013-7.

Shi, M., Chu, F., Zhu, F., Zhu, J., 2024. Peripheral blood amyloid-β involved in the pathogenesis of Alzheimer's disease via impacting on peripheral innate immune cells. J. Neuroinflammation. 1742-2094. 21 (1). doi:10.1186/s12974-023-03003-5.

Shi, Y., Yamada, K., Liddelow, S.A., Smith, S.T., Zhao, L., Luo, W., Tsai, R.M., Spina, S., Grinberg, L.T., Rojas, J.C., Gallardo, G., Wang, K., Roh, J., Robinson, G., Finn, M.B., Jiang, H., Sullivan, P.M., Baufeld, C., Wood, M.W., Sutphen, C., McCue, L., Xiong, C., Del-Aguila, J.L., Morris, J.C.,

Cruchaga, C., Fagan, A.M., Miller, B.L., Boxer, A.L., Seeley, W.W., Butovsky, O., Barres, B.A., Paul, S.M., Holtzman, D.M., 2017. ApoE4 markedly exacerbates tau-mediated neurodegeneration in a mouse model of tauopathy. Nature. 14764687. 549 (7673), 523–527. doi:10.1038/nature24016.

Shirvan, A., Reshef, A., Yogev-Falach, M., Ziv, I., 2009. Molecular imaging of neurodegeneration by a novel cross-disease biomarker. Exp. Neurol. 00144886. 219 (1), 274–283. doi:10.1016/j.expneurol.2009.05.032.

Silva, J.C., Shen, Y., Chan, J., Kwan, P., Jones, N.C., 2022. Anti-epileptogenic effects of synaptic vesicle protein 2A modulation in a mouse model of Alzheimer's disease. Epilepsy Res. 09201211. 186, 106994. doi:10.1016/j.eplepsyres.2022.106994.

Silva, M.V.F., Loures, C.D.M.G., Alves, L.C.V., De Souza, L.C., Borges, K.B.G., Carvalho, M.D.G., 2019. Alzheimer's disease: risk factors and potentially protective measures. J. Biomed. Sci. 14230127. 26 (1). doi:10.1186/s12929-019-0524-y.

Smith, D.E., Roberts, J., Gage, F.H., Tuszynski, M.H., 1999. Age-associated neuronal atrophy occurs in the primate brain and is reversible by growth factor gene therapy. Proc. Natl. Acad. Sci. U. S. A. 0027-8424. 96 (19), 10893–10898. doi:10.1073/pnas.96.19.10893.

Spilman, P.R., Corset, V., Gorostiza, O., Poksay, K.S., Galvan, V., Zhang, J., et al., 2016. Netrin-1 Interrupts Amyloid-beta Amplification, Increases sA-betaPPalpha in vitro and in vivo, and Improves Cognition in a Mouse Model of Alzheimer's Disease. J. Alzheimers Dis. 52 (1), 223–242. doi:10.3233/JAD-151046.

Spires, T.L., Orne, J.D., SantaCruz, K., Pitstick, R., Carlson, G.A., Ashe, K.H., Hyman, B.T., 2006. Region-specific dissociation of neuronal loss and neurofibrillary pathology in a mouse model of tauopathy. Am. J. Pathol. 00029440. 168 (5), 1598–1607. doi:10.2353/ajpath.2006.050840.

Steiner, H., 2004. Uncovering gamma-secretase. Curr. Alzheimer Res. 1 (3), 175–181. doi:10.2174/1567205043332081.

Sullivan, P.M., Mace, B.E., Maeda, N., Schmechel, D.E., 2004. Marked regional differences of brain human apolipoprotein e expression in targeted replacement mice. Neuroscience. 03064522. 124 (4), 725–733. doi:10.1016/j.neuroscience.2003.10.011.

Sun, Y., Guo, Y., Feng, X., Jia, M., Ning, A., Dong, Y., Zheng, Y., Fu, L., Bin, Y., Zhang, H., Jiaxin, W., Xianghui, Y., Hui, W., Kong, W., 2020. The behavioural and neuropathologic sexual dimorphism and absence of MIP-3α in tau P301S mouse model of Alzheimer's disease. J. Neuroinflammation. 1742-2094. 17 (1). doi:10.1186/s12974-020-01749-w.

Swanson, C.J., Zhang, Y., Dhadda, S., Wang, J., Kaplow, J., Lai, R.Y.K., Lannfelt, L., Bradley, H., Rabe, M., Koyama, A., Reyderman, L., Berry, D.A., Berry, S., Gordon, R., Kramer, L.D., Cummings, J.L., 2021. A randomized, double-blind, phase 2b proof-of-concept clinical trial in early Alzheimer's disease with lecanemab, an anti-Aβ protofibril antibody. Alzheimers Res. Ther. 17589193. 13 (1). doi:10.1186/s13195-021-00813-8.

Tai, L.M., Balu, D., Avila-Munoz, E., Abdullah, L., Thomas, R., Collins, N., Valencia-Olvera, A.C., LaDu, M.J., 2017. EFAD transgenic mice as a human APOE relevant preclinical model of Alzheimer's disease. J. Lipid Res. 15397262. 58 (9), 1733–1755. doi:10.1194/jlr.R076315.

Takeuchi, A., Irizarry, M.C., Duff, K., Saido, T.C., Ashe, K.H., Hasegawa, M., Mann, D.M.A., Hyman, B.T., Iwatsubo, T., 2000. Age-related amyloid β deposition in transgenic mice overexpressing both Alzheimer mutant presenilin 1 and amyloid β precursor protein Swedish mutant is not associated with global neuronal loss. Am. J. Pathol. 00029440. 157 (1), 331–339. doi:10.1016/S0002-9440(10)64544-0.

Takeuchi, H., Iba, M., Inoue, H., Higuchi, M., Takao, K., Tsukita, K., Karatsu, Y., Iwamoto, Y., Miyakawa, T., Suhara, T., Trojanowski, J.Q., Lee, V.M.-Y., Takahashi, R., Bush, A.I., 2011. P301S Mutant Human Tau Transgenic Mice Manifest Early Symptoms of Human Tauopathies with Dementia and Altered Sensorimotor Gating. PLoS ONE. 1932-6203. 6 (6), e21050. doi:10.1371/journal.pone.0021050.

Tan, V.T.Y., Mockett, B.G., Ohline, S.M., Parfitt, K.D., Wicky, H.E., Peppercorn, K., Schoderboeck, L., Yahaya, M.F.B., Tate, W.P., Hughes, S.M., Abraham, W.C., 2018. Lentivirus-mediated expression of human secreted amyloid precursor protein-alpha prevents development of memory and plasticity deficits in a mouse model of Alzheimer's disease. Mol. Brain. 17566606. 11 (1). doi:10.1186/s13041-018-0348-9.

Tanzi, R.E., Bertram, L., 2005. Twenty years of the Alzheimer's disease amyloid hypothesis: a genetic perspective. Cell. 00928674. 120 (4), 545–555. doi:10.1016/j.cell.2005.02.008.

TCW, J., Qian, L., Pipalia, N.H., Chao, M.J., Liang, S.A., Shi, Y., Jain, B.R., Bertelsen, S.E., Kapoor, M., Marcora, E., Sikora, E., Andrews, E.J., Martini, A.C., Karch, C.M., Head, E., Holtzman, D.M., Zhang, B., Wang, M., Maxfield, F.R., Poon, W.W., Goate, A.M., 2022. Cholesterol and matrisome pathways dysregulated in astrocytes and microglia. Cell. 10974172. 185 (13), 2213. doi:10.1016/j.cell.2022.05.017.

Terwel, D., Lasrado, R., Snauwaert, J., Vandeweert, E., Van Haesendonck, C., Borghgraef, P., Van Leuven, F., 2005. Changed Conformation of Mutant Tau-P301L Underlies the Moribund Tauopathy, Absent in Progressive, Nonlethal Axonopathy of Tau-4R/2N Transgenic Mice. J. Biol. Chem. 00219258. 280 (5), 3963–3973. doi:10.1074/jbc.m409876200.

Thinakaran, G., Koo, E.H., 2008. Amyloid precursor protein trafficking, processing, and function. J. Biol. Chem. 1083351X. 283 (44), 29615–29619. doi:10.1074/jbc.R800019200.

Toljan, K., Homolak, J., 2021. Circadian changes in Alzheimer's disease: neurobiology, clinical problems, and therapeutic opportunities. Handb. Clin. Neurol. 179, 285–300. doi:10.1016/B978-0-12-819975-6.00018-2.

Turnbull, I.R., Gilfillan, S., Cella, M., Aoshi, T., Miller, M., Piccio, L., Hernandez, M., Colonna, M., 2006. Cutting edge: TREM-2 attenuates macrophage activation. J. Immunol. 00221767. 177 (6), 3520–3524. doi:10.4049/jimmunol.177.6.3520.

Tuszynski, M.H., Smith, D.E., Roberts, J., McKay, H., Mufson, E., 1998. Targeted intraparenchymal delivery of human NGF by gene transfer to the primate basal forebrain for 3 months does not accelerate β-amyloid plaque deposition. Exp. Neurol. 00144886. 154 (2), 573–582. doi:10.1006/exnr.1998.6956.

Tuszynski, M.H., Thal, L., Pay, M., Salmon, D.P., Hoi Sang, U., Bakay, R., Patel, P., Armin Blesch, H.L.V., Gilbert, H., Tong, G., Potkin, S.G., Fallon, J., Hansen, L., Mufson, E.J., Kordower, J.H., Gall, C., Conner, J., 2005. A phase 1 clinical trial of nerve growth factor gene therapy for Alzheimer disease. Nat. Med. 1078-8956. 11 (5), 551–555. doi:10.1038/nm1239.

Tuszynski, M.H., Yang, J.H., Barba, D., Hoi-Sang, U., Bakay, R.A.E., Pay, M.M., Masliah, E., Conner, J.M., Kobalka, P., Roy, S., Nagahara, A.H., 2015. Nerve growth factor gene therapy activation of neuronal responses in Alzheimer disease. JAMA Neurol. 21686157. 72 (10), 1139–1147. doi:10.1001/jamaneurol.2015.1807.

Ulland, T.K., Song, W.M., Huang, S.C.C., Ulrich, J.D., Sergushichev, A., Beatty, W.L., Loboda, A.A., Zhou, Y., Cairns, N.J., Kambal, A., Loginicheva, E., Gilfillan, S., Cella, M., Virgin, H.W., Unanue, E.R., Wang, Y., Artyomov, M.N., Holtzman, D.M., Colonna, M., 2017. TREM2 Maintains Microglial Metabolic Fitness in Alzheimer's Disease. Cell. 10974172. 170 (4), 649. doi:10.1016/j.cell.2017.07.023.

van der Spek, S.J.F., Gonzalez-Lozano, M.A., Koopmans, F., Miedema, S.S.M., Paliukhovich, I., Smit, A.B., Li, K.W., 2021. Age-dependent hippocampal proteomics in the app/ps1 Alzheimer mouse model: a comparative analysis with classical swath/dia and directdia approaches. Cells. 20734409. 10 (7). doi:10.3390/cells10071588.

Volianskis, A., Køstner, R., Mølgaard, M., Hass, S., Jensen, M.S., 2010. Episodic memory deficits are not related to altered glutamatergic synaptic transmission and plasticity in the CA1 hippocampus of the APPswe/PS1ΔE9-deleted transgenic mice model of β-amyloidosis. Neurobiol. Aging. 01974580. 31 (7), 1173–1187. doi:10.1016/j.neurobiolaging.2008.08.005.

Wakabayashi, T., De Strooper, B., 2008. Presenilins: members of the γ-Secretase Quartets, But Part-Time Soloists Too. Physiology. 1548-9213. 23 (4), 194–204. doi:10.1152/physiol.00009.2008.

Wan, H.L., Hong, X.Y., Zhao, Z.H., Li, T., Zhang, B.G., Liu, Q., Wang, Q., Zhao, S., Wang, J.Z., Shen, X.F., Liu, G.P., 2021. STAT3 ameliorates cognitive deficits via regulation of NMDAR expression in an Alzheimer's disease animal model. Theranostics. 18387640. 11 (11), 5511–5524. doi:10.7150/thno.56541.

Wang, F., Wei, X.X., Chang, L.S., Dong, L., Wang, Y.L., Li, N.N., 2021. Ultrasound Combined With Microbubbles Loading BDNF Retrovirus to Open Blood–Brain Barrier for Treatment of Alzheimer's Disease. Front. Pharmacol. 16639812. 12. doi:10.3389/fphar.2021.615104.

Wang, F., Wei, X.X., Chang, L.S., Dong, L., Wang, Y.L., Li, N.N., 2021. Ultrasound Combined With Microbubbles Loading BDNF Retrovirus to Open BloodBrain Barrier for Treatment of Alzheimer's Disease. Front Pharmacol. 5 (12), 615104. doi:10.3389/fphar.2021.615104.

Wang, P., Yang, P., Qian, K., Li, Y., Shuting, X., Meng, R., Guo, Q., Cheng, Y., Cao, J., Minjun, X., Wei, L., Zhang, Q., 2022. Precise gene delivery systems with detachable albumin shell remodeling dysfunctional microglia by TREM2 for treatment of Alzheimer's disease. Biomaterials. 01429612. 281, 121360. doi:10.1016/j.biomaterials.2021.121360.

Wang, Q.H., Wang, Y.R., Zhang, T., Jiao, S.S., Liu, Y.H., Zeng, F., Li, J., Yao, X.Q., Zhou, H.D., Zhou, X.F., Wang, Y.J., 2016. Intramuscular delivery of p75NTR ectodomain by an AAV vector attenuates cognitive deficits and Alzheimer's disease-like pathologies in APP/PS1 transgenic mice. J. Neurochem. 14714159. 163–173. doi:10.1111/jnc.13616.

Wang, Y., Cella, M., Mallinson, K., Ulrich, J.D., Young, K.L., Robinette, M.L., Gilfillan, S., Krishnan, G.M., Sudhakar, S., Zinselmeyer, B.H., Holtzman, D.M., Cirrito, J.R., Colonna, M., 2015. TREM2 lipid sensing sustains the microglial response in an Alzheimer's disease model. Cell. 10974172. 160 (6), 1061–1071. doi:10.1016/j.cell.2015.01.049.

Wang, Y.J., Gao, C.Y., Yang, M., Liu, X.H., Sun, Y., Pollard, A., Dong, X.Y., Wu, X.B., Zhong, J.H., Zhou, H.D., Zhou, X.F., 2010. Intramuscular delivery of a single chain antibody gene prevents brain Aβ deposition and cognitive impairment in a mouse model of Alzheimer's disease. Brain, Behavior, and Immunity. 08891591. 24 (8), 1281–1293. doi:10.1016/j.bbi.2010.05.010.

Watamura, N., Sato, K., Shiihashi, G., Iwasaki, A., Kamano, N., Takahashi, M., Sekiguchi, M., Mihira, N., Fujioka, R., Nagata, K., Hashimoto, S., Saito, T., Ohshima, T., Saido, T.C., Sasaguri, H., 2022. An isogenic panel of App knock-in mouse models: profiling β-secretase inhibition and endosomal abnormalities. Sci. Adv. 2375-2548. 8 (23). doi:10.1126/sciadv.abm6155.

Weeraratna, A.T., Kalehua, A., DeLeon, I., Bertak, D., Maher, G., Wade, M.S., Lustig, A., Becker, K.G., Wood, W., Walker, D.G., Beach, T.G., Taub, D.D., 2007. Alterations in immunological and neurological gene expression patterns in Alzheimer's disease tissues. Exp. Cell Res. 00144827. 313 (3), 450–461. doi:10.1016/j.yexcr.2006.10.028.

Wen, X., Wang, L., Liu, Z., Liu, Y., Hu, J., 2014. Intracranial injection of PEG-PEI/ROCK II-siRNA improves cognitive impairment in a mouse model of Alzheimer's disease. Int. J. Neurosci. 15635279. 124 (9), 697–703. doi:10.3109/00207454.2013.877014.

Westmark, C.J., Westmark, P.R., Malter, J.S., 2010. Alzheimer's disease and down syndrome rodent models exhibit audiogenic seizures. J. Alzheimers Dis. 13872877. 20 (4), 1009–1013. doi:10.3233/JAD-2010-100087.

Whitehouse, P.J., Price, D.L., Struble, R.G., Clark, A.W., Coyle, J.T., Mahlon, R.D., 1982. Alzheimer's Disease and Senile Dementia: loss of Neurons in the Basal Forebrain. Science. 0036-8075. 215 (4537), 1237–1239. doi:10.1126/science.7058341.

Wisor, J.P., Edgar, D.M., Yesavage, J., Ryan, H.S., McCormick, C.M., Lapustea, N., Murphy, G.M., 2005. Sleep and circadian abnormalities in a transgenic mouse model of Alzheimer's disease: a role for cholinergic transmission. Neuroscience. 03064522. 131 (2), 375–385. doi:10.1016/j.neuroscience.2004.11.018.

Woodhouse, A., Vickers, J.C., Adlard, P.A., Dickson, T.C., 2009. Dystrophic neurites in TgCRND8 and Tg2576 mice mimic human pathological brain aging. Neurobiol. Aging. 01974580. 30 (6), 864–874. doi:10.1016/j.neurobiolaging.2007.09.003.

Wu, K., Meyers, C.A., Guerra, N.K., King, M.A., Meyer, E.M., 2004. The effects of rAAV2-mediated NGF gene delivery in adult and aged rats. Mol. Ther. 15250016. 9 (2), 262–269. doi:10.1016/j.ymthe.2003.11.010.

Xia, D., Lianoglou, S., Sandmann, T., Calvert, M., Suh, J.H., Thomsen, E., Dugas, J., Pizzo, M.E., DeVos, S.L., Earr, T.K., Lin, C.C., Davis, S., Ha, C., Leung, A.W.S., Nguyen, H., Chau, R., Yulyaningsih, E., Lopez, I., Solanoy, H., Masoud, S.T., Liang, C.C., Lin, K., Astarita, G., Khoury, N., Zuchero, J.Y., Thorne, R.G., Shen, K., Miller, S., Palop, J.J., Garceau, D., Sasner, M., Whitesell, J.D., Harris, J.A., Hummel, S., Gnörich, J., Wind, K., Kunze, L., Zatcepin, A., Brendel, M., Willem, M., Haass, C., Barnett, D., Zimmer, T.S., Orr, A.G., Scearce-Levie, K., Lewcock, J.W., Di Paolo, G., Sanchez, P.E., 2022. Novel App knock-in mouse model shows key features of amyloid pathology and reveals profound metabolic dysregulation of microglia. Mol. Neurodegener. 17501326. 17 (1). doi:10.1186/s13024-022-00547-7.

Xiao, N.A., Zhang, J., Zhou, M., Wei, Z., Wu, X.L., Dai, X.M., Zhu, Y.G., Chen, X.C., 2015. Reduction of glucose metabolism in olfactory bulb is an earlier Alzheimer's disease-related biomarker in 5XFAD mice. Chin. Med. J. 25425641. 128 (16), 2220–2227. doi:10.4103/0366-6999.162507.

Yamazaki, Y., Zhao, N., Caulfield, T.R., Liu, C.C., Bu, G., 2019. Apolipoprotein E and Alzheimer disease: pathobiology and targeting strategies. Nat. Rev. Neurol. 17594766. 15 (9), 501–518. doi:10.1038/s41582-019-0228-7.

Yan, Q., Rosenfeld, R.D., Matheson, C.R., Hawkins, N., Lopez, O.T., Bennett, L., Welcher, A.A., 1997. Expression of brain-derived neurotrophic factor protein in the adult rat central nervous system. Neuroscience. 03064522. 78 (2), 431–448. doi:10.1016/s0306-4522(96)00613-6.

Yokoyama, A.S., Rutledge, J.C., Medici, V., 2017. DNA methylation alterations in Alzheimer's disease. Environ. Epigenet. 20585888. 3 (2). doi:10.1093/eep/dvx008.

Yu, L., Chibnik, L.B., Srivastava, G.P., Pochet, N., Yang, J., Jishu, X., Kozubek, J., Obholzer, N., Leurgans, S.E., Schneider, J.A., Meissner, A., De Jager, P.L., Bennett, D.A., 2015. Association of Brain DNA Methylation in SORL1, ABCA7, HLA-DRB5, SLC24A4, and BIN1 with Pathological Diagnosis of Alzheimer Disease . JAMA Neurology. 2168-6149. 72 (1), 15–24. doi:10.1001/jamaneurol.2014.3049.

Yue, M., Hanna, A., Wilson, J., Roder, H., Janus, C., 2011. Sex difference in pathology and memory decline in rTg4510 mouse model of tauopathy. Neurobiol. Aging. 01974580. 32 (4), 590–603. doi:10.1016/j.neurobiolaging.2009.04.006.

Zeng, C.Y., Yang, T.T., Zhou, H.J., Zhao, Y., Kuang, X., Duan, W., Du, J.R., 2019. Lentiviral vector–mediated overexpression of Klotho in the brain improves Alzheimer's disease–like pathology and cognitive deficits in mice. Neurobiol. Aging. 15581497. 78, 18–28. doi:10.1016/j.neurobiolaging.2019.02.003.

Zhang, B., Carroll, J., Trojanowski, J.Q., Yao, Y., Iba, M., Potuzak, J.S., Hogan, A.M.L., Xie, S.X., Ballatore, C., Smith, A.B., Lee, V.M.Y., Brunden, K.R., 2012. The microtubule-stabilizing agent, epothilone D, reduces axonal dysfunction, neurotoxicity, cognitive deficits, and Alzheimer-like pathology in an interventional study with aged tau transgenic mice. J. Neurosci. 15292401. 32 (11), 3601–3611. doi:10.1523/JNEUROSCI.4922-11.2012.

Zhang, H., Wei, W., Zhao, M., Ma, L., Jiang, X., Pei, H., Cao, Y., Hao, L., 2021. Interaction between Aβ and Tau in the Pathogenesis of Alzheimer's Disease. Int. J. Biol. Sci. 1449-2288. 17 (9), 2181–2192. doi:10.7150/ijbs.57078.

Zhang, Yu, Lai, Li, Sun, 2013. Effects of small interfering RNA targeting sphingosine kinase-1 gene on the animal model of Alzheimer's disease. J. Huazhong Univ. Sci. Technol. Med. Sci. 33 (3), 427–432. doi:10.1007/s11596-013-1136-5.

Zhao, L., Gottesdiener, A.J., Parmar, M., Li, M., Kaminsky, S.M., Chiuchiolo, M.J., Sondhi, D., Sullivan, P.M., Holtzman, D.M., Crystal, R.G., Paul, S.M., 2016. Intracerebral adeno-associated virus gene delivery of apolipoproteinE2markedly reduces brain amyloid pathology in Alzheimer's disease mouse models. Neurobiol. Aging. 15581497. 44, 159–172. doi:10.1016/j.neurobiolaging.2016.04.020.

Zheng, J.Y., Sun, J., Ji, C.M., Shen, L., Chen, Z.J., Xie, P., Sun, Y.Z., Yu, R.T., 2017. Selective deletion of apolipoprotein E in astrocytes ameliorates the spatial learning and memory deficits in Alzheimer's disease (APP/PS1) mice by inhibiting TGF-β/Smad2/STAT3 signaling. Neurobiol. Aging. 15581497. 54, 112–132. doi:10.1016/j.neurobiolaging.2017.03.002.

Zhong, M.Z., Peng, T., Duarte, M.L., Wang, M., Cai, D., 2024. Updates on mouse models of Alzheimer's disease. Mol. Neurodegener. 17501326. 19 (1). doi:10.1186/s13024-024-00712-0.

Zhou, Y., Lai, C.S.W., Bai, Y., Li, W., Zhao, R., Yang, G., Frank, M.G., Gan, W.B., 2020. REM sleep promotes experience-dependent dendritic spine elimination in the mouse cortex. Nat. Commun. 20411723. 11 (1). doi:10.1038/s41467-020-18592-5.

Zhu, L., Zhong, M., Elder, G.A., Sano, M., Holtzman, D.M., Gandy, S., Cardozo, C., Haroutunian, V., Robakis, N.K., Cai, D., 2015. Phospholipid dysregulation contributes to apoe4-associated cognitive deficits in Alzheimer's disease pathogenesis. Proc. Natl. Acad. Sci. U. S. A. 10916490. 112 (38), 11965–11970. doi:10.1073/pnas.1510011112.

Chapter 11

Gene therapy in Parkinson's disease (PD)

Samaneh Toutounchian and Iman Kiani
Tehran University of Medical Sciences, Medicine Faculty, Tehran, Iran

1 Introduction

Parkinson's disease (PD) is a progressive neurodegenerative disorder that presents with the selective loss of dopaminergic neurons in the substantia nigra pars compacta (SNpc) and the subsequent depletion of striatal dopamine (DA) (Bloem et al., 2021). Clinically, PD manifests with cardinal motor features, bradykinesia, resting tremor, rigidity, and postural instability, as well as a broad range of nonmotor symptoms, including cognitive decline, autonomic dysfunction, mood disorders, and sleep disturbances (Sveinbjornsdottir, 2016). Although current pharmacologic and surgical approaches, such as levodopa therapy and deep brain stimulation (DBS), provide symptomatic benefits, none have demonstrated the capacity to modify the underlying disease process or stop neurodegeneration (Lee and Yankee, 2022). Over time, many patients experience treatment-related complications, diminishing efficacy, and fluctuating responsiveness, which highlights the need for novel treatment strategies (Axelsen and Woldbye, 2018). PD currently affects >8–9 million individuals worldwide, and its prevalence is projected to more than double by 2050 as populations age in both high-income and low- to middle-income countries (Rocca, 2018; Dongning et al., 2025).

Gene therapy is a promising frontier in the management of PD, which has shown the potential to restore dopaminergic function, provide neuroprotection, and address molecular mechanisms underlying pathogenesis (Uwishema et al., 2022). Advances in vector design, delivery techniques, and target selection have transformed the field from experimental proof-of-concept studies into early-phase clinical trials with encouraging safety and efficacy profiles (Winston et al., 2025). Contemporary gene-based approaches in PD generally fall into three interrelated categories: enzyme replacement strategies to enhance DA synthesis, modulation of dysfunctional neural circuits through inhibitory or excitatory gene expression, and neuroprotective/neurorestorative interventions aimed at promoting neuronal survival and limiting disease progression (Szunyogh et al., 2025).

As gene therapy continues to evolve, it occupies a unique intersection between neurology, molecular medicine, and neurosurgical innovation. Long-term efficacy, optimal vector dosing, delivery accuracy, immunogenicity, and ethical considerations remain active areas of investigation. Nevertheless, the trajectory of research to date suggests that gene-based interventions may eventually complement or even surpass existing treatments.

This chapter provides an overview of the scientific rationale, vector technologies, molecular targets, clinical trial landscape, and future directions of gene therapy in PD. We aim to evaluate preclinical discoveries and trial data to demonstrate the current state of the field.

2 Pathogenesis

Understanding PD's pathogenesis is crucial for developing targeted gene therapies that address the root causes of neuronal death instead of just the symptoms. PD results from the interaction of circuit dysfunction in the nigrostriatal system, α-synuclein-driven proteinopathy, genetic susceptibilities (e.g., *SNCA, LRRK2, GBA1, PRKN, PINK1*), and downstream failures in mitochondria, proteostasis, and immune homeostasis. These processes combine to kill dopaminergic neurons in the SNpc and deprive the dorsal striatum of DA, leading to the motor symptoms. Mapping these failure points helps identify actionable nodes for gene therapy.

Gene Therapy in Neurodegenerative Disorders. https://doi.org/10.1016/B978-0-443-40467-2.00004-X

2.1 The nigrostriatal system: The epicenter of motor decline

The nigrostriatal pathway links DA-producing neurons in the SNpc to the dorsal striatum (putamen and caudate). Striatal medium spiny neurons (MSNs) segregate into D1 receptor-expressing cells of the direct pathway, which facilitate movement, and D2 receptor-expressing cells of the indirect pathway, which constrain it. Tonic and phasic DA signaling keeps these pathways in balance and enables the initiation and scaling of voluntary actions (Kordower et al., 2013).

DA biogenesis in nigrostriatal neurons occurs via a tightly linked synthesis–packaging–recycling process. In the cytosol, tyrosine is converted to L-DOPA by tyrosine hydroxylase (TH), the rate-limiting step. Then, L-DOPA is decarboxylated by aromatic L-amino acid decarboxylase (AADC/DDC) to DA. TH requires tetrahydrobiopterin (BH4), which is produced de novo under the control of GTP cyclohydrolase I (GCH1) (Fig. 11.1) (Christenson et al., 1970; Fanet et al., 2021). Newly synthesized DA needs to be stored in synaptic vesicles by VMAT2, a proton-gradient-driven antiporter (Eiden and Weihe, 2011). Vesicle fusion at dopaminergic release sites is triggered by Ca^{2+} and coordinated by active-zone scaffolds

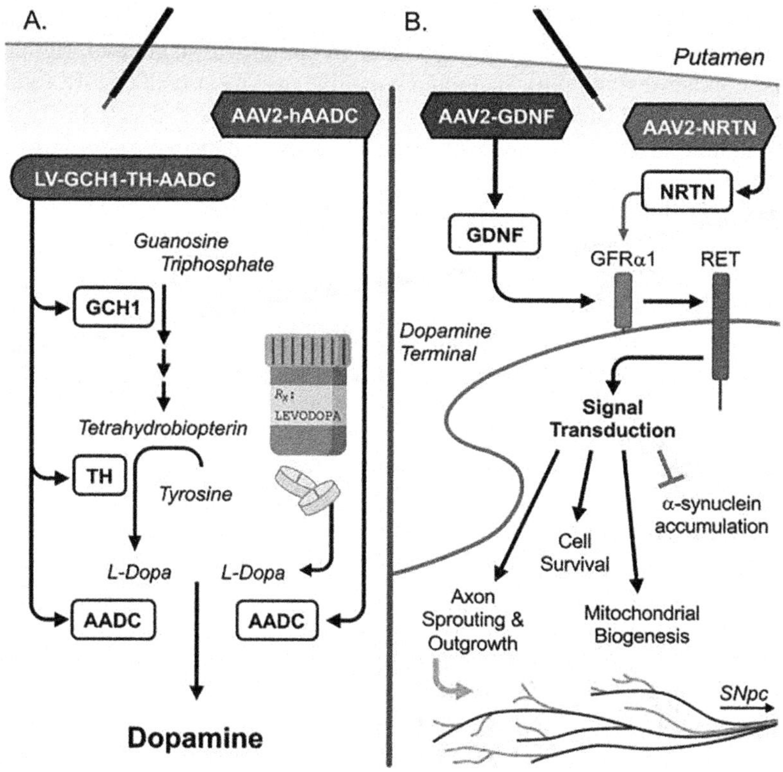

FIG. 11.1 Dopamine synthesis and methods of action for current gene therapies. (A) Enhancement of dopamine production. LV-GCH1-TH-AADC transduction of putaminal neurons restores key enzymes of the DA production pathway, leading to increased production of the TH cofactor tetrahydrobiopterin (via GCH1), increased production of levodopa from tyrosine (via TH), and enhanced conversion of levodopa (L-DOPA) to readily available DA (via AADC). AAV2-hAADC transduction of putaminal neurons leads to the increased local production of AADC to enhance the conversion of L-DOPA to readily available DA. Both therapies durably enhance the amount and consistent production of DA, from both endogenously produced and medication-derived L-DOPA, within the putamen with the goal of reducing "Off" time symptoms. (B) Restoration of neurotrophic signaling. Transduction of putaminal neurons by AAV2-GDNF or AAV2-NRTN leads to increased expression of glial cell line-derived neurotrophic factor (GDNF) and neurturin (NRTN), respectively, both of which are decreased in PD brains. These neurotrophic factors exert their effects by binding to GDNF family receptor α (GFRα) members on the surface of the DA neuron terminals. GDNF has a high affinity for GFRα1, which is highly expressed on DA neurons. NRTN can also bind to GFRα1, though with a lower affinity. The receptor/ligand complex attracts and activates the transmembrane receptor RET, a receptor tyrosine kinase, triggering a cell survival signaling cascade within the DA neurons. Evidence from animal models of PD has shown that enhanced neurotrophic factor expression in the striatum can protect against nigrostriatal DA neuron loss, reduce α-synuclein accumulation in DA neurons, improve mitochondrial biogenesis and function, and encourage sprouting and growth of DA axons. *Based on Van Laar, A.D., Van Laar, V.S., Sebastian, W.S., Merola, A., Elder, J.B., Lonser, R.R., Bankiewicz, K.S., Björklund, A., Bloem, B.R., Brundin, P., Federoff, H., 2021. An update on gene therapy approaches for Parkinson's disease: restoration of dopaminergic function. J. Parkinsons Dis. 18777171. 11 (s2). https://doi.org/10.3233/JPD-212724*

in the presynaptic terminal (e.g., RIM/Munc13). After release, DA is primarily cleared by DA transporter (DAT)-mediated reuptake into terminals, with additional inactivation through monoamine oxidase-B (MAO-B) and catechol-O-methyltransferase (COMT) pathways (Silm et al., 2019; Banerjee et al., 2022). Furthermore, cytosolic DA can form o-quinones and aminochrome, and MAO generates the toxic aldehyde 3,4-Dihydroxyphenylacetaldehyde (DOPAL), which is detoxified by aldehyde dehydrogenase and aldo-keto reductases. Moreover, glia convert extracellular DA to homovanillic acid (Inyushin et al., 2012).

Neuropathologically, PD features progressive loss of SNpc neurons and striatal DA depletion. Once signaling falls below a critical threshold, bradykinesia, rigidity, and rest tremor emerge. Reduced striatal DA weakens D1-MSNs and disinhibits D2-MSNs, driving excessive basal ganglia output that suppresses thalamocortical motor drive (Kordower et al., 2013). Subthalamic nucleus (STN) hyperactivity further elevates globus pallidus internus (GPi) and substantia nigra pars reticulata (SNpr) firing, reinforcing a GABAergic brake on the thalamus and producing the core motor syndrome (Galvan et al., 2015). Therefore restoring DA synthesis or signaling in the putamen is anatomically and physiologically sound.

2.2 Molecular hallmarks: Pathological genes and proteins

While the degeneration of the nigrostriatal pathway explains the primary motor symptoms of PD, modern research has identified several core pathological processes, centered on protein misfolding and genetic mutations. These molecular hallmarks not only offer crucial insights into why dopaminergic neurons die but also identify druggable pathways for gene knockdown, augmentation, or enzyme regulation replacement.

2.2.1 *Alpha-synuclein aggregation*

The central pathology of PD is the misfolding and aggregation of the protein alpha-synuclein (α-synuclein), encoded by the SNCA gene. In its native state, α-synuclein is a largely unfolded protein, though it can form aggregation-resistant tetramers (Bartels et al., 2011). However, in PD, it undergoes a conformational shift to a β-sheet-rich structure that is highly prone to aggregation. This pathological transformation is heavily promoted by cellular stressors and biochemical changes, most notably posttranslational modifications like the Ser129 phosphorylation (pS129) (Barrett and Greenamyre, 2015).

These misfolded proteins form a toxic hierarchy of species, ranging from soluble oligomers and protofibrils to large, insoluble filaments that make up the core of Lewy bodies (LBs) and can spread through a prion-like process, transferring the pathology from cell to cell and throughout the nervous system (Danzer et al., 2009). A key shift from previous views is that the most neurotoxic forms are the early, soluble oligomeric versions of α-synuclein, rather than the mature, insoluble fibrils found in LBs (Karpinar et al., 2009).

2.2.2 *The genetic blueprint*

The discovery of genes responsible for rare, monogenic forms of PD is essential for understanding the molecular pathways of the disease. These genetic findings create a direct link between specific molecular dysfunctions and neurodegeneration, offering valuable targets for gene therapy.

LRRK2 mutations

Mutations in the leucine-rich repeat kinase 2 (LRRK2) gene are the most common cause of autosomal dominant PD, with p.G2019S the most prevalent worldwide (Baba et al., 1998). Such mutations result in a hyperactive LRRK2 kinase enzyme, disrupting various cellular processes, including vesicular trafficking, cytoskeletal organization, autophagy, and inflammatory signaling, ultimately increasing neuronal vulnerability and causing cell death. Notably, LRRK2 can inhibit chaperone-mediated autophagy (CMA), which reduces α-synuclein degradation and increases its toxic buildup (Pan et al., 2017).

GBA mutations

Mutations in GBA1 are among the strongest genetic risk factors for PD (Billingsley et al., 2018). While inheriting two mutated copies of this gene causes the severe lysosomal storage disorder known as Gaucher disease, it is now shown that a heterozygous state is enough to create a significant risk for PD (Hruska et al., 2008). The GBA1 gene encodes glucocerebrosidase (GCase), a crucial enzyme responsible for breaking down the glycolipid glucosylceramide within the lysosome. GBA1 mutations lead to a reduction in GCase enzymatic activity, impairing the function of the entire autophagy-lysosome pathway (Magalhaes et al., 2016). Heterozygous variants reduce GCase activity, impair lysosomal flux, and raise α-syn burden. α-syn also impairs GCase trafficking, forming a bidirectional loop (Magalhaes et al., 2016).

Parkin and PINK1 mutations

Mutations in the Parkin RBR E3 Ubiquitin Protein Ligase (PARK) gene, which encodes the protein Parkin, and the PTEN induced kinase 1 (PINK1) gene, are common causes of autosomal recessive, early-onset PD (Valente et al., 2001).

PINK1 acts as a sensor that accumulates on the surface of damaged mitochondria, activating Parkin, which then tags the damaged organelle with ubiquitin, marking it for degradation by the autophagy system, a process named mitophagy. Loss-of-function mutations in these genes prevent mitophagy of depolarized mitochondria, causing reactive oxygen species (ROS) and bioenergetic failure, and ultimately neuronal death (Billingsley et al., 2018).

2.3 Downstream cellular consequences

The pathogenesis of PD is not limited to the accumulation of α-synuclein aggregates; however, it is also critically linked to downstream cellular disruptions, which together amplify the neurodegenerative process.

2.3.1 *Mitochondrial dysfunction*

A breakdown in mitochondrial function is a central and recurring theme in both idiopathic and familial PD. Specifically, a marked deficiency in the activity of mitochondrial complex I (CI) is found in PD, which is a critical enzyme in the electron transport chain responsible for ATP production (Parker et al., 2008).

The 1-methyl-4-phenyl-1,2,3,6-tetrahydropyridine (MPTP) model supports causality. MPTP is converted to MPP^+, which is transported into dopaminergic neurons via the DAT and inhibits CI, reproducing selective nigrostriatal degeneration in vivo (Meredith and Rademacher, 2011).

Moreover, PINK1 and Parkin preserve mitochondrial quality through mitophagy; α-synuclein oligomers target mitochondria and bind to TOM20, disrupting protein import and reducing CI-dependent respiration, thereby increasing ROS (Luth et al., 2014). These interactions place mitochondria as a point where genetic, environmental, and proteopathic stressors converge.

2.3.2 *Failure of proteostasis*

Neuronal survival relies on proteostasis maintained by the ubiquitin–proteasome system (UPS) and the autophagy–lysosome pathway (ALP), including macroautophagy and CMA (Hou et al., 2020). Genetics further implicates UPS components through PARK2 (Parkin, an E3 ligase) and PARK5 (UCH-L1) (Nishikawa et al., 2003).

By contrast, ALP function shows converging deficits. Nigral neurons accumulate autophagic vacuoles with elevated LC3-II and display reduced lysosomal membrane proteins, LAMP1 and LAMP2A. Mutations in GBA1 directly impair lysosomes, and defects in PINK1 and Parkin disrupt mitophagy (Alvarez-Erviti et al., 2010). Together, these failures promote α-synuclein accumulation and damaged organelles, creating a self-reinforcing cycle of cellular stress.

2.3.3 *Neuroinflammation*

Concurrently, neuroinflammation actively drives disease progression. PD brains exhibit microglial activation, T-cell infiltration, and increased proinflammatory cytokines (Hirsch and Hunot, 2009). Pathologic α-synuclein strongly activates microglia, which further amplifies damage (Xiaomin et al., 2008). Therefore impaired proteostasis and chronic inflammation together accelerate nigrostriatal degeneration.

2.4 The missing support: Loss of neurotrophic factors

The survival of dopaminergic neurons also depends on a supportive extracellular environment rich in neurotrophic factors. Dopaminergic neurons are particularly dependent on specific neurotrophic factors for their long-term survival and function, such as glial cell line-derived neurotrophic factor (GDNF) and neurturin (NRTN). In PD, these pathways are diminished, and α-syn pathology may further weaken TrkB signaling.

GDNF-family ligands bind GFRα coreceptors on dopaminergic terminals. The ligand–GFRα complex activates the RET tyrosine kinase and downstream survival pathways. GDNF shows high affinity for GFRα1. Similarly, NRTN interacts with the same axis but with lower affinity (Kramer and Liss, 2015).

Furthermore, brain-derived neurotrophic factor (BDNF), which promotes neuronal health through its interaction with the TrkB receptor, is disrupted in PD. Pathological α-synuclein directly impairs BDNF/TrkB signaling, effectively blocking a crucial prosurvival pathway and contributing to dopaminergic cell death (Fang et al., 2017; Kang et al., 2017). This concept of "missing support" provides a clear and compelling reason for many gene therapies focused on neuroprotection.

Taken together, SNpc neurons feature very extensive axonal trees, high energy demands, accumulation of iron and neuromelanin, and high DAT expression, all factors that increase oxidative and proteostatic stress. These traits help explain why these neurons are the first to die and guide vector targeting to the putamen and midbrain.

3 Current therapies

Levodopa remains the cornerstone of symptomatic management in PD. As a DA precursor administered with a peripheral decarboxylase inhibitor (carbidopa or benserazide), it restores striatal dopaminergic tone and provides the greatest improvements in bradykinesia and rigidity. It is delivered via immediate-, controlled-, or extended-release formulations (Hadjiconstantinou and Neff, 2008). Gastrointestinal kinetics is influenced by dietary protein and fat, which can reduce bioavailability (Rusch et al., 2023). With advancing disease, the short pharmacokinetic half-life and variable absorption contribute to wearing-off, ON–OFF fluctuations, and dyskinesia despite peripheral decarboxylase inhibition (Cenci, 2014).

Adjunct pharmacotherapies extend benefits and mitigate fluctuations. COMT inhibitors (entacapone, opicapone, tolcapone) prolong levodopa availability and reduce end-of-dose OFF (Müller, 2010). Monoamine oxidase-B inhibitors (selegiline, rasagiline) stabilize synaptic DA (Nayak and Henchcliffe, 2008), while safinamide additionally modulates glutamatergic transmission and increases ON time without troublesome dyskinesia (Borgohain et al., 2009). Nonergoline DA agonists (pramipexole, ropinirole, rotigotine) are useful as early monotherapy or fluctuation "smoothers," though they are less potent than levodopa and are limited by somnolence, edema, hallucinations, and impulse control disorders (Jing et al., 2023). Amantadine, anticholinergics, and adenosine A2A antagonists (e.g., istradefylline) may also alleviate symptoms (Tarsy, 2006). Detailed data for each category are reviewed in Table 11.1.

When oral regimens plateau, device-aided continuous dopaminergic (Table 11.1) delivery and circuit-level interventions (e.g., DBS or selective lesioning) stabilize motor control. Among device-aided strategies, intrajejunal levodopa–carbidopa gel (LCIG/LECIG) (Olanow, et al., 2014) and continuous subcutaneous options (apomorphine infusion or 24-hour foslevodopa/foscarbidopa) (Katzenschlager et al., 2018) reduce OFF time and extend good ON time. When fluctuations or dyskinesia persist, circuit-level interventions are considered. DBS of the STN or GPi improves motor outcomes and quality of life, with target selection individualized; yet DBS is not disease-modifying, and axial or cognitive features may remain (Weaver et al., 2012; Petersen et al., 2024). For patients unsuitable for neuromodulation, MRI-guided focused ultrasound offers incisionless lesioning but is unilateral and irreversible (Krishna et al., 2023; Martínez-Fernández et al., 2020). Overall, these modalities improve symptoms but do not alter disease biology.

Nonmotor symptoms in PD require personalized care. Depression and anxiety often respond to selective serotonin reuptake inhibitors or serotonin–norepinephrine reuptake inhibitors; apathy benefits from behavioral support (Richard et al., 2012). For psychosis, reducing dopaminergic medications and considering pimavanserin or clozapine is useful (Cummings et al., 2014; Weiner, 2009). Rivastigmine can improve cognition (Emre, et al., 2004). Sleep issues, including rapid eye movement sleep behavior disorder and excessive daytime sleepiness, reflect nocturnal hypodopaminergia and can be controlled by controlled-release agents, along with melatonin or clonazepam (Lauterbach, 2004). These nonmotor burdens often progress independently of motor symptoms; treatment must be adjusted individually over time.

Despite advances, current PD care still has several ongoing gaps. First, according to pharmacokinetics, synaptic DA remains oscillatory, leading to nonphysiologic receptor activation and dyskinesia. Second, precise anatomical targeting is crucial. Systemic drugs are widespread, and neuromodulation does not regenerate the degenerating putaminal terminal field. Third, there is a mismatch in pathobiology. No approved therapy reduces α-synuclein, normalizes lysosomal flux, rescues mitochondrial quality control, or restores trophic support. As a result, symptoms often worsen. These gaps drive the development of targeted gene therapies, which the next section explains.

TABLE 11.1 Current therapies in PD.

Category	Class	Core mechanism	Representative agents/procedure	Typical dosing/ programming	Clinical use and strengths	Common adverse effects/key risks	Special notes	Ref (#)
Oral dopaminergic	**Levodopa + peripheral AADC inhibitor**	DA precursor	Levodopa/carbidopa; levodopa/ benserazide	Typical total 300–1200 mg/day in 3–12 divided doses; space from protein; consider ER at night	Greatest effect on bradykinesia and rigidity; QoL benefit; rapid onset in early PD	Nausea, orthostasis; motor fluctuations, LID; hallucinations	Dose with protein separation; fractionate when wearing-off emerges; avoid abrupt withdrawal (NMS-like)	Dhall and Kreitzman (2016)
Adjunct to levodopa	**COMT inhibitors**	Inhibit peripheral levodopa O-methylation	Entacapone, tolcapone, opicapone	Entacapone 200 mg with each levodopa dose; tolcapone 100–200 mg TID; opicapone 50 mg qHS	Decreases end-of-dose OFF	Diarrhea, urine discoloration; tolcapone hepatotoxicity	LFTs with tolcapone	Müller (2010, 2009)
	MAO-B inhibitors	Inhibit DA metabolism in synapse	Rasagiline, selegiline, safinamide	Rasagiline 1 mg qd; selegiline 5 mg bid (or ODT lower); safinamide 50–100 mg qd	Modest symptomatic benefit; add-on increases ON time without troublesome dyskinesia (safinamide)	Insomnia (selegiline), nausea; serotonergic interactions; tyramine effect at higher doses	Drug interactions (SSRIs/SNRIs, tramadol, meperidine)	Nayak and Henchcliffe (2008)
Dopamine receptor stimulation	**Nonergoline dopamine agonists**	Direct D2/ D3 receptor agonism	Pramipexole, ropinirole, rotigotine patch	Pramipexole 0.125 mg TID, then 0.5–1.5 mg TID (ER 0.375–4.5 mg qd); ropinirole up to 24 mg/day ER; rotigotine 2–16 mg/24 h patch	Early monotherapy in younger patients; fluctuation "smoothers"; an option when limiting levodopa	Somnolence, edema, hallucinations; ICDs; sleep attacks	Screen regularly for ICDs; taper slowly to avoid DAWS; avoid ergot agonists (fibrosis)	Jing et al. (2023)
Dyskinesia control	**Amantadine**	NMDA antagonism; indirect dopaminergic effects	Amantadine IR/ER	IR 100 mg bid–tid; ER 274 mg qHS (dyskinesia)	Reduces LID; can improve OFF-related dyskinesia	Hallucinations, confusion; livedo reticularis, edema; insomnia	Renal dose adjust	Müller et al. (2003)
Tremor-predominant	**Anticholinergics**	Muscarinic antagonism	Trihexyphenidyl, benztropine	Trihexyphenidyl 1–2 mg bid/tid; benztropine 0.5–2 mg qd–bid	Useful in younger patients with tremors	Cognitive worsening, dry mouth, urinary retention, constipation, blurred vision	Avoid in cognitive vulnerability, glaucoma, BPH	Tarsy (2006)

Adjunct to levodopa	**A2A receptor antagonists**	Block striatal A2A on indirect pathway MSNs → decrease OFF	Istradefylline	20–40 mg qd	Decrease in daily OFF time in fluctuators	Dyskinesia, insomnia, nausea	CYP3A4 interactions; consider when MAO-B/COMT insufficient	LeWitt et al. (2008)
Device-aided continuous delivery	**LCIG/LECIG (intestinal levodopa gel)**	Continuous jejunal levodopa infusion → stable brain exposure	PEG-J pump; morning bolus + continuous rate (daytime)	Daytime 16-h infusion; titrate to symptoms; LECIG lowers gel volume (entacapone added)	Decreases OFF; increases good ON vs. optimized oral therapy	Tube/device complications, weight loss; polyneuropathy reported	B12/folate monitoring	Olanow et al. (2014)
	Subcutaneous apomorphine infusion	Continuous D1/D2 agonism	SC pump (with PRN rescue possible)	Continuous 1–6 mg/h (individualized); antiemetic pretreatment	Meaningful OFF reduction in optimized fluctuators	Nausea, nodules at the infusion site, hypotension; somnolence	QT and drug interactions; skin care for nodules	Katzenschlager et al. (2018)
Nonmotor symptoms	**5-HT2A inverse agonist; atypical antipsychotic**	Reduce hallucinations and delusions with minimal DA blockade	Pimavanserin; clozapine	Pimavanserin 34 mg qd; clozapine 12.5–50 mg/day (very low)	Improves psychosis without worsening motor symptoms	QT prolongation (pimavanserin); agranulocytosis (clozapine), sedation	ANC monitoring for clozapine; CYP interactions	Cummings et al. (2014)
	SSRIs/SNRIs	Serotonergic and noradrenergic modulation	Paroxetine; venlafaxine XR	Paroxetine 10–40 mg/day; venlafaxine XR 75–225 mg/day	Class I evidence; no motor worsening	GI upset, sexual dysfunction; increase in BP (venlafaxine)	Interactions with MAO-B; start low, go slow	Richard et al. (2012)
	Cholinesterase inhibitor	Increase cortical cholinergic tone for better cognition	Rivastigmine	PO 6–12 mg/day, divided; patch 9.5–13.3 mg/24 h	Modest cognitive and behavioral gains in PDD	Nausea, weight loss; bradycardia	Patch improves GI tolerance; monitor weight/HR	Emre et al. (2004)
	Melatonin; clonazepam; dopaminergic optimization; wake-promoting agent	Normalize REM atonia; improve sleep continuity; reduce EDS	Melatonin IR; clonazepam; evening CR levodopa or rotigotine; modafinil	Melatonin 3–6 mg qHS; clonazepam 0.25–1 mg qHS; modafinil 100–200 mg am	RBD control; insomnia relief; EDS reduction	Daytime sedation/falls (clonazepam); vivid dreams; headache (modafinil)	Prioritize safety in RBD (bedroom measures); treat nocturnal hypodopaminergia	Howell et al. (2023) and Adler et al. (2003)

Abbreviations: *5-HT2A*, serotonin (5-hydroxytryptamine) 2A receptor; *A2A*, adenosine A2A receptor; *AADC*, aromatic L-amino acid decarboxylase; *AM*, morning (ante meridiem); *ANC*, absolute neutrophil count; *BP*, blood pressure; *BPH*, benign prostatic hyperplasia; *COMT*, catechol-O-methyltransferase; *CR*, controlled-release; *CYP*, cytochrome P450; *CYP3A4*, cytochrome P450 3A4; *DA*, dopamine; *DAWS*, dopamine agonist withdrawal syndrome; *DBS*, deep brain stimulation; *D1*, dopamine D1 receptor; *EDS*, excessive daytime sleepiness; *ER*, extended-release; *GI*, gastrointestinal; *GPi*, globus pallidus internus; *HR*, heart rate; *ICDs*, impulse-control disorders; *IR*, immediate-release; *LCIG*, levodopa–carbidopa intestinal gel; *LECIG*, levodopa–entacapone–carbidopa intestinal gel; *LFTs*, liver function tests; *LID*, levodopa-induced dyskinesia; *MAO-B*, monoamine oxidase-B; *MSNs*, medium spiny neurons; *NMDA*, N-methyl-D-aspartate; *NMS*, neuroleptic malignant syndrome; *ODT*, orally disintegrating tablet; *OFF*, periods of poor motor control; *ON*, periods of good motor control; *PD*, Parkinson's disease; *PDD*, Parkinson's disease dementia; *PEG-J*, percutaneous endoscopic gastrojejunostomy; *PO*, by mouth (per os); *PRN*, as needed (pro re nata); *QoL*, quality of life; qd, once daily; bid, twice daily; *tid/TID*, three times daily; *qHS*, at bedtime; *RBD*, REM sleep behavior disorder; *REM*, rapid eye movement; *SC*, subcutaneous; *SNRIs*, serotonin–norepinephrine reuptake inhibitors; *SSRIs*, selective serotonin reuptake inhibitors; *STN*, subthalamic nucleus; *QT*, QT interval (electrocardiogram); *XR*, extended-release; *pts*, patients.

4 Gene therapy strategies

Gene therapy in PD depends on delivering therapeutic genes with high anatomical accuracy, strong expression, and minimal off-target effects. Gene delivery strategies for PD combine vector type, promoter, and delivery route to achieve targeted, lasting expression (Fig. 11.2). Adeno-associated viruses (AAVs) are favored for their neuronal affinity and low pathogenicity. For example, AAV2 is commonly used for putaminal infusion, while AAV9 allows for blood–brain barrier (BBB)-permissive systemic approaches currently under evaluation (Björklund et al., 2021). In contrast, lentiviral vectors provide larger cargo capacity for multigene cassettes with stable integration but carry a risk of insertional mutagenesis. Nonviral systems (such as lipids and polymers) may reduce immunogenicity but are still in early development and have so far been explored only in preclinical PD settings (Van Laar et al., 2021). Promoter design involves balancing strength and specificity. Broad promoters (e.g., cytomegalovirus [CMV], chicken β-actin [CBA]) maximize expression but increase off-target risks, whereas cell-specific elements (e.g., TH for dopaminergic neurons; Glial fibrillary acidic protein [GFAP] for astrocytes) improve safety (Matsuzaki et al., 2024). Different therapeutic objectives also affect promoter selection. Neurotrophic factors benefit from sustained, moderate levels, while enzyme replacement requires high activity (Van Laar et al., 2021).

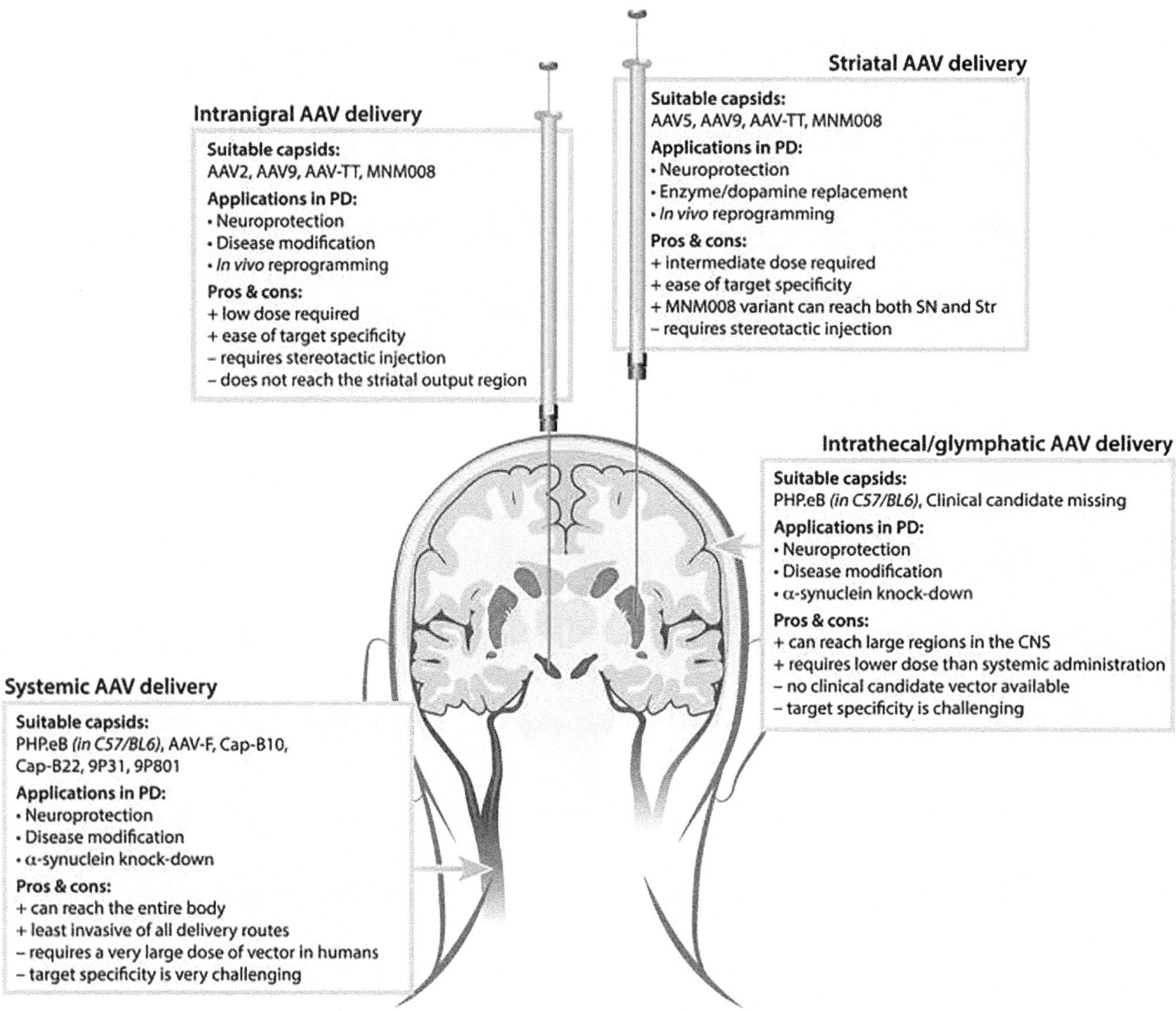

FIG. 11.2 Delivery routes and suitable AAV variants for gene therapy in Parkinson's disease. To target the basal ganglia, AAV vectors can be delivered through four different routes: intranigral, striatal, intrathecal, or systemic. All routes have advantages but also present challenges. Fortunately, we now have engineered AAV-vector capsids suitable for most routes. *Based on Björklund, T., Davidsson, M., Björklund, A., Bloem, B.R., Brundin, P., Federoff, H., 2021. Next-generation gene therapy for Parkinson's disease using engineered viral vectors. J. Parkinsons Dis. 18777171. 11 (s2), S209. https://doi.org/10.3233/JPD-212674.*

Delivery usually involves stereotactic, MRI-guided infusion into the posterior putamen, with optional cotargeting of the SNpc for neuroprotection. Convection-enhanced delivery (CED) applies positive pressure to improve volume spread and reduce reflux (Bobo et al., 1994). Real-time MRI with gadolinium enhances visualization during delivery (Richardson et al., 2020). In addition, MRI-guided low-intensity focused ultrasound combined with microbubbles transiently disrupts the BBB, enabling noninvasive, region-specific delivery of systemic vectors such as AAV9 for widespread gene silencing (e.g., antiSNCA shRNA) or neurotrophic factor delivery without intracranial surgery.

Efficacy of each therapy is assessed using clinical outcomes and objective biomarkers. The Unified Parkinson's Disease Rating Scale (UPDRS) remains central. Reductions, particularly in Part III, indicate motor improvement (Goetz et al., 2008). Patient motor diaries quantify ON and OFF time, linking longer ON and shorter OFF to meaningful function and reduced medication needs (Hauser et al., 2004). Neuroimaging complements these measures. Dopaminergic positron emission tomography (PET) (^{18}F-DOPA, FMT) indexes presynaptic DA synthesis and AADC activity. Postgene transfer increases signal target engagement and may precede clinical gains due to circuit adaptation or dosing (Eberling et al., 2008). Concordant changes across UPDRS, ON/OFF time, dyskinesia burden, and PET provide mechanistic validation. Longitudinal assessment in defined ON/OFF states guides dosing, delivery optimization, and trial progression.

5 Approaches mapped to PD pathology

This section connects gene therapy strategies with the fundamental biology of PD, shifting from symptom management to mechanism-based treatment. Each approach targets a specific aspect of the disease process (Fig. 11.3). For each strategy, the rationale,

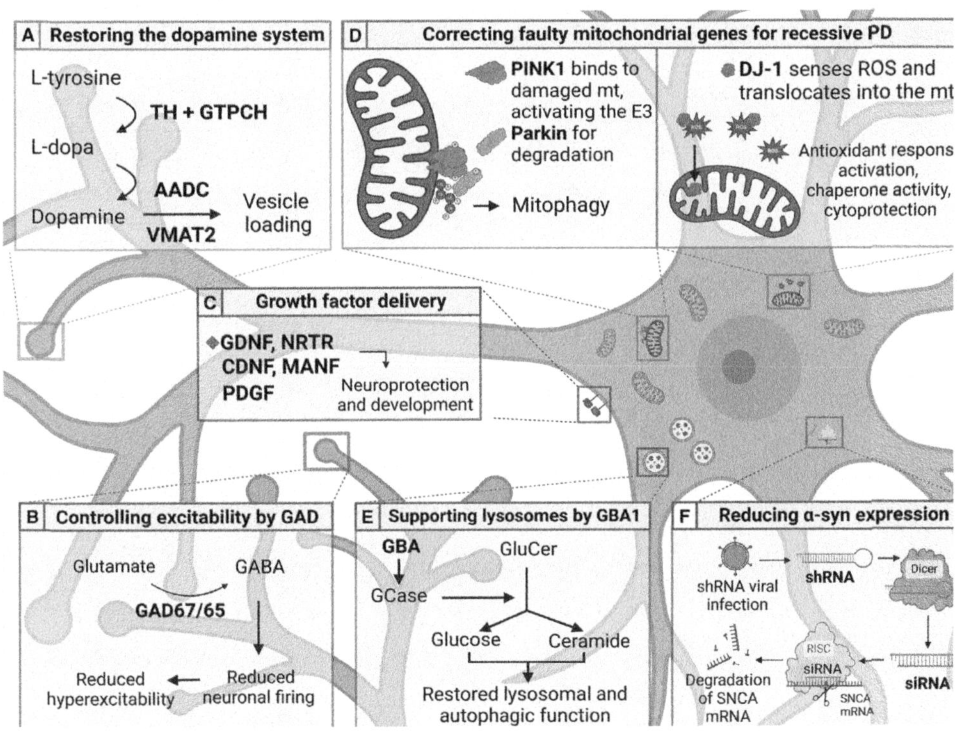

FIG. 11.3 Gene therapy approaches in PD. (A) Restoring the dopamine system through the delivery of genes involved in dopamine biosynthesis and dopamine release, including tyrosine hydroxylase (TH), GTP cyclohydrolase I (GTPCH), aromatic amino acid decarboxylase (AADC), and vesicular monoamine transporter 2 (VMAT2). (B) Reinforcing inhibitory pathways in the basal ganglia through the delivery of genes encoding glutamic acid decarboxylase isoforms (GAD65/67), which convert glutamate to GABA to reduce neuronal firing. (C) Promoting the survival and development of dopaminergic neurons through the delivery of genes encoding growth factors, including glial cell line-derived neurotrophic factor (GDNF), neurturin (NRTN), cerebral dopaminergic neurotrophic factor (CDNF), mesencephalic astrocyte-derived neurotrophic factor (MANF), and platelet-derived growth factor (PDGF). (D) Delivery of genes encoding PINK1, PRKN, and DJ-1 to address mitochondrial dysfunction resulting from recessive familial PD mutations. (E) Supporting lysosomal function through the delivery of GBA1 to increase glucocerebrosidase activity (GCase) and promote the breakdown of glucosylceramide (GluCer) into glucose and ceramide in the lysosome. (F) Reducing α-synuclein expression through the delivery of short hairpin RNA (shRNA). shRNA is processed by dicer to form short interfering RNA (siRNA), which targets the SNCA gene and promotes mRNA degradation. *Based on Szunyogh, S., Carroll, E., Wade-Martins, R., 2025. Recent developments in gene therapy for Parkinson's disease. Mol. Ther. 15250024. 33 (5), 2052–2064. Available from: https://www.sciencedirect.com/science/journal/15250016. https://doi.org/10.1016/j.ymthe.2025.03.030*

TABLE 11.2 Gene therapy approaches.

Approach	Therapeutic goal	Cargo/construct	Vector and promoter	Target anatomy and delivery	Primary efficacy readouts	Strengths	Limitations	Ref
DA restoration, AADC augmentation	Resensitize putamen to levodopa	AADC	AAV2	Postcommissural putamen; MRI-guided CED; reflux-resistant cannula; gadolinium tracking	Improving levodopa response; decrease OFF; PET (FMT/FDOPA) target engagement	Simple payload; fits AAV; durable	Cytosolic DA diffusion in GABAergic MSNs; surgical risks	(Christine et al., 2009; Eberling et al., 2008; Nutt et al., 2020), (Björklund et al., 2021; Christine et al., 2019; Hwu et al., 2021)
DA restoration, de novo **biosynthesis**	Local, continuous DA production	TH + AADC + GCH1 (± VMAT2)	LV (ProSavin, OXB-102); HSV-based 4-gene (preclinical)	Putamen; MRI-CED; coverage goals	Motor scales; PET DA synthesis	Less dependent on surviving terminals; VMAT2 → vesicular release, less cytosolic DA	Integration risk (LV); dyskinesia if excessive DA; surgical risks	(Palfi et al., 2018, Palfi et al., 2014; Badin et al., 2019; Sun et al., 2004)
Neurotrophic,GDNF/NRTN	Neuroprotection	GDNF, NRTN	AAV2	Putamen ± SN; MRI-CED; dual-site in some designs	Improve PET (FDOPA/FMT); motor scales	Mechanism-based survival signaling (RET/Nurr1)	Mixed clinical signals; distribution-dependent; timing/RET expression matters	(Zheng et al., 2005; Lapchak et al., 1997; Kells et al., 2010)
Neurotrophic,CDNF/MANF/PDGF	UPR modulation; antiinflammatory	CDNF, MANF, PDGF-BB	AAV2/AAV8 (CDNF); LV (CDNF + MANF)	Nigra vs. striatum (site-dependent efficacy); MRI-CED	Motor recovery; TH markers; PET	ER-stress and inflammation targeting; potential synergy (CDNF + MANF)	Site and stage sensitive; vector risks	(Nam et al., 2024; Wang et al., 2017; Li et al., 2022; Othberg et al., 1995; Pietz et al., 1996)
Proteinopathy, SNCA lowering (RNAi)	Reduce α-syn load	AAV/LV shRNA; miR-embedded guides	AAV2; miR-30 backbones	SNpc (primary); regional upstream sites in seeding paradigms; MRI-CED	Decrease α-syn; preserved terminals; behavioral gains	Dicer saturation; inflammation; TH downmodulation if overdone	Use miR backbones; limit knockdown; allele-specific guides (e.g., A53T)	(Zharikov et al., 2015; Zharikov et al., 2019; Zharikov et al., 2015; Khodr et al., 2011; Sapru et al., 2006)
Proteinopathy, transcriptional repression	Locus-level *SNCA* suppression	dCas9-DNMT3A epigenetic editing; AAV ZFR	LV (dCas9-DNMT3A); AAV (ZFR)	Intraparenchymal or systemic (preclinical)	Decrease *SNCA* mRNA/protein; cell viability	Sustained repression; DNA-encoded specificity	Off-target methylation	(Kantor et al., 2018)

Circuit rebalancing, STN GAD	Inhibit STN → lower GPi/SNpr output; restore motor flow	GAD65/67	AAV2 (e.g., JDK promoter > CMV)	STN; MRI-CED; DBS-style targeting	Motor scales; FDG-PET cortical patterns	Hardware-free DBS-like physiology; autoregulated synapses	Not disease-modifying; speech/gait may persist; surgical risks	(Alharbi et al., 2024; Winston et al., 2025; Luo et al., 2002; Emborg et al., 2007)
Lysosomal restoration, GBA1 augmentation	Increase GCase → normalize lipid flux; lower α-syn	GBA1 cDNA	AAV9, AAV2/5, AAV1; PHP.B	Striatum ± SN (parenchymal); CSF/IV for broad access	Decrease insoluble α-syn and gliosis; DA preservation	Immune responses; systemic exposure with BBB-permissive capsids	Benefit may extend beyond GBA carriers; capsid drives spread; monitor coverage	(Sardi et al., 2011; Rockenstein et al., 2016; Rocha et al., 2015; Morabito et al., 2017)
Mitochondrial QC, PINK1/Parkin augmentation	Restore mitophagy; reduce ROS	PINK1, Parkin (± aMTD-Parkin)	AAV9 and related; neuron-biased promoters	Striatum ± SN; stereotactic infusion	Motor benefit; TH preservation; stress markers	Overexpression effects; off-target ubiquitination	Pair with lysosomal repair; target high-stress regions (posterior putamen/SNpc)	(Narendra and Youle, 2024; Petit et al., 2005)

Abbreviations: *DA*, dopamine; *AADC*, aromatic L-amino acid decarboxylase; *AAV*, adeno-associated virus; *AAV2/5/9*, serotypes 2/5/9 of AAV; *LV*, lentiviral vector; *HSV*, herpes simplex virus; *CED*, convection-enhanced delivery; *MRI*, magnetic resonance imaging; *PET*, positron emission tomography; *FDOPA*, 6-[^{18}F]fluoro-L-DOPA; *FMT*, 6-[^{18}F]fluoro-m-tyrosine; *MSN(s)*, medium spiny neuron(s); *GABA*, gamma-aminobutyric acid; *TH*, tyrosine hydroxylase; *GCH1*, GTP cyclohydrolase-1; *VMAT2*, vesicular monoamine transporter 2; *GDNF*, glial cell line-derived neurotrophic factor; *NRTN*, neurturin; *SN*, substantia nigra; *SNpc*, substantia nigra pars compacta; *SNpr*, substantia nigra pars reticulata; *RET*, rearranged during transfection receptor tyrosine kinase; *Nurr1*, nuclear receptor-related 1 protein (NR4A2); *UPR*, unfolded protein response; *ER*, endoplasmic reticulum; *CDNF*, cerebral dopamine neurotrophic factor; *MANF*, mesencephalic astrocyte-derived neurotrophic factor; *PDGF-BB*, platelet-derived growth factor-BB; *SNCA*, alpha-synuclein gene; *α-syn*, alpha-synuclein; *RNAi*, RNA interference; *shRNA*, short hairpin RNA; *miR*, microRNA; *ZFR*, zinc-finger repressor; *dCas9*, catalytically inactive Cas9; *DNMT3A*, DNA methyltransferase 3A; *STN*, subthalamic nucleus; *GPi*, globus pallidus internus; *DBS*, deep brain stimulation; *FDG-PET*, [^{18}F] fluorodeoxyglucose PET; *CMV*, cytomegalovirus (immediate-early promoter); *JDK promoter*, program-specific synthetic promoter; *GCase*, glucocerebrosidase; *GBA1*, glucocerebrosidase beta acid gene; *PHP.B*, BBB-permissive AAV9 capsid variant (mouse); *CSF*, cerebrospinal fluid; *IV*, intravenous; *BBB*, blood–brain barrier; *QC*, quality control; *PINK1*, PTEN-induced kinase 1; *ROS*, reactive oxygen species; *aMTD*, advanced macromolecule transduction domain (a cell-penetrating peptide); *cDNA*, complementary DNA; *GAD*, glutamic acid decarboxylase (isoforms 65/67); *mRNA*, messenger RNA.

vector and promoter choices, anatomical targets, and delivery methods are examined alongside effectiveness signals. Collectively, these complementary approaches demonstrate how aligning payload and delivery with the underlying defect can move beyond drug replacement toward lasting modification of the disease course. The details of each strategy are illustrated in Table 11.2.

5.1 DA restoration strategies

The goal of DA restoration gene therapy is to reestablish striatal DA despite ongoing nigrostriatal degeneration. There are two complementary strategies for this: (1) enhancing residual levodopa responsiveness by increasing AADC in the putamen and (2) creating new DA biosynthesis within striatal neurons through multigene cassettes (usually TH + AADC + GTP cyclohydrolase-1 [GCH1], and in newer designs, adding VMAT2 to support vesicular storage and activity-dependent release). These methods use AAV or lentiviral vectors delivered intraparenchymally. The most common route of administration is into the postcommissural putamen to maximize on-target transduction (Van Laar et al., 2021; Sun et al., 2004).

5.1.1 Putaminal AADC augmentation

In advanced PD, reduced AADC activity in the denervated putamen limits the conversion of exogenous levodopa to DA. Using AAV to transfer AADC to striatal neurons enhances local enzymatic conversion capacity and can "resensitize" the putamen to levodopa input, and target engagement can be demonstrated with ^{18}F-DOPA or FMT PET (Christine et al., 2009; Eberling et al., 2008; Nutt et al., 2020).

The AADC coding sequence easily fits into AAV. AAV2 remains a standard choice for putaminal delivery because of its neuronal preference and extensive use in the basal ganglia. Promoter selection (pan-neuronal versus more specific) and new capsid engineering offer additional options for cell-type targeting and controlling expression (Björklund et al., 2021).

To resolve the problem of undercoverage while infusing, modern neurosurgical systems use MRI-guided CED with reflux-resistant cannulae and gadolinium coinfusion to visualize and improve intraputaminal coverage in real time (Christine et al., 2019).

However, since MSNs are GABAergic and do not naturally release DA, cytosolic DA can potentially diffuse abnormally. This requires adjusting dosage, promoter strength, and volumetric spread (Hwu et al., 2021).

5.1.2 De novo striatal DA biosynthesis (multigene cassettes)

Lentiviral constructs that coexpress TH + AADC + GCH1 enable local DA synthesis by providing both enzymatic steps and the BH4 cofactor. This method aims for continuous restoration of DA in the area of greatest deficiency, the posterior putamen, without relying on surviving nigral terminals (ProSavin (Palfi et al., 2018, 2014) and OXB-102 (Badin et al., 2019)).

Including VMAT2 enhances vesicular packaging and activity-dependent release, decreases cytosolic DA exposure, and more accurately mimics physiological signaling. VMAT2 transports DA into synaptic vesicles, supporting regulated release and relieving TH from DA feedback inhibition. Preclinical studies show normalization of extracellular DA and K^+-evoked release with four-gene designs compared to TH/AADC/GCH1 alone using HSV type 1 (Sun et al., 2004).

As with AADC augmentation, intraputaminal MRI-guided CED is used to achieve sufficient volumetric coverage of the postcommissural putamen.

5.2 Neurotrophic factor delivery

Mechanistic axis and preclinical foundation: Gene transfer can restore trophic signaling to nigrostriatal circuits via GDNF-family ligands. The most developed ligands of this approach are GDNF and NRTN. Moreover, there are emerging factors, including CDNF, MANF, and PDGF (Paul and Sullivan, 2019).

As mentioned in Section 2.4, ligand binding to GFRα coreceptors on striatal dopaminergic terminals activates RET, driving survival programs and dopaminergic gene expression through Nurr1.

CDNF and MANF originate from the ER to regulate the unfolded protein response (UPR), inflammation, and oxidative stress. Intrastriatal injection of recombinant CDNF has been shown to improve motor function and protect dopaminergic neurons in animal models of PD. Similarly, MANF has been able to reduce neuron death by shielding them from stress-related damage (88–90). In contrast, PDGF-BB activates membrane survival pathways. It is shown to improve the survival of embryonic TH-positive neurons in vitro (Li et al., 2022; Othberg et al., 1995; Pietz et al., 1996)

Supporting the role of neurotrophic factors, a double-blind, randomized Phase 1 trial evaluated intraputaminal infusion of recombinant GDNF in 41 patients. Despite a significant increase in ^{18}F-FDOPA uptake seen in the GDNF group, there were no statistically significant improvements in the OFF-state UPDRS motor scores or any secondary and exploratory motor endpoints compared to placebo (Whone et al., 2019). This highlights the need for durable, focal expression that can be achieved through gene delivery.

Vectorization: AAV2 is the canonical vector for putaminal delivery of GDNF or NRTN because of its neuronal tropism and established manufacturing and surgical experience (Heiss et al., 2019; Marks et al., 2010; Warren Olanow et al., 2015; Marks et al., 2008). For CDNF, both AAV8 and AAV2 have been used preclinically (Bäck et al., 2013; Ren et al., 2013; Wang et al., 2017). Moreover, a single lentiviral cassette has enabled combined CDNF and MANF expression when larger payloads or dual factors are desired (Cordero-Llana et al., 2015).

Target anatomy and delivery: The posterior putamen is prioritized to engage remaining terminals. However, some designs also target the substantia nigra to address transport constraints and support cell bodies (Bartus et al., 2013). Similar to previous approaches, MRI-guided CED with reflux-resistant cannulae and gadolinium coinfusion monitors coverage. Coverage is a critical efficacy determinant: restricted expression has been linked to limited putaminal and nigral coverage (Bartus et al., 2015).

Delivering to both the putamen and substantia nigra increases the fraction of NRTN-positive neurons compared to putamen-only approaches, guiding current anatomical strategies. Postmortem analysis conducted 10 years after viral delivery showed NRTN expression in 1.3%–5.9% of the putamen and 11% of the substantia nigra, along with a threefold rise in TH levels. Dual delivery of AAV2-NRTN into both regions resulted in more NRTN-positive neurons than injection into the putamen alone, indicating that gene therapy targeting both the axon terminals and cell bodies of nigrostriatal neurons may be more effective for treating PD (Chu and Kordower, 2023).

In addition, AAV2-GDNF shows anterograde distribution in basal ganglia circuits, suggesting exposure to additional networks beyond the directly transduced terminals. Notably, this transport occurs independently of degenerated nigrostriatal dopaminergic neurons (Kells et al., 2012; Salegio et al., 2013).

Preclinical foundation: Across toxin-based models, striatal overexpression of **GDNF** consistently protects dopaminergic neurons, lowers α-synuclein burden, enhances mitochondrial biogenesis, and stimulates axonal sprouting (101–103).

In rodent PD models, intrastriatal GDNF gene transfer consistently normalizes dopaminergic markers and behavior: in 6-OHDA rats, it reduces apomorphine rotations, increases spontaneous locomotion, and raises striatal DA/DOPAC levels, with independent studies also showing more TH-positive nigral neurons and higher DA (Zheng et al., 2005; Lapchak et al., 1997). In a proteasome lesion model, AAV2-GDNF preserved nigral and striatal DA and stimulated SVZ and dentate gyrus neurogenesis with more newly born dopaminergic neurons (Yunlan et al., 2013). Furthermore, brief AAV5-GDNF dosing was enough to restore long-term striatal DA and protect nigrostriatal neurons (Tereshchenko et al., 2014). Similarly, **NRTN** provides RET-dependent neuroprotection in 6-OHDA rats comparable to GDNF.

In nonhuman primates (NHPs), intraputaminal AAV2-GDNF delivered by convection-enhanced infusion after a stable MPTP lesion (with initiation deliberately delayed until the lesioning was complete) produced durable neurorestoration: bilateral increases in FMT-PET uptake at 6 months (around 18% in the moderately lesioned putamen; around 54% in the severely lesioned putamen) that continued to rise by 22 months (about 37% and about 138%, respectively). PET gains closely matched sustained improvements on a blinded clinical rating scale over 6–24 months. In addition, histology showed extensive regrowth of TH-positive fibers in the putamen, and ELISA plus mapping confirmed putaminal expression with significant transport to the substantia nigra. Importantly, midbrain exposure occurred even when nigrostriatal axons on the infused side were severely degenerated, which indicates anterograde spread through striatonigral connections (Kells et al., 2010). For NRTN, AAV2-NRTN studies in monkeys showed reliable striatal coverage with robust retrograde transport to nigral cell bodies and high induction of nigral TH. Coverage in representative cases ranged from approximately 4%–25% of the striatal volume in aged animals (and about 2%–6% in young monkeys), averaging roughly 13% in MPTP-lesioned monkeys, while still achieving clear nigral exposure. These results supported dual-site (putamen plus substantia nigra) targeting and strict coverage goals in later-stage designs (Bartus et al., 2011).

Mechanistically, delivering human **CDNF** (hCDNF) using AAV-hCDNF in MPTP mice lessened motor and gait deficits while decreasing glial IL-1β/C3 and ER stress markers (e.g., CHOP, GRP78), supporting neuroprotection through modulation of UPR and inflammation (Nam et al., 2024).

Gene delivery of CDNF restores dopaminergic function in toxin models, depending on the stage of the lesion: AAV8-CDNF improved motor performance and increased TH levels in mild 6-OHDA lesions, whereas CDNF + AADC was necessary to recover function in severe lesions (Wang et al., 2017). Similarly, striatal AAV2-CDNF led to behavioral recovery along with the restoration of TH-positive neurons and fibers, as well as PET evidence of dopaminergic activity (Ren et al., 2013), achieving durable expression with reduced rotations but only partial TH preservation in the nigra and striatum (Bäck et al., 2013). Also, location plays an important role: striatal overexpression of CDNF or **MANF** alone was ineffective. Whereas nigral CDNF reduced rotations and increased striatal TH fibers, nigral MANF preserved TH-positive cells, and combined nigral CDNF + MANF produced synergistic improvements in behavior, fiber density, and cell survival (Cordero-Llana et al., 2015).

5.3 Proteinopathy-targeted approaches

Mechanistic axis: As mentioned earlier, excess α-synuclein drives vulnerable neurons toward oligomer formation, propagation, and damage. The practical implication is subtle: physiological α-synuclein aids presynaptic function, but the aim is to carefully reduce it to prevent nucleation and spread, while still maintaining its essential role at synapses.

Vectorization methods and preclinical foundation: Several methods have been proposed to lower the ultimate α-synuclein level inside neurons: (1) RNA interference (RNAi) cassettes (AAV-shRNA/miRNA), (2) transcription-level repression (epigenetic editing), and (3) synthetic transcriptional repressors (zinc-finger repressors [ZFRs]).

In RNAi cassettes, AAV is used to express short hairpin or microRNA scaffolds targeting SNCA, resulting in durable knockdown in nigral neurons. In rotenone-sensitized rats, AAV2-shRNA targeting SNCA achieved about 35% protein reduction, along with decreased motor decline and restoration of dopaminergic terminals and dendrites (Zharikov et al., 2015). However, some levels of toxicity were observed in high-concentration shRNA neurons (Khodr et al., 2011). Similar lentiviral vectors of shRNA in the SN reduced limb use deficits and preserved dopaminergic markers (Sapru et al., 2006). In addition, embedding the guide in a miR-30 backbone reduces Dicer loading and off-target effects but can still cause inflammation and TH downregulation (Khodr et al., 2014). Importantly, long-term (12-month) studies with AAV2-shRNA achieved over 70% protein reduction without evident neurodegeneration, although TH downmodulation also occurred with control vectors, suggesting that vector exposure and shRNA burden are contributing factors (Zharikov et al., 2019). Building on this approach, a 21-nt site in SNCA supports strong RNAi, including allele-specific silencing of A53T. Using plasmids and a dual-cassette lentivirus that coexpresses an SNCA-targeting shRNA with EGFP, they achieved significant suppression of endogenous α-synuclein in SH-SY5Y cells and of experimentally expressed human α-synuclein in vivo in the rat brain (Sapru et al., 2006). This approach shows that mutation-constrained synucleinopathy can be targeted with selective guides, potentially enhancing safety when a pathogenic allele is the main cause.

Collectively, these data support choosing appropriate doses and vector designs (e.g., miR-embedded guides, weaker promoters, capsid serotype) that limit maximum knockdown and prevent saturation of the RNAi machinery.

Beyond RNAi-based strategies, gene repression can also be achieved at the transcriptional level using programmable DNA-binding proteins.

In transcription-level repression, epigenetic editing is proposed using an all-in-one lentiviral vector containing a combination of CRISPR–deactivated Cas9 (dCas9) and the catalytic domain of DNA methyltransferase 3A (DNMT3A). Targeting CpG sites in SNCA intron 1 induces hypermethylation and decreases SNCA mRNA and protein levels in triplication iPSC-derived dopaminergic neurons, along with enhanced cellular viability (Kantor et al., 2018).

In addition, synthetic transcriptional repressors, such as AAV-delivered ZFRs, have been shown to reduce SNCA mRNA levels in human iPSC-neurons. Similarly, following systemic AAV delivery in a humanized PAC-SNCA mouse model, they decreased expression in the midbrain and thalamus at 6 weeks (Andrew Young et al., 2024).

Target anatomy and delivery: Most RNAi programs have targeted the SNpc directly to lower α-synuclein, where degeneration initiates (Zharikov et al., 2015; Khodr et al., 2011; Sapru et al., 2006). In prion-like seeding paradigms, AAV1-miR-SNCA placed in the hippocampus curtailed pathology spread, illustrating that regional lowering upstream of propagation routes can be disease-modifying (Menon et al., 2021). For synthetic repressors, systemic AAV has been explored preclinically (tail-vein), with measurable knockdown in the midbrain and thalamus (Andrew Young et al., 2024)

When using intraparenchymal AAV, stereotactic CED and slow infusion rates are standard to optimize volumetric spread while minimizing reflux; vector choice (e.g., AAV2 versus AAV2/5/8) and infusate parameters significantly influence both biodistribution and host response, again linking delivery physics to safety signals (Zharikov et al., 2019).

For specificity controls, dopaminergic-leaning promoters or microRNA target-site detuning can restrict excessive silencing outside the nigrostriatal system. Epigenetic editors inherently operate at transcriptional regulatory DNA, offering locus specificity. Nonetheless, guide selection and payload exposure remain critical to minimizing off-target methylation (Kantor et al., 2018).

5.4 Circuit-level rebalancing

Mechanistic axis: Beyond proteinopathy and trophic repair, PD also presents as a circuit disorder: loss of dopaminergic tone causes the STN to become hyperactive and bursty, overdriving the Gpi and SNpr, which suppress thalamocortical output and movement. If we can restore inhibition locally, we can silence this harmful loop. GAD-gene delivery implements this approach: by equipping STN neurons with the enzyme GAD65/67, they are able to convert glutamate into GABA. Therefore the STN's output becomes more inhibitory, reducing Gpi and SNpr firing and restoring normal function to downstream motor networks (Alharbi et al., 2024).

GAD delivery does not replace DA or clear aggregates; instead, it rebalances the basal ganglia network internally, potentially offering DBS-like benefits without hardware and with autoregulatory synaptic physiology. Two main principles guide its translation: first, prioritize coverage and promoter selection for consistent expression; second, combine circuit-level biomarkers with clinical outcomes to demonstrate mechanism-linked benefits (Winston et al., 2025).

Preclinical foundation and vectorization: A series of rodent and primate studies demonstrated that AAV-mediated GAD delivery to the STN can restore inhibitory balance and improve motor function, providing the mechanistic basis for translation. Unilateral AAV-GAD65/67 to the STN in 6-OHDA rats increased GABA release in the SNpr during STN stimulation and improved motor function (Luo et al., 2002). Later work emphasized that promoter engineering matters: an AAV2-JDK-GAD65 cassette drove higher expression and more consistent firing-rate normalization than AAV2-CMV-GAD65, tracking with stronger behavioral rescue (Lee et al., 2005).

Furthermore, in MPTP macaques, STN AAV-GAD was well-tolerated and led to sustained motor improvement for about a year, along with FDG-PET changes in the motor cortex that correlated with clinical ratings (Emborg et al., 2007).

Target anatomy: Gene transfer employs a "biologic DBS" approach: stereotactic, intraparenchymal infusion into the STN using MRI-guided CED and reflux-resistant cannulae to maximize coverage and keep expression localized. Because the STN projects to the SNpr and GPi, increasing local GABA production enhances inhibition along the specific pathways that are pathologically overactive (Winston et al., 2025; Luo et al., 2002).

5.5 Lysosomal restoration

Mechanistic axis: After rebalancing circuits and trimming α-synuclein at its source, we now focus on the lysosome, where impaired GCase activity disrupts proteolipid turnover and increases α-synuclein stress. Heterozygous GBA1 mutations lower GCase levels, impair autophagy and lysosomes, and create a feedback loop in which α-synuclein further disrupts GCase trafficking and maturation (Menozzi et al., 2023). The therapeutic approach is simple: restore GCase to normalize lipid breakdown and, secondarily, reduce α-synuclein accumulation.

Vectorization and target anatomy: Because enzyme replacement does not cross the BBB and requires lifelong dosing, AAV-mediated GBA1 delivery provides long-lasting expression (Stojkovska et al., 2018). Programs have packaged GBA1 cDNA into AAV serotypes suitable for brain delivery, such as AAV9, AAV2/5, and AAV1, with promoter options (e.g., CBA, GUSB, synapsin) selected to adjust expression levels and target specific cell types (Sardi et al., 2011; Rockenstein et al., 2016; Rocha et al., 2015; Morabito et al., 2017).

Two complementary strategies have been identified: (1) targeted intraparenchymal delivery to the nigrostriatal system (including the striatum and sometimes the SN or hippocampus) to replenish GCase at the intersection of α-synuclein pathology and DA vulnerability; and (2) approaches through CSF-facing or intravenous routes to access widespread cortical–subcortical regions involved in GBA1-related PD (Sardi et al., 2011). Notably, BBB-permissive AAV9 variants (e.g., PHP.B) allow systemic delivery with widespread neural transduction in mice, raising the possibility of treating diffuse synucleinopathy noninvasively. However, the trade-offs of broad exposure must be considered (Morabito et al., 2017).

Preclinical foundation: Across GD and PD models, increasing GCase through AAV-GBA1 consistently lowers insoluble α-synuclein, reduces gliosis and neuroinflammation, and maintains dopaminergic structure, with the promoter, serotype, and route chosen to meet coverage goals (Menozzi et al., 2023). In mice, AAV-PHP.B-GBA1 delivered intravenously produced widespread CNS expression, prevented α-synuclein deposits across multiple corticostriatal regions, and improved behavior in an A53T-SNCA model (Morabito et al., 2017). Similarly, in α-synuclein-overexpressing rodents, intracerebral AAV-GBA1 reduced nigral and striatal aggregates and protected nigrostriatal neurons, with accompanying normalization of autophagy markers (Rocha et al., 2015). Mechanistic counterpoints strengthen causality: partial GCase inhibition worsens synucleinopathy and related deficits, while GCase enhancement delays disease progression and preserves striatal markers, even without GBA1 mutations, indicating broad therapeutic potential (Rockenstein et al., 2016). Collectively, these findings support a unified model: GCase restoration normalizes lysosomal lipid flow, reduces α-synuclein load, and stabilizes vulnerable dopaminergic circuits.

5.6 Mitochondrial quality control

As discussed earlier, PINK1 accumulates on depolarized mitochondria, recruiting Parkin to ubiquitinate outer-membrane substrates and trigger mitophagy. When this process fails, defective organelles persist, increasing oxidative stress and the vulnerability of dopaminergic neurons (Narendra and Youle, 2024). Gene transfer aims to restore this system to maintain healthy mitochondria in the nigrostriatal system.

In practice, both cargoes are compatible with AAV, indicating that the coding sequences easily fit within the AAV payload. Programs have focused on AAV9 and related capsids, with neuron-specific promoters to bias expression in the basal ganglia. One approach involves adding a hydrophobic cell-penetrating peptide (aMTD) to the Parkin construct (AAV-aMTD-Parkin) to improve tissue permeability at low doses in 6-OHDA models. As a result, Parkin reintroduction restored motor performance and increased TH expression, supporting a genuine neuroprotective effect (Lee et al., 2024). Anatomically, the delivery has focused on the striatum, with or without the SN, regions under the greatest oxidative stress. This has been done through standard stereotactic infusion. This targeted approach matches the underlying biology, as the highest mitochondrial turnover happens where DA neurons fire continuously and have long, highly branched axons.

On the complementary arm, PINK1 overexpression decreases both basal and toxin-induced apoptosis and reduces α-syn-driven neurotoxicity in neuronal systems, while PINK1 loss worsens dopaminergic degeneration, highlighting that increasing mitophagy is mechanistically linked to PD vulnerability, and decreasing it can alleviate the symptoms of PD (Petit et al., 2005). In conclusion, targeting PINK1/Parkin adds a third pillar to PD gene therapy, alongside proteinopathy and lysosomal repair, by restoring the organelle function that dopaminergic neurons rely on.

6 Clinical trials

Gene therapy has emerged as a novel approach capable of directly modifying neural circuits, restoring neurotransmitter synthesis, or delivering neurotrophic factors for neuroprotection. Recent advances in vector engineering, stereotactic delivery, and imaging-guided monitoring have enabled multiple human clinical trials targeting different pathogenic axes. This review summarizes trial evidence across three main categories: DA enzyme replacement, neuroprotective growth factor delivery, and thalamic/subthalamic circuit modulation (Table 11.3).

6.1 DA enzyme replacement

Given the centrality of DA depletion in PD, one strategy has aimed to restore striatal DA metabolism through viral vector-mediated expression of synthetic enzymes. Three main enzymes have been targeted in clinical trials: AADC, TH, and GCH1 (Duan et al., 2005; Nagatsu, 2024).

The earliest clinical studies used AAV2-AADC infusion into the postcommissural putamen to enhance local conversion of levodopa to DA. Christine et al. (2009) treated ten patients with escalating doses and reported a 31% improvement in UPDRS motor scores, with increased FMT-PET uptake confirming target engagement (Christine et al., 2009). Long-term follow-up showed persistence of PET signal elevation but gradual regression of motor benefits toward baseline over 4 years.

Later dose escalation studies employed MRI-guided infusions with larger coverage (up to 42% of the putamen). These trials demonstrated dose-dependent increases in FMT-PET signal (up to 79%), reduced levodopa-equivalent dose requirements (up to 42%), and improvements in both on- and off-state UPDRS scores. Adverse events were generally mild and transient, such as dyskinesia, highlighting the tolerability of this approach (Muramatsu et al., 2010; Christine et al., 2019).

To overcome AAV cargo limitations, ProSavin, a lentiviral vector encoding TH, AADC, and GTP cyclohydrolase 1 (GTPCH1), was developed. In a Phase ½ trial (n = 15), bilateral putaminal delivery resulted in UPDRS motor improvements of 7–15 points depending on dose, alongside reduced levodopa burden. Long-term follow-up up to 6 years confirmed durable tolerability and sustained symptomatic improvement, with no severe drug-related adverse events. However, efficacy was variable, and moderate adverse events, such as dyskinesia, were reported (Palfi et al., 2018, 2014). Notably, despite the promising findings, Sio Gene Therapies (formerly Axovant) halted development of AXO-Lenti-PD, and the

TABLE 11.3 Clinical trial for gene therapy of PD.

Study	NCT	Name of product	Design and phase	Vector/ target	Delivery site and dose	Eligibility criteria	N (patients)	Follow-up and monitoring	Motor outcomes (UPDRS/ MDS-UPDRS part III, "off" state)	Safety
Subthalamic										
Kaplitt et al. (2007)	NCT00195143	–	Phase 1, open-label, single center	AAV2-GAD (1×10^{11}–1×10^{12} vg/mL)	Unilateral STN, 50 μL	Eligible patients were 25–70 years old with PD ≥ 5 years, Hoehn and Yahr ≥ 3, and UPDRS III (off) ≥ 30, with levodopa-related motor complications and stable medications for ≥ 3 months. Exclusion criteria included significant cognitive impairment, surgical contraindications, atypical/secondary parkinsonism, and major psychiatric illness.	12 (4 per dose group)	12 months	~28% reduction in UPDRS score across all cohorts†	No vector-related adverse events
LeWitt et al. (2011)	NCT00643890	–	Phase 2, randomized, double-blind, multicenter	AAV2-GAD (1×10^{12} vg/mL)	Bilateral STN, ~ 34.5 μL per side	Progressive, levodopa-responsive PD per UK PDS criteria; stable dopaminergic regimen ≥ 4 weeks; UPDRS III off-medication ≥ 25; age 30–75; symptom duration ≥ 5 years; levodopa responsiveness ≥ 12 months. Excluded: prior brain surgery, dopamine-blocking drugs, focal neurological deficits, abnormal MRI, atypical PD pattern on FDG-PET, cognitive impairment (Mattis DRS < 130).	45 (22 active, 23 sham)	6 months	8.1-point decrease vs. 4.7 of UPDRS score in sham†	No serious treatment-related adverse events
Enzyme modification										
Christine et al. (2009)	NCT00229736	NBIb-1817	Phase 1, open-label, single center	AAV2-AADC (9×10^{10}–3×10^{11} vg/mL)	Bilateral postcommissural putamen, 100 μL/side	PD with intractable motor fluctuations despite optimized therapy; H&Y stage III–IV off medication; documented improvement with dopaminergic challenge; stable antiPD regimen ≥ 2 months. Exclusions: atypical parkinsonism, violent dyskinesias (past 6 months), prior stereotactic surgery, MMSE ≤ 26, hallucinations/delusions (past 6 months), major psychiatric illness, GDS > 10 (>5 if on antidepressants), recent malignancy (≤5 years except BCC), or high AAV-2 neutralizing antibody titer (≥1:1200).	10 (5 low, 5 high doses)	6 months	31% reduction of UPDRS †	One intracranial bleed (not vector-related)

Continued

TABLE 11.3 Clinical trial for gene therapy of PD—Cont'd

Study	NCT	Name of product	Design and phase	Vector/ target	Delivery site and dose	Eligibility criteria	N (patients)	Follow-up and monitoring	Motor outcomes (UPDRS/ MDS-UPDRS part III, "off" state)	Safety
Christine et al. (2019)	NCT01973543	NBIb-1817	Phase 1, open-label, multicenter	AAV2-AADC (8.3×10^{11}–2.6×10^{12} vg/mL)	Bilateral postcommissural putamen, 450–900 μL/side	Eligibility criteria included age 40–70 years; moderately advanced PD; responsiveness to dopaminergic therapy; disabling motor fluctuations despite optimal medical treatment; disease duration ≥ 5 years; mH&Y stage ≥ 2.5 in the off-medication state; stable PD medication regimen ≥ 4 weeks before screening.	15 (5 per cohort)	12 months	15.6 (low), 17.8 (medium), 13.6 (high) point decreases of UPDRS‡	No major treatment-related complications
Christine et al. (2022)	NCT01973543	NBIb-1817	Phase 1, open-label, multicenter	AAV2-AADC (8.3×10^{11}–2.6×10^{12} vg/mL)	Bilateral postcommissural putamen, 450–900 μL/side	Eligibility criteria included age 40–70 years; moderately advanced PD; responsiveness to dopaminergic therapy; disabling motor fluctuations despite optimal medical treatment; disease duration ≥ 5 years; mH&Y stage ≥ 2.5 in the off-medication state; stable PD medication regimen ≥ 4 weeks before screening.	15 (5 per cohort)	3 years	UPDRS III "off" scores showed sustained improvement across all cohorts, with mean decreases of 19.0 points (high dose), 12.2 points (medium dose), and 10.2 points (low dose)	No major treatment-related complications
Muramatsu et al. (2010)	–	–	Phase 1, open-label, single center	AAV2-AADC (1.5×10^{12} vg/mL)	Bilateral putamen, 100 μL/side	Eligible participants were 45–75 years old with moderate–advanced PD (Hoehn and Yahr stage IV), ≥ 5 years of levodopa therapy, OFF-state UPDRS III ≥ 20, ≥ 8-point levodopa response, and motor complications despite optimized medical treatment. Exclusion criteria included atypical parkinsonism, dementia (MMSE < 20), and any prior neurosurgical intervention for PD.	6	6 months	46% reduction in UDPRS †	One transient venous infarct (not therapy-related)

Palfi et al. (2014)	NTC00627588/ NCT01856439	ProSavin	Phase ½, open-label, multicenter	Lentiviral vector encoding TH + AADC + GCH1 (1.9×10^7–1×10^8 TU)	Bilateral putamen	Patients aged 48–65 years with ≥ 5 years of disease duration, Hoehn and Yahr stages III–IV (off), UPDRS III (off) 20–60, L-Dopa-responsive motor complications, stable medications ≥ 4 weeks, and ≥ 50% improvement in UPDRS III after an acute L-Dopa challenge.	15 (3 low, 6 medium, 6 high doses)	12 months	7.7, 14.8, and 13.2-point decreases in UPDRS in low, medium, and high groups, respectively	No major treatment-related complications
Palfi et al. (2018)	NTC00627588/ NCT01856439	ProSavin	Phase ½, open-label, multicenter	Lentiviral vector encoding TH + AADC + GCH1 (1.9×10^7–1×10^8 TU)	Bilateral putamen	Patients aged 48–65 years with ≥ 5 years of disease duration, Hoehn and Yahr stages III–IV (off), UPDRS III (off) 20–60, L-Dopa-responsive motor complications, stable medications ≥ 4 weeks, and ≥ 50% improvement in UPDRS III after an acute L-Dopa challenge.	15 (3 low, 6 medium, 6 high doses)	Minimum of 2 years	Significant improvement in UPDRS III (off state) versus baseline: the mean score improved from 38.4 to 29.2 at 2 years ($n = 14$, $p < 0.05$), with sustained benefit at 4 years in 8 of 15 patients	No major treatment-related complications
Neurodegeneration										
Marks et al. (2008)	NCT00252850	CERE-120	Phase 1, open-label, multicenter	AAV2-NRTN (1.3×10^{11}–5.4×10^{11} vg)	Bilateral putamen, 40 µL/side	Adults aged 35–75 years with moderate–severe PD (Hoehn and Yahr stages III–IV; UPDRS III ≥ 30 in the off state) for ≥ 5 years, good levodoparesponse, motor complications inadequately controlled medically, ≥ 3 h/day of off-time, stable PD medications ≥ 1 month, and medically fit for surgery. Key exclusions: atypical parkinsonism, dementia/ MMSE < 25, recent investigational or contraindicated drugs, previous PD neurosurgery, MRI abnormalities, major medical/ psychiatric illness, recent immunotherapy/chemotherapy/ vaccination, prior gene therapy, or substance abuse	12 (6 low, 6 high dose)	12 months	14-point reduction in UPDRS at 12 months; no PET signal change was observed	No major treatment-related complications

Continued

TABLE 11.3 Clinical trial for gene therapy of PD—Cont'd

Study	NCT	Name of product	Design and phase	Vector/ target	Delivery site and dose	Eligibility criteria	N (patients)	Follow-up and monitoring	Motor outcomes (UPDRS/ MDS-UPDRS part III, "off" state)	Safety
Marks et al. (2010)	NCT00400634	CERE-120	Phase 2, randomized, double-blind, multicenter	AAV2-NRTN (5.4×10^{11} vg)	Bilateral putamen	Eligible patients were 35–75 years old with idiopathic PD ≥ 5 years, good levodopa responsiveness, and ≥ 2 h/day of off-time. Required UPDRS III (off) ≥ 30 and stable antiparkinsonian therapy for ≥ 1 month. Exclusion criteria included atypical/secondary parkinsonism, MMSE ≤ 26, prior neurosurgical PD treatment, and significant medical or psychiatric comorbidity.	58 (38 active, 20 sham)	12 months	7.2 vs. 6.9-point reduction in UPDRS at 12 months; having no significant difference	37 SAEs in 17 patients, none clearly vector-related
Bartus et al. (2013)	NCT00985517	CERE-120	Phase 1, open-label	AAV2-NRTN (5.4×10^{11} vg) to Putamen, 4.0×10^{11} vg to substantia nigra	Putamen and substantia nigra	Eligible patients were 35–75 years old with idiopathic PD ≥ 5 years, good levodopa responsiveness, and ≥ 2 h/day of off-time. Required UPDRS III (off) ≥ 30 and stable antiparkinsonian therapy for ≥ 1 month. Exclusion criteria included atypical/secondary parkinsonism, MMSE ≤ 26, prior neurosurgical PD treatment, and significant medical or psychiatric comorbidity.	6	2 years	UPDRS decreased from 38.2 (62.6) to 32.7 (63.6)	No major treatment-related complications

Marks et al. (2016)	–	CERE-120	Long-term follow-up (3–5 years) of patients completing Phase 1, open-label (putamen), Phase 2a randomized sham-controlled (putamen), and revised Phase 1 (putamen + SN)	AAV2-NRTN (5.4×10^{11} vg)	Bilateral putamen (Phase 1 and Phase 2a) or bilateral putamen + substantia nigra (revised Phase 1); doses: 1.3×10^{11} vg (6 pts), 5.4×10^{11} vg (41 pts), 9.4×10^{11} vg (3 pts), 24×10^{11} vg (3 pts)	–	53 enrolled (12 Phase 1, 35 Phase 2a, 6 revised Phase 1); 41 completed the full 5-year protocol	5.3 (range 3.0–7.75) years	Mean OFF-state UPDRS motor score at 5 years: Phase 1 cohort ↓14.9% (41.8→33); Phase 2a cohort ↑3.2% (38.6→38.8); no formal 5-year scores for revised Phase 1, but symptoms similar to 24-month outcomes	184 AEs in 44 subjects (29 severe in 17); only 9 AEs in 6 subjects "possibly related" to therapy. 33 SAEs in 18 subjects, mostly unrelated. Two AEs (neck pain, cervical spine stenosis) in revised Phase 1 deemed unrelated. No intracranial or laboratory abnormalities attributable to treatment
Olanow et al. (2015)	NCT00985517	CERE-120	Phase 2, randomized, double-blind, multicenter	AAV2-NRTN	Bilateral putamen (1×10^{12} vg/side) + SNpc (2×10^{12} vg/side)	Eligible participants were 35–75 years old with idiopathic PD per UK Brain Bank criteria, demonstrated a good L-dopa response, and had motor complications inadequately controlled by medication despite stable therapy for ≥ 6 weeks. Key exclusions included atypical or secondary parkinsonism, Mattis Dementia Rating Scale < 130, prior neurosurgical PD treatment, and any clinically significant medical, psychiatric, or laboratory abnormality.	51 (23 active, 25 sham)	Mean of 18.9 and 18.1 months for sham-surgery groups, respectively.	7.0 vs. 5.2-point reduction in UPDRS at 15 months	Two transient intracranial hemorrhages

Continued

TABLE 11.3 Clinical trial for gene therapy of PD—Cont'd

Study	NCT	Name of product	Design and phase	Vector/ target	Delivery site and dose	Eligibility criteria	N (patients)	Follow-up and monitoring	Motor outcomes (UPDRS/ MDS-UPDRS part III, "off" state)	Safety
Heiss et al. (2019)	NCT01621581	–	Phase 1, open-label, single center	AAV2-GDNF (9×10^{10}–9×10^{11} vg)	Bilateral putamen	Patients older than 18, with advanced PD not well-controlled with medication	13 (6 low, 6 medium, 1 high dose)	18 months	+2.6 (low dose), – 7.1 (medium dose) at 18 months	No serious drug-related adverse events
Van Laar et al. (2025)	NCT04167540	–	Phase 1b, open-label	AAV2-GDNF (3.3×10^{12} vg/mL)	Bilateral putamen	Adults aged 35–75 years with mild–moderate idiopathic PD (UK Brain Bank criteria), mH&Y stages I–III (off medication), and clear levodopa responsiveness (≥30% MDS-UPDRS III improvement). Key exclusions were atypical parkinsonism, pathogenic PD-related gene mutations, UDysRS > 30, prior brain surgery, and significant cognitive, psychiatric, or substance-use comorbidity.	11 (mild PD n = 6; moderate PD n = 5)	18 months	Mild PD: stable; Moderate cohort: mean reduction of −20.4 ± 4.5 points (47.8 % ± 8.5% improvement) in UPDRS at 18 months off	no vector-related SAEs
Rocco et al. (2022)	NCT01621581	–	Phase 1, open-label, single center	AAV2-GDNF (9×10^{10} vg [n = 6]; 3×10^{11} vg [n = 6]; 9×10^{11} vg [n = 1])	Bilateral anterior + posterior putamen, coinfusion with gadoteridol; total 450 μL per putamen under real-time MRI guidance	Adults aged > 18 years with idiopathic PD for > 5 years, Hoehn and Yahr stages III–IV (off), UPDRS motor ≥ 30, clear levodopa responsiveness (≥30% improvement), and disability despite optimized therapy, with normal laboratory and coagulation profiles, and no atypical parkinsonism, dementia (MoCA ≤ 25), major psychiatric or substance-use disorders, significant medical comorbidity, prior brain surgery, MRI contraindications, genetic PD, normal PET uptake, active infection, or recent investigational treatment.	13 (10 men, 3 women)	Up to 5 years	UPDRS "off" scores stabilized over 36–60 months; increased ^{18}F-DOPA uptake at 18 months in 92% of participants	All infusions well tolerated; no treatment-related serious AEs; 1 death 3.5 years post-gene therapy (unrelated)

Abbreviations: *AADC*, aromatic L-amino acid decarboxylase; *AAV*, adeno-associated virus; *AAV2-GDNF*, adeno-associated virus serotype 2 encoding glial cell line-derived neurotrophic factor; *AAV-GAD*, adeno-associated virus encoding glutamic acid decarboxylase; *AAV9*, adeno-associated virus serotype 9; *AE*, adverse event; *AXO-Lenti-PD*, lentiviral gene therapy vector for Parkinson's disease; *BDI-II*, Beck Depression Inventory II; *BOLD-MRI*, blood oxygen level-dependent magnetic resonance imaging; *CAP-003*, investigational GBA1-targeted gene therapy; *CGI*, Clinical Global Impression; *CGI-C*, Clinical Global Impression—Change; *CGI-I*, Clinical Global Impression—Improvement; *CSF*, cerebrospinal fluid; *CTCAE*, Common Terminology Criteria for Adverse Events; *DLT*, dose-limiting toxicity; *DTI*, diffusion tensor imaging; F-Dopa, [^{18}F]-fluorodopa; *FMT-PET*, [^{18}F]-fluoromethyltyrosine positron emission tomography; *GAD*, glutamic acid decarboxylase; *GBA1*, glucocerebrosidase 1 gene; *GCase*, glucocerebrosidase enzyme; *GluSph*, glucosylsphingosine; *MDS-UPDRS*, Movement Disorder Society—Unified Parkinson's Disease Rating Scale; *MoCA*, Montreal Cognitive Assessment; *MRI*, magnetic resonance imaging; *MSA-P*, multiple system atrophy, parkinsonian subtype; *NCS*, nerve conduction study; *NfL*, neurofilament light chain; *NM-MRI*, neuromelanin-sensitive magnetic resonance imaging; *NMSS*, Nonmotor Symptoms Scale; *PD*, Parkinson's disease; *PDQ-39*, 39-item Parkinson's Disease Questionnaire; *PDSS-2*, Parkinson's Disease Sleep Scale-2; *PET*, positron emission tomography; *PGI*, *Patient Global Impression*; *PGI-C*, Patient Global Impression—Change; *PGI-I*, *Patient Global Impression—Improvement*; *QoL*, quality of life; *SAE*, serious adverse event; *STN*, subthalamic nucleus; *TEAE*, treatment-emergent adverse event; *UDysRS*, Unified Dyskinesia Rating Scale; *UPDRS*, Unified Parkinson's Disease Rating Scale.†: Denotes statistical significance.

corresponding clinical trial (NCT03720418) is now recorded as "terminated" on ClinicalTrials.gov. Highlighting such terminations and nonprogression beyond early phases is important, as they reveal not only safety, manufacturing, or efficacy hurdles, but also strategic and financial barriers that can prevent promising biologically targeted therapies from reaching Phase 3 trials and eventual clinical implementation.

7 Neuroprotective approaches

While DA enzyme strategies offer symptomatic relief, neuroprotection aims to modify disease progression by preserving vulnerable nigrostriatal neurons.

7.1 Neurturin

AAV2-mediated delivery of NRTN, a GDNF analog, was among the first attempts. A Phase 1 study (n = 12) demonstrated safety and modest motor improvements at 12 months, but no changes in PET dopaminergic activity (Marks et al., 2008). A subsequent Phase 2 trial (n = 58) found no significant difference between treatment and sham groups, raising doubts about efficacy (Marks et al., 2010). Later attempts combining putaminal and substantia nigra delivery also failed to meet endpoints, though postmortem data confirmed persistent NRTN expression for up to 10 years (Bartus et al., 2013). To address possible issues with axonal transport, a subsequent trial involved delivering AAV2-NRTN not only to the putamen but also to the substantia nigra. In this double-blind, randomized controlled study, 51 patients with advanced PD received bilateral injections of AAV2-NRTN at doses of 2×10^{11} vg in the substantia nigra and 1×10^{12} vg in the putamen. However, after 15 months of follow-up, the trial found no significant differences in either the primary or secondary outcomes (Warren Olanow et al., 2015). A long-term follow-up study involving 53 patients, conducted 5 years after receiving bilateral AAV2-NRTN injections either in the putamen alone or in both the putamen and substantia nigra, reported no deaths. However, one case of psychosis and three cases of dyskinesia were observed (Marks et al., 2016). Postmortem analysis performed a decade after the gene therapy revealed NRTN expression in 1.3%–5.9% of the putamen and in 11% of the substantia nigra, alongside a threefold increase in TH expression (Chu and Kordower, 2023).

7.2 Glial cell line-derived neurotrophic factor

Continuous intraputaminal infusion of recombinant GDNF had earlier demonstrated safety but inconsistent efficacy. More recently, AAV2-GDNF was infused into the putamen under intraoperative MRI guidance. A Phase 1 trial (n = 13) showed stabilization of UPDRS scores and increased [18 F]-DOPA uptake in some patients, though clinical improvements were modest. Importantly, long-term follow-up up to 5 years confirmed sustained GDNF expression and dopaminergic fiber preservation in postmortem analyses, providing biological proof-of-concept (Whone et al., 2019). Due to the challenges of managing continuous brain infusion systems and the potential advantages of using the body's own cells to produce growth factors, researchers explored new gene therapy approaches involving a single infusion of vectors to achieve localized, sustained GDNF expression. In a Phase 1 trial, AAV2-GDNF was infused bilaterally into the putamen, with gadoteridol added to the vector to enable real-time MRI monitoring during surgery (Heiss et al., 2019). About 26% of the putamen's radiographic volume was covered. Although UPDRS scores did not improve, they remained stable, and no treatment-related adverse effects were noted over a 5-year follow-up. The trial also used ^{18}F-6-fluoro-L-DOPA PET imaging to observe bilateral tracer uptake in the posterior putamen, supporting ongoing investigations (NCT04167540) (Rocco et al., 2022). In addition, postmortem analysis nearly 4 years after treatment revealed sustained GDNF expression in the putamen, along with increased TH-positive fibers and elevated local catecholamine levels on HPLC, which correlated with the patient's earlier F-DOPA PET findings 18 months posttreatment (Heiss et al., 2024).

Lastly, a recent Phase 1b open-label trial evaluated the safety and preliminary efficacy of a single high-dose bilateral intraputaminal infusion of AAV2-GDNF (AB-1005) in eleven patients with mild to moderate PD. The vector was administered under real-time MRI guidance, achieving a mean putaminal coverage of 63%. At 18 months posttreatment, the gene therapy was well-tolerated, with no AAV2-GDNF-related serious adverse events. While patients with mild disease showed numerical stability in clinical measures, those with moderate disease demonstrated improvements in MDS-UPDRS Part III "off" scores (−20.4 ± 4.5), reduced "off" time on motor diaries (−1.7 ± 1.1 h), and lower levodopa equivalent daily dose (−257.6 ± 162.2 mg) (Van Laar et al., 2025).

7.3 Glucocerebrosidase (GBA1)

Heterozygous GBA mutations are a common genetic risk factor for PD, associated with earlier onset and faster progression. The PROPEL trial (NCT04127578) is testing AAV9-GBA delivery via intracisternal infusion in GBA-associated PD. Though results are pending, this trial represents the first gene therapy targeting a causative mutation in PD. Safety and biomarker analyses (e.g., CSF GCase activity, glycosphingolipid levels) will provide critical translational insights.

8 Thalamic and subthalamic circuit modulation

Another strategy leverages gene therapy to normalize aberrant basal ganglia circuitry by altering neurotransmitter balance.

8.1 Glutamic acid decarboxylase

The first-in-human trial, initiated in 2002 (NCT00195143), was an open-label, dose-escalation study assessing the safety and tolerability of unilateral rAAV-glutamic acid decarboxylase (GAD) infusion into the STN of twelve patients with PD. Participants were allocated to receive low (1.0×10^{11} vg/mL), medium (3.0×10^{11} vg/mL), or high (1.0×10^{12} vg/mL) vector doses, infused at a controlled rate of 0.5 μL/min for 100 min. The viral construct expressed both GAD65 and GAD67 isoforms under the CMV/CBA promoter in a 1:1 ratio. Notably, patients exhibited significant improvements in motor UPDRS scores by 3 months postinjection, with benefits sustained for up to 12 months. The clinical gains were most evident contralateral to the treated hemisphere, consistent with the lateralized action of the intervention (Kaplitt et al., 2007). A larger Phase 2 sham-controlled trial (n = 45) showed significantly greater UPDRS motor improvement in the AAV2-GAD group (−8.1 vs −4.7 points) and reduced levodopa-induced dyskinesias. FDG-PET analyses confirmed normalization of network metabolism. A subsequent randomized, double-blind, placebo-controlled Phase 2 trial enrolled 44 patients with advanced PD, who received bilateral STN infusions of rAAV-GAD (1.0×10^{12} vg/mL) at 0.23 μL/min for 2.5 h. Among patients assessed in the off-medication state, motor UPDRS scores improved by 23.1% at 6 months, with effects persisting for at least 1 year. However, the placebo group also showed a 12.7% improvement, yielding a net therapeutic gain of approximately 10%, which, although meaningful, was less pronounced than the improvement typically achieved with DBS (LeWitt et al., 2011).

More recently, a Phase 1/2 double-blind, placebo-controlled study (NCT05894343), launched in 2022, further evaluated the safety and efficacy of rAAV-GAD at two dose levels (7.0×10^{10} vg and 2.1×10^{11} vg) compared with sham treatment. The trial included 14 PD patients and assessed safety, tolerability, and motor performance at baseline, 12 weeks, and 26 weeks. Completed in September 2024, preliminary reports indicated that rAAV-GAD was well tolerated, and significant motor improvement was observed in the high-dose group only. Long-term follow-up extending to 5 years posttreatment is ongoing to evaluate durability, safety, and sustained clinical benefit.

Moreover, a list of unpublished clinical trials is shown in Table 11.4.

9 Preclinical advance

9.1 Animal models

No single animal model replicates the entire clinical and pathological spectrum of PD; instead, each captures a manageable aspect of the biology. For gene therapy, that is an advantage, not a flaw: we select models whose lesion mechanisms, time courses, and readouts align with the payload's goal, such as DA restoration, proteostasis repair, mitochondrial quality control, or circuit rebalancing, and with the delivery methods we can practically implement. Below, we organize the main platforms by the biology they model and how they support gene therapy research. The models are summarized in Table 11.5.

9.1.1 Neurotoxin platforms

Models created by neurotoxins provide quick, targeted stressors for aligning "mechanism-readout." The 6-hydroxydopamine (6-OHDA) hemilesion remains the primary neurotoxin model for PD because it produces a clear, unilateral loss of dopaminergic neurons with quick and measurable motor outcomes (Deumens et al., 2002). Mechanistically, 6-OHDA enters catecholaminergic neurons through DAT (and the noradrenaline transporter), auto-oxidizes to create ROS, inhibits mitochondrial complex I and IV, collapses the membrane potential, depletes ATP, and causes oxidative damage to lipids, proteins, and DNA, ultimately leading to the death of dopaminergic neurons in the SNpc (Puspita et al., 2017). Since it cannot cross the BBB, the toxin must be delivered stereotactically into areas like the SNpc, medial forebrain bundle (MFB), or

TABLE 11.4 The list of ongoing clinical trials without published results on PD.

NCT number	Study title	Status	Interventions	Primary outcome measures (brief)	Key secondary outcome measures (brief)	Collaborators/ sponsor	Phase	Study type	Planned/ Actual completion
NCT07011771	A Clinical Trial of CAP-003 Gene Therapy in Adult Patients With GBA1 Associated Parkinson's Disease	Suspended	CAP-003 gene therapy (single IV infusion)	TEAEs/safety laboratories, ECG, vitals, physical examination over 2 years	Plasma and CSF GluSph and GCase levels; change in MDS-UPDRS II–III scores over 24 months	–	Phase ½	Interventional	2029-11
NCT06285643	A Study of AAV2-GDNF in Adults With Moderate Parkinson's Disease (REGENERATE-PD)	Recruiting	AAV2-GDNF gene therapy vs. control surgery	Change from baseline to Month 18 in PD motor diary "good ON" and "OFF" time (Hauser diary)	–	Bayer	Phase 2	Interventional	2027-11-30
NCT07085195	A Clinical Study of the Safety and Efficacy of Chemogenetics Therapy in the Treatment of Parkinson's Disease	Not yet recruiting	STP-001 AAV chemogenetic vector + low-dose clozapine	Incidence of AEs/ SAEs; change in AAV binding/neutralizing antibody titers over 12 months	Change in antiPD drug use; MDS-UPDRS; multiple motor, nonmotor, sleep, cognitive, mood, and QoL scales; serial MRI (BOLD, DTI, QSM, NM-MRI)	–	Early Phase 1	Interventional	2025-12-30
NCT02418598	AADC Gene Therapy for Parkinson's Disease	Terminated	AAV-hAADC-2 putaminal infusion (two cohorts)	Safety of intraputaminal AAV-hAADC-2 (AEs, laboratory abnormalities) at 6 months	Change in PD symptoms (diaries, clinical assessment, levodopa dose); FMT-PET assessment of intraputaminal AADC expression	Takara Bio Inc.; Gene Therapy Research Institute	Phase ½	Interventional	2018-03-31
NCT05894343	Long-term Follow-up of Glutamic Acid Decarboxylase (GAD) Gene Transfer in Parkinson's Disease	Active, not recruiting	AAV-GAD (long-term follow-up only)	Number of study-drug-related AEs/SAEs up to 60 months	–	–	Phase ½	Interventional	2029-10
NCT05603312	A Double-blind Study to Evaluate the Safety of Glutamic Acid Decarboxylase Gene Transfer in Parkinson's Participants	Completed	AAV-GAD low dose; AAV-GAD high dose; sham surgery	Incidence of treatment-related AEs and SAEs (week 26)	–	–	Phase ½	Interventional	2024-09-06

Continued

TABLE 11.4 The list of ongoing clinical trials without published results on PD—Cont'd

NCT number	Study title	Status	Interventions	Primary outcome measures (brief)	Key secondary outcome measures (brief)	Collaborators/ sponsor	Phase	Study type	Planned/ Actual completion
NCT05822739	Safety and Efficacy Study of Parkinson's Disease Gene Therapy Drug (BBM-P002)	Not yet recruiting	BBM-P002 gene therapy	Incidence of AEs/ SAEs; MRI evidence of toxicity at/around the infusion site (up to week 52)	Change in MDS-UPDRS (efficacy) at week 52	–	Early Phase 1	Interventional	2028-12-30
NCT03720418	Study of OXB-102 (AXO-Lenti-PD) in Patients With Bilateral, Idiopathic Parkinson's Disease	Terminated	OXB-102 (AXO-Lenti-PD) vs. imitation surgical procedure	TEAEs/SAEs; clinically significant changes in laboratories, vitals, MRI, and physical examination (3-month timepoint)	Change in UPDRS scores ("OFF" and "ON"); change in OFF time; change in dyskinesia rating scale (up to 6 months)	–	Phase ½	Interventional	2022-04-12
NCT04127578	Phase 1/2a Clinical Trial of PR001 (LY3884961) in Patients With Parkinson's Disease With at Least One GBA1 Mutation (PROPEL)	Recruiting	LY3884961 (AAV9-GCase) + immunosuppression	TEAEs/SAEs; procedure- or treatment-emergent AEs (MRI, spine MRI, NCS); immunogenicity of AAV9 and GCase (blood/CSF)	Changes in glycolipid and GCase levels and enzyme activity (blood/CSF), over 12 months	Eli Lilly and Company	Phase ½	Interventional	2030-12-31
NCT01301573	Long Term Follow-Up Study for rAAV-GAD Treated Subjects	Terminated	rAAV-GAD (STN; observational follow-up)	Long-term safety: AEs/ SAEs over 5 years	–	–	–	Observational	–
NCT05819359	Efficacy, Safety, Tolerability, Pharmacodynamics, and Pharmacokinetics of BIA 28-6156 in GBA-PD	Active, not recruiting	BIA 28-6156 10 mg; BIA 28-6156 60 mg; placebo	Time to clinically meaningful motor progression (MDS-UPDRS II ≥ 2-point ↑ plus no improvement in III) over 78 weeks	Time to progression on MDS-UPDRS III; CGI-C, PGI-C; change in MDS-UPDRS total, modified Hoehn and Yahr, PDQ-39, and other clinical endpoints	–	Phase 2	Interventional	2026-07-31

NCT07081841	AB-1005 Long-Term Follow-up Study	Enrolling by invitation	AB-1005 (prior gene therapy recipients only)	Long-term safety and tolerability: serious and nonserious AEs up to 10 years post-surgery	–	–	–	Observational	2038-03-30
NCT03562494	VY-AADC02 for Parkinson's Disease With Motor Fluctuations (RESTORE-1)	Completed	VY-AADC02 vs sham surgery	Number of TEAEs up to 5 years after surgery	—(Registry focused on long-term safety; efficacy outcomes reported in parent protocol)	Voyager Therapeutics	Phase 1	Interventional	2024-10-30
NCT07195825	A Clinical Study to Evaluate the Safety, and Tolerability of BBM-P002 in the Treatment of Parkinson's Disease	Not yet recruiting	BBM-P002 gene therapy (bilateral putaminal injection)	Incidence of dose-limiting toxicity (DLT) and AEs/SAEs within 4 weeks	Long-term AEs/SAEs; antibody titers; MRI changes; MDS-UPDRS I–IV; PGI-I, CGI-I; 18 F-DOPA PET activity; levodopa dose and ON-time over 5 years	–	Phase 1	Interventional	2031-12-30
NCT03733496	Observational, Long-term Extension Study for Participants of Prior VY-AADC01 Studies	Completed	VY-AADC01 (extension follow-up)	Incidence of TEAEs/SAEs (long-term safety)	F-DOPA PET AADC activity; PD medication changes; UPDRS motor function; ON/OFF diary states over up to 8 years	Voyager Therapeutics	–	Observational	2023-07-18
NCT03065192	Safety and Efficacy Study of VY-AADC01 for Advanced Parkinson's Disease	Completed	VY-AADC01 gene transfer	Safety: AEs/SAEs (CTCAE 4.03), MRI abnormalities, physical examination, laboratories, C-SSRS over 3 years	Changes in PD medications; motor function (PD diaries, stand-walk-sit test, Modified Hoehn and Yahr, UPDRS, UDysRS); mood, cognition, sleep, QoL scales	Voyager Therapeutics	Phase 1	Interventional	2021-08-10

TABLE 11.5 Animal models in PD gene therapy research.

Category	Platform/model	Core lesion biology	Induction and target	Key readouts	Typical gene therapy use
Neurotoxin	**6-OHDA (rat/mouse, unilateral hemilesion)**	Catecholaminergic toxin; oxidative stress; mitochondrial complex I/IV dysfunction; SNpc DA neuron loss	Stereotactic injection to the MFB, SNpc, or striatum; DAT uptake; desipramine cotreatment to spare NE	Amphetamine/apomorphine rotations, cylinder/stepping tests; TH immunostaining; striatal DA/DOPAC	Dose/payload titration for AADC, TH/AADC/GCH1 (±VMAT2); GDNF/NRTN rescue; STN-GAD circuit work; delivery physics
	MPTP (mouse)	Systemic toxin; astrocytic MAO-B to MPP^+; mitochondrial complex I inhibition	Repeated i.p. dosing (≤20 mg/kg); systemic	Open-field, pole test; TH counts; DA biochemistry	Mitochondrial-axis therapies; imaging pharmacodynamics
	MPTP (nonhuman primate)	Same mechanism; bilateral or hemi-parkinsonism	Systemic or intracarotid dosing	Clinical rating scales; levodopa responsiveness; dyskinesia; PET (FDOPA/FMT)	Translational validation; CED scale-up; surgical workflow rehearsal
	Pesticides (rotenone, paraquat, maneb)	Environmental toxicants; oxidative and mitochondrial stress; chronic rotenone may show α-syn-like inclusions	Systemic exposure (varies by agent/protocol)	Motor tests; pathology; inflammatory markers	Mitochondrial restoration; peripheral–central interaction studies
Proteinopathy	**rAAV/LV α-syn overexpression (WT, A53T/A30P)**	α-Syn proteinopathy; mitochondrial and autophagy stress; neuroinflammation	Viral delivery to SNpc (rodent, NHP); capsid/promoter/titer configurable	Motor tests; TH loss; α-syn inclusions; microglial markers	SNCA-lowering (RNAi/ASO/ZF-repressor); GBA1 augmentation; delivery tuning
	α-Syn preformed fibrils (PFF) seeding	Pathology seeding of endogenous α-syn; network propagation	Intrastriatal PFF injection; sonicated recombinant or brain-derived fibrils	Phospho-α-syn pathology; motor deficits on a delayed timeline	SNCA-lowering, GBA1 gene therapy; route/coverage studies (including gut–brain)

Genetic models	**PINK1/PRKN loss (mouse)**	Mitophagy failure; mitochondrial stress; neuroinflammation	Knockout/knock-in	Mitochondrial assays; inflammatory markers; motor tasks	PINK1/Parkin augmentation; mitophagy enhancers
	PINK1 gene-edited NHP	Translational mitophagy deficiency	CRISPR-based editing	Primatology scales; imaging; histology	Late-stage validation for mitophagy restoration
	DJ-1 loss (mouse/rat)	Oxidative stress defense deficit	Knockout; often combined with MPTP	Stress vulnerability; mitochondrial assays	Antioxidant/mitophagy payload testing
	MitoPark mouse (DAT-Cre TFAM deletion)	Mitochondrial transcription failure in DA neurons	Conditional TFAM loss in DA neurons	Motor decline; nonmotor features; TH/DA markers	Mitochondrial rescue; chronic efficacy; symptom progression
	CHCHD2 loss	Cristae maintenance defect; oxidative stress	Knockout	Mitochondrial function; DA measures	Bioenergetic/proteostasis interventions
	HTRA2 deficiency	Mitochondrial protease deficit; protein stress	Knockout/deficiency lines	Proteostasis and DA markers	Proteostasis-targeted payloads
	GBA1 variants (L444P, N370S; ± SNCA/PFF)	Lysosomal/GCase deficit; lipid dysregulation; DA vulnerability	Knock-in/transgenic; crosses with SNCA/PFF	Lysosomal flux; α-syn biochemistry; motor tests	AAV-GBA1; small-molecule chaperones
	LRRK2 (G2019S, R1441C/G)	Endolysosomal traffic defect; kinase hyperactivity	Transgenic/knock-in; fly and mouse lines	Rab10 phosphorylation; synaptic and tau phenotypes; motor tests	LRRK2 kinase inhibitors; ASO programs
	ATP13A2 loss	Lysosomal and polyamine homeostasis defect	Knockout	ALP markers; DA structure/function	Lysosome-stabilizing strategies; polyamine/metal homeostasis
	Parkin knockout	UPS/mitophagy cross-talk; E3 ligase loss	Knockout	Stress-induced rescue; UPS/autophagy markers	Parkin augmentation; pathway activators
	UCH-L1 models	Deubiquitination defect	Transgenic/knockout	UPS markers; motor tasks	UPS-targeted therapies
	FBXO7 models	SCF-E3 ligase pathway defect; mitophagy interface	Transgenic/knockout	UPS/mitophagy markers; motor behavior	UPS–autophagy interface modulation

Abbreviations: *6-OHDA*, 6-hydroxydopamine; *SNpc*, substantia nigra pars compacta; *DA*, dopamine; *MFB*, medial forebrain bundle; *DAT*, dopamine transporter; *NE*, norepinephrine; *TH*, tyrosine hydroxylase; *DOPAC*, 3,4-dihydroxyphenylacetic acid; *AADC*, aromatic L-amino acid decarboxylase; *GCH1*, GTP cyclohydrolase-1; *VMAT2*, vesicular monoamine transporter 2; *GDNF*, glial cell line-derived neurotrophic factor; *NRTN*, neurturin; *STN*, subthalamic nucleus; *GAD*, glutamate decarboxylase; *MPTP*, 1-methyl-4-phenyl-1,2,3,6-tetrahydropyridine; *MAO-B*, monoamine oxidase-B; *MPP*$^+$, 1-methyl-4-phenylpyridinium; *i.p.*, intraperitoneal; *PET*, positron emission tomography; *FDOPA*, 6-[^{18}F]fluoro-L-DOPA; *FMT*, 6-[^{18}F]fluoro-m-tyrosine; *NHP*, nonhuman primate; *rAAV*, recombinant adeno-associated virus; *LV*, lentiviral vector; *WT*, wild type; *α-syn*, alpha-synuclein; *SNCA*, alpha-synuclein gene; *RNAi*, RNA interference; *ASO*, antisense oligonucleotide; *ZF*, zinc finger; *PFF*, preformed fibrils; *CRISPR*, clustered regularly interspaced short palindromic repeats; *TFAM*, mitochondrial transcription factor A; *CHCHD2*, coiled-coil-helix-coiled-coil-helix domain containing 2; *HTRA2*, HtrA serine peptidase 2; *GBA1*, glucosylceramidase beta gene; *GCase*, glucocerebrosidase; *LRRK2*, leucine-rich repeat kinase 2; *ALP*, autophagy–lysosome pathway; *UPS*, ubiquitin–proteasome system; *UCH-L1*, ubiquitin C-terminal hydrolase L1; *FBXO7*, F-box only protein 7; *SCF*, SKP1-CUL1-F-box (E3 ubiquitin ligase) complex; *CED*, convection-enhanced delivery.

striatum (Kirik et al., 1998). However, coadministration of desipramine is often used to protect noradrenergic neurons. The site of injection is important: MFB or SNpc injections lead to rapid, severe degeneration, while intrastriatal injections cause a slower, retrograde degeneration that more closely resembles the progression of human disease (Blum et al., 2001). The unilateral approach produces contralateral motor deficits and robust rotation tests, but also reveals nonmotor features such as depressive-like behavior, delayed gastric emptying, and hyposmia, depending on the dose and injection site (Kuruvilla et al., 2013). However, these models are largely acute, do not show LB-like α-synuclein pathology, and have a narrower range of behavioral symptoms compared to clinical PD. This profile makes 6-OHDA ideal for optimizing symptomatic gene therapies (such as AADC or multigene DA cassettes), trophic support delivery, and circuit rebalancing strategies (e.g., STN-GAD) (Gubellini and Kachidian, 2015), while approaches targeting aggregation or propagation require complementary α-synuclein-based platforms.

MPTP, an organic compound with analgesic properties, crosses the BBB. It is converted to MPP^+ by astrocytic MAO-B and inhibits complex I, directly linking the model to mitochondrial dysfunction (Nicklas et al., 1987; Przedborski et al., 2001). In mice, four acute i.p. injections (≤20 mg/kg) cause about 90% loss of striatal DA with approximately 70% of SNpc neurons losing function within days (Jackson-Lewis and Przedborski, 2007). In NHPs, systemic or intracarotid MPTP induces bilateral or hemiparkinsonism with human-like symptoms, levodopa responsiveness, and dyskinesia on chronic dosing (Fox and Brotchie, 2010; Bankiewicz et al., 1986). Therefore it is the standard platform for scaling surgical delivery (e.g., convection-enhanced putaminal infusions), pharmacodynamic imaging (FDOPA/FMT-PET), and confirming efficacy before first-in-human trials.

Environmental toxicants, especially pesticides like rotenone, paraquat, and maneb, are strongly linked to the onset of PD development (Tanner et al., 2011; Costello et al., 2009). These models emphasize oxidative and mitochondrial stress and, in chronic rotenone paradigms, can display inclusions with α-syn features alongside motor and some nonmotor phenotypes (Cannon et al., 2009; Bonneh-Barkay et al., 2005). They are valuable when a therapy aims to normalize mitochondrial function or peripheral–central interactions, but variability and systemic toxicity limit their widespread preclinical use.

Neurotoxin models are ideal when you need a clean dopaminergic deficit and fast, quantitative motor readouts to titrate dose, promoter strength, or vector spread. They are also the right place to prototype delivery physics (e.g., intraparenchymal convection vs. CSF routes) before moving to NHP MPTP for translational confirmation.

9.1.2 Proteinopathy platforms

Among "protein-first" models, viral overexpression of human α-synuclein (WT or mutants such as A53T/A30P) using AAV or lentiviral vectors remains the most versatile method to induce synucleinopathy in the adult brain, offering control over brain region, titer, and transduced cell populations (Kirik et al., 2002a; Ip et al., 2017). Beyond nigrostriatal circuits, these vectors have also been used to explore potential peripheral-to-central spread of pathology, enabling transmission studies that traditional toxin models cannot address (Recasens and Dehay, 2014).

When expressed in the SNpc, α-syn drives progressive loss of dopaminergic neurons and striatal projections over weeks to months, though the effect size varies with serotype, promoter, dose, age, sex, strain, and species (Koprich et al., 2010; Volpicelli-Daley et al., 2016). In rodents, intranigral α-syn overexpression usually causes about 40% loss of SNpc DA cells and motor deficits by around 8 weeks (Decressac et al., 2012). Transgene and capsid choices are important: WT and A53T cause broadly similar damage in rat rAAV models, while A30P is weaker and occurs later (Van der Perren et al., 2014; Kirik et al., 2002b). In mice, rAAV2/7 delivery can overcome generally mild phenotypes, leading to dose-dependent nigral degeneration by 8 weeks and motor problems by 12 weeks (Oliveras-Salvá et al., 2013). In NHPs, intranigral AAV (WT or A53T) results in approximately 30% nigral loss by 8 weeks with worsening motor decline by about 16 weeks, though with more variability between animals than in rodents, and it allows for longer observation periods. Macaques seem especially susceptible to A53T-encoded vectors (Kirik et al., 2003; Yang et al., 2015; Koprich et al., 2018).

Based on pathology, these viral platforms exhibit several hallmark features: mitochondrial dysfunction, autophagy impairment, early and persistent neuroinflammation, and LB-like inclusions (McKinnon et al., 2020; Chung et al., 2009). In practice, these proteinopathy platforms are crucial for testing α-syn-lowering strategies (such as RNAi, ASO, or synthetic repressors) and lysosomal-axis gene therapies (e.g., GBA1 augmentation) because they sustain a consistent α-syn burden in adult-onset, cell-intrinsic α-syn stress with adjustable severity (Recasens and Dehay, 2014; Oliveras-Salvá et al., 2013).

PFF paradigms complement viral overexpression by seeding pathology rather than overproducing α-syn. Sonicated recombinant (or brain-derived) α-syn fibrils enter host circuits and convert endogenous α-syn into a phosphorylated, aggregation-prone state, driving prion-like propagation across connected nodes (Luk et al., 2012; Recasens et al., 2014). After intrastriatal injection in wild-type mice, pathology appears within approximately 1 month, followed by striatal denervation, SNpc cell loss, and motor deficits on a slower, progressive timeline, features that mirror idiopathic spread dynamics (Chung et al., 2019). Importantly for gene therapy, PFF models impose a sustained, endogenous α-syn burden that is ideal for testing SNCA-lowering vectors and lysosomal restoration (AAV-GBA1) and for probing route and coverage questions, including gut-to-brain seeding to evaluate systemic or CSF-facing delivery strategies (Ahn et al., 2020; Challis et al., 2020). They also reproduce mitochondrial and autophagic stress and neuroinflammation, offering mechanism-linked readouts (Guiney et al., 2020). In NHPs, LB-enriched or fibrillar inocula produce nigrostriatal pathology with variable motor expression, supporting translational biomarker development (Wang et al., 2019).

9.1.3 Genetic platforms

Preclinical gene therapy requires models that emphasize the targeted pathway, whether it's mitochondrial quality control, lysosome and autophagy, or the UPS, and allow for measuring on-target rescue. Below, we highlight concise platforms commonly used to evaluate payloads such as AAV-GBA1, PINK1/Parkin augmentation, LRRK2 inhibition, and UPS modulators.

Key mitochondrial quality control genes and models include PINK1, PRKN/Parkin, DJ-1, the MitoPark mouse (mitochondrial transcription factor A [TFAM] deletion), CHCHD2, and HTRA2. Invertebrates display strong mitochondrial and motor phenotypes useful for screening (Park et al., 2006; Cooper et al., 2017), while mouse PINK1/PRKN deletions model neuroinflammation and mitochondrial stress with slow or age-related nigral changes. Therefore they are ideal for long-term rescue studies but time-consuming for testing efficacy (Noda et al., 2020; Dawson et al., 2010). A PINK1 gene-edited NHP using the CRISPR system improves translation, showing behavioral changes around 1.5 years and SNpc neuron loss by about 3 years (Yang et al., 2019). Moreover, DJ-1 models emphasize oxidative stress defenses. Knouckout (KO) mice and rats are more vulnerable to mitochondrial and inflammatory challenges and are often combined with MPTP to speed up phenotypes for therapeutic testing (Xu et al., 2017; Ji et al., 2020). The MitoPark mouse is another model that features conditional loss of TFAM in DA neurons. It exhibits a progressive, age-related, levodopa-responsive phenotype, valuable for testing mitochondrial rescue strategies against both motor and early nonmotor symptoms (Langley et al., 2021; Ekstrand et al., 2007; Beckstead and Howell, 2021). Other mitochondrial regulators include CHCHD2, which is essential for maintaining cristae. Its knockout leads to oxidative stress and DA loss (Meng et al., 2017; Wang et al., 2025; Sato et al., 2021). HTRA2 is a mitochondrial protease; deficiencies result in misfolded protein buildup and DA damage. These insights allow for testing interventions that target bioenergetics and proteostasis (Chung et al., 2020) (reviewed in Peng et al., 2025).

In the ALP pathway and endolysosomal traffic, genes such as GBA1, LRRK2, and ATP13A2 are engaged. GBA1 models (L444P, N370S, and combinations with SNCA/PFF) demonstrate lysosomal stress, mitochondrial dysfunction, and DA vulnerability (Farfel-Becker et al., 2019; Hongyu et al., 2019; Migdalska-Richards et al., 2017; Abul Khair et al., 2018). Therefore they serve as the main platform for AAV-GBA1 and small-molecule chaperones (ambroxol, pioglitazone) (Shola-Dare et al., 2021; Mishra and Krishnamurthy, 2020; Sardi et al., 2013). LRRK2 mutations (G2019S, R1441C/G) disrupt endolysosomal traffic. In Drosophila, human LRRK2 overexpression yields DA degeneration, as well as mitochondrial and autophagic defects (Bardai et al., 2018). In addition, mouse knock-in or transgenic lines exhibit synaptic and Tau-related changes, with late nigrostriatal loss in promoter-targeted lines, making them ideal for testing LRRK2 kinase inhibitors or ASOs and using Rab10 phosphorylation as a pharmacodynamic marker (Volta et al., 2017; Kozina et al., 2018; Xiong et al., 2018). Based on this mechanism, tool compounds (e.g., DNL201, GNE-7915) and ASO41-1 show pathway reversibility in LRRK2 models (Jennings et al., 2022; Hu et al., 2023; Korecka et al., 2020).

ATP13A2 (P5-ATPase) maintains lysosomal and polyamine homeostasis. Loss of ATP13A2 in the brains of mice or NHPs causes progressive nigrostriatal degeneration with ALP defects, making it a useful model for lysosome-stabilizing payloads and strategies for polyamine and metal homeostasis (Erb et al., 2024; Sikora et al., 2024; van Veen et al., 2020).

Genetic alterations in the UPS and mitophagy cross-talk, including Parkin, UCH-L1, and FBXO7, also introduce animal models of PD. Although rodent Parkin KO is often subtle at baseline, it is commonly used to test Parkin enhancement and to study stress-induced rescue, including nicotine and other UPS-autophagy activators (Ge et al., 2020; Ullah et al., 2024). UCH-L1 (Cerqueira et al., 2020; Tran et al., 2018) and FBXO7 models explore deubiquitination and SCF-E3 ligase biology, respectively. Despite limitations (late or nonnigral phenotypes, developmental confounds), they can be combined with α-syn overexpression or toxin hits to reveal targetable deficits (Stott et al., 2019; Spagnol et al., 2021).

10 Limitations and future direction

Gene therapy for PD has advanced from an idea to credible clinical trials, with shared strategies that restore DA production, repair lysosomal and mitochondrial functions, and rebalance basal ganglia circuits. However, progress still depends on two interconnected challenges: delivering enough vector to the correct cells and doing so safely and consistently across different patients.

(1) Delivery and coverage remain the main obstacles. Intraparenchymal infusion works but lacks standardization; infusion rates, catheter technology, and real-time imaging differ across studies, leading to inconsistent target coverage and dose-response relationships. Using gadolinium coinfusion more widely to visualize spread during CED should improve reliability, while determining the best dose and serotype requires systematic research. Crossing the BBB with systemic AAV9 has been shown in mice, but long-term safety, effectiveness, and immune responses in humans are still unresolved. In addition, disease-related axonal transport issues could complicate vector trafficking.
(2) Vector size limits and payload design restrict what we can deliver. Most vectors only carry cDNAs, which restricts multigene or regulatory payloads. HSV-1 amplicons, capable of holding up to approximately 150 kb, allow delivery of entire genomic regions and complex enzyme pathways, an attractive approach for tackling complex PD targets and combination therapies (like support for both trophic factors and enzymatic support). The field is rapidly adopting engineered AAVs with improved targeting to reduce the dose needed and lower immune responses. Advances in the control of gene expression using enhancers and cell-specific promoters should help refine regional and cellular targeting.
(3) New platform innovations are broadening options beyond DNA replacement. RNA editing tools (such as CasRx) offer high specificity without permanent changes to the genome and have even been used to reprogram striatal glia into dopaminergic neurons in vivo, which could be a promising route for circuit repair if confirmed across models. Engineered AAVs might also work with cell therapies to adjust graft output, reduce adverse effects, and decrease rejection. Long-term vectors could correct pathways that increase risk or enhance resilience (like macroautophagy, aging, and rejuvenation), and, if proven safe, could someday be used for presymptomatic interventions in at-risk populations.
(4) Clinical development remains at an early stage. Although gene therapies are approved for certain neurodegenerative and neurometabolic disorders, such as spinal muscular atrophy and specific leukodystrophies, none has yet been approved for PD. Most PD trials still enroll small cohorts. Future studies require sufficiently large cohorts, standardized surgical and infusion procedures, predefined imaging and clinical endpoints, and extended immunosurveillance to satisfy regulatory and ethical standards. Surgical procedures are also evolving toward shorter, image-guided methods, with bedside MRI and improved catheters expected to expand access and ensure consistent results.

11 Conclusion

Overall, the opportunity is clear: better capsids, smarter promoters, higher-capacity vectors, and image-guided delivery can align mechanisms, anatomy, and durability. The challenges are equally clear: coverage, dosage, immune responses, and variability. Overcoming these with careful trial design and next-generation tools is the pathway from promising signs to approved, disease-modifying treatments.

AI disclosure

During the preparation of this work, the author(s) used Perplexity to discover papers. After using this tool/service, the author(s) reviewed and edited the content as needed and take(s) full responsibility for the content of the publication.

References

Abul Khair, S.B., Dhanushkodi, N.R., Ardah, M.T., Chen, W., Yang, Y., Haque, M.E., 2018. Silencing of glucocerebrosidase gene in drosophila enhances the aggregation of Parkinson's disease associated α-synuclein mutant A53T and affects locomotor activity. Front. Neurosci. 1662-453X. 12. doi:10.3389/fnins.2018.00081.

Adler C.H., Caviness J.N., Hentz J.G., Lind M., Tiede J., 2003. Randomized trial of modafinil for treating subjective daytime sleepiness in patients with Parkinson's disease. Mov. Disord. 18(3), 287–293. doi:10.1002/mds.10390.

Ahn, E.H., Kang, S.S., Liu, X., Chen, G., Zhang, Z., Chandrasekharan, B., Alam, A.M., Neish, A.S., Cao, X., Ye, K., 2020. Initiation of Parkinson's disease from gut to brain by δ-secretase. Cell Res. 17487838. 30 (1), 70–87. Available from: http://www.nature.com/cr/index.html. doi:10.1038/s41422-019-0241-9.

Alharbi, B., Al-kuraishy, H.M., Al-Gareeb, A.I., Elekhnawy, E., Alharbi, H., Alexiou, A., Papadakis, M., Batiha, G.E.-S., 2024. Role of GABA pathway in motor and non-motor symptoms in Parkinson's disease: a bidirectional circuit. Eur. J. Med. Res. 2047-783X. 29 (1). doi:10.1186/s40001-024-01779-7.

Alvarez-Erviti, L., Rodriguez-Oroz, M.C., Cooper, J.M., Caballero, C., Ferrer, I., Obeso, J.A., Schapira, A.H.V., 2010. Chaperone-mediated autophagy markers in Parkinson disease brains. Arch. Neurol. 15383687. 67 (12), 1464–1472. Available from: http://archneur.ama-assn.org/cgi/reprint/67/12/1464. doi:10.1001/archneurol.2010.198.

Andrew Young, K.K., Cifrese, J., Hatami, A., Nguyen, H.-O., et al., SNCA Gene Repression Mediated by Zinc Finger Repressors (ZFRs) as a Therapeutic Approach for Parkinson's Disease. 2024.

Axelsen, T.M., Woldbye, D.P.D., 2018. Gene therapy for Parkinson's disease, an update. J. Parkinsons Dis. 1877718X. 8 (2), 195–215. Available from: http://www.iospress.nl/journal/journal-of-parkinsons-disease/. doi:10.3233/JPD-181331.

Baba, M., Nakajo, S., Tu, P.H., Tomita, T., Nakaya, K., Lee, V.M.Y., Trojanowski, J.Q., Iwatsubo, T., 1998. Aggregation of α-synuclein in Lewy bodies of sporadic Parkinson's disease and dementia with Lewy bodies. Am. J. Pathol. 00029440. 152 (4), 879–884.

Bäck, S., Peränen, J., Galli, E., Pulkkila, P., Lonka-Nevalaita, L., Tamminen, T., Voutilainen, M.H., Raasmaja, A., Saarma, M., Männistö, P.T., Tuominen, R.K., 2013. Gene therapy with AAV 2- CDNF provides functional benefits in a rat model of P arkinson's disease. Brain Behav. 2162-3279. 3 (2), 75–88. doi:10.1002/brb3.117.

Badin, R.A., Binley, K., Van Camp, N., Jan, C., Gourlay, J., Robert, C., Gipchtein, P., Fayard, A., Stewart, H., Ralph, G.S., Lad, Y., Kelleher, M., Loader, J., Hosomi, K., Palfi, S., Mitrophanous, K.A., Hantraye, P., 2019. Gene Therapy for Parkinson's Disease: preclinical Evaluation of Optimally Configured TH:CH1 Fusion for Maximal Dopamine Synthesis. Mol. Ther. Methods Clin. Dev. 23290501. 14, 206–216. Available from: https://www.journals.elsevier.com/molecular-therapy-methods-and-clinical-development/. doi:10.1016/j.omtm.2019.07.002.

Banerjee, A., Imig, C., Balakrishnan, K., Kershberg, L., Lipstein, N., Uronen, R.-L., Wang, J., Cai, X., Benseler, F., Rhee, J.S., Cooper, B.H., Liu, C., Wojcik, S.M., Brose, N., Kaeser, P.S., 2022. Molecular and functional architecture of striatal dopamine release sites. Neuron. 08966273. 110 (2), 248. doi:10.1016/j.neuron.2021.10.028.

Bankiewicz, K.S., Oldfield, E.H., Chiueh, C.C., Doppman, J.L., Jacobowitz, D.M., Kopin, I.J., 1986. Hemiparkinsonism in monkeys after unilateral internal carotid artery artery infusion of 1-methyl-4-phenyl-1,2,3,6-tetrahydropyridine (MPTP). Life Sci. 00243205. 39 (1), 7–16. doi:10.1016/0024-3205(86)90431-5.

Bardai, F.H., Ordonez, D.G., Bailey, R.M., Hamm, M., Lewis, J., Feany, M.B., Polleux, F., 2018. Lrrk promotes tau neurotoxicity through dysregulation of actin and mitochondrial dynamics. PLoS Biol. 1545-7885. 16 (12), e2006265. doi:10.1371/journal.pbio.2006265.

Barrett, P.J., Greenamyre, J.T., 2015. Post-translational modification of α-synuclein in Parkinson's disease. Brain Res. 18726240. 1628, 247–253. Available from: https://www.elsevier.com/locate/bri. doi:10.1016/j.brainres.2015.06.002.

Bartels, T., Choi, J.G., Selkoe, D.J., 2011. α-Synuclein occurs physiologically as a helically folded tetramer that resists aggregation. Nature. 14764687. 477 (7362), 107–111. doi:10.1038/nature10324.

Bartus, R.T., Baumann, T.L., Siffert, J., Herzog, C.D., Alterman, R., Boulis, N., Turner, D.A., Stacy, M., Lang, A.E., Lozano, A.M., Olanow, C.W., 2013. Safety/feasibility of targeting the substantia nigra with AAV2-neurturin in Parkinson patients. Neurology. 0028-3878. 80 (18), 1698–1701. doi:10.1212/WNL.0b013e3182904faa.

Bartus, R.T., Herzog, C.D., Chu, Y., Wilson, A., Brown, L., Siffert, J., Johnson, E.M., Olanow, C.W., Mufson, E.J., Kordower, J.H., 2011. Bioactivity of AAV2-neurturin gene therapy (CERE-120): differences between Parkinson's disease and nonhuman primate brains. Mov. Disord. 15318257. 26 (1), 27–36. doi:10.1002/mds.23442.

Bartus, R.T., Kordower, J.H., Johnson, E.M., Brown, L., Kruegel, B.R., Chu, Y., Baumann, T.L., Lang, A.E., Olanow, C.W., Herzog, C.D., 2015. Post-mortem assessment of the short and long-term effects of the trophic factor neurturin in patients with α-synucleinopathies. Neurobiol. Dis. 09699961. 78, 162–171. doi:10.1016/j.nbd.2015.03.023.

Beckstead, M.J., Howell, R.D., 2021. Progressive parkinsonism due to mitochondrial impairment: lessons from the MitoPark mouse model. Exp. Neurol. 00144886. 341, 113707. doi:10.1016/j.expneurol.2021.113707.

Billingsley, K.J., Bandres-Ciga, S., Saez-Atienzar, S., Singleton, A.B., 2018. Genetic risk factors in Parkinson's disease. Cell Tissue Res. 0302-766X. 373 (1), 9–20. doi:10.1007/s00441-018-2817-y.

Björklund, T., Davidsson, M., Björklund, A., Bloem, B.R., Brundin, P., Federoff, H., 2021. Next-Generation Gene Therapy for Parkinson's Disease Using Engineered Viral Vectors. J. Parkinsons Dis. 18777171. 11 (s2), S209. doi:10.3233/jpd-212674.

Bloem, B.R., Okun, M.S., Klein, C., 2021. Parkinson's disease. Lancet. 1474547X. 397 (10291), 2284–2303. Available from: http://www.journals.elsevier.com/the-lancet/. doi:10.1016/S0140-6736(21)00218-X.

Blum, D., Torch, S., Lambeng, N., Nissou, M.-F., Benabid, A.-L., Sadoul, R., Verna, J.-M., 2001. Molecular pathways involved in the neurotoxicity of 6-OHDA, dopamine and MPTP: contribution to the apoptotic theory in Parkinson's disease. Prog. Neurobiol. 03010082. 65 (2), 135–172. doi:10.1016/s0301-0082(01)00003-x.

Bobo, R.H., Laske, D.W., Akbasak, A., Morrison, P.F., Dedrick, R.L., Oldfield, E.H., 1994. Convection-enhanced delivery of macromolecules in the brain. Proc. Natl. Acad. Sci. U. S. A. 0027-8424. 91 (6), 2076–2080. doi:10.1073/pnas.91.6.2076.

Bonneh-Barkay, D., Reaney, S.H., Langston, W.J., Di Monte, D.A., 2005. Redox cycling of the herbicide paraquat in microglial cultures. Mol. Brain Res. 0169328X. 134 (1), 52–56. doi:10.1016/j.molbrainres.2004.11.005.

Borgohain, R., et al, 2009. Efficacy and safety of Safinamide in patients with Parkinson's disease experiencing motor fluctuations: results of a 6-month Phase III, randomized, double-blind, placebo-controlled study. In: 13th International Congress of the Movement Disorder Society, Paris, France.

Cannon, J.R., Tapias, V., Na, H.M., Honick, A.S., Drolet, R.E., Greenamyre, J.T., 2009. A highly reproducible rotenone model of Parkinson's disease. Neurobiol. Dis. 09699961. 34 (2), 279–290. doi:10.1016/j.nbd.2009.01.016.

Cenci, M.A., 2014. Presynaptic Mechanisms of l-DOPA-Induced Dyskinesia: the Findings, the Debate, and the Therapeutic Implications. Front. Neurol. 1664-2295. 5. doi:10.3389/fneur.2014.00242.

Cerqueira, F.M., von Stockum, S., Giacomello, M., Goliand, I., Kakimoto, P., Marchesan, E., De Stefani, D., Kowaltowski, A.J., Ziviani, E., Shirihai, O.S., 2020. A new target for an old DUB: UCH-L1 regulates mitofusin-2 levels, altering mitochondrial morphology, function and calcium uptake. Redox Biol. 22132317. 37, 101676. doi:10.1016/j.redox.2020.101676.

Challis, C., Hori, A., Sampson, T.R., Yoo, B.B., Challis, R.C., Hamilton, A.M., Mazmanian, S.K., Volpicelli-Daley, L.A., Gradinaru, V., 2020. Gut-seeded α-synuclein fibrils promote gut dysfunction and brain pathology specifically in aged mice. Nat. Neurosci. 1097-6256. 23 (3), 327–336. doi:10.1038/s41593-020-0589-7.

Christenson, J.G., Dairman, W., Udenfriend, S., 1970. Preparation and properties of a homogeneous aromatic l-amino acid decarboxylase from hog kidney. Arch. Biochem. Biophys. 10960384. 141 (1), 356–367. doi:10.1016/0003-9861(70)90144-X.

Christine, C.W., et al., 2022. Safety of AADC gene therapy for moderately advanced Parkinson disease: three-year outcomes from the PD-1101 trial. Neurology 98 (1), e40–e50. doi:10.1212/WNL.0000000000012952.

Christine, C.W., Bankiewicz, K.S., Van Laar, A.D., Richardson, R.M., Ravina, B., Kells, A.P., Boot, B., Martin, A.J., Nutt, J., Thompson, M.E., Larson, P.S., 2019. Magnetic resonance imaging–guided phase 1 trial of putaminal AADC gene therapy for Parkinson's disease. Ann. Neurol. 15318249. 85 (5), 704–714. Available from: http://onlinelibrary.wiley.com/journal/10.1002/(ISSN)1531-8249. doi:10.1002/ana.25450.

Christine, C.W., Starr, P.A., Larson, P.S., Eberling, J.L., Jagust, W.J., Hawkins, R.A., Vanbrocklin, H.F., Wright, J.F., Bankiewicz, K.S., Aminoff, M.J., 2009. Safety and tolerability of putaminal AADC gene therapy for Parkinson disease. Neurology. 1526632X. 73 (20), 1662–1669. Available from: http://www.neurology.org. doi:10.1212/WNL.0b013e3181c29356.

Chu, Y., Kordower, J.H., 2023. Post-mortem studies of neurturin gene therapy for Parkinson's disease: two subjects with 10 years CERE120 delivery. Mov. Disord. 15318257. 38 (9), 1728–1736. Available from: http://onlinelibrary.wiley.com/journal/10.1002/(ISSN)1531-8257. doi:10.1002/mds.29518.

Chung, C.Y., Koprich, J.B., Siddiqi, H., Isacson, O., 2009. Dynamic changes in presynaptic and axonal transport proteins combined with striatal neuroinflammation precede dopaminergic neuronal loss in a rat model of AAV α-synucleinopathy. J. Neurosci. 15292401. 29 (11), 3365–3373. Available from: http://www.jneurosci.org/cgi/reprint/29/11/3365. doi:10.1523/JNEUROSCI.5427-08.2009.

Chung, H.J., Jamal, M.A.H.M., Hong, S.T., 2020. The function of bacterial HtrA is evolutionally conserved in mammalian HtrA2/Omi. Sci. Rep. 20452322. 10 (1). Available from: https://www.nature.com/srep/index.html. doi:10.1038/s41598-020-62309-z.

Chung, H.K., Ho, H.A., Pérez-Acuña, D., Lee, S.J., 2019. Modeling α-synuclein propagation with preformed Fibril injections. J. Mov. Disord. 20934939. 12 (3), 139–151. Available from: https://www.e-jmd.org/upload/jmd-19046.pdf. doi:10.14802/jmd.19046.

Cooper, J.F., Machiela, E., Dues, D.J., Spielbauer, K.K., Senchuk, M.M., Van Raamsdonk, J.M., 2017. Activation of the mitochondrial unfolded protein response promotes longevity and dopamine neuron survival in Parkinson's disease models. Sci. Rep. 20452322. 7 (1). Available from: https://www.nature.com/srep/index.html. doi:10.1038/s41598-017-16637-2.

Cordero-Llana, Ó., Houghton, B.C., Rinaldi, F., Taylor, H., Yáñez-Muñoz, R.J., Uney, J.B., Wong, L.-F., Caldwell, M.A., 2015. Enhanced efficacy of the CDNF/MANF family by Combined intranigral overexpression in the 6-OHDA Rat Model of Parkinson's disease. Mol. Ther. 15250016. 23 (2), 244–254. doi:10.1038/mt.2014.206.

Costello, S., Cockburn, M., Bronstein, J., Zhang, X., Ritz, B., 2009. Parkinson's disease and residential exposure to Maneb and Paraquat from agricultural applications in the Central Valley of California. Am. J. Epidemiol. 0002-9262. 169 (8), 919–926. doi:10.1093/aje/kwp006.

Cummings, J., Isaacson, S., Mills, R., Williams, H., Chi-Burris, K., Corbett, A., Dhall, R., Ballard, C., 2014. Pimavanserin for patients with Parkinson's disease psychosis: a randomised, placebo-controlled phase 3 trial. Lancet 383 (9916), 533–540. doi:10.1016/S0140-6736(13)62106-6.

Danzer,, et al, 2009. Seeding induced by alpha-synuclein oligomers provides evidence for spreading of alpha-synuclein pathology. J. Neurochem. 111 (1), 192–203.

Dawson, T.M., Ko, H.S., Dawson, V.L., 2010. Genetic animal models of Parkinson's Disease. Neuron. 08966273. 66 (5), 646–661. doi:10.1016/j.neuron.2010.04.034.

Decressac, M., Mattsson, B., Lundblad, M., Weikop, P., Björklund, A., 2012. Progressive neurodegenerative and behavioural changes induced by AAV-mediated overexpression of α-synuclein in midbrain dopamine neurons. Neurobiol. Dis. 09699961. 45 (3), 939–953. doi:10.1016/j.nbd.2011.12.013.

Deumens, R., Blokland, A., Prickaerts, J., 2002. Modeling Parkinson's disease in rats: an evaluation of 6-OHDA lesions of the nigrostriatal pathway. Exp. Neurol. 00144886. 175 (2), 303–317. Available from: http://www.elsevier.com/inca/publications/store/6/2/2/8/2/8/index.htt. doi:10.1006/exnr.2002.7891.

Dhall, R., Kreitzman, D.L., 2016. Advances in levodopa therapy for Parkinson disease. Neurology 86 (14_supplement_1). doi:10.1212/WNL.0000000000002510.

Dongning, S., Cui, Y., Chengzhang, H., Yin, P., Bai, R., Zhu, J., Lam, J.S.T., Zhang, J., Yan, R., Zheng, X., Jiayi, W., Zhao, D., Wang, A., Zhou, M., Feng, T., 2025. Projections for prevalence of Parkinson's disease and its driving factors in 195 countries and territories to 2050: modelling study of Global Burden of Disease Study 2021. BMJ. 1756-1833. 388, e080952. doi:10.1136/bmj-2024-080952.

Duan, C.-L., Yue, S., Zhao, C.-L., Ling-Ling, L., Qun-Yuan, X., Yang, H., 2005. The assays of activities and function of TH, AADC, and GCH1 and their potential use in ex vivo gene therapy of PD. Brain Res. Protoc. 1385299X. 16 (1-3), 37–43. doi:10.1016/j.brainresprot.2005.10.005.

Eberling, J.L., Jagust, W.J., Christine, C.W., Starr, P., Larson, P., Bankiewicz, K.S., Aminoff, M.J., 2008. Results from a phase I safety trial of hAADC gene therapy for Parkinson disease. Neurology. 0028-3878. 70 (21), 1980–1983. doi:10.1212/01.wnl.0000312381.29287.ff.

Eiden, L.E., Weihe, E., 2011. VMAT2: a dynamic regulator of brain monoaminergic neuronal function interacting with drugs of abuse. Ann. N. Y. Acad. Sci. 17496632. 1216 (1), 86–98. Available from: http://onlinelibrary.wiley.com/journal/10.1111/(ISSN)1749-6632. doi:10.1111/j.1749-6632.2010.05906.x.

Ekstrand, M.I., Terzioglu, M., Galter, D., Zhu, S., Hofstetter, C., Lindqvist, E., Thams, S., Bergstrand, A., Hansson, F.S., Trifunovic, A., Hoffer, B., Cullheim, S., Mohammed, A.H., Olson, L., Larsson, N.-G., 2007. Progressive parkinsonism in mice with respiratory-chain-deficient dopamine neurons. Proc. Natl. Acad. Sci. U. S. A. 0027-8424. 104 (4), 1325–1330. doi:10.1073/pnas.0605208103.

Emborg, M.E., Carbon, M., Holden, J.E., During, M.J., Ma, Y., Tang, C., Moirano, J., Fitzsimons, H., Roitberg, B.Z., Tuccar, E., Roberts, A., Kaplitt, M.G., Eidelberg, D., 2007. Subthalamic glutamic acid decarboxylase gene therapy: changes in motor function and cortical metabolism. J. Cereb. Blood Flow Metab. 0271678X. 27 (3), 501–509. doi:10.1038/sj.jcbfm.9600364.

Emre, M., et al., 2004. Rivastigmine for dementia associated with Parkinson's disease. N. Engl. J. Med. 351 (24), 2509–2518. doi:10.1056/NEJMoa041470.

Erb, M.L., Sipple, K., Levine, N., Chen, X., Moore, D.J., 2024. Adult-onset deletion of ATP13A2 in mice induces progressive nigrostriatal pathway dopaminergic degeneration and lysosomal abnormalities. NPJ Parkinsons Dis. 2373-8057. 10 (1). doi:10.1038/s41531-024-00748-5.

Fanet, H., Capuron, L., Castanon, N., Calon, F., Vancassel, S., 2021. Tetrahydrobioterin (BH4) pathway: from metabolism to neuropsychiatry. Curr. Neuropharmacol. 1570159X. 19 (5), 591–609. doi:10.2174/1570159x18666200729103529.

Fang, F., Yang, W., Florio, J.B., Rockenstein, E., Spencer, B., Orain, X.M., Dong, S.X., Li, H., Chen, X., Sung, K., Rissman, R.A., Masliah, E., Ding, J., Wu, C., 2017. Synuclein impairs trafficking and signaling of BDNF in a mouse model of Parkinson's disease. Sci. Rep. 20452322. 7 (1). Available from: https://www.nature.com/srep/index.html. doi:10.1038/s41598-017-04232-4.

Farfel-Becker, T., Jenny, D., Tayebi, N., Sidransky, E., 2019. Can GBA1-associated Parkinson disease be modeled in the mouse? Trends Neurosci. 01662236. 42 (9), 631–643. doi:10.1016/j.tins.2019.05.010.

Fox, S.H., Brotchie, J.M., 2010. The MPTP-lesioned non-human primate models of Parkinson's disease. Past, present, and future. Prog. Brain Res. 184, 133–157. Available from: http://www.sciencedirect.com/science/bookseries/00796123. doi:10.1016/S0079-6123(10)84007-5.

Galvan, A., Devergnas, A., Wichmann, T., 2015. Alterations in neuronal activity in basal ganglia-thalamocortical circuits in the Parkinsonian State. Front. Neuroanat. 16625129. 9, 1–21. Available from: http://journal.frontiersin.org/Article/10.3389/fnana.2015.00005/pdf. doi:10.3389/fnana.2015.00005.

Ge, P., Dawson, V.L., Dawson, T.M., 2020. PINK1 and Parkin mitochondrial quality control: a source of regional vulnerability in Parkinson's disease. Mol. Neurodegener. 17501326. 15 (1). Available from: http://www.molecularneurodegeneration.com/. doi:10.1186/s13024-020-00367-7.

Goetz, C.G., Tilley, B.C., Shaftman, S.R., Stebbins, G.T., Fahn, S., Martinez-Martin, P., Poewe, W., Sampaio, C., Stern, M.B., Dodel, R., Dubois, B., Holloway, R., Jankovic, J., Kulisevsky, J., Lang, A.E., Lees, A., Leurgans, S., LeWitt, P.A., Nyenhuis, D., Olanow, C.W., Rascol, O., Schrag, A., Teresi, J.A., van Hilten, J.J., LaPelle, N., Agarwal, P., Athar, S., Bordelan, Y., Bronte-stewart, H.M., Camicioli, R., Chou, K., Cole, W., Dalvi, A., Delgado, H., Diamond, A., Dick, J.P., Duda, J., Elble, R.J., Evans, C., Evidente, V.G., Fernandez, H.H., Fox, S., Friedman, J.H., Fross, R.D., Gallagher, D., Goetz, C.G., Hall, D., Hermanowicz, N., Hinson, V., Horn, S., Hurtig, H., Kang, U.J., Kleiner-Fisman, G., Klepitskaya, O., Kompoliti, K., Lai, E.C., Leehey, M.L., Leroi, I., Lyons, K.E., McClain, T., Metzer, S.W., Miyasaki, J., Morgan, J.C., Nance, M., Nemeth, J., Pahwa, R., Parashos, S.A., Schneider, J.S., Sethi, K., Shulman, L.M., Siderowf, A., Silverdale, M., Simuni, T., Stacy, M., Stern, M.B., Stewart, R.M., Sullivan, K., Swope, D.M., Wadia, P.M., Walker, R.W., Walker, R., Weiner, W.J., Wiener, J., Wilkinson, J., Wojcieszek, J.M., Wolfrath, S., Wooten, F., Wu, A., Zesiewicz, T.A., Zweig, R.M., 2008. Movement disorder society-sponsored revision of the unified Parkinson's disease rating scale (MDS-UPDRS): scale presentation and clinimetric testing results. Mov. Disord. 15318257. 23 (15), 2129–2170. Available from: http://www3.interscience.wiley.com/cgi-bin/fulltext/121536316/PDFSTART. doi:10.1002/mds.22340.

Gubellini, P., Kachidian, P., 2015. Animal models of Parkinson's disease: AN updated overview. Revue Neurol. 00353787. 171 (11), 750–761. doi:10.1016/j.neurol.2015.07.011.

Guiney, S.J., Adlard, P.A., Lei, P., Mawal, C.H., Bush, A.I., Finkelstein, D.I., Ayton, S., 2020. Fibrillar α-synuclein toxicity depends on functional lysosomes. J. Biol. Chem. 1083351X. 295 (51), 17497–17513. Available from: https://www.jbc.org/content/295/51/17497.full.pdf. doi:10.1074/jbc.RA120.013428.

Hadjiconstantinou, M., Neff, N.H., 2008. Enhancing aromatic L-amino acid decarboxylase activity: implications for L-DOPA treatment in Parkinson's disease. CNS Neurosci. Ther. 17555949. 14 (4), 340–351. doi:10.1111/j.1755-5949.2008.00058.x.

Hauser, R.A., Deckers, F., Lehert, P., 2004. Parkinson's disease home diary: further validation amd implications for clinical trials. Mov. Disord. 08853185. 19 (12), 1409–1413. doi:10.1002/mds.20248.

Heiss, J.D., Lungu, C., Hammoud, D.A., Herscovitch, P., Ehrlich, D.J., Argersinger, D.P., Sinharay, S., Scott, G., Wu, T., Federoff, H.J., Zaghloul, K.A., Hallett, M., Lonser, R.R., Bankiewicz, K.S., 2019. Trial of magnetic resonance–guided putaminal gene therapy for advanced Parkinson's disease. Mov. Disord. 15318257. 34 (7), 1073–1078. Available from: http://onlinelibrary.wiley.com/journal/10.1002/(ISSN)1531-8257. doi:10.1002/mds.27724.

Heiss, J.D., Ray-Chaudhury, A., Kleiner, D.E., Ehrlich, D.J., Scott, G., Edwards, N.A., Goldstein, D.S., Hammoud, D.A., Hadaczek, P., Van Laar, V.S., Graff, S.A., Herscovitch, P., Lungu, C., Hallett, M., Lonser, R.R., Zaghloul, K.A., Bankiewicz, K.S., 2024. Persistent GDNF expression 45 Months after Putaminal Infusion of AAV2-GDNF in a patient with Parkinson's disease. Mov. Disord. 15318257. 39 (8), 1412–1417. Available from: http://onlinelibrary.wiley.com/journal/10.1002/(ISSN)1531-8257. doi:10.1002/mds.29820.

Hirsch, E.C., Hunot, S., 2009. Neuroinflammation in Parkinson's disease: a target for neuroprotection? Lancet Neurol. 14744422. 8 (4), 382–397. doi:10.1016/S1474-4422(09)70062-6.

Hongyu, L., Ham, A., Thong Chi, M., Kuo, S.-H., Kanter, E., Kim, D., Han Seok, K., Quan, Y., Sardi, S.P., Aiqun, L., Arancio, O., Kang, U.J., Sulzer, D., Tang, G., 2019. Mitochondrial dysfunction and mitophagy defect triggered by heterozygous GBA mutations. Autophagy. 1554-8627. 15 (1), 113–130. doi:10.1080/15548627.2018.1509818.

Hou, X., Watzlawik, J.O., Fiesel, F.C., Springer, W., 2020. Autophagy in Parkinson's disease. J. Mol. Biol. 00222836. 432 (8), 2651–2672. doi:10.1016/j.jmb.2020.01.037.

Howell, M., et al., 2023. Management of REM sleep behavior disorder: an American Academy of Sleep Medicine clinical practice guideline. J. Clin. Sleep. Med. 19 (4), 759–768. doi:10.5664/jcsm.10424.

Hruska, K.S., LaMarca, M.E., Scott, C.R., Sidransky, E., 2008. Gaucher disease: mutation and polymorphism spectrum in the glucocerebrosidase gene (GBA). Hum. Mutat. 10597794. 29 (5), 567–583. doi:10.1002/humu.20676.

Hu, J., Zhang, D., Tian, K., Ren, C., Heng, L., Lin, C., Huang, X., Liu, J., Mao, W., Zhang, J., 2023. Small-molecule LRRK2 inhibitors for PD therapy: current achievements and future perspectives. Eur. J. Med. Chem. 02235234. 256, 115475. doi:10.1016/j.ejmech.2023.115475.

Hwu, P.W.L., Kiening, K., Anselm, I., Compton, D.R., Nakajima, T., Opladen, T., Pearl, P.L., Roubertie, A., Roujeau, T., Muramatsu, S.I., 2021. Gene therapy in the putamen for curing AADC deficiency and Parkinson's disease. EMBO Mol. Med. 17574684. 13 (9). Available from: http://embomolmed.embopress.org/. doi:10.15252/emmm.202114712.

Inyushin, M.Y., Huertas, A., Kucheryavykh, Y.V., Kucheryavykh, L.Y., Tsydzik, V., Sanabria, P., Eaton, M.J., Skatchkov, S.N., Rojas, L.V., Wessinger, W.D., 2012. L-DOPA uptake in astrocytic endfeet enwrapping blood vessels in rat brain. Parkinsons Dis. 2090-8083. 2012, 1–8. doi:10.1155/2012/321406.

Ip, C.W., Klaus, L.C., Karikari, A.A., Visanji, N.P., Brotchie, J.M., Lang, A.E., Volkmann, J., Koprich, J.B., 2017. AAV1/2-induced overexpression of A53T-α-synuclein in the substantia nigra results in degeneration of the nigrostriatal system with Lewy-like pathology and motor impairment: a new mouse model for Parkinson's disease. Acta Neuropathol. Commun. 20515960. 5 (1), 11. doi:10.1186/s40478-017-0416-x.

Jackson-Lewis, V., Przedborski, S., 2007. Protocol for the MPTP mouse model of Parkinson's disease. Nat. Protoc. 17502799. 2 (1), 141–151. doi:10.1038/nprot.2006.342.

Jennings, D., Huntwork-Rodriguez, S., Henry, A.G., Sasaki, J.C., Meisner, R., Diaz, D., Solanoy, H., Wang, X., Negrou, E., Bondar, V.V., Ghosh, R., Maloney, M.T., Propson, N.E., Zhu, Y., Maciuca, R.D., Harris, L., Kay, A., LeWitt, P., King, T.A., Kern, D., Ellenbogen, A., Goodman, I., Siderowf, A., Aldred, J., Omidvar, O., Masoud, S.T., Davis, S.S., Arguello, A., Estrada, A.A., de Vicente, J., Sweeney, Z.K., Astarita, G., Borin, M.T., Wong, B.K., Wong, H., Nguyen, H., Scearce-Levie, K., Carole, H., Troyer, M.D., 2022. Preclinical and clinical evaluation of the LRRK2 inhibitor DNL201 for Parkinson's disease. Sci. Transl. Med. 1946-6234. 14 (648). doi:10.1126/scitranslmed.abj2658.

Ji, Y.J., Wang, H.L., Yin, B.L., Ren, X.Y., 2020. Down-regulation of DJ-1 Augments Neuroinflammation via Nrf2/Trx1/NLRP3 Axis in MPTP-induced Parkinson's Disease Mouse Model. Neuroscience. 18737544. 442, 253–263. Available from: https://www.elsevier.com/locate/neuroscience. doi:10.1016/j.neuroscience.2020.06.001.

Jing, X.Z., Yang, H.J., Taximaimaiti, R., Wang, X.P., 2023. Advances in the therapeutic use of non-ergot dopamine agonists in the treatment of motor and non-motor symptoms of Parkinson's disease. Curr. Neuropharmacol. 18756190. 21 (5), 1224–1240. Available from: http://www.eurekaselect.com/603/journal/current-neuropharmacology. doi:10.2174/1570159X20666220915091022.

Kang, S.S., Zhang, Z., Liu, X., Manfredsson, F.P., Benskey, M.J., Cao, X., Xu, J., Sun, Y.E., Ye, K., 2017. TrkB neurotrophic activities are blocked by α-synuclein, triggering dopaminergic cell death in Parkinson's disease. Proc. Natl. Acad. Sci. U. S. A. 10916490. 114 (40), 10773–10778. Available from: http://www.pnas.org/content/114/40/10773.full.pdf. doi:10.1073/pnas.1713969114.

Kantor, B., Tagliafierro, L., Jeffrey, G., Zamora, M.E., Ilich, E., Grenier, C., Huang, Z.Y., Murphy, S., Chiba-Falek, O., 2018. Downregulation of SNCA expression by targeted editing of DNA methylation: a potential strategy for precision therapy in PD. Mol. Ther. 15250016. 26 (11), 2638–2649. doi:10.1016/j.ymthe.2018.08.019.

Kaplitt, M.G., Feigin, A., Tang, C., Fitzsimons, H.L., Mattis, P., Lawlor, P.A., Bland, R.J., Young, D., Strybing, K., Eidelberg, D., During, M.J., 2007. Safety and tolerability of gene therapy with an adeno-associated virus (AAV) borne GAD gene for Parkinson's disease: an open label, phase I trial. Lancet. 01406736. 369 (9579), 2097–2105. Available from: http://www.journals.elsevier.com/the-lancet/. doi:10.1016/S0140-6736(07)60982-9.

Karpinar, D.P., Balija, M.B.G., Kügler, S., Opazo, F., Rezaei-Ghaleh, N., Wender, N., Kim, H.Y., Taschenberger, G., Falkenburger, B.H., Heise, H., Kumar, A., Riedel, D., Fichtner, L., Voigt, A., Braus, G.H., Giller, K., Becker, S., Herzig, A., Baldus, M., Jäckle, H., Eimer, S., Schulz, J.B., Griesinger, C., Zweckstetter, M., 2009. Pre-fibrillar α-synuclein variants with impaired B-structure increase neurotoxicity in Parkinson's disease models. EMBO J. 14602075. 28 (20), 3256–3268. doi:10.1038/emboj.2009.257.

Katzenschlager, R., Poewe, W., Rascol, O., Trenkwalder, C., Günther Deuschl, K.R.C., Henriksen, T., van Laar, T., Spivey, K., Vel, S., Staines, H., Lees, A., 2018. Apomorphine subcutaneous infusion in patients with Parkinson's disease with persistent motor fluctuations (Toledo): a multicentre, double-blind, randomised, placebo-controlled trial. Lancet Neurol. 14744422. 17 (9), 749–759. doi:10.1016/s1474-4422(18)30239-4.

Kells, A.P., Eberling, J., Su, X., Pivirotto, P., Bringas, J., Hadaczek, P., Narrow, W.C., Bowers, W.J., Federoff, H.J., Forsayeth, J., Bankiewicz, K.S., 2010. Regeneration of the MPTP-lesioned dopaminergic system after convection-enhanced delivery of AAV2-GDNF. J. Neurosci. 0270-6474. 30 (28), 9567–9577. doi:10.1523/jneurosci.0942-10.2010.

Kells, A.P., Forsayeth, J., Bankiewicz, K.S., 2012. Glial-derived neurotrophic factor gene transfer for Parkinson's disease: anterograde distribution of AAV2 vectors in the primate brain. Neurobiol. Dis. 1095953X. 48 (2), 228–235. doi:10.1016/j.nbd.2011.10.004.

Khodr, C.E., Becerra, A., Han, Y., Bohn, M.C., 2014. Targeting alpha-synuclein with a microRNA-embedded silencing vector in the rat substantia nigra: positive and negative effects. Brain Res. 18726240. 1550, 47–60. doi:10.1016/j.brainres.2014.01.010.

Khodr, C.E., Sapru, M.K., Pedapati, J., Han, Y., West, N.C., Kells, A.P., Bankiewicz, K.S., Bohn, M.C., 2011. An alpha-synuclein AAV gene silencing vector ameliorates a behavioral deficit in a rat model of Parkinson's disease, but displays toxicity in dopamine neurons. Brain Res. 00068993. 1395, 94–107. doi:10.1016/j.brainres.2011.04.036.

Kirik, D., Annett, L.E., Burger, C., Muzyczka, N., Mandel, R.J., Björklund, A., 2003. Nigrostriatal α-synucleinopathy induced by viral vector-mediated overexpression of human α-synuclein: a new primate model of Parkinson's disease. Proc. Natl. Acad. Sci. U. S. A. 0027-8424. 100 (5), 2884–2889. doi:10.1073/pnas.0536383100.

Kirik, D., Rosenblad, C., Björklund, A., 1998. Characterization of behavioral and neurodegenerative changes following partial lesions of the nigrostriatal dopamine system induced by intrastriatal 6-hydroxydopamine in the rat. Exp. Neurol. 00144886. 152 (2), 259–277. Available from: http://www.elsevier.com/inca/publications/store/6/2/2/8/2/8/index.htt. doi:10.1006/exnr.1998.6848.

Kirik, D., Rosenblad, C., Burger, C., Lundberg, C., Johansen, T.E., Muzyczka, N., Mandel, R.J., Björklund, A., 2002a. Parkinson-like neurodegeneration induced by targeted overexpression of α-synuclein in the nigrostriatal system. J. Neurosci. 02706474. 22 (7), 2780–2791. Available from: https://www.jneurosci.org. doi:10.1523/jneurosci.22-07-02780.2002.

Kirik, D., Rosenblad, C., Burger, C., Lundberg, C., Johansen, T.E., Muzyczka, N., Mandel, R.J., Björklund, A., 2002b. Parkinson-like neurodegeneration induced by targeted overexpression of α-synuclein in the nigrostriatal system. J. Neurosci. 0270-6474. 22 (7), 2780–2791. doi:10.1523/jneurosci.22-07-02780.2002.

Koprich, J.B., Johnston, T.H., Reyes, M.G., Sun, X., Brotchie, J.M., 2010. Expression of human A53T alpha-synuclein in the rat substantia nigra using a novel AAV1/2 vector produces a rapidly evolving pathology with protein aggregation, dystrophic neurite architecture and nigrostriatal degeneration with potential to model the pathology of Parkinson's disease. Mol. Neurodegener. 1750-1326. 5 (1), 43. doi:10.1186/1750-1326-5-43.

Koprich, J., Johnston, T., Howson, P., Reyes, G., Omana, V., Brotchie, J., 2018. Characterization and reproducibility of a macaque model of Parkinson's disease alpha-synucleinopathy. Parkinsonism Relat. Disord. 13538020. 46, e3. doi:10.1016/j.parkreldis.2017.11.012.

Kordower, J.H., Olanow, C.W., Dodiya, H.B., Chu, Y., Beach, T.G., Adler, C.H., Halliday, G.M., Bartus, R.T., 2013. Disease duration and the integrity of the nigrostriatal system in Parkinson's disease. Brain. 14602156. 136 (8), 2419–2431. Available from: http://brain.oxfordjournals.org/. doi:10.1093/brain/awt192.

Korecka, J.A., Thomas, R., Hinrich, A.J., Moskites, A.M., Macbain, Z.K., Hallett, P.J., Isacson, O., Hastings, M.L., 2020. Splice-switching antisense oligonucleotides reduce LRRK2 kinase activity in human LRRK2 transgenic mice. Mol. Ther. Nucleic Acids. 21622531. 21, 623–635. Available from: https://www.journals.elsevier.com/molecular-therapy-nucleic-acids. doi:10.1016/j.omtn.2020.06.027.

Kozina, E., Sadasivan, S., Jiao, Y., Dou, Y., Zhijun, M., Tan, H., Kodali, K., Shaw, T., Peng, J., Smeyne, R.J., 2018. Mutant LRRK2 mediates peripheral and central immune responses leading to neurodegeneration in vivo. Brain. 0006-8950. 141 (6), 1753–1769. doi:10.1093/brain/awy077.

Kramer, E.R., Liss, B., 2015. GDNF-Ret signaling in midbrain dopaminergic neurons and its implication for Parkinson disease. FEBS Lett. 18733468. 589 (24), 3760–3772. Available from: https://www.elsevier.com/locate/febslet. doi:10.1016/j.febslet.2015.11.006.

Krishna, V., Fishman, P.S., Eisenberg, H.M., Kaplitt, M., Baltuch, G., Chang, J.W., Chang, W.-C., Fernandez, R.M., Del Alamo, M., Halpern, C.H., Ghanouni, P., Eleopra, R., Cosgrove, R., Guridi, J., Gwinn, R., Khemani, P., Lozano, A.M., McDannold, N., Fasano, A., Constantinescu, M., Schlesinger, I., Dalvi, A., Elias, W.J., 2023. Trial of globus pallidus focused ultrasound ablation in Parkinson's disease. N. Engl. J. Med. 0028-4793. 388 (8), 683–693. doi:10.1056/nejmoa2202721.

Kuruvilla, K.P., Nandhu, M.S., Paul, J., Paulose, C.S., 2013. Oxidative stress mediated neuronal damage in the corpus striatum of 6-hydroxydopamine lesioned Parkinson's rats: neuroprotection by Serotonin, GABA and Bone Marrow Cells Supplementation. J. Neurol. Sci. 18785883. 331 (1-2), 31–37. Available from: https://www.elsevier.com/locate/jns. doi:10.1016/j.jns.2013.04.020.

Langley, M.R., Ghaisas, S., Palanisamy, B.N., Ay, M., Jin, H., Anantharam, V., Kanthasamy, A., Kanthasamy, A.G., 2021. Characterization of nonmotor behavioral impairments and their neurochemical mechanisms in the MitoPark mouse model of progressive neurodegeneration in Parkinson's disease. Exp. Neurol. 00144886. 341, 113716. doi:10.1016/j.expneurol.2021.113716.

Lapchak, P.A., Araujo, D.M., Hilt, D.C., Sheng, J., Jiao, S., 1997. Adenoviral vector-mediated GDNF gene therapy in a rodent lesion model of late stage Parkinson's disease. Brain Res. 00068993. 777 (1-2), 153–160. doi:10.1016/S0006-8993(97)01100-1.

Lauterbach, E.C., 2004. The neuropsychiatry of Parkinson's disease and related disorders. Psychiatr. Clin. North Am. 0193953X. 27 (4), 801–825. doi:10.1016/j.psc.2004.07.001.

Lee, B., Lee, H., Nam, Y.R., Oh, J.H., Cho, Y.H., Chang, J.W., 2005. Enhanced expression of glutamate decarboxylase 65 improves symptoms of rat parkinsonian models. Gene Ther. 0969-7128. 12 (15), 1215–1222. doi:10.1038/sj.gt.3302520.

Lee, S., Kang, M., Lee, S., Yoon, S., Cho, Y., Min, D., Ann, D., Shin, J., Paik, Y.-K., Daewoong, J., 2024. AAV-aMTD-Parkin, a therapeutic gene delivery cargo, enhances motor and cognitive functions in Parkinson's and Alzheimer's diseases. Pharmacol. Res. 10436618. 208, 107326. doi:10.1016/j.phrs.2024.107326.

Lee, T.K., Yankee, E.L., 2022. A review on Parkinson's disease treatment. Neuroimmunol. Neuroinflamm. 23478659. 8, 222. doi:10.20517/2347-8659.2020.58.

LeWitt, P.A., Guttman, M., Tetrud, J.W., Tuite, P.J., Mori, A., Chaikin, P., Sussman, N.M., 2008. Adenosine A2A receptor antagonist istradefylline (KW-6002) reduces "off" time in Parkinson's disease: a double-blind, randomized, multicenter clinical trial (6002-US-005). Ann. Neurol. 63 (3), 295–302. doi:10.1002/ana.21315.

LeWitt, P.A., Rezai, A.R., Leehey, M.A., Ojemann, S.G., Flaherty, A.W., Eskandar, E.N., Kostyk, S.K., Thomas, K., Sarkar, A., Siddiqui, M.S., Tatter, S.B., Schwalb, J.M., Poston, K.L., Henderson, J.M., Kurlan, R.M., Richard, I.H., Van Meter, L., Sapan, C.V., During, M.J., Kaplitt, M.G., Feigin, A., 2011. AAV2-GAD gene therapy for advanced Parkinson's disease: a double-blind, sham-surgery controlled, randomised trial. Lancet Neurol. 14744422. 10 (4), 309–319. doi:10.1016/S1474-4422(11)70039-4.

Li, S.Y.S., Johnson, R., Smyth, L.C.D., Dragunow, M., 2022. Platelet-derived growth factor signalling in neurovascular function and disease. Int. J. Biochem. Cell Biol. 13572725. 145, 106187. doi:10.1016/j.biocel.2022.106187.

Luk, K.C., Kehm, V., Carroll, J., Zhang, B., O'Brien, P., Trojanowski, J.Q., Lee, V.M.Y., 2012. Pathological α-synuclein transmission initiates Parkinson-like neurodegeneration in nontransgenic mice. American Association for the Advancement of Science, United States Science. 10959203. 338 (6109), 949–953. Available from: http://www.sciencemag.org/content/338/6109/949.full.pdf. doi:10.1126/science.1227157.

Luo, J., Kaplitt, M.G., Fitzsimons, H.L., Zuzga, D.S., Liu, Y., Oshinsky, M.L., During, M.J., 2002. Subthalamic GAD gene therapy in a Parkinson's disease rat model. Science. 0036-8075. 298 (5592), 425–429. doi:10.1126/science.1074549.

Luth, E.S., Stavrovskaya, I.G., Bartels, T., Kristal, B.S., Selkoe, D.J., 2014. Soluble, prefibrillar α-synuclein oligomers promote complex I-dependent, Ca2+-induced mitochondrial dysfunction. J. Biol. Chem. 1083351X. 289 (31), 21490–21507. Available from: http://www.jbc.org/content/289/31/21490.full.pdf+html. doi:10.1074/jbc.M113.545749.

Magalhaes, J., Gegg, M.E., Migdalska-Richards, A., Doherty, M.K., Whitfield, P.D., Schapira, A.H.V., 2016. Autophagic lysosome reformation dysfunction in glucocerebrosidase deficient cells: relevance to Parkinson disease. Hum. Mol. Genet. 0964-6906. 25 (16), 3432–3445. doi:10.1093/hmg/ddw185.

Marks, W.J., Bartus, R.T., Siffert, J., Davis, C.S., Lozano, A., Boulis, N., Vitek, J., Stacy, M., Turner, D., Verhagen, L., Bakay, R., Watts, R., Guthrie, B., Jankovic, J., Simpson, R., Tagliati, M., Alterman, R., Stern, M., Baltuch, G., Starr, P.A., Larson, P.S., Ostrem, J.L., Nutt, J., Kieburtz, K., Kordower, J.H., Olanow, C.W., 2010. Gene delivery of AAV2-neurturin for Parkinson's disease: a double-blind, randomised, controlled trial. Lancet Neurol. 14744422. 9 (12), 1164–1172. doi:10.1016/S1474-4422(10)70254-4.

Marks, W.J., Baumann, T.L., Bartus, R.T., 2016. Long-term safety of patients with Parkinson's disease receiving rAAV2-neurturin (CERE-120) gene transfer. Hum. Gene Ther. 15577422. 27 (7), 522–527. Available from: https://www.liebertonline.com/hum. doi:10.1089/hum.2015.134.

Marks, W.J., Ostrem, J.L., Verhagen, L., Starr, P.A., Larson, P.S., Bakay, R.A., Taylor, R., Cahn-Weiner, D.A., Stoessl, A.J., Olanow, C.W., Bartus, R.T., 2008. Safety and tolerability of intraputaminal delivery of CERE-120 (adeno-associated virus serotype 2-neurturin) to patients with idiopathic Parkinson's disease: an open-label, phase I trial. Lancet Neurol. 14744422. 7 (5), 400–408. doi:10.1016/S1474-4422(08)70065-6.

Martínez-Fernández, R., Jorge, U.M.-M., Rodríguez-Rojas, R., Del Álamo, M., Shah, B.B., Hernández-Fernández, F., Pineda-Pardo, J.A., Monje, M.H.G., Fernández-Rodríguez, B., Sperling, S.A., Mata-Marín, D., Guida, P., Alonso-Frech, F., Obeso, I., Gasca-Salas, C., Vela-Desojo, L., Elias, W.J., Obeso, J.A., 2020. Randomized trial of focused ultrasound subthalamotomy for Parkinson's disease. N. Engl. J. Med. 0028-4793. 383 (26), 2501–2513. doi:10.1056/nejmoa2016311.

Matsuzaki, Y., Fukai, Y., Konno, A., Hirai, H., 2024. Optimal different adeno-associated virus capsid/promoter combinations to target specific cell types in the common marmoset cerebral cortex. Mol. Ther. Methods Clin. Dev. 23290501. 32 (4), 101337. doi:10.1016/j.omtm.2024.101337.

McKinnon, C., De Snoo, M.L., Gondard, E., Neudorfer, C., Chau, H., Ngana, S.G., O'Hara, D.M., Brotchie, J.M., Koprich, J.B., Lozano, A.M., Kalia, L.V., Kalia, S.K., 2020. Early-onset impairment of the ubiquitin-proteasome system in dopaminergic neurons caused by α-synuclein. Acta Neuropathol. Commun. 2051-5960. 8 (1). doi:10.1186/s40478-020-0894-0.

Meng, H., Yamashita, C., Shiba-Fukushima, K., Inoshita, T., Funayama, M., Sato, S., Hatta, T., Natsume, T., Umitsu, M., Takagi, J., Imai, Y., Hattori, N., 2017. Loss of Parkinson's disease-associated protein CHCHD2 affects mitochondrial crista structure and destabilizes cytochrome c. Nat. Commun. 2041-1723. 8 (1). doi:10.1038/ncomms15500.

Menon, S., Kofoed, R.H., Nabbouh, F., Xhima, K., Al-Fahoum, Y., Langman, T., Mount, H.T.J., Shihabuddin, L.S., Sardi, S.P., Fraser, P.E., Watts, J.C., Aubert, I., Tandon, A., 2021. Viral alpha-synuclein knockdown prevents spreading synucleinopathy. Brain Commun. 2632-1297. 3 (4). doi:10.1093/braincomms/fcab247.

Menozzi, E., Toffoli, M., Schapira, A.H.V., 2023. Targeting the GBA1 pathway to slow Parkinson disease: insights into clinical aspects, pathogenic mechanisms and new therapeutic avenues. Pharmacol. Ther. 01637258. 246, 108419. doi:10.1016/j.pharmthera.2023.108419.

Meredith, G.E., Rademacher, D.J., 2011. MPTP mouse models of Parkinson's disease: an update. J. Parkinsons Dis. 1877718X. 1 (1), 19–33. doi:10.3233/JPD-2011-11023.

Migdalska-Richards, A., Wegrzynowicz, M., Rusconi, R., Deangeli, G., Di Monte, D.A., Spillantini, M.G., Schapira, A.H.V., 2017. The L444P Gba1 mutation enhances alpha-synuclein induced loss of nigral dopaminergic neurons in mice. Brain. 0006-8950. 140 (10), 2706–2721. doi:10.1093/brain/awx221.

Mishra, A., Krishnamurthy, S., 2020. Neurorestorative effects of sub-chronic administration of ambroxol in rodent model of Parkinson's disease. Naunyn Schmiedebergs Arch. Pharmacol. 14321912. 393 (3), 429–444. Available from: https://link.springer.de/link/service/journals/00210/index.htm. doi:10.1007/s00210-019-01737-9.

Morabito, G., Giannelli, S.G., Ordazzo, G., Bido, S., Castoldi, V., Indrigo, M., Cabassi, T., Cattaneo, S., Luoni, M., Cancellieri, C., Sessa, A., Bacigaluppi, M., Taverna, S., Leocani, L., Lanciego, J.L., Broccoli, V., 2017. AAV-PHP.B-mediated global-scale expression in the mouse nervous system enables GBA1 gene therapy for wide protection from synucleinopathy. Mol. Ther. 15250016. 25 (12), 2727–2742. doi:10.1016/j.ymthe.2017.08.004.

Müller, T., 2009. Levodopa/carbidopa and entacapone in the treatment of Parkinson's disease: efficacy, safety and patient preference. Patient Prefer Adherence 3, 51–59. doi:10.2147/ppa.s4084.

Müller, T., 2010. Catechol-O-methyltransferase enzyme. Cofactor S-adenosyl-l-methionine and related mechanisms. Int. Rev. Neurobiol. 95, 49–71. Available from: http://www.elsevier.com/wps/find/bookdescription.cws_home/704657/description#description. doi:10.1016/B978-0-12-381326-8.00004-1.

Müller, T., Kuhn, W., Schulte, T., Przuntek, H., 2003. Intravenous amantadine sulphate application improves the performance of complex but not simple motor tasks in patients with Parkinson's disease. Neurosci. Lett. 339 (1), 25–28. doi:10.1016/s0304-3940(02)01462-3.

Muramatsu, S.I., Fujimoto, K.I., Kato, S., Mizukami, H., Asari, S., Ikeguchi, K., Kawakami, T., Urabe, M., Kume, A., Sato, T., Watanabe, E., Ozawa, K., Nakano, I., 2010. A phase i study of aromatic l-amino acid decarboxylase gene therapy for Parkinson's disease. Mol. Ther. 15250024. 18 (9), 1731–1735. Available from: https://www.journals.elsevier.com/molecular-therapy. doi:10.1038/mt.2010.135.

Nagatsu, T., 2024. Catecholamines and Parkinson's disease: tyrosine hydroxylase (TH) over tetrahydrobiopterin (BH4) and GTP cyclohydrolase I (GCH1) to cytokines, neuromelanin, and gene therapy: a historical overview. J. Neural Transm. 0300-9564. 131 (6), 617–630. doi:10.1007/s00702-023-02673-y.

Nam, J., Richie, C.T., Harvey, B.K., Voutilainen, M.H., 2024. Delivery of CDNF by AAV-mediated gene transfer protects dopamine neurons and regulates ER stress and inflammation in an acute MPTP mouse model of Parkinson's disease. Sci. Rep. 2045-2322. 14 (1). doi:10.1038/s41598-024-65735-5.

Narendra, D.P., Youle, R.J., 2024. The role of PINK1–Parkin in mitochondrial quality control. Nat. Cell Biol. 14764679. 26 (10), 1639–1651. Available from: https://www.nature.com/ncb/. doi:10.1038/s41556-024-01513-9.

Nayak, L., Henchcliffe, C., 2008. Rasagiline in treatment of Parkinson's disease. Neuropsychiatr. Dis. Treat. 4 (1), 23–32.

Nicklas, W.J., Youngster, S.K., Kindt, M.V., Heikkila, R.E., 1987. IV. MPTP, MPP+ and mitochondrial function. Life Sci. 00243205. 40 (8), 721–729. doi:10.1016/0024-3205(87)90299-2.

Nishikawa, K., Hang, L., Kawamura, R., Osaka, H., Wang, Y.-L., Hara, Y., Hirokawa, T., Manago, Y., Amano, T., Noda, M., Aoki, S., Wada, K., 2003. Alterations of structure and hydrolase activity of parkinsonism-associated human ubiquitin carboxyl-terminal hydrolase L1 variants. Biochem. Biophys. Res. Commun. 0006291X. 304 (1), 176–183. doi:10.1016/s0006-291x(03)00555-2.

Noda, S., Sato, S., Fukuda, T., Tada, N., Uchiyama, Y., Tanaka, K., Hattori, N., 2020. Loss of Parkin contributes to mitochondrial turnover and dopaminergic neuronal loss in aged mice. Neurobiol. Dis. 09699961. 136, 104717. doi:10.1016/j.nbd.2019.104717.

Nutt, J.G., Curtze, C., Hiller, A., Anderson, S., Larson, P.S., Van Laar, A.D., Richardson, R.M., Thompson, M.E., Sedkov, A., Leinonen, M., Ravina, B., Bankiewicz, K.S., Christine, C.W., 2020. Aromatic L-amino acid decarboxylase gene therapy enhances levodopa response in Parkinson's Disease. Mov. Disord. 15318257. 35 (5), 851–858. Available from: http://onlinelibrary.wiley.com/journal/10.1002/(ISSN)1531-8257. doi:10.1002/mds.27993.

Olanow, C.W., et al., 2014. Continuous intrajejunal infusion of levodopa-carbidopa intestinal gel for patients with advanced Parkinson's disease: a randomised, controlled, double-blind, double-dummy study. Lancet Neurol. 13 (2), 141–149. doi:10.1016/S1474-4422(13)70293-X.

Olanow, C.W., Bartus, R.T., Baumann, T.L., Factor, S., Boulis, N., Stacy, M., Turner, D.A., Marks, W., Larson, P., Starr, P.A., Jankovic, J., Simpson, R., Watts, R., Guthrie, B., Poston, K., Henderson, J.M., Stern, M., Baltuch, G., Goetz, C.G., Herzog, C., Kordower, J.H., Alterman, R., Lozano, A.M.,

Lang, A.E., 2015. Gene delivery of neurturin to putamen and substantia nigra in P arkinson disease: a double-blind, randomized, controlled trial. Ann. Neurol. 0364-5134. 78 (2), 248–257. doi:10.1002/ana.24436.

Oliveras-Salvá, M., Van der Perren, A., Casadei, N., Stroobants, S., Nuber, S., D'Hooge, R., Van den Haute, C., Baekelandt, V., 2013. rAAV2/7 vector-mediated overexpression of alpha-synuclein in mouse substantia nigra induces protein aggregation and progressive dose-dependent neurodegeneration. Mol. Neurodegener. 1750-1326. 8 (1). doi:10.1186/1750-1326-8-44.

Othberg, A., Odin, P., Ballagi, A., ÅAhgren, A., Funa, K., Lindvall, O., 1995. Specific effects of platelet derived growth factor (PDGF) on fetal rat and human dopaminergic neurons in vitro. Exp. Brain Res. 14321106. 105 (1), 111–122. doi:10.1007/BF00242187.

Palfi, S., Gurruchaga, J.M., Lepetit, H., Katy Howard, G.S.R., Mason, S., Gouello, G., Domenech, P., Buttery, P.C., Hantraye, P., Tuckwell, N.J., Barker, R.A., Mitrophanous, K.A., 2018. Long-term follow-up of a phase I/II study of ProSavin, a lentiviral vector gene therapy for Parkinson's disease. Hum. Gene Ther. Clin. Dev. 2324-8637. 29 (3), 148–155. doi:10.1089/humc.2018.081.

Palfi, S., Jean Marc Gurruchaga, G.S.R., Lepetit, H., Lavisse, S., Buttery, P.C., Watts, C., Miskin, J., Kelleher, M., Deeley, S., Iwamuro, H., Lefaucheur, J.P., Thiriez, C., Fenelon, G., Lucas, C., Brugières, P., Gabriel, I., Abhay, K., Drouot, X., Tani, N., Kas, A., Ghaleh, B., Corvoisier, P.L., Dolphin, P., Breen, D.P., Mason, S., Guzman, N.V., Mazarakis, N.D., Radcliffe, P.A., Harrop, R., Kingsman, S.M., Rascol, O., Naylor, S., Barker, R.A., Hantraye, P., Remy, P., Cesaro, P., Mitrophanous, K.A., 2014. Long-term safety and tolerability of ProSavin, a lentiviral vector-based gene therapy for Parkinson's disease: a dose escalation, open-label, phase 1/2 trial. Lancet. 01406736. 383 (9923), 1138–1146. doi:10.1016/s0140-6736(13)61939-x.

Pan, P.Y., Li, X., Wang, J., Powell, J., Wang, Q., Zhang, Y., Chen, Z., Wicinski, B., Hof, P., Ryan, T.A., Yue, Z., 2017. Parkinson's disease-associated LRRK2 hyperactive kinase mutant disrupts synaptic vesicle trafficking in ventral midbrain neurons. J. Neurosci. 15292401. 37 (47), 11366–11376. Available from: http://www.jneurosci.org/content/jneuro/37/47/11366.full.pdf. doi:10.1523/JNEUROSCI.0964-17.2017.

Park, J., Lee, S.B., Lee, S., Kim, Y., Song, S., Kim, S., Bae, E., Kim, J., Shong, M., Kim, J.-M., Chung, J., 2006. Mitochondrial dysfunction in Drosophila PINK1 mutants is complemented by parkin. Nature. 0028-0836. 441 (7097), 1157–1161. doi:10.1038/nature04788.

Parker, W.D., Parks, J.K., Swerdlow, R.H., 2008. Complex I deficiency in Parkinson's disease frontal cortex. Brain Res. 00068993. 1189 (1), 215–218. doi:10.1016/j.brainres.2007.10.061.

Paul, G., Sullivan, A.M., 2019. Trophic factors for Parkinson's disease: where are we and where do we go from here? Eur. J. Neurosci. 14609568. 49 (4), 440–452. Available from: http://onlinelibrary.wiley.com/journal/10.1111/(ISSN)1460-9568. doi:10.1111/ejn.14102.

Peng, Y., Jiang, D.-Y., Yao, S.-Y., Zhang, X., Kazuo, S., Liu, J., Miao-qiao, D., Lin, L.-X., Chen, Q., Jin, H., 2025. Gene-modified animal models of Parkinson's disease. Exp. Neurol. 00144886. 390, 115287. doi:10.1016/j.expneurol.2025.115287.

Petersen, J.J., Kamp, C.B., Faltermeier, P., Juul, S., Løkkegaard, A., Gluud, C., Jakobsen, J.C., 2024. Deep brain stimulation for Parkinson's disease: systematic review with meta-analysis and trial sequential analysis. BMJ Med. 2754-0413. 3 (1), e000705. doi:10.1136/bmjmed-2023-000705.

Petit, A., Kawarai, T., Paitel, E., Sanjo, N., Maj, M., Scheid, M., Chen, F., Yongjun, G., Hasegawa, H., Salehi-Rad, S., Wang, L., Rogaeva, E., Fraser, P., Robinson, B., St George-hyslop, P., Tandon, A., 2005. Wild-type PINK1 prevents basal and induced neuronal apoptosis, a protective effect abrogated by Parkinson disease-related mutations. J. Biol. Chem. 00219258. 280 (40), 34025–34032. doi:10.1074/jbc.m505143200.

Pietz, K., Odin, P., Funa, K., Lindvall, O., 1996. Protective effect of platelet-derived growth factor against 6-hydroxydopamine-induced lesion of rat dopaminergic neurons in culture. Neurosci. Lett. 03043940. 204 (1-2), 101–104. doi:10.1016/0304-3940(96)12326-0.

Przedborski, S., Jackson-Lewis, V., Naini, A.B., Jakowec, M., Petzinger, G., Miller, R., Akram, M., 2001. The parkinsonian toxin 1-methyl-4-phenyl-1,2,3,6-tetrahydropyridine (MPTP): a technical review of its utility and safety. J. Neurochem. 0022-3042. 76 (5), 1265–1274. doi:10.1046/j.1471-4159.2001.00183.x.

Puspita, L., Chung, S.Y., Shim, J.W., 2017. Oxidative stress and cellular pathologies in Parkinson's disease. Mol. Brain. 17566606. 10 (1). Available from: http://www.molecularbrain.com/. doi:10.1186/s13041-017-0340-9.

Recasens, A., Dehay, B., 2014. Alpha-synuclein spreading in Parkinson's disease. Front. Neuroanat. 16625129. 8. Available from: http://journal.frontiersin.org/Journal/10.3389/fnana.2014.00159/pdf. doi:10.3389/fnana.2014.00159.

Recasens, A., Dehay, B., Bové, J., Carballo-Carbajal, I., Dovero, S., Pérez-Villalba, A., Fernagut, P., Blesa, J., Parent, A., Perier, C., Fariñas, I., Obeso, J.A., Bezard, E., Vila, M., 2014. Lewy body extracts from Parkinson disease brains trigger α-synuclein pathology and neurodegeneration in mice and monkeys. Ann. Neurol. 0364-5134. 75 (3), 351–362. doi:10.1002/ana.24066.

Ren, X., Zhang, T., Gong, X., Guanzheng, H., Ding, W., Wang, X., 2013. AAV2-mediated striatum delivery of human CDNF prevents the deterioration of midbrain dopamine neurons in a 6-hydroxydopamine induced parkinsonian rat model. Exp. Neurol. 00144886. 248, 148–156. doi:10.1016/j.expneurol.2013.06.002.

Richard, I.H., McDermott, M.P., Kurlan, R., Lyness, J.M., Como, P.G., Pearson, N., Factor, S.A., Juncos, J., Ramos, C.S., Brodsky, M., Manning, C., Marsh, L., Shulman, L., Fernandez, H.H., Black, K.J., Panisset, M., Christine, C.W., Jiang, W., Singer, C., Horn, S., Pfeiffer, R., Rottenberg, D., Slevin, J., Elmer, L., Press, D., Hyson, H.C., McDonald, W., 2012. A randomized, double-blind, placebo-controlled trial of antidepressants in Parkinson disease. Neurology. 0028-3878. 78 (16), 1229–1236. doi:10.1212/WNL.0b013e3182516244.

Richardson, R.M., Bankiewicz, K.S., Christine, C.W., Van Laar, A.D., Gross, R.E., Lonser, R., Factor, S.A., Kostyk, S.K., Kells, A.P., Ravina, B., Larson, P.S., 2020. Data-driven evolution of neurosurgical gene therapy delivery in Parkinson's disease. Journal of Neurology, Neurosurgery and Psychiatry. 1468330X. 91 (11), 1210–1218. Available from: http://jnnp.bmj.com/. doi:10.1136/jnnp-2020-322904.

Rocca, W.A., 2018. The burden of Parkinson's disease: a worldwide perspective. Lancet Neurol. 14744465. 17 (11), 928–929. Available from: http://www.journals.elsevier.com/the-lancet-neurology/. doi:10.1016/S1474-4422(18)30355-7.

Rocco, M.T., Akhter, A.S., Ehrlich, D.J., Scott, G.C., Lungu, C., Munjal, V., Aquino, A., Lonser, R.R., Fiandaca, M.S., Hallett, M., Heiss, J.D., Bankiewicz, K.S., 2022. Long-term safety of MRI-guided administration of AAV2-GDNF and gadoteridol in the putamen of individuals with Parkinson's disease. Mol. Ther. 15250024. 30 (12), 3632–3638. Available from: https://www.journals.elsevier.com/molecular-therapy. doi:10.1016/j.ymthe.2022.08.003.

Rocha, E.M., Smith, G.A., Park, E., Cao, H., Brown, E., Hayes, M.A., Beagan, J., McLean, J.R., Izen, S.C., Perez-Torres, E., Hallett, P.J., Isacson, O., 2015. Glucocerebrosidase gene therapy prevents α-synucleinopathy of midbrain dopamine neurons. Neurobiol. Dis. 1095953X. 82, 495–503. Available from: http://www.elsevier.com/inca/publications/store/6/2/2/9/2/3/index.htt. doi:10.1016/j.nbd.2015.09.009.

Rockenstein, E., Clarke, J., Viel, C., Panarello, N., Treleaven, C.M., Kim, C., Spencer, B., Adame, A., Park, H., Dodge, J.C., Cheng, S.H., Shihabuddin, L.S., Masliah, E., Sardi, S.P., 2016. Glucocerebrosidase modulates cognitive and motor activities in murine models of Parkinson's disease. Hum. Mol. Genet. 14602083. 25 (13), 2645–2660. Available from: http://hmg.oxfordjournals.org/. doi:10.1093/hmg/ddw124.

Rusch, C., Flanagan, R., Suh, H., Subramanian, I., 2023. To restrict or not to restrict? Practical considerations for optimizing dietary protein interactions on levodopa absorption in Parkinson's disease. NPJ Parkinsons Dis. 2373-8057. 9 (1). doi:10.1038/s41531-023-00541-w.

Salegio, E.A., Samaranch, L., Kells, A.P., Mittermeyer, G., Sebastian, W.S., Zhou, S., Beyer, J., Forsayeth, J., Bankiewicz, K.S., 2013. Axonal transport of adeno-associated viral vectors is serotype-dependent. Gene Ther. 0969-7128. 20 (3), 348–352. doi:10.1038/gt.2012.27.

Sapru, M.K., Yates, J.W., Hogan, S., Jiang, L., Halter, J., Bohn, M.C., 2006. Silencing of human α-synuclein in vitro and in rat brain using lentiviral-mediated RNAi. Exp. Neurol. 10902430. 198 (2), 382–390. doi:10.1016/j.expneurol.2005.12.024.

Sardi, S.P., Clarke, J., Kinnecom, C., Tamsett, T.J., Li, L., Stanek, L.M., Passini, M.A., Grabowski, G.A., Schlossmacher, M.G., Sidman, R.L., Cheng, S.H., Shihabuddin, L.S., 2011. CNS expression of glucocerebrosidase corrects α-synuclein pathology and memory in a mouse model of Gaucher-related synucleinopathy. Proc. Natl. Acad. Sci. U. S. A. 10916490. 108 (29), 12101–12106. Available from: http://www.pnas.org/content/108/29/12101.full.pdf+html. doi:10.1073/pnas.1108197108.

Sardi, S.P., Clarke, J., Viel, C., Chan, M., Tamsett, T.J., Treleaven, C.M., Bu, J., Sweet, L., Passini, M.A., Dodge, J.C., Yu, W.H., Sidman, R.L., Cheng, S.H., Shihabuddin, L.S., 2013. sAugmenting CNS glucocerebrosidase activity as a therapeutic strategy for parkinsonism and other Gaucher-related synucleinopathies. Proc. Natl. Acad. Sci. U. S. A. 0027-8424. 110 (9), 3537–3542. doi:10.1073/pnas.1220464110.

Sato, S., Noda, S., Torii, S., Amo, T., Ikeda, A., Funayama, M., Yamaguchi, J., Fukuda, T., Kondo, H., Tada, N., Arakawa, S., Watanabe, M., Uchiyama, Y., Shimizu, S., Hattori, N., 2021. Homeostatic p62 levels and inclusion body formation in CHCHD2 knockout mice. Hum. Mol. Genet. 0964-6906. 30 (6), 443–453. doi:10.1093/hmg/ddab057.

Shola-Dare, O., Bailess, S., Flores, C.C., Vanderheyden, W.M., Gerstner, J.R., 2021. Glitazone treatment rescues phenotypic deficits in a fly model of Gaucher/Parkinson's Disease. Int. J. Mol. Sci. 1422-0067. 22 (23), 12740. doi:10.3390/ijms222312740.

Sikora, J., Dovero, S., Kinet, R., Arotcarena, M.-L., Bohic, S., Bezard, E., Fernagut, P.-O., Dehay, B., 2024. Nigral ATP13A2 depletion induces Parkinson's disease-related neurodegeneration in a pilot study in non-human primates. NPJ Parkinsons Dis. 2373-8057. 10 (1). doi:10.1038/s41531-024-00757-4.

Silm, K., Yang, J., Marcott, P.F., Asensio, C.S., Eriksen, J., Guthrie, D.A., Newman, A.H., Ford, C.P., Edwards, R.H., 2019. Synaptic vesicle recycling pathway determines neurotransmitter content and release properties. Neuron. 08966273. 102 (4), 786. doi:10.1016/j.neuron.2019.03.031.

Spagnol, V., Oliveira, C.A.B., Randle, S.J., Passos, P.M.S., Correia, C.R.S.T.B., Simaroli, N.B., Oliveira, J.S., Mevissen, T.E.T., Medeiros, A.C., Gomes, M.D., Komander, D., Laman, H., Teixeira, F.R., 2021. TheE3ubiquitin ligase SCF(Fbxo7) mediates proteasomal degradation of UXT isoform 2 (UXT-V2) to inhibit the NF-κB signaling pathway. Biochim. Biophys. Acta Gen. Subj. 03044165. 1865 (1), 129754. doi:10.1016/j.bbagen.2020.129754.

Stojkovska, I., Krainc, D., Mazzulli, J.R., 2018. Molecular mechanisms of α-synuclein and GBA1 in Parkinson's disease. Cell Tissue Res. 14320878. 373 (1), 51–60. Available from: https://link.springer.de/link/service/journals/00441/index.htm. doi:10.1007/s00441-017-2704-y.

Stott, S.R.W., Randle, S.J., Rawi, S.A., Rowicka, P.A., Harris, R., Mason, B., Xia, J., Dalley, J.W., Barker, R.A., Laman, H., 2019. Loss of FBXO7 results in a Parkinson's-like dopaminergic degeneration via an RPL23–MDM2–TP53 pathway. J. Pathol. 0022-3417. 249 (2), 241–254. doi:10.1002/path.5312.

Sun, M., Kong, L., Wang, X., Holmes, C., Gao, Q., Zhang, G.-R., Pfeilschifter, J., Goldstein, D.S., Geller, A.I., 2004. Coexpression of tyrosine hydroxylase, GTP cyclohydrolase I, aromatic amino acid decarboxylase, and vesicular monoamine transporter 2 from a helper virus-free herpes simplex virus type 1 vector supports high-level, long-term biochemical and behavioral correction of a rat model of Parkinson's disease. Hum. Gene Ther. 1043-0342. 15 (12), 1177–1196. doi:10.1089/hum.2004.15.1177.

Sveinbjornsdottir, S., 2016. The clinical symptoms of Parkinson's disease. J. Neurochem. 0022-3042. 139 (S1), 318–324. doi:10.1111/jnc.13691.

Szunyogh, S., Carroll, E., Wade-Martins, R., 2025. Recent developments in gene therapy for Parkinson's disease. Mol. Ther. 15250024. 33 (5), 2052–2064. Available from: https://www.sciencedirect.com/science/journal/15250016. doi:10.1016/j.ymthe.2025.03.030.

Tanner, C.M., Kame, F., Ross, G.W., Hoppin, J.A., Goldman, S.M., Korell, M., Marras, C., Bhudhikanok, G.S., Kasten, M., Chade, A.R., Comyns, K., Richards, M.B., Meng, C., Priestley, B., Fernandez, H.H., Cambi, F., Umbach, D.M., Blair, A., Sandler, D.P., Langston, J.W., 2011. Rotenone, paraquat, and Parkinson's disease. Environ. Health Perspect. 15529924. 119 (6), 866–872. Available from: http://ehp03.niehs.nih.gov/article/fetchObjectAttachment.action?uri=info%3Adoi%2F10.1289%2Fehp.1002839&representation=PDF. doi:10.1289/ehp.1002839.

Tarsy, D., 2006. Initial treatment of Parkinson's disease. Curr. Treat. Options Neurol. 1092-8480. 8 (3), 224–235. doi:10.1007/s11940-006-0013-y.

Tereshchenko, J., Maddalena, A., Bähr, M., Kügler, S., 2014. Pharmacologically controlled, discontinuous GDNF gene therapy restores motor function in a rat model of Parkinson's disease. Neurobiol. Dis. 09699961. 65, 35–42. doi:10.1016/j.nbd.2014.01.009.

Tran, H.H., Dang, S.N.A., Nguyen, T.T., Huynh, A.M., Dao, L.M., Kamei, K., Yamaguchi, M., Dang, T.T.P., 2018. Drosophila Ubiquitin C-Terminal Hydrolase Knockdown Model of Parkinson's Disease. Sci. Rep. 20452322. 8 (1). Available from: https://www.nature.com/srep/index.html. doi:10.1038/s41598-018-22804-w.

Ullah, I., Uddin, S., Zhao, L., Wang, X., Hongyu, L., 2024. Autophagy and UPS pathway contribute to nicotine-induced protection effect in Parkinson's disease. Exp. Brain Res. 0014-4819. 242 (4), 971–986. doi:10.1007/s00221-023-06765-9.

Uwishema, O., Onyeaka, H., Badri, R., Nazlı Yücel, A., Korkusuz, A.K., Ajagbe, A.O., Abuleil, A., Chaaya, C., Alhendawi, B.H.M., 2022. Elie Chalhoub, The understanding of Parkinson's disease through genetics and new therapies. Brain Behav. 2162-3279. 12 (5). doi:10.1002/brb3.2577.

Valente, E.M., Bentivoglio, A.R., Dixon, P.H., Ferraris, A., Lalongo, T., Frontali, M., Albanese, A., Wood, N.W., 2001. Localization of a novel locus for autosomal recessive early-onset parkinsonism, PARK6, on human chromosome 1p35-p36. Am. J. Hum. Genet. 00029297. 68 (4), 895–900. doi:10.1086/319522.

Van der Perren, A., Van den Haute, C., Baekelandt, V., 2014. Viral vector-based models of Parkinson's disease. Curr. Top. Behav. Neurosci. 18663389. 22, 271–301. Available from: http://www.springerlink.com/content/1866-3370/. doi:10.1007/7854_2014_310.

Van Laar, A.D., Christine, C.W., Phielipp, N., Larson, P.S., Elder, J.B., Merola, A., Sebastian, W.S., Fiandaca, M.S., Kells, A.P., Wisniewski, M.E., Bankiewicz, K.S., 2025. Intraputaminal delivery of adeno-associated virus serotype 2–glial cell line–derived neurotrophic factor in mild or moderate Parkinson's disease. Mov. Disord. 15318257. 40 (7), 1297–1306. Available from: http://onlinelibrary.wiley.com/journal/10.1002/(ISSN)1531-8257. doi:10.1002/mds.30193.

Van Laar, A.D., Van Laar, V.S., Sebastian, W.S., Merola, A., Elder, J.B., Lonser, R.R., Bankiewicz, K.S., Björklund, A., Bloem, B.R., Brundin, P., Federoff, H., 2021. An update on gene therapy approaches for Parkinson's disease: restoration of dopaminergic function. J. Parkinsons Dis. 18777171. 11 (s2), S173. doi:10.3233/jpd-212724.

van Veen, S., Martin, S., Van den Haute, C., Benoy, V., Lyons, J., Vanhoutte, R., Kahler, J.P., Decuypere, J.-P., Gelders, G., Lambie, E., Zielich, J., Swinnen, J.V., Annaert, W., Agostinis, P., Ghesquière, B., Verhelst, S., Baekelandt, V., Eggermont, J., Vangheluwe, P., 2020. ATP13A2 deficiency disrupts lysosomal polyamine export. Nature. 0028-0836. 578 (7795), 419–424. doi:10.1038/s41586-020-1968-7.

Volpicelli-Daley, L.A., Kirik, D., Stoyka, L.E., Standaert, D.G., Harms, A.S., 2016. How can rAAV -α-synuclein and the fibril α-synuclein models advance our understanding of Parkinson's disease? J. Neurochem. 0022-3042. 139 (S1), 131–155. doi:10.1111/jnc.13627.

Volta, M., Beccano-Kelly, D.A., Paschall, S.A., Cataldi, S., MacIsaac, S.E., Kuhlmann, N., Kadgien, C.A., Tatarnikov, I., Fox, J., Khinda, J., Mitchell, E., Bergeron, S., Melrose, H., Farrer, M.J., Milnerwood, A.J., 2017. Initial elevations in glutamate and dopamine neurotransmission decline with age, as does exploratory behavior, in LRRK2 G2019S knock-in mice. eLife. 2050-084X. 6. doi:10.7554/elife.28377.

Wang, F., Liu, X., Chen, M., Xiaoxin, X., Yang, Y., Qiuhong, X., Zhu, H., Anding, X., Pouladi, M.A., Xu, X., 2025. Neuroprotective role of CHCHD2 in Parkinson's disease: insights into the GPX4-related ferroptosis pathway. Free Radic. Biol. Med. 08915849. 226, 348–363. doi:10.1016/j.freeradbiomed.2024.11.034.

Wang, L., Wang, Z., Zhu, R., Jinpeng, B., Feng, X., Liu, W., Jiaxin, W., Zhang, H., Hui, W., Kong, W., Bin, Y., Xianghui Yu, R.L.M., 2017. Therapeutic efficacy of AAV8-mediated intrastriatal delivery of human cerebral dopamine neurotrophic factor in 6-OHDA-induced parkinsonian rat models with different disease progression. PLOS ONE. 1932-6203. 12 (6), e0179476. doi:10.1371/journal.pone.0179476.

Wang, X., Becker, K., Levine, N., Zhang, M., Lieberman, A.P., Moore, D.J., Ma, J., 2019. Pathogenic alpha-synuclein aggregates preferentially bind to mitochondria and affect cellular respiration. Acta Neuropathol. Commun. 20515960. 7 (1), 41. doi:10.1186/s40478-019-0696-4.

Weaver, F.M., Follett, K.A., Stern, M., Luo, P., Harris, C.L., Hur, K., Marks, W.J., Rothlind, J., Sagher, O., Moy, C., Pahwa, R., Burchiel, K., Hogarth, P., Lai, E.C., Duda, J.E., Holloway, K., Samii, A., Horn, S., Bronstein, J.M., Stoner, G., Starr, P.A., Simpson, R., Baltuch, G., De Salles, A., Huang, G.D., Reda, D.J., 2012. Randomized trial of deep brain stimulation for Parkinson disease: thirty-six-month outcomes. Neurology. 1526632X. 79 (1), 55–65. Available from: http://www.neurology.org. doi:10.1212/WNL.0b013e31825dcdc1.

Weiner, W.J., 2009. What do clinical trials tell us about treating patients? Parkinsonism Relat. Disord. 13538020. 15 (3), S34. doi:10.1016/S1353-8020(09)70776-X.

Whone, A., Luz, M., Boca, M., Woolley, M., Mooney, L., Dharia, S., Broadfoot, J., Cronin, D., Schroers, C., Barua, N.U., Lara Longpre, C.L.B., Boiko, C., Johnson, G.A., Fibiger, H.C., Harrison, R., Lewis, O., Pritchard, G., Howell, M., Irving, C., Johnson, D., Kinch, S., Marshall, C., Lawrence, A.D., Blinder, S., Vesna Sossi, A.J.S., Skinner, P., Mohr, E., Gill, S.S., 2019. Randomized trial of intermittent intraputamenal glial cell line-derived neurotrophic factor in Parkinson's disease. Brain. 0006-8950. 142 (3), 512–525. doi:10.1093/brain/awz023.

Winston, G., Kharas, N., Svenningsson, P., Jha, A., Kaplitt, M.G., 2025. Gene therapy for Parkinson's disease: trials and technical advances. Lancet Neurol. 14744422. 24 (6), 548–556. doi:10.1016/s1474-4422(25)00125-5.

Xiaomin, S., Maguire-Zeiss, K.A., Giuliano, R., Prifti, L., Venkatesh, K., Federoff, H.J., 2008. Synuclein activates microglia in a model of Parkinson's disease. Neurobiol. Aging. 01974580. 29 (11), 1690–1701. doi:10.1016/j.neurobiolaging.2007.04.006.

Xiong, Y., Neifert, S., Karuppagounder, S.S., Liu, Q., Stankowski, J.N., Lee, B.D., Ko, H.S., Lee, Y., Grima, J.C., Mao, X., Jiang, H., Kang, S.-U., Swing, D.A., Iacovitti, L., Tessarollo, L., Dawson, T.M., Dawson, V.L., 2018. Robust kinase- and age-dependent dopaminergic and norepinephrine neurodegeneration in LRRK2 G2019S transgenic mice. Proc. Natl. Acad. Sci. U. S. A. 0027-8424. 115 (7), 1635–1640. doi:10.1073/pnas.1712648115.

Xu, C.Y., Kang, W.Y., Chen, Y.M., Jiang, T.F., Zhang, J., Zhang, L.N., Ding, J.Q., Liu, J., Chen, S.D., 2017. DJ-1 inhibits α-synuclein aggregation by regulating chaperone-mediated autophagy. Front. Aging Neurosci. 16634365. 9. Available from: https://www.frontiersin.org/articles/10.3389/fnagi.2017.00308/full. doi:10.3389/fnagi.2017.00308.

Yang, W., Li, S., Li, X.J., 2019. A CRISPR monkey model unravels a unique function of PINK1 in primate brains. Mol. Neurodegener. 17501326. 14 (1). Available from: http://www.molecularneurodegeneration.com/. doi:10.1186/s13024-019-0321-9.

Yang, W., Wang, G., Wang, C.E., Guo, X., Yin, P., Gao, J., Tu, Z., Wang, Z., Wu, J., Hu, X., Li, S., Li, X.J., 2015. Mutant alpha-synuclein causes age-dependent neuropathology in monkey brain. J. Neurosci. 15292401. 35 (21), 8345–8358. Available from: http://www.jneurosci.org/content/35/21/8345.full.pdf. doi:10.1523/JNEUROSCI.0772-15.2015.

Yunlan, D., Zhang, X., Tao, Q., Chen, S., Weidong, L., 2013. Adeno-associated virus Type 2 vector-mediated glial cell line-derived neurotrophic factor gene transfer induces neuroprotection and neuroregeneration in a ubiquitin-proteasome system impairment animal model of Parkinson's disease. Neurodegener. Dis. 1660-2854. 11 (3), 113–128. doi:10.1159/000334527.

Zharikov, A., Bai, Q., De Miranda, B.R., Laar, A.V., Greenamyre, J.T., Edward, A.B., 2019. Long-term RNAi knockdown of α-synuclein in the adult rat substantia nigra without neurodegeneration. Neurobiol. Dis. 09699961. 125, 146–153. doi:10.1016/j.nbd.2019.01.004.

Zharikov, A.D., Cannon, J.R., Tapias, V., Bai, Q., Horowitz, M.P., Shah, V., Ayadi, A.E., Hastings, T.G., Greenamyre, J.T., Burton, E.A., 2015. ShRNA targeting α-synuclein prevents neurodegeneration in a Parkinson's disease model. J. Clin. Invest. 15588238. 125 (7), 2721–2735. Available from: http://www.jci.org/articles/view/64502/pdf. doi:10.1172/JCI64502.

Zheng, J.S., Tang, L.L., Zheng, S.S., Zhan, R.Y., Zhou, Y.Q., Goudreau, J., Kaufman, D., Chen, A.F., 2005. Delayed gene therapy of glial cell line-derived neurotrophic factor is efficacious in a rat model of Parkinson's disease. Mol. Brain Res. 0169328X. 134 (1), 155–161. Available from: https://www.elsevier.com/locate/molbrainres. doi:10.1016/j.molbrainres.2004.06.029.

Chapter 12

Gene therapy in Huntington's disease (HD)

Fatemeh Tirgar[a,b], Mina Hosseini[c], Sakineh Alijanpour[d], Mohamad-Reza Aghanoori[e], Sahar Alijanpour[f], and Hami Mahdavinataj[b]

[a]*Iranian National Center for Addiction Studies (INCAS), Tehran University of Medical Sciences, Tehran, Iran*

[b]*Department of Neuroscience and Addiction Studies, School of Advanced Technologies in Medicine, Tehran University of Medical Sciences, Tehran, Iran*

[c]*Department of Pharmaceutical Biotechnology, School of Pharmacy, Tehran University of Medical Sciences, Tehran, Iran*

[d]*Department of Biology, Faculty of Science, Gonbad Kavous University, Gonbad Kavous, Iran*

[e]*Department of Molecular Medicine, National Institute of Genetic Engineering and Biotechnology (NIGEB), Tehran, Iran*

[f]*Department of Genetics, Faculty of Medicine, Alborz University of Medical Sciences, Karaj, Iran*

1 Introduction

Huntington's disease (HD), among neurodegenerative disorders, holds a distinctive place due to its well-established genetic source, predictable clinical course, and profound motor and cognitive impairments. The earliest scientific definition of this condition was provided in 1872 by the American physician George Huntington, whose observations laid the foundation for what is now a widely studied and clinically recognized disorder Huntington (1872). In his seminal paper on what he referred to as "hereditary chorea," Huntington clearly identified three hallmark features of the disease: its hereditary nature, the presence of neuropsychiatric and cognitive disturbances, and its progressive, unrelenting course. Huntington's miserable statement "when once it begins it clings to the bitter end" Huntington (1872) underscores the irreversible and ultimately fatal nature of the disease (McGarry et al., 2024). These hallmarks have made HD a crucial model for investigating the underlying mechanisms of neurodegeneration and for advancing the development of targeted therapeutic strategies.

1.1 The worldwide prevalence of HD

Although HD is a worldwide occurring neurodegenerative disease, significant disparities in its prevalence exist across ancestral groups, reflecting underlying genetic variability. Typically, HD has a higher incidence rate in Europe and North America than in Asia and Africa (Medina et al., 2022; Kay et al., 2017; Bates et al., 2015; Jiang et al., 2023a). In Western countries, the prevalence of HD ranges approximately from 10.6 to 13.7 cases per 100,000 population (Bates et al., 2015; Jiang et al., 2023a; McColgan and Tabrizi, 2018). In contrast, significantly lower rates have been documented in Asian populations, including Japan, Taiwan, and Hong Kong, with prevalence estimates ranging from 1 to 7 per million population (Bates et al., 2015; Jiang et al., 2023a). Likewise, in South Africa, HD is less common among individuals of Black African ancestry compared to those of White or mixed ancestry (McColgan and Tabrizi, 2018; Baine et al., 2016). The observed variations in disease prevalence are primarily explained by ethnic differences in the length of CAG repeats within the huntingtin (*HTT*) gene. Populations with increased disease rates commonly exhibit higher average CAG repeats. Particularly, individuals of European ancestry display an average of 18.4 to 18.7 repeats, in contrast to East Asian populations, whose averages range from 17.5 to 17.7 (Bates et al., 2015; Jiang et al., 2023a).

Emerging evidence suggests that the prevalence of the HD mutation in the general population is likely higher than earlier reports have estimated. This apparent increase in prevalence is likely driven by an assortment of factors, including improved access to diagnostic testing, earlier and more accurate diagnoses, increased life expectancy, and the occurrence of de novo mutations. Furthermore, the rising prevalence of HD is accompanied by a demographic change, marked by an increasing number of late-onset cases and a progressively later average age of onset (Medina et al., 2022; Kay et al., 2017).

Gene Therapy in Neurodegenerative Disorders. https://doi.org/10.1016/B978-0-443-40467-2.00011-7

1.2 Common clinical signs and symptoms of HD

The symptoms of HD are typically categorized into three principal domains: motor, psychiatric, and cognitive symptoms (Bates et al., 2015; Jiang et al., 2023a; Palaiogeorgou et al., 2022; Ajitkumar et al., 2025; Ross et al., 2014; Liu et al., 2015; Nance, 1998), as demonstrated in Fig. 12.1. The following sections provide a detailed exploration of these symptoms.

1.2.1 Motor symptoms

The first domain involves hyperkinetic motor disturbances, most notably chorea, characterized by involuntary, abrupt, and purposeless movements, which tend to progress over time into more severe and disabling motor impairments, including bradykinesia, akathisia, dystonia, myoclonus, and rigidity (Bates et al., 2015; Jiang et al., 2023a; McColgan and Tabrizi, 2018; Ajitkumar et al., 2025; Nance, 1998; Karagas et al., 2020; Roos, 2010). Chorea, from the Greek word for dance (Burgunder, 2013; Fazio et al., 2018), is characterized by involuntary, arrhythmic motor activity that initially manifests in the distal musculature, often presenting with low amplitude. As the disorder progresses, these hyperkinetic movements may involve the orofacial area and later spread to the proximal limb and axial muscles (Ajitkumar et al., 2025; Roos, 2010). This rostrocaudal pattern is commonly accompanied by an increase in movement amplitude, reflecting the progressive nature of the underlying neuropathology. Bradykinesia, defined as slowness in the initiation and execution of voluntary movement, is a less conspicuous but clinically significant motor symptom in HD, particularly in advanced stages and in the juvenile-onset form (Bakels et al., 2022). Unlike chorea, which dominates the early disease course, bradykinesia often emerges as a marker of disease progression, contributing to impaired fine motor control, decreased facial expressivity, and profound functional disability (Garcia Ruiz et al., 2000). Akathisia, characterized by an inner sense of restlessness and an urgent need to move, is an underrecognized but distressing motor and neuropsychiatric complication in HD (Karagas et al., 2020).

As HD progresses, dysarthria and dysphagia commonly emerge, increasing the risk of aspiration events and predisposing patients to aspiration pneumonia, a frequent cause of mortality (Jiang et al., 2023; Ajitkumar et al., 2025; Roos, 2010). Additional involuntary motor phenomena such as tics and cerebellar signs, including ataxia, may also occur. Pyramidal tract involvement is apparent in some patients, as demonstrated by signs like the Babinski reflex. Over time, motor dysfunction

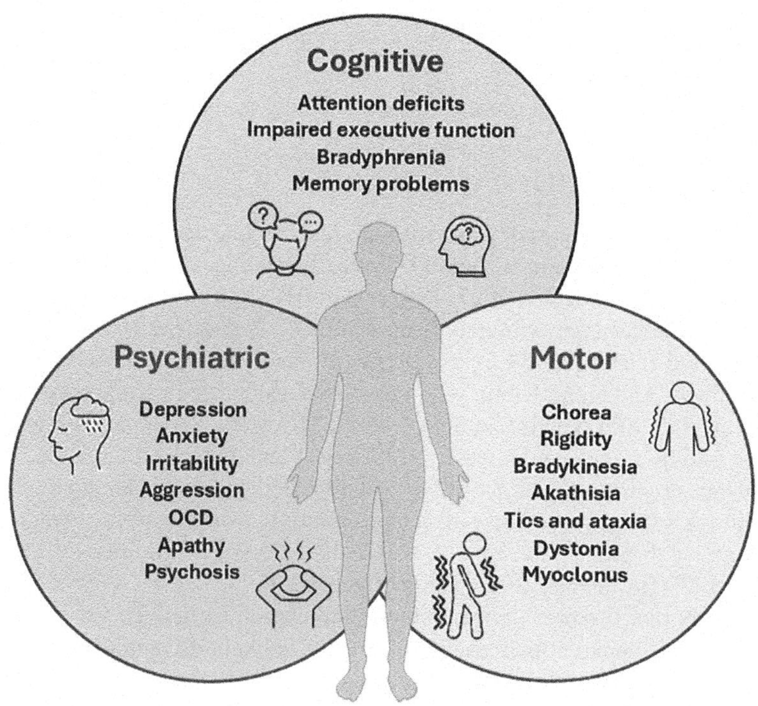

FIG. 12.1 Common signs and symptoms of HD. HD is characterized by a combination of motor, cognitive, and psychiatric manifestations. Motor symptoms include chorea, rigidity, bradykinesia, dystonia, and other movement abnormalities. Cognitive impairments often involve executive dysfunction, along with memory and attention deficits. Psychiatric and behavioral changes may include depression, irritability, anxiety, aggression, psychosis, and compulsive behaviors. The progressive nature of HD leads to increasing functional impairment over time.

significantly impairs daily functioning, manifesting as gait instability, impaired postural control, and an increased risk of falls (Ajitkumar et al., 2025; Roos, 2010).

1.2.2 Psychiatric and behavioral symptoms

The second domain pertains to psychiatric disturbances, which may manifest as depression, anxiety, irritability, obsessive-compulsive disorder, emotional lability, including aggression, apathy, and in more severe cases, psychotic features (Jiang et al., 2023a; McColgan and Tabrizi, 2018; Snowden, 2017; Teixeira et al., 2016; Paulsen et al., 2017). These symptoms may reduce quality of life and increase caregiver burden, necessitating targeted pharmacological intervention (Paoli et al., 2017).

1.2.3 Cognitive symptoms

The third domain encompasses progressive cognitive deficits, including early impairments in attention, emotion recognition, memory retrieval, decision-making, goal-directed behavior, and other executive functions, with patients exhibiting difficulty in organizing tasks, multitasking, cognitive flexibility, and planning (Bates et al., 2015; Jiang et al., 2023a; Snowden, 2017; Teixeira et al., 2016; Paulsen et al., 2017; Vinther-Jensen et al., 2014). In addition, impairments in verbal learning and visuospatial processing are common, eventually leading to more global memory dysfunction as the disease evolves (Jiang et al., 2023a; Nance, 1998; Roos, 2010; Snowden, 2017). As the disease advances, these deficits intensify and culminate in a pattern consistent with subcortical dementia rather than cortical dementia (McColgan and Tabrizi, 2018; Ajitkumar et al., 2025; Teixeira et al., 2016; Martinez-Horta et al., 2024). Notably, hallmark features of cortical involvement such as apraxia, aphasia, and agnosia are generally absent in HD (Ajitkumar et al., 2025; Teixeira et al., 2016).

Longitudinal studies of individuals with HD have demonstrated that subtle cognitive impairments and neuropsychiatric symptoms often precede the onset of motor symptoms and show gradual progression over time (Bates et al., 2015; McColgan and Tabrizi, 2018; Ajitkumar et al., 2025; Paulsen et al., 2017). In addition, a lack of insight into the nature and impact of their symptoms is frequently observed among patients (Ajitkumar et al., 2025).

1.2.4 Other clinical symptoms

Other less common symptoms include slow eye movement with dysfunctional gaze and seizures, which only occur in approximately 30%–50% of juvenile-onset HD cases, particularly among patients diagnosed before the age of 10 years (Ajitkumar et al., 2025). Sleep disturbances are also noticed and are characterized by considerable disruptions in circadian rhythm, including decreased sleep efficiency, delayed sleep onset, and a shortened duration of rapid eye movement sleep (Arnulf et al., 2008; Morton, 2013). These irregularities reflect broader dysregulation within neural systems managing sleep–wake cycles and homeostasis.

In addition, patients with HD often exhibit peripheral symptoms such as metabolic dysfunction, including weight loss, as well as skeletal muscle atrophy, cardiac failure, testicular atrophy, osteoporosis, and other manifestations of autonomic nervous system dysfunction (Jiang et al., 2023a; van der Burg et al., 2009; Carroll et al., 2015; Sassone et al., 2009). These pathological alterations are not merely a consequence of central nervous system (CNS) dysfunction; instead, a substantial portion seems to arise directly from the expression of the mutant protein in peripheral tissues (van der Burg et al., 2009). Consequently, therapeutic strategies aimed at targeting peripheral organs and systems may offer substantial improvements in quality of life for individuals with HD, even in the absence of interventions that alter the course of the neurodegenerative process (Carroll et al., 2015).

1.3 The prognosis of HD

Symptom onset in HD typically occurs between the ages of 35 and 46, though this range may vary across different ethnic populations (Nance, 1998; Carroll et al., 2015; Sun et al., 2022; Stanisławska-Sachadyn et al., 2024; Kim et al., 2016; Hwang et al., 2024). Notably, the median age of onset tends to be higher in women compared to men (Stanisławska-Sachadyn et al., 2024). In a subset of cases, symptom onset may deviate significantly from the typical age range, emerging either much earlier or later in life. Juvenile-onset HD, characterized by symptom emergence before the age of 21, accounts for approximately 1% to 5% of all cases (Bakels et al., 2022; Quarrell et al., 2012; Achenbach and Saft, 2021; Bakels et al., 2024). In contrast, late-onset HD, which begins after the age of 60, is observed in an estimated 4% to 12% of affected individuals (Koutsis et al., 2014; Cornejo-Olivas et al., 2015; Chaganti et al., 2017).

Following the onset of symptoms, the disease typically progresses over 15–20 years (Bates et al., 2015; Jiang et al., 2023a; Sun et al., 2022). However, some studies have reported shorter median survival times in specific populations, for instance, approximately 14.5 years in Korean patients (Kim et al., 2016) and 12 years in cohorts from the United Kingdom

(Furby et al., 2022). In contrast, longer median survival durations have been observed in European populations, with some reports indicating survival of up to 35 years from symptom onset (Rodrigues et al., 2017). In a recent study, it has been revealed that American individuals aged 65 and older exhibited the highest age-adjusted mortality rate due to HD, followed by middle-aged individuals (45–64 years) and younger adults (25–44 years) (Bin Kashif et al., 2024). The median age at death is estimated to range approximately between 54 and 64 years, with variations observed across different racial and ethnic groups (Sun et al., 2022; Kim et al., 2016; Hwang et al., 2024; Rodrigues et al., 2017; Solberg et al., 2018). Although mortality rates are slightly higher in men than in women (Bin Kashif et al., 2024), no significant sex-related variations have been observed in the age at death among individuals with HD (Solberg et al., 2018).

Notably, the length of the CAG repeat shows an inverse relationship with both the age of symptom onset and age at death; however, it does not appear to be associated with overall mortality status and duration of the disease (Sun et al., 2022; Stanisławska-Sachadyn et al., 2024; Keum et al., 2016).

2 HD pathogenesis and genetic patterns

HD pathogenesis is primarily explained by two key hypotheses: a toxic gain-of-function from the mutant huntingtin (mHTT) protein, which disrupts various cellular processes, and a dominant-negative loss-of-function, where mHTT impairs the typical roles of wild-type huntingtin (wtHTT) protein (Bardai et al., 2013; Finkbeiner, 2011). In the following section, we discuss the physiological functions of wtHTT and the dysfunctions of mHTT that contribute to the pathogenesis of HD.

2.1 Key functional roles of wild-type huntingtin

The evolutionarily conserved *HTT* gene encodes the 348-kDa HTT protein comprising 3144 amino acids (Jurcau, 2022). The HTT protein contains an N-terminal polyglutamine (polyQ) tract beginning at residue 18 (Ersoy Tunalı, 2021) downstream into the HEAT (Huntingtin, Elongation factor 3, protein phosphatase 2A, and yeast kinase Target of rapamycin 1) repeat motif that mediates protein–protein interactions (Wanker et al., 2019; Jung et al., 2020). The first 17 amino acids enable subcellular localization with organelles such as the nucleus (Didiot et al., 2018), mitochondria, endoplasmic reticulum (ER), and Golgi (Jurcau, 2022; Schulte, 2011; Hoffner et al., 2002; Strehlow et al., 2007). This protein is also found in neurites and synaptic terminals, where it participates in processes such as vesicular trafficking, microtubule transport, and synaptic function (Hoffner et al., 2002; Velier et al., 1998). The HTT protein adopts an α-helical conformation and forms a stable complex with huntingtin-associated protein 40 (HAP40), which stabilizes its structure (Guo et al., 2018).

HTT is ubiquitously expressed, with the highest levels in neurons (Jurcau, 2022; Schulte, 2011; Han et al., 2010) and is also found in glial cells (Jurcau, 2022; Schulte, 2011; Jansen et al., 2017) and peripheral tissues like muscle (Sassone et al., 2009; Jurcau, 2022; Schulte, 2011; Kojer et al., 2019). Posttranslational modifications, including acetylation, phosphorylation, ubiquitination, sumoylation, and palmitoylation, play a critical role in modulating the subcellular distribution, structural stability, turnover, and functional dynamics of the HTT protein (Jurcau, 2022).

As demonstrated in Fig. 12.2, wtHTT is implicated in a wide range of cellular processes, particularly within the nervous system. Key functional roles include the following:

2.1.1 Cell survival

HTT plays a critical role in cell survival (Dragatsis et al., 2018; Tong et al., 2011; Reiner et al., 2003). HTT exerts antiapoptotic effects through modulation of caspase activity. For instance, HTT inhibits apoptosis directly by inhibiting caspase-3 (Zhang et al., 2006; Leavitt et al., 2006) and caspase-9 activity (Tong et al., 2011; Rigamonti et al., 2000, 2001), and indirectly by sequestering Hip-1, thereby blocking the formation of the Hip-1/Hippi (Hip-1 protein interactor) complex, leading to the inhibition of the recruitment and activation of procaspase-8 (Rigamonti et al., 2000; Gervais et al., 2002). Through its interaction with p21-activated kinase 2 (Pak2), HTT also indirectly inhibits apoptosis by limiting the ability of caspase-3 and caspase-8 to cleave Pak2 into a death-promoting fragment (Luo and Rubinsztein, 2009). Besides its role in conserving the integrity of the PI3K/AKT/GSK3β prosurvival signaling pathway (Hansen et al., 2019), HTT also regulates members of the Bcl-2 family, highlighting its multifaceted antiapoptotic functions (Rigamonti et al., 2000; Sassone et al., 2013).

2.1.2 Intracellular trafficking

HTT is a key regulator of axonal transport, facilitating the trafficking of organelles and vesicles along microtubules (Jurcau, 2022; Reiner et al., 2003; White et al., 2015; Vitet et al., 2020; Caviston and Holzbaur, 2009; Harjes and Wanker, 2003). Kinesin protein mediates anterograde transport, and dynein protein mediates retrograde transport (Sleigh et al., 2019). HTT

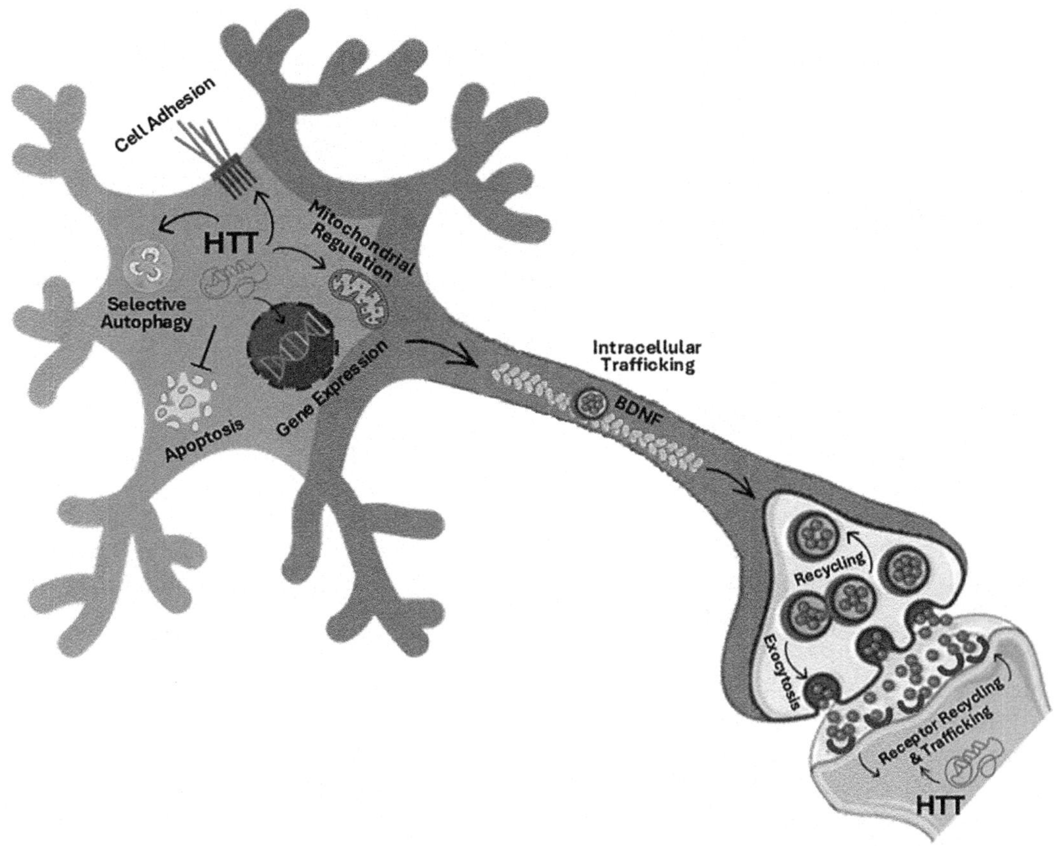

FIG. 12.2 Physiological functions of wtHTT within the nervous system. wtHTT plays essential roles in neuronal survival during development and adulthood, vesicular trafficking and axonal transport, modulation of transcriptional pathways, support of synaptic function and plasticity, facilitation of selective autophagy, cell adhesion, and mitochondrial regulation. These integrated functions highlight why the widespread suppression of the *HTT gene* requires careful therapeutic balancing to avoid loss-of-function consequences.

modulates both proteins. For instance, HTT interacts with the dynactin complex to modulate dynein activity (Gunawardena and Goldstein, 2005; Caviston et al., 2007). On the contrary, phosphorylation of HTT enhances the anterograde transport of brain-derived neurotrophic factor (BDNF)-containing vesicles, while dephosphorylation favors retrograde movement (Vitet et al., 2020; Gauthier et al., 2004; Ehinger et al., 2020; Colin et al., 2008). In addition, HTT phosphorylation promotes the anterograde transport of synaptic vesicle precursors, supporting synaptic vesicle replenishment and neurotransmission (Vitet et al., 2020).

2.1.3 Embryonic development

During embryogenesis, HTT is essential for proper neuronal cytoarchitecture, regulation of cell division, and cell survival via antiapoptotic pathways (Dragatsis et al., 2018; Saudou and Humbert, 2016; Barnat et al., 2017). Deletion of the *Htt* gene in mouse models results in early embryonic lethality, marked by excessive apoptosis in the embryonic ectoderm, impaired neuronal proliferation, increased neuroepithelial cell death, and impaired neuronal migration to the proper laminar position (Dragatsis et al., 2018; Tong et al., 2011; Reiner et al., 2003; Zeitlin et al., 1995; Duyao et al., 1995; Nasir et al., 1995; Woda et al., 2005).

2.1.4 Cell adhesion

Interactions between HTT and various cytoskeletal components highlight its essential role in regulating cell adhesion (O'Regan et al., 2020). HTT, through its interaction with α-actinin, is an important regulator of cell adhesion, morphology, and actin-dependent processes (Tousley et al., 2019). In addition, HTT plays a critical role in N-cadherin-mediated adhesion and distribution of the tight junction protein ZO-1 (Saudou and Humbert, 2016; Lo Sardo et al., 2012).

2.1.5 Synaptic activity

HTT plays a critical role in maintaining synaptic integrity by regulating both pre- and postsynaptic functions. At the presynaptic terminal, HTT supports synaptic vesicle trafficking, including exocytosis (Barron et al., 2021) and vesicle recycling (Barron et al., 2021; McAdam et al., 2020). At the postsynaptic site, HTT mediates the proper transport, recycling, and localization of receptors, for instance, inhibitory gamma-aminobutyric acid (GABA)-A (Barron et al., 2021; Twelvetrees et al., 2010) and excitatory α-amino-3-hydroxy-5-methyl-4-isoxazolepropionic acid (AMPA) (Barron et al., 2021; Mandal et al., 2011) receptors to postsynaptic sites. HTT also binds directly to PSD95, anchoring glutamate receptors at the postsynaptic density (Smith et al., 2005; Sun et al., 2001; Zuccato et al., 2010). Impaired HTT function may lead to aberrant receptor distribution, such as extrasynaptic clustering of N-methyl-D-aspartate (NMDA) receptors (Kang et al., 2019). HTT protein partners are involved in neurotransmitter release, particularly glutamate, and reducing HTT levels can disrupt the normal function of these complexes (Barron et al., 2021).

2.1.6 Mitochondrial regulation and cellular energy

HTT plays a crucial role in mitochondrial function. Through other proteins, it facilitates the transport of mitochondria along axons, ensuring proper distribution to energy-demanding sites (Trushina and McMurray, 2007). In addition, HTT contributes to mitochondrial dynamics and cellular energy homeostasis by regulating fission/fusion processes and modulating the activity of the respiratory chain to maintain ATP production. Furthermore, HTT is involved in the regulation of mitochondrial calcium, which is essential for both metabolism and cell survival (Jurcau, 2022 Shirendeb et al., 2011a).

2.1.7 Transcriptional regulation

HTT has a critical place in transcriptional regulation through multiple pathways (Jurcau, 2022). wtHTT promotes transcription via the cAMP-responsive element-binding (CREB)/CREB-binding protein pathway (Jiang et al., 2023a; Wyttenbach et al., 2001; Li and Li, 2004; Landles and Bates, 2004; Choi et al., 2009; Steffan et al., 2000), supports specificity protein 1 activity (Li et al., 2002; Zhai et al., 2005; Xiang et al., 2018), and prevents the nuclear entry of the repressor-element-1 transcription factor–neuron-restrictive silencer factor complex, thereby inhibiting its subsequent repressive action on the transcription of Neuron-Restrictive Silencer Element (NRSE)-containing genes, including BDNF (Jurcau, 2022; Landles and Bates, 2004; Zuccato et al., 2001, 2003). It also facilitates nuclear factor κ light-chain-enhancer of activated B cells (NF-κB) translocation to the nucleus, maintaining its levels to drive target gene expression (Marcora and Kennedy, 2010). In addition, HTT serves as a cofactor for nuclear receptors, including liver X receptor, promoting the expression of genes such as ABCA1 (Futter et al., 2009). Notably, HTT can directly interact with DNA and alter its conformation. mHTT enhances HTT–DNA binding, particularly at recognition sites of transcription factors, interfering with their binding and ultimately contributing to gene expression dysregulation (Benn et al., 2008).

2.1.8 Autophagy

HTT exhibits strong interactions with components of the autophagy machinery, which has recently emerged as a key regulator of neurotransmission and synaptic homeostasis (Barron et al., 2021). Notably, the C-terminal domain of HTT shares structural and binding similarities with the yeast scaffold protein Atg11, raising the possibility that HTT normally serves as a scaffold protein coordinating different forms of selective autophagy, the elimination of targeted cytoplasmic components instead of bulk nonspecific degradation (Ochaba et al., 2014). HTT contributes critically to autophagy by engaging in interactions with the cargo adaptor and autophagy initiation kinase proteins, thereby facilitating both the recognition of selective autophagy cargo and the initiation of the autophagic process (Barron et al., 2021; Martinez-Vicente et al., 2010).

2.2 Molecular mechanisms associated with HD

mHTT triggers multiple pathogenic mechanisms that likely act in concert to promote neurodegeneration, and these processes provide the foundation for current therapeutic strategies. Neurodegeneration occurs mainly in subcortical brain regions, including the striatum, with a marked vulnerability of medium spiny neurons (MSNs) (Finkbeiner, 2011; Ehrlich, 2012; Bergonzoni et al., 2021). GABAergic MSNs constitute approximately 95% of the neuronal population in the striatum. These neurons serve as the principal output cells of the striatum and maintain extensive intrastriatal connectivity with other MSNs and local interneurons (Ehrlich, 2012; Bergonzoni et al., 2021). In addition, there has been growing interest in cortical pathology, particularly within layers V and VI, which house the corticostriatal projection neurons (Ehrlich, 2012;

Johnson et al., 2021; C. Estevez-Fraga et al., 2023a; Hickman et al., 2022). The following section is a summary of the key underlying mHTT-induced neurodegeneration:

2.2.1 Excitotoxicity-mediated neurodegeneration

Glutamate-mediated excitotoxicity has long been proposed as a key mechanism underlying striatal neurodegeneration in HD (Jiang et al., 2023a; Zuccato et al., 2010; Jiang et al., 2024; Sepers and Raymond, 2014). Excessive glutamate transmission in striatal synapses due to increased release and reduced clearance by astrocytes (via decreased glutamate transporter GLT1 and GLAST expression) (Jurcau, 2022; Sepers and Raymond, 2014; Shin et al., 2005; Miller and Bezprozvanny, 2010; Lee et al., 2013; Faideau et al., 2010; Estrada-Sánchez et al., 2009; Liévens et al., 2001), leads to overstimulation of glutamate receptors and, in particular, NMDA receptors. This results in prolonged depolarization, calcium influx, oxidative stress, and initiation of death pathways (Jiang et al., 2023, Jiang et al., 2024; Sepers and Raymond, 2014; Estrada-Sánchez et al., 2009; Fernandes and Raymond, 2008). Experimental models have shown that the expression and activation of these receptors induce degeneration of striatal MSNs (Milnerwood et al., 2010; Heng et al., 2009), which are particularly vulnerable due to their high density of ionotropic glutamate receptors (Jurcau, 2022; Ehrlich, 2012; Jiang et al., 2024; Rigby et al., 1996).

2.2.2 Dopaminergic system disruption

Disruption of the dopamine system and alterations in dopamine levels, receptor expression, and receptor binding are observed in HD, contributing to motor symptoms (Cepeda et al., 2014; Chen et al., 2013; Koch and Raymond, 2019). In the early hyperkinetic stage, elevated dopamine levels and reduced expression of striatal D1 and D2 dopamine receptors are associated with chorea, while later stages show decreased dopamine in the caudate and striatum, correlating with rigidity and akinesia (Jiang et al., 2023a; Cepeda et al., 2014; Chen et al., 2013; Koch and Raymond, 2019; Yang et al., 2022). Therefore it has been suggested that dopamine levels in HD exhibit a biphasic, time-dependent pattern, characterized by an initial increase during early stages, followed by a decline in later stages (Ortiz et al., 2011). Reduced striatal expression of dopamine transporter and vesicular monoamine transporter 2 (VMAT2) has also been observed in HD, potentially reflecting the degeneration of dopaminergic nigrostriatal terminals (Chen et al., 2013). However, it remains unclear whether these changes are a cause or a consequence of neurodegeneration (Jiang et al., 2023b).

On the contrary, DA metabolism induces oxidative stress in HD. Enzymatic degradation by monoamine oxidases (MAOs) produces 3,4-dihydroxyphenylacetate and hydrogen peroxide, the latter of which can generate hydroxyl radicals through reactions with iron. In addition, DA can undergo nonenzymatic autooxidation, forming semiquinones and superoxide radicals, further exacerbating oxidative-induced striatal neurodegeneration (Jurcau, 2022; Kumar et al., 2010).

2.2.3 Mitochondrial dysfunction

Mitochondrial malfunction plays a starring role in HD development, contributing to neurodegeneration through impaired bioenergetics, disrupted calcium homeostasis, increased oxidative stress, and defective mitochondrial dynamics and trafficking. Evidence indicates a significant reduction in mitochondrial respiratory chain activity and bioenergetic impairment in HD (Jurcau, 2022; Vos et al., 2012). Systemic administration of 3-nitropropionic acid (3-NP), an irreversible inhibitor of complex II, reproduces HD-like symptoms in animal models (Beal et al., 1993; Brouillet et al., 1995). In addition, deficits in ATP production and altered ATP/ADP ratios are consistently observed in HD (Mochel et al., 2012). On the contrary, several studies reveal early impairments in the glycolytic pathway (Kedaigle et al., 2021) and elevated lactate levels (Zuccato et al., 2010; Han et al., 2021; Solís-Maldonado et al., 2018), even before clinical symptoms appear.

Moreover, impaired electron flow in dysfunctional mitochondria leads to the inordinate production of several reactive oxygen species (ROS), including superoxide, hydroxyl radicals, and peroxynitrite (Ribeiro et al., 2014; Addabbo et al., 2009; Guo et al., 2013). These ROS contribute to oxidative damage and promote apoptotic pathways in HD (Dash et al., 2025; Essa et al., 2019). As mentioned earlier, DA metabolism further exacerbates oxidative stress-induced neurodegeneration in HD (Jurcau, 2022; Kumar et al., 2010).

It appears that mitochondrial fission and fusion, two processes that require supplying the necessary energy demands, are dysregulated in HD (Jurcau, 2022, Jurcau (2021)). In addition, proper mitochondrial positioning is essential for neuronal function due to high energy demands at synapses (Jiang et al., 2023a; Jurcau and Nunkoo, 2021; Sheng, 2017). mHTT impairs mitochondrial trafficking by disrupting interactions among HTT, huntingtin-associated protein 1 (HAP1), kinesin, dynein, and Miro, a protein anchored to the outer mitochondrial membrane (U. P. Shirendeb et al., 2011b; Franco-Iborra et al., 2018; Chang et al., 2006; López-Doménech et al., 2018; Fenton et al., 2021). An increased cytosolic level of Ca^{2+}

downregulates mitochondrial trafficking, and subsequently, disrupted trafficking leads to energy deficits in vulnerable regions, including the striatum (Jurcau and Nunkoo, 2021).

Affected neurons by direct mHTT binding exert heightened sensitivity of inositol 1,4,5-trisphosphate receptors, leading to increased cytosolic Ca^{2+} levels (Bezprozvanny, 2009; Mackay et al., 2018). Subsequently, an elevated mitochondrial Ca^{2+} level is observed due to compensatory increased influx via the mitochondrial Ca^{2+} uniporter (Mammucari et al., 2018). Excess mitochondrial Ca^{2+} promotes the opening of the mitochondrial permeability transition pore, triggering the release of cytochrome *c* and apoptosis (Bezprozvanny, 2009; Briston et al., 2019).

2.2.4 Neuroinflammation

Neuroinflammation is a hallmark of HD, characterized by the progressive activation of glial cells, especially microglia and astrocytes (Guzman-Martinez et al., 2019; Saba et al., 2022; Adamu et al., 2024). While these cells initially respond to neuronal injury by clearing debris and providing neuroprotection (Harry and Kraft, 2008; Palpagama et al., 2019; Crotti and Glass, 2015), chronic activation results in the sustained release of proinflammatory mediators (Crotti and Glass, 2015; Jia et al., 2022; Wilton and Stevens, 2020)—such as cytokines and nitric oxide—which exacerbate neuronal damage (Palpagama et al., 2019; Crotti and Glass, 2015; Silvestroni et al., 2009; Yang et al., 2017a; Möller, 2010; Hsiao et al., 2013; Bradford et al., 2010). The progressive accumulation of reactive glia in the brains of individuals with HD has been documented over the course of the disease (Crotti and Glass, 2015; Tai et al., 2007; Sapp et al., 2001; Pavese et al., 2006; Pagano et al., 2016), and notably, neuroinflammatory changes have been identified in presymptomatic HD mutant carriers and may be detectable as early as 15 years prior to the estimated onset of clinical symptoms (Tai et al., 2007; Björkqvist et al., 2008).

Microglia in HD exhibit morphological changes and shift from a supporting cell to an inflammatory, phagocytic phenotype (Hickman et al., 2022; Palpagama et al., 2019; Hickman et al., 2018). Evidence reveals that mHTT may drive microglial activation through NF-κB (Palpagama et al., 2019 Yang et al., 2017a; Khoshnan and Patterson, 2011; Träger et al., 2014), kynurenine (Palpagama et al., 2019 Yang et al., 2017a; Möller, 2010; Schwarcz et al., 2010), and cannabinoid receptor (Palpagama et al., 2019 Yang et al., 2017a) signaling pathways. Elevated levels of proinflammatory cytokines, such as interleukin 1β, interleukin 6, interleukin 8, and tumor necrosis factor alpha, have been detected in HD brains as a result of microglial activation (Palpagama et al., 2019; Crotti and Glass, 2015 Yang et al., 2017a; Björkqvist et al., 2008).

Astrocytes also exhibit reactive changes in both neurotoxic A1 (Palpagama et al., 2019; Liddelow and Barres, 2017; Liddelow et al., 2017) and neuroprotective A2 (Diaz-Castro et al., 2019) subtypes. A1 astrocytes may contribute to neuronal dysfunction by releasing inflammatory cytokines (Hsiao et al., 2013; Lee et al., 2022), impairing potassium buffering (Crotti and Glass, 2015; Liddelow et al., 2017; Lee et al., 2022; Khakh et al., 2017; Tong et al., 2014), and reducing glutamate uptake, leading to the neurodegeneration seen in HD (Liévens et al., 2001; Crotti and Glass, 2015; Liddelow and Barres, 2017; Lee et al., 2022; Khakh et al., 2017; Allaman et al., 2011; Guttenplan and Liddelow, 2018). In contrast, A2 astrocytes appear to support the upregulation of neurotrophic factors and neuronal survival by producing antioxidants and promoting repair (Liddelow and Barres, 2017; Zamanian et al., 2012). Although A2 astrocytes are present in HD, there is an imbalance favoring the A1 phenotype, which is associated with heightened neuronal damage (Palpagama et al., 2019). Interestingly, in HD, activated microglia may induce neurotoxic reactive A1 astrocytes, leading to inflammation-induced neurodegeneration (Liddelow et al., 2017).

2.2.5 Abnormal protein degradation

Autophagy, particularly lysosome-mediated macroautophagy, plays a crucial role in clearing misfolded proteins like mHTT (Croce and Yamamoto, 2019). In HD, multiple aspects of autophagy are impaired, including cargo recognition and substrate degradation (Jurcau, 2022; Croce and Yamamoto, 2019). Rhes, a small GTPase enriched in the striatum, normally promotes autophagy. However, Rhes is inactivated upon becoming sequestered in mHTT aggregates, potentially contributing to the selective vulnerability of striatal neurons (Mealer et al., 2014; Harding and Tong, 2018). Conversely, mHTT aggregates may also stimulate autophagy by sequestering the mammalian target of rapamycin, a negative regulator of autophagy (Wanker et al., 2019; Ravikumar et al., 2004). Although increased numbers of autophagic vesicles have been reported in HD (Kaliszewski et al., 2015; Valionyte et al., 2020), many appear empty, likely due to impaired cargo recognition, contributing to the accumulation of misfolded mHTT and damaged organelles, and eventually cellular toxicity (Martinez-Vicente et al., 2010; Croce and Yamamoto, 2019; Kaliszewski et al., 2015; Cortes and Spada, 2014; Yang et al., 2021). In addition, posttranslational acetylation of mHTT facilitates its autophagosomal trafficking and degradation (Son et al., 2021).

On the contrary, the ER is a key organelle in HTT-mediated toxicity. mHTT interferes with the ER-associated degradation system, leading to the accumulation of misfolded proteins, and eventually triggers ER stress and activates the unfolded

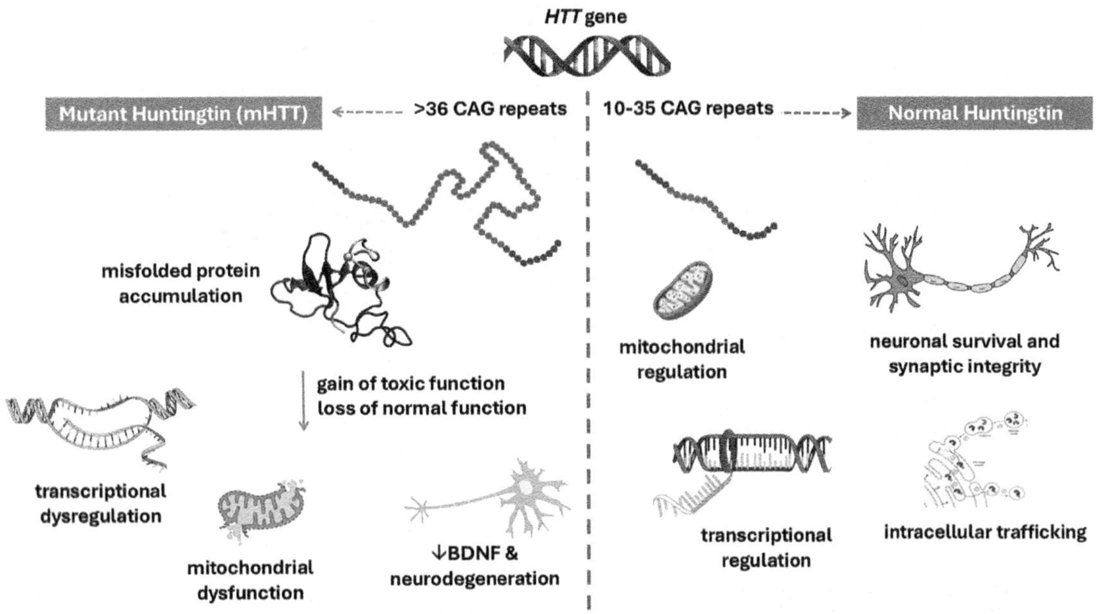

FIG. 12.3 Comparative overview of wtHTT and mHTT functions. Normal HTT contributes to neuronal health by supporting intracellular trafficking, axonal transport, transcriptional regulation, synaptic integrity, mitochondrial regulation, and cell survival. In contrast, mHTT exhibits toxic gain-of-function and loss of normal properties, including aberrant aggregation, disruption of intracellular transport, transcriptional dysregulation, mitochondrial dysfunction, and eventual neurodegeneration. The balance between wtHTT function and mHTT toxicity underlies disease pathogenesis and informs therapeutic strategies aimed at selectively reducing mHTT while preserving wtHTT activity.

protein response (UPR) (Cortes and Spada, 2014; Duennwald and Lindquist, 2008; Noh et al., 2009). ER stress and UPR activation have been linked to the induction of autophagy, suggesting a compensatory relationship between these pathways, with autophagy acting as a protective mechanism to mitigate ER stress Cortes and Spada (2014).

2.2.6 *Disrupted intracellular trafficking*

mHTT perturbs vesicular and receptor trafficking through aberrant protein interactions, implicating intracellular transport deficits in HD pathogenesis. HAP1, a central regulator of microtubule-dependent transport, is crucial for maintaining proper intracellular trafficking. An abnormally tight interaction between HAP1 and mHTT, compared with wtHTT, leads to the disruption of vesicular and receptor trafficking (Chen et al., 2023). Notably, mHTT weakens the association of HAP1 with motor proteins. This reduction impairs the intracellular trafficking of tropomyosin-related kinase A receptor tyrosine kinase (TrkA) (Rong et al., 2006), GABA-A (Twelvetrees et al., 2010; Yuen et al., 2012), and GluR2-containing AMPA (Mandal et al., 2011) receptors.

Fig. 12.3 summarizes some important functions of wtHTT while also demonstrating the pathological activities associated with mHTT. Gaining insight into these genetic factors is essential for elucidating HD mechanisms, refining diagnostic accuracy, discovering biomarkers of progression, and ultimately paving the way for novel therapeutic strategies.

2.3 Genetic patterns and associated genes

In 1993 the genetic mutation underlying HD was discovered as an abnormal expansion of a CAG (cytosine–adenine–guanine) trinucleotide repeat within exon 1 of the *HTT*gene, mapped to the short arm (p arm) of chromosome 4 (4p16.3) (MacDonald et al., 1993; Capiluppi et al., 2020). This mutation leads to the production of the HTT protein with an abnormally expanded polyglutamine (polyQ) tract, rendering it highly prone to aggregation and deposition (Finkbeiner, 2011; Bates, 2003; Tabrizi et al., 2019b). HD is inherited in an autosomal dominant manner, meaning that the presence of a single mutated copy of the gene is sufficient to cause the disease. Therefore there is a 50% likelihood that a child born to an affected parent will inherit the disease (Parsons and Raymond, 2023; Quaid et al., 2010). In rare reports, the condition may occur in an individual with no family history due to a de novo mutation (Houge et al., 2013; Semaka et al., 2010, 2015; Warby et al., 2009; Lin et al., 2022; Myers et al., 1993; Ratna et al., 2025).

The number of CAG repeats in the *mHTT* gene often increases across generations, particularly through the paternal line, due to CAG repeat instability during spermatogenesis (Roos, 2010; Semaka et al., 2010, 2015; Pengo and Squitieri, 2024). This expansion is associated with anticipation, a phenomenon in which symptoms appear at an earlier age in subsequent generations (McColgan and Tabrizi, 2018; Ajitkumar et al., 2025; Roos, 2010; Testa and Jankovic, 2019; Fusilli et al., 2018). A CAG repeat length of 26 or fewer in the *HTT* gene is considered normal. HD does not develop in people with 27–35 repeats (intermediate mutable alleles) (De Luca et al., 2021), although they may transmit expanded repeats to their offspring due to meiotic instability (Ajitkumar et al., 2025; Gatto et al., 2020; Migliore et al., 2019). Repeats of 36–39 (low- or incomplete-penetrance alleles) may potentially lead to disease in some cases (Ajitkumar et al., 2025; Capiluppi et al., 2020; De Luca et al., 2021; Gatto et al., 2020; Migliore et al., 2019). Repeat lengths of 40 or more (full- or complete-penetrance alleles) are fully penetrant and consistently result in HD (De Luca et al., 2021; Gatto et al., 2020; Shiwal et al., 2025). Rare expansions beyond 60–80 repeats are associated with pediatric-onset HD, characterized by early onset, rapid progression, atypical symptoms, and reduced life expectancy (Ajitkumar et al., 2025; Fusilli et al., 2018; De Luca et al., 2021; Migliore et al., 2019). There is a well-established inverse association indicating that individuals with longer repeat expansions tend to experience earlier onset of symptoms and reduced lifespan (Keum et al., 2016).

Beyond the well-established role of CAG repeat expansion, emerging studies have uncovered additional genetic factors contributing to HD pathogenesis (Pengo and Squitieri, 2024; Gatto et al., 2020; Chao et al., 2017). These factors, known as HD genetic modifiers, play a significant role in modulating disease onset and progression. A substantial proportion of these modifiers is implicated in DNA repair mechanisms and is thought to promote somatic expansion of the CAG repeat tract. This modulation occurs through both cis- and trans-acting elements, as well as mechanisms involving RNA toxicity, mitochondrial DNA mutations, and epigenetic alterations. Notably, recent findings have identified loci on chromosomes 8 (Ribonucleotide Reductase Regulatory TP53 Inducible Subunit M2B: RRM2B and Ubiquitin Protein Ligase E3 Component N-Recognin 5: UBR5 genes) and 15 (Myotubularin Related Protein 10: MTMR10 and FANCD2/FANCI Associated Nuclease 1: FAN1 genes, and HERC2 Pseudogene 10: HERC2P10) that are associated with DNA repair and maintenance pathways and therefore modulate HD pathogenicity (Pengo and Squitieri, 2024; Gatto et al., 2020; Holmans et al., 2017; Jain and Chen-Plotkin, 2018; Lee et al., 2015). For instance, FAN1 restricts CAG repeat expansion, leading to a protective effect in HD (Goold et al., 2019). Along with these loci, there are new loci on other chromosomes, such as chromosomes 2 (PMS1 Homolog 1, DNA Mismatch Repair System Component: PMS1 gene), 3 (MutL Homolog 1, DNA Mismatch Repair Gene: MLH1 gene), 5 (MutS Homolog 3: MSH3, Dihydrofolate Reductase: DHFR, Transcription Elongation Regulator 1: TCERG1), 7 (PMS1 Homolog 2, Mismatch Repair System Component: PMS2), 11 (Coiled-Coil Domain Containing 82: CCDC82), and 19 (DNA Ligase 1: LIG1). MSH3, a component of the mismatch repair (MMR) pathway, forms complexes with MSH2 and MLH1 to recognize and repair DNA mismatches. However, in HD, it promotes erroneous loop extrusion during repair, thereby driving CAG expansions. For review, see Lee et al. (2019)).

In addition to modifier DNA repair and maintenance genes, a number of other genetic factors contribute to the disease, including polymorphisms in glutamic acid, glutamate receptors, and mitochondrial-related genes. Finally, it has been revealed that certain genes that colocalize with the expanded CAG repeats appear to have a higher likelihood of undergoing replication (Jurcau, 2022).

3 Current approved therapeutic approaches for HD

Despite over a century since HD was first described, and decades of molecular insight into its genetic basis, there is still no disease-modifying treatment for this progressive neurodegenerative disorder (Saade and Mestre, 2024). While substantial effort has been directed toward developing disease-modifying or gene-targeting therapies, the only currently approved treatments are those that provide symptomatic relief, especially for the hallmark hyperkinetic movement disorder known as chorea (Shin et al., 2024). Alongside pharmacological management, nonpharmacological and supportive care interventions play a critical role in mitigating the multidimensional burden of the disease (Mendizabal et al., 2023).

3.1 Pharmacological management of motor symptoms

3.1.1 Chorea

The only drugs specifically approved for the most prominent motor symptom, chorea, are inhibitors of VMAT2, which reduce dopaminergic transmission in the basal ganglia (Saade and Mestre, 2024). Tetrabenazine was the first agent approved by the US Food and Drug Administration (FDA) in 2008, specifically for chorea associated with HD. It acts by reversibly inhibiting VMAT2, thereby depleting monoamines, particularly dopamine, from presynaptic vesicles (Scott, 2011).

Clinical trials have shown that tetrabenazine significantly reduces choreiform movements, as assessed by the Unified Huntington's Disease Rating Scale (UHDRS) (Frank, 2009). However, its relatively short half-life necessitates multiple daily doses, and its side-effect profile, particularly depression, anxiety, sedation, and Parkinsonism, has limited its broader utility (Frank, 2010).

To address these limitations, deutetrabenazine, a deuterated analog of tetrabenazine, was developed and received FDA approval in 2017 (Dean and Sung, 2018). The strategic substitution of hydrogen atoms with deuterium results in slower metabolic degradation, more stable plasma concentrations, and reduced peak–trough fluctuations (Dean and Sung, 2018). Clinical trials, such as (First Time Use of SD-809) FIRST-HD and Alternatives for Reducing Chorea (ARC)-HD, have demonstrated deutetrabenazine's superior tolerability and noninferiority in chorea reduction compared to tetrabenazine (Claassen et al., 2017; Frank et al., 2021, Frank et al. (2016)). Furthermore, the extended-release formulation, deutetrabenazine XR (Austedo XR), approved in 2023, allows for once-daily dosing, potentially improving adherence and patient quality of life (Moondra and Jimenez-Shahed, 2024).

More recently, valbenazine, another VMAT2 inhibitor initially developed and approved for tardive dyskinesia, received regulatory approval for the treatment of HD-related chorea in a sprinkle formulation (Hauser et al., 2017; Patino et al., 2025). Valbenazine exhibits selectivity for VMAT2, minimal interaction with MAO, and a long half-life conducive to once-daily administration (Patino et al., 2025). Though its development in HD-specific populations has been less extensive than that of other agents, its inclusion broadens the therapeutic arsenal and offers alternative dosing options for patients with dysphagia or those unable to tolerate other VMAT2 inhibitors.

Notably, while these agents provide significant symptomatic relief, none target the underlying disease process. In addition, all VMAT2 inhibitors carry the risk of neuropsychiatric side effects, particularly in a population already vulnerable to depression and suicidality (Wong et al., 2025). Consequently, their use necessitates individualized risk–benefit assessments and careful clinical monitoring, particularly in patients with preexisting psychiatric comorbidities.

3.1.2 Dystonia and myoclonus

These symptoms often prove more refractory and disabling than chorea, complicating clinical management. Pharmacological strategies for dystonia primarily rely on the off-label use of anticholinergics (e.g., trihexyphenidyl), benzodiazepines, and GABAergic agents such as baclofen, which offer modest benefits by reducing muscle overactivity (Cloud and Jinnah, 2010). Botulinum toxin injections can be useful for focal or segmental dystonia, particularly when limb or cervical involvement causes pain or functional impairment (Hammoud and Jankovic, 2022). Myoclonus, while less common, may emerge in Westphal variant HD or during disease progression (Gonzalez Rojas et al., 2022). Treatment often includes levetiracetam, valproic acid, or clonazepam, chosen for their efficacy in suppressing cortical and subcortical myoclonic discharges. However, these agents may exacerbate sedation, cognitive slowing, or mood disturbances, necessitating careful titration (Saft et al., 2006; Bachoud-Lévi et al., 2019). Importantly, the overlapping presentation of motor symptoms in HD requires individualized therapeutic strategies, balancing efficacy against neuropsychiatric burden and functional decline. As no pharmacological agent is currently approved specifically for dystonia or myoclonus in HD, management remains symptomatic and guided by clinical experience, with an urgent need for targeted therapies validated in controlled trials (Shin et al., 2024).

3.1.3 Akathisia

Akathisia may occur intrinsically as part of the disease spectrum or, more commonly, emerge secondary to pharmacologic treatments, particularly dopamine-depleting agents (e.g., tetrabenazine, deutetrabenazine) and antipsychotics used for chorea or behavioral symptoms (Karagas et al., 2020; Mathews et al., 2005). Clinically, akathisia can be difficult to distinguish from anxiety or agitation, especially in advanced stages where communication is impaired, leading to frequent misdiagnosis or under-treatment (Karagas et al., 2020). Pharmacological management is empirical and often based on strategies borrowed from treatment-induced akathisia in other neuropsychiatric disorders. Beta-blockers, particularly propranolol, are considered first-line due to their central and peripheral effects on adrenergic tone (Sharma et al., 2005). Benzodiazepines such as clonazepam may offer symptom relief through anxiolytic and muscle-relaxant properties, though their sedating effects limit long-term use. In some cases, dose reduction or substitution of the offending agent may be necessary (Pringsheim et al., 2018). Mirtazapine or low-dose anticholinergics have also been explored anecdotally (Chidiac et al., 2024; Rathbone and Soares-Weiser, 2006). As no standardized treatment guidelines exist, clinical management must be individualized, carefully balancing motor, cognitive, and psychiatric considerations. Increased awareness of akathisia as a potential iatrogenic burden in HD is essential to avoid exacerbating patient distress and functional impairment.

3.1.4 *Bradykinesia*

Bradykinesia may coexist with rigidity and gait disturbance, complicating the clinical picture and requiring nuanced therapeutic strategies (Delval et al., 2006). Pharmacological management is limited, as dopamine-depleting agents such as tetrabenazine and antipsychotics may exacerbate bradykinesia by further suppressing striatal dopaminergic tone. Conversely, in select juvenile-onset cases exhibiting Parkinsonian features, levodopa and other dopaminergic therapies have been trialed with modest and variable benefit (Askenasy et al., 1990; Jongen et al., 1980). However, their use is constrained by concerns over neuropsychiatric side effects and inconsistent efficacy. Amantadine, with its combined dopaminergic and NMDA-antagonist properties, has also been considered an off-label option, though supporting evidence remains limited (Lucetti et al., 2002). As with other hypokinetic features in HD, bradykinesia is largely refractory to pharmacological intervention and often requires multidisciplinary support, including occupational therapy and adaptive strategies to preserve functional independence. Further research is needed to clarify its mechanistic underpinnings and identify targeted therapeutic approaches.

3.1.5 *Gait and balance*

Unlike chorea, gait and balance disturbances are often less responsive to direct pharmacological intervention and remain a major contributor to falls and functional decline. While dopamine-depleting agents such as tetrabenazine and deutetrabenazine may reduce excessive movements that interfere with gait, they can paradoxically worsen balance by unmasking rigidity or hypokinesia, particularly in mid-to-late stages (Claassen et al., 2017). Similarly, antipsychotics and benzodiazepines, occasionally used for behavioral or dystonic symptoms, may exacerbate postural instability due to sedation or extrapyramidal effects (Unti et al., 2017; Díaz-Gutiérrez et al., 2017). Currently, there is no pharmacological agent specifically approved to target gait or balance dysfunction in HD. In selected cases, low-dose amantadine has been explored for its potential to improve bradykinetic features and axial stability, although robust evidence is lacking (Magnet et al., 2004). Ultimately, pharmacological approaches must be cautiously individualized, aiming to optimize motor control without impairing alertness, strength, or coordination. Given the limited efficacy of drugs in this domain, pharmacotherapy is best viewed as an adjunct to physical therapy, gait training, and fall prevention strategies, which remain the cornerstone of managing balance-related disability in HD (Quinn et al., 2020). Further research is urgently needed to define targeted interventions for these disabling motor deficits.

3.2 Pharmacological management of psychiatric and behavioral symptoms

Selective serotonin reuptake inhibitors (SSRIs) are considered first-line for depression and anxiety, given their tolerability and potential benefits on irritability and obsessive-compulsive symptoms (Zadegan et al., 2024). Serotonin–norepinephrine reuptake inhibitors (SNRIs) and bupropion may be considered in cases with prominent apathy or low energy, though bupropion's dopaminergic activity warrants caution in individuals with chorea (Bogdan et al., 2020). Antipsychotics, particularly atypical agents such as olanzapine, risperidone, and quetiapine, are often used off-label to manage severe irritability, aggression, psychosis, or comorbid motor symptoms, although their risk of metabolic side effects and exacerbation of Parkinsonian features must be weighed carefully (Eaton and Claassen, 2024). Mood stabilizers like valproic acid or carbamazepine may be employed in cases of affective lability or bipolar-like presentations (Andriessen et al., 2022). Importantly, treatment should be individualized based on the dominant psychiatric phenotype, comorbid motor features, and side effect profiles, with regular reassessment over the disease course. Nonpharmacological interventions, including psychotherapy and caregiver support, should be integrated where feasible, though access remains limited in many settings (Anderson et al., 2018).

3.3 Pharmacological management of cognitive symptoms

There are currently no pharmacological agents approved specifically for cognitive symptoms in HD (Dickey and Spada, 2018). Cholinesterase inhibitors, including donepezil and rivastigmine, and the NMDA receptor antagonist memantine have been trialed with inconsistent results (Vattakatuchery and Kurien, 2013). Some studies have suggested that memantine may offer modest cognitive benefits or neuroprotective effects, but robust, controlled data are lacking, and its routine use is not currently recommended (Ogilvie and Schultz, 2023). Psychostimulants like modafinil have occasionally been used off-label to address apathy or attention deficits, but these agents may exacerbate irritability or insomnia (Blackwell et al., 2008). Given the overlap between cognitive dysfunction and functional decline, pharmacological efforts to enhance cognition have also sought to preserve independence in daily activities, though evidence remains sparse (Huynh et al.,

2023). As such, current pharmacological management remains symptomatic and largely supportive, emphasizing treatment of comorbid psychiatric symptoms that may exacerbate cognitive deficits. A multidisciplinary approach incorporating cognitive rehabilitation, environmental structuring, and caregiver support remains critical in addressing the broader impact of cognitive and functional impairment in HD (Clare et al., 2019).

3.4 Nonpharmacological and multidisciplinary interventions

Nonpharmacological and multidisciplinary interventions are fundamental to comprehensive care in HD, complementing pharmacological treatments to address its complex symptoms. As HD progresses, tailored rehabilitation programs involving physical, occupational, speech, and cognitive therapies help maintain function and quality of life (Bartlett et al., 2020). Physical therapy focuses on gait, balance, and coordination, while occupational therapy addresses daily living and environmental adjustments (Fritz et al., 2017). Speech therapy manages common issues such as dysarthria and swallowing difficulties (Burnip et al., 2020). Cognitive rehabilitation, though still developing, aims to support executive functions and compensate for deficits (Yhnell et al., 2020). Psychological support for patients and caregivers is also vital to alleviate emotional distress and reduce caregiver burden (Røthing et al., 2015). Multidisciplinary teams, including neurologists, psychiatrists, therapists, social workers, and dietitians, work collaboratively to provide personalized care. Evidence demonstrates that such integrated approaches improve functional outcomes, reduce hospitalizations, and enhance well-being, making nonpharmacological strategies an essential component of HD management (Piira et al., 2014; Mühlbäck et al., 2023).

In summary, while symptomatic treatments dominate the current therapeutic landscape for HD, several recent advances have expanded the choices available to clinicians. Table 12.1 summarizes the principal FDA-approved pharmacological therapies specifically indicated for HD, alongside commonly used agents for off-label symptomatic management.

4 Gene therapy in HD

Current therapeutic approaches for HD are primarily focused on symptomatic management, and an effective disease-modifying treatment remains unavailable. Gene therapy has recently emerged as a promising strategy, being actively explored not only for HD but also for other neurodegenerative disorders. In contrast to polygenic neurodegenerative conditions like Alzheimer's or Parkinson's, HD's etiology is monogenic and precisely delineated, with the CAG repeat length inversely correlating with the age of onset. This genetic clarity renders HD a prime target for gene therapy. The core therapeutic premise is to attenuate or amend *mHTT* gene expression, thereby potentially altering disease trajectory, unlike symptomatic palliatives, such as tetrabenazine for chorea or antipsychotics for behavioral issues, that dominate current management and offer no halt to neurodegeneration (Cheng et al., 2024; Piao et al., 2025).

4.1 Targets for gene therapy and mechanisms of action

Therapeutic paradigms in HD gene therapy primarily focus on diminishing mHTT levels or directly ameliorating the causal mutation. Diverse strategies are under investigation, each characterized by unique molecular mechanisms and translational hurdles, ranging from RNA-level interference to DNA editing. These approaches aim to address the gain-of-function toxicity of mHTT while ideally preserving the essential functions of wtHTT.

4.1.1 Antisense oligonucleotide-based approach

Antisense oligonucleotides (ASOs) represent one of the most advanced modalities. ASOs are synthetic, chemically stabilized DNA analogs, often modified with phosphorothioate backbones for nuclease resistance and enhanced cellular uptake, that anneal to complementary *HTT* mRNA sequences via Watson–Crick base pairing. Upon hybridization, they can trigger RNase H-mediated degradation of *HTT* transcripts, thereby abrogating protein translation (Alkanli et al., 2023a; Seo et al., 2023a; Shin et al., 2022a), or modulate splicing to destabilize transcripts or excise mutant exons, effectively reducing mHTT production at the posttranscriptional level (Caron et al., 2024). The RNase H mechanism involves the recruitment of the endogenous RNase H enzyme, which cleaves the RNA strand in the RNA–DNA hybrid duplex formed by the ASO and target mRNA, leading to mRNA degradation by cellular exonucleases and preventing translation into protein (Rook and Southwell, 2022; Khorkova and Wahlestedt, 2017). For splicing modulation, ASOs can bind to pre-mRNA at splice sites or enhancers, altering exon inclusion or exclusion to produce nonproductive transcripts that are degraded or encode truncated, nontoxic proteins (Scoles et al., 2019; Rinaldi and Wood, 2018).

Two principal ASO categories are in development: nonallele-selective ASOs, which suppress both *wtHTT* and *mHTT*, and allele-selective variants. Nonselective ASOs broadly hybridize to shared sequences in *HTT* mRNA, reducing overall

TABLE 12.1 Current pharmacological options for common symptoms in HD.

Symptom	Medication	Mechanism of action	Indication	Regulatory status
Chorea	Tetrabenazine	VMAT2 inhibitor (dopamine depletion)	Chorea in HD	FDA-approved (2008)
	Deutetrabenazine	VMAT2 inhibitor (deuterated analog of tetrabenazine)	Chorea in HD	FDA-approved (2017)
	Valbenazine	Selective VMAT2 inhibitor	Chorea in HD	FDA-approved (2024)
	Atypical antipsychotics (e.g., olanzapine, risperidone, quetiapine)	Dopamine D2 receptor antagonism	Off-label for chorea with behavioral symptoms	Off-label
Dystonia and myoclonus	Baclofen	GABA-B receptor agonist	Spasticity, dystonia	Off-label
	Trihexyphenidyl	Anticholinergic	Focal/segmental dystonia	Off-label
	Clonazepam	GABA-A receptor modulator	Myoclonus, dystonia	Off-label
	Levetiracetam	Modulates synaptic neurotransmission	Myoclonus	Off-label
	Valproic acid	GABAergic and antiepileptic	Myoclonus, mood symptoms	Off-label
	Botulinum toxin	Inhibits acetylcholine release	Focal dystonia (e.g., cervical)	Off-label
Gait and balance	Amantadine	NMDA receptor antagonist, dopaminergic	Bradykinesia, axial rigidity	Off-label
Rigidity and bradykinesia	Levodopa/carbidopa	Dopamine precursor + decarboxylase inhibitor	Juvenile-onset HD (Westphal variant)	Off-label
	Amantadine	NMDA antagonist, dopaminergic	Bradykinesia, rigidity	Off-label
Akathisia	Propranolol	Nonselective beta-blocker	First-line for akathisia	Off-label
	Clonazepam	GABA-A modulator	Adjunct for akathisia	Off-label
	Mirtazapine, anticholinergics	Noradrenergic/serotonergic modulation/antimuscarinic	Anecdotal use for refractory cases	Off-label
Psychiatric symptoms	SSRIs (e.g., sertraline, fluoxetine)	Serotonin reuptake inhibition	Depression, anxiety, irritability, OCD traits	Off-label
	SNRIs (e.g., venlafaxine)	Serotonin-norepinephrine reuptake inhibition	Depression, apathy	Off-label
	Bupropion	Norepinephrine–dopamine reuptake inhibition	Apathy, fatigue (caution: may worsen chorea)	Off-label
	Atypical antipsychotics	Dopamine D2 antagonism, serotonin modulation	Irritability, aggression, psychosis	Off-label
	Valproic acid, carbamazepine	Mood stabilization, antiepileptic	Affective lability, bipolar-like features	Off-label
	Methylphenidate, modafinil	Central nervous system stimulants	Apathy[S1], attention deficits	Off-label
Cognitive decline	Memantine	NMDA receptor antagonist	Cognitive symptoms (unproven efficacy)	Off-label
	Donepezil, rivastigmine	Cholinesterase inhibitors	Cognitive symptoms (limited benefit)	Off-label

HTT gene expression but risking the loss of wtHTT function (Farag et al., 2024; McColgan et al., 2023). Tominersen (Ionis/Roche), a nonselective ASO, pioneered large-scale clinical evaluation, exhibiting dose-dependent reductions in cerebrospinal fluid (CSF) mHTT levels in early trials. Nonetheless, the Phase III GENERATION-HD1 trial was terminated in 2021 due to a lack of clinical efficacy and safety issues at elevated doses, including elevated CSF protein levels indicative of inflammation (Farag et al., 2024; McColgan et al., 2023). Post hoc analyses suggested potential benefits in subgroups with lower disease burden, leading to ongoing exploratory studies with modified regimens.

Allele-selective ASOs, in contrast, are engineered to preferentially inhibit the mutant allele by targeting single-nucleotide polymorphisms (SNPs) in linkage disequilibrium with the expanded CAG repeat, allowing discrimination between alleles through enhanced binding affinity to the mutant transcript. WVE-003 (Wave Life Sciences; ClinicalTrials.gov: NCT05032196) exemplifies this, incorporating stereopure chemistry to enhance potency and selectivity. Interim clinical data from the SELECT-HD trial indicate selective mHTT reduction of up to 30% in CSF while preserving wtHTT levels, with no serious adverse events reported as of 2024. This selectivity is paramount, as wtHTT facilitates vesicular trafficking, embryogenesis, and neuronal viability; complete knockdown in animal models has shown developmental lethality and exacerbated neurodegeneration (Iwamoto et al., 2024; Gulumkar and Dowdy, 2024). Challenges for ASOs include the need for repeated intrathecal administrations and potential immune activation, but their reversibility offers a safety advantage over permanent edits (Gagliardi and Ashizawa, 2021).

4.1.2 RNA interference-based approach

RNA interference (RNAi) leverages intrinsic cellular silencing pathways to achieve similar HTT-lowering effects. Small interfering RNAs (siRNAs) or synthetic microRNAs (miRNAs) are vectored, typically via adeno-associated viruses (AAVs), to degrade *HTT* transcripts via the RNA-induced silencing complex (RISC), which cleaves target mRNA in a sequence-specific manner. AAV vectors, particularly serotype 5 or 9, are favored for their nonpathogenic nature, low immunogenicity, and ability to transduce nondividing neurons (Seyhan, 2011; Davidson and Boudreau, 2007). The mechanism begins with the introduction of double-stranded RNA (dsRNA) precursors, which are processed by the enzyme Dicer into short siRNAs or miRNAs (typically 21–23 nucleotides long). These short RNAs are then loaded into the RISC, where the Argonaute protein unwinds the duplex, retaining the guide strand that is complementary to the target *HTT* mRNA (Germain et al., 2023; Harper et al., 2005). The activated RISC binds to the mRNA via base pairing, leading to endonucleolytic cleavage of the target transcript (for perfect complementarity, as in siRNAs) or translational repression and mRNA destabilization (for partial complementarity, as in miRNAs). Allele-selective RNAi can be designed by incorporating SNPs into the guide sequence, ensuring preferential silencing of mutant transcripts (Germain et al., 2023; Harper et al., 2005). AMT-130 (uniQure) leads RNAi candidates, utilizing an AAV5 vector to express an artificial miRNA targeting exon 1 of *HTT* mRNA. Administered via bilateral intrastriatal infusion, a minimally invasive stereotactic surgery, it pursues sustained silencing post single-dose, with expression persisting for years in preclinical models. Interim Phase I/II data as of 2025 reveal a favorable safety profile, albeit with transient neurofilament light chain (NfL) spikes indicative of procedural neuronal perturbation. Notably, high-dose cohorts demonstrated slowed disease progression, with a 20–30% reduction in mHTT and stabilization on the composite UHDRS (cUHDRS) over 36 months (Estevez-Fraga et al., 2022b, Estevez-Fraga et al., 2023b). Relative to ASOs, RNAi affords protracted suppression via one-time delivery, though neurosurgical administration poses a barrier, including risks of hemorrhage or infection.

4.1.3 Gene-editing approach

Gene-editing approaches, such as clustered regularly interspaced short palindromic repeats (CRISPR)/CRISPR-associated protein 9 (Cas9) and its derivatives, have revolutionized HD gene therapy by enabling precise *HTT* genomic modifications at the DNA level, offering potential for a permanent cure. Strategies encompass excision of the expanded CAG repeat using dual guide RNAs to truncate the polyglutamine domain, allele-selective inactivation of *mHTT* while sparing the wild-type via SNP-targeted nucleases, and advanced techniques like base editing (e.g., adenine or cytosine base editors) or prime editing for nucleotide-precision corrections without inducing double-strand breaks, thereby minimizing genomic instability (Duan et al., 2023; Liao et al., 2024). For allele-selectivity, SNPs create mutant-specific PAM sites or gRNA binding sequences, enabling targeted DSBs only on the expanded allele (Shin et al., 2022b). Cytosine and adenine base editors—engineered as Cas9 nickases (nCas9) fused to deaminases—have been used to convert individual CAG codons into CAA within the repeat tract. Because both codons encode glutamine but only CAG is susceptible to pathological expansion, this targeted substitution has the potential to enhance repeat stability and slow the underlying disease processes (Choi et al., 2024; Liao et al., 2024b; Matuszek et al., 2025a). Base editing has also been applied to splice-site modulation; selective alteration of canonical donor or acceptor motifs can enforce exon 13 skipping, producing an HTT isoform that lacks the caspase-6 cleavage site and thereby generates fewer toxic N-terminal fragments. This intervention has been shown to reduce aggregation and attenuate brain atrophy in HD models (Seo et al., 2023b). Prime editing, which relies on a Cas9 nickase tethered to a reverse transcriptase and directed by a prime editing guide RNA (pegRNA), enables the installation or replacement of short sequences with high precision. This has made it possible to perturb specific regulatory motifs, such as NF-κB binding sites in the *HTT* promoter, which in turn leads to measurable reductions in *HTT* transcript levels in vivo (Alkanli et al., 2023b; Seo et al., 2023c; Shin et al., 2022c). Taken together, these emerging approaches offer a more nuanced therapeutic framework for HD by enabling stabilization of the repeat tract, controlled alteration of splicing patterns, and fine-scale tuning of *HTT* expression. Preclinical studies in HD models have shown up to a 50% reduction in mHTT aggregates, improved motor function, and extended survival. Although confined to preclinical stages as of 2025, these hold promise for enduring effects, with recent allele-selective CRISPR variants demonstrating high specificity in human induced pluripotent stem cell (iPSC)-derived neurons (Pinto et al., 2025). Obstacles include off-target effects (e.g., unintended edits at homologous sites), CNS-wide delivery (which requires widespread vector distribution), and the irreversible nature of genomic alterations, raising concerns about long-term oncogenic risks or compensatory mutations (Duan et al., 2023; Alkanli et al., 2023c).

4.1.4 *Adjunctive approaches*

Beyond direct *HTT* suppression, alternative and adjunctive targets are explored to complement primary strategies. These include DNA repair enzymes like MSH3 and FAN1, which modulate somatic CAG expansions (Tabrizi et al., 2022). Targeting MSH3 with ASOs or small molecules inhibits erroneous loop extrusion during repair, thereby stabilizing repeats by preventing MMR initiation at CAG slips (Farag et al., 2024; Salman et al., 2025). FAN1 acts as a nuclease that resolves DNA interstrand crosslinks and competes with MSH3 in MMR, sequestering MLH1 to suppress expansion-prone repair; enhancing FAN1 activity or expression could mitigate instability (Kim et al., 2024).

In addition, neurotrophic augmentation via viral delivery of BDNF, whose levels are depleted in HD, aims to bolster neuronal resilience, promoting survival and synaptic integrity through activation of TrkB receptors and downstream signaling cascades like MAPK/ERK and PI3K/Akt, which enhance neuronal plasticity, antiapoptotic pathways, and dendritic arborization (Pengo and Squitieri, 2024; Tabrizi et al., 2022; Kim et al., 2024; McLean et al., 2024). These adjunctive approaches may synergize with HTT-lowering therapies for enhanced efficacy.

4.2 Delivery and monitoring

Efficacious delivery is pivotal to gene therapy success in HD, given the protected environment of the CNS. Intrathecal infusion, commonly used for ASOs, enables CSF-mediated dissemination across the brain and spinal cord, leveraging natural fluid flow for broad distribution without surgery. In contrast, intrastriatal injection for AAV-based RNAi yields focal, concentrated exposure to the most affected regions but necessitates stereotactic neurosurgery under magnetic resonance imaging (MRI) guidance, with risks mitigated by real-time imaging. Evolving vectors, such as engineered AAV capsids (e.g., AAV.PHP.eB) with enhanced blood–brain barrier (BBB) traversal via intravenous administration, are in preclinical appraisal and could revolutionize noninvasive delivery, achieving up to 80% neuronal transduction in nonhuman primates (Sampaio and Disease, 2024; Ling et al., 2025).

Monitoring therapeutic progress relies on a multimodal framework. Biomarkers such as CSF mHTT levels provide direct evidence of target engagement, while NfL, a marker of axonal damage, gauges neuroaxonal integrity, with elevations signaling acute injury but declines indicating stabilization. Advanced neuroimaging, including MRI volumetrics to track striatal atrophy (typically 1–2% annual loss in HD) and positron emission tomography (PET) with ligands like [18 F]FDG for metabolic changes, quantifies structural and functional alterations (Cheng et al., 2024; Wang et al., 2025; Farag et al., 2025). Clinical metrics, such as the UHDRS for motor, cognitive, and behavioral domains, and total functional capacity (TFC) for daily independence, assess outcomes, albeit with gradual sensitivity requiring long-term follow-up. Composite scores like cUHDRS integrate these for powered evaluations, essential for detecting subtle modifications in a slowly progressive disease (Cheng et al., 2024; Estevez-Fraga et al., 2024a).

4.3 Current role of gene therapy in HD treatment

Gene therapy has advanced rapidly from substantiation to interrogation, marking the most significant strides toward HD modification since the gene's discovery in 1993. As of mid-2025, investigations focus on HTT-lowering with encouraging biomarker data. Endpoints integrate biomarkers (CSF mHTT for engagement, NfL for injury), clinical evaluations (UHDRS subscales, TFC for functionality), and neuroimaging (MRI/PET for atrophy and metabolism). These multifaceted metrics are indispensable, given HD's protracted evolution over decades (Farag et al., 2024). Fig. 12.4 demonstrates the three main concepts of gene therapy currently under investigation for HD.

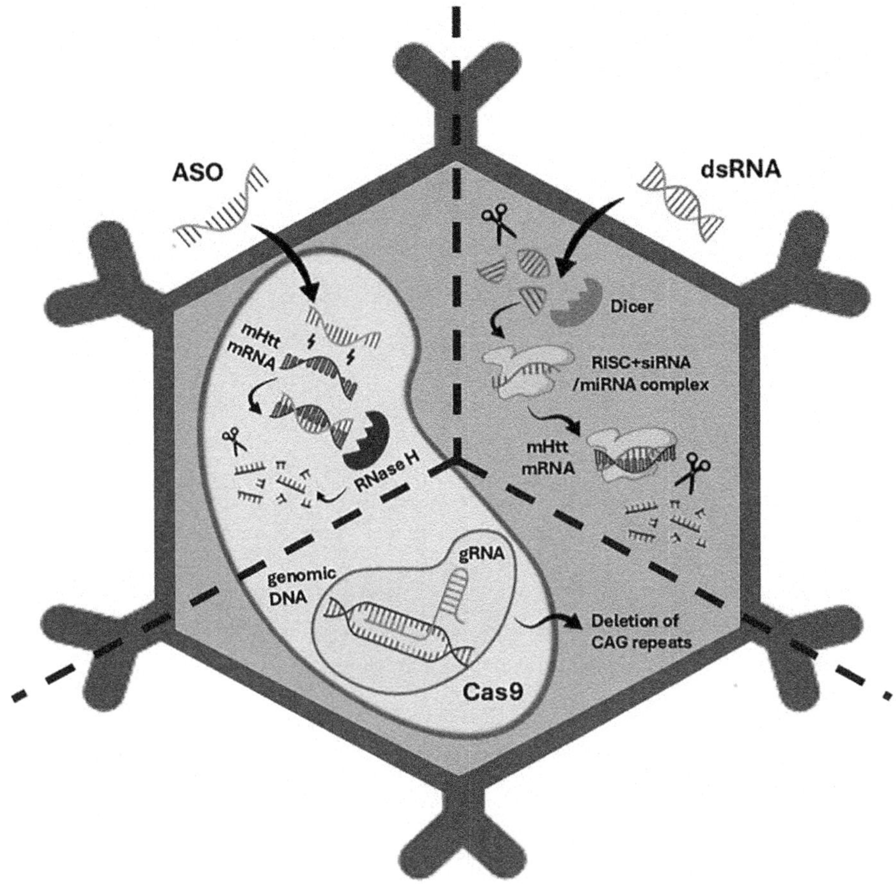

FIG. 12.4 Main gene therapy approaches currently under investigation for HD. There are three principal gene-targeting strategies currently being investigated in HD, including ASOs, RNAi platforms such as siRNA and miRNA, and gene-editing approaches such as CRISPR/Cas9. These modalities aim to reduce *mHTT* expression or correct pathogenic variants at the genomic or transcript level.

5 Recent and ongoing clinical trials of gene therapy in HD

Over the last decade, clinical trials of gene therapy in HD have matured into a diverse pipeline of experimental approaches, defined by their gene-silencing modalities and delivery strategies. Below, we will discuss trials by mechanistic class, moving from early ASOs to splicing modulators, RNAi approaches, and neuroprotective gene delivery across a chronological framework, as summarized in Table 12.2.

5.1 ASO-based clinical trials

The first human trials of ASO-based *HTT* suppression began in 2015 with tominersen (IONIS-HTTRx/RG6042), Roche/Ionis's nonallele-selective ASO targeting both *wtHTT* and *mHTT* transcripts. Its Phase I/II (NCT02519036) established intrathecal dosing feasibility and target engagement, demonstrating dose-dependent reduction of mHTT levels in CSF samples of HD patients (Tabrizi et al., 2019a). This led to GENERATION-HD1 and GENERATION-HD2 (Phase II/III, NCT03761849/NCT03842969), enrolling hundreds of manifest HD patients. Despite initial promise, the Phase III study was halted in early 2021 due to a lack of efficacy and a potential signal of worsening in some subgroups (McColgan et al., 2023; Reilmann, 2024). In fact, an ad hoc analysis of data obtained at week 69 revealed that participants who received tominersen every 8 weeks experienced a significantly worse cUHDRS score than that observed in the placebo group (−0.54 points; adjusted $P = 0.001$). However, changes in the TFC score were not significantly different (−0.40 points; adjusted $P = 0.09$). For those treated every 16 weeks, mean changes in both cUHDRS and TFC scores were comparable to placebo (−0.16 points; adjusted $P = 0.84$ and − 0.04 points; adjusted $P = 1.00$, respectively). Moreover, the incidence of adverse events was higher among participants in the every-8-week group compared with both placebo and the every-16-week group (McColgan et al., 2023; Reilmann, 2024).

TABLE 12.2 Recent and ongoing clinical trials of gene therapy in HD.

Class	Agent name	Mechanism of action	Trial identifier	Trial start–(estimated) end year	Route of administration, and dosing	Population (number)	Primary outcome(s)	Follow-up duration	Blinding/randomization	Phase	Status
Antisense oligonucleotides	RO7234292 (ISIS 443139/ tominersen)	Antisense oligonucleotide targeting *HTT* mRNA	NCT02519036	2015–2017	Intrathecal (10, 30, 60, 90, and 120 mg vs. placebo, monthly injections)	Manifest HD (46)	Number of participants with treatment-related AEs[1]	28 weeks	Double-blind, randomized	I/II	Completed
			NCT03342053	2018–2019	Intrathecal (dosing N/A, monthly vs. bimonthly injections)	Manifest HD (46)	Percentage of participants with AEs	18 months	Randomized, no blinding	II	Completed
			NCT04000594 (GEN-PEAK)	2019–2022	Intrathecal (3 dosing levels (N/A) vs. each other, monthly injections)	Manifest HD (12)	Drug concentration in CSF[2] and plasma, mHTT[3] concentration in CSF	7 months	No randomization, no blinding	I	Completed
			NCT03842969	2019–2022	Intrathecal (dosing N/A, bimonthly vs. quadrimonthly injections)	Manifest HD (236)	Percentage of participants with treatment-emergent AEs, number of participants with suicidal ideation/behavior, change from baseline in cognition	3 years	Randomized, no blinding	III	Completed
			NCT03761849 (Generation HD1)	2019–2022	Intrathecal (120 mg, bimonthly vs. quadrimonthly vs. placebo injections)	Manifest HD (899)	Change from baseline in cUHDRS[4] and TFC[5] scores	6 years	Double-blind, randomized	III	Completed
			NCT05686551 (Generation HD2)	2023–2027	Intrathecal (60 mg vs. 100 mg vs. placebo, quadrimonthly injections)	Early-manifest/ prodroma HD (301)	Incidence and severity of AEs, change from baseline in: CSF WBC[6], CSF protein, MRI[7] abnormalities, CSF mHTT, cUHDRS, and TFC scores	24 months	Double-blind, randomized	II	Active, not recruiting

	WVE-120101	Antisense oligonucleotide targeting *HTT* mRNA	NCT03225833 (PRECISION-HD1)	2017–2021	Intrathecal (5 dosing levels (N/A) vs. each other, monthly injections)	HD (61)	Number of patients with (severe/serious) treatment-emergent AEs, number of patients who withdraw *due to treatment*-emergent AEs	7 months	Double-blind, randomized	I/II	Terminated
			NCT04617847	2020–2021	Intrathecal (single-arm, dosing N/A, monthly injections)	HD (27)	Number of patients with (severe/serious) treatment-emergent AEs, number of patients who withdraw due to treatment-emergent AEs	12 months	No randomization, no blinding	I/II	Terminated
	WVE-120102	Antisense oligonucleotide targeting *HTT* mRNA	NCT03225846 (PRECISION-HD2)	2017–2021	Intrathecal (2, 4, 8, 16, and 32 mg vs. placebo, monthly injections)	HD (88)	Number of patients with (severe/serious) treatment-emergent AEs, number of patients who withdraw due to treatment-emergent AEs	7 months	Double-blind, randomized	I/II	Terminated
			NCT04617860	2019–2021	Intrathecal (single-arm, dosing N/A, monthly injections)	Early-manifest HD (36)	Number of patients with (severe/serious) treatment-emergent AEs, number of patients who withdraw due to treatment-emergent AEs	12 months	No randomization, no blinding	I/II	Terminated
	WVE-003	Antisense oligonucleotide targeting *HTT* mRNA	NCT05032196 (SELECT-HD)	2021–2024	Intrathecal (30, 60, and 90 mg vs. placebo, bimonthly injections)	Early-manifest HD (47)	Number of patients with treatment-emergent AEs	28 weeks	Double-blind, randomized	I/II	Completed
	VO659	Antisense oligonucleotide targeting *HTT* mRNA	NCT05822908	2023–2025	Intrathecal (10, 20, 40, and 60 mg vs. each other, monthly injections)	SCA1, SCA3, and early-manifest HD (65)	AEs, vital signs, body weight, ECG[8], laboratory parameters, MRI abnormalities, percentage of participants with suicidal ideation/behavior	58 weeks	No randomization, no blinding	I/II	Recruiting

Continued

Table 12.2 Recent and ongoing clinical trials of gene therapy in HD—cont'd

Class	Agent name	Mechanism of action	Trial identifier	Trial start–(estimated) end year	Route of administration, and dosing	Population (number)	Primary outcome(s)	Follow-up duration	Blinding/randomization	Phase	Status
Splicing modulators	LMI070 (Branaplam)	*HTT* mRNA splicing modifier	NCT05111249 (VIBRANT-HD)	2021–2023	Oral (56 mg or 112 mg vs. placebo, weekly)	Early-manifest HD (26)	Baseline change in CSF mHTT (up to week 17), number of participants with (serious) AEs	69 weeks	Double-blind, randomized	II	Terminated
	PTC518 (Votoplam)	*HTT* mRNA splicing modulator	NCT05358717 (PIVOT-HD)	2022–2025	Oral (5, 10, and 20 mg vs. placebo, daily)	HD (252)	Number of patients with AEs, baseline change in blood mHTT protein	18 months	Double-blind, randomized	II	Active, not recruiting
			NCT06254482	2023–2027	Oral (5, 10, and 20 mg vs. each other daily)	HD (250)	Number of patients with treatment-emergent AEs, baseline change in blood mHTT protein	54 months	Double-blind, randomized	II	Enrolling by invitation
	SKY-0515	*HTT* mRNA splicing modulator	NCT06873334 (Falcon-HD)	2025–2027	Oral (3, 6, and 9 mg vs. placebo, daily)	HD (120)	Changes in blood mHTT, changes in brain volume measured through MRI, changes in UHDRS	20 months	Double-blind, randomized	II/III	Recruiting
Gene silencing miRNA/siRNAs	AMT-130	*HTT* mRNA-silencing miRNA delivered in AAV5 vector	NCT04120493	2019–2029	Intrastriatal (low (6×10^{12} gc/subject) and high (6×10^{13} gc/subject) doses vs. sham, single injection)	Early-manifest HD (43)	Number and type of AEs	72 months	Double-blind, randomized	I/II	Active, not recruiting
			NCT05243017	2021–2029	Intrastriatal (low (6×10^{12} gc/subject) vs. high (6×10^{13} gc/subject) dose, single injection)	Early-manifest HD (14)	AEs, vital signs, ECG, neurological/physical examinations, serum/CSF biomarkers, MRI, cognitive scores	60 months	Randomized, no blinding	I/II	Active, not recruiting

	VY-HTT01	AAV vector carrying *HTT*-targeting miRNA	NCT04885114 (VYTAL)	2021–2024	Intraparenchymal (unilateral low (3×10^9 vg[11]/ml) dose vs. bilateral low and mid (1.7×10^{10} vg/ml) dose vs. bilateral high (1.7×10^{10} vg/ml) dose, single injection)	Early-stage HD (0)	Incidence and type of AEs	24 months	Randomized, no blinding	I	Withdrawn
	ER2001	*HTT*-targeting siRNA	NCT06024265	2023–2024	Intravenous (ascending doses of 0.04, 0.08, 0.16, and 0.32 mg/kg, 8 injections)	Early-manifest HD (15)	Incidence and severity of (serious) AEs	6.5 months	No randomization, no blinding	I	Recruiting
	ALN-HTT02	*HTT* exon 1-targeting siRNA	NCT06585449	2024–2028	Intrathecal (drug vs. placebo [dosing N/A], single injection)	HD (54)	Frequency of AEs	12 months	Double-blind, randomized	I	Recruiting
	SPK-10001	AAV vector carrying *HTT*-targeting miRNA	NCT06826612	2025–2035	Intraparenchymal (bilateral injection of drug vs. placebo surgery, dosing and intervals N/A)	HD (53)	Number and severity of treatment-emergent AEs, baseline changes in UHDRS and TFC[10] scores	5 years	Double-blind, randomized	I/II	Recruiting
Neuroprotective gene therapy	AB-1001 (BV-101)	AAV vector encoding CYP46A1 gene	NCT05541627	2022–2028	Intrastriatal (bilateral low (4×10^8 vg/subject) and high (1.1×10^9 vg/subject) doses, single injection)	Early-manifest HD (5)	Incidence of dose-limiting toxicities (DLTs), treatment-emergent, and serious AEs	5 years	No randomization, no blinding	I/II	Active, not recruiting

[1]Adverse events; [2]Cerebrospinal fluid; [3]Mutant huntingtin; [4]Composite Unified Huntington's Disease Rating Scale Score; [5]Total Functional Capacity Score; [6]White blood cells; [7]Magnetic Resonance Imaging; [8]Electro cardiogram; [9]Genome copies; [10]Total functional capacity; [11]Vector genomes.

Post hoc subgroup analyses indicated that younger participants, who typically have lower baseline disease burden, might derive greater benefit from treatment (McColgan et al., 2023; Reilmann, 2024). This hypothesis is now being prospectively evaluated in the ongoing Phase II study, GENERATION HD2 (NCT05686551). This study, enrolling early manifest or prodromal patients, is now underway but not yet recruiting, testing either the 60 or 100 mg dosing of tominersen, which is lower in its highest dose compared to the GENERATION HD1 trial (120 mg). Supplementing this, public sources note that tominersen remains important as the benchmark that clarified the necessity of dosing regimen, allele selectivity, and patient stratification (Tabrizi et al., 2022).

Wave Life Sciences pursued allele-selective ASOs (WVE-120101, WVE-120102). After Phase I/II studies (NCT03225846, NCT03225833) showed insufficient pharmacodynamic response or biomarker lowering, the programs were discontinued (Estevez-Fraga et al., 2023a). WVE-003, the most recent allele-selective ASO developed by Wave Life Sciences for HD, is currently in a Phase I/II trial (NCT05032196/SELECT-HD). In the initial single-dose phase of this trial, 47 participants received one intrathecal administration of WVE-003 (30, 60, or 90 mg) or placebo and were monitored longitudinally. The treatment was well tolerated across all dose groups, with no serious adverse events reported. Interim analyses revealed a 22% mean reduction in CSF mHTT levels after 85 days, corresponding to a 35% greater decrease than in placebo, while wtHTT concentrations remained unchanged, confirming allele selectivity. In the subsequent multiple-dose phase, 23 participants received three intrathecal doses of 30 mg WVE-003 or placebo at 8-week intervals. No serious safety concerns emerged. By week 24 (8 weeks after the final dose), mHTT levels were reduced by 46% compared with placebo, and this effect persisted through week 28 (44% reduction). Imaging analyses indicated that greater mHTT lowering was associated with reduced caudate nucleus atrophy, and exploratory clinical assessments suggested less progression of motor symptoms among treated participants (Farag et al., 2024). Collectively, these interim findings demonstrate that WVE-003 achieves selective mHTT reduction with a favorable safety profile, supporting continued evaluation of its therapeutic potential in HD (Farag et al., 2024).

Vico Therapeutics' VO659 is an allele-preferential ASO designed to target expanded CAG repeats within the *HTT* transcript, thereby reducing *mHTT* expression while largely sparing the wild-type allele (Estevez-Fraga et al., 2024a). The agent is currently being evaluated in a multicenter, first-in-human Phase I/IIa study (NCT05822908) enrolling individuals with early-manifest HD and those with spinocerebellar ataxia types 1 and 3 (Estevez-Fraga et al., 2024b; Stahl et al., 2024). Interim results presented at the 2024 European Huntington's Disease Network (EHDN) and Enroll-HD meeting demonstrated a 28% mean reduction in CSF mHTT at Day 85 without evidence of sustained increases in NfL and with a tolerable safety profile. Recruitment remains ongoing as of mid-2025 (Estevez-Fraga et al., 2024b). By targeting the toxic repeat expansion directly, VO659 broadens the ASO therapeutic landscape beyond the nonallele-selective tominersen program and the SNP-restricted allele-selective strategies pursued by Wave Life Sciences, offering a potentially more widely applicable allele-preferential approach for HTT lowering (Tabrizi et al., 2022).

5.2 Splicing modulation-based clinical trials

Splicing modulation offers a small-molecule alternative to RNA-based therapies by altering *HTT* pre-mRNA splicing and triggering mRNA decay pathways, thus reducing *HTT* expression (Tabrizi et al., 2022). Branaplam, a splicing modulator initially developed for spinal muscular atrophy, was rapidly repurposed for *HTT* mRNA suppression in the VIBRANT-HD trial (NCT05111249) (Estevez-Fraga et al., 2022b; Tabrizi et al., 2022). While it showed CSF mHTT lowering, the trial was terminated early due to safety concerns, including peripheral neuropathy (Estevez-Fraga et al., 2022a).

PTC Therapeutics' PTC518, an oral splicing modulator targeting *HTT* pre-mRNA, commenced the PIVOT-HD Phase II trial (NCT05358717) in 2022, enrolling approximately 252 early HD patients (Farag et al., 2024). Interim results at 12 months, reported in June 2024, revealed dose-dependent lowering of mHTT in blood and CSF, favorable trends in motor and cUHDRS outcomes, and no safety signals or increases in CSF NfL (Estevez-Fraga et al., 2024b). A proposed Phase III efficacy study is now under development, signaling high regulatory and clinical priority for this oral small-molecule approach.

Skyhawk Therapeutics' SKY-0515 represents another orally bioavailable splicing modulator designed to selectively reduce *HTT* expression. Early-phase studies in healthy volunteers and individuals with HD demonstrated dose-dependent reductions of up to ~ 72% in blood *HTT* mRNA at the 9 mg dose after 12 weeks of daily consumption, while preserving wtHTT levels. The compound was well-tolerated across tested doses, with no serious adverse events reported to date. Based on these encouraging biomarker outcomes, SKY-0515 has advanced into the Phase II/III FALCON-HD trial (NCT06873334), which will further evaluate its long-term safety and clinical efficacy in early manifest HD (Farag et al., 2024).

5.3 RNAi-based clinical trials

While ASOs and splicing modulators are repeat-dosed intrathecal or oral agents, viral vector delivery offers the potential for durable, sometimes one-time, suppression. The most advanced program in this class is AMT-130, an AAV5-based miRNA targeting exon 1 of *HTT* delivered via direct, single-dose intrastriatal infusion (Stimming et al., 2022). Phase I/II trials in the United States (NCT04120493, 2019) and Europe/UK (NCT05243017, 2021) began in 2019 and resumed after a recruitment pause in 2022. As of mid-2025, combined enrollment totals approximately 57 early-manifest HD patients across low-dose (6 trillion vector genomes) and high-dose (60 trillion vector genomes) cohorts; the US trial is randomized versus sham surgery, and the European study is primarily open-label. Long-term follow-up is planned through 2029. AMT-130 has been granted Breakthrough Therapy and Regenerative Medicine Advanced Therapy designations by the FDA ("A phase I/II, randomized, double-blind, sham control and open-label study to explore safety, tolerability, and efficacy signals of multiple doses of striatally-administered rAAV5-miHTT total huntingtin gene (HTT) lowering therapy (AMT-130," 2019; "A phase Ib/II randomized, double-blind study to explore safety, tolerability, and efficacy signals of multiple doses of striatally-administered rAAV5-miHTT total huntingtin gene (HTT," 2021). Interim 36-month data from uniQure's ongoing Phase I/II study of AMT-130 indicate that its intrastriatal administration may slow clinical disease progression. Among 24 evaluable patients, the high-dose cohort demonstrated a 75% reduction in disease progression as measured by cUHDRS, showing a mean decline of –0.38 compared with –1.52 in a propensity score-matched external control group ($p = 0.003$). Secondary endpoints supported this trend, including TFC with 60% slowing (–0.36 vs –0.88; $p = 0.033$), symbol digit modalities test (SDMT) with 88% slowing (–0.44 vs –3.73; $p = 0.057$), and stroop word reading test (SWRT) showing 113% slowing (+0.88 vs –6.98; $p = 0.0021$). Improvements were also observed in total motor score (59% slowing; 2.01 vs 4.88; $p = 0.1741$) (Stimming et al., 2022). Biologically, CSF NfL levels decreased by 8.2% from baseline at 36 months, contrasting with the expected increase in untreated natural history cohorts, suggesting reduced neuronal injury. A dose-response relationship was noted, with the high-dose group showing greater consistency across endpoints than the low-dose cohort. Safety was acceptable overall, with no new drug-related serious adverse events reported since 2022; most adverse events were procedure-related and transient (Stimming et al., 2022).

While these findings represent a major milestone for gene therapy in HD and have been described as "the most convincing to date," several caveats merit caution. First, the analysis is based on a small cohort: only 12 participants completed the high-dose regimen, and no concurrent placebo control was included, relying instead on a historical comparator, which may inflate apparent treatment effects. Second, the magnitude of change on cUHDRS (–0.38 vs. –1.52 over 3 years) corresponds to an annual difference of roughly 0.13 points, which is below clinically perceptible thresholds for most patients. Thus confirmation in a randomized, placebo-controlled Phase III study remains essential before claims of disease modification can be substantiated.

Additional gene therapy candidates using viral vector-mediated RNAi approaches include VY-HTT01, SPK-10,001, ER2001, and ALN-HTT02. VY-HTT01 was developed by Voyager Therapeutics as an AAV-based gene-therapy candidate, which delivered an HTT-lowering miRNA construct to the brain. While demonstrating robust preclinical *Htt* knockdown, clinical testing in the Phase I VYTAL trial never enrolled patients, and the program has since been withdrawn due to the company's focus on other noninvasive, systemic gene therapies (Ferguson et al., 2022). Similarly, SPK-10,001 (Spark Therapeutics/Roche) employs an AAV vector to deliver a miRNA targeting *HTT* mRNA nonallele-specifically through a one-time bilateral intraparenchymal infusion into the striatum. Early-phase clinical testing commenced in 2025, with promising preclinical data demonstrating durable Htt lowering and favorable safety profiles in nonhuman primates, though human safety and efficacy data are still pending (Matuszek et al., 2025a).

ER2001 (Engensis, developed by uniQure) is an AAV5-mediated gene therapy encoding both an siRNA targeting *HTT* mRNA and a neuron-targeting rabies virus glycoprotein tag. This platform, known as IVSA (in-vivo self-assembled exosome), allows peripheral cells to assemble the plasmid into exosomes, which can carry the siRNA to neurons by crossing the BBB. A Phase I/II clinical trial (NCT06024265) investigating safety, tolerability, and preliminary efficacy in early-manifest HD patients began in 2023 and is ongoing. A preliminary report presented at the 2024 EHDN/Enroll-HD meeting revealed ER2001 to be safe and tolerable, with signs of a decrease in CSF NfL (Pei et al., 2024). Furthermore, ALN-HTT02 (Alnylam Pharmaceuticals), an siRNA designed to selectively silence *HTT* mRNA through intrathecal delivery, targets the conserved exon 1 of *HTT*, thereby covering multiple pathogenic HTT isoforms (Sloan and on behalf of the ALN-HTT02 program team, 2024). While detailed clinical trial identifiers are not publicly disclosed, studies on NHPs demonstrated evidence of dose-dependent HTT lowering in CSF, with acceptable tolerability reported (Sloan and on behalf of the ALN-HTT02 program team, 2024). These approaches illustrate the ongoing diversification of RNAi-based therapies, balancing the challenges of delivery methods, safety, and sustained mHTT reduction in HD.

5.4 Neuroprotective gene therapy-based clinical trials

Beyond mHTT-lowering strategies, emerging neuroprotective gene-therapy approaches aim to restore neuronal function and counteract metabolic disturbances contributing to disease progression. Among these, interventions targeting cholesterol homeostasis and neurotrophic support, such as AB-1001, represent a promising direction in HD therapy. AB-1001 is an AAV-based CYP46A1 gene augmentation therapy, given as a one-time bilateral striatal infusion to restore neuronal cholesterol metabolism in HD. A first-in-human, Phase I/II open-label, dose escalation study is underway to evaluate its safety, tolerability, and preliminary efficacy in adults with early-manifest HD (NCT05541627). Participants receive a single intracerebral administration of AB-1001 using an AAVrh10 vector at one of two dose levels (4.0×10^8 or 1.1×10^9 vg/μL), followed by up to 5 years of monitoring. The primary objectives focus on dose-limiting toxicities and treatment-emergent adverse events, while secondary measures assess MRI-based neuroanatomical changes, cUHDRS performance, and magnetic resonance spectroscopy metabolic profiles (Schultz et al., 2023). If successful, this strategy could establish a novel therapeutic paradigm focused on neuronal restoration rather than gene silencing, complementing existing HTT-lowering approaches in HD.

6 Current advances in preclinical gene therapy for HD

Selecting an appropriate animal model is crucial for understanding HD pathogenesis, assessing potential therapeutics, and facilitating drug development. Animal models of HD can be broadly categorized as toxin-induced (nongenetic) or genetic, encompassing species ranging from invertebrates to vertebrates. Each model provides unique advantages for investigating neurobehavioral mechanisms and testing novel interventions. Given the limited treatment options for HD, it is essential to employ preclinical models that accurately recapitulate the disease's diverse features. The choice between nongenetic and genetic models should be guided by the specific research objectives Yu et al. (2024). This section offers an overview of commonly used HD animal models and outlines their characteristics, followed by a discussion of recent advances in gene therapy evaluated in these models.

6.1 Animal models for HD

HD animal models range from simple invertebrate systems to sophisticated mammalian models, each offering unique strengths and limitations. These models can be broadly categorized into invertebrate, nongenetic, genetic (fragment, full-length, and knock-in), and large-animal systems (Upadhayay et al., 2023; Han et al., 2024).

6.1.1 Invertebrate models

Caenorhabditis elegans and *Drosophila melanogaster* are widely used invertebrate models due to their short life cycles, cost-effectiveness, and genetic tractability. In *C. elegans*, the expression of polyglutamine repeats in neurons or muscle enables the study of protein aggregation and toxicity, supporting large-scale genetic and drug screens. *Drosophila* transgenic models carrying human *HTT* fragments with expanded repeats (e.g., Q93, Q120) allow the investigation of mHTT toxicity, small-molecule screening, and genetic modifiers. While highly efficient for high-throughput studies, their simple nervous systems limit their ability to fully replicate human HD pathology.

6.1.2 Nongenetic models

Neurotoxic models in rodents and primates replicate specific HD-related mechanisms, especially mitochondrial dysfunction and excitotoxicity. Mitochondrial toxins, such as 3-NP and malonic acid, inhibit succinate dehydrogenase, impair oxidative phosphorylation, and deplete ATP. 3-NP, which crosses the BBB, is commonly used in various dosing regimens to model metabolic stress. Excitotoxins like quinolinic acid (QA) and kainic acid act as NMDA receptor agonists, leading to glutamate-mediated neuronal death. QA is particularly valuable because direct intrastriatal injection selectively destroys MSNs and induces cognitive deficits, resembling human HD. Although less commonly used today, these models remain important for dissecting isolated pathological processes, such as excitotoxicity and mitochondrial impairment.

6.1.3 Genetic models

N-terminal fragment transgenics

Mouse models such as R6/1 and R6/2 are among the most widely used. Both express exon 1 of the human *HTT* gene but differ in CAG repeat length and disease severity. R6/2 mice (125–150 repeats) show early-onset, rapidly progressing symptoms, including motor and cognitive deficits, but have very short lifespans (~13 weeks). R6/1 mice (115 repeats) display milder, slower disease progression, making them useful for early-stage studies. Other fragment models include N171-82Q and N586-82Q, which capture progressive neurodegeneration, motor deficits, and Htt aggregation. Transgenic rat models carrying human *HTT* mutations (e.g., 51 repeats) resemble late-onset HD, allowing long-term studies of disease progression and therapeutic interventions.

Full-length transgenics

These models express the entire human *mHTT* gene using yeast or bacterial artificial chromosomes (YAC or BAC, respectively). They preserve the genomic context, leading to gradual phenotypes that better reflect human disease. YAC128 mice (128 repeats) show progressive striatal/cortical atrophy, hyperactivity, and cognitive decline over 12–18 months. BACHD mice (97 repeats) and BAC226Q mice (226 repeats) display progressive motor, cognitive, and psychiatric-like deficits, with slower but more clinically relevant pathology. Full-length BACHD rats similarly exhibit motor, cognitive, and psychiatric abnormalities, but without the weight gain seen in mice.

Knock-in models

Knock-in (KI) mice carry expanded human CAG repeats directly inserted into the endogenous *Htt* locus, providing the closest genetic match to human HD. In the replacement strategy, endogenous repeats are replaced with expanded sequences (e.g., HdhQ72, HdhQ150, HdhQ200, HdhQ250). These lines show progressive pathology, including cognitive decline, motor deficits, striatal/cortical atrophy, and Htt aggregation, while maintaining a normal lifespan. In the chimeric strategy, insertion of humanized exon 1 with expanded repeats occurs (e.g., CAG140, zQ175). These mice display early behavioral and neuropathological changes, including hyperactivity, anxiety-like behavior, and MSN dysfunction. Notably, zQ175 mice exhibit early-onset motor impairments, cortical/striatal atrophy, and shorter lifespan, while CAG140 mice show early mHtt aggregation, psychiatric-like features, and progressive neuropathology without premature death.

6.1.4 *Large-animal models*

Pigs represent a valuable large-animal system due to their neuroanatomical similarity to humans and suitability for advanced genetic manipulation. Transgenic minipigs expressing mHtt fragments or full-length gene variants show age-dependent motor, cognitive, and behavioral deficits. Transgenic HD sheep exhibit mild motor and cognitive deficits. Their long lifespan and large brains make them ideal for studying early disease stages, though phenotype development is slow and maintenance is costly. Nonhuman primates like Rhesus macaque models exhibit striatal pathology and cognitive decline, closely resembling human HD, making them valuable for translational research despite ethical concerns and high costs. These large-animal models are particularly important for translational studies, as they allow long-term assessment of disease mechanisms and therapeutic interventions not feasible in rodents. Although large-animal models closely mimic human physiology and are essential for testing gene therapies, they are more expensive and less amenable to genetic manipulation (Morton and Howland, 2013).

6.2 Therapeutic strategies in preclinical studies

Preclinical gene therapy for HD focuses on four main strategies: ASOs, RNAi, gene editing, and adjunctive neuroprotective interventions. Each approach targets different aspects of HD pathology, from reducing toxic protein levels to mitigating neuronal loss. Preclinical models have shown that suppressing *mHtt* expression can reverse neuronal inclusions and improve symptoms like motor function (Yamamoto et al., 2000). In BACHD mice, reducing cortical mHtt partially improves motor and psychiatric-like deficits without preventing neurodegeneration, whereas mHtt reduction in both cortical and striatal neurons ameliorates all behavioral deficits and brain atrophy (Wang et al., 2014).

To achieve these therapeutic effects, efficient delivery of gene-modifying molecules is essential, and viral vectors have become the primary tools for this purpose. AAV and lentivirus (LV) vectors are commonly used to deliver components in preclinical studies. Both vectors are replication-deficient and considered nonpathogenic; however, they differ in persistence and integration. AAV vectors remain largely episomal and do not integrate into the host genome; therefore expression of therapeutic RNA gradually diminishes over time due to cell division, although this is minimal in neurons. In contrast, LV vectors integrate into the host genome, providing more stable expression but with a potential risk of insertional mutagenesis, which is mitigated by using self-inactivating LV constructs. Overall, viral vector-based delivery enables prolonged and targeted expression of therapeutic RNA molecules, reducing the need for repeated invasive administration into the CNS (Wang et al., 2024; Srivastava et al., 2021).

6.2.1 ASO-based preclinical studies

ASOs, synthetic DNA-like molecules, have also shown promise as a research area that binds to *mHTT* mRNA and reduces mHTT protein synthesis. ASOs are typically delivered directly into the CNS, allowing lower doses and reduced toxicity risk (Southwell et al., 2012). While rodent studies show even CNS distribution, larger-brain models reveal lower ASO concentrations in deeper brain regions (Kordasiewicz et al., 2012). Kordasiewicz et al. revealed that in BACHD and YAC128 mouse models, intracerebroventricular (ICV) infusion of ASOs targeting human huntingtin (HuASO) achieved broad brain distribution and dose-dependent suppression of human *mHTT* RNA and protein, lasting 2–3 months and reversing disease phenotypes (Kordasiewicz et al., 2012). The ASO selectively targeted *mHtt* in mice, but in humans, it would affect both *wtHTT* and *mHTT*. To assess the safety and efficacy of attenuating both forms, a second ASO targeting both mouse *wtHtt* and human *mHTT* was tested, effectively reducing both. This team also found that intrathecal infusion of an ASO targeting both Rhesus monkey and human huntingtin mRNA (MkHuASO) (4 mg/day for 21 days) led to widespread cortical ASO distribution and reduced Htt protein levels to 47% and 63% of normal in the prefrontal and occipital cortices, respectively (Kordasiewicz et al., 2012). ICV bolus injection of allele-specific ASOs was tested in Hu97/18 and Hu18/18 mouse models of HD, where both early and late interventions led to improvements in cognitive and behavioral performance, mitigated forebrain atrophy, and preserved levels of striatal DARPP-32, a dopamine- and cyclic AMP-regulated phosphoprotein. To evaluate this approach in a larger primate brain, the researchers delivered ASOs through lumbar puncture in cynomolgus monkeys, resulting in widespread suppression of *Htt* expression across cortical and limbic regions (the hippocampus, amygdala, and cingulate and entorhinal cortices), suggesting the use of ASOs as a therapeutic potential for HD (Southwell et al., 2018). As wtHtt is essential for neuronal health and synaptic transmission, reducing its expression may be detrimental. Allele-specific ASOs targeting only the *mHtt* could mitigate this risk, using strategies such as binding to the expanded CAG repeat or specific SNPs (Southwell et al., 2013). In a recent study, Aly et al., to enhance CNS delivery of ASOs, evaluated apolipoprotein A-I nanodisks (apoA-I NDs) as a novel intranasal delivery platform for mHtt-lowering ASOs (Aly et al., 2023). They found that apoA-I NDs can bypass the BBB following intranasal administration in the BACHD mouse model of HD. Notably, a single intranasal dose of apoA-I ASO-NDs reduced mHtt levels in the brain regions most affected in HD, including the cortex and striatum. These results suggest that improved ASO delivery via apoA-I NDs may have broad applications for treating CNS disorders (Aly et al., 2023). These findings are especially important, as they pave the way for future clinical applications aimed at modifying disease progression through targeted gene silencing.

6.2.2 RNAi-based preclinical studies

RNAi-based strategies utilize miRNAs, short hairpin RNAs (shRNAs), or siRNAs to selectively bind *mHtt* transcripts. These molecules recruit the RISC, which mediates posttranscriptional gene silencing through targeted degradation of the *mHtt* mRNA, thereby reducing the synthesis of the toxic Htt protein. Spronck et al. found that AAV5-mi*HTT* (miRNA targeting human *HTT*) delivery to the striatum of Q175 KI mice produced a dose-dependent, sustained reduction in Htt protein and suppressed mHtt aggregates in both the striatum and cortex for at least 12 months (Spronck et al., 2019). In R6/2 transgenic mice, this treatment improved motor coordination, extended median survival by 4 weeks, and reduced neuronal dysfunction and HD-like symptoms. These findings demonstrate long-term efficacy and support further development of Htt-lowering gene therapies for HD (Spronck et al., 2019). AAV-miR132 infusion into the striatum of R6/2 mice effectively improved motor function, significantly prolonged lifespan, slowed disease progression, and enhanced neuronal function despite having minimal impact on *mHtt* mRNA levels and inclusion body formation. These findings suggest that HD symptoms can potentially be alleviated without directly targeting the disease-causing *Htt* gene or its products (Fukuoka et al., 2018). Intrastriatal infusion of an AAV9-GFP-mi*RHTT* vector in Q140/Q140 KI mice resulted in an approximately 50% reduction of *Htt* mRNA expression in the striatum and enhanced striatal neuron survival (Keeler et al., 2016). Using an acute HD rat model, Miniarikova et al. demonstrated that therapeutic miRNAs delivered via AAV5 can achieve both total and allele-selective *Htt* silencing in the brain (Miniarikova et al., 2017). Intracerebral administration of mi*HTT* (AAV5-miHTT-155/AAV5-miHTT-451) or allele-selective candidates (miSNP50T/miSNP67T) suppressed mHtt aggregates and prevented DARPP-32-related neuronal dysfunction in rat striata, with AAV5-mi*HTT*-451 showing the strongest in vivo efficacy (Miniarikova et al., 2017). For assessment of the gene-silencing approach in larger brains, HD transgenic sheep expressing human *mHTT* were treated by intrastriatal injection of AAV9-miRNA. The therapy reduced *mHTT* mRNA and protein levels by 50–80% in the caudate and putamen, which sustained up to 6 months posttreatment. Importantly, the intervention spared normal HTT, preserved neuronal markers (DARPP-32 and NeuN), and did not trigger microglial activation (Pfister et al., 2018). In another study, injection of the highest dose of AAV5-mi*HTT* into the striatum of transgenic HD minipigs produced a dose-dependent reduction of human *mHTT* mRNA and protein levels in the putamen, caudate nucleus, thalamus, and cortex regions in a manner (Evers et al., 2018). It should be noted that AAVs are widely used to deliver genes and siRNAs to neurons, and their minimal immunogenicity makes them well-suited for brain-targeted gene therapy (Bravo-Hernandez et al., 2020; Wang et al., 2020).

Using siRNA or ASOs to reduce *mHtt* mRNA requires repeated dosing, can cause off-target effects, and may be toxic to the kidneys and liver. In contrast, a single AAV-delivered shRNA dose can provide long-lasting *Htt* knockdown with lower toxicity and fewer off-target effects, though controlling shRNA expression remains challenging (Aguiar et al., 2017). Recent advancements include the development of divalent siRNAs (di-siRNA), which achieved robust brain and spinal cord silencing with minimal off-target effects or toxicity following a single injection into the CSF of nonhuman primates, highlighting the potential of di-siRNAs for CNS-targeted RNAi therapies (Alterman et al., 2019).

6.2.3 Gene-editing-based preclinical studies

Zinc finger protein (ZFP) repressors and CRISPR–Cas9-based strategies aim to reduce *mHtt* expression at the transcriptional level by directly targeting DNA (Wild and Tabrizi, 2017). Effective treatment of HD requires that CRISPR–Cas9 or other gene-editing tools be efficiently delivered to the specific neuronal populations affected by the disease. In preclinical models, targeted delivery of CRISPR–Cas components to affected cells has been achieved using AAV, lipid nanoparticles (LNPs), and virus-like particles (Raguram et al., 2022). Among these, AAV vectors are currently the most effective and ideal gene delivery vectors (Wang et al., 2019). Preclinical studies have used CRISPR/Cas9 to excise the CAG repeat region in mouse models. For example, permanent suppression of *mHtt* via AAV-mediated delivery of CRISPR/Cas9 in the striatum of HD140Q-KI mice effectively depleted Htt aggregates, even after the formation of extensive Htt aggregates. Reducing *Htt* expression in striatal neuronal cells does not affect the survival of the adult HD140Q-KI mice but attenuates their motor impairment and neurological symptoms for 3 months (Yang et al., 2017b). Ekman et al. indicated that the small Cas9 nuclease, which can be packaged with a single gRNA into a single AAV vector, effectively disrupts *mHtt* expression in the striatum of R6/2 HD mice (Ekman et al., 2019). This treatment resulted in a decrease of ~50% in neuronal inclusions, protected striatal neurons from degeneration, and significantly improved lifespan and certain motor deficits (Ekman et al., 2019). Using a CRISPR/Cas9 system, *mHtt* and *wtHtt* expression were reduced in striatal neurons of heterozygous zQ175 mice. Longitudinal assessment with iVASO MRI, behavioral tests, and structural MRI showed that *Htt* silencing at the premanifest stage restored arteriolar cerebral blood volumes (CBVa), delayed disease onset, and slowed motor decline and striatal atrophy (Liu et al., 2021). These results thus illustrate that although this genome-editing strategy was not allele-specific, CRISPR–Cas9 technology can be a potential approach to treat HD (Ekman et al., 2019). Monteys and colleagues developed an allele-specific genome-editing approach targeting the *mHtt* gene using CRISPR/Cas9 technology (Monteys et al., 2017). Their strategy resulted in a reduction of *mHtt* expression by approximately 40% both in vitro and in vivo, achieved through intracerebral delivery of recombinant adeno-associated virus (rAAV) vectors carrying the CRISPR–Cas9 components (Monteys et al., 2017). This work demonstrated the potential of allele-specific gene editing for selective suppression of *mHtt* in HD models.

Shin et al. applied two CRISPR/Cas9 gRNAs targeting PAM sites created by SNPs on the *mHtt* chromosome to excise ~ 44 kb encompassing the promoter, transcription start site, and CAG expansion (Shin et al., 2016). This approach permanently inactivated the mutant allele, eliminating *mHtt* mRNA and protein without affecting the normal allele. Its perfect allele selectivity suggests broad potential for precision medicine in other gain-of-function disorders (Shin et al., 2016). ZFP repressors are engineered transcription factors that bind specific DNA sequences within the *Htt* gene promoter to block transcriptional machinery. In a study by Zeitler et al. (2019) on patient-derived fibroblasts and neurons, they observed that ZFP transcription factors (ZFP-TFs) repressed over 99% of HD-causing alleles while maintaining >86% of normal allele expression, with minimal impact on other CAG-containing genes. In HdhQ50, R6/2, and zQ175 het mice, one-time striatal AAV-ZFP infusion improved molecular, histopathological, electrophysiological, and functional outcomes, supporting further development of allele-selective ZFP-TFs for HD therapy (Zeitler et al., 2019). Bilateral injection of AAV-ZFP into the striatum of R6/2 mice repressed *mHtt* expression without affecting *wtHtt* or other genes and alleviated HD-related symptoms in a dose-dependent manner (Garriga-Canut et al., 2012). Lopes et al. investigated mitochondrial dysfunction in HD patient-derived iPSCs and neural stem cells (NSCs) compared with controls, and after CRISPR/Cas9-mediated deletion of the expanded CAG repeats in the *mHtt* gene (Lopes et al., 2020). HD cells showed mitochondrial and metabolic impairment, including reduced complex III and PDH activity, decreased oxygen consumption and ATP production, increased mitochondrial ROS, and fragmented mitochondrial morphology. CRISPR/Cas9 CAG deletion partially rescued these mitochondrial and metabolic defects, highlighting early transcription-linked mitochondrial dysfunction in HD pathogenesis (Lopes et al., 2020).

6.2.4 Adjunctive neuroprotection-based preclinical studies

Neuroprotective gene therapies aim to mitigate mHtt-induced neuronal damage by delivering trophic factors or modulating downstream pathways. Previous studies have demonstrated decreased BDNF levels in the hippocampus of R6/2 mice, and systemic administration of BDNF treatment effectively restores these levels to normal physiological values (Paldino, 2019). In addition, Connor and colleagues (2016) found that an intrastriatal $AAV_{1/2}$/BDNF treatment led to increased BDNF protein levels within the striatum in HOMO HD rats. Bilateral expression of BDNF significantly reduced deficits in both motor and cognitive functions. Notably, a gender difference was observed, with female transgenic HD rats showing milder functional impairments than their male counterparts. Furthermore, transgenic HD rats receiving AAV1/2-BDNF treatment displayed increased striatal volume and a higher number of NeuN-positive cells (Connor et al., 2016). These findings suggest that BDNF has promising therapeutic potential to mitigate behavioral deficits and neuronal loss in HD. In a QA rat model of HD, unilateral injections of AAV/BDNF and AAV/glial cell line-derived neurotrophic factor (GDNF) into the striatum markedly reduced QA-induced death of GABAergic medium spiny projection neurons. Notably, BDNF and GDNF provided distinct neuroprotective effects on different GABAergic interneuron subtypes within the striatum. BDNF preserved neurons expressing nitric oxide synthase (NOS), while GDNF mitigated the loss of parvalbumin-expressing neurons in the QA-damaged striatum (Kells et al., 2004).

In conclusion, preclinical gene therapy for HD has advanced significantly, utilizing sophisticated delivery systems and targeted therapeutic strategies to address mHtt and its downstream pathological effects. Adeno-associated and lentiviral vectors, along with emerging nonviral platforms such as LNPs, have improved CNS targeting. Concurrently, RNAi, ASOs, and CRISPR-based gene-editing techniques have shown robust *mHtt* suppression in animal models. Table 12.3 summarizes preclinical studies indicated for HD. Continued innovation in vector design, gene-editing precision, and large animal modeling will be crucial for translating these therapies into clinical trials, offering promising avenues toward effective treatments for HD.

7 Current challenges and limitations in gene therapy for HD

The therapeutic landscape for HD gene therapy has evolved from repeated-dosing oligonucleotide approaches toward longer-acting strategies, including viral vector-mediated RNAi and gene editing. Over the past 5 years, for the first time in HD, gene therapies have demonstrated the potential to modify disease progression, marking a significant milestone in the field. Despite advances, substantial challenges remain in preclinical and clinical gene therapy.

Immune responses to AAV vectors are commonly observed in animal models and can limit therapeutic efficacy, prompting the inclusion of immunosuppressive strategies in preclinical and clinical protocols (Prasad et al., 2022; Jiang and Dalby, 2023). Furthermore, major concerns with CRISPR-based therapies include the risk of off-target effects that may disrupt other genes or activate oncogenes, preexisting immunity to commonly used Cas9 orthologs, CNS CRISPR delivery, limitations in delivery efficiency and specificity, and significant challenges in large-scale vector manufacturing (Charlesworth et al., 2019; Guo et al., 2023; Xiong et al., 2025). Unlike RNAi, gene editing permanently alters gene expression, so insufficient specificity could have serious long-term consequences. Another challenge in the CRISPR approach is its delivery to deep brain regions. Thus current preclinical HD studies focus on improving allele specificity, minimizing off-target effects, and reducing immune stimulation (Colella et al., 2018; Rabinowitz et al., 2019). This strategy further complements the aforementioned approaches by helping to maintain neuronal function.

On the contrary, clinical translation requires robust integration of biomarkers, imaging, and clinical outcome measures, alongside continuous refinement of delivery methods and safety protocols. Matching each therapeutic modality to the optimal stage of disease remains a critical determinant of success (Duan et al., 2023).

Several persistent challenges complicate the pathway toward effective gene therapy in HD. These include:

- **Safety concerns**: immune responses to viral vectors or delivery vehicles may reduce efficacy and increase adverse effects. Irreversible vector exposure necessitates long-term monitoring for genotoxicity and immunoreactivity.
- **Ethical considerations:** rigorous informed consent in cognitively impaired individuals must be addressed through robust designs and patient registries.
- **Efficacy risks**: unintended lowering of wtHTT could worsen cognitive outcomes, complicating the therapeutic index.
- **Delivery limitations**: invasive procedures and the burden of repeated interventions restrict scalability.
- **Equitable access**: given the rarity of HD, cost and availability pose significant barriers (Sampaio and Disease, 2024; Stimming et al., 2022).

To date (2025), no gene therapy has received regulatory approval for HD. Nevertheless, key milestones include validated mHTT-lowering in human studies and evidence of slowed disease progression in subsets of patients.

TABLE 12.3 Recent preclinical gene therapy in HD.

Strategy	Intervention	Animal model	Mechanism of action	Route of administration	Effect	Reference
ASOs	HuASO	BACHD, YAC128 mice	Bind *mHtt* mRNA → reduce mHtt protein synthesis	ICV injection	Dose-dependent suppression of *mHtt* RNA/protein; sustained reversal of HD-like phenotypes	(Kordasiewicz et al., 2012)
	MkHuASO	Rhesus monkeys	Reduce *Htt* mRNA (*wt* and *mHtt*)	Intrathecal infusion	Htt protein reduction in the cortex	(Kordasiewicz et al., 2012)
	Allele-specific *mHtt* ASO	Hu97/18 and Hu18/18 mice	Selective silencing of mHtt	ICV bolus injection	Reduced cognitive/behavioral deficits while sparing wtHtt, mitigated forebrain atrophy, and preserved levels of striatal DARPP-32	(Southwell et al., 2018)
	Allele-specific ASO	Cynomolgus monkeys	Selective silencing of *mHtt*	Lumbar puncture	Suppression of Htt expression across cortical and limbic regions	(Southwell et al., 2013)
	apoA-I nanodisk ASO	BACHD mice	Enhanced CNS delivery of mHtt-lowering ASOs	Intranasal	Reduction of mHtt in the cortex and striatum	(Aly et al., 2023)
RNAi	AAV5-miHTT	Q175 KI mice; R6/2 mice	miRNA → RISC-mediated degradation of *mHtt* mRNA	Intrastriatal injection	Dose-dependent, sustained Htt reduction; reduction in aggregation; improved motor function and survival observed only in R6/2 mice	(Spronck et al., 2019)
	AAV-miR132	R6/2 mice	miRNA modulation of neuronal pathways (neuroprotective)	Intrastriatal injection	Improved motor function, prolonged lifespan, slowed progression (despite minimal change in *mHtt* mRNA)	(Fukuoka et al., 2018)
	AAV9-GFP-miRHTT	Q140/Q140 KI mice	miRNA/RNAi → *Htt* mRNA silencing	Intrastriatal injection	~50% reduction in *Htt* mRNA in the striatum; enhanced striatal neuron survival	(Keeler et al., 2016)
	AAV5-miHTT-155/ AAV5-miHTT-451/allele-selective miRNAs (miSNP50T)	Acute, lentiviral HD rat model	Total or allele-selective *Htt* silencing via miRNA	ICV	Suppressed mHtt aggregates and prevented DARPP-32 neuronal dysfunction	(Miniarikova et al., 2017)
	AAV9-miRNA	transgenic HD sheep	Allele-selective *Htt* silencing via miRNA	Intrastriatal injection	50–80% decrease in *mHtt* mRNA/protein in the caudate and putamen, lasting 6 months	(Pfister et al., 2018)
	AAV5-miHTT	transgenic HD minipig	*mHtt* mRNA silencing	Intrastriatal injection	Reduction of *mHtt* mRNA and protein levels in the brain following the highest dose of treatment	(Evers et al., 2018)
	di-siRNA (divalent siRNA)	Cynomolgus macaques	Divalent siRNA → robust *Htt* mRNA degradation	CSF injection	Robust silencing across the brain and spinal cord after a single dose; minimal off-target/toxicity	(Alterman et al., 2019)

Continued

Table 12.3 Recent preclinical gene therapy in HD—con'd

Strategy	Intervention	Animal model	Mechanism of action	Route of administration	Effect	Reference
Gene editing	AAV-HTT-gRNA/Cas9	HD140Q-KI mice	Excise/disrupt CAG repeat region → permanent suppression of *mHtt*	Intrastriatal injection	Depletion of Htt aggregates, improvement of motor/neurological symptoms for months	(Su Yang et al., 2017b)
	Small Cas9 nuclease + gRNA	R6/2 mice	Genome editing to disrupt the *mHtt* coding sequence	Intrastriatal injection	~50% reduction in neuronal inclusions; protection of striatal neurons; improved lifespan and motor deficits	(Ekman et al., 2019)
	CRISPR/Cas9 (nonallele-specific)	zQ175 heterozygous mice	*Htt* gene disruption/silencing at the DNA level	Intrastriatal injection	Restored CBVa, delayed disease onset, slowed motor decline, and striatal atrophy	(Liu et al., 2021)
	AAV-ZFP transcription factors (ZFP-TFs)	HdhQ50, R6/2, zQ175 mice	Engineered ZFPs bind the *Htt* promoter → transcriptional repression of the mHtt allele	Intrastriatal injection	>99% repression of the HD allele, ~ 86% preservation of wtHtt; improved molecular, histopathological, and functional readouts	(Zeitler et al., 2019)
Neuroprotection	AAV1/2-BDNF	HOMO HD rats	Restore/increase BDNF levels → trophic support and synaptic maintenance	Intrastriatal injection	Increased striatal volume, improved motor and cognitive function; increased NeuN + cell counts; sex differences noted (females milder)	(Connor et al., 2016)
	AAV-BDNF + AAV-GDNF	QA-lesioned rat model	Neurotrophic support (BDNF and GDNF act on different interneuron subtypes)	Intrastriatal injection	Marked reduction of QA-induced GABAergic neuron death; BDNF preserved NOS + neurons, GDNF preserved PV + neurons	(Kells et al., 2004)

Near-term prospects center on allele-selective ASOs and AAV-mediated RNAi, which could potentially achieve approval by 2027–2030 (Sampaio and Disease, 2024). Looking further ahead, emerging technologies such as nonviral nanoparticles and focused ultrasound-mediated BBB opening may expand therapeutic scalability and accessibility. Patient advocacy groups, such as the Huntington's Disease Society of America (HDSA), further emphasize the importance of quality-of-life outcomes beyond traditional motor endpoints, underscoring the need to align therapeutic innovation with patient-centered values (Cheng et al., 2024; Sampaio and Disease, 2024).

In summary, while challenges remain substantial, gene therapy represents the most rational disease-modifying avenue for HD. If current trajectories hold, the prospect of slowing or even intervening prior to clinical onset may soon move from aspiration to reality, reshaping prognosis and offering new hope to affected families.

AI disclosure

During the preparation of this work, the author(s) used Chat GPT-4 to assist in improving their writing style. All text edited by AI has been double-checked by the authors to ensure clarity and consistency. After using this tool/service, the author(s) reviewed and edited the content as needed and take(s) full responsibility for the content of the publication.

References

Achenbach, J., Saft, C., 2021. Data from ENROLL-HD: is the prevalence of juvenile and pediatric Huntington's disease overestimated? Parkinsonism Relat. Disord. 18735126. 88, 1–2. doi: 10.1016/j.parkreldis.2021.05.012.

Adamu, A., Shuo, L., Gao, F., 2024. Guofang Xue, The role of neuroinflammation in neurodegenerative diseases: current understanding and future therapeutic targets. Front. Aging Neurosci. 1663-4365. 16. doi: 10.3389/fnagi.2024.1347987.

Addabbo, F., Montagnani, M., Goligorsky, M.S., 2009. Mitochondria and reactive oxygen species. Hypertension. 0194911X. 53 (6), 885–892. doi: 10.1161/HYPERTENSIONAHA.109.130054.

Aguiar, S., van der Gaag, B., Cortese, F.A.B., 2017. RNAi mechanisms in Huntington's disease therapy: siRNA versus shRNA. Transl. Neurodegener. 2047-9158. 6 (1). doi: 10.1186/s40035-017-0101-9.

Ajitkumar, Lui, De Jesus, O., 2025. Huntington Disease. StatPearls. Available from: http://www.ncbi.nlm.nih.gov/books/NBK559166/.

Alkanli, S.S., Alkanli, N., Ay, A., Albeniz, I., 2023. CRISPR/Cas9 Mediated Therapeutic Approach in Huntington's Disease. Mol. Neurobiol. 15591182. 60 (3), 1486–1498. doi: 10.1007/s12035-022-03150-5.

Allaman, I., Bélanger, M., Magistretti, P.J., 2011. Astrocyte-neuron metabolic relationships: for better and for worse. Trends Neurosci. 01662236. 34 (2), 76–87. doi: 10.1016/j.tins.2010.12.001.

Alterman, J.F., Godinho, B.M.D.C., Hassler, M.R., Ferguson, C.M., Echeverria, D., Sapp, E., Haraszti, R.A., Coles, A.H., Conroy, F., Miller, R., Roux, L., Yan, P., Knox, E.G., Turanov, A.A., King, R.M., Gernoux, G., Mueller, C., Gray-Edwards, H.L., Moser, R.P., Bishop, N.C., Jaber, S.M., Gounis, M.J., Sena-Esteves, M., Pai, A.A., DiFiglia, M., Aronin, N., Khvorova, A., 2019. A divalent siRNA chemical scaffold for potent and sustained modulation of gene expression throughout the central nervous system. Nat. Biotechnol. 15461696. 37 (8), 884–894. doi: 10.1038/s41587-019-0205-0.

Aly, A.E.E., Caron, N.S., Black, H.F., Schmidt, M.E., Anderson, C., Ko, S., Baddeley, H.J.E., Anderson, L., Casal, L.L., Rahavi, R.S.M., Martin, D.D.O., Hayden, M.R., 2023. Delivery of mutant huntingtin-lowering antisense oligonucleotides to the brain by intranasally administered apolipoprotein A-I nanodisks. J. Control. Release. 18734995. 360, 913–927. doi: 10.1016/j.jconrel.2023.07.027.

Anderson, K.E., Van Duijn, E., Craufurd, D., Drazinic, C., Edmondson, M., Goodman, N., Van Kammen, D., Loy, C., Priller, J., Goodman, L.V.V., 2018. Clinical management of neuropsychiatric symptoms of huntington disease: expert-based consensus guidelines on agitation, anxiety, apathy, psychosis and sleep disorders. J. Huntingtons Dis. 18796400. 7 (4), 355–366. doi: 10.3233/JHD-180293.

Andriessen, R.L., Oosterloo, M., Hollands, A., Linden, D.E.J., de Greef, B.T.A., Leentjens, A.F.G., 2022. Psychotropic medication use in Huntington's disease: a retrospective cohort study. Parkinsonism Relat. Disord. 18735126. 105, 69–74. doi: 10.1016/j.parkreldis.2022.11.004.

Arnulf, I., Nielsen, J., Lohmann, E., Schieffer, J., Wild, E., Jennum, P., Konofal, E., Walker, M., Oudiette, D., Tabrizi, S., Durr, A., 2008. Rapid eye movement sleep disturbances in Huntington disease. Arch. Neurol. 15383687. 65 (4), 482–488. doi: 10.1001/archneur.65.4.482France.

Askenasy, J.J.M., Mendelson, L., Keren, O., Braun, Z., 1990. Juvenile Parkinson's disease and its response to L-dopa therapy. J. Neural Transm. 14351463. 2 (1), 23–30. doi: 10.1007/BF02251243.

Bachoud-Lévi, A.C., Ferreira, J., Massart, R., Youssov, K., Rosser, A., Busse, M., Craufurd, D., Reilmann, R., De Michele, G., Rae, D., Squitieri, F., Seppi, K., Perrine, C., Scherer-Gagou, C., Audrey, O., Verny, C., Burgunder, J.M., 2019. International guidelines for the treatment of Huntington's disease. Front. Neurol. 16642295. 10. doi: 10.3389/fneur.2019.00710.

Baine, F.K., Krause, A., Greenberg, L.J., 2016. The frequency of huntington disease and huntington disease-like 2 in the South African population. Neuroepidemiology. 14230208. 46 (3), 198–202. doi: 10.1159/000444020.

Bakels, H.S., Feleus, S., Rodríguez-Girondo, M., Losekoot, M., Bijlsma, E.K., Roos, R.A.C., de Bot, S.T., 2024. Prevalence of juvenile-onset and pediatric Huntington's disease and their availability and ability to participate in trials: a Dutch population and Enroll-HD observational study. J. Huntingtons Dis. 18796400. 13 (3), 357–368. doi: 10.3233/JHD-240034.

Bakels, H.S., Roos, R.A.C., van Roon-mom, W.M.C., de Bot, S.T., 2022. Juvenile-Onset Huntington Disease Pathophysiology and Neurodevelopment: a Review. Mov. Disord. 15318257. 37 (1), 16–24. doi: 10.1002/mds.28823.

Bardai, F.H., Verma, P., Smith, C., Rawat, V., Wang, L., D'Mello, S.R., 2013. Disassociation of histone deacetylase-3 from normal huntingtin underlies mutant huntingtin neurotoxicity. J. Neurosci. 15292401. 33 (29), 11833–11838. doi:10.1523/JNEUROSCI.5831-12.2013.

Barnat, M., Friec, J.L., Benstaali, C., Humbert, S., 2017. Huntingtin-Mediated Multipolar-Bipolar Transition of Newborn Cortical Neurons Is Critical for Their Postnatal Neuronal Morphology. Neuron. 10974199. 93 (1), 99–114. doi:10.1016/j.neuron.2016.11.035.

Barron, J.C., Hurley, E.P., Parsons, M.P., 2021. Huntingtin and the Synapse. Front. Cell. Neurosci. 16625102. 15. doi:10.3389/fncel.2021.689332.

Bartlett, D.M., Govus, A., Rankin, T., Lampit, A., Feindel, K., Poudel, G., Teo, W.-P., Johnny, L., Georgiou-Karistianis, N., Ziman, M.R., Cruickshank, T.M., 2020. The effects of multidisciplinary rehabilitation on neuroimaging, biological, cognitive and motor outcomes in individuals with premanifest Huntington's disease. J. Neurol. Sci. 0022510X. 416, 117022,. doi:10.1016/j.jns.2020.117022.

Bates, G., 2003. Huntingtin aggregation and toxicity in Huntington's disease. Lancet. 01406736. 361 (9369), 1642–1644. doi:10.1016/S0140-6736(03)13304-1.

Bates, G.P., Dorsey, R., Gusella, J.F., Hayden, M.R., Kay, C., Leavitt, B.R., Nance, M., Ross, C.A., Scahill, R.I., Wetzel, R., Wild, E.J., Tabrizi, S.J., 2015. Huntington disease. Nat. Rev. Dis. Primers. 2056676X. 1. doi:10.1038/nrdp.2015.5.

Beal, M.F., Brouillet, E., Jenkins, B.G., Ferrante, R.J., Kowall, N.W., Miller, J.M., Storey, E., Srivastava, R., Rosen, B.R., Hyman, B.T., 1993. Neurochemical and histologic characterization of striatal excitotoxic lesions produced by the mitochondrial toxin 3-nitropropionic acid. The Journal of Neuroscience. 0270-6474. 13 (10), 4181–4192. doi:10.1523/jneurosci.13-10-04181.1993.

Benn, C.L., Sun, T., Sadri-Vakili, G., McFarland, K.N., DiRocco, D.P., Yohrling, G.J., Clark, T.W., Bouzou, B., Cha, J.H.J., 2008. Huntingtin modulates transcription, occupies gene promoters in vivo, and binds directly to DNA in a polyglutamine-dependent manner. J. Neurosci. 02706474. 28 (42), 10720–10733. doi:10.1523/JNEUROSCI.2126-08.2008.

Bergonzoni, G., Döring, J., Biagioli, M., 2021. D1R- and D2R-Medium-Sized Spiny Neurons Diversity: insights Into Striatal Vulnerability to Huntington's Disease Mutation. Front. Cell. Neurosci. 16625102. 15. doi: 10.3389/fncel.2021.628010.

Bezprozvanny, I., 2009. Calcium signaling and neurodegenerative diseases. Trends Mol. Med. 14714914. 15 (3), 89–100. doi:10.1016/j.molmed.2009.01.001.

Björkqvist, M., Wild, E.J., Thiele, J., Silvestroni, A., Andre, R., Lahiri, N., Raibon, E., Lee, R.V., Benn, C.L., Soulet, D., Magnusson, A., Woodman, B., Landles, C., Pouladi, M.A., Hayden, M.R., Khalili-Shirazi, A., Lowdell, M.W., Brundin, P., Bates, G.P., Leavitt, B.R., Möller, T., Tabrizi, S.J., A novel pathogenic pathway of immune activation detectable before clinical onset in Huntington's disease. J. Exp. Med. 00221007. 205 (8) (2008), 1869–1877. doi:10.1084/jem.20080178.

Blackwell, A.D., Paterson, N.S., Barker, R.A., Robbins, T.W., Sahakian, B.J., 2008. The effects of modafinil on mood and cognition in Huntington's disease. Psychopharmacology. 00333158. 199 (1), 29–36. doi:10.1007/s00213-008-1068-0.

Bogdan, A., Manera, V., Koenig, A., David, R., 2020. Pharmacologic Approaches for the Management of Apathy in Neurodegenerative Disorders. Front. Pharmacol. 1663-9812. 10. doi:10.3389/fphar.2019.01581.

Bradford, J., Shin, J.Y., Roberts, M., Wang, C.E., Sheng, G., Li, S., Li, X.J., 2010. Mutant huntingtin in glial cells exacerbates neurological symptoms of huntington disease mice. J. Biol. Chem. 1083351X. 285 (14), 10653–10661. doi:10.1074/jbc.M109.083287.

Bravo-Hernandez, M., Tadokoro, T., Navarro, M.R., Platoshyn, O., Kobayashi, Y., Marsala, S., Miyanohara, A., Juhas, S., Juhasova, J., Skalnikova, H., Tomori, Z., Vanicky, I., Studenovska, H., Proks, V., Chen, P., Govea-Perez, N., Ditsworth, D., Ciacci, J.D., Gao, S., Zhu, W., Ahrens, E.T., Driscoll, S.P., Glenn, T.D., McAlonis-Downes, M., Cruz, S.D., Pfaff, S.L., Kaspar, B.K., Cleveland, D.W., Marsala, M., 2020. Spinal subpial delivery of AAV9 enables widespread gene silencing and blocks motoneuron degeneration in ALS. Nat. Med. 1078-8956. 26 (1), 118–130. doi:10.1038/s41591-019-0674-1.

Briston, T., Selwood, D.L., Szabadkai, G., Duchen, M.R., 2019. Mitochondrial Permeability Transition: a Molecular Lesion with Multiple Drug Targets. Trends Pharmacol. Sci. 18733735. 40 (1), 50–70. doi:10.1016/j.tips.2018.11.004.

Brouillet, E., Hantraye, P., Ferrante, R.J., Dolan, R., Leroy-Willig, A., Kowall, N.W., Beal, M.F., 1995. Chronic mitochondrial energy impairment produces selective striatal degeneration and abnormal choreiform movements in primates. Proc. Natl. Acad. Sci. 0027-8424. 92 (15), 7105–7109. doi:10.1073/pnas.92.15.7105.

Burgunder, J.M., 2013. Recent advances in the management of choreas. Therapeutic Advances in Neurological Disorders. 17562864. 6 (2), 117–127. doi:10.1177/1756285612471700.

Burnip, E., Wallace, E., Gozdzikowska, K., Huckabee, M.L., 2020. A Systematic Review of Rehabilitation for Corticobulbar Symptoms in Adults with Huntington's Disease. J. Huntingtons Dis. 18796400. 9 (1), 1–12. doi: 10.3233/JHD-190384.

Capiluppi, E., Romano, L., Rebora, P., Nanetti, L., Castaldo, A., Gellera, C., Mariotti, C., Antonella Macerollo, M.G.C., 2020. Late-onset Huntington's disease with 40–42 CAG expansion. Neurol. Sci. 1590-1874. 41 (4), 869–876. doi:10.1007/s10072-019-04177-8.

Caron, N.S., Aly, A.-E.-E., Black, H.F., Martin, D.D.O., Schmidt, M.E., Ko, S., Anderson, C., Harvey, E.M., Casal, L.L., Anderson, L.M., Rahavi, S.M.R., Reid, G.S.D., Oda, M.N., Stanimirovic, D., Abulrob, A., McBride, J.L., Leavitt, B.R., Hayden, M.R., 2024. Systemic delivery of mutant huntingtin lowering antisense oligonucleotides to the brain using apolipoprotein A-I nanodisks for Huntington disease. J. Control. Release. 01683659. 367, 27–44. doi:10.1016/j.jconrel.2024.01.011.

Carroll, J.B., Bates, G.P., Steffan, J., Saft, C., Tabrizi, S.J., 2015. Treating the whole body in Huntington's disease. Lancet Neurol. 14744465. 14 (11), 1135–1142. doi:10.1016/S1474-4422(15)00177-5.

Caviston, J.P., Holzbaur, E.L.F., 2009. Huntingtin as an essential integrator of intracellular vesicular trafficking. Trends in Cell Biology. 09628924. 19 (4), 147–155. doi:10.1016/j.tcb.2009.01.005.

Caviston, J.P., Ross, J.L., Antony, S.M., Tokito, M., Holzbaur, E.L.F., 2007. Huntingtin facilitates dynein/dynactin-mediated vesicle transport. Proc. Natl. Acad. Sci. U. S. A. 0027-8424. 104 (24), 10045–10050. doi:10.1073/pnas.0610628104.

Cepeda, C., Murphy, K.P.S., Parent, M., Levine, M.S., The role of dopamine in huntington's disease. Progress in Brain Research. 211 (2014), 235–254. doi:10.1016/B978-0-444-63425-2.00010-6.

Chaganti, S.S., McCusker, E.A., Loy, C.T., 2017. What do we know about late onset Huntington's disease?J. Huntingtons Dis. 18796400. 6 (2), 95–103. doi:10.3233/JHD-170247.

Chang, D.T.W., Rintoul, G.L., Pandipati, S., Reynolds, I.J., 2006. Mutant huntingtin aggregates impair mitochondrial movement and trafficking in cortical neurons. Neurobiol. Dis. 09699961. 22 (2), 388–400. doi:10.1016/j.nbd.2005.12.007.

Chao, T.K., Hu, J., Pringsheim, T., 2017. Risk factors for the onset and progression of Huntington disease. NeuroToxicology. 18729711. 61, 79–99. doi:10.1016/j.neuro.2017.01.005.

Charlesworth, C.T., Deshpande, P.S., Dever, D.P., Camarena, J., Lemgart, V.T., Cromer, M.K., Vakulskas, C.A., Collingwood, M.A., Zhang, L., Bode, N.M., Behlke, M.A., Dejene, B., Cieniewicz, B., Romano, R., Lesch, B.J., Gomez-Ospina, N., Mantri, S., Pavel-Dinu, M., Weinberg, K.I., Porteus, M.H., 2019. Identification of preexisting adaptive immunity to Cas9 proteins in humans. Nat. Med. 1546170X. 25 (2), 249–254. doi:10.1038/s41591-018-0326-x.

Chen, J.Y., Wang, E.A., Cepeda, C., Levine, M.S., 2013. Dopamine imbalance in Huntington's disease: a mechanism for the lack of behavioral flexibility. Front. Neurosci. 1662453X. (7). doi:10.3389/fnins.2013.00114.

Chen, X., He, E., Su, C., Zeng, Y., Xu, J., 2023. Huntingtin-associated protein 1-associated intracellular trafficking in neurodegenerative diseases. Front. Aging Neurosci. 16634365. 15. doi:10.3389/fnagi.2023.1100395.

Cheng, Y., Zhang, S., Shang, H., 2024. Latest advances on new promising molecular-based therapeutic approaches for Huntington's disease. J. Transl. Intern. Med. 2224-4018. 12 (2), 134–147. doi:10.2478/jtim-2023-0142.

Chidiac, M., Elhusein, B., Gajebasia, N., 2024. Reassessing mirtazapine and akathisia: a case report on its efficacy in treating severe, treatment-resistant akathisia and a review of the evidence. SAGE Open Med. Case Rep. 2050-313X. 12. doi:10.1177/2050313x241299947.

Choi, D.E., Shin, J.W., Zeng, S., Hong, E.P., Jang, J.H., Loupe, J.M., Wheeler, V.C., Stutzman, H.E., Kleinstiver, B., Lee, J.M., 2024. Base editing strategies to convert CAG to CAA diminish the disease-causing mutation in Huntington's disease. eLife. 2050084X. doi:10.7554/eLife.89782.2.

Choi, Y.S., Lee, B., Cho, H.Y., Reyes, I.B., Pu, X.A., Saido, T.C., Hoyt, K.R., Obrietan, K., 2009. CREB is a key regulator of striatal vulnerability in chemical and genetic models of Huntington's disease. Neurobiol. Dis. 09699961. 36 (2), 259–268. doi:10.1016/j.nbd.2009.07.014.

Claassen, D.O., Carroll, B., De Boer, L.M., Wu, E., Ayyagari, R., Gandhi, S., Stamler, D., 2017. Indirect tolerability comparison of Deutetrabenazine and Tetrabenazine for Huntington disease. J. Clin. Mov. Disord. 2054-7072. 4 (1). doi:10.1186/s40734-017-0051-5.

Clare, L., Teale, J.C., Toms, G., Kudlicka, A., Evans, I., Abrahams, S., Goldstein, L.H., Hindle, J.V., Ho, A.K., Jahanshahi, M., Langdon, D., Morris, R., Snowden, J.S., Davies, R., Markova, I., Busse, M., Thompson-Coon, J., 2019. Cognitive rehabilitation, self-management, psychotherapeutic and caregiver support interventions in progressive neurodegenerative conditions: a scoping review. NeuroRehabilitation. 18786448. 43 (4), 443–471. doi:10.3233/NRE-172353.

Cloud, L.J., Jinnah, H.A., 2010. Treatment strategies for dystonia. Expert Opin. Pharmacother. 14656566. 11 (1), 5–15. doi:10.1517/14656560903426171.

Colella, P., Ronzitti, G., Mingozzi, F., 2018. Emerging Issues in AAV-Mediated In Vivo Gene Therapy. Mol. Ther. Methods Clin. Dev. 23290501. 8, 87–104. doi:10.1016/j.omtm.2017.11.007.

Colin, E., Zala, D., Liot, G., Rangone, H., Borrell-Pagès, M., Xiao-Jiang, L., Saudou, F., Humbert, S., 2008. Huntingtin phosphorylation acts as a molecular switch for anterograde/retrograde transport in neurons. EMBO J. 0261-4189. 27 (15), 2124–2134. doi:10.1038/emboj.2008.133.

Connor, B., Sun, Y., von Hieber, D., Tang, S.K., Jones, K.S., Maucksch, C., 2016. AAV1/2-mediated BDNF gene therapy in a transgenic rat model of Huntington's disease. Gene Ther. 0969-7128. 23 (3), 283–295. doi:10.1038/gt.2015.113.

Cornejo-Olivas, M.R., Inca-Martinez, M.A., Espinoza-Huertas, K., Veliz-Otani, D., Velit-Salazar, M.R., Marca, V., Ortega, O., Cornejo-Herrera, I.F., Lindo-Samanamud, S., Mora-Alferez, P., Mazzetti, P., 2015. Clinical and Molecular Features of Late Onset Huntington Disease in a Peruvian Cohort. J. Huntingtons Dis. 18796400. 4 (1), 99–105. doi: 10.3233/JHD-140119.

Cortes, C.J., Spada, A.R.L., 2014. The many faces of autophagy dysfunction in Huntington's disease: from mechanism to therapy. Drug Discov. Today. 18785832. 19 (7), 963–971. doi:10.1016/j.drudis.2014.02.014.

Croce, K.R., Yamamoto, A., 2019. A role for autophagy in Huntington's disease . Neurobiol. Dis. 1095953X. 122, 16–22. doi:10.1016/j.nbd.2018.08.010.

Crotti, A., Glass, C.K., 2015. The choreography of neuroinflammation in Huntington's disease. Trends Immunol. 14714981. 36 (6), 364–373. doi:10.1016/j.it.2015.04.007.

Dash, U.C., Bhol, N.K., Swain, S.K., Samal, R.R., Nayak, P.K., Raina, V., Panda, S.K., Kerry, R.G., Duttaroy, A.K., Jena, A.B., 2025. Oxidative stress and inflammation in the pathogenesis of neurological disorders: mechanisms and implications. Acta Pharm. Sin. B. 22113843. 15 (1), 15–34. doi:10.1016/j.apsb.2024.10.004.

Davidson, B.L., Boudreau, R.L., 2007. RNA Interference: a Tool for Querying Nervous System Function and an Emerging Therapy. Neuron. 08966273. 53 (6), 781–788. doi:10.1016/j.neuron.2007.02.020.

De Luca, A., Morella, A., Consoli, F., Fanelli, S., Thibert, J.R., Statt, S., Latham, G.J., Squitieri, F., 2021. A novel triplet-primed pcr assay to detect the full range of trinucleotide cag repeats in the huntingtin gene (Htt). Int. J. Mol. Sci. 14220067. 22 (4), 1–13. doi:10.3390/ijms22041689.

Dean, M., Sung, V.W., 2018. Review of deutetrabenazine: a novel treatment for chorea associated with Huntington's disease. Drug Design, Development and Therapy. 11778881. 12, 313–319. doi:10.2147/DDDT.S138828.

Delval, A., Krystkowiak, P., Blatt, J.L., Labyt, E., Dujardin, K., Destée, A., Derambure, P., Defebvre, L., 2006. Role of hypokinesia and bradykinesia in gait disturbances in Huntington's disease: a biomechanical study. J. Neurol. 14321459. 253 (1), 73–80. doi:10.1007/s00415-005-0929-2.

Diaz-Castro, B., Gangwani, M.R., Yu, X., Coppola, G., Khakh, B.S., 2019. Astrocyte molecular signatures in Huntington's disease. Sci. Transl. Med. 19466242. 11 (514). doi:10.1126/scitranslmed.aaw8546.

Díaz-Gutiérrez, M.J., Martínez-Cengotitabengoa, M., de Adana, E.S., Cano, A.I., Martínez-Cengotitabengoa, M.T., Besga, A., Segarra, R., González-Pinto, A., 2017. Relationship between the use of benzodiazepines and falls in older adults: a systematic review. Maturitas. 18734111. 101, 17–22. doi:10.1016/j.maturitas.2017.04.002.

Dickey, A.S., Spada, A.R.L., 2018Therapy development in Huntington disease: from current strategies to emerging opportunities. American Journal of Medical Genetics, Part A. 15524833. 176 (4), 842–861. doi:10.1002/ajmg.a.38494.

Didiot, M.C., Ferguson, C.M., Ly, S., Coles, A.H., Smith, A.O., Bicknell, A.A., Hall, L.M., Sapp, E., Echeverria, D., Pai, A.A., DiFiglia, M., Moore, M.J., Hayward, L.J., Aronin, N., Khvorova, A., 2018. Nuclear Localization of Huntingtin mRNA Is Specific to Cells of Neuronal Origin. Cell Rep. 22111247. 24 (10), 2553. doi:10.1016/j.celrep.2018.07.106.

Dragatsis, I., Dietrich, P., Ren, H., Deng, Y.P., Mar, N.D., Wang, H.B., Johnson, I.M., Jones, K.R., Reiner, A., 2018Effect of early embryonic deletion of huntingtin from pyramidal neurons on the development and long-term survival of neurons in cerebral cortex and striatum. Neurobiol. Dis. 1095953X. 111, 102–117. doi:10.1016/j.nbd.2017.12.015.

Duan, W., Urani, E., Mattson, M.P., 2023. The potential of gene editing for Huntington's disease. Trends Neurosci. 1878108X. 46 (5), 365–376. doi:10.1016/j.tins.2023.02.005.

Duennwald, M.L., Lindquist, S., 2008. Impaired ERAD and ER stress are early and specific events in polyglutamine toxicity. Genes Dev. 0890-9369. 22 (23), 3308–3319. doi:10.1101/gad.1673408.

Duyao, M.P., Auerbach, A.B., Ryan, A., Persichetti, F., Barnes, G.T., McNeil, S.M., Ge, P., Vonsattel, J.P., Gusella, J.F., Joyner, A.L., MacDonald, M.E., 1995. Inactivation of the mouse huntington's disease gene homolog Hdh. Science. 00368075. 269 (5222), 407–410. doi:10.1126/science.7618107.

Eaton, J.E., Claassen, D.O., 2024. Guidance on antipsychotic selection for agitation and aggressive behavior in persons with Huntington's disease. Expert Rev. Neurother. 17448360. 24 (10), 937–940. doi:10.1080/14737175.2024.2376836.

Ehinger, Y., Bruyère, J., Panayotis, N., Abada, Y., Borloz, E., Matagne, V., Scaramuzzino, C., Vitet, H., Delatour, B., Saidi, L., Villard, L., Saudou, F., Roux, J., 2020. Huntingtin phosphorylation governs BDNF homeostasis and improves the phenotype of Mecp2 knockout mice. EMBO Mol. Med. 1757-4676. 12 (2). doi:10.15252/emmm.201910889.

Ehrlich, M.E., 2012. Huntington's Disease and the Striatal Medium Spiny Neuron: cell-Autonomous and Non-Cell-Autonomous Mechanisms of Disease. Neurotherapeutics. 18787479. 9 (2), 270–284. doi:10.1007/s13311-012-0112-2.

Ekman, F.K., Ojala, D.S., Adil, M.M., Lopez, P.A., Schaffer, D.V., Gaj, T., 2019. CRISPR-Cas9-Mediated Genome Editing Increases Lifespan and Improves Motor Deficits in a Huntington's Disease Mouse Model. Mol. Ther. Nucleic Acids. 21622531. 17, 829–839. doi:10.1016/j.omtn.2019.07.009.

Essa, M.M., Moghadas, M., Ba-Omar, T., Qoronfleh, M.W., Guillemin, G.J., Manivasagam, T., Justin-Thenmozhi, A., Ray, B., Bhat, A., Chidambaram, S.B., Fernandes, A.J., Song, B.J., Akbar, M., 2019. Protective Effects of Antioxidants in Huntington's Disease: an Extensive Review. Neurotoxicity Research. 14763524. 35 (3), 739–774. doi: 10.1007/s12640-018-9989-9.

Estevez-Fraga, C., Altmann, A., Parker, C.S., Scahill, R.I., Costa, B., Chen, Z., Manzoni, C., Zarkali, A., Durr, A., Roos, R.A.C., Landwehrmeyer, B., Leavitt, B.R., Rees, G., Tabrizi, S.J., Mccolgan, P., 2023a. Genetic topography and cortical cell loss in Huntington's disease link development and neurodegeneration. Brain. 14602156. 146 (11), 4532–4546. doi:10.1093/brain/awad275.

Estevez-Fraga, C., Rodrigues, F.B., Tabrizi, S.J., Wild, E.J., 2022a. Huntington's Disease Clinical Trials Corner: April 2022. J. Huntingtons Dis. 18796397. 11 (2), 105–118. doi:10.3233/jhd-229002.

Estevez-Fraga, C., Tabrizi, S.J., Wild, E.J., 2022b. Huntington's Disease Clinical Trials Corner: November 2022. J. Huntingtons Dis. 18796397. 11 (4), 351–367. doi:10.3233/jhd-229006.

Estevez-Fraga, C., Tabrizi, S.J., Wild, E.J., 2024. Huntington's Disease Clinical Trials Corner: March 2024. J. Huntingtons Dis. 18796397. 13 (1), 1–14. doi:10.3233/jhd-240017.

Estevez-Fraga, C., Tabrizi, S.J., Wild, E.J., Morton, J., 2023b. Huntington's Disease Clinical Trials Corner: August 2023. J. Huntingtons Dis. 1879-6397. 12 (2), 169–185. doi:10.3233/jhd-239001.

Estrada-Sánchez, A.M., Montiel, T., Segovia, J., Massieu, L., 2009. Glutamate toxicity in the striatum of the R6/2 Huntington's disease transgenic mice is age-dependent and correlates with decreased levels of glutamate transporters. Neurobiol. Dis. 09699961. 34 (1), 78–86. doi:10.1016/j.nbd.2008.12.017.

Evers, M.M., Miniarikova, J., Juhas, S., Vallès, A., Bohuslavova, B., Juhasova, J., Skalnikova, H.K., Vodicka, P., Valekova, I., Brouwers, C., Blits, B., Lubelski, J., Kovarova, H., Ellederova, Z., van Deventer, S.J., Petry, H., Motlik, J., Konstantinova, P., 2018. AAV5-miHTT Gene Therapy Demonstrates Broad Distribution and Strong Human Mutant Huntingtin Lowering in a Huntington's Disease Minipig Model. Mol. Ther. 15250024. 26 (9), 2163–2177. doi:10.1016/j.ymthe.2018.06.021.

Faideau, M., Kim, J., Cormier, K., Gilmore, R., Welch, M., Auregan, G., Dufour, N., Guillermier, M., Brouillet, E., Hantraye, P., Deglon, N., Ferrante, R.J., Bonvento, G., 2010. In vivo expression of polyglutamine-expanded huntingtin by mouse striatal astrocytes impairs glutamate transport: a correlation with Huntington's disease subjects. Hum. Mol. Genet. 0964-6906. 19 (15), 3053–3067. doi: 10.1093/hmg/ddq212.

Farag, M., Tabrizi, S.J., Wild, E.J., 2024. Huntington's Disease Clinical Trials Update: september 2024. J. Huntingtons Dis. 1879-6397. 13 (4), 409–418. doi:10.1177/18796397241293955.

Farag, M., Tabrizi, S.J., Wild, E.J., 2025. Huntington's disease clinical trials update: march 2025. J. Huntingtons Dis. 1879-6397. 14 (2), 191–206. doi:10.1177/18796397251337000.

Fazio, P., Paucar, M., Svenningsson, P., Varrone, A., 2018. Novel Imaging Biomarkers for Huntington's Disease and Other Hereditary Choreas. Curr. Neurol. Neurosci. Rep. 1528-4042. 18 (12). doi:10.1007/s11910-018-0890-y.

Fenton, A.R., Jongens, T.A., Holzbaur, E.L.F., 2021. Mitochondrial adaptor TRAK2 activates and functionally links opposing kinesin and dynein motors. Nat. Commun. 20411723. 12 (1). doi:10.1038/s41467-021-24862-7.

Ferguson, M.W., Kennedy, C.J., Palpagama, T.H., Waldvogel, H.J., Faull, R.L.M., Kwakowsky, A., 2022. Current and Possible Future Therapeutic Options for Huntington's Disease. J. Cent. Nerv. Syst. Dis. 1179-5735. 14. doi:10.1177/11795735221092517.

Fernandes, H.B., Raymond, L.A., 2008. NMDA Receptors and Huntington's Disease Biology of the NMDA Receptor. CRC Press, Canada, 17–40. Available from: https://www.taylorfrancis.com/books/e/9781420044157.

Finkbeiner, S., 2011. Huntington's Disease. Cold Spring Harb. Perspect. Biol. 3 (6), a007476.

Franco-Iborra, S., Vila, M., Perier, C., 2018. Mitochondrial quality control in neurodegenerative diseases: focus on Parkinson's disease and Huntington's disease. Front. Neurosci. 1662453X. 12 (MAY). doi:10.3389/fnins.2018.00342.

Frank, S., 2009. Tetrabenazine as anti-chorea therapy in Huntington Disease: an open-label continuation study. Huntington Study Group/TETRA-HD Investigators. BMC Neurol. 14712377. 9, doi:10.1186/1471-2377-9-62.

Frank, S., 2010. Tetrabenazine: the first approved drug for the treatment of chorea in US patients with Huntington disease. Neuropsychiatric Disease and Treatment. 1176-6328. 6, 657. doi:10.2147/NDT.S6430.

Frank, S., Testa, C.M., Stamler, D., Kayson, E., Davis, C., Edmondson, M.C., Kinel, S., Leavitt, B., Oakes, D., O'Neill, C., Vaughan, C., Goldstein, J., Herzog, M., Snively, V., Whaley, J., Wong, C., Suter, G., Jankovic, J., Jimenez-Shahed, J., Hunter, C., Claassen, D.O., Roman, O.C., Sung, V., Smith, J., Janicki, S., Clouse, R., Saint-Hilaire, M., Hohler, A., Turpin, D., James, R.C., Rodriguez, R., Rizer, K., Anderson, K.E., Heller, H., Carlson, A., Criswell, S., Racett, B.A., Revilla, F.J., Nucifora, F., Margolis, R.L., Ong, M.J., Mendis, T., Mendis, N., Singer, C., Quesada, M., Paulsen, J.S., Brashers-Krug, T., Miller, A., Kerr, J., Dubinsky, R.M., Gray, C., Factor, S.A., Sperin, E., Molho, E., Eglow, M., Evans, S., Kumar, R., Reeves, C., Samii, A., Chouinard, S., Beland, M., Scott, B.L., Hickey, P.T., Esmail, S., Fung, W.L.A., Gibbons, C., Qi, L., Colcher, A., Hackmyer, C., McGarry, A., Klos, K., Gudesblatt, M., Fafard, L., Graffitti, L., Schneider, D.P., Dhall, R., Wojcieszek, J.M., Faver, K.L., Duker, A., Neefus, E., Wilson-Perez, H., Shprecher, D., Wall, P., Blindauer, K.A., Wheeler, L., Boyd, J.T., Houston, E., Farbman, E.S., Agarwal, P., Eberly, S.W., Watts, A., Tariot, P.N., Feigin, A., Evans, S., Beck, C., Orme, C., Edicola, J., Christopher, E., 2016. Effect of deutetrabenazine on chorea among patients with huntington disease: A randomized clinical trial. JAMA. 15383598. 316 (1), 40–50. doi:10.1001/jama.2016.8655.

Frank, S., Testa, C.M., Stamler, D., Kayson, E., Oakes, D., Vaughan, C., Goldstein, J., Whaley, J., Davis, M.D., Gordon, M.F., Savola, J.-M., 2021. Long-Term Efficacy and Safety of Deutetrabenazine for Chorea in Huntington's Disease: results From the ARC-HD Open-label Study. CNS Spectr. 1092-8529. 26 (2), 164–165. doi:10.1017/s1092852920002655.

Fritz, N.E., Rao, A.K., Kegelmeyer, D., Kloos, A., Busse, M., Hartel, L., Carrier, J., Quinn, L., 2017. Physical Therapy and Exercise Interventions in Huntington's Disease: a Mixed Methods Systematic Review. J. Huntingtons Dis. 18796397. 6 (3), 217–235. doi:10.3233/jhd-170260.

Fukuoka, M., Takahashi, M., Fujita, H., Tomoko Chiyo, H.A.P., Watanabe, S., Furuya, H., Murata, M., Wada, K., Okada, T., Nagai, Y., Hohjoh, H., 2018. Supplemental Treatment for Huntington's Disease with miR-132 that Is Deficient in Huntington's Disease Brain. Mol. Ther. Nucleic Acids. 21622531. 11, 79–90. doi:10.1016/j.omtn.2018.01.007.

Furby, H., Siadimas, A., Rutten-Jacobs, L., Rodrigues, F.B., Wild, E.J., 2022. Natural history and burden of Huntington's disease in the UK: a population-based cohort study. Eur. J. Neurol. 14681331. 29 (8), 2249–2257. doi:10.1111/ene.15385.

Fusilli, C., Migliore, S., Mazza, T., Consoli, F., De Luca, A., Barbagallo, G., Ciammola, A., Gatto, E.M., Cesarini, M., Etcheverry, J.L., Parisi, V., Al-Oraimi, M., Al-Harrasi, S., Al-Salmi, Q., Marano, M., Vonsattel, J.P.G., Sabatini, U., Landwehrmeyer, G.B., Squitieri, F., 2018. Biological and clinical manifestations of juvenile Huntington's disease: a retrospective analysis. Lancet Neurol. 14744465. 17 (11), 986–993. doi:10.1016/S1474-4422(18)30294-1.

Futter, M., Diekmann, H., Schoenmakers, E., Sadiq, O., Chatterjee, K., Rubinsztein, D.C., 2009. Wild-type but not mutant huntingtin modulates the transcriptional activity of liver X receptors. J. Med. Genet. 0022-2593. 46 (7), 438–446. doi:10.1136/jmg.2009.066399.

Gagliardi, M., Ashizawa, A.T., 2021. The challenges and strategies of antisense oligonucleotide drug delivery. Biomedicines. 22279059. 9 (4). doi: 10.3390/biomedicines9040433.

Garriga-Canut, M., Agustín-Pavón, C., Herrmann, F., Sánchez, A., Dierssen, M., Fillat, C., Isalan, M., 2012. Synthetic zinc finger repressors reduce mutant huntingtin expression in the brain of R6/2 mice. Proc. Natl. Acad. Sci. U. S. A. 0027-8424. 109 (45). doi: 10.1073/pnas.1206506109.

Gatto, E.M., Rojas, N.G., Persi, G., Etcheverry, J.L., Cesarini, M.E., Perandones, C., 2020. Huntington disease: advances in the understanding of its mechanisms. Clin. Park. Relat. Disord. 25901125. 3. doi: 10.1016/j.prdoa.2020.100056.

Gauthier, L.R., Charrin, B.C., Borrell-Pagès, M., Dompierre, J.P., Rangone, H., Cordelières, F.P., De Mey, J., MacDonald, M.E., Leßmann, V., Humbert, S., Saudou, F., 2004. Huntingtin controls neurotrophic support and survival of neurons by enhancing BDNF vesicular transport along microtubules. Cell. 00928674. 118 (1), 127–138. doi: 10.1016/j.cell.2004.06.018.

Germain, N.D., Chung, W.K., Sarmiere, P.D., 2023. RNA interference (RNAi)-based therapeutics for treatment of rare neurologic diseases. Mol. Aspects Med. 00982997. 91, 101148. doi:10.1016/j.mam.2022.101148.

Gervais, F.G., Singaraja, R., Xanthoudakis, S., Gutekunst, C.A., Leavitt, B.R., Metzler, M., Hackam, A.S., Tam, J., Vaillancourt, J.P., Houtzager, V., Rasper, D.M., Roy, S., Hayden, M.R., Nicholson, D.W., 2002. Recruitment and activation of caspase-8 by the Huntingtin-interacting protein Hip-1 and a novel partner Hippi. Nat. Cell Biol. 14657392. 4 (2), 95–105. doi:10.1038/ncb735.

Goold, R., Flower, M., Moss, D.H., Medway, C., Wood-Kaczmar, A., Andre, R., Farshim, P., Bates, G.P., Holmans, P., Jones, L., Tabrizi, S.J., 2019. FAN1 modifies Huntington's disease progression by stabilizing the expanded HTT CAG repeat. Hum. Mol. Genet. 14602083. 28 (4), 650–661. doi:10.1093/hmg/ddy375.

Gulumkar, V.N., Dowdy, S.F., 2024. Stereopure ASOs: an unanticipated increase in selectivity for targeting mutant HTT. Mol. Ther. Nucleic Acids. 21622531. 35 (3), 102312. doi:10.1016/j.omtn.2024.102312.

Gunawardena, S., Goldstein, L.S.B., 2005Polyglutamine diseases and tansport problems. Arch. Neurol. 0003-9942. 62 (1) 46. doi: 10.1001/archneur.62.1.46.

Guo, C., Xiaoteng, M., Gao, F., Guo, Y., 2023. Off-target effects in CRISPR/Cas9 gene editing. Front. Bioeng. Biotechnol. 2296-4185. 11. doi:10.3389/fbioe.2023.1143157.

Guo, C.Y., Sun, L., Chen, X.P., Zhang, D.S., 2013. Oxidative stress, mitochondrial damage and neurodegenerative diseases. Neural Regen. Res. 18767958. 8 (21), 2003–2014. doi:10.3969/j.issn.1673-5374.2013.21.009.

Guo, Q., Huang, B., Cheng, J., Seefelder, M., Engler, T., Pfeifer, G., Oeckl, P., Otto, M., Moser, F., Maurer, M., Pautsch, A., Baumeister, W., Fernández-Busnadiego, R., Kochanek, S., 2018. The cryo-electron microscopy structure of huntingtin. Nature. 0028-0836. 555 (7694), 117–120. doi: 10.1038/nature25502.

Guttenplan, K.A., Liddelow, S.A., 2018. Play it again, SAM: macrophages control peripheral fat metabolism. Trends Immunol. 14714981. 39 (2), 81–82. doi: 10.1016/j.it.2017.12.004.

Guzman-Martinez, L., Maccioni, R.B., Andrade, V., Navarrete, L.P., Pastor, M.G., Ramos-Escobar, N., 2019. Neuroinflammation as a common feature of neurodegenerative disorders. Front. Pharmacol. 16639812, 10. doi: 10.3389/fphar.2019.01008.

Hammoud, N., Jankovic, J.2022. Botulinum Toxin in the Treatment of Cervical Dystonia: Evidence-Based Review . Dystonia. 2813-2106. 1. doi:10.3389/dyst.2022.10655.

Han, B., Liang, W., Xiao-Jiang, L., Shihua, L., Yan, S., Zhuchi, T., 2024. Large animal models for Huntington's disease research. Zoolog. Res. 45 (2), 275–283. doi:10.24272/j.issn.2095-8137.2023.199.

Han, I., You, Y., Kordower, J.H., Brady, S.T., Morfini, G.A., 2010. Differential vulnerability of neurons in Huntington's disease: the role of cell type-specific features. J. Neurochem. 14714159. 113 (5), 1073–1091. doi:10.1111/j.1471-4159.2010.06672.x.

Han, R., Liang, J., Zhou, B., 2021. Glucose Metabolic Dysfunction in Neurodegenerative Diseases—New Mechanistic Insights and the Potential of Hypoxia as a Prospective Therapy Targeting Metabolic Reprogramming. Int. J. Mol. Sci. 1422-0067. 22 (11), 5887. doi:10.3390/ijms22115887.

Hansen, T., Thant, C., White, J.A., Banerjee, R., Thuamsang, B., Gunawardena, S., 2019. Excess active P13K rescues huntingtin-mediated neuronal cell death but has no effect on axonal transport defects. Apoptosis. 1573675X. 24 (3-4), 341–358. doi:10.1007/s10495-019-01520-4.

Harding, R.J., Tong, Y.F., 2018. Proteostasis in Huntington's disease: disease mechanisms and therapeutic opportunities. Acta Pharmacol. Sin. 17457254. 39 (5), 754–769. doi:10.1038/aps.2018.11.

Harjes, P., Wanker, E.E., 2003. The hunt for huntingtin function: interaction partners tell many different stories. Trends Biochem. Sci. 09680004. 28 (8), 425–433. doi:10.1016/S0968-0004(03)00168-3.

Harper, S.Q., Staber, P.D., He, X., Eliason, S.L., Martins, I.H., Mao, Q., Yang, L., Kotin, R.M., Paulson, H.L., Davidson, B.L., 2005. RNA interference improves motor and neuropathological abnormalities in a Huntington's disease mouse model. Proc. Natl. Acad. Sci. U. S. A. 102 (16), 5820–5825. doi:10.1073/pnas.0501507102.

Harry, G.J., Kraft, A.D., 2008. Neuroinflammation and microglia: considerations and approaches for neurotoxicity assessment. Expert Opin. Drug Metab. Toxicol. 17425255. 4 (10), 1265–1277 doi:10.1517/17425255.4.10.1265.

Hauser, R.A., Factor, S.A., Marder, S.R., Knesevich, M.A., Ramirez, P.M., Jimenez, R., Burke, J., Liang, G.S., O'Brien, C.F., 2017. KINECT 3: a phase 3 randomized, double-blind, placebo-controlled trial of valbenazine for tardive dyskinesia. Am. J. Psychiatry. 15357228 174(5), 476–484. doi:10.1176/appi.ajp.2017.16091037.

Heng, M.Y., Detloff, P.J., Wang, P.L., Tsien, J.Z., Albin, R.L., 2009. In vivo evidence for NMDA receptor-mediated excitotoxicity in a murine genetic model of huntington disease. J. Neurosci. 15292401. 29 (10), 3200–3205. doi:10.1523/JNEUROSCI.5599-08.2009.

Hickman, R.A., Faust, P.L., Marder, K., Yamamoto, A., Vonsattel, J.P., 2022. The distribution and density of Huntingtin inclusions across the Huntington disease neocortex: regional correlations with Huntingtin repeat expansion independent of pathologic grade. Acta Neuropathol. Commun. 20515960. 10 (1). doi:10.1186/s40478-022-01364-1.

Hickman, S., Izzy, S., Sen, P., Morsett, L., Khoury, J.E., 2018. Microglia in neurodegeneration. Nat. Neurosci. 1097-6256. 21 (10), 1359–1369. doi:10.1038/s41593-018-0242-x.

Hoffner, G., Kahlem, P., Djian, P., 2002. Perinuclear localization of huntingtin as a consequence of its binding to microtubules through an interaction with β-tubulin: relevance to Huntington's disease. J. Cell Sci. 00219533. 115 (5), 941–948. doi:10.1242/jcs.115.5.941.

Holmans, P.A., Massey, T.H., Jones, L., 2017. Genetic modifiers of Mendelian disease: huntington's disease and the trinucleotide repeat disorders. Hum. Mol. Genet. 14602083. 26 (2), R83. doi: 10.1093/hmg/ddx261.

Houge, G., Bruland, O., Bjørnevoll, I., Hayden, M.R., Semaka, A., 2013. De novo huntington disease caused by 26-44 CAG repeat expansion on a low-risk haplotype. Neurology. 1526632X. 81 (12), 1099–1100. doi:10.1212/WNL.0b013e3182a4a4af.

Hsiao, H.Y., Chen, Y.C., Chen, H.M., Tu, P.H., Chern, Y., 2013. A critical role of astrocyte-mediated nuclear factor-κB-dependent inflammation in huntington's disease. Hum. Mol. Genet. 14602083. 22 (9), 1826–1842. doi:10.1093/hmg/ddt036.

Huntington, G., 1872. On Chorea. In: The Medical and Surgical Reporter. pp. 317–321.

Huynh, K., Nategh, L., Jamadar, S., Stout, J., Georgiou-Karistianis, N., Lampit, A., 2023. Cognition-oriented treatments and physical exercise on cognitive function in Huntington's disease: a systematic review. J. Neurol. 0340-5354. 270 (4), 1857–1879. doi:10.1007/s00415-022-11516-x.

Hwang, Y.S., Sungyang, J., Kim, G.-H., Lee, J.-Y., Ryu, H.-S., Eungseok, O., Lee, S.-H., Kim, Y.S., Chung, S.J., 2024. Clinical and Genetic Characteristics Associated With Survival Outcome in Late-Onset Huntington's Disease in South Korea. J. Clin. Neurol. 1738-6586. 20 (4), 394. doi:10.3988/jcn.2023.0329.

Iwamoto, N., Liu, Y., Frank-Kamenetsky, M., Maguire, A., Tseng, W.C., Taborn, K., Kothari, N., Akhtar, A., Bowman, K., Shelke, J.D., Lamattina, A., Xiao Shelley, H., Jang, H.G., Kandasamy, P., Liu, F., Longo, K., Meena, R.L., Metterville, J., Pan, Q., Purcell-Estabrook, E., Shimizu, M., Prakasha, P.S., Standley, S., Upadhyay, H., Yang, H., Yin, Y., Zhao, A., Francis, C., Byrne, M., Dale, E., Verdine, G.L., Vargeese, C., 2024. Preclinical evaluation of stereopure antisense oligonucleotides for allele-selective lowering of mutant HTT. Mol. Ther. Nucleic Acids. 21622531. 35 (3), 102246. doi:10.1016/j.omtn.2024.102246.

Jain, N., Chen-Plotkin, A.S., 2018. Genetic Modifiers in Neurodegeneration. Curr. Genet. Med. Rep. 2167-4876. 6 (1), 11–19. doi:10.1007/s40142-018-0133-1.

Jansen, A.H.P., van Hal, M., op den Kelder, I.C., Meier, R.T., de Ruiter, A.A., Schut, M.H., Smith, D.L., Grit, C., Brouwer, N., Kamphuis, W., Boddeke, H.W.G.M., den Dunnen, W.F.A., van Roon, W.M.C., Bates, G.P., Hol, E.M., Reits, E.A., Frequency of nuclear mutant huntingtin inclusion formation in neurons and glia is cell-type-specific. GLIA. 10981136. 65 (1) (2017), 50–61. doi:10.1002/glia.23050.

Jia, Q., Li, S., Li, X.J., Yin, P., 2022Neuroinflammation in Huntington's disease: from animal models to clinical therapeutics. Front. Immunol. 16643224. 13. doi:10.3389/fimmu.2022.1088124.

Jiang, A., Handley, R.R., Lehnert, K., Snell, R.G., 2023. From Pathogenesis to Therapeutics: a Review of 150 Years of Huntington's Disease Research. Int. J. Mol. Sci. 1422-0067. 24 (16), 13021. doi:10.3390/ijms241613021.

Jiang, A., You, L., Handley, R.R., Hawkins, V., Reid, S.J., Jacobsen, J.C., Patassini, S., Rudiger, S.R., McLaughlan, C.J., Kelly, J.M., Verma, P.J., Bawden, C.S., Gusella, J.F., Macdonald, M.E., Waldvogel, H.J., Faull, R.L.M., Lehnert, K., Snell, R.G., 2024Single nuclei RNA-seq reveals a medium spiny neuron glutamate excitotoxicity signature prior to the onset of neuronal death in an ovine Huntington's disease model. Hum. Mol. Genet. 14602083. 33 (17), 1524–1539. doi:10.1093/hmg/ddae087.

Jiang, Z., Dalby, P.A., Challenges in scaling up AAV-based gene therapy manufacturing. Trends Biotechnol. 18793096. 41 (10) (2023), 1268–1281. doi:10.1016/j.tibtech.2023.04.002.

Johnson, E.B., Ziegler, G., Penny, W., Rees, G., Tabrizi, S.J., Scahill, R.I., Gregory, S.2021, Dynamics of Cortical Degeneration Over a Decade in Huntington's Disease. Biol. Psychiatry. 18732402. 89 (8), 807–816. doi:10.1016/j.biopsych.2020.11.009.

Jongen, P.J.H., Renier, W.O., Gabreëls, F.J.M., 1980. Seven cases of huntington's disease in childhood and levodopa induced improvement in the hypokinetic — rigid form. Clin. Neurol. Neurosurg. 03038467. 82 (4), 251–261. doi:10.1016/0303-8467(80)90017-7.

Jung, T., Shin, B., Tamo, G., Kim, H., Vijayvargia, R., Leitner, A., Marcaida, M.J., Astorga-Wells, J., Jung, R., Aebersold, R., Peraro, M.D., Hebert, H., Seong, I.S., Song, J.J., 2020. The Polyglutamine Expansion at the N-Terminal of Huntingtin Protein Modulates the Dynamic Configuration and Phosphorylation of the C-Terminal HEAT Domain. Structure. 18784186. 28 (9, 1035. doi:10.1016/j.str.2020.06.008.

Jurcau, A., 2021. Insights into the Pathogenesis of Neurodegenerative Diseases: focus on Mitochondrial Dysfunction and Oxidative Stress. Int. J. Mol. Sci. 1422-0067. 22 (21), 11847. doi:10.3390/ijms222111847.

Jurcau, A., 2022. Molecular Pathophysiological Mechanisms in Huntington's Disease. Biomedicines. 2227-9059. 10 (6), 1432. doi:10.3390/biomedicines10061432.

Jurcau, A., Nunkoo, V.S., 2021. Clinical Markers May Identify Patients at Risk for Early Parkinson's Disease Dementia: a Prospective Study. Am. J. Alzheimers Dis. Other Demen. 19382731. 36. doi:10.1177/15333175211021369.

Kaliszewski, M., Knott, A.B., Bossy-Wetzel, E., 2015. Primary cilia and autophagic dysfunction in Huntington's disease. Cell Death Differ. 1350-9047. 22 (9), 1413–1424. doi:10.1038/cdd.2015.80.

Kang, R., Wang, L., Sanders, S.S., Zuo, K., Hayden, M.R., Raymond, L.A., 2019Altered regulation of striatal neuronal N-methyl-D-aspartate receptor trafficking by palmitoylation in Huntington disease mouse model. Front. Synaptic Neurosci. 16633563. 11. doi:10.3389/fnsyn.2019.00003.

Karagas, N.E., Rocha, N.P., Stimming, E.F., 2020. Irritability in Huntington's Disease. J. Huntingtons Dis. 18796400. 9 (2), 107–113. doi:10.3233/JHD-200397.

Kashif, M.A.B., Mahmood, S., Saqib, T., Waheed, S.T., Kumar, P., Javaid, A., Riaz, M.A., Fatima, U., Nadeem, Z.A., Nasir, S.A., Hassan, A., 2024. Huntington's Disease-Related Mortality Patterns: a Two-Decade Analysis of Mortality Trends in the United States, from 1999–2019. J. Huntingtons Dis. 18796400. 13 (3), 329–338. doi:10.3233/JHD-240037.

Kay, C., Hayden, M.R., Leavitt, B.R., 2017. Epidemiology of Huntington disease. Handb. Clin. Neurol. 144, 31–46. doi:10.1016/B978-0-12-801893-4.00003-1.

Kedaigle, A.J., Fraenkel, E., Atwal, R.S., Wu, M., Gusella, J.F., MacDonald, M.E., Kaye, J.A., Finkbeiner, S., Mattis, V.B., Tom, C.M., Svendsen, C., King, A.R., Chen, Y., Stocksdale, J.T., Lim, R.G., Casale, M., Wang, P.H., Thompson, L.M., Akimov, S.S., Ratovitski, T., Arbez, N., Ross, C.A., 2021Bioenergetic deficits in Huntington's disease iPSC-derived neural cells and rescue with glycolytic metabolites. Hum. Mol. Genet. 14602083. 29 (11), 1757–1771. doi:10.1093/HMG/DDY430.

Keeler, A.M., Sapp, E., Chase, K., Sottosanti, E., Danielson, E., Pfister, E., Stoica, L., DiFiglia, M., Aronin, N., Sena-Esteves, M., 2016Cellular Analysis of Silencing the Huntington's Disease Gene Using AAV9 Mediated Delivery of Artificial Micro RNA into the Striatum of Q140/Q140 Mice. J. Huntingtons Dis. 18796400. 5 (3), 239–248. doi:10.3233/JHD-160215.

Kells, A.P., Fong, D.M., Dragunow, M., During, M.J., Young, D., Connor, B., 2004. AAV-mediated gene delivery of BDNF or GDNF is neuroprotective in a model of Huntington disease. Mol. Ther. 15250016. 9 (5), 682–688. doi:10.1016/j.ymthe.2004.02.016.

Keum, J.W., Shin, A., Gillis, T., Mysore, J.S., Elneel, K.A., Lucente, D., Hadzi, T., Holmans, P., Jones, L., Orth, M., Kwak, S., Macdonald, M.E., Gusella, J.F., Lee, J.M., 2016. The HTT CAG-Expansion Mutation Determines Age at Death but Not Disease Duration in Huntington Disease. Am. J. Hum. Genet. 15376605. 98 (2), 287–298. doi:10.1016/j.ajhg.2015.12.018.

Khakh, B.S., Beaumont, V., Cachope, R., Munoz-Sanjuan, I., Goldman, S.A., Grantyn, R., 2017. Unravelling and Exploiting Astrocyte Dysfunction in Huntington's Disease. Trends Neurosci. 1878108X. 40 (7), 422–437. doi:10.1016/j.tins.2017.05.002.

Khorkova, O., Wahlestedt, C., 2017. Oligonucleotide therapies for disorders of the nervous system. Nat. Biotechnol. 1087-0156. 35 (3), 249–263. doi:10.1038/nbt.3784.

Khoshnan, A., Patterson, P.H., 2011. The role of IκB kinase complex in the neurobiology of Huntington's disease. Neurobiol. Dis. 1095953X. 43 (2), 305–311. doi:10.1016/j.nbd.2011.04.015.

Kim, H.-J., Shin, C.-W., Jeon, B., Park, H., 2016. Survival of Korean Huntington's Disease Patients. Journal of Movement Disorders. 2005-940X. 9 (3), 166–170. doi:10.14802/jmd.16022.

Kim, K.H., Hong, E.P., Lee, Y., McLean, Z.L., Elezi, E., Lee, R., Kwak, S., McAllister, B., Massey, T.H., Lobanov, S., Holmans, P., Orth, M., Ciosi, M., Monckton, D.G., Long, J.D., Lucente, D., Wheeler, V.C., MacDonald, M.E., Gusella, J.F., Lee, J.M., 2024. Posttranscriptional regulation of FAN1 by miR-124-3p at rs3512 underlies onset-delaying genetic modification in Huntington's disease. Proc. Natl. Acad. Sci. U. S. A. 10916490. 121 (16). doi:10.1073/pnas.2322924121.

Koch, E.T., Raymond, L.A., 2019. Dysfunctional striatal dopamine signaling in Huntington's disease. Journal of Neuroscience Research. 10974547. 97 (12), 1636–1654. doi:10.1002/jnr.24495.

Kojer, K., Hering, T., Bazenet, C., Weiss, A., Herrmann, F., Taanman, J.W., Orth, M., 2019. Huntingtin Aggregates and Mitochondrial Pathology in Skeletal Muscle but not Heart of Late-Stage R6/2 Mice. J. Huntingtons Dis. 18796400. 8 (2), 145–159. doi:10.3233/JHD-180324.

Kordasiewicz, H.B., Stanek, L.M., Wancewicz, E.V., Mazur, C., McAlonis, M.M., Pytel, K.A., Artates, J.W., Weiss, A., Cheng, S.H., Shihabuddin, L.S., Hung, G., Bennett, C.F., Cleveland, D.W., 2012. Sustained Therapeutic Reversal of Huntington's Disease by Transient Repression of Huntingtin Synthesis. Neuron. 10974199. 74 (6), 1031–1044. doi:10.1016/j.neuron.2012.05.009.

Koutsis, G., Karadima, G., Kladi, A., Panas, M., 2014. Late-onset Huntington's disease: diagnostic and prognostic considerations. Parkinsonism Relat. Disord. 13538020. 20 (7), 726–730. doi:10.1016/j.parkreldis.2014.03.017.

Kumar, P., Kalonia, H., Kumar, A., 2010. Huntington's disease: pathogenesis to animal models. Pharmacological Reports. 22995684. 62 (1), 1–14. doi:10.1016/S1734-1140(10)70238-3.

Landles, C., Bates, G.P., 2004. Huntingtin and the molecular pathogenesis of Huntington's disease. EMBO Rep. 1469-221X. 5 (10), 958–963. doi: 10.1038/sj.embor.7400250.

Leavitt, B.R., Van Raamsdonk, J.M., Shehadeh, J., Fernandes, H., Murphy, Z., Graham, R.K., Wellington, C.L., Raymond, L.A., Hayden, M.R., 2006. Wild-type huntingtin protects neurons from excitotoxicity. J. Neurochem. 14714159. 96 (4), 1121–1129. doi: 10.1111/j.1471-4159.2005.03605.x.

Lee, H.G., Wheeler, M.A., Quintana, F.J., 2022Function and therapeutic value of astrocytes in neurological diseases. Nat. Rev. Drug Discov. 14741784. 21 (5), 339–358. doi: 10.1038/s41573-022-00390-x.

Lee, J.-M., Wheeler, V.C., Chao, M.J., Vonsattel, J.P.G., Pinto, R.M., Lucente, D., Abu-Elneel, K., Ramos, E.M., Mysore, J.S., Gillis, T., MacDonald, M.E., Gusella, J.F., Harold, D., Stone, T.C., Escott-Price, V., Han, J., Vedernikov, A., Holmans, P., Jones, L., Kwak, S., Mahmoudi, M., Michael Orth, G.B.L., Jane, S.P., Dorsey, E.R., Shoulson, I., Richard, H.M., 2015. Identification of genetic factors that modify clinical onset of huntington's disease. Cell. 00928674. 162 (3), 516–526. doi: 10.1016/j.cell.2015.07.003.

Lee, J.M., Correia, K., Loupe, J., Kim, K.H., Barker, D., Hong, E.P., Chao, M.J., Long, J.D., Lucente, D., Vonsattel, J.P.G., Pinto, R.M., Elneel, K.A., Ramos, E.M., Mysore, J.S., Gillis, T., Wheeler, V.C., MacDonald, M.E., Gusella, J.F., McAllister, B., Massey, T., Medway, C., Stone, T.C., Hall, L., Jones, L., Holmans, P., Kwak, S., Ehrhardt, A.G., Sampaio, C., Ciosi, M., Maxwell, A., Chatzi, A., Monckton, D.G., Orth, M., Landwehrmeyer, G.B., Paulsen, J.S., Dorsey, E.R., Shoulson, I., Myers, R.H., 2019. CAG repeat, not polyglutamine length, determines timing of Huntington's disease onset. Cell. 10974172. 178 (4), 887. doi: 10.1016/j.cell.2019.06.036.

Lee, W., Reyes, R.C., Gottipati, M.K., Lewis, K., Lesort, M., Parpura, V., Gray, M., 2013. Enhanced Ca2+-dependent glutamate release from astrocytes of the BACHD Huntington's disease mouse model. Neurobiol. Dis. 1095953X. 58, 192–199. doi: 10.1016/j.nbd.2013.06.002.

Li, S.H., Cheng, A.L., Zhou, H., Lam, S., Rao, M., Li, H., Li, X.J., 2002. Interaction of Huntington disease protein with transcriptional activator Sp1. Mol. Cell. Biol. 22 (5), 1277–1287. doi: 10.1128/MCB.22.5.1277-1287.2002.

Li, S.H., Li, X.J., 2004. Huntingtin-protein interactions and the pathogenesis of Huntington's disease. Trends Genet. 01689525. 20 (3), 146–154. doi: 10.1016/j.tig.2004.01.008.

Liao, H., Wu, J., VanDusen, N.J., Li, Y., Zheng, Y., 2024. CRISPR-Cas9-mediated homology-directed repair for precise gene editing. Mol. Ther. Nucleic Acids. 21622531. 35 (4), 102344. doi: 10.1016/j.omtn.2024.102344.

Liddelow, S.A., Barres, B.A., 2017. Reactive astrocytes: production, function, and therapeutic potential. Immunity. 10974180. 46 (6), 957–967. doi: 10.1016/j.immuni.2017.06.006.

Liddelow, S.A., Guttenplan, K.A., Clarke, L.E., Bennett, F.C., Bohlen, C.J., Schirmer, L., Bennett, M.L., Münch, A.E., Chung, W.S., Peterson, T.C., Wilton, D.K., Frouin, A., Napier, B.A., Panicker, N., Kumar, M., Buckwalter, M.S., Rowitch, D.H., Dawson, V.L., Dawson, T.M., Stevens, B., Barres, B.A., 2017. Neurotoxic reactive astrocytes are induced by activated microglia. Nature. 14764687. 541 (7638), 481–487. doi: 10.1038/nature21029.

Liévens, J.-C., Woodman, B., Mahal, A., Spasic-Boscovic, O., Samuel, D., Goff, -L.K.-L., Bates, G.P., 2001. Impaired glutamate uptake in the R6 Huntington's disease transgenic mice. Neurobiol. Dis. 09699961. 8 (5), 807–821. doi: 10.1006/nbdi.2001.0430.

Lin, L., Su, F., Chen, D., Pei, Z., 2022. The first patient with sporadic Huntington's disease due to a de novo (CAG)n expansion in China. J. Transl. Intern. Med. 22244018. 10 (1), 76–78. doi: 10.2478/jtim-2022-0006.

Ling, S., Zhang, X., Dai, Y., Jiang, Z., Zhou, X., Sicong, L., Qian, X., Liu, J., Selfjord, N., Satir, T.M., Lundin, A., Touza, J.L., Firth, M., Van Zuydam, N., Bilican, B., Akcakaya, P., Hong, J., Cai, Y., 2025. Customizable virus-like particles deliver CRISPR–Cas9 ribonucleoprotein for effective ocular neovascular and Huntington's disease gene therapy. Nat. Nanotechnol. 1748-3387. 20 (4), 543–553. doi: 10.1038/s41565-024-01851-7.

Liu, D., Long, J.D., Zhang, Y., Raymond, L.A., Marder, K., Rosser, A., McCusker, E.A., Mills, J.A., Paulsen, J.S., 2015. Motor onset and diagnosis in Huntington disease using the diagnostic confidence level. J. Neurol. 14321459. 262 (12), 2691–2698. doi: 10.1007/s00415-015-7900-7.

Liu, H., Zhang, C., Jiadi, X., Jin, J., Cheng, L., Miao, X., Qian, W., Wei, Z., Liu, P., Hanzhang, L., van Zijl, P.C.M., Ross, C.A., Hua, J., Duan, W., 2021. Huntingtin silencing delays onset and slows progression of Huntington's disease: a biomarker study. Brain. 0006-8950. 144 (10), 3101–3113. doi: 10.1093/brain/awab190.

Lo Sardo, V., Zuccato, C., Gaudenzi, G., Vitali, B., Ramos, C., Tartari, M., Myre, M.A., Walker, J.A., Pistocchi, A., Conti, L., Valenza, M., Drung, B., Schmidt, B., Gusella, J., Zeitlin, S., Cotelli, F., Cattaneo, E., 2012. An evolutionary recent neuroepithelial cell adhesion function of huntingtin implicates ADAM10-Ncadherin. Nat. Neurosci. 15 (5), 713–721. doi: 10.1038/nn.3080.

Lopes, C., Tang, Y., Anjo, S.I., Manadas, B., Onofre, I., de Almeida, L.P., Daley, G.Q., Schlaeger, T.M., Rego, A.C.C., 2020. Mitochondrial and Redox Modifications in Huntington Disease Induced Pluripotent Stem Cells Rescued by CRISPR/Cas9 CAGs Targeting. Front. Cell Dev. Biol. 2296-634X. 8, doi: 10.3389/fcell.2020.576592.

López-Doménech, G., Covill-Cooke, C., Ivankovic, D., Halff, E.F., Sheehan, D.F., Norkett, R., Birsa, N., Kittler, J.T., 2018. Miro proteins coordinate microtubule- and actin-dependent mitochondrial transport and distribution. EMBO J. 14602075. 37 (3), 321–336. doi: 10.15252/embj.201696380.

Lucetti, C., Gambaccini, G., Bernardini, S., Dell'Agnello, G., Petrozzi, L., Rossi, G., 2002. Amantadine in Huntington's disease: open-label video-blinded study. Neurol. Sci. 23 (2), S83–S84.

Luo, S., Rubinsztein, D.C.2009, Huntingtin promotes cell survival by preventing Pak2 cleavage. J. Cell Sci. 00219533. 122 (6), 875–885. doi: 10.1242/jcs.050013.

MacDonald, M.E., Ambrose, C.M., Duyao, M.P., Myers, R.H., Lin, C., Srinidhi, L., Barnes, G., Taylor, S.A., James, M., Groot, N., MacFarlane, H., Jenkins, B., Anderson, M.A., Wexler, N.S., Gusella, J.F., Bates, G.P., Baxendale, S., Hummerich, H., Kirby, S., North, M., Youngman, S., Mott, R., Zehetner, G., Sedlacek, Z., Poustka, A., Frischauf, A.M., Lehrach, H., Buckler, A.J., Church, D., Doucette-Stamm, L., O'Donovan, M.C., Riba-Ramirez, L., Shah, M., Stanton, V.P., Strobel, S.A., Draths, K.M., Wales, J.L., Dervan, P., Housman, D.E., Altherr, M., Shiang, R., Thompson, L., Fielder, T., Wasmuth, J.J., Tagle, D., Valdes, J., Elmer, L., Allard, M., Castilla, L., Swaroop, M., Blanchard, K., Collins, F.S., Snell, R., Holloway, T., Gillespie, K., Datson, N., Shaw, D., Harper, P.S., 1993. A novel gene containing a trinucleotide repeat that is expanded and unstable on Huntington's disease chromosomes. Cell. 00928674. 72 (6), 971–983. doi:10.1016/0092-8674(93)90585-E.

Mackay, J.P., Nassrallah, W.B., Raymond, L.A., 2018. Cause or compensation?—Altered neuronal Ca2+ handling in Huntington's disease. CNS Neurosci. Ther. 17555949. 24 (4), 301–310. doi:10.1111/cns.12817.

Magnet, M.K., Bonelli, R.M., Kapfhammer, H.P., 2004. Amantadine in the akinetic-rigid variant of Huntington's disease. Ann. Pharmacother. 10600280. 38 (7-8), 1194–1196. doi:10.1345/aph.1E004.

Mammucari, C., Raffaello, A., Reane, D.V., Gherardi, G., De Mario, A., Rizzuto, R., 2018. Mitochondrial calcium uptake in organ physiology: from molecular mechanism to animal models. Pflugers Arch. 0031-6768. 470 (8), 1165–1179. doi:10.1007/s00424-018-2123-2.

Mandal, M., Wei, J., Zhong, P., Cheng, J., Duffney, L.J., Liu, W., Yuen, E.Y., Twelvetrees, A.E., Li, S., Li, X.J., Kittler, J.T., Yan, Z., 2011. Impaired α-amino-3-hydroxy-5-methyl-4-isoxazolepropionic acid (AMPA) receptor trafficking and function by mutant Huntingtin. J. Biol. Chem. 1083351X. 286 (39), 33719–33728. doi:10.1074/jbc.M111.236521.

Marcora, E., Kennedy, M.B., 2010. The Huntington's disease mutation impairs Huntingtin's role in the transport of NF-κB from the synapse to the nucleus. Hum. Mol. Genet. 14602083. 19 (22), 4373–4384. doi: 10.1093/hmg/ddq358.

Martinez-Vicente, M., Talloczy, Z., Wong, E., Tang, G., Koga, H., Kaushik, S., de Vries, R., Arias, E., Harris, S., Sulzer, D., Cuervo, A.M., 2010. Cargo recognition failure is responsible for inefficient autophagy in Huntington's disease. Nat. Neurosci. 1097-6256. 13 (5), 567–576. doi: 10.1038/nn.2528.

Martinez-Horta, S., Perez-Perez, J., Perez-Gonzalez, R., Sampedro, F., Horta-Barba, A., Campolongo, A., Rivas-Asensio, E., Puig-Davi, A., Pagonabarraga, J., Kulisevsky, J., 2024. Cognitive phenotype and neurodegeneration associated with Tau in Huntington's disease. Ann. Clin. Transl. Neurol. 2328-9503. 11 (5), 1160–1171. doi: 10.1002/acn3.52031.

Mathews, G., Adetunji, G., Mathews, B., 2005. Antipsychotic-induced movement disorders: evaluation and treatment. Psychiatry (Edgmont (Pa: Townsh)) 2 (3), 36–41.

Matuszek, Z., Arbab, M., Kesavan, M., Hsu, A., Roy, J.C.L., Zhao, J., Yu, T., Weisburd, B., Newby, G.A., Doherty, N.J., Wu, M., Shibata, S., Cristian, A., Tao, Y.A., Fearnley, L.G., Bahlo, M., Rehm, H.L., Xie, J., Gao, G., Pinto, R.M., Liu, D.R., 2025a. Base editing of trinucleotide repeats that cause Huntington's disease and Friedreich's ataxia reduces somatic repeat expansions in patient cells and in mice. Nat. Genet. 15461718. doi: 10.1038/s41588-025-02172-8.

McAdam, R.L., Morton, A., Gordon, S.L., Alterman, J.F., Khvorova, A., Cousin, M.A., Smillie, K.J., 2020. Loss of huntingtin function slows synaptic vesicle endocytosis in striatal neurons from the httQ140/Q140 mouse model of Huntington's disease. Neurobiol. Dis. 1095953X. 134. doi:10.1016/j.nbd.2019.104637.

McColgan, P., Tabrizi, S.J., 2018. Huntington's disease: a clinical review. Eur. J. Neurol. 14681331. 25 (1), 24–34. doi:10.1111/ene.13413.

McColgan, P., Thobhani, A., Boak, L., Schobel, S.A., Nicotra, A., Palermo, G., Trundell, D., Zhou, J., Schlegel, V., Ducray, P.S., Hawellek, D.J., Dorn, J., Simillion, C., Lindemann, M., Wheelock, V., Durr, A., Anderson, K.E., Long, J.D., Wild, E.J., Landwehrmeyer, G.B., Leavitt, B.R., Tabrizi, S.J., Doody, R., 2023. Tominersen in Adults with Manifest Huntington's Disease. N. Engl. J. Med. 0028-4793. 389 (23), 2203–2205. doi:10.1056/nejmc2300400.

McGarry, A., Biglan, K., Marshall, F., 2024. Huntington's disease. In: Rosenberg's Molecular and Genetic Basis of Neurological and Psychiatric Disease, vol. 2, seventh ed. Academic Press, pp. 171–181. doi:10.1016/B978-0-443-19176-3.00012-1.

McLean, Z.L., Gao, D., Correia, K., Roy, J.C.L., Shibata, S., Farnum, I.N., Valdepenas-Mellor, Z., Kovalenko, M., Rapuru, M., Morini, E., Ruliera, J., Gillis, T., Lucente, D., Kleinstiver, B.P., Lee, J.M., MacDonald, M.E., Wheeler, V.C., Pinto, R.M., Gusella, J.F., 2024. Splice modulators target PMS1 to reduce somatic expansion of the Huntington's disease-associated CAG repeat. Nat. Commun. 20411723. 15 (1), doi: 10.1038/s41467-024-47485-0.

Mealer, R.G., Murray, A.J., Shahani, N., Subramaniam, S., Snyder, S.H., 2014. Rhes, a Striatal-selective Protein Implicated in Huntington Disease, Binds Beclin-1 and activates autophagy. J. Biol. Chem. 1083351X. 289 (6), 3547–3554. doi: 10.1074/jbc.M113.536912.

Medina, A., Mahjoub, Y., Shaver, L., Pringsheim, T., 2022Prevalence and Incidence of Huntington's Disease: an Updated Systematic Review and Meta-Analysis. Mov. Disord. 0885-3185. 37 (12), 2327–2335. doi: 10.1002/mds.29228.

Mendizabal, A., Diaz, J.M., Bustamante, A.V., Bordelon, Y., 2023. Health services in Huntington disease: a systematic literature review. Neurol. Clin. Pract. 21630933. 13 (1), doi: 10.1212/CPJ.0000000000200108.

Migliore, S., Jankovic, J., Squitieri, F., 2019. Genetic counseling in Huntington's disease: potential new challenges on the horizon? Front. Neurol. 1664-2295. 10, doi: 10.3389/fneur.2019.00453.

Miller, B.R., Bezprozvanny, I., 2010. Corticostriatal circuit dysfunction in Huntingtons disease: intersection of glutamate, dopamine and calcium. Future Neurol. 14796708. 5 (5), 735–756. doi: 10.2217/fnl.10.41.

Milnerwood, A.J., Gladding, C.M., Pouladi, M.A., Kaufman, A.M., Hines, R.M., Boyd, J.D., Ko, R.W.Y., Vasuta, O.C., Graham, R.K., Hayden, M.R., Murphy, T.H., Raymond, L.A., 2010. Early increase in extrasynaptic NMDA receptor signaling and expression contributes to phenotype onset in Huntington's disease mice. Neuron. 08966273. 65 (2), 178–190. doi: 10.1016/j.neuron.2010.01.008.

Miniarikova, J., Zimmer, V., Martier, R., Brouwers, C.C., Pythoud, C., Richetin, K., Rey, M., Lubelski, J., Evers, M.M., van Deventer, S.J., Petry, H., Déglon, N., Konstantinova, P., 2017. AAV5-miHTT gene therapy demonstrates suppression of mutant huntingtin aggregation and neuronal dysfunction in a rat model of Huntington's disease. Gene Ther. 0969-7128. 24 (10), 630–639. doi:10.1038/gt.2017.71.

Mochel, F., Durant, B., Meng, X., O'Callaghan, J., Yu, H., Brouillet, E., Wheeler, V.C., Humbert, S., Schiffmann, R., Durr, A., 2012. Early alterations of brain cellular energy homeostasis in huntington disease models. J. Biol. Chem. 1083351X. 287 (2), 1361–1370. doi:10.1074/jbc.M111.309849France.

Möller, T., 2010. Neuroinflammation in Huntington's disease. J. Neural Transm. 0300-9564. 117 (8), 1001–1008. doi:10.1007/s00702-010-0430-7.

Monteys, A.M., Ebanks, S.A., Keiser, M.S., Davidson, B.L., 2017. CRISPR/Cas9 editing of the mutant huntingtin allele in vitro and in vivo. Mol. Ther. 15250024. 25 (1), 12–23. doi:10.1016/j.ymthe.2016.11.010.

Moondra, P., Jimenez-Shahed, J., 2024. Profiling deutetrabenazine extended-release tablets for tardive dyskinesia and chorea associated with Huntington's disease. Expert Rev. Neurother. 1473-7175. 24 (9), 849–863. doi:10.1080/14737175.2024.2376107.

Morton, A.J., 2013. Circadian and sleep disorder in Huntington's disease. Exp. Neurol. 10902430. 243, 34–44. doi:10.1016/j.expneurol.2012.10.014.

Morton, A.J., Howland, D.S., 2013. Large genetic animal models of Huntington's disease. J. Huntingtons Dis. 18796397. 2 (1), 3–19. doi:10.3233/jhd-130050.

Mühlbäck, A., Walsem, M., Nance, M., Arnesen, A., Page, K., Fisher, A., 2023. What we don't need to prove but need to do in multidisciplinary treatment and care in Huntington's disease: a position paper. Orphanet J. Rare Dis. 18 (1), 19.

Myers, R.H., Macdonald, M.E., Koroshetz, W.J., Duyao, M.P., Ambrose, C.M., Taylor, S.A.M., Barnes, G., Srinidhi, J., Lin, C.S., Whaley, W.L., Lazzarini, A.M., Schwarz, M., Wolff, G., Bird, E.D., Vonsattel, J.P.G., Gusella, J.F., 1993. De novo expansion of a (CAG)n repeat in sporadic Huntington's disease. Nat. Genet. 15461718. 5 (2), 168–173. doi:10.1038/ng1093-168.

Nance, M.A., 1998. Huntington disease: clinical, genetic, and social aspects. J. Geriatr. Psychiatry Neurol. 08919887. 11 (2), 61–70. doi:10.1177/089198879801100204.

Nasir, J., Floresco, S.B., O'Kusky, J.R., Diewert, V.M., Richman, J.M., Zeisler, J., Borowski, A., Marth, J.D., Phillips, A.G., Hayden, M.R., 1995. Targeted disruption of the Huntington's disease gene results in embryonic lethality and behavioral and morphological changes in heterozygotes. Cell. 00928674. 81 (5), 811–823. doi:10.1016/0092-8674(95)90542-1.

NCT04120493, 2019. A phase 1/2, randomized, double-blind, sham control and open-label study to explore safety, tolerability, and efficacy signals of multiple doses of striatally-administered rAAV5-miHTT total huntingtin gene (HTT) lowering therapy (AMT-130). Early Manifest Huntington's Disease. Available from: https://clinicaltrials.gov/study/NCT04120493.

NCT05243017, 2021. A phase ib/II randomized, double-blind study to explore safety, tolerability, and efficacy signals of multiple doses of striatally-administered rAAV5-miHTT total huntingtin gene (HTT). Lowering Therapy (AMT 130) in Early Manifest Huntington's Disease. Available from: https://clinicaltrials.gov/study/NCT05243017.

Noh, J.Y., Lee, H., Song, S., Kim, N.S., Im, W., Kim, M., Seo, H., Chung, C.W., Chang, J.W., Ferrante, R.J., Yoo, Y.J., Ryu, H., Jung, Y.K., 2009. SCAMP5 links endoplasmic reticulum stress to the accumulation of expanded polyglutamine protein aggregates via endocytosis inhibition. J. Biol. Chem. 1083351X. 284 (17), 11318–11325. doi:10.1074/jbc.M807620200.

O'Regan, G.C., Farag, S.H., Ostroff, G.R., Tabrizi, S.J., Andre, R., 2020. Wild-type huntingtin regulates human macrophage function. Sci. Rep. 20452322. 10 (1). doi: 10.1038/s41598-020-74042-8.

Ochaba, J., Lukacsovich, T., Csikos, G., Zheng, S., Margulis, J., Salazar, L., Mao, K., Lau, A.L., Yeung, S.Y., Humbert, S., Saudou, F., Klionsky, D.J., Finkbeiner, S., Scott, O.Z., Marsh, J.L., Housman, D.E., Thompson, L.M., Steffan, J.S., 2014. Potential function for the Huntingtin protein as a scaffold for selective autophagy. Proc. Natl. Acad. Sci. U. S. A. 0027-8424. 111 (47), 16889–16894. doi: 10.1073/pnas.1420103111.

Ogilvie, A.C., Schultz, J.L., 2023. Memantine use and cognitive decline in Huntington's disease: an Enroll-HD study. Mov. Disord. Clin. Pract. 23301619. 10 (7), 1120–1125. doi: 10.1002/mdc3.13763.

Ortiz, A.N., Kurth, B.J., Osterhaus, G.L., Johnson, M.A., 2011. Impaired dopamine release and uptake in R6/1 Huntington's disease model mice. Neurosci. Lett. 03043940. 492 (1), 11–14. doi: 10.1016/j.neulet.2011.01.036.

Pagano, G., Niccolini, F., Politis, M., 2016. Current status of PET imaging in Huntington's disease. Eur. J. Nucl. Med. Mol. Imaging. 1619-7070. 43 (6), 1171–1182. doi: 10.1007/s00259-016-3324-6.

Palaiogeorgou, A., Papakonstantinou, E., Golfinopoulou, R., Sigala, M., Mitsis, T., Papageorgiou, L., Diakou, I., Pierouli, K., Dragoumani, K., Spandidos, D., Bacopoulou, F., Chrousos, G., Eliopoulos, E., Vlachakis, D., 2022. Recent approaches on Huntington's disease. Biomed. Rep. 18 (1), 5. doi: 10.3892/br.2022.1587.

Paldino, E., 2019. Modulation of phospho-CREB by systemically administered recombinant BDNF in the hippocampus of the R6/2 mouse model of huntington's disease. Neurosci J. 2019 (1). 8363274.

Palpagama, T.H., Waldvogel, H.J., Faull, R.L.M., Kwakowsky, A., 2019. The role of microglia and astrocytes in Huntington's disease. Front. Mol. Neurosci. 16625099. 12. doi: 10.3389/fnmol.2019.00258.

Paoli, R.A., Botturi, A., Ciammola, A., Silani, V., Prunas, C., Lucchiari, C., Zugno, E., Caletti, E., 2017. Neuropsychiatric burden in huntington's disease. Brain Sci. 20763425. 7 (6). doi: 10.3390/brainsci7060067.

Parsons, M.P., Raymond, L.A., 2023. Huntington Disease. Elsevier BV, 275–292. doi: 10.1016/b978-0-323-85654-6.00042-3.

Patino, J., Stimming, E.F., Testa, C.M., Mehanna, R., 2025. Valbenazine for the treatment of chorea associated with Huntington's disease. Expert Opin. Pharmacother. 17447666. 26 (2), 127–132. doi: 10.1080/14656566.2024.2445728.

Paulsen, J.S., Miller, A.C., Hayes, T., Shaw, E., 2017. Cognitive and behavioral changes in Huntington disease before diagnosis. Handb. Clin. Neurol. 144, 69–91. doi: 10.1016/B978-0-12-801893-4.00006-7.

Pavese, N., Gerhard, A., Tai, Y.F., Ho, A.K., Turkheimer, F., Barker, R.A., Brooks, D.J., Piccini, P., 2006. Microglial activation correlates with severity in Huntington disease. Neurology. 0028-3878. 66 (11), 1638–1643. doi: 10.1212/01.wnl.0000222734.56412.17.

Pei, Z., Lin, L., Chen, X., Tengteng, W., Lin, L., Pingyi, X., Meng, X., 2024. J001 Preliminary results from the first-in-human study of ER2001, an innovative treatment for Huntington's disease using in vivo self-assembled sirna. J. Neurol. Neurosurg. Psychiatry. 95 (Suppl 1), A157 . doi: 10.1136/jnnp-2024-EHDN.322.

Pengo, M., Squitieri, F., 2024. Beyond CAG repeats: the multifaceted role of genetics in Huntington disease. Genes. 2073-4425. 15 (6), 807. doi: 10.3390/genes15060807.

Pfister, E.L., Dinardo, N., Mondo, E., Borel, F., Conroy, F., Fraser, C., Gernoux, G., Han, X., Hu, D., Johnson, E., Kennington, L., Liu, P., Reid, S.J., Sapp, E., Vodicka, P., Kuchel, T., Morton, A.J., Howland, D., Moser, R., Sena-Esteves, M., Gao, G., Mueller, C., Difiglia, M., Aronin, N., Artificial miRNAs Reduce Human Mutant Huntingtin Throughout the Striatum in a Transgenic Sheep Model of Huntington's Disease. Hum. Gene Ther. 15577422. 29 (6) (2018), 663–673, Available from: www.liebertonline.com/hum. doi: 10.1089/hum.2017.199.

Piao, X., Dan, L., Liu, H., Guo, Q., Yang, Y., Advances in gene and cellular therapeutic approaches for Huntington's disease. Protein Cell. 1674-800X. 16 (5) (2025), 307–337, doi: 10.1093/procel/pwae042.

Piira, A., van Walsem, M.R., Mikalsen, G., Øie, L., Frich, J.C., Knutsen, S., Effects of a Two-Year Intensive Multidisciplinary Rehabilitation Program for Patients with Huntington's Disease: a Prospective Intervention Study. PLoS Curr. 2157-3999. 6 (2014), doi: 10.1371/currents.hd.2c56ceef7f9f8e239a59ecf2d94cddac.

Pinto, R.M., Murtha, R., Azevedo, A., Douglas, C., Kovalenko, M., Ulloa, J., Crescenti, S., Burch, Z., Oliver, E., Kesavan, M., Shibata, S., Vitalo, A., Mota-Silva, E., Riggs, M.J., Correia, K., Elezi, E., Demelo, B., Carroll, J.B., Gillis, T., Gusella, J.F., MacDonald, M.E., Wheeler, V.C., In vivo CRISPR–Cas9 genome editing in mice identifies genetic modifiers of somatic CAG repeat instability in Huntington's disease. Nat. Genet. 1061-4036. 57 (2) (2025), 314–322, doi: 10.1038/s41588-024-02054-5.

Prasad, S., Dimmock, D.P., Greenberg, B., Walia, J.S., Sadhu, C., Tavakkoli, F., Lipshutz, G.S., Immune Responses and Immunosuppressive Strategies for Adeno-Associated Virus-Based Gene Therapy for Treatment of Central Nervous System Disorders: current Knowledge and Approaches. Hum. Gene Ther. 1043-0342. 33 (23-24) (2022), 1228–1245, doi: 10.1089/hum.2022.138.

Pringsheim, T., Gardner, D., Addington, D., Martino, D., Morgante, F., Ricciardi, L., Poole, N., Remington, G., Edwards, M., Carson, A., Barnes, T.R.E., The Assessment and Treatment of Antipsychotic-Induced Akathisia. Can. J. Psychiatry. 14970015. 63 (11) (2018), 719–729, Available from: https://journals.sagepub.com/home/cpa. doi: 10.1177/0706743718760288.

Quaid, K.A., Swenson, M.M., Sims, S.L., Harrison, J.M., Moskowitz, C., Stepanov, N., Suter, G.W., Westphal, B.J., What were you thinking? Individuals at risk for huntington disease talk about having children. J. Genet. Couns. 15733599. 19 (6) (2010), 606–617, doi: 10.1007/s10897-010-9312-2.

Quarrell, O., O'Donovan, K.L., Bandmann, O., Strong, M., The Prevalence of Juvenile Huntington's Disease: a Review of the Literature and Meta-Analysis. PLoS Curr. 4 (2012), e4f8606b742ef3.

Quinn, L., Kegelmeyer, D., Kloos, A., Rao, A.K., Busse, M., Fritz, N.E., Clinical recommendations to guide physical therapy practice for Huntington disease. Neurology. 1526632X. 94 (5) (2020), 217–228, Available from: http://www.neurology.org. doi: 10.1212/WNL.0000000000008887.

Rabinowitz, J., Chan, Y.K., Samulski, R.J., Adeno-associated Virus (AAV) versus immune response. Viruses. 19994915. 11 (2) (2019), Available from: https://www.mdpi.com/1999-4915/11/2/102/pdf. doi: 10.3390/v11020102.

Raguram, A., Banskota, S., Liu, D.R., Therapeutic in vivo delivery of gene editing agents. Cell. 10974172. 185 (15) (2022), 2806–2827, Available from: https://www.sciencedirect.com/journal/cell. doi: 10.1016/j.cell.2022.03.045.

Rathbone, J., Soares-Weiser, K., 2006. Anticholinergics for neuroleptic-induced acute akathisia. Cochrane Database Syst. Rev. 4, CD003727. doi:10.1002/14651858.CD003727.pub3.

Ratna, N., Venkatesh, S.D., Pasupulati, S., Murari, G., Kamble, N., Viswanath, B., Yadav, R., Varghese, M., Pal, P.K., Jain, S., Purushottam, M., Allelic Diversity, De Novo CAG Expansions and Intergenerational Instability at the HTT Locus in a Sample from India. Can. J. Neurol. Sci. 0317-1671. (2025), 1–7, doi: 10.1017/cjn.2025.10092.

Ravikumar, B., Vacher, C., Berger, Z., Davies, J.E., Luo, S., Oroz, L.G., Scaravilli, F., Easton, D.F., Duden, R., O'Kane, C.J., Rubinsztein, D.C., Inhibition of mTOR induces autophagy and reduces toxicity of polyglutamine expansions in fly and mouse models of Huntington disease. Nat. Genet. 36 (6) (2004), 585–595, doi: 10.1038/ng1362.

Reilmann, R., Concern about Tominersen in Patients with Huntington's Disease. N. Engl. J. Med. 15334406. 390 (11) (2024), 1058–1059, doi: 10.1056/NEJMc2400161.

Reiner, A., Dragatsis, I., Zeitlin, S., Goldowitz, D., Wild-Type Huntingtin Plays a Role in Brain Development and Neuronal Survival. Mol. Neurobiol. 0893-7648. 28 (3) (2003), 259–276, doi: 10.1385/MN:28:3:259.

Ribeiro, M., Rosenstock, T.R., Oliveira, A.M., Oliveira, C.R., Rego, A.C., Insulin and IGF-1 improve mitochondrial function in a PI-3K/Akt-dependent manner and reduce mitochondrial generation of reactive oxygen species in Huntington's disease knock-in striatal cells. Free Radical Biology and Medicine. 18734596. 74 (2014), 129–144, Available from: www.elsevier.com/locate/freeradbiomed. doi: 10.1016/j.freeradbiomed.2014.06.023.

Rigamonti, D., Bauer, J.H., De-Fraja, C., Conti, L., Sipione, S., Sciorati, C., Clementi, E., Hackam, A., Hayden, M.R., Li, Y., Cooper, J.K., Ross, C.A., Govoni, S., Vincenz, C., Cattaneo, E., Wild-type huntingtin protects from apoptosis upstream of caspase-3. J. Neurosci. 02706474. 20 (10) (2000), 3705–3713, Available from: www.jneurosci.org. doi: 10.1523/jneurosci.20-10-03705.2000.

Rigamonti, D., Sipione, S., Goffredo, D., Zuccato, C., Fossale, E., Cattaneo, E., Huntingtin's Neuroprotective Activity Occurs via Inhibition of Procaspase-9 Processing. J. Biol. Chem. 00219258. 276 (18) (2001), 14545–14548, doi: 10.1074/jbc.c100044200.

Rigby, M., Bourdellès, B.L., Heavens, R.P., Kelly, S., Smith, D., Butler, A., Hammans, R., Hills, R., Xuereb, J.H., Hill, R.G., Whiting, P.J., Sirinathsinghji, D.J.S., The messenger RNAs for the N-methyl- d-aspartate receptor subunits show region-specific expression of different subunit composition in the human brain. Neuroscience. 03064522. 73 (2) (1996), 429–447, doi: 10.1016/0306-4522(96)00089-9.

Rinaldi, C., Wood, M.J.A., Antisense oligonucleotides: the next frontier for treatment of neurological disorders. Nat. Rev. Neurol. 1759-4758. 14 (1) (2018), 9–21, doi: 10.1038/nrneurol.2017.148.

Rodrigues, F.B., Abreu, D., Damásio, J., Goncalves, N., Correia-Guedes, L., Coelho, M., Ferreira, J.J., Survival, M., Causes and Places of Death in a European Huntington's Disease Prospective Cohort. Mov. Disord. Clin. Pract. 23301619. 4 (5) (2017), 737–742, Available from: http://onlinelibrary.wiley.com/journal/10.1002/(ISSN)2330-1619. doi: 10.1002/mdc3.12502.

Rojas, N.G., Cesarini, M.E., Peker, G., Prat, G.A.D., Etcheverry, J.L., Gatto, E.M., Review of Huntington's Disease: from Basics to Advances in Diagnosis and Treatment. Journal of Neurology Research. 1923-2845. 12 (3) (2022), 93–113, doi: 10.14740/jnr721.

Rong, J., McGuire, J.R., Fang, Z.H., Sheng, G., Shin, J.Y., Li, S.H., Li, X.J., Regulation of intracellular trafficking of huntingtin-associated protein-1 is critical for TrkA protein levels and neurite outgrowth. J. Neurosci. 02706474. 26 (22) (2006), 6019–6030, Available from: http://www.jneurosci.org/cgi/reprint/26/22/6019. doi: 10.1523/JNEUROSCI.1251-06.2006.

Rook, M.E., Southwell, A.L., Antisense Oligonucleotide Therapy: from Design to the Huntington Disease Clinic. BioDrugs. 1179190X. 36 (2) (2022), 105–119, Available from: http://rd.springer.com/journal/40259. doi: 10.1007/s40259-022-00519-9.

Roos, R.A.C., Huntington's disease: a clinical review. Orphanet J. Rare Dis. 17501172. 5 (1) (2010), doi: 10.1186/1750-1172-5-40.

Ross, C.A., Aylward, E.H., Wild, E.J., Langbehn, D.R., Long, J.D., Warner, J.H., Scahill, R.I., Leavitt, B.R., Stout, J.C., Paulsen, J.S., Reilmann, R., Unschuld, P.G., Wexler, A., Margolis, R.L., Tabrizi, S.J., Huntington disease: natural history, biomarkers and prospects for therapeutics. Nat. Rev. Neurol. 17594766. 10 (4) (2014), 204–216, Available from: http://www.nature.com/nrneurol/archive/index.html. doi: 10.1038/nrneurol.2014.24.

Røthing, M., Malterud, K., Frich, J.C., Balancing needs as a family caregiver in Huntington's disease: a qualitative interview study. Health Soc. Care Community. 13652524. 23 (5) (2015), 569–576, Available from: http://onlinelibrary.wiley.com/journal/10.1111/(ISSN)1365-2524. doi: 10.1111/hsc.12174Norway.

Ruiz, P.J.G., Tortosa, E.G., Bernados, V.S., Rojo, A., Fontán, A., De Yebenes, J.G., Bradykinesia in Huntington's disease. Clinical Neuropharmacology. 03625664. 23 (1) (2000), 50–52, doi: 10.1097/00002826-200001000-00009.

Saade, J., Mestre, T.A., Huntington's Disease: latest Frontiers in Therapeutics. Curr. Neurol. Neurosci. Rep. 15346293. 24 (8) (2024), 255–264, Available from: https://www.springer.com/journal/11910. doi: 10.1007/s11910-024-01345-y.

Saba, J., Couselo, F.L., Bruno, J., Carniglia, L., Durand, D., Lasaga, M., Caruso, C., Neuroinflammation in Huntington's Disease: a Starring Role for Astro-cyte and Microglia. Current Neuropharmacology. 18756190. 20 (6) (2022), 1116–1143, Available from: http://www.eurekaselect.com/603/journal/current-neuropharmacology. doi: 10.2174/1570159X19666211201094608.

Saft, C., Lauter, T., Kraus, P.H., Przuntek, H., Andrich, J.E., Dose-dependent improvement of myoclonic hyperkinesia due to valproic acid in eight Huntington's Disease patients: a case series. BMC Neurol. 14712377. 6 (2006), Available from: http://www.biomedcentral.com/1471-2377/6/11. doi: 10.1186/1471-2377-6-11.

Salman, A., Eid, A.H., Khalaf, S.S., El-Dessouki, A.M., El-Shiekh, R.A., Aly, S.H., Natural products proposed for the management of Huntington's disease (HD): a comprehensive review. Naunyn Schmiedebergs Arch. Pharmacol. 0028–1298. (2025), doi: 10.1007/s00210-025-04444-w.

Sampaio, C., Disease, H., Update on ongoing therapeutic developments and a look toward the future. Park Relat Disord. 122 (2024), 106049.

Sapp, E., Kegel, K.B., Aronin, N., Hashikawa, T., Uchiyama, Y., Tohyama, K., Bhide, P.G., Vonsattel, J.P., Difiglia, M., Early and Progressive Accumulation of Reactive Microglia in the Huntington Disease Brain. J. Neuropathol. Exp. Neurol. 0022-3069. 60 (2) (2001), 161–172, doi: 10.1093/jnen/60.2.161.

Sassone, J., Colciago, C., Cislaghi, G., Silani, V., Ciammola, A., Huntington's disease: the current state of research with peripheral tissues. Exp. Neurol. 10902430. 219 (2) (2009), 385–397, doi: 10.1016/j.expneurol.2009.05.012.

Sassone, J., Maraschi, A., Sassone, F., Silani, V., Ciammola, A., Defining the role of the Bcl-2 family proteins in Huntington's disease. Cell Death Dis. 2041-4889. 4 (8) (2013), e772, doi: 10.1038/cddis.2013.300.

Saudou, F., Humbert, S., The Biology of Huntingtin. Neuron. 10974199. 89 (5) (2016), 910–926, Available from: www.neuron.org. doi: 10.1016/j.neuron.2016.02.003.

Schulte, L., 2011. The biological function of the Huntingtin protein and its relevance to Huntington's Disease pathology. Curr. Trends Neur. 5, 65–78.

Schultz, M., Durr, A., Karachi, C., O'Callaghan, M., Alves, S., Lamontagne, A., Hurmic, H., Cartier, N., Investigating the Effects of Brain Cholesterol Metabolism Using CYP46A1 Gene Therapy in Subjects with Huntington's Disease (P1-11.016). Neurology. 100 (17_supplement_2) (2023), 3739. doi: 10.1212/WNL.0000000000203493.

Schwarcz, R., Guidetti, P., Sathyasaikumar, K.V., Muchowski, P.J., Of mice, rats and men: revisiting the quinolinic acid hypothesis of Huntington's disease. Progress in Neurobiology. 03010082. 90 (2) (2010), 230–245, doi: 10.1016/j.pneurobio.2009.04.005.

Scoles, D.R., Minikel, E.V., Pulst, S.M., Antisense oligonucleotides: a primer. Neurol. Genet. 23767839. 5 (2) (2019), Available from: http://ng.neurology.org/. doi: 10.1212/NXG.0000000000000323.

Scott, L.J., Tetrabenazine: for chorea associated with huntingtons disease. CNS Drugs. 11791934. 25 (12) (2011), 1073–1085, doi: 10.2165/11208330-000000000-00000.

Semaka, A., Collins, J.A., Hayden, M.R., Unstable familial transmissions of Huntington disease alleles with 27–35 CAG repeats (intermediate alleles). Am. J. Med. Genet. B Neuropsychiatr. Genet. 1552-4841. 153B (1) (2010), 314–320, doi: 10.1002/ajmg.b.30970.

Semaka, A., Kay, C., Belfroid, R.D.M., Bijlsma, E.K., Losekoot, M., van Langen, I.M., van Maarle, M.C., Oosterloo, M., Hayden, M.R., van Belzen, M.J., A new mutation for Huntington disease following maternal transmission of an intermediate allele. European Journal of Medical Genetics. 18780849. 58 (1) (2015), 28–30, Available from: http://www.elsevier.com/wps/find/journaldescription.cws_home/705239/description#description. doi: 10.1016/j.ejmg.2014.11.005.

Seo, J.H., Shin, J.H., Lee, J., Kim, D., Hwang, H.Y., Nam, B.G., Lee, J., Kim, H.H., Cho, S.R., DNA double-strand break-free CRISPR interference delays Huntington's disease progression in mice. Commun. Biol. 23993642. 6 (1) (2023), Available from: https://www.springer.com/journal/42003. doi: 10.1038/s42003-023-04829-8.

Sepers, M.D., Raymond, L.A., Mechanisms of synaptic dysfunction and excitotoxicity in Huntington's disease. Drug Discov. Today. 18785832. 19 (7) (2014), 990–996, Available from: www.elsevier.com/locate/drugdiscov. doi: 10.1016/j.drudis.2014.02.006.

Seyhan, A.A., RNAi: a potential new class of therapeutic for human genetic disease. Hum. Genet. 14321203. 130 (5) (2011), 583–605, doi: 10.1007/s00439-011-0995-8.

Sharma, A., Madaan, V., Petty, F., Propranolol treatment for neuroleptic-Induced akathisia [2]. Prim. Care Companion J. Clin. Psychiatry. 15235998. 7 (4) (2005), 202–203, Available from: http://www.psychiatrist.com/pcc/. doi: 10.4088/pcc.v07n0412b.

Sheng, Z.H., The Interplay of Axonal Energy Homeostasis and Mitochondrial Trafficking and Anchoring. Trends in Cell Biology. 18793088. 27 (6) (2017), 403–416, Available from: www.elsevier.com/locate/tcb. doi: 10.1016/j.tcb.2017.01.005.

Shin, J.H., Yang, H.J., Ahn, J.H., Jo, S., Chung, S.J., Lee, J.Y., Kim, H.S., Kim, M., Evidence-Based Review on Symptomatic Management of Huntington's Disease. Journal of Movement Disorders. 20934939. 17 (4) (2024), 369–386, Available from: https://www.e-jmd.org/upload/jmd-24140.pdf. doi: 10.14802/jmd.24140.

Shin, J.W., Hong, E.P., Park, S.S., Choi, D.E., Zeng, S., Chen, R.Z., Lee, J.M., PAM-altering SNP-based allele-specific CRISPR-Cas9 therapeutic strategies for Huntington's disease. Mol. Ther. Methods Clin. Dev. 23290501. 26 (2022), 547–561, Available from: https://www.journals.elsevier.com/molecular-therapy-methods-and-clinical-development/. doi: 10.1016/j.omtm.2022.08.005.

Shin, J.W., Kim, K.H., Chao, M.J., Atwal, R.S., Gillis, T., MacDonald, M.E., Gusella, J.F., Lee, J.M., Permanent inactivation of Huntington's disease mutation by personalized allele-specific CRISPR/Cas9. Hum. Mol. Genet. 14602083. 25 (20) (2016), 4566–4576, Available from: 10.1093/hmg/ddw286. doi: 10.1093/hmg/ddw286.

Shin, J.Y., Fang, Z.H., Yu, Z.X., Wang, C.E., Li, S.H., Li, X.J., Expression of mutant huntingtin in glial cells contributes to neuronal excitotoxicity. J. Cell Biol. 00219525. 171 (6) (2005), 1001–1012, Available from: http://www.jcb.org/cgi/reprint/171/6/1001. doi: 10.1083/jcb.200508072.

Shirendeb, U.P., Calkins, M.J., Manczak, M., Anekonda, V., Dufour, B., McBride, J.L., Mao, P., Reddy, P.H., Mutant huntingtin's interaction with mitochondrial protein Drp1 impairs mitochondrial biogenesis and causes defective axonal transport and synaptic degeneration in Huntington's disease. Hum. Mol. Genet. 0964-6906. 21 (2) (2011b), 406–420, doi: 10.1093/hmg/ddr475.

Shirendeb, U., Reddy, A.P., Manczak, M., Calkins, M.J., Mao, P., Tagle, D.A., Reddy, P.H., Abnormal mitochondrial dynamics, mitochondrial loss and mutant huntingtin oligomers in Huntington's disease: implications for selective neuronal damage. Hum. Mol. Genet. 14602083. 20 (7) (2011a), 1438–1455, doi: 10.1093/hmg/ddr024.

Shiwal, A., Nibrad, D., Tadas, M., Katariya, R., Kale, M., Wankhede, N., Kotagale, N., Umekar, M., Taksande, B., Polyamines signalling pathway: a key player in unveiling the molecular mechanisms underlying Huntington's disease. Neuroscience. 18737544. 570 (2025), 213–224, Available from: https://www.sciencedirect.com/science/journal/03064522. doi: 10.1016/j.neuroscience.2025.02.042.

Silvestroni, A., Faull, R.L.M., Strand, A.D., Möllera, T., Distinct neuroinflammatory profile in post-mortem human Huntington's disease. NeuroReport. 09594965. 20 (12) (2009), 1098–1103, doi: 10.1097/WNR.0b013e32832e34ee.

Sleigh, J.N., Rossor, A.M., Fellows, A.D., Tosolini, A.P., Schiavo, G., Axonal transport and neurological disease. Nat. Rev. Neurol. 17594766. 15 (12) (2019), 691–703, Available from: http://www.nature.com/nrneurol/archive/index.html. doi: 10.1038/s41582-019-0257-2.

Sloan, K., on behalf of the ALN-HTT02 program team, 2024. J004 ALN-HTT02, a novel C16-sirna conjugate for HTT-lowering in the CNS. J. Neurol. Neurosurg. Psychiatry. 95 (Suppl 1), A158. Available from: http://jnnp.bmj.com/content/95/Suppl_1/A158.3.abstract. doi:10.1136/jnnp-2024-EHDN.325.

Smith, R., Brundin, P., Li, J.-Y., Synaptic dysfunction in Huntington's disease: a new perspective. Cell. Mol. Life Sci. 1420-682X. 62 (17) (2005), 1901–1912, doi: 10.1007/s00018-005-5084-5.

Snowden, J.S., The Neuropsychology of Huntington's Disease. Arch. Clin. Neuropsychol. 18735843. 32 (7) (2017), 876–887, Available from: http://acn.oxfordjournals.org/archive/. doi: 10.1093/arclin/acx086.

Solberg, O.K., Filkuková, P., Frich, J.C., Billaud Feragen, K.J., Age at Death and Causes of Death in Patients with Huntington Disease in Norway in 1986–2015. J. Huntingtons Dis. 18796397. 7 (1) (2018), 77–86, doi: 10.3233/jhd-170270.

Solís-Maldonado, M., Miró, M.P., Acuña, A.I., Covarrubias-Pinto, A., Loaiza, A., Mayorga, G., Beltrán, F.A., Cepeda, C., Levine, M.S., Concha, I.I., Bátiz, L.F., Carrasco, M.A., Castro, M.A., Altered lactate metabolism in Huntington's disease is dependent on GLUT3 expression. CNS Neurosci. Ther. 17555949. 24 (4) (2018), 343–352, Available from: http://onlinelibrary.wiley.com/journal/10.1111/(ISSN)1755-5949. doi: 10.1111/cns.12837.

Son, S.M., Park, S.J., Fernandez-Estevez, M., Rubinsztein, D.C., Autophagy regulation by acetylation—implications for neurodegenerative diseases. Exp. Mol. Med. 20926413. 53 (1) (2021), 30–41, Available from: https://www.nature.com/emm/. doi: 10.1038/s12276-021-00556-4.

Southwell, A.L., Kordasiewicz, H.B., Langbehn, D., Skotte, N.H., Parsons, M.P., Villanueva, E.B., Caron, N.S., Ostergaard, M.E., Anderson, L.M., Xie, Y., Cengio, L.D., Findlay-Black, H., Doty, C.N., Fitsimmons, B., Swayze, E.E., Seth, P.P., Raymond, L.A., Bennett, C.F., Hayden, M.R., Huntingtin suppression restores cognitive function in a mouse model of Huntington's disease. Sci. Transl. Med. 19466242. 10 (461) (2018), Available from: http://stm.sciencemag.org/content/scitransmed/10/461/eaar3959.full.pdf. doi: 10.1126/scitranslmed.aar3959.

Southwell, A.L., Skotte, N.H., Bennett, C.F., Hayden, M.R., Antisense oligonucleotide therapeutics for inherited neurodegenerative diseases. Trends Mol. Med. 1471499X. 18 (11) (2012), 634–643, doi: 10.1016/j.molmed.2012.09.001.

Southwell, A.L., Warby, S.C., Carroll, J.B., Doty, C.N., Skotte, N.H., Zhang, W., Villanueva, E.B., Kovalik, V., Xie, Y., Pouladi, M.A., Collins, J.A., Yang, X.W., Franciosi, S., Hayden, M.R., A fully humanized transgenic mouse model of Huntington disease. Hum. Mol. Genet. 14602083. 22 (1) (2013), 18–34, doi: 10.1093/hmg/dds397.

Spronck, E.A., Brouwers, C.C., Vallès, A., de Haan, M., Petry, H., van Deventer, S.J., Konstantinova, P., Evers, M.M., AAV5-miHTT Gene Therapy Demonstrates Sustained Huntingtin Lowering and Functional Improvement in Huntington Disease Mouse Models. Mol. Ther. Methods Clin. Dev. 23290501. 13 (2019), 334–343, Available from: https://www.journals.elsevier.com/molecular-therapy-methods-and-clinical-development/. doi: 10.1016/j.omtm.2019.03.002.

Srivastava, A., Mallela, K.M.G., Deorkar, N., Brophy, G., Manufacturing Challenges and Rational Formulation Development for AAV Viral Vectors. J. Pharm. Sci. 00223549. 110 (7) (2021), 2609–2624, doi: 10.1016/j.xphs.2021.03.024.

Stahl, F., Evert, B.O., Han, X., Breuer, P., Wüllner, U., Spinocerebellar Ataxia Type 3 Pathophysiology—Implications for Translational Research and Clinical Studies. Int. J. Mol. Sci. 1422-0067. 25 (7) (2024), 3984, doi: 10.3390/ijms25073984.

Stanisławska-Sachadyn, A., Krzemiński, M., Zielonka, D., Krygier, M., Ziętkiewicz, E., Sławek, J., Limon, J., Sex contribution to average age at onset of Huntington's disease depends on the number of (CAG)n repeats. Sci. Rep. 2045-2322. 14 (1) (2024), doi: 10.1038/s41598-024-64105-5.

Steffan, J.S., Kazantsev, A., Spasic-Boskovic, O., Greenwald, M., Zhu, Y.-Z., Gohler, H., Wanker, E.E., Bates, G.P., Housman, D.E., Thompson, L.M., The Huntington's disease protein interacts with p53 and CREB-binding protein and represses transcription. Proc. Natl. Acad. Sci. U. S. A. 0027-8424. 97 (12) (2000), 6763–6768, doi: 10.1073/pnas.100110097.

Stimming, E.F., Sung, V., Testa, C., Kostyk, C.A.R., Samii, A., et al., J05 interim results from cohort 1 of the double-blind, dose-escalation phase I/II clinical trial of amt-130 (HD-genetrx-1) for early-stage huntington's disease (HD). J. Neurol. Neurosurg. Psychiatry. 93 (2022), A95

Strehlow, A.N.T., Li, J.Z., Myers, R.M., Wild-type huntingtin participates in protein trafficking between the Golgi and the extracellular space. Hum. Mol. Genet. 14602083. 16 (4) (2007), 391–409, doi: 10.1093/hmg/ddl467.

Sun, E., Kang, M., Wibawa, P., Tsoukra, Z.C., Farrand, S., et al., Huntington's disease: mortality and risk factors in an Australian cohort. J. Neurol. Sci. 442 (2022), 120437.

Sun, Y., Savanenin, A., Reddy, P.H., Liu, Y.F., Polyglutamine-expanded Huntingtin Promotes Sensitization of N-Methyl-D-aspartate Receptors via Post-synaptic Density 95. J. Biol. Chem. 276 (27) (2001), 24713–24718, doi: 10.1074/jbc.M103501200.

Tabrizi, S.J., Estevez-Fraga, C., van Roon-mom, W.M.C., Flower, M.D., Scahill, R.I., Wild, E.J., Muñoz-Sanjuan, I., Sampaio, C., Rosser, A.E., Leavitt, B.R., Potential disease-modifying therapies for Huntington's disease: lessons learned and future opportunities. Lancet Neurol. 14744465. 21 (7) (2022), 645–658, Available from: http://www.journals.elsevier.com/the-lancet-neurology/. doi: 10.1016/S1474-4422(22)00121-1.

Tabrizi, S.J., Ghosh, R., Leavitt, B.R., Huntingtin Lowering Strategies for Disease Modification in Huntington's Disease. Neuron. 10974199. 101 (5) (2019b), 801–819, Available from: http://www.cell.com/neuron/home. doi: 10.1016/j.neuron.2019.01.039.

Tabrizi, S.J., Leavitt, B.R., Landwehrmeyer, G.B., Wild, E.J., Saft, C., Barker, R.A., Blair, N.F., Craufurd, D., Priller, J., Rickards, H., Rosser, A., Kordasiewicz, H.B., Czech, C., Swayze, E.E., Norris, D.A., Baumann, T., Gerlach, I., Schobel, S.A., Paz, E., Smith, A.V., Bennett, C.F., Lane, R.M., Targeting huntingtin expression in patients with Huntington's disease. N. Engl. J. Med. 15334406. 380 (24) (2019a), 2307–2316, Available from: http://www.nejm.org/medical-index. doi: 10.1056/NEJMoa1900907.

Tai, Y.F., Pavese, N., Gerhard, A., Tabrizi, S.J., Barker, R.A., Brooks, D.J., Piccini, P., Microglial activation in presymptomatic Huntington's disease gene carriers. Brain. 14602156. 130 (7) (2007), 1759–1766, Available from: http://brain.oxfordjournals.org/. doi: 10.1093/brain/awm044.

Teixeira, A.L., de Souza, L.C., Rocha, N.P., Furr-Stimming, E., Lauterbach, E.C., Revisiting the neuropsychiatry of Huntington's disease. Dementia & Neuropsychologia. 1980-5764. 10 (4) (2016), 261–266, doi: 10.1590/s1980-5764-2016dn1004002.

Testa, C.M., Jankovic, J., Huntington disease: a quarter century of progress since the gene discovery. J. Neurol. Sci. 18785883. 396 (2019), 52–68, Available from: www.elsevier.com/locate/jns. doi: 10.1016/j.jns.2018.09.022.

Tong, X., Ao, Y., Faas, G.C., Nwaobi, S.E., Xu, J., Haustein, M.D., Anderson, M.A., Mody, I., Olsen, M.L., Sofroniew, M.V., Khakh, B.S., Astrocyte Kir4.1 ion channel deficits contribute to neuronal dysfunction in Huntington's disease model mice. Nat. Neurosci. 15461726. 17 (5) (2014), 694–703, Available from: http://www.nature.com/neuro/index.html. doi: 10.1038/nn.3691.

Tong, Y., Ha, T.J., Liu, L., Nishimoto, A., Reiner, A., Goldowitz, D., Spatial and temporal requirements for huntingtin (Htt) in neuronal migration and survival during brain development. J. Neurosci. 15292401. 31 (41) (2011), 14794–14799, Available from: http://www.jneurosci.org/content/31/41/14794.full.pdf+html. doi: 10.1523/JNEUROSCI.2774-11.2011.

Tousley, A., Iuliano, M., Weisman, E., Sapp, E., Richardson, H., Vodicka, P., Alexander, J., Aronin, N., DiFiglia, M., Kegel-Gleason, K.B., Frédéric Saudou, Huntingtin associates with the actin cytoskeleton and α-actinin isoforms to influence stimulus dependent morphology changes. PLOS ONE. 1932-6203. 14 (2) (2019), e0212337, doi: 10.1371/journal.pone.0212337.

Träger, U., Andre, R., Lahiri, N., Magnusson-Lind, A., Weiss, A., Grueninger, S., McKinnon, C., Sirinathsinghji, E., Kahlon, S., Pfister, E.L., Moser, R., Hummerich, H., Antoniou, M., Bates, G.P., Luthi-Carter, R., Lowdell, M.W., Björkqvist, M., Ostroff, G.R., Aronin, N., Tabrizi, S.J., HTT-lowering reverses Huntington's disease immune dysfunction caused by NFκB pathway dysregulation. Brain. 14602156. 137 (3) (2014), 819–833, Available from: http://brain.oxfordjournals.org/. doi: 10.1093/brain/awt355.

Trushina, E., McMurray, C.T., Oxidative stress and mitochondrial dysfunction in neurodegenerative diseases. Neuroscience. 03064522. 145 (4) (2007), 1233–1248, doi: 10.1016/j.neuroscience.2006.10.056.

Tunalı, N.E., Molecular Mechanisms of Polyglutamine Pathology and Lessons Learned from Huntington's Disease. IntechOpen, (2021), doi: 10.5772/intechopen.93508.

Twelvetrees, A.E., Yuen, E.Y., Arancibia-Carcamo, I.L., MacAskill, A.F., Rostaing, P., Lumb, M.J., Humbert, S., Triller, A., Saudou, F., Yan, Z., Kittler, J.T., Delivery of GABAARs to Synapses Is Mediated by HAP1-KIF5 and Disrupted by Mutant Huntingtin. Neuron. 08966273. 65 (1) (2010), 53–65, doi: 10.1016/j.neuron.2009.12.007.

Unti, E., Mazzucchi, S., Palermo, G., Bonuccelli, U., Ceravolo, R., Antipsychotic drugs in Huntington's disease. Expert Rev. Neurother. 17448360. 17 (3) (2017), 227–237, doi: 10.1080/14737175.2016.1226134.

Upadhayay, S., Jamwal, S., Kumar, P., Animal models of Huntington's disease and their applicability to novel drug discovery and development. Expert Opin. Drug Discov. 18 (5) (2023), 527–538, doi: 10.1080/17460441.2023.2201493.

Valionyte, E., Yang, Y., Roberts, S.L., Kelly, J., Lu, B., Luo, S., Lowering Mutant Huntingtin Levels and Toxicity: autophagy-Endolysosome Pathways in Huntington's Disease. Journal of Molecular Biology. 10898638. 432 (8) (2020), 2673–2691, Available from: https://www.journals.elsevier.com/journal-of-molecular-biology. doi: 10.1016/j.jmb.2019.11.012.

van der Burg, J.M., Björkqvist, M., Brundin, P., Beyond the brain: widespread pathology in Huntington's disease. Lancet Neurol. 14744422. 8 (8) (2009), 765–774, Available from: http://www.journals.elsevier.com/the-lancet-neurology/. doi: 10.1016/S1474-4422(09)70178-4.

Vattakatuchery, K., Acetylcholinesterase inhibitors in cognitive impairment in Huntington's disease: a brief review. World J. Psychiatry. 3 (3) (2013), 62.

Velier, J., Kim, M., Schwarz, C., Kim, T.W., Sapp, E., Chase, K., Aronin, N., DiFiglia, M., Wild-type and mutant huntingtins function in vesicle trafficking in the secretory and endocytic pathways. Exp. Neurol. 00144886. 152 (1) (1998), 34–40, Available from: http://www.elsevier.com/inca/publications/store/6/2/2/8/2/8/index.htt. doi: 10.1006/exnr.1998.6832.

Vinther-Jensen, T., Larsen, I.U., Hjermind, L.E., Budtz-Jørgensen, E., Nielsen, T.T., Nørremølle, A., Nielsen, J.E., Vogel, A., A clinical classification acknowledging neuropsychiatric and cognitive impairment in Huntington's disease. Orphanet J. Rare Dis. 17501172. 9 (1) (2014), Available from: http://www.ojrd.com/. doi: 10.1186/s13023-014-0114-8.

Vitet, H., Brandt, V., Saudou, F., Traffic signaling: new functions of huntingtin and axonal transport in neurological disease. Curr. Opin. Neurobiol. 18736882. 63 (2020), 122–130, Available from: http://www.elsevier.com/locate/conb. doi: 10.1016/j.conb.2020.04.001.

Vos, T., Flaxman, A.D., Naghavi, M., Lozano, R., Michaud, C., Ezzati, M., Shibuya, K., Salomon, J.A., Abdalla, S., Aboyans, V., Abraham, J., Ackerman, I., Aggarwal, R., Ahn, S.Y., Ali, M.K., AlMazroa, M.A., Miriam Alvarado, H.R.A., Anderson, L.M., Andrews, K.G., Atkinson, C., Baddour, L.M., Bahalim, A.N., Barker-Collo, S., Barrero, L.H., Bartels, D.H., Basáñez, M.-G., Baxter, A., Bell, M.L., Benjamin, E.J., Bennett, D., Bernabé, E., Bhalla, K., Bhandari, B., Bikbov, B., Abdulhak, A.B., Birbeck, G., Black, J.A., Blencowe, H., Blore, J.D., Blyth, F., Bolliger, I., Bonaventure, A., Boufous, S., Bourne, R., Boussinesq, M., Braithwaite, T., Brayne, C., Bridgett, L., Brooker, S., Brooks, P., Brugha, T.S., Bryan-Hancock, C., Bucello, C., Buchbinder, R., Buckle, G., Budke, C.M., Burch, M., Burney, P., Burstein, R., Calabria, B., Campbell, B., Canter, C.E., Carabin, H., Carapetis, J., Carmona, L., Cella, C., Charlson, F., Chen, H., Cheng, A.T.-A., Chou, D., Chugh, S.S., Coffeng, L.E., Colan, S.D., Samantha Colquhoun, K.E.C., Condon, J., Connor, M.D., Cooper, L.T., Corriere, M., Cortinovis, M., de Vaccaro, K.C., Couser, W., Cowie, B.C., Criqui, M.H., Cross, M., Dabhadkar, K.C., Dahiya, M., Dahodwala, N., Damsere-Derry, J., Danaei, G., Davis, A., De Leo, D., Degenhardt, L., Dellavalle, R., Delossantos, A., Denenberg, J., Derrett, S., Jarlais, D.C.D., Dharmaratne, S.D., Dherani, M., Diaz-Torne, C., Helen Dolk, E.R.D., Driscoll, T., Duber, H., Ebel, B., Edmond, K., Elbaz, A., Ali, S.E., Erskine, H., Erwin, P.J., Espindola, P., Ewoigbokhan, S.E., Farzadfar, F., Feigin, V., Felson, D.T., Ferrari, A., Ferri, C.P., Fèvre, E.M., Finucane, M.M., Flaxman, S., Flood, L., Foreman, K., Forouzanfar, M.H., Fowkes, F.G.R., Franklin, R., Fransen, M., Freeman, M.K., Gabbe, B.J., Gabriel, S.E., Gakidou, E., Ganatra, H.A., Garcia, B., Gaspari, F., Gillum, R.F., Gmel, G., Gosselin, R., Grainger, R., Groeger, J., Guillemin, F., Gunnell, D., Gupta, R., Haagsma, J., Hagan, H., Halasa, Y.A., Hall, W., Haring, D., Haro, J.M., Harrison, J.E., Havmoeller, R., Hay, R.J., Higashi, H., Hill, C., Hoen, B., Hoffman, H., Hotez, P.J., Hoy, D., Huang, J.J., Ibeanusi, S.E., Jacobsen, K.H., James, S.L., Jarvis, D., Jasrasaria, R., Jayaraman, S., Johns, N., Jonas, J.B., Karthikeyan, G., Kassebaum, N., Kawakami, N., Keren, A., Khoo, J.-P., King, C.H., Knowlton, L.M., Kobusingye, O., Koranteng, A., Krishnamurthi, R., Lalloo, R., Laslett, L.L., Lathlean, T., Leasher, J.L., Lee, Y.Y., Leigh, J., Lim, S.S., Limb, E., Lin, J.K., Lipnick, M., Lipshultz, S.E., Liu, W., Loane, M., Ohno, S.L., Lyons, R., Jixiang, M., Mabweijano, J., MacIntyre, M.F., Malekzadeh, R., Mallinger, L., Manivannan, S., Marcenes, W., March, L., Margolis, D.J., Marks, G.B., Marks, R., Matsumori, A., Matzopoulos, R., Mayosi, B.M., McAnulty, J.H., McDermott, M.M., McGill, N., McGrath, J., Medina-Mora, M.E., Meltzer, M., Memish, Z.A., Mensah, G.A., Merriman, T.R., Meyer, A.-C., Miglioli, V., Miller, M., Miller, T.R., Mitchell, P.B., Mocumbi, A.O., Moffitt, T.E., Mokdad, A.A., Monasta, L., Montico, M., Moradi-Lakeh, M., Moran, A., Morawska, L., Mori, R., Murdoch, M.E., Mwaniki, M.K., Kovin Naidoo, M.N.N., Luigi Naldi, K.M.V.N., Nelson, P.K., Nelson, R.G., Nevitt, M.C., Newton, C.R., Nolte, S., Norman, P., Norman, R., O'Donnell, M., O'Hanlon, S., Olives, C., Omer, S.B., Ortblad, K., Osborne, R., Ozgediz, D., Page, A., Pahari, B., Pandian, J.D., Rivero, A.P., Patten, S.B., Pearce, N., Padilla, R.P., Perez-Ruiz, F., Perico, N., Pesudovs, K., Phillips, D., Phillips, M.R., Pierce, K., Pion, S., Polanczyk, G.V., Suzanne Polinder, C.A.P., Popova, S., Porrini, E., Pourmalek, F., Prince, M., Pullan, R.L., Kapa, D., Ramaiah, D.R., Razavi, H., Regan, M., Rehm, J.T., Rein, D.B., Remuzzi, G., Richardson, K., Rivara, F.P., Roberts, T., Robinson, C., De Leòn, F.R., Ronfani, L., Room, R., Rosenfeld, L.C., Rushton, L., Sacco, R.L., Saha, S., Sampson, U., Sanchez-Riera, L., Sanman, E., Schwebel, D.C., Scott, J.G., Segui-Gomez, M., Shahraz, S., Shepard, D.S., Shin, H., Shivakoti, R., Silberberg, D., Singh, D., Singh, G.M., Singh, J.A., Singleton, J., Sleet, D.A., Sliwa, K., Smith, E., Smith, J.L., Stapelberg, N.J.C., Steer, A., Steiner, T., Stolk, W.A., Stovner, L.J., Sudfeld, C., Syed, S., Tamburlini, G., Tavakkoli, M., Taylor, H.R., Taylor, J.A., Taylor, W.J., Thomas, B., Thomson, W.M., Thurston, G.D., Tleyjeh, I.M., Tonelli, M., Towbin, J.A., Truelsen, T., Tsilimbaris, M.K., Ubeda, C., Undurraga, E.A., van der Werf, M.J., van Os, J., Vavilala, M.S., Venketasubramanian, N., Wang, M., Wang, W., Watt, K., Weatherall, D.J., Weinstock, M.A., Weintraub, R., Weisskopf, M.G., Weissman, M.M., White, R.A., Whiteford, H., Wiersma, S.T., Wilkinson, J.D., Williams, H.C., Williams, S.R.M., Witt, E., Wolfe, F., Woolf, A.D., Wulf, S., Yeh, P.-H., Zaidi, A.K.M., Zheng, Z.-J., Zonies, D., Lopez, A.D., Murray, C.J.L., Years lived with disability (YLDs) for 1160 sequelae of 289 diseases and injuries 1990–2010: a systematic analysis for the Global Burden of Disease Study 2010. Lancet. 01406736. 380 (9859) (2012), 2163–2196, doi: 10.1016/s0140-6736(12)61729-2.

Wang, D., Tai, P.W.L., Gao, G., Adeno-associated virus vector as a platform for gene therapy delivery. Nat. Rev. Drug Discov. 1474-1776. 18 (5) (2019), 358–378, doi: 10.1038/s41573-019-0012-9.

Wang, D., Zhang, F., Gao, G., CRISPR-Based Therapeutic Genome Editing: strategies and In Vivo Delivery by AAV Vectors. Cell. 00928674. 181 (1) (2020), 136–150, doi: 10.1016/j.cell.2020.03.023.

Wang, J.H., Gessler, D.J., Zhan, W., Gallagher, T.L., Gao, G., Adeno-associated virus as a delivery vector for gene therapy of human diseases. Signal Transduction and Targeted Therapy. 20593635. 9 (1) (2024), Available from: https://www.nature.com/sigtrans. doi: 10.1038/s41392-024-01780-w.

Wang, L., Lin, M., Gao, Z., Wang, Y., Qiu, J., Significance of gene therapy in neurodegenerative diseases. Front. Neurosci. 1662-453X. 19 (2025), doi: 10.3389/fnins.2025.1515255.

Wang, N., Gray, M., Xiao-Hong, L., Cantle, J.P., Holley, S.M., Greiner, E., Xiaofeng, G., Shirasaki, D., Cepeda, C., Yuqing, L., Dong, H., Levine, M.S., Yang, X.W., Neuronal targets for reducing mutant huntingtin expression to ameliorate disease in a mouse model of Huntington's disease. Nat. Med. 1078-8956. 20 (5) (2014), 536–541, doi: 10.1038/nm.3514.

Wanker, E.E., Ast, A., Schindler, F., Trepte, P., Schnoegl, S., The pathobiology of perturbed mutant huntingtin protein–protein interactions in Huntington's disease. J. Neurochem. 14714159. 151 (4) (2019), 507–519, Available from: http://onlinelibrary.wiley.com/journal/10.1111/(ISSN)1471-4159. doi: 10.1111/jnc.14853.

Warby, S.C., Montpetit, A., Hayden, A.R., Carroll, J.B., Butland, S.L., Visscher, H., Collins, J.A., Semaka, A., Hudson, T.J., Hayden, M.R., CAG Expansion in the Huntington Disease Gene Is Associated with a Specific and Targetable Predisposing Haplogroup. Am. J. Hum. Genet. 00029297. 84 (3) (2009), 351–366, doi: 10.1016/j.ajhg.2009.02.003.

White, J.A., Anderson, E., Zimmerman, K., Zheng, K.H., Rouhani, R., Gunawardena, S., Huntingtin differentially regulates the axonal transport of a sub-set of Rab-containing vesicles in vivo. Hum. Mol. Genet. 14602083. 24 (25) (2015), 7182–7195, Available from: http://hmg.oxfordjournals.org/. doi: 10.1093/hmg/ddv415.

Wild, E.J., Tabrizi, S.J., Therapies targeting DNA and RNA in Huntington's disease. Lancet Neurol. 14744465. 16 (10) (2017), 837–847, Available from: http://www.journals.elsevier.com/the-lancet-neurology/. doi: 10.1016/S1474-4422(17)30280-6.

Wilton, D.K., Stevens, B., The contribution of glial cells to Huntington's disease pathogenesis. Neurobiol. Dis. 1095953X. 143 (2020), Available from: http://www.elsevier.com/inca/publications/store/6/2/2/9/2/3/index.htt. doi: 10.1016/j.nbd.2020.104963.

Woda, J.M., Calzonetti, T., Hilditch-Maguire, P., Duyao, M.P., Conlon, R.A., MacDonald, M.E., Inactivation of the Huntington's disease gene (Hdh) impairs anterior streak formation and early patterning of the mouse embryo. BMC Dev. Biol. 1471-213X. 5 (1) (2005), doi: 10.1186/1471-213x-5-17.

Wong, S., Le, G.H., Kwan, A.T.H., Rhee, T.G., Teopiz, K.M., Ho, R.C., Cao, B., Rosenblat, J.D., Mansur, R., McIntyre, R.S., Risk of VMAT2 inhibitors on suicidality and parkinsonism: report utilizing the United States Food and Drug Administration adverse event reporting system. Int. Clin. Psychopharmacol. 14735857. 40 (3) (2025), 176–181, Available from: https://journals.lww.com/intclinpsychopharm/pages/default.aspx. doi: 10.1097/YIC.0000000000000553.

Wyttenbach, A., Swartz, J., Kita, H., Thykjaer, T., Carmichael, J., Bradley, J., Brown, R., Maxwell, M., Schapira, A., Orntoft, T.F., Kato, K., Rubinsztein, D.C., Polyglutamine expansions cause decreased CRE-mediated transcription and early gene expression changes prior to cell death in an inducible cell model of Hungtington's disease. Hum. Mol. Genet. 09646906. 10 (17) (2001), 1829–1845, Available from: http://hmg.oxfordjournals.org/.

Xiang, C., Zhang, S., Dong, X., Ma, S., Cong, S., Transcriptional dysregulation and post-translational modifications in polyglutamine diseases: from pathogenesis to potential therapeutic strategies. Front. Mol. Neurosci. 16625099. 11 (2018), Available from: https://www.frontiersin.org/articles/10.3389/fnmol.2018.00153/pdf. doi: 10.3389/fnmol.2018.00153.

Xiong, K., Wang, X., Feng, C., Zhang, K., Chen, D., Yang, S., Vectors in CRISPR Gene Editing for Neurological Disorders: challenges and Opportunities. Adv. Biol. 2701-0198. 9 (3) (2025), doi: 10.1002/adbi.202400374.

Yamamoto, A., Lucas, J.J., Hen, R., Reversal of neuropathology and motor dysfunction in a conditional model of Huntington's disease. Cell. 00928674. 101 (1) (2000), 57–66, Available from: https://www.sciencedirect.com/journal/cell. doi: 10.1016/S0092-8674(00)80623-6.

Yang, H.M., Yang, S., Huang, S.S., Tang, B.S., Guo, J.F., Microglial activation in the pathogenesis of Huntington's Disease. Front. Aging Neurosci. 16634365. 9 (2017a), Available from: http://journal.frontiersin.org/article/10.3389/fnagi.2017.00193/full. doi: 10.3389/fnagi.2017.00193.

Yang, J., Chen, X., Huilin, X., SQSTM1/p62 droplet -mediated autophagosome formation:insights into Huntington disease. Autophagy. 1554-8627. 17 (10) (2021), 3256–3259, doi: 10.1080/15548627.2021.1953820.

Yang, S., Chang, R., Yang, H., Zhao, T., Hong, Y., Kong, H.E., Sun, X., Qin, Z., Jin, P., Shihua, L., Xiao-Jiang, L., CRISPR/Cas9-mediated gene editing ameliorates neurotoxicity in mouse model of Huntington's disease. Journal of Clinical Investigation. 0021-9738. 127 (7) (2017b), 2719–2724, doi: 10.1172/jci92087.

Yang, S.J., Del Bonis O'donnell, J.T., Giordani, F., Wang, J., Lui, A., Piekarski, D., Irrinki, A., Schaffer, D.V., Landry, M.P., Synaptic scale dopamine disruption in Huntington's Disease model mice imageds with near infrared catecholamine nanosensors. bioRxiv. 26928205. (2022), Available from: https://www.biorxiv.org. doi: 10.1101/2022.09.19.508617.

Yhnell, E., Furby, H., Lowe, R.S., Brookes-Howell, L.C., Drew, C.J.G., Playle, R., Watson, G., Metzler-Baddeley, C., Rosser, A.E., Busse, M.E., A randomised feasibility study of computerised cognitive training as a therapeutic intervention for people with Huntington's disease (CogTrainHD). Pilot and Feasibility Studies. 20555784. 6 (1) (2020), Available from: https://pilotfeasibilitystudies.biomedcentral.com/. doi: 10.1186/s40814-020-00623-z.

Yu, Z., Yang, W., Yan, S., 2024. Advantages and differences among various animal models of Huntington's disease. Ageing Neurodegener. Dis. 2769-5301. 4 (3). doi: 10.20517/and.2024.13.

Yuen, E.Y., Wei, J., Zhong, P., Yan, Z., Disrupted GABAAR trafficking and synaptic inhibition in a mouse model of Huntington's disease. Neurobiol. Dis. 1095953X. 46 (2) (2012), 497–502, doi: 10.1016/j.nbd.2012.02.015.

Zadegan, S.A., Ramirez, F., Reddy, K.S., Sahin, O., Rocha, N.P., Teixeira, A.L., Stimming, E.F., Treatment of Depression in Huntington's Disease: a Systematic Review. The Journal of Neuropsychiatry and Clinical Neurosciences. 15457222. 36 (4) (2024), 283–299, doi: 10.1176/appi.neuropsych.20230120.

Zamanian, J.L., Xu, L., Foo, L.C., Nouri, N., Zhou, L., Giffard, R.G., Barres, B.A., Genomic analysis of reactive astrogliosis. J. Neurosci. 15292401. 32 (18) (2012), 6391–6410, Available from: http://www.jneurosci.org/content/32/18/6391.full.pdf+html. doi: 10.1523/JNEUROSCI.6221-11.2012.

Zeitler, B., Froelich, S., Marlen, K., Shivak, D.A., Yu, Q., Davis, L., Pearl, J.R., Miller, J.C., Zhang, L., Paschon, D.E., Hinkley, S.J., Ankoudinova, I., Lam, S., Guschin, D., Kopan, L., Cherone, J.M., Nguyen, H.-O.B., Qiao, G., Ataei, Y., Mendel, M.C., Amora, R., Surosky, R., Josee Laganiere, B.J.V., Narayanan, A., Sedaghat, Y., Tillack, K., Thiede, C., Gärtner, A., Kwak, S., Bard, J., Mrzljak, L., Park, L., Heikkinen, T., Lehtimäki, K.K., Svedberg, M.M., Häggkvist, J., Tari, L., Tóth, M., Varrone, A., Halldin, C., Kudwa, A.E., Ramboz, S., Day, M., Jyothisri Kondapalli, D.J.S., Urnov, F.D., Gregory, P.D., Rebar, E.J., Ignacio Muñoz-Sanjuán, H.S.Z., Allele-selective transcriptional repression of mutant HTT for the treatment of Huntington's disease. Nat. Med. 1078-8956. 25 (7) (2019), 1131–1142, doi: 10.1038/s41591-019-0478-3.

Zeitlin, S., Liu, J.P., Chapman, D.L., Papaioannou, V.E., Efstratiadis, A., Increased apoptosis and early embryonic lethality in mice nullizygous for the Huntington's disease gene homologue. Nat. Genet. 15461718. 11 (2) (1995), 155–163, doi: 10.1038/ng1095-155.

Zhai, W., Jeong, H., Cui, L., Krainc, D., Tjian, R., In vitro analysis of huntingtin-mediated transcriptional repression reveals multiple transcription factor targets. Cell. 00928674. 123 (7) (2005), 1241–1253, Available from: https://www.sciencedirect.com/journal/cell. doi: 10.1016/j.cell.2005.10.030.

Zhang, Y., Leavitt, B.R., Van Raamsdonk, J.M., Dragatsis, I., Goldowitz, D., MacDonald, M.E., Hayden, M.R., Friedlander, R.M., Huntingtin inhibits caspase-3 activation. EMBO J. 14602075. 25 (24) (2006), 5896–5906, doi: 10.1038/sj.emboj.7601445.

Zuccato, C., Ciammola, A., Rigamonti, D., Leavitt, B.R., Goffredo, D., Conti, L., MacDonald, M.E., Friedlander, R.M., Silani, V., Hayden, M.R., Timmusk, T., Sipione, S., Cattaneo, E., Loss of huntingtin-mediated BDNF gene transcription in Huntington's disease. Science. 293 (5529) (2001), 493–498, doi: 10.1126/science.1059581.

Zuccato, C., Tartari, M., Crotti, A., Goffredo, D., Valenza, M., Conti, L., Cataudella, T., Leavitt, B.R., Hayden, M.R., Timmusk, T., Rigamonti, D., Cattaneo, E., Huntingtin interacts with REST/NRSF to modulate the transcription of NRSE-controlled neuronal genes. Nat. Genet. 1061-4036. 35 (1) (2003), 76–83, doi: 10.1038/ng1219.

Zuccato, C., Valenza, M., Cattaneo, E., Molecular mechanisms and potential therapeutical targets in Huntington's disease. Physiol. Rev. 15221210. 90 (3) (2010), 905–981, Available from: http://physrev.physiology.org/cgi/reprint/90/3/905. doi: 10.1152/physrev.00041.2009Italy.

Chapter 13

Gene therapy in spinocerebellar ataxia (SCA)

Feyza Sule Aslan[a] and Enes Akyuz[b]
[a]*School of Medicine, Marmara University, Istanbul, Turkey*
[b]*Department of Pediatrics, School of Medicine, University of Wisconsin-Madison, Madison, WI, United States*

1 Introduction

Spinocerebellar ataxia (SCA) is a group of genetically inherited, progressive neurodegenerative disorders that follow an autosomal dominant inheritance pattern. This disorder primarily affects the cerebellum, which is responsible for balance, and the spinal cord, which controls vital reflexes. SCA manifests through progressive impairments in balance and coordination (ataxia), speech disturbances (dysarthria), abnormal eye movements, and cognitive decline (Klockgether et al., 2019; Schöls et al., 2004; Ando et al., 2018). More than 40 subtypes of SCA have been identified, each resulting from different genetic mutations. Although the clinical manifestations are heterogeneous and not pathognomonic, they assist in distinguishing among subtypes.

SCA is not restricted to the cerebellum and spinal cord. For instance, SCA1 may affect peripheral nerves and the optic nerve, while SCA2 can lead to retinal involvement. Among these, SCA3 (also known as Machado–Joseph disease) is the most prevalent subtype, predominantly affecting the cerebellum and resulting in symptoms such as speech impairment, uncontrolled muscle contractions (dystonia), muscle stiffness (spasticity), rigidity, tremors, proptosis, and diplopia (Sullivan et al., 2019; Paulson and Shakkottai, 1998). SCA4 has been associated with epilepsy.

The incidence of SCA is estimated to be approximately 2.7 per 100,000 individuals, with a prevalence of 5.6 per 100,000. SCA3 accounts for 25%–50% of all SCA cases, followed by SCA2 (13%–18%) and SCA6 (13%–15%). However, the prevalence of SCA subtypes varies based on geographic and ethnic factors. Notably, SCA3 is remarkably common in Portugal, followed by Germany, France, and the Netherlands (van Prooije et al., 2021; Orozco-Gutiérrez et al., 2017; Fujioka et al., 2013; Ruano et al., 2014).

At present, there is no curative treatment for SCA. Many subtypes lead to progressive neurodegeneration, often resulting in premature mortality. Some forms, such as SCA6 and SCA11, progress slowly over decades, whereas others, such as SCA1, SCA2, and SCA3, deteriorate rapidly within months. Most patients require a wheelchair within 10–20 years after diagnosis. While no definitive treatment exists, symptom management remains the primary therapeutic approach. Physiotherapy interventions aimed at strengthening muscles are crucial for maintaining daily functional abilities (Sullivan et al., 2019; Jacobi et al., 2023; Matilla-Dueñas, 2012; Chien et al., 2022).

A deeper understanding of the genetic and pathogenic mechanisms underlying SCA, the identification of causative genes, and the elucidation of its molecular pathways hold promise for novel therapeutic strategies. In this book chapter, we explore the pathogenesis and genetic basis of SCA, emphasizing molecular mechanisms. We summarize approved treatment approaches and highlight emerging gene therapy applications that offer potential avenues for definitive treatment.

1.1 SCA pathogenesis and genetic patterns

SCA subtypes are classified into repeat expansion mutations and nonrepeat mutations, which include missense, nonsense, insertion, or deletion mutations (Klockgether et al., 2019) (Table 13.1). The pathogenesis of SCA is generally associated with proteotoxicity, RNA toxicity, mitochondrial dysfunction, oxidative stress, and impaired proteasomal pathways due to trinucleotide repeat expansions. SCA1, 2, 3, 6, 7, and 17 result from cytosine–adenine–guanine (CAG) trinucleotide

Gene Therapy in Neurodegenerative Disorders. https://doi.org/10.1016/B978-0-443-40467-2.00001-4

TABLE 13.1 SCA types and associated genes, loci, and genetic mechanisms.

SCA type	Gene	Locus	Genetic mechanism	References
SCA1	*ATXN1*	6p22.3	CAG repeat extension	(Opal and Ashizawa, 1998)
SCA2	*ATXN2*	12q24.12	CAG repeat extension	(Scoles and Pulst, 2018)
SCA3	*ATXN3*	14q32.12	CAG repeat extension	(Bettencourt et al., 2010)
SCA4	*ZFHX3*	16q22.1	Unknown	(Flanigan et al., 1996)
SCA5	*SPTBN2*	11q13.4	Missense mutation	(Dick et al., 2012)
SCA6	*CACNA1A*	19p13.13	CAG repeat extension	(Du and Gomez, 2018)
SCA7	*ATXN7*	3p14.1	CAG repeat extension	(Bouwman et al., 2024)
SCA8	*ATXN8OS*	13q21	CTG repeat extension	(Subramony and Vig, 2001)
SCA10	*ATXN10*	22q13.31	ATTCT repeat extension	(Morato Torres et al., 2022)
SCA11	*TTBK2*	3p21.1	Missense mutation	(Felício and Santos, 2024)
SCA12	*PPP2R2B*	5q31.2	CAG repeat extension	(Bahl et al., 2005)
SCA13	*KCNC3*	19q13.33	Missense mutation	(Waters, 2006)
SCA14	*PRKCG*	19q13.42	Missense mutation	(Chen et al., 2005)
SCA15/16	*ITPR1*	3p26.1	Deletion	(Iwaki et al., 2008)
SCA17	*TBP*	6q27	CAG/CAA repeat extension	(Zühlke et al., 2003)
SCA18	–	7q22-q32	Unknown	(Echeveste et al., 2017)
SCA19/22	*KCND3*	1p13.2	Missense mutation	(Duarri et al., 2015)
SCA20	–	11q12.2-q12.3	Deletion	(Knight et al., 2008)
SCA21	*TMEM240*	1p36.33	Missense mutation	(Delplanque et al., 2014)
SCA23	*PDYN*	20p13	Missense mutation	(Verbeek et al., 2004)
SCA25	*PNPT1*	2p15-p21	Unknown	(Whaley et al., 2011)
SCA26	*ELOVL5*	6p12.1	Missense mutation	(Rudaks et al., 2024)
SCA27	*FGF14*	13q33.1	Missense mutation	(Coebergh et al., 2014)
SCA28	*AFG3L2*	18p11.21	Missense mutation	(Brussino et al., 2011)
SCA29	*ITPR1*	3p26.1	Missense mutation	(Huang et al., 2012)
SCA30	–	4q34.3	Unknown	(Fujioka et al., 2013)
SCA31	*BEAN1, TK2*	16q22.1	TGGAA repeat extension	(Ishiguro et al., 2021)
SCA32	–	7q32-q33	Unknown	(Müller, 2021)
SCA34	*ELOVL4*	6q14.1	Missense mutation	(Yeboah et al., 2021)
SCA35	*TGM6*	20p13	Missense mutation	(Maass et al., 2023)
SCA36	*NOP56*	20p13	GGC repeat extension	(Kobayashi et al., 2011)
SCA37	*DAB1*	1p32.2	ATTTC repeat extension	(Seixas et al., 2017)
SCA38	*ELOVL5*	6p12.1	Missense mutation	(Brusco and Gregorio, 2019)

repeat expansions and are categorized as polyglutamine (PolyQ) disorders. The increase in CAG trinucleotide repeats leads to toxic PolyQ chains, forming intracellular aggregates (Cui et al., 2024; Kang and Hong, 2009; Koeppen, 2005). When these aggregates are not adequately cleared by the proteasomal system and autophagy mechanisms, they induce toxicity and promote apoptosis (Lee et al., 2020). This toxicity triggers the endoplasmic reticulum stress response and unfolded protein response by upregulating transcription factors such as CHOP, ATF6, and XBP1 (Chen et al., 2023; Guo et al., 2018; Matilla-Dueñas and Volpini, 2019). The increased cellular stress contributes to mitochondrial dysfunction, leading to reduced ATP production and an increase in reactive oxygen species, which cause oxidative stress (Wang et al., 2011). Mutations in ataxin proteins involved in RNA processing disrupt gene expression by interacting with transcription

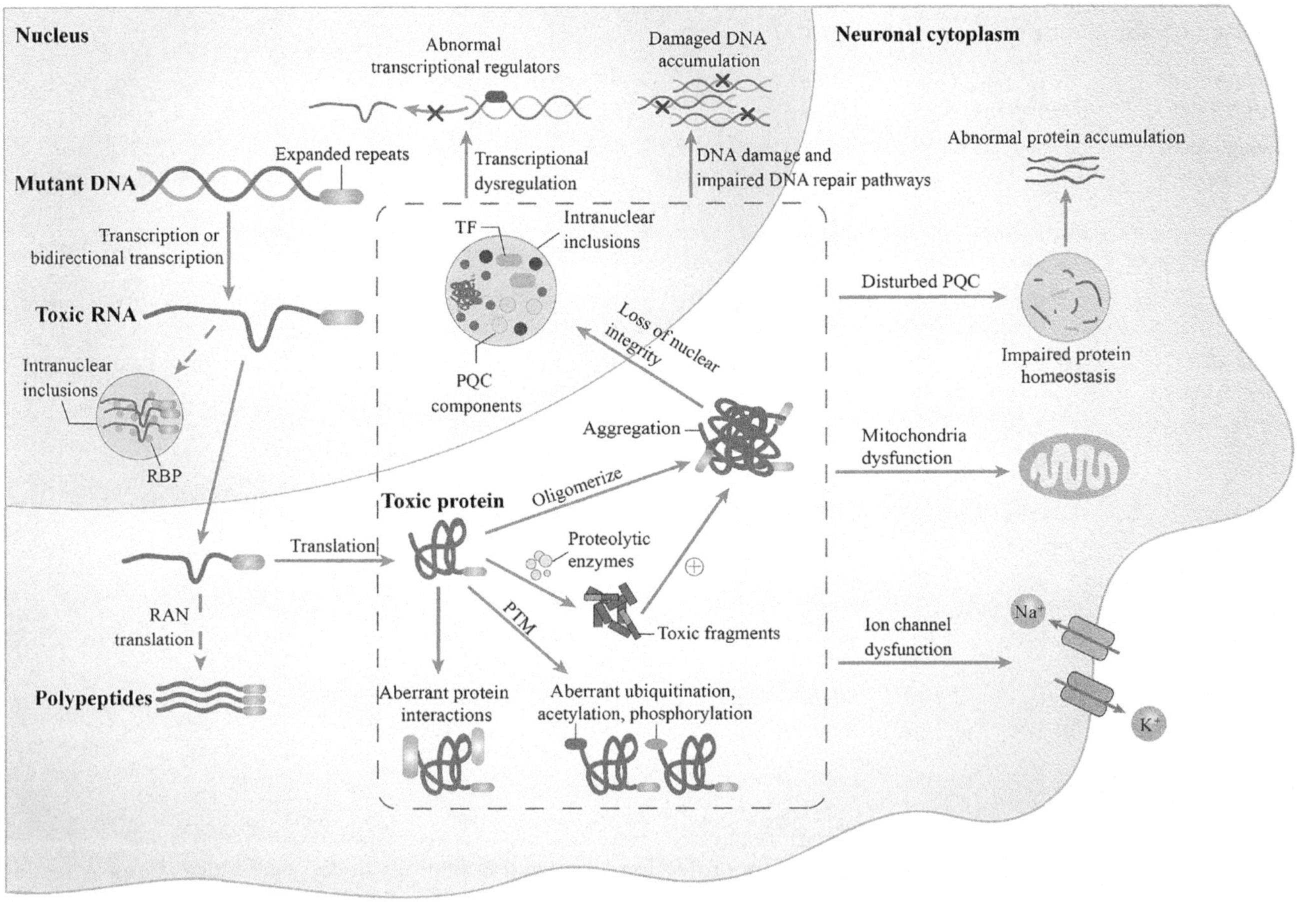

FIG. 13.1 Possible pathogenic mechanisms of spinocerebellar ataxias. Solid arrows indicate possible mechanisms for polyQ SCAs, and dashed arrows indicate possible mechanisms for SCAs caused by noncoding repeat expansions. RAN, non-AUG translation; RBP, RNA-binding protein; PQC, protein quality control; PTM, posttranslational modifier; TF, transcription factor. *(From Cui, Z.T., Mao, Z.T., Yang, R., Li, J.J., Jia, S.S., Zhao, J.L., Zhong, F.T., Yu, P., Dong, M., 2024. Spinocerebellar ataxias: from pathogenesis to recent therapeutic advances. Front. Neurosci. 18. https://doi.org/10.3389/fnins.2024.1422442.)*

factors, further contributing to neuronal damage (Dueñas et al., 2006; Vig et al., 2000). Ultimately, these aggregates impair neuronal function and integrity by causing ion channel dysfunction, mitochondrial dysfunction, transcriptional dysregulation, and DNA damage, leading to ataxia in the brain. This degeneration particularly affects Purkinje cells responsible for cerebellar motor output, as well as neurons in the inferior olivary nuclei, cerebellar dentate nuclei, and red nuclei, resulting in symptoms such as ataxia and dysarthria in SCA (Koeppen, 2005) (Fig. 13.1).

1.2 Currently approved therapeutic approaches for SCA

Although the etiology of SCA is well understood, no treatment has been found to completely halt the disease due to its genetic and heterogeneous neurodegenerative nature. Currently, no pharmacological treatment has been approved for routine use in SCA patients. However, various therapeutic approaches have been developed to alleviate symptoms and slow disease progression. Nonpharmacological approaches include physical therapy, occupational therapy, and speech therapy, which aim to improve balance, coordination, posture, and muscle strength. Pharmacological treatments primarily focus on symptom management, with drugs such as baclofen and levodopa being commonly used (Klockgether et al., 2019; Cui et al., 2024; Braga Neto et al., 2016) (Table 13.2).

Baclofen is a medication used to treat muscle spasticity. A study examining the experiences of four patients with chronic progressive hereditary ataxia undergoing intrathecal baclofen therapy reported improvements in body posture, movement, sleep, and pain management. However, the disease and treatment process also introduced challenges, including uncertainty

TABLE 13.2 Current treatment options for SCA.

Therapeutic approach	Drug(s)/strategy	Effect	Targeted SCA types	Approval status	References
Symptomatic treatments	Baclofen, Botox	Reduces muscle spasms and dystonia	Different SCA types	Approved	(Berntsson et al., 2017; Bushart et al., 2021; Thirumurugesan et al., 2022)
	Propranolol, primidone	Reduces tremors	Different SCA types	Approved	(Zhou et al., 2022; Naderi et al., 2012)
	Levodopa	Improves motor function	SCA2, SCA3	Approved	(Miranda and Cubo, 2022; Ferrara et al., 2010)
Disease-modifying therapies	Riluzole	Neuroprotective effects	SCA1, SCA2, SCA3	On progress	(Ayala et al., 2022; Coarelli et al., 2022)
	Coenzyme Q10	Supports mitochondrial function	SCA2, SCA3	Approved	(Wu et al., 2022)
	Varenicline	Improves gait and balance	SCA3	Approved	(Zesiewicz et al., 2012)
Experimental and other approaches	Stem cell therapy	Aims to regenerate nerve cells	Different SCA types	On progress	(Appelt et al., 2021)
	Immunomodulatory therapies	Modulates immune response	Different SCA types	On progress	(Egorova and Bezprozvanny, 2019)

about the future, the necessity of withdrawing from professional life, and limitations in daily activities (Berntsson et al., 2017). A combination of chlorzoxazone and baclofen has been shown to improve Purkinje neuron firing regularity and ameliorate motor deficits in SCA1 mouse models. Chlorzoxazone is an FDA-approved activator of calcium-activated potassium channels that normalizes Purkinje cell electrophysiology and action potential firing. In a mouse model, this combination improved motor performance without affecting muscle strength, and optimal dose levels were determined to minimize toxicity while improving ataxia symptoms, paving the way for clinical trials (Bushart et al., 2021). This treatment option may be beneficial for preventing muscle spasticity in SCA.

The underlying cause of cervical dystonia, hyperkinetic dysarthria, intentional hand tremors, progressive torticollis, and positional lumbar scoliosis in a 14-year-old boy with a normal developmental history was investigated. Genetic analysis identified a variant in the *TGM6* gene, leading to a diagnosis of SCA35. The patient showed no improvement with levodopa, baclofen, trihexyphenidyl, or clonazepam, and botulinum toxin (Botox) injection was administered to alleviate cervical dystonia. A significant improvement in torticollis was observed following the injection (Thirumurugesan et al., 2022). Botox could therefore be considered a treatment option for dystonia and muscle spasms in SCA.

Tremor, the most common movement disorder, is also observed in SCA patients and is associated with the cerebello-thalamo-cortical circuit. While medications can reduce tremor severity, no drug completely eliminates it, and in refractory cases, surgical interventions such as deep brain stimulation may be beneficial (Mukherjee and Pandey, 2024). Propranolol, a β-adrenergic receptor blocker, regulates cerebellar circuit activity and reduces both physiological and pathophysiological tremors. Studies in mouse models have demonstrated that propranolol alters Purkinje cell and cerebellar nucleus neuron activity, reducing pathological tremors in SCA models without correcting ataxia (Zhou et al., 2022). These findings suggest that propranolol could play an important role in tremor management in SCA patients. Similarly, primidone, a barbiturate used to treat partial and generalized seizures and essential tremors, has been shown to be effective and well-tolerated in reducing severe cerebellar tremors (Naderi et al., 2012).

Given that SCA3 presents with both cerebellar ataxia and Parkinsonian symptoms, two patients with a levodopa-responsive Parkinsonian phenotype were successfully treated using carbidopa–levodopa intestinal gel infusion pumps. These cases suggest that SCA3 should be considered in patients with early-onset Parkinsonian symptoms, and such patients may benefit from anti-Parkinson treatments (Miranda and Cubo, 2022). Similarly, in an SCA2 patient with Parkinsonian symptoms such as rigidity and bradykinesia, carbidopa–levodopa treatment was initiated. While initially effective, the treatment later led to levodopa-induced dyskinesias and motor fluctuations (Ferrara et al., 2010). This suggests that levodopa treatment may be effective in SCA patients before reaching a therapeutic threshold but requires careful dose adjustments to avoid adverse effects.

In addition to Parkinson's treatment, riluzole—used for amyotrophic lateral sclerosis and other motor neuron diseases—has also been investigated for SCA. Among three clinical trials, two reported statistically significant improvements, while one did not show any benefit; however, riluzole was found to be safe in all studies, supporting its potential use in hereditary ataxia treatment (Ayala et al., 2022). A study in France evaluated the efficacy of riluzole in SCA2 patients, randomly assigning 45 participants to receive riluzole (50 mg twice daily) or placebo for 12 months. The study found that riluzole did not clinically or radiologically improve disease progression but provided valuable data on symptom progression in SCA2 (Coarelli et al., 2022).

A study investigating the therapeutic effects of the antioxidant coenzyme Q10 (CoQ10) in an SCA3 mouse model found that a CoQ10-supplemented diet improved locomotor function, enhanced cerebellar cell survival, and facilitated autophagy-mediated clearance of mutant ATXN-3 proteins (Wu et al., 2022). These findings suggest that CoQ10 may help prevent muscle atrophy and could be a potential therapeutic option for SCA3. In addition, the nicotinic receptor partial agonist varenicline was evaluated for the symptomatic treatment of SCA3. In a study involving 20 patients, varenicline was administered for 8 weeks, resulting in significant improvements in axial symptoms (gait and posture) and rapid alternating movements. However, no improvement was observed in appendicular function (arm movements) (Zesiewicz et al., 2012). Varenicline may thus be beneficial for improving gait and balance in SCA patients.

Alternative approaches such as stem cell therapy, immunomodulatory treatments, and gene therapy are also being explored to reduce the overall medication burden and achieve more meaningful outcomes in SCA patients, while pharmacological treatments primarily aim to alleviate symptoms. Stem cells have been shown to function as immunomodulators and neuroprotectors in SCA patients and may help regenerate affected neurons (Egorova and Bezprozvanny, 2019). Although studies investigating stem cell therapy have demonstrated clinical improvement, no statistically significant difference has been found. This highlights the need for further large-scale, placebo-controlled studies to evaluate the efficacy of stem cell therapy for SCA (Appelt et al., 2021).

1.3 Gene therapy in SCA

Gene therapy is an emerging method aimed at eliminating various genetic diseases by targeting the underlying genetic mutations. It can be categorized into three main approaches: gene silencing using small interfering RNA (siRNA) and microRNA (miRNA), gene modification through the direct delivery of desired genes into the cell via plasmids and viral vectors, and gene editing using specific nucleases such as clustered regularly interspaced short palindromic repeats (CRISPR)/CRISPR-associated protein (Cas) nucleases (Sayed et al., 2022; Liu et al., 2024; Kohn et al., 2023).

Gene therapy is considered a promising approach for treating SCA, with its primary goal being to mitigate the toxic effects of expanded polyQ repeats in affected genes such as *ATXN1* (SCA1), *ATXN2* (SCA2), and *ATXN3* (SCA3). The main mechanisms of action include antisense oligonucleotides (ASOs), RNA interference (RNAi), and CRISPR/Cas-based gene editing, which aim to reduce mutant protein expression, regulate disease-associated pathways, or restore normal gene function (Cui et al., 2024; Ghanekar et al., 2022).

ASOs are molecules that target RNA transcripts specifically to slow or eliminate the progression of rare genetic diseases by binding to single-stranded mRNA and regulating gene expression. This method allows for personalized treatment options by selectively targeting specific mRNAs for patient groups or individual cases. ASOs are used to degrade mRNA and prevent the production of abnormal proteins (Rinaldi and Wood, 2018; Lauffer et al., 2024). However, not all genetic disorders are suitable for ASO-based therapies. In SCA, ASOs are being developed to reduce the production of mutant ataxin proteins caused by CAG trinucleotide repeat expansions in ATXN genes (Teive et al., 2018). By selectively targeting only the mutant mRNA without completely silencing normal gene expression, ASOs offer a safer therapeutic option. Similarly, RNAi provides a simple and rapid method for gene silencing by binding to mRNA. In this approach, small RNA molecules (siRNA, miRNA) associate with the RNA-induced silencing complex to target mRNA and mediate gene silencing. There are two main types of small RNA molecules used in RNAi: siRNA and miRNA. siRNAs are highly specific and are primarily used to decrease the translation of specific proteins, while miRNAs, which originate from single-stranded RNA precursors with characteristic hairpin structures, act as more general repressors (Hannon, 2002; Agrawal et al., 2003; Pushparaj et al., 2008). Similar to ASOs, RNAi therapies are being developed to target the mutant ATXN gene and reduce toxic protein levels in neurons.

CRISPR–Cas is one of the most powerful gene-editing tools used to directly correct genetic mutations that cause diseases. The Cas enzyme, guided by RNA, targets specific genomic regions and introduces double-strand breaks in DNA.

TABLE 13.3 Types of gene therapy methods and mechanisms in SCA.

Treatment strategy	Mechanism	Targeted types of SCA	Development stage
ASO	Binding to single-stranded mRNA to regulate gene expression and prevent abnormal protein production	SCA1, 2, 3, 36, 13	Preclinical stage
RNAi	siRNA and miRNA bind to the RISC complex to facilitate the destruction of target mRNA and prevent abnormal protein production	SCA1, 3, 7	Preclinical stage
CRISPR–Cas9	Direct regulation of mutated genes	SCA1, 3	Preclinical stage

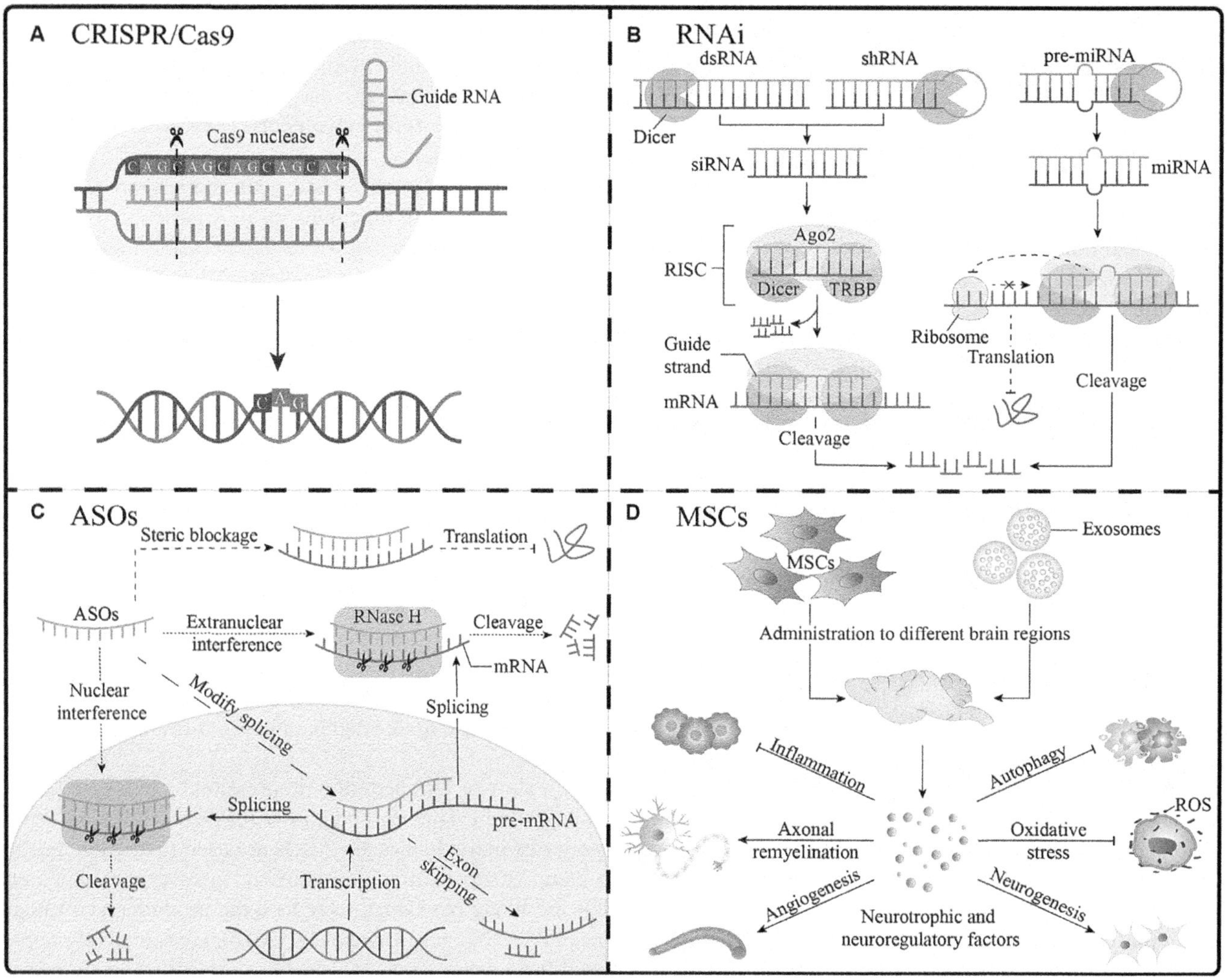

FIG. 13.2 Gene therapy and stem cell therapy for spinocerebellar ataxias. (A) The CRISPR/Cas9 platform enables gene editing by guiding RNA strands to bind to target DNA, as shown in the figure for the knockdown of the expanded CAG repeats. (B) Three RNAi molecules mediate the process of mRNA cleavage and inhibition of translation outside the nucleus. (C) The ASOs inhibit gene expression through mRNA cleavage, interference with splicing, and steric blockage. (D) After entering different regions of the brain, MSCs or their exosomes secrete a variety of neuroprotective or neuroregulatory factors, which may have the effects of promoting neurogenesis processes (e.g., neuronal proliferation, differentiation, and myelination), improving angiogenesis, acting as antioxidants, reducing inflammation, and inhibiting apoptosis, which is beneficial to neuronal cell survival and neurological function recovery. Ago2, argonaute2; ASOs, antisense oligonucleotides; CRISPR, clustered regularly interspaced short palindromic repeats; dsRNA, double-stranded RNA; miRNA, microRNA; MSCs, mesenchymal stem cells; RISC, RNA-induced silencing complexes; RNAi, RNA interference; ROS, reactive oxygen species; shRNA, short hairpin RNA; siRNA, small interfering RNA; TRBP, TAR RNA binding proteins. *(From Cui, Z.T., Mao, Z.T., Yang, R., Li, J.J., Jia, S.S., Zhao, J.L., Zhong, F.T., Yu, P., Dong, M., 2024. Spinocerebellar ataxias: from pathogenesis to recent therapeutic advances. Front. Neurosci. 18. https://doi.org/10.3389/fnins.2024.1422442.)*

This damage triggers the cell's repair mechanisms, leading to either random repair (resulting in gene mutation or silencing) or precise insertion of a desired DNA sequence, which can correct the mutation and provide a permanent cure (Li et al., 2023; Chavez et al., 2023; Wang and Doudna, 2023). In SCA, CRISPR–Cas9 is being explored for its potential to directly correct faulty genetic mutations. By excising or reducing CAG trinucleotide repeat expansions, CRISPR–Cas9 can mitigate the effects of the mutation and suppress mutant gene expression through epigenetic modifications, ultimately slowing neurodegeneration and enabling genetic correction in SCA (Miclăuş and Balmus, 2024) (Table 13.3 and Fig. 13.2).

Currently, gene therapy strategies for SCA remain in preclinical and early clinical stages. Studies on animal models, particularly mice, have demonstrated that reducing mutant ataxin protein levels can slow neurodegeneration and improve motor function. Although various gene therapy strategies have been tested in these models, no gene therapy has been approved for clinical use yet. However, ongoing in vitro and in vivo research continues to investigate various genetic approaches for the treatment of SCA.

2 Current advances in preclinical gene therapy for SCA

ASO-based preclinical studies for the treatment of SCAs have been tested on cellular or animal models of SCA1, 2, 3, 36, and 13. In SCA1, ASO-mediated suppression of *ATXN1* has been shown not to have adverse effects on tumor suppressor Capicua activity, BACE1 protease activation, or hippocampal neuronal precursor cells. The results support the idea that targeting *ATXN1* therapeutically for SCA1 could be safe and effective (O'Callaghan et al., 2020). ASO-(CUG)7 has been found to significantly reduce mutant *ATXN3* and *ATXN1* levels in patient-derived cell lines and mouse models. Furthermore, in the SCA3 model, it promoted the formation of a non-polyQ-containing, potentially less toxic form of *ATXN3* (Kourkouta et al., 2019).

In SCA2, ASO7 treatment has been found to reduce *ATXN2* mRNA and protein levels. Intraventricular injection of ASO7 reduced *ATXN2* levels in Purkinje cells to below 75%, improved motor functions without microglial activation, and normalized the firing impairment in these cells (Scoles et al., 2017). Similarly, ASO-1 treatment for SCA2 was observed to reduce *ATXN2* expression by up to 80% in both induced pluripotent stem cells (iPSCs) and mice, while also penetrating Purkinje cells. No increase in microglial activation (Iba1 expression) was observed after ASO-1 treatment, suggesting it as a reliable therapeutic candidate (Scoles et al., 2014).

A single ASO treatment can suppress both mutant and wild-type *ATXN3* levels by over 90%. However, to selectively target the mutant allele, specific ASOs have been developed utilizing the c.987G > C SNP in the *ATXN3* gene. Among them, ASOmut4 reduced mutant *ATXN3* levels by 80% while largely preserving the wild-type protein, with no observed neurotoxicity in cells (Hauser et al., 2022). Another study showed that in the SCA3 mouse model, intraventricular injection reduced disease-related ATXN3 protein levels by >50%. ASO treatment was effective in brain regions most affected by the disease, such as the cerebellum and diencephalon, with no signs of cell toxicity (Moore et al., 2017). SCA3-hESC (human embryonic stem cells) is a preferred cell line to study the molecular hallmarks of the disease and differentiation processes. In a study, ASO treatment in SCA3-hESC was shown to prevent p62-positive aggresome formation (Moore et al., 2019). This new SCA3-hESC line could be a powerful model for understanding disease mechanisms and testing potential treatments. In addition, ASOs that induce exon 10 skipping in *ATXN3* pre-mRNA have been shown to produce an ATXN3 form that lacks the toxic polyQ expansion but retains its ubiquitin-binding and cleavage functions. This strategy reduced insoluble ATXN3 and nuclear accumulation, indicating a potential therapeutic effect against disease pathology. These findings suggest that exon skipping could be a promising approach for SCA3 treatment (Toonen et al., 2017). In SCA3 mouse models, mutant ATXN3 protein disrupts oligodendrocyte maturation, but ASO treatment was shown to improve this process. Thus in addition to neuronal targeting, glial cells should also be considered in therapeutic strategies (Schuster et al., 2024). A study examining the effects of ASO on neurochemical abnormalities in SCA3 found that ASO treatment partially or completely corrected neurochemical changes in the cerebellum and brainstem. Increased total choline levels were associated with improved oligodendrocyte function. Corrected neurochemical levels correlated with improved motor function and reduced disease-associated protein accumulation (McLoughlin et al., 2023). ASO treatment appears to be a promising gene-silencing strategy for SCA3.

In SCA36, an expansion of a hexanucleotide repeat (GGCCTG) in the noncoding region of the NOP56 gene leads to RNA foci formation. In SCA36-iPSC cells, ASO treatment targeting NOP56 pre-mRNA reduced RNA foci by approximately 50% without altering mRNA levels (Matsuzono et al., 2017). In addition, since some SCA types result from ion channel dysfunction, targeting ion channels emerges as a therapeutic option. In SCA13, the $K_V3.3$-G592R mutation leads to cerebellar degeneration and motor dysfunction by increasing Tank-binding kinase 1 (TBK1) activation and disrupting Hax-1 protein. Experiments in mouse models have shown that ASO targeting $K_V3.3$ reduces TBK1 activation, restores Hax-1 levels, and improves motor function in mutant mice. This treatment has reversed the effects of the SCA13 mutation. Consequently, ASO therapy targeting the $K_V3.3$ channel may be a promising strategy for SCA13 (Zhang et al., 2021). The results suggest that a single ASO targeting CAG repeats could be applicable to multiple polyQ diseases, while ion channel targeting could also be effective.

RNAi-shRNA targeting polyQ-CAG repeats has been shown to specifically reduce mutant protein levels in patient-derived cells. This method offers a universal and allele-selective therapeutic strategy for polyQ diseases. The results indicate that RNAi triggers could serve as potential therapeutic agents by suppressing the expression of disease-associated mutant genes (Kotowska-Zimmer et al., 2020).

In SCA1 mouse models, recombinant adeno-associated virus (rAAV) vectors have been used to suppress mutant ATXN-1 expression via shRNA. This resulted in preserved motor coordination, cerebellar morphology, and dissolution of disease-specific ATXN-1 inclusions (Xia et al., 2004). In addition to partially suppressing mutant *ATXN1* gene expression, rAAV treatment was effective even after symptom onset, slowing and reversing neurodegeneration. These findings suggest that RNAi therapy in SCA1 could both prevent and reverse symptoms (Keiser, 2016). Overexpression of ATXN1-like (ATXN1L) via rAAV vectors also improved motor function and histological phenotypes, reducing the toxic effects of mutant ATXN1 (Keiser et al., 2013). Similarly, a combined approach using RNAi targeting ATXN1 and overexpression of ATXN1L showed that ATXN1L overexpression improved motor function and, when combined with RNAi against mutant *ATXN1*, restored disease-related gene expression changes. These findings suggest that a two-component gene therapy approach could be effective for SCA1, alleviating symptoms and providing genetic-level recovery (Carrell et al., 2022). Injection of AAV vectors targeting mutant ATXN1 expression into deep cerebellar nuclei in mice prevented neuronal degeneration and improved motor function. RNAi treatment preserved environmental adaptation performance, prevented transcriptional changes, and corrected gait abnormalities. The results suggest that targeting only the deep cerebellar nuclei with RNAi may be sufficient for SCA1 treatment and could guide human clinical applications (Keiser et al., 2014).

In SCA3, a large-scale RNAi screen in a transgenic *Caenorhabditis elegans* model identified 387 disease-related transcription factors. Inactivation of FKH-2/FOXG1 worsened movement impairment and neurodegeneration, while its overexpression reversed these effects. This suggests that FKH-2/FOXG1 has neuroprotective properties and could be a potential target for SCA3 treatment (Fardghassemi and Alex Parker, 2021). An AAV encoding miRATXN3 effectively and durably suppressed ATXN3 in the cerebellum of SCA3 mice. However, motor impairments did not improve, and survival was not extended. These findings suggest that targeting only the cerebellum may be insufficient, and broader brain-wide gene silencing strategies should be explored (Do Carmo Costa et al., 2013).

SCA7 is characterized by motor coordination loss and retinal degeneration. In a mouse model, RNAi-mediated reduction of both mutant and wild-type ATXN7 expression preserved retinal function. Reducing ATXN-7 by ≥50% prevented retinal degeneration without causing toxicity. This suggests that RNAi-based therapy could be a potential approach to prevent vision loss in SCA7 patients (Ramachandran et al., 2014). These results highlight the potential of RNAi therapies for both symptom relief and genetic correction in clinical applications.

In SCA1, Cas9 application in the B05 mouse model and SCA1-iPSC neurons reduced ATXN1 levels by 20% without causing inflammation, leading to behavioral improvement. These results, also validated in human iPSC-derived neurons, are promising for clinical applications (Fagan et al., 2024). Another study screened 10 different sgRNAs targeting exon 8 of the *ATXN1* gene, identifying the most effective ones, which were validated in fibroblasts from SCA1 patients. This confirmed that CRISPR/Cas9 successfully reduced ATXN1 expression in patient-derived fibroblasts (Pappadà et al., 2022).

For SCA3, CRISPR/Cas9-mediated homologous recombination corrected the expanded CAG repeats in SCA3-iPSC cells, preventing mutant ATXN3 accumulation and correcting disease-associated cellular abnormalities. After correction, cells retained pluripotency and successfully differentiated into healthy neurons with preserved electrophysiological properties (He et al., 2021). Similarly, CRISPR/Cas9 reversed disease-related phenotypes in cerebellar neurons during differentiation in SCA3-iPSC cells, preventing ATXN3 accumulation (Song et al., 2022). These results suggest that CRISPR/Cas9 could be a promising genetic therapy for SCA1 and SCA3 in the future. However, further testing in human-iPSC and animal models is necessary before clinical application (Table 13.4).

2.1 Current challenges and limitations in gene therapy for SCA

Gene therapy approaches, including ASO, RNAi, and CRISPR–Cas systems, hold great promise for the treatment of SCA. However, like all therapeutic strategies, these approaches come with challenges and limitations.

ASO- and RNAi-based therapies work by reducing the expression of mutant genes at the mRNA level, thereby lowering abnormal protein accumulation. However, these therapies may cause adverse side effects, including neurotoxicity and immune responses. In addition, since these gene therapies are susceptible to degradation by cellular mechanisms, their effects are temporary, requiring repeated administrations. This raises concerns regarding long-term treatment adherence.

TABLE 13.4 The summary of preclinical gene therapy, including ASO, RNAi, and CRISPR/Cas9 in SCA.

Gene therapy	Study type	Study material	Observation	Remarks	References
ASO	In vivo	Atxn1−/− mice	ASO-mediated suppression of *ATXN1* in SCA1 does not have adverse effects on tumor suppressor CIC activity, BACE1 protease activation, or hippocampal neuronal precursor cells.	Therapeutic targeting of *ATXN1* for SCA1 may be safe and effective.	(O'Callaghan et al., 2020)
	In vitro; in vivo	Patient-derived SCA1 and SCA3 fibroblasts; MJD84.2 and SCA1154Q/2Q mouse model	ASO-(CUG)7 significantly reduces mutant *ATXN-3* and *ATXN-1* levels.	Also, a less toxic form of *ATXN-3* that lacks polyglutamine was formed with ASO-(CUG)7.	(Kourkouta et al., 2019)
	In vivo	ATXN2-Q127 mice	ASO7 treatment in SCA2 reduces mRNA and protein levels of ATXN2.	ASO7 reduced ATXN2 levels in Purkinje cells by 75% and improved motor functions without microglial activation.	(Scoles et al., 2017)
	In vitro; in vivo	HepG2 cells; ATXN2-Q172 and BAC-ATXN2-Q72 transgenic mice	ASO-1 treatment for SCA2 reduced both human and mouse *ATXN2* expression by 80% and penetrated into Purkinje cells.	No increase in microglial activation (Iba1 expression) was observed after ASO-1 treatment, suggesting it as a reliable therapeutic candidate.	(Scoles et al., 2014)
	In vitro	SCA3-iPSC cells	A single ASO treatment can suppress both mutant and wild-type *ATXN-3* levels by >90%.	ASOmut4 reduces mutant ataxin-3 levels by 80% while largely preserving the wild-type protein.	(Hauser et al., 2022)
	In vivo	MJD-Q84.2 and MJD-Q135 mice	ASO reduces disease-associated ATXN3 protein levels by >50%.	ASO treatment is effective in the brain regions, including the cerebellum and diencephalon.	(Moore et al., 2017)
	In vitro	SCA3-hESC cells	ASO treatment in SCA3-hESC prevents p62-positive aggresome formation.	The SCA3-hESC line may be a powerful model for understanding disease mechanisms and testing potential treatments.	(Moore et al., 2019)
	In vivo	MJD84.2 SCA3 mouse model	Exon 10 skipping ASO therapy reduces insoluble ATXN-3 and nuclear accumulation.	Exon skipping strategy may be promising in SCA3 treatment.	(Toonen et al., 2017)
	In vivo	MJD-Q84 mice	ASO treatment suppresses mutant ATXN3 protein and improves oligodendrocyte maturation.	In addition to neuronal targeting, glial cells should also be included in treatment strategies.	(Schuster et al., 2024)
	In vivo	SCA3 mouse: YACMJD84.2	ASO treatment partially or completely corrects neurochemical changes in the cerebellum and brainstem.	Remediated neurochemical levels correlate with improved motor function and decreased disease-associated protein accumulation.	(McLoughlin et al., 2023)
	In vitro	SCA36-iPSCs cells and SCA36-iPSCs-derived neurons	ASO treatment targeting NOP56 pre-mRNA in SCA36-iPSC cells reduces RNA foci by approximately 50% without altering mRNA levels.	ASOs may be useful for preventing GGCCTG repeat expansion in SCA36.	(Matsuzono et al., 2017)
	In vitro; in vivo	Mouse hippocampal neurons; Kv3.3-G592R knock-in mice	ASO targeting Kv3.3 reduces Tbk1 activation in mutant mice, normalizes Hax-1 levels, and improves motor function.	ASO reversed the effects of the SCA13 mutation.	(Zhang et al., 2021)

Continued

TABLE 13.4 The summary of preclinical gene therapy, including ASO, RNAi, and CRISPR/Cas9 in SCA—Cont'd

Gene therapy	Study type	Study material	Observation	Remarks	References
RNAi	In vivo	SCA1 mouse	Mutant *ATXN-1* expression is suppressed with shRNA in SCA1.	Motor coordination and cerebellar morphology are preserved with shRNA.	(Xia et al., 2004)
	In vivo	SCA1-B05 mice	rAAV partially suppressed the expression of the mutant *ATXN1* gene and was effective even after symptom onset.	RNAi therapy in SCA1 may both prevent and reverse symptoms.	(Keiser, 2016)
	In vitro; in vivo	HEK293; SCA1-B05 *mice*	Overexpression of ATXN1-like also improves motor functions and histological phenotypes.	Overexpression of ATXN-1-like reduced the toxic effects of mutant *ATXN-1*.	(Keiser et al., 2013)
	In vivo	SCA1-B05 mice	Combined use of RNAi targeting mutant ATXN1 and ATXN1L improves motor function and normalizes disease-associated gene expression changes.	A two-component gene therapy approach may be effective for SCA1.	(Carrell et al., 2022)
	In vivo	SCA1-B05 mice	Injecting AAV vectors that reduce mutant *ATXN-1* expression into the deep cerebellar nuclei of mice prevents neuronal degeneration and improves motor function.	Targeting only the deep cerebellar nuclei with RNAi may be sufficient for the treatment of SCA1.	(Keiser et al., 2014)
	In vivo	ATXN3 transgenic *C. elegans* model	Inactivation of the *FKH-2/FOXG1* gene aggravates movement disorders and neurodegeneration.	*FKH-2/FOXG1* has neuroprotective properties and may be a potential target for SCA3 treatment.	(Fardghassemi and Alex Parker, 2021)
	In vivo	MJD-Q84 mouse	Despite the effective and long-lasting suppression of *ATXN3* in the cerebellum of SCA3 mice by an AAV encoding a microRNA called miRATXN3, motor impairment was not improved, and survival was not prolonged.	This may suggest that targeting the cerebellum alone may not be sufficient and that gene-silencing strategies targeting a wider brain region should be investigated.	(Do Carmo Costa et al., 2013)
	In vivo	BAC-PrP-SCA7-92Q transgenic mice	Reduction of *ATXN-7* by ≥50% prevented retinal degeneration without causing toxicity.	Indicating that RNAi-based therapy may be a potential method to prevent vision loss in SCA7 patients.	(Ramachandran et al., 2014)
CRISPR/ Cas9	In vitro; in vivo	SCA1-iPSCs cells, SCA1-iPSCs-derived neurons; SCA1-B05 mice	It has been determined that reducing ATXN1 levels by 20% provides behavioral improvement without causing inflammation.	The results, which have also been confirmed in human iPSC-derived neurons, can be considered promising for clinical application and SCA1 treatment.	(Fagan et al., 2024)
	In vitro	SCA1-human fibroblasts	10 different sgRNAs targeting Exon 8 of the ATXN1 gene were screened, and the most effective ones were determined and validated in fibroblasts obtained from SCA1 patients.	It was determined that CRISPR/Cas9 successfully reduced *ATXN1* expression in fibroblasts obtained from patients.	(Pappadà et al., 2022)
	In vitro	SCA3-iPSC	CRISPR/Cas9-mediated homologous recombination strategy corrected disease-causing expanded CAG repeats in SCA3-iPSC cells, preventing the accumulation of mutant *ATXN-3*.	After correction, cells were able to differentiate into healthy neuronal cells, maintaining pluripotency, and retaining electrophysiological properties.	(He et al., 2021)
	In vitro	SCA3 iPSC	CRISPR/Cas9 application in SCA3 reverses disease-associated phenotypes during differentiation into cerebellar neurons.	No accumulation of ATXN-3 was observed in CRISPR/Cas9-treated SCA3-iPSC cells.	(Song et al., 2022)

Moreover, delivering ASO or RNAi molecules effectively to deep brain regions, including the cerebellum—one of the most affected areas in SCA—remains a major challenge. Current delivery methods, such as intrathecal or intraventricular administration, require invasive procedures and may not achieve homogeneous distribution across neurons in the targeted regions. CRISPR–Cas9 gene editing aims to permanently correct SCA mutations by targeting disease-causing genes. However, given that CRISPR–Cas9 induces permanent genetic changes, unintended modifications could lead to irreversible defects in patients. The Cas enzyme, guided to the target site via guide RNA, can cause off-target genetic alterations, leading to disease-unrelated mutations or undesirable cellular effects. Similar to ASO and RNAi therapies, efficient delivery to cerebellar neurons remains a significant hurdle. The use of viral vectors presents additional limitations, such as triggering immune responses and having restricted cargo capacity (Lauffer et al., 2024; Hou et al., 2023; Yang et al., 2021; Aagaard and Rossi, 2007).

In gene therapy research, the prioritization of polyQ-repeat SCA types (such as SCA1, SCA2, and SCA3) in preclinical studies may delay the transition of non-polyQ SCA types into clinical trials. This could limit therapeutic options for patients with non-polyQ SCA, further underscoring the need for broader research efforts.

3 Conclusion

In conclusion, while ASO-, RNAi-, and CRISPR–Cas-based strategies are promising candidates for treating SCAs and other genetic disorders, their transition into clinical therapies requires addressing challenges related to delivery, specificity, and safety. Further preclinical studies and trials across different SCA subtypes are essential to ensure their reliability and efficacy.

References

Aagaard, L., Rossi, J.J., 2007. RNAi therapeutics: principles, prospects and challenges. Adv. Drug Deliv. Rev. 0169409X. 59 (2–3), 75–86. doi:10.1016/j.addr.2007.03.005.

Agrawal, N., Dasaradhi, P.V.N., Mohmmed, A., Malhotra, P., Bhatnagar, R.K., Mukherjee, S.K., 2003. RNA interference: biology, mechanism, and applications. Microbiol. Mol. Biol. Rev. 67 (4), 657–685. doi:10.1128/MMBR.67.4.657-685.2003.

Ando, H., Hirose, M., Mikoshiba, K., 2018. Aberrant IP3 receptor activities revealed by comprehensive analysis of pathological mutations causing spinocerebellar ataxia 29. Proc. Natl. Acad. Sci. USA. 10916490. 115 (48), 12259–12264. Available from: http://www.pnas.org/content/pnas/115/48/12259.full.pdf. doi:10.1073/pnas.1811129115.

Appelt, P.A., Comella, K., de Souza, L.A.P.S., Luvizutto, G.J., 2021. Effect of stem cell treatment on functional recovery of spinocerebellar ataxia: systematic review and meta-analysis. Cerebellum Ataxias. 20538871. 8 (1). Available from: http://www.cerebellumandataxias.com/. doi:10.1186/s40673-021-00130-8.

Ayala, I.N., Aziz, S., Argudo, J.M., Yepez, M., Camacho, M., Ojeda, D., Aguirre, A.S., Oña, S., Andrade, A.F., Vasudhar, A., Moncayo, J.A., Hassen, G., Ortiz, J.F., Tambo, W., 2022. Use of riluzole for the treatment of hereditary ataxias: a systematic review. Brain Sci. 20763425. 12 (8). Available from: http://www.mdpi.com/journal/brainsci. doi:10.3390/brainsci12081040.

Bahl, S., Virdi, K., Mittal, U., Sachdeva, M.P., Kalla, A.K., Holmes, S.E., O'Hearn, E., Margolis, R.L., Jain, S., Srivastava, A.K., Mukerji, M., 2005. Evidence of a common founder for SCA12 in the Indian population. Ann. Hum. Genet. 0003-4800. 69 (5), 528–534. doi:10.1046/j.1529-8817.2005.00173.x.

Berntsson, S.G., Landtblom, A.M., Flensner, G., 2017. Cerebellar ataxia and intrathecal baclofen therapy: focus on patients' experiences. PLoS ONE. 19326203. 12 (6). Available from: http://journals.plos.org/plosone/article/file?id=10.1371/journal.pone.0180054&type=printable. doi:10.1371/journal.pone.0180054.

Bettencourt, C., Santos, C., Montiel, R., Kay, T., Vasconcelos, J., Maciel, P., Lima, M., 2010. The (CAG)n tract of Machado–Joseph Disease gene (ATXN3): a comparison between DNA and mRNA in patients and controls. Eur. J. Hum. Genet. 1018-4813. 18 (5), 621–623. doi:10.1038/ejhg.2009.215.

Bouwman, L.F., Joosen, M.E.M., Buijsen, R.A.M., van der Graaf, L.M., Pepers, B.A., Voesenek, B.J.B., Brosens, E., van de Warrenburg, B.P.C., van Roon-mom, W.M.C., 2024. Generation of human induced pluripotent stem cell lines (LUMCi051-A,B and LUMCi052-A,B,C) of two patients with Spinocerebellar ataxia type 7. Stem Cell Res. 18767753. 78. Available from: https://www.sciencedirect.com/science/journal/18735061. doi:10.1016/j.scr.2024.103462.

Brusco, A., Gregorio, B.B., 2019. Spinocerebellar Ataxia Type 38 ., GeneReviews®, University of Washington.

Brussino, A., Brusco, A., Durr, A., Mancini, C., et al., 2011. Spinocerebellar Ataxia Type 28 ., GeneReviews®. University of Washington.

Bushart, D.D., Huang, H., Man, L.J., Morrison, L.M., Shakkottai, V.G., 2021. A chlorzoxazone-baclofen combination improves cerebellar impairment in spinocerebellar ataxia type 1. Mov. Disord. 15318257. 36 (3), 622–631. Available from: http://onlinelibrary.wiley.com/journal/10.1002/(ISSN)1531-8257. doi:10.1002/mds.28355.

Carrell, E.M., Keiser, M.S., Robbins, A.B., Davidson, B.L., 2022. Combined overexpression of ATXN1L and mutant ATXN1 knockdown by AAV rescue motor phenotypes and gene signatures in SCA1 mice. Mol. Ther. Methods Clin. Dev. 23290501. 25, 333–343. Available from: https://www.journals.elsevier.com/molecular-therapy-methods-and-clinical-development/. doi:10.1016/j.omtm.2022.04.004.

Chavez, M., Chen, X., Finn, P.B., Qi, L.S., 2023. Advances in CRISPR therapeutics. Nat. Rev. Nephrol. 1759507X. 19 (1), 9–22. Available from: https://www.nature.com/nrneph/. doi:10.1038/s41581-022-00636-2.

Chen, D.H., Bird, T.D., Raskind, W.H., 2005. Spinocerebellar Ataxia Type 14 ., GeneReviews®, University of Washington.

Chen, X., Shi, C., He, M., Xiong, S., Xia, X., 2023. Endoplasmic reticulum stress: molecular mechanism and therapeutic targets. Signal Transduct. Target. Ther. 2059-3635. 8 (1). doi:10.1038/s41392-023-01570-w.

Chien, H.F., Zonta, M.B., Chen, J., Diaferia, G., Viana, C.F., Teive, H.A.G., Pedroso, J.L., Barsottini, O.G.P., 2022. Rehabilitation in patients with cerebellar ataxias. Arquivos de Neuro-Psiquiatria. 16784227. 80 (3), 306–315. Available from: https://www.scielo.br/j/anp/i/2022.v80n3/. doi:10.1590/0004-282X-ANP-2021-0065.

Coarelli, G., Heinzmann, A., Ewenczyk, C., Fischer, C., Chupin, M., Monin, M.L., Hurmic, H., Calvas, F., Calvas, P., Goizet, C., Thobois, S., Anheim, M., Nguyen, K., Devos, D., Verny, C., Ricigliano, V.A.G., Mangin, J.F., Brice, A., du Montcel, S.T., Durr, A., 2022. Safety and efficacy of riluzole in spinocerebellar ataxia type 2 in France (ATRIL): a multicentre, randomised, double-blind, placebo-controlled trial. Lancet Neurol. 14744465. 21 (3), 225–233. Available from: http://www.journals.elsevier.com/the-lancet-neurology/. doi:10.1016/S1474-4422(21)00457-9.

Coebergh, J.A., van de Putte, D.E.F., Snoeck, I.N., Ruivenkamp, C., van Haeringen, A., Smit, L.M., 2014. A new variable phenotype in spinocerebellar ataxia 27 (SCA 27) caused by a deletion in the FGF14 gene. Eur. J. Paediatr. Neurol. 10903798. 18 (3), 413–415. doi:10.1016/j.ejpn.2013.10.006.

Cui, Z.T., Mao, Z.T., Yang, R., Li, J.J., Jia, S.S., Zhao, J.L., Zhong, F.T., Yu, P., Dong, M., 2024. Spinocerebellar ataxias: from pathogenesis to recent therapeutic advances. Front. Neurosci. 1662453X. 18. Available from: https://www.frontiersin.org/journals/neuroscience#. doi:10.3389/fnins.2024.1422442.

Delplanque, J., Devos, D., Huin, V., Genet, A., Sand, O., Moreau, C., Goizet, C., Charles, P., Anheim, M., Monin, M.L., Buée, L., Destée, A., Grolez, G., Delmaire, C., Dujardin, K., Dellacherie, D., Brice, A., Stevanin, G., Strubi-Vuillaume, I., Dürr, A., Sablonnière, B., 2014. TMEM240 mutations cause spinocerebellar ataxia 21 with mental retardation and severe cognitive impairment. Brain. 14602156. 137 (10), 2657–2663. Available from: http://brain.oxfordjournals.org/. doi:10.1093/brain/awu202.

Dick, K.A., Ikeda, Y., Day, J.W., Ranum, L.P.W., 2012. Spinocerebellar ataxia type 5. Handb. Clin. Neurol. 103, 451–459. Available from: http://www.sciencedirect.com/science/journal/00729752. doi:10.1016/B978-0-444-51892-7.00028-0.

Do Carmo Costa, M., Luna-Cancalon, K., Fischer, S., Ashraf, N.S., Ouyang, M., Dharia, R.M., Martin-Fishman, L., Yang, Y., Shakkottai, V.G., Davidson, B.L., Rodríguez-Lebrón, E., Paulson, H.L., 2013. Toward RNAi therapy for the polyglutamine disease Machado-Joseph disease. Mol. Ther. 15250024. 21 (10), 1898–1908. Available from: https://www.journals.elsevier.com/molecular-therapy. doi:10.1038/mt.2013.144.

Du, X., Gomez, C.M., 2018. Spinocerebellum ataxia type 6: molecular mechanisms and calcium channel genetics. Adv. Exp. Med. Biol. 1049, 147–173. Available from: http://www.springer.com/series/5584. doi:10.1007/978-3-319-71779-1_7.

Duarri, A., Lin, M.C.A., Fokkens, M.R., Meijer, M., Smeets, C.J.L.M., Nibbeling, E.A.R., Boddeke, E., Sinke, R.J., Kampinga, H.H., Papazian, D.M., Verbeek, D.S., 2015. Spinocerebellar ataxia type 19/22 mutations alter heterocomplex Kv4.3 channel function and gating in a dominant manner. Cell. Mol. Life Sci. 14209071. 72 (17), 3387–3399. Available from: http://link.springer.de/link/service/journals/00018/index.htm. doi:10.1007/s00018-015-1894-2.

Dueñas, A.M., Goold, R., Giunti, P., 2006. Molecular pathogenesis of spinocerebellar ataxias. Brain. 14602156. 129 (6), 1357–1370. Available from: http://brain.oxfordjournals.org/. doi:10.1093/brain/awl081.

Echeveste, B., Covarrubias, E., Cabello, J.P., Luquin, M.R., 2017. Isolated palatal tremor as unique clinical manifestation of SCA 18 due to a new mutation of IFRD1. Parkinsonism Relat. Disord. 18735126. 42, 100–101. Available from: www.elsevier.com/locate/parkreldis. doi:10.1016/j.parkreldis.2017.05.026.

Egorova, P.A., Bezprozvanny, I.B., 2019. Molecular mechanisms and therapeutics for spinocerebellar ataxia type 2. Neurotherapeutics. 18787479. 16 (4), 1050–1073. Available from: http://www.springerlink.com/content/1933-7213/. doi:10.1007/s13311-019-00777-6.

Fagan, K.J., Chillon, G., Carrell, E.M., Waxman, E.A., Davidson, B.L., 2024. Cas9 editing of ATXN1 in a spinocerebellar ataxia type 1 mice and human iPSC-derived neurons. Mol. Ther. Nucleic Acids. 21622531. 35 (4). Available from: https://www.sciencedirect.com/science/journal/21622531. doi:10.1016/j.omtn.2024.102317.

Fardghassemi, Y., Alex Parker, J., 2021. Overexpression of FKH-2/FOXG1 is neuroprotective in a C. elegans model of Machado-Joseph disease. Exp. Neurol. 00144886. 337, 113544. doi:10.1016/j.expneurol.2020.113544.

Felício, D., Santos, M., 2024. Spinocerebellar ataxia type 11 (SCA11): TTBK2 variants, functions and associated disease mechanisms. Cerebellum. 1473-4230. 23 (2), 678–687. doi:10.1007/s12311-023-01540-6.

Ferrara, J.M., Adam, O.R., Ondo, W.G., 2010. Levodopa-induced dyskinesias in spinocerebellar ataxia type 2. Arch. Neurol. 0003-9942. 67 (1), 114–115. doi:10.1001/archneurol.2009.291.

Flanigan, K., Gardner, K., Alderson, K., Galster, B., Otterud, B., Leppert, M.F., Kaplan, C., Ptáček, L.J., 1996. Autosomal dominant spinocerebellar ataxia with sensory axonal neuropathy (SCA4): clinical description and genetic localization to chromosome 16q22.1. Am. J. Hum. Genet. 00029297. 59 (2), 392–399. Available from: http://www.elsevier.com/wps/find/journaldescription.cws_home/713561/description#description.

Fujioka, S., Sundal, C., Wszolek, Z.K., 2013. Autosomal dominant cerebellar ataxia type III: a review of the phenotypic and genotypic characteristics. Orphanet J. Rare Dis. 17501172. 8. doi:10.1186/1750-1172-8-14.

Ghanekar, S.D., Kuo, S.H., Staffetti, J.S., Zesiewicz, T.A., 2022. Current and emerging treatment modalities for spinocerebellar ataxias. Expert Rev. Neurother. 17448360. 22 (2), 101–114. Available from: http://www.tandfonline.com/loi/iern20. doi:10.1080/14737175.2022.2029703.

Guo, J., Cui, Y., Liu, Q., Yang, Y., Li, Y., Weng, L., Tang, B., Jin, P., Li, X.J., Yang, S., Li, S., 2018. Piperine ameliorates SCA17 neuropathology by reducing ER stress. Mol. Neurodegener. 17501326. 13 (1). Available from: http://www.molecularneurodegeneration.com/. doi:10.1186/s13024-018-0236-x.

Hannon, G.J., 2002. RNA interference. Nature 418 (6894), 244–251. doi:10.1038/418244a.

Hauser, S., Helm, J., Kraft, M., Korneck, M., Hübener-Schmid, J., Schöls, L., 2022. Allele-specific targeting of mutant ataxin-3 by antisense oligonucleotides in SCA3-iPSC-derived neurons. Mol. Ther. Nucleic Acids. 21622531. 27, 99–108. Available from: https://www.journals.elsevier.com/molecular-therapy-nucleic-acids. doi:10.1016/j.omtn.2021.11.015.

He, L., Wang, S., Peng, L., Zhao, H., Shuai, L., Han, X., de Dieu Habimana, J., Chen, Z., Wang, C., Peng, Y., Peng, H., Xie, Y., Lei, L., Deng, Q., Wan, L., Wan, N., Yuan, H., Gong, Y., Zou, G., Zhiyuan, L., Tang, B., Jiang, H., 2021. CRISPR/Cas9 mediated gene correction ameliorates abnormal phenotypes in spinocerebellar ataxia type 3 patient-derived induced pluripotent stem cells. Transl. Psychiatry. 2158-3188. 11 (1). doi:10.1038/s41398-021-01605-2.

Hou, B., Hai, Y., Buyin, B., Hasi, S., 2023. Research progress and limitation analysis of RNA interference in Haemonchus contortus in China. Front. Vet. Sci. 2297-1769. 10. doi:10.3389/fvets.2023.1079676.

Huang, L., Chardon, J.W., Carter, M.T., Friend, K.L., Dudding, T.E., Schwartzentruber, J., Zou, R., Schofield, P.W., Douglas, S., Bulman, D.E., Boycott, K.M., 2012. Missense mutations in ITPR1 cause autosomal dominant congenital nonprogressive spinocerebellar ataxia. Orphanet J. Rare Dis. 17501172. 7 (1). Available from: http://www.ojrd.com/. doi:10.1186/1750-1172-7-67.

Ishiguro, T., Nagai, Y., Ishikawa, K., 2021. Insight into spinocerebellar ataxia type 31 (SCA31) from drosophila model. Front. Neurosci. 1662-453X. 15. doi:10.3389/fnins.2021.648133.

Iwaki, A., Kawano, Y., Miura, S., Shibata, H., Matsuse, D., Li, W., Furuya, H., Ohyagi, Y., Taniwaki, T., Kira, J., Fukumaki, Y., 2008. Heterozygous deletion of ITPR1, but not SUMF1, in spinocerebellar ataxia type 16. J. Med. Genet. 00222593. 45 (1), 32–35. doi:10.1136/jmg.2007.053942.

Jacobi, H., Schaprian, T., Schmitz-Hübsch, T., Schmid, M., Klockgether, T., 2023. Disease progression of spinocerebellar ataxia types 1, 2, 3 and 6 before and after ataxia onset. Ann. Clin. Transl. Neurol. 23289503. 10 (10), 1833–1843. Available from: http://onlinelibrary.wiley.com/journal/10.1002/(ISSN)2328-9503. doi:10.1002/acn3.51875.

Kang, S., Hong, S., 2009. Molecular pathogenesis of spinocerebellar ataxia type 1 disease. Mol. Cells. 10168478. 27 (6), 621–627. doi:10.1007/s10059-009-0095-y.

Keiser, M.S., Geoghegan, J.C., Boudreau, R.L., Lennox, K.A., Davidson, B.L., 2013. RNAi or overexpression: alternative therapies for spinocerebellar ataxia type 1. Neurobiol. Dis. 1095953X. 56, 6–13. doi:10.1016/j.nbd.2013.04.003.

Keiser, M.S., Boudreau, R.L., Davidson, B.L., 2014. Broad therapeutic benefit after RNAi expression vector delivery to deep cerebellar nuclei: implications for spinocerebellar ataxia type 1 therapy. Mol. Ther. 15250024. 22 (3), 588–595. Available from: https://www.journals.elsevier.com/molecular-therapy. doi:10.1038/mt.2013.279.

Keiser, M.S., 2016. Translating RNAi therapy for Spinocerebellar Ataxia 1 to the clinic. Mol. Ther. doi:10.1016/S1525-0016(16)33391-3.

Klockgether, T., Mariotti, C., Paulson, H.L., 2019. Spinocerebellar ataxia. Nat. Rev. Dis. Primers. 2056676X. 5 (1). Available from: http://www.nature.com/nrdp/. doi:10.1038/s41572-019-0074-3.

Knight, M.A., Hernandez, D., Diede, S.J., Dauwerse, H.G., Rafferty, I., van de Leemput, J., Forrest, S.M., Gardner, R.J.M., Storey, E., van Ommen, G.-J.B., Tapscott, S.J., Fischbeck, K.H., Singleton, A.B., 2008. A duplication at chromosome 11q12.2-11q12.3 is associated with spinocerebellar ataxia type 20. Hum. Mol. Genet. 0964-6906. 17 (24), 3847–3853. doi:10.1093/hmg/ddn283.

Kobayashi, H., Abe, K., Matsuura, T., Ikeda, Y., Hitomi, T., Akechi, Y., Habu, T., Liu, W., Okuda, H., Koizumi, A., 2011. Expansion of intronic GGCCTG hexanucleotide repeat in NOP56 causes SCA36, a type of spinocerebellar ataxia accompanied by motor neuron involvement. Am. J. Hum. Genet. 15376605. 89 (1), 121–130. Available from: http://www.elsevier.com/wps/find/journaldescription.cws_home/713561/description#description. doi:10.1016/j.ajhg.2011.05.015.

Koeppen, A.H., 2005. The pathogenesis of spinocerebellar ataxia. Cerebellum 4 (1), 62–73. doi:10.1080/14734220510007950.

Kohn, D.B., Chen, Y.Y., Spencer, M.J., 2023. Successes and challenges in clinical gene therapy. Gene Ther. 14765462. 30 (10–11), 738–746. Available from: https://www.nature.com/gt/. doi:10.1038/s41434-023-00390-5.

Kotowska-Zimmer, A., Ostrovska, Y., Olejniczak, M., 2020. Universal RNAi Triggers for the Specific Inhibition of Mutant Huntingtin, Atrophin-1, Ataxin-3, and Ataxin-7 Expression. Mol. Ther. Nucleic Acids. 21622531. 19, 562–571. doi:10.1016/j.omtn.2019.12.012.

Kourkouta, E., Weij, R., González-Barriga, A., Mulder, M., Verheul, R., Bosgra, S., Groenendaal, B., Puoliväli, J., Toivanen, J., van Deutekom, J.C.T., Datson, N.A., 2019. Suppression of mutant protein expression in SCA3 and SCA1 Mice Using a CAG Repeat-Targeting Antisense Oligonucleotide. Mol. Ther. Nucleic Acids. 21622531. 17, 601–614. Available from: https://www.journals.elsevier.com/molecular-therapy-nucleic-acids. doi:10.1016/j.omtn.2019.07.004.

Lauffer, M.C., van Roon-mom, W., Aartsma-Rus, A., 2024. Possibilities and limitations of antisense oligonucleotide therapies for the treatment of monogenic disorders. Commun. Med. 2730664X. 4 (1). Available from: https://www.nature.com/commsmed/. doi:10.1038/s43856-023-00419-1.

Lee, D., Lee, Y.I., Lee, Y.S., Lee, S.B., 2020. The Mechanisms of Nuclear Proteotoxicity in Polyglutamine Spinocerebellar Ataxias. Front. Neurosci. 1662453X. 14. Available from: https://www.frontiersin.org/journals/neuroscience#. doi:10.3389/fnins.2020.00489.

Li, T., Yang, Y., Qi, H., Cui, W., Zhang, L., Fu, X., He, X., Liu, M., Li, P.F., Yu, T., 2023. CRISPR/Cas9 therapeutics: progress and prospects. Signal Transduct. Target. Ther. 20593635. 8 (1). Available from: https://www.nature.com/sigtrans. doi:10.1038/s41392-023-01309-7.

Liu, F., Li, R., Zhu, Z., Yang, Y., Lu, F., 2024. Current developments of gene therapy in human diseases. MedComm. 26882663. 5 (9). Available from: https://onlinelibrary.wiley.com/journal/26882663. doi:10.1002/mco2.645.

Maass, F., Jamous, A., Biskup, S., Eisenberg, H., D'Hedouville, Z., Bähr, M., van Riesen, C., 2023. Spinocerebellar Ataxia Type 35 Caused by a New TGM6 Variant: video Documentation of a German Family. Mov. Disord. Clin. Pract. 23301619. 10 (6), 1016–1019. Available from: http://onlinelibrary.wiley.com/journal/10.1002/(ISSN)2330-1619. doi:10.1002/mdc3.13717.

Matilla-Dueñas, A., 2012. Machado-Joseph disease and other rare spinocerebellar ataxias. Adv. Exp. Med. Biol. 00652598. 724, 172–188. doi:10.1007/978-1-4614-0653-2_14.

Matilla-Dueñas, A., Volpini, V., 2019. Spinocerebellar Ataxia Type 37.

Matsuzono, K., Imamura, K., Murakami, N., Tsukita, K., Yamamoto, T., Izumi, Y., Kaji, R., Ohta, Y., Yamashita, T., Abe, K., Inoue, H., 2017. Antisense Oligonucleotides Reduce RNA Foci in Spinocerebellar Ataxia 36 Patient iPSCs. Mol. Ther. Nucleic Acids. 21622531. 8, 211–219. doi:10.1016/j.omtn.2017.06.017.

McLoughlin, H.S., Gundry, K., Rainwater, O., Schuster, K.H., Wellik, I.G., Zalon, A.J., Benneyworth, M.A., Eberly, L.E., Öz, G., 2023. Antisense Oligonucleotide Silencing Reverses Abnormal Neurochemistry in Spinocerebellar Ataxia 3 Mice. Ann. Neurol. 15318249. 94 (4), 658–671. Available from: http://onlinelibrary.wiley.com/journal/10.1002/(ISSN)1531-8249. doi:10.1002/ana.26713.

Miclăuş, M., Balmus, G., 2024. CRISPR-Cas9-directed gene therapy for spinocerebellar ataxia type 1. Mol. Ther. Nucleic Acids. 21622531. 35 (4), 102377. doi:10.1016/j.omtn.2024.102377.

Miranda, J., Cubo, E., 2022. Spinocerebellar ataxia type 3: response to levodopa infusion in two cases. Neurol. Sci. 1590-1874. 43 (5), 3423–3425. doi:10.1007/s10072-022-05962-8.

Moore, L.R., Rajpal, G., Dillingham, I.T., Qutob, M., Blumenstein, K.G., Gattis, D., Hung, G., Kordasiewicz, H.B., Paulson, H.L., McLoughlin, H.S., 2017. Evaluation of Antisense Oligonucleotides Targeting ATXN3 in SCA3 Mouse Models. Mol. Ther. Nucleic Acids. 21622531. 7, 200–210. Available from: https://www.journals.elsevier.com/molecular-therapy-nucleic-acids. doi:10.1016/j.omtn.2017.04.005.

Moore, L.R., Keller, L., Bushart, D.D., Delatorre, R.G., Li, D., McLoughlin, H.S., Do Carmo Costa, M., Shakkottai, V.G., Smith, G.D., Paulson, H.L., 2019. Antisense oligonucleotide therapy rescues aggresome formation in a novel spinocerebellar ataxia type 3 human embryonic stem cell line. Stem Cell Res. 18767753. 39. Available from: http://www.elsevier.com/wps/find/journaldescription.cws_home/711630/description#description. doi:10.1016/j.scr.2019.101504.

Mukherjee, A., Pandey, S., 2024. Tremor in Spinocerebellar Ataxia: a Scoping Review. Tremor Other Hyperkinet. Mov. 2160-8288. 14 (1). doi:10.5334/tohm.911.

Müller, U., 2021. Spinocerebellar ataxias (SCAs) caused by common mutations. Neurogenetics. 1364-6745. 22 (4), 235–250. doi:10.1007/s10048-021-00662-5.

Naderi, F., Javadi, S.A., Motamedi, M., Sahraian, M.A., 2012. The efficacy of primidone in reducing severe cerebellar tremors in patients with multiple sclerosis. Clin. Neuropharmacol. 1537162X. 35 (5), 224–226. doi:10.1097/WNF.0b013e31826249bb.

Neto, P.B., Luiz Pedroso, J., Sheng-han Kuo, C.F.M.J., Afonso Ghizoni Teive, H., Barsottini, O.G.P., 2016. Current concepts in the treatment of hereditary ataxias. Arquivos de Neuro-Psiquiatria. 0004-282X. 74 (3), 244–252. doi:10.1590/0004-282X20160038.

O'Callaghan, B., Hofstra, B., Handler, H.P., Kordasiewicz, H.B., Cole, T., Duvick, L., Friedrich, J., Rainwater, O., Yang, P., Benneyworth, M., Nichols-Meade, T., Heal, W., Haar, R.T., Henzler, C., Orr, H.T., 2020. Antisense Oligonucleotide Therapeutic Approach for Suppression of Ataxin-1 Expression: a Safety Assessment. Mol. Ther. Nucleic Acids. 21622531. 21, 1006–1016. Available from: https://www.journals.elsevier.com/molecular-therapy-nucleic-acids. doi:10.1016/j.omtn.2020.07.030.

Opal, Ashizawa, 1998. Spinocerebellar Ataxia Type 1.

Orozco-Gutiérrez, M.H., Cervantes-Aragón, I., García-Cruz, D., 2017. Ethical considerations in presymptomatic diagnosis of autosomal dominant spinocerebellar ataxias. Neurologia. 15781968. 32 (7), 469–475. Available from: http://www.sciencedirect.com/science/journal/02134853. doi:10.1016/j.nrl.2015.06.004.

Pappadà, M., Bonuccelli, O., Buratto, M., Fontana, R., Sicurella, M., Caproni, A., Fuselli, S., Benazzo, A., Bertorelli, R., De Sanctis, V., Cavallerio, P., Simioni, V., Tugnoli, V., Salvatori, F., Marconi, P., 2022. Suppressing gain-of-function proteins via CRISPR/Cas9 system in SCA1 cells. Sci. Rep. 2045-2322. 12 (1). doi:10.1038/s41598-022-24299-y.

Paulson, H., Shakkottai, V., 1998. Spinocerebellar Ataxia Type 3.

Pushparaj, P.N., Aarthi, J.J., Manikandan, J., Kumar, S.D., 2008. siRNA, miRNA, and shRNA: in vivo Applications. J. Dent. Res. 0022-0345. 87 (11), 992–1003. doi:10.1177/154405910808701109.

Ramachandran, P.S., Bhattarai, S., Singh, P., Boudreau, R.L., Thompson, S., LaSpada, A.R., Drack, A.V., Davidson, B.L., 2014. RNA interference-based therapy for spinocerebellar ataxia type 7 retinal degeneration. PLoS ONE. 19326203. 9 (4). Available from: http://www.plosone.org/article/fetchObject.action?uri=info%3Adoi%2F10.1371%2Fjournal.pone.0095362&representation=PDF. doi:10.1371/journal.pone.0095362.

Rinaldi, C., Wood, M.J.A., 2018. Antisense oligonucleotides: the next frontier for treatment of neurological disorders. Nat. Rev. Neurol. 17594766. 14 (1), 9–22. Available from: http://www.nature.com/nrneurol/archive/index.html. doi:10.1038/nrneurol.2017.148.

Ruano, L., Melo, C., Silva, M.C., Coutinho, P., 2014. The global epidemiology of hereditary ataxia and spastic paraplegia: a systematic review of prevalence studies. Neuroepidemiology. 14230208. 42 (3), 174–183. Available from: http://www.karger.com/journals/ned/ned_jh.htm. doi:10.1159/000358801.

Rudaks, L.I., Yeow, D., Ng, K., Deveson, I.W., Kennerson, M.L., Kumar, K.R., 2024. An Update on the Adult-Onset Hereditary Cerebellar Ataxias: novel Genetic Causes and New Diagnostic Approaches. Cerebellum. 14734230. 23 (5), 2152–2168. Available from: https://www.springer.com/journal/12311. doi:10.1007/s12311-024-01703-z.

Sayed, N., Allawadhi, P., Khurana, A., Singh, V., Navik, U., Pasumarthi, S.K., Khurana, I., Banothu, A.K., Weiskirchen, R., Bharani, K.K., 2022. Gene therapy: comprehensive overview and therapeutic applications. Life Sci. 18790631. 294. Available from: www.elsevier.com/locate/lifescie. doi:10.1016/j.lfs.2022.120375.

Schöls, L., Bauer, P., Schmidt, T., Schulte, T., Riess, O., 2004. Autosomal dominant cerebellar ataxias: clinical features, genetics, and pathogenesis. Lancet Neurol. 14744422. 3 (5), 291–304. doi:10.1016/s1474-4422(04)00737-9.

Schuster, K.H., Zalon, A.J., DiFranco, D.M., Putka, A.F., Stec, N.R., Jarrah, S.I., Naeem, A., Haque, Z., Zhang, H., Guan, Y., McLoughlin, H.S., 2024. ASOs are an effective treatment for disease-associated oligodendrocyte signatures in premanifest and symptomatic SCA3 mice. Mol. Ther. 15250024. 32 (5), 1359–1372. Available from: https://www.sciencedirect.com/science/journal/15250016. doi:10.1016/j.ymthe.2024.02.033.

Scoles, D., Hung, G., Pflieger, L., Thai, K., Bennett, F., Pulst, S., 2014. Treatment Of Spinocerebellar Ataxia Type 2 (SCA2) with MOE Antisense Oligonucleotides (S47.006). Neurology. 0028-3878. 82 (10_supplement). doi:10.1212/wnl.82.10_supplement.s47.006.

Scoles, D.R., Meera, P., Schneider, M.D., Paul, S., Dansithong, W., Figueroa, K.P., Hung, G., Rigo, F., Bennett, C.F., Otis, T.S., Pulst, S.M., 2017. Antisense oligonucleotide therapy for spinocerebellar ataxia type 2. Nature. 14764687. 544 (7650), 362–366. Available from: http://www.nature.com/nature/index.html. doi:10.1038/nature22044.

Scoles, D.R., Pulst, S.M., 2018. Spinocerebellar ataxia type 2. Adv. Exp. Med. Biol. 1049, 175–195. Available from: http://www.springer.com/series/5584. doi:10.1007/978-3-319-71779-1_8.

Seixas, A.I., Loureiro, J.R., Costa, C., Ordóñez-Ugalde, A., Marcelino, H., Oliveira, C.L., Loureiro, J.L., Dhingra, A., Brandão, E., Cruz, V.T., Timóteo, A., Quintáns, B., Rouleau, G.A., Rizzu, P., Carracedo, Á., Bessa, J., Heutink, P., Sequeiros, J., Sobrido, M.J., Coutinho, P., Silveira, I., 2017. A Pentanucleotide ATTTC Repeat Insertion in the Non-coding Region of DAB1, Mapping to SCA37, Causes Spinocerebellar Ataxia. Am. J. Hum. Genet. 15376605. 101 (1), 87–103. Available from: http://www.elsevier.com/wps/find/journaldescription.cws_home/713561/description#description. doi:10.1016/j.ajhg.2017.06.007.

Song, G., Ma, Y., Gao, X., Zhang, X., Zhang, F., Tian, C., Hou, J., Liu, Z., Zhao, Z., Tian, Y., 2022. CRISPR/Cas9-mediated genetic correction reverses spinocerebellar ataxia 3 disease-associated phenotypes in differentiated cerebellar neurons. Life Med. 2755-1733. 1 (1), 27–44. doi:10.1093/lifemedi/lnac020.

Subramony, S.H., Vig, P.J.S., 2001. Spinocerebellar Ataxia Type 3. Cambridge University Press (CUP), pp. 428–439. doi:10.1017/cbo9780511666469.030.

Sullivan, R., Yau, W.Y., O'Connor, E., Houlden, H., 2019. Spinocerebellar ataxia: an update. J. Neurol. 14321459. 266 (2), 533–544. Available from: https://link.springer.com/journal/415. doi:10.1007/s00415-018-9076-4.

Teive, H.A.G., Camargo, C.H.F., Munhoz, R.P., 2018. Antisense Oligonucleotide Therapy for Spinocerebellar Ataxias: good News for Terrible Diseases. Mov. Disord. Clin. Pract. 23301619. 5 (4), 402–403. Available from: http://onlinelibrary.wiley.com/journal/10.1002/(ISSN)2330-1619. doi:10.1002/mdc3.12627.

Thirumurugesan, S.L., Delhikumar, C.G., Karunakar, P., Sugumaran, R., Biswal, N., Kalatheeswaran, S., 2022. Cervical Dystonia—A Rare Presentation of Spinocerebellar Ataxia Type 35. Indian J. Pediatr. 09737693. 89 (6), 626. Available from: http://www.springerlink.com/content/0019-5456. doi:10.1007/s12098-022-04158-4.

Toonen, L.J.A., Rigo, F., van Attikum, H., van Roon-mom, W.M.C., 2017. Antisense Oligonucleotide-Mediated Removal of the Polyglutamine Repeat in Spinocerebellar Ataxia Type 3 Mice. Mol. Ther. Nucleic Acids. 21622531. 8, 232–242. Available from: https://www.journals.elsevier.com/molecular-therapy-nucleic-acids. doi:10.1016/j.omtn.2017.06.019.

Torres, C.A.M., Zafar, F., Tsai, Y.-C., Vazquez, J.P., Gallagher, M.D., McLaughlin, I., Hong, K., Lai, J., Lee, J., Chirino-Perez, A., Romero-Molina, A.O., Torres, F., Fernandez-Ruiz, J., Ashizawa, T., Ziegle, J., Javier Jiménez Gil, F., Schüle, B., 2022. ATTCT and ATTCC repeat expansions in the ATXN10 gene affect disease penetrance of spinocerebellar ataxia type 10. Hum. Genet. Genomics Adv. 26662477. 3 (4), 100137. doi:10.1016/j.xhgg.2022.100137.

van Prooije, T., Ibrahim, N.M., Azmin, S., van de Warrenburg, B., 2021. Spinocerebellar ataxias in Asia: prevalence, phenotypes and management. Parkinsonism Relat. Disord. 18735126. 92, 112–118. Available from: www.elsevier.com/locate/parkreldis. doi:10.1016/j.parkreldis.2021.10.023.

Verbeek, D.S., Van De Warrenburg, B.P., Wesseling, P., Pearson, P.L., Kremer, H.P., Sinke, R.J., 2004. Mapping of the SCA23 locus involved in autosomal dominant cerebellar ataxia to chromosome region 20p13-12.3. Brain 127 (11), 2551–2557. doi:10.1093/brain/awh276.

Vig, P.J.S., Subramony, S.H., Qin, Z., McDaniel, D.O., Fratkin, J.D., 2000. Relationship between ataxin-1 nuclear inclusions and Purkinje cell specific proteins in SCA-1 transgenic mice. J. Neurol. Sci. 0022510X. 174 (2), 100–110. doi:10.1016/S0022-510X(00)00262-8.

Wang, J.Y., Doudna, J.A., 2023. CRISPR technology: a decade of genome editing is only the beginning. Science. 10959203. 379 (6629). Available from: https://www.science.org/doi/10.1126/science.add8643. doi:10.1126/science.add8643.

Wang, Y.C., Lee, C.M., Lee, L.C., Tung, L.C., Hsieh-Li, H.M., Lee-Chen, G.J., Su, M.T., 2011. Mitochondrial dysfunction and oxidative stress contribute to the pathogenesis of spinocerebellar ataxia type 12 (SCA12). J. Biol. Chem. 1083351X. 286 (24), 21742–21754. Available from: http://www.jbc.org/content/286/24/21742.full.pdf+html. doi:10.1074/jbc.M110.160697.

Waters, M.F., 2006. Spinocerebellar Ataxia Type 13.

Whaley, N.R., Fujioka, S., Wszolek, Z.K., 2011. Autosomal dominant cerebellar ataxia type I: a review of the phenotypic and genotypic characteristics. Orphanet J. Rare Dis. 17501172. 6 (1). doi:10.1186/1750-1172-6-33.

Wu, Y.L., Chang, J.C., Sun, H.L., Cheng, W.L., Yen, Y.P., Lin, Y.S., Chao, Y.C., Liu, K.H., Huang, C.S., Liu, K.L., Liu, C.S., 2022. Coenzyme Q10 Supplementation Increases Removal of the ATXN3 Polyglutamine Repeat, Reducing Cerebellar Degeneration and Improving Motor Dysfunction in Murine Spinocerebellar Ataxia Type 3. Nutrients. 20726643. 14 (17). Available from: http://www.mdpi.com/journal/nutrients/. doi:10.3390/nu14173593.

Xia, H., Mao, Q., Eliason, S.L., Harper, S.Q., Martins, I.H., Orr, H.T., Paulson, H.L., Yang, L., Kotin, R.M., Davidson, B.L., 2004. RNAi suppresses polyglutamine-induced neurodegeneration in a model of spinocerebellar ataxia. Nat. Med. 10 (8), 816–820. doi:10.1038/nm1076.

Yang, Y., Xu, J., Ge, S., Lai, L., 2021. CRISPR/Cas: advances, Limitations, and Applications for Precision Cancer Research. Front. Med. 2296-858X. 8. doi:10.3389/fmed.2021.649896.

Yeboah, G.K., Lobanova, E.S., Brush, R.S., Agbaga, M.P., 2021. Very long chain fatty acid-containing lipids: a decade of novel insights from the study of ELOVL4. J. Lipid Res. 15397262. 62. Available from: https://linkinghub.elsevier.com/retrieve/pii/S0022227521000109. doi:10.1016/J.JLR.2021.100030.

Zesiewicz, T.A., Greenstein, P.E., Sullivan, K.L., Wecker, L., Miller, A., Jahan, I., Chen, R., Perlman, S.L., 2012. A randomized trial of varenicline (Chantix) for the treatment of spinocerebellar ataxia type 3. Neurology. 0028-3878. 78 (8), 545–550. doi:10.1212/WNL.0b013e318247cc7a.

Zhang, Y., Quraishi, I.H., McClure, H., Williams, L.A., Cheng, Y.C., Kale, S., Dempsey, G.T., Agrawal, S., Gerber, D.J., McManus, O.B., Kaczmarek, L.K., 2021. Suppression of Kv3.3 channels by antisense oligonucleotides reverses biochemical effects and motor impairment in spinocerebellar ataxia type 13 mice. FASEB J. 15306860. 35 (12). Available from: https://onlinelibrary.wiley.com/journal/15306860. doi:10.1096/fj.202101356R.

Zhou, J., Van der Heijden, M.E., Leon, L.E.S., Lin, T., Miterko, L.N., Kizek, D.J., Perez, R.M., Pavešković, M., Brown, A.M., Sillitoe, R.V., 2022. Propranolol Modulates Cerebellar Circuit Activity and Reduces Tremor. Cells. 20734409. 11 (23). Available from: https://www.mdpi.com/journal/cells. doi:10.3390/cells11233889.

Zühlke, C.H., Spranger, M., Spranger, S., Voigt, R., Lanz, M., Gehlken, U., Hinrichs, F., Schwinger, E., 2003. SCA17 caused by homozygous repeat expansion in TBP due to partial isodisomy 6. Eur. J. Hum. Genet. 11 (8), 629–632. doi:10.1038/sj.ejhg.5201018.

Index

Note: Page numbers followed by "*f*" indicate figures, "*t*" indicate tables.

O

P

R

S

T

U

V

W

X

Z

Printed and bound by CPI Group (UK) Ltd, Croydon, CR0 4YY
17/09/2026
02218345-0001